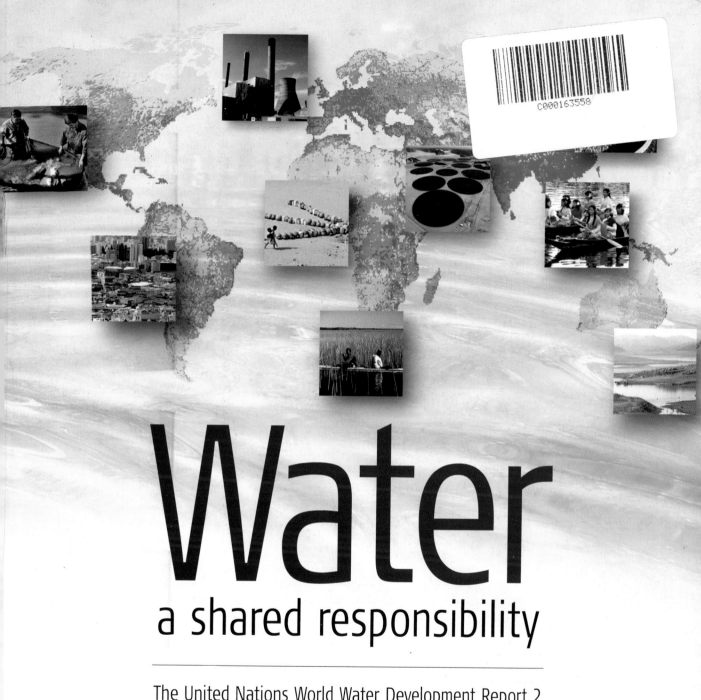

Water

a shared responsibility

The United Nations World Water Development Report 2

World **Water**
Assessment Programme

UNESCO
United Nations
Educational, Scientific and
Cultural Organization

BERGHAHN BOOKS

UN WATER

Published in 2006 jointly by the
**United Nations Educational, Scientific
and Cultural Organization (UNESCO)**
7, place de Fontenoy, 75007 Paris, France,
and **Berghahn Books**, 150 Broadway, Suite 812, New
York, NY 10038, United States of America.

This Report has been published on behalf of the partners
of the United Nations World Water Assessment Programme
(WWAP) with the support of the following countries and
organizations:

Argentina, Bolivia, Brazil, Denmark, Estonia, Ethiopia,
France, Germany, International Commission for the
Protection of the Danube River Basin (ICPDR) in
cooperation with the countries of the Danube River Basin
District, Japan, Kenya, Mali, Mexico, Mongolia, Paraguay,
Peru, Russian Federation, South Africa, Spain, Sri Lanka,
Thailand, Turkey, Uganda, Uruguay, United Kingdom.

United Nations Funds and Programmes
United Nations Centre for Human Settlements (UN-
HABITAT)
United Nations Children's Fund (UNICEF)
United Nations Department of Economic and Social Affairs
(UNDESA)
United Nations Development Programme (UNDP)
United Nations Environment Programme (UNEP)
United Nations High Commissioner for Refugees (UNHCR)
United Nations University (UNU)

Specialized UN Agencies
Food and Agriculture Organization (FAO)
International Atomic Energy Agency (IAEA)
International Bank for Reconstruction and Development
(World Bank)
World Health Organization (WHO)
World Meteorological Organization (WMO)
United Nations Educational, Scientific and Cultural
Organization (UNESCO)
International Fund for Agricultural Development (IFAD)
United Nations Industrial Development Organization
(UNIDO)

United Nations Regional Commissions
Economic Commission for Europe (ECE)
Economic and Social Commission for Asia and the Pacific
(ESCAP)
Economic Commission for Africa (ECA)
Economic Commission for Latin America and the Caribbean
(ECLAC)
Economic Commission for Western Asia (ESCWA)

**Secretariat of United Nations Conventions
and Decades**
Secretariat of the Convention to Combat Desertification
(CCD)
Secretariat of the Convention on Biological Diversity (CBD)
Secretariat of the United Nations Framework Convention
on Climate Change (CCC)
Secretariat of the International Strategy for Disaster
Reduction (ISDR)

Library of Congress Cataloging-in-Publication Data
A catalogue record for this book is available from the
Library of Congress.

British Library Cataloguing in Publication Data
A catalogue record for this book is available from the
Library of Congress.

ISBN UNESCO: 92-3-104006-5
ISBN Berghahn: 1-84545-177-5

The designations employed and the presentation of
material throughout this publication do not imply the
expression of any opinion whatsoever on the part of
UNESCO and WWAP concerning the legal status of any
country, territory, city or area or of its authorities, or the
delimitation of its frontiers or boundaries.

UNESCO Publishing: http://upo.unesco.org/
Berghahn Books: www.berghahnbooks.com

Printed in Barcelona.

Design & production
Andrew Esson, Baseline Arts Ltd, Oxford, UK

All websites accessed in February 2006.

Contents

*Water distribution during
a drought in Mandera,
Kenya*

Foreword

Water is an essential life-sustaining element. It pervades our lives and is deeply embedded in our cultural backgrounds. The basic human needs of a secure food supply and freedom from disease depend on it. Social development – endeavours such as the smooth functioning of hospitals – likewise relies on the availability of clean water. Economic development requires energy resources and industrial activities, and both are in turn water-dependent. The provision of sanitation for girls in schools offers yet another example of water's broader links – it has positive effects on hygiene and health, keeps girls in school, and helps to safeguard the natural environment. For these reasons and many more, access to safe drinking water and sanitation is both a development target in its own right and integrally linked to achieving all the Millennium Development Goals.

The United Nations *World Water Development Report* is the flagship publication of UN-Water, the inter-agency mechanism established to coordinate the activities of all United Nations agencies and entities working in the area of freshwater resources. First published in 2003 as a contribution to the International Year of Freshwater, the Report is produced by UN-Water's World Water Assessment Programme (WWAP). Working closely with governments, non-governmental organizations, civil society groups and the private sector, WWAP monitors water problems, provides recommendations for meeting future demand, and develops case studies in order to promote informed discussion of freshwater issues.

This second edition of the *World Water Development Report – Water, A Shared Responsibility –* shows that collective responsibility is essential for assessing and monitoring progress and for meeting internationally-agreed targets and goals. As we move further into the International Decade for Action, 'Water for Life' (2005–2015), I urge all partners to work more closely together to promote respect for the natural ecosystems on which we depend, and to ensure that all people enjoy access to safe water and the benefits it makes possible.

Kofi A. Annan
UN Secretary General

Fishing on the Mekong River in Viet Nam

Prologue

In March 2003, at the Third World Water Forum held in Kyoto, Japan, I had the pleasure of introducing the first *World Water Development Report*, which is now being used as an educational tool and as a guide for decision-makers in many countries around the world. Its impact was such that it created a momentum at the international level for the creation of the International Decade for Action, 'Water for Life' (2005-2015).

It gives me great pleasure, therefore, to introduce the second in this series of World Water Development Reports – *Water, A Shared Responsibility.* Its publication is most timely, coming just one year after the launch of the Decade and in time for the fourth World Water Forum in Mexico City in March 2006. Subsequent editions of the Report are scheduled for production in 2009, 2012 and 2015, and will provide substantive content for the Decade's agenda. They will assist in monitoring progress towards achieving the targets set at the Millennium Summit and the World Summit for Sustainable Development, many of which have timelines culminating in 2015.

Water, of course, is everyone's business. Hardly a day goes by when we do not hear of another flood, another drought or another pollution spill into surface waters or groundwaters. Each of these issues has a direct or indirect impact not only on human security but also on livelihoods and development. The issues involved range from those of basic human well-being (food security and health), to those of economic development (industry and energy), to essential questions about the preservation of natural ecosystems on which ultimately we all depend. These issues are inter-related and have to be considered together in a holistic manner.

It is thus entirely appropriate that some twenty-four agencies and entities within the United Nations system are involved, with a shared purpose, in producing a comprehensive and objective global report on water issues and the measures being taken to address the related challenges that beset humanity worldwide.

I am very proud that UNESCO, by housing the Secretariat for the World Water Assessment Programme and providing a trust fund to help underwrite the costs of the production of the Report, is facilitating the process of bringing the UN agencies together in common cause. I firmly believe that understanding the many systems that underlie water issues – scientific and cultural, economic and social – will enhance our ability to better manage this precious resource and will help lead to poverty elimination and world peace.

Koïchiro Matsuura
UNESCO Director General

Perito Moreno glacier,
Argentina

Preface

In the three years since the launch of the first *World Water Development Report* at the Third World Water Forum in Kyoto (March 2003), the world has witnessed considerable change. There have been many instances of major water-related disasters: the 2004 Indian Ocean tsunami; the 2004 and 2005 hurricanes in the Caribbean, the west Pacific and the United States; the 2005 floods in central and eastern Europe as well as in many other regions; and the extensive droughts in Niger, Mali, Spain and Portugal. These are a constant reminder of both the destructive power of water and the misery deriving from lack of it in so many regions of the world.

These extreme events are the most prominent illustrations of fundamental changes that are affecting water resources worldwide. In many cases, this evolution is most probably linked to slow but persistent changes in the global climate, a phenomenon supported by a growing body of evidence. The combination of lower precipitation and higher evaporation in many regions is diminishing water quantities in rivers, lakes and groundwater storage, while increased pollution is damaging ecosystems and the health, lives and livelihoods of those without access to adequate, safe drinking water and basic sanitation.

Major demographic changes are also seriously affecting the quality and quantity of available freshwater on the planet. While the more developed countries enjoy relatively stable populations, the less-developed regions of the world are generally experiencing rapid growth and population shifts, particularly in towns, small cities and mega-cities. In many rapidly growing urban areas, it is proving difficult to build the infrastructure necessary to deliver water supply and sanitation facilities to service the population, leading to poor health, low quality of life and, in many cases, to social unrest. To the urban demands for water must be added the increasing demands on water for food production, energy creation and industrial uses.

Large shifts in the geographic distribution of populations occur in various contexts, often adding to water supply problems and social tension. In areas, such as Darfur, there are both internally displaced persons and transboundary refugees. Legal and illegal economic migrants are swelling populations in parts of the United States, and Western Europe, as elsewhere. Increasing tourism to many holiday destinations often exerts a strain on the water supplies of these regions. Whether the result of continued unrest and warfare, terrorist activities or economic instability, population movement is a factor that has a substantial impact on water availability in the world.

It is against these changes in the global situation – some rapid and very noticeable, others insidious and yet persistent – that the governance of water resources must be assessed. This second Report, *Water, A Shared Responsibility*, sets water issues against this evolving background and places greater emphasis on governance issues.

It is proving extremely difficult for many governments to effectively confront the many intertwined issues concerning water. Not only is it difficult for departments within national governments to collaborate effectively, but problems are compounded when many management decisions have to be taken at sub-national and community levels, as the linkage and cooperation between different levels of government is often tenuous at best. The challenges for government agencies to link to NGOs and the private sector for resolving water issues further complicate management and decision-making. The task of managing water becomes even more complex when rivers flow from one country to another. The building of cooperative upstream-downstream relationships is becoming increasingly important with close to half of the world's people living in river basins or above aquifers that cross international borders.

An important goal of the World Water Assessment Programme – founded in 2000 at the request of governments within the Commission on Sustainable Development – is therefore to assist governments in developing their national water management plans. Thus, a number of case studies have been developed and included in the Report. In the first Report, 7 case studies involving 12 countries were included to illustrate the variety of circumstances in different regions of the world. Since then, the number of case studies has grown to 17 involving 41 countries. In a single volume, it is not possible to describe all case studies in detail. Thus we choose to summarize the case studies in the Report and publish the details of each study on our website. This strategy also allows us to make all the necessary updates as new data and information become available.

As we move through the International Decade for Action, 'Water for Life', 2005-2015, the World Water Development Reports will provide a series of assessments that will facilitate the monitoring of change in the water sector, both on a global basis and within a growing number of case study countries and river basins. The purpose of the Decade is to focus on the implementation of water-related programmes and projects, while striving to ensure cooperation at all levels, including the participation of women, to achieve the internationally-agreed water-related goals.

A number of issues identified by UN-Water as priorities for the Decade include coping with water scarcity, access to drinking water, sanitation and hygiene, and disaster risk reduction, particularly in Africa. The Decade aims to support countries in addressing the challenges and achieving the water-related goals of Agenda 21, the UN Millennium Declaration and the Johannesburg Plan of Implementation, as well as those of the 12th and 13th sessions of the Commission on Sustainable Development.

The triennial World Water Development Reports will provide substantive content for the Decade's agenda (subsequent editions of the Report are scheduled for production in 2009, 2012 and 2015) and lay the foundation for a continuous, global monitoring system, pooling the unique perspectives and expertise of the 24 UN agencies that comprise UN-Water, in partnership with governments and other entities concerned with freshwater issues.

We trust that you will find this Report both informative and stimulating.

Gordon Young
WWAP Coordinator

Acknowledgements

This report would not have been possible without the generous and varied contributions of many individuals and organizations from around the world. In addition to the twenty-four agencies that make up UN-Water, numerous other UN organizations, universities, institutes, NGOs and national governments have contributed invaluable input. We would especially like to thank the Government of Japan for its generous support and the publishers, Berghahn Books and UNESCO Publishing.

Team for the preparation of WWDR2: *Water, A Shared Responsibility*

WWAP coordination

Gordon Young	*Coordinator*
Carlos Fernández-Jáuregui	*Deputy coordinator*

WWAP editorial team

Engin Koncagül	*Programme officer, case studies*
Deanna Donovan	*Programme officer, indicators*
Janine Treves-Habar	*Editor-in-chief*
Sean Lee	*Editor*
Isabelle Brugnon	*Photo editor*
Kristin Pittman	*Editorial assistant and CD preparation*
Alejandra Núñez-Luna	*Research assistant*
Casey Walther	*Publicity and editorial support*
Alia Hassan	*Assistant*

WWAP communications and administration

Cornelia Hauke	*Assistant project coordinator*
Georgette Gobina	*Programme secretary*
Pilar González Meyaui	*Communications officer*
Saskia Castelein	*Project officer*
Toshihiro Sonoda	*Liaison officer*
Maria Rosa Cárdenas	*Communications assistant*
Mustapha Boutegrabet	*Technical assistant*

External Contributors

Tony Milburn	*Scientific editor*
Marie-Aude Bodin	*Proofreader*
David McDonald	*Assistant editor*

Baseline Arts

Andrew Esson	*Design, typography and layout*
Nicki Averill and Shirley Bolton	*Design, illustration and layout*
Sue Bushell	*Typesetter*

Berghahn Books

Mark Stanton and Jen Cottrill	*Proofreaders*
Caroline Richards	*Proofreader*
Jim Henderson	*Indexer*

Chapter 1: Living in a Changing World

This chapter was drafted by Tony Milburn (consultant) and the WWAP editorial team, with contributions from UN-ECE and IOC.

Chapter 2: The Challenges of Governance

This chapter was coordinated and drafted by Håkan Tropp (UNDP Water Governance Facility, Stockholm International Water Institute).

The following individuals contributed as authors, reviewers, editors, working group members and/or workshop and meeting participants: Nighisty Ghezae, Karin Krchnak, Joakim Harlin, Melvyn Kay, Alan Hall, Sebastian Silva Leander, David Trouba, Rudolph Cleveringa and Thord Palmlund.

Chapter 3: Water and Human Settlements in an Urbanizing World

This chapter draws from the second issue of UN-HABITAT's Water and Sanitation in the World's Cities (2006), which is currently under print, to be published by Earthscan Publications, London. UN-HABITAT gratefully acknowledges the contribution of David Satterthwaite of IIED, London, and UN-HCR for their contribution on refugees.

Chapter 4: The State of the Resource

This chapter was drafted by Keith Kennedy (consultant) and supervised by Alice Aureli (UNESCO) and Avinash Tyagi (WMO).

The following individuals contributed to the chapter either as working group members and/or workshop and meeting participants, authors, reviewers or editors: Tommaso Abrate, Pradeep Aggarwal, Bo Appelgren, Roger Barry, Åse Eliasson, Andy Fraser, Regula Frauenfelder, Lindsey Higgs, Hege Hisdal, Regine Hock, Jippe Hoogeveen, Kshitij Kulkarni, Annukka Lipponen, Jean Margat, Datius Rutashobya, Joop Steenvoorden, Mohammed Tawfik, Jac van der Gun, Jaroslav Vrba, Bruce Webb and Gary Wright.

Chapter 5: Coastal and Freshwater Ecosystems

The chapter was coordinated by S. Diop and P. M'mayi (UNEP) and drafted by C. Revenga (The Nature Conservancy-TNC), R. D. Robarts (UNEP-GEMS /Water) and C. Zöckler (UNEP-WCMC).

The following individuals contributed to the chapter either as working group members and/or workshop and meeting participants, authors, reviewers or editors: M. Adriaanse, K. Ambrose, N. Ash, S. Barker, C. Bene, S. Butchart, W. Darwall, N. Davidson, R. Davis, M. Diamond, N. Dudley, P. Dugan, M. Dyhr-Nielsen, J. M. Faures, M. Finlayson, D. Gerten, M. Hatziolos, R. Hirji, H. Hoff, N. Holmes, C. Lacambra, B. Lankford, C. Leveque, E. McManus, Muchina-Kreutzberg, C. Nilsson, S. Oppenheimer, C. A. Reidy, M. Schomaker, K. Schuyt, D. Stroud and S. Tomkins. The following individuals contributed to the chapter as reviewers, contributors and/or participants in meetings: A. Calcagno, G. Carr, M. Cheatle, N. Cox, D. Daler, H. Drammeh, T. Goverse, J. Heppeler, K. Hodgson, R. Johnstone (Editor), E. Khaka, S. Koeppel, H. M. Lindblom, P. Manyara, F. Masai, C. Ouma, W. Rast, M. Scheffer, D. Smith, K. Vervuurt and R. G. Witt.

Chapter 6: Protecting and Promoting Human Health

This chapter was coordinated by Robert Bos (WHO) and drafted by Wim van der Hoek (IWMI) and Rolf Luyendijk (UNICEF).

Chapter 7: Water for Food, Agriculture and Rural Livelihoods

This chapter was coordinated and drafted by Wulf Klohn and Jean-Marc Faurès (FAO).

The following individuals contributed as authors, reviewers, editors, working group members and/or workshop and meeting participants: Melvyn Kay, Karen Frenken, Rudolph Cleveringa, Cécile Brugère, Jake Burke, Carlos Garces, Paul van Hofwegen, Sasha Koo-Oshima, Audrey Nepveu de Villemarceau, Åse Eliasson, David Molden, Daniel Renault, Uwe Barg, Leon Hermans, Pasquale Steduto and Michael Wales.

Chapter 8: Water and Industry

This chapter was coordinated and drafted by Ania Grobicki (consultant) on behalf of UNIDO.

The following individuals contributed as authors, reviewers, editors, working group members and/or workshop and meeting participants: Pablo Huidobro, Susanna Galloni, Takashi Asano and Karen Franz Delgau.

Chapter 9: Water and Energy

This chapter was drafted by Ania Grobicki (consultant) and coordinated by Robert Williams (UNIDO).

The following individuals contributed as authors, reviewers, editors, working group members and/or workshop and meeting participants: Richard Taylor, Pravin Karki, Margaret McMorrow, Vestal Tutterow, Michael Brown, Tong Jiandong, Lucille Langlois, Ferenc Toth, John Topper, Gordon Couch and Drona Upadhyay.

Chapter 10: Managing Risks: Securing the Gains of Development

This chapter was coordinated and drafted by Bastien Affeltranger (UNU-EHS).

The following individuals contributed as working group members: Wolfgang Grabs (WMO) and Yuichi Ono (UN-ISDR). And the following individuals contributed as authors or reviewers: Mohammed Abchir, Reid Basher, Janos Bogardi, Salvano Briceno, Xiaotao Cheng, Ken Davidson, John Harding, Tarek Merabtene, Tony Milburn, Buruhani Nyenzi, Pascal Peduzzi, Erich Plate, Rajib Shaw, Slobodan Simonovic, Caroline Sullivan, Mohamed Tawfik, Avinash Tyagi and Junichi Yoshitani.

Chapter 11: Sharing Water

This chapter was drafted by Evan Vlachos (Colorado State University) and coordinated by Léna Salamé (UNESCO).

The following individuals contributed as authors, reviewers, editors, working group members and/or workshop and meeting participants: Shammy Puri, P. B. Anand, Aaron T. Wolf, Joshua T. Newton, Houria

Tazi Sadeq, Raya Stefan, Volker Böge, Lars Wirkus, Arjen Y. Hoekestra, Dipak Gyawali, Bruce Hooper, Monica Porto, Eugene Stakhiv, and Waleed El Zubeiri, Bozena Blix, András Szöllösi-Nagy, Alberto Tejada-Guibert and Alice Aureli.

Chapter 12: Valuing and Charging for Water

This chapter was drafted by Robert A. Young (consultant) and M. Aslam Chaudhry (UNDESA) and coordinated by M. Aslam Chaudhry (UNDESA).

The following individuals contributed as reviewers: Manuel Dengo, Claude Sauveplane, Jean-Michel Chene and Leon Hermans. The following individuals provided background material and reports: Marcia Brewster and Jacob Burke.

Chapter 13: Enhancing Knowledge and Capacity

This chapter was coordinated and drafted by Jan Luijendijk, Roland Price and Kyle Robertson (UNESCO-IHE).

The following individuals contributed as authors, reviewers, editors, working group members and/or workshop and meeting participants: Diego Mejia-Velez, Guy J. F. R. Alaerts, Paul W. J. van Hofwegen, Saba Bokhari, Claudio Caponi, Ralph Daley, Jac van der Gun, Keith Kennedy, Kees Leendertse, Wouter Lincklaen Arriens, Annukka Lipponen, Paul Taylor, Alexey Volynets, Charles Vörösmarty and Jan Yap.

Chapter 14: Case Studies: Moving Towards an Integrated Approach

This chapter was coordinated by Engin Koncagül. Jean-Marie Barrat contributed to the African case studies, and the chapter is based on information provided by the following case study partners:

The Autonomous Community of the Basque Country
Ana Isabel Oregi Bastarrika, Tomas Epalza Solano, José María Sanz de Galdeano Equiza, Iñaki Arrate Jorrín, Jasone Unzueta, Mikel Mancisidor and Iñaki Urrutia Garayo.

The Danube River Basin
Ursula Schmedtje, Igor Liska and Michaela Popovici.

Ethiopia
Abera Mekonen, Teshome Workie, Tesfaye Woldemihret, Michael Abebe, Mesfin Amare, Zeleke Chafamo and Teshome Afrassa.

France
The Water Directorate at the French Ministry of Ecology and Sustainable Development, its regional services and the French Water Agencies.

Japan
Kouji Ikeuchi, Masaru Kunitomo, Satoru Ohtani, Takashi Nimura, Hiroki Ishikawa, Junichi Yoshitani, Tarek Merabtene, Daisuke Kuribayashi, Masato Toyama, Katsutoshi Koga and Ken Yoneyama.

Kenya
George O. Krhoda, Simeon Ochieng, George K. Chesang, Samuel Mureithi Kioni, Patrick Opondo Oloo, Zablon N. Isaboke Oonge, Francis J. Edalia, Bernard Imbambi Kasabuli, Andy Tola Maro, Josiah W. Kaara, Peterson Nyaga Njiru, Evelyn M. Mbatia, Simon Kariuki Mugera, Peter Musuva, Patrick O. Hayombe, Daniel M. Mbithi, John Gachuki Kariuki and Helen Musyoki.

Lake Peipsi/Chudskoe-Pskovskoe Basin
Estonia: Ago Jaani, Harry Liiv, Margus Korsjukov.
Russian Federation: Natalia P. Alexeeva, Vladimir F. Budarin, Alla A. Sedova.

Lake Titicaca
Alberto Crespo Milliet, Jorge Molina Carpio and Julio Sanjinez-Goytia.

Mali
Malick Alhousseni, Adama Tiémoko Diarra, Sidi Toure, Housseini Maiga and Karaba Traore.

State of Mexico
Enrique Peña Nieto, Benjamín Fournier Espinosa, José Manuel Camacho Salmón, Edgardo Castañeda Espinosa, José Raúl Millán López, Mario Gerardo Macay Lim, José Luis Luege Tamargo and Mónica Salazar Balderrama.

Mongolia
Dr. Basandorj, G. Davaa, N. Jadambaa, N. Batsukh, Z. Batbayar and Ramasamy Jayakumar.

La Plata
Victor Pochat, Silvia González, Elena Benítez, Carlos Díaz, Miguel Giraut, Julio Thadeu Kettelhut, Luis Loureiro, Ana Mugetti, Silvia Rafaelli, Roberto Torres, Helio de Macedo Soares and the technical and administrative staff of Comité Intergubernamental Coordinador de los Países de la Cuenca del Plata.

South Africa
Fred Van Zyl and Eberhard Braune.

Sri Lanka
Maithripala Sirisena, B.V.R. Punyawardane, Ananda Wijesuriya, B.J.P. Mendis, K.S.R. de Silva, M.H. Abeygunawardane, A.P.R. Jayasinghe, W.A.N. Somaweera, Amara Satharasinghe, Badra Kamaladasa, T.J. Meegastenne, T.M. Abayawickrama, Tissa Warnasuriya, A.D.S. Gunawardane, H.P.S. Somasiri, B.M.S. Samarasekara, K.Athukorale, M.S. Wickramarachchi, U.S. Imbulana, M. Wickramage, Dayantha S. Wijeyesekara, Malik Ranasinghe, L.T. Wijesuriya, P.P.G. Dias, C.R. Panabokke, A.P.G.R.L. Perera, H.M. Jayatillake, L. Chandrapala, B.R.S.B. Basnayake, G.H.P. Dharmaratne, R.S.C. George, K.W. Nimal Rohana, C.K. Shanmugarajah, S.L. Weerasena, A.R.M. Mahrouf, N.Senanayake, G.A.M.S. Emitiyagoda, K.D.N. Weerasinghe, M.P. de Silva, U. Wickramasinghe, R.N. Karunaratne, B.K.C.C. Seneviratne, T.D. Handagama, S. Senaratne, U.R. Ratnayake, G. Herath, M.M. Ariyabandu, B.M.S. Batagoda, N.K. Atapattu, R.W.F. Ratnayake, N.T.S. Wijesekara and B.R. Neupane.

Thailand
Department of Water Resources, Surapol Pattanee and Sukontha Aekaraj.

Uganda
Patrick Kahangire, Nsubuga Senfuma, Sottie Bomukama, Fred Kimaite, Justin Ecaat, Henry Bidasala, Disan Ssozi, Mohammed Badaza, Abushen Majugu, Joseph Epitu, Nicholas Azza, Joyce Ikwaput, Joel Okonga, Callist Tindimugaya, Patrick Okuni and Ben Torach.

Chapter 15: Conclusions and Recommendations for Action

This chapter was drafted by the WWAP editorial team.

SECTION 1
Changing Contexts

The key challenges of water management can only be understood within the context of water's role in the world today. Many of the world's socio-economic systems are becoming linked at an unprecedented rate. Fast developing communications and transportation systems – including television, Internet and mobile phones – enable many of us to see first hand, and often in real time, what is happening in the world and even take us there should we wish. We are witnessing the impact of extreme climates in floods and drought conditions as well as that of poverty, warfare and disease, which still bedevil so many people of the world, often in increasingly crowded urban conditions.

It is within this setting that the world's water managers have to manage what is an increasingly scarce resource. The pressures and complexity that they face, in what is often a fast changing setting where the available resources can vary greatly in time and space, are huge. This section gives an overview of this and the increasingly refined techniques necessary to secure the equitable management of one of the planet's most precious resources.

Global Map 1: *Index of Non-sustainable Water Use*
Global Map 2: *Urban Population Growth*

Chapter 1 – **Living in a Changing World**

Emphasizing the central role of water use and allocation in poverty alleviation and socio-economic development, this chapter discusses some of the many ways in which demographic and technological change, globalization and trade, climate variability, HIV/AIDS, warfare, etc., affect and are impacted by water. Key concepts of water management, sustainability and equity are introduced, as is the pivotal role of the many activities of the UN system in the water sector.

Chapter 2 – **The Challenges of Governance** (UNDP with IFAD)

Recognizing that the water crisis is largely a crisis of governance, this chapter outlines many of the leading obstacles to sound and sustainable water management: sector fragmentation, poverty, corruption, stagnated budgets, declining levels of development assistance and investment in the water sector, inadequate institutions and limited stakeholder participation. While the progress towards reforming water governance remains slow, this chapter provides recommendations for balancing the social, economic, political and environmental dimensions of water.

Chapter 3 – **Water and Human Settlements in an Urbanizing World** (UN-HABITAT)

Increasing population growth is creating major problems worldwide. Growing urban water supply and sanitation needs, particularly in lower- and middle-income countries, face increasing competition with other sectors. Rising incomes in other portions of the world population fuel demand for manufactured goods and environmental services and amenities, all of which require water. This chapter emphasizes the scale of the growing urban water challenges, pointing out that nearly one-third of urban dwellers worldwide live in slums.

Index of Non-sustainable Water Use

In general, human society has positioned itself in areas with locally sustainable water supplies, in the form of runoff, and/or river and stream flows (Postel et al., 1996; Vörösmarty et al., 2005b). This map illustrates where human water use (domestic, industrial and agricultural) exceeds average water supplies annually. Areas of high water overuse (highlighted in red to brown tones) tend to occur in regions that are highly dependent on irrigated agriculture, such as the Indo-Gangetic Plain in South Asia, the North China Plain and the High Plains in North America. Urban concentration of water demand adds a highly localized dimension to these broader geographic trends. Where water use exceeds local supplies society is dependent on infrastructure that transports water over long distances (i.e., pipelines and canals) or on groundwater extraction – an unsustainable practice over the long-term. Both the map and the graph below understate the problem, as the impact of seasonal shortages are not reflected. The consequences of overuse include diminished river flow, depletion of groundwater reserves, reduction of environmental flows needed to sustain aquatic ecosystems, and potential societal conflict.

Water use in excess of natural supply (average annual)

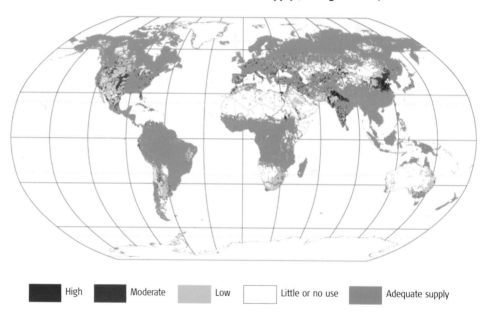

High | Moderate | Low | Little or no use | Adequate supply

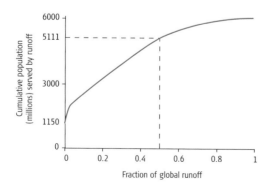

The graph (left) shows that in 2000, of the world's total population 20% had no appreciable natural water supply, 65% (85% minus the 20% with no appreciable water supply mentioned above) shared low-to-moderate supplies (≤50% of global runoff) and only 15% enjoyed relative abundance (>50% of global runoff).

Source: Water Systems Analysis Group, University of New Hampshire. Datasets available for download at http://wwdrii.sr.unh.edu/

Urban Population Growth

In 1950, the world's population was about 2.5 billion people; by 2000, global population was just over 6 billion, an increase of nearly 150 percent in only 50 years. During this time, the proportion of the global population living in urban areas increased from 29 to 47 percent and it is estimated that by 2010, more than 50 percent of the global population will be urban dwellers (UN, 2003).

In less developed regions of the world, this increase has been even more dramatic: in Africa and Asia the fraction of urban population has nearly tripled in the last 50 years (see graph below). Between 2000 and 2030, most population growth is expected to occur within the urban areas of less developed countries, while overall, rural population is expected to decline slightly.

Global Population Density, 2000

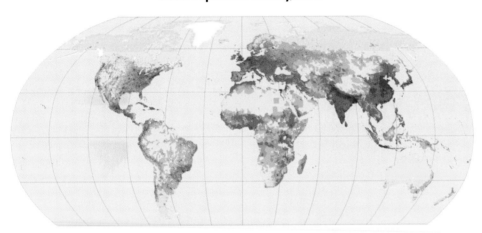

Global Rural Urban Mapping Project (GRUMP) alpha Centre for International Earth Science Information Network (CIESIN) Columbia University in the City of New York

Persons per square km

| <1 | 1–4 | 5–24 | 25–249 | 250–999 | 1000+ | No data |

Proportion of total population that resides in urban areas by region

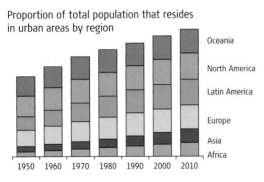

Oceania
North America
Latin America
Europe
Asia
Africa

1950 1960 1970 1980 1990 2000 2010

Roughly 3% of the earth's land surface is occupied by urban areas, with the highest concentrations occurring along the coasts and waterways. The historical importance of water as a means of transport as well as a resource has meant that inland water and river corridors have been important in determining the spatial organization and distribution of human settlements.

Sources: Center for International Earth Science Information Network, Columbia University, Water Systems Analysis Group, University of New Hampshire
Data available for download at http://wwdrii.sr.unh.edu/

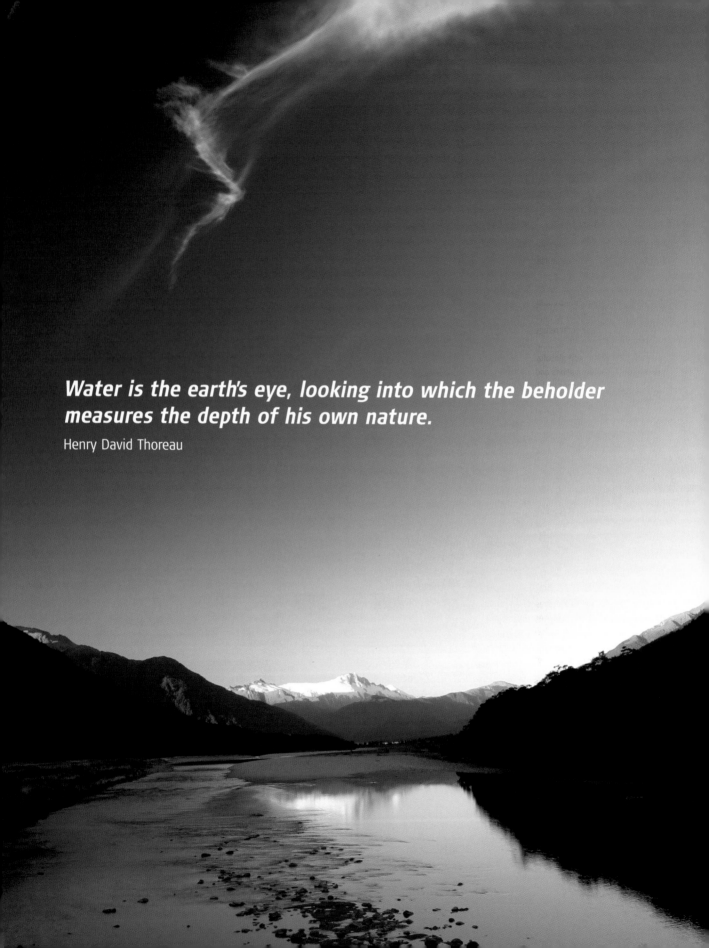

Water is the earth's eye, looking into which the beholder measures the depth of his own nature.

Henry David Thoreau

CHAPTER 1

Living in a Changing World

The confluence of two rivers, New Zealand

Key messages:

At present, our society has not yet attained a level of sustainability whereby humanity honours and respects life upon this planet and uses fairly and equitably the resources it provides. The UN system has taken on a lead role in addressing this challenge through the setting of the Millennium Development Goals and water has a crucial role to play in this. Forty percent of the world's population live in situations of extreme poverty and efforts are underway to lift them out of the poverty trap and to offer greater protection from the natural hazards that still prevail. This introductory chapter will give a flavour of some of the critical challenges involved in managing freshwater to enable poverty alleviation and socio-economic development, within an environmentally-sound integrated framework.

- Water is the primary life-giving resource. Its availability is an essential component in socio-economic development and poverty reduction. Today, a number of significant factors have an impact both on this resource and on managing water in an integrated, sustainable and equitable manner. These include widespread poverty, malnutrition, the dramatic impacts of demographic change, growing urbanization, the effects of globalization – with the threats and opportunities this brings – and the recent manifestations of climate change. All these factors impinge on the water sector in increasingly complex ways.

- The defining characteristic of today's world is change. In almost every sector, including the natural world, the pace of change is unprecedented in recent history. Technological change, especially in information and communications technology, facilitates 'globalization', which in turn affects virtually all aspects of our lives as physical and cultural products move ever more easily around the world. As internationalization and rapid economic growth in many societies alter traditional socio-economic structures, it is clear that change, although virtually pervasive, is not entirely positive. Many people, especially in the developing world and especially those on urban margins and in rural areas, are left behind in poverty and mired in preventable disease. All the chapters in this Report address this issue in one form or another.

- Exacerbating the challenge of economic development is the issue of climate change, which strongly influences the hydrological cycle. Droughts and floods, intensified by climate change, can lead to famine, loss of resources and contamination of water supplies. Population pressure on forest resources can accelerate land degradation and compromise watershed functions, increasing the vulnerability of the poorest communities. Warming temperatures, rising sea levels, uncertain effects on ecosystems, and increased climatic variability are just some of the changes expected to have a disproportionate and significant impact on developing countries. While climate may reduce poor people's assets, increased climate variability will increase their vulnerability and undermine their resilience and coping ability. Thus, climate variability and change present a fundamental challenge to the long-term development prospects of many developing countries, and will make it difficult to meet and sustain the Millennium Development Goals (MDGs).

- In short, water is fundamental to our way of life, at whatever point in the socio-economic spectrum a community may be situated. It is likewise crucial to the preservation of the essential ecosystems upon which our lives depend. Whatever development initiatives are proposed over and above the provision of secure access to water – and such initiatives are many and varied – unless the requisite water services are secured and provided, these initiatives will not succeed. Access to secure water supplies is essential. This seems self-evident. Yet, as this Report shows, it is clear that the central role of water in development is neither well understood nor appreciated. Much more needs to be done by the water sector to educate the world at large and decision-makers in particular.

Part 1. Changing Socio-economic Contexts

Poverty impinges on individual households and families. In aggregate it affects approximately 1 billion people worldwide. This represents one-sixth of the total world population who, through sickness, hunger, thirst, destitution and marginalization, find it nearly impossible to climb out of the pit of extreme poverty. Very poor people struggle to pay for adequate food and water, for housing, for medicines and drugs to treat sick family members, for transport to get to places of work or carry sick family members to treatment centres, and for the education of their children.

The rural poor are often at the end of irrigation systems, and at the whim of richer upstream users for water...

1a. Poverty, water and development

The extreme poor live hand to mouth – what they earn, in an urban area on a good day, will buy food and water for the family for that day. Very often, the quantity of water needed for good personal and domestic hygiene is too expensive to buy from street water vendors, too far to carry in the case of distant water sources, and often necessitates the use of polluted water from nearby, heavily used, rivers and streams. Rarely do they have access to improved sanitation and, where this may be available from a public facility in towns and cities, the cost to the whole family may be prohibitive (see **Chapter 6**). Many poor families occupy land over which they have no formal legal rights – in a squatter community or slum, often with little flood protection infrastructure (see **Chapters 3** and **10**). Many also farm on marginal lands owned by others with limited access to reliable water (see **Chapter 7**). Drainage systems for urban rain and storm water are frequently inadequate, no formal systems for solid waste collection are provided, and there is a lack of paved areas, such as footpaths and roads. The latter are important and not just for movement; they also provide a location for the installation and ready operation and maintenance of network utility services such as water, drainage and electricity. The payment structure for many utility services (e.g. water, electricity), with their up-front connection charges and monthly consumption charges (see **Chapter 12**), are often too high for the poor to pay them. On top of all of this, debt frequently adds to the burdens of poor households.

In rural areas, the food and water needed by families will be taken largely from the natural environment. Water is carried from a distant spring or pool, some

not very nutritious food may be grown on marginally productive land, or collected from forests and is most often insufficient to satisfy hunger and provide needed nourishment. The rural poor are often at the end of irrigation systems, and at the whim of richer upstream users for water, or pushed out onto land dependent totally on what may be, with growing climate variability, increasingly erratic rainfall. Deep well water abstractions by richer farmers and water-using industries can lower water tables to the extent that poorer families and communities cannot then access the groundwater. Untreated municipal and industrial effluents also pollute the surface and groundwater sources relied on by the poor for their water supplies, without redress.

Unbridled competition from richer farmers and industry, productive land, agriculture and fisheries often puts the poor at a serious disadvantage. The implementation of national food policies (through subsidies, taxes, tariffs, food aid etc.) can distort markets and marginalize the rural poor; and inadequately organized and non pro-poor international trade liberalization can exacerbate this. Because of the difficulties poor families face in accumulating any surpluses – food or financial – they find it difficult to maintain consumption when their incomes are interrupted or their crops fail. The poor are excluded from many life-saving and livelihood opportunities, either because of inadequate provision of basic community services by local authorities – health care, transport, education and training, emergency services (e.g. fire-fighting) and law enforcement – or their inability to pay for these services.

Perhaps at no stage in the poverty relief process is water and sanitation more critical than at the

BOX 1.1: THE PARTICULAR PROBLEMS OF AFRICA

Africa is subject to famine, beset by disease and mostly isolated from contemporary international trade. Bad governance and rapid population growth add to the problems. Its very narrow range of exports, restricted to agricultural commodities, some mineral resources and hydrocarbons – all of which are subject to the price vagaries of international markets – further restricts economic development. Africa is not dissimilar to Asia around forty years ago. In Asia, however, high-yield staple crops (rice and wheat), introduced during the Asian green revolution, enabled food production to grow rapidly as the continent's vast plains were well suited to irrigation. Agricultural income grew rapidly, and production diversified. Surplus labour migrated to the cities, leading to rapid urbanization and industrialization. However, Africa does not have the same sort of irrigation potential as Asia. Malaria and the growing AIDs crisis have added to its burdens. Unlike Asia, which hosts coastal cities with good access to ports and sea routes, most African people live inland, too far from ports to enable growth from industrial exports. Road, rail and inland water transport infrastructure are poor and air travel is cumbersome between African countries.

Source: Sachs, 2005.

Child carrying water across an open drain, Ghana

beginning of movement out of abject poverty. Access to a reliable nearby source of water provides relief from the burden of carrying water from distant springs and wells, freeing up time for livelihood activities and, in the case of girls, for school attendance. Having enough water to cover drinking and domestic hygiene needs promotes better health and well-being. Sanitation facilities help to ensure the safe disposal of human waste and reduce disease and death. Adequate water supplies improve the prospects of new livelihood activities, which are otherwise denied, and are often a key step out of poverty. In many lower-income countries, large parts of the population depend on agriculture for their basic livelihood. Others, living in great poverty on marginal land, struggle just to survive. Access to reliable water sources, under the control of the people concerned, reduces crop-loss risks and leads to the possibility of accumulating surpluses and the chance to invest in more intensive agriculture (Moench et al., 2003). Diversification into other activities becomes possible, education opportunities for children grow and, maybe, transition out of agriculture to more profitable enterprises. Industry at all scales needs reliable water resources to prosper and encourage investment in industrial growth. Available water resources and freedom from water-related disease also encourage inward foreign investment.

1b. Demographic changes

The present global population is around 6.4 billion and growing at some 70 million per year, mostly in low-income countries. What has been called the global demographic transition, from populations with short life expectancy and large families, to longer life expectation and smaller families, is very far from complete. Approximately one-third of all countries are still in the early stages of the process, all of which are low-income countries. In fact, of the projected population growth to 8.1 billion by 2030 and some 8.9 billion by 2050, almost all will be in low-income countries (Browne, 2005).

A growing problem, as covered in Section 3 of the Report, is the increasing competition for freshwater between agriculture and urban and industrial uses, causing tension between rural and urban areas and possibly threatening regional or national food security. In fact nearly all malnutrition and low food yield problems are found in low-income countries in the tropics, where water scarcity in relation to food, people and the environment are at their greatest. The four principal demographic risk factors currently challenging humanity – (1) the increasing percentages of young adults, (2) rapid urbanization, (3) reduced availability of freshwater and cropland for food production, (4) HIV/AIDs – rarely occur singly. More usually they occur in combination and coupled with other obstacles, such as weak institutions, unresponsive governments and historic ethnic tensions. The resulting challenges to the leadership of governments can reduce the ability of countries to function effectively (Worldwatch Institute, 2005), as explored in **Chapter 2**.

A particular problem of recent rapid population growth is the so-called 'youth bulge', where young people between the ages of 15 and 30 represent over 40 percent of the total adult population. Many people in this age group have no jobs and even the educated can struggle to find meaningful work. As things stand now, 85 percent of the world's young people are in low-income countries and the average unemployment rate is four times the overall adult rate. The potential consequences of this situation are significant social and political unrest.

The problem is often worst in rural areas where young men cannot inherit land because plot sizes, subdivided through successive generations of inheritance to large families, have become so small that they are no longer viable. Thus, the men lack a secure livelihood and reduced prospects of marriage – a socially destabilizing situation. While the youth bulge will decline as fertility rates continue to fall, some countries (in sub-Saharan Africa and the Middle East) are still experiencing rapid growth in young adult populations. It is likely that these countries will pose a challenge to the development of their region and to international security (Worldwatch Institute, 2005).

Population growth and urbanization

Back in 1970, about two-thirds of the total world population lived in rural areas. By 2001, this had dropped to just over 50 percent. On current predictions, by 2020 this will have fallen to 44 percent with 56 percent of the population living in urban areas. Until recently, Africa was considered the least urbanized continent. This has changed. By 2020, Africa's urban population is estimated to reach 500 million – up from 138 million in 1990. Malawi is the current fastest urbanizing nation due to population flight from severe flooding. Nigeria has also seen tremendous urban growth while huge slums are found in Johannesburg and Nairobi (Worldwatch Institute, 2005).[1]

Urbanization can be a force for good, in terms of economic growth and global integration. However, some of the factors that helped to create wealth in industrialized countries, such as youthful populations, a middle class, nearness to political power and ethnic/religious diversity, can be potential sources of conflict in fast-growing but

Kibera slum, Nairobi, Kenya

1. www.worldwatch.org

BOX 1.2: ENVIRONMENTAL REFUGEES

It is estimated that there are presently some 30 million environmental refuges and a further 17 million other refugees and displaced persons from wars, persecution and other causes. The former have fled from resource scarcity, from deforestation and environmental degradation, climate change impacts, overpopulation, displacement by development projects, etc. Large displacements of population can cause instability or conflict in the host country, country of origin, or within a region. They entail depletion of scarce resources, overcrowding, shortage of potable water and unsanitary conditions that can lead to disease epidemics.

It has been suggested that the number of environmental refugees could rise to 150 million by 2050 as one of the results of climate change.

Source: Worldwatch Institute, 2005.

BOX 1.3: TOURISM AND GLOBALIZATION

An increasing number of low-income countries have actively promoted a large increase in tourism activities to foster their economic development. While there are clear economic benefits, there is also a downside. Problems of excessive water consumption in tourist complexes in water-scarce areas, especially where golf courses are involved, an increase in marine pollution in coastal areas from inadequate wastewater treatment and loss of crucial marine biodiversity, including coral reef destruction, have all occurred. Competition by tourism for scarce water supplies has led to instances of water diversion from farmers, effectively driving them out of agriculture and their livelihoods.

Source: www.uneptie.org/pc/tourism/sust-tourism/env-3main.htm

---

Otash camp for Internally Displaced Persons (IDP), Sudan

poor cities in developing countries. Statistics show that countries with urban growth exceeding 4 percent per year are twice as likely as others to experience civil disturbances (Population Action International, 2003). **Chapter 3** looks at the issues of urbanization and human settlements in much more detail, especially the challenges of providing water and sanitation to improve both health and livelihood activities. The food and water implications related to demographic change and urbanization are reviewed in **Chapter 7**.

1c. Political and economic changes
We are living in a period of rapid and significant geopolitical change. Previously established empires and countries have broken up (e.g. the Soviet Union and Yugoslavia) while neighbouring groups of countries seek closer economic collaboration or consolidation (e.g. the European Union). The former centrally controlled economy of the Soviet Union is now a collection of nation states trying to enter the global economy, without the experience or institutions to cope effectively. Ethnic tensions suppressed under former political systems within the former Yugoslavia have erupted into armed conflict in the Balkans. New nation states, wary of sharing the water resources of transboundary rivers and aquifers, become very defensive about their perceived sovereignty over such resources, especially as resources are pressured by increased demand and deteriorating water quality.

Warfare and conflicts
The Post Conflict Assessment Unit of the United Nations Environment Programme (UNEP) has shown that conflicts are almost always followed by environmental crises: chemicals leaching into waterways, damage to irrigation systems, deforestation, the destruction of infrastructure and

collapses of governance systems – local and national. Rebuilding economies, damaged lives, shattered infrastructure including water and power systems, rebuilding and restoring damaged irrigation systems, removing landmines in post-conflict situations, absorb 27 percent of all Overseas Development Assistance (Worldwatch Institute, 2005). The Convention on the Prohibition of Military or any Other Hostile Use of Environmental Modification Techniques (ENMOD Convention)[2] seeks to prohibit acts such as weather modification and harmful flood creation. The threat posed by the release of toxic chemicals into the environment has further prompted calls for a new convention. It is no coincidence that many of the countries yet to make progress on debt relief are those recently emerged from conflict situations (World in 2005). **Chapters 2, 3** and **11** cover these and other issues of warfare and refugees, and point to the need for introduction of agreements and conventions to address these problems.

Globalization
At present, the world is going through an unprecedented process of integrating finance, trade, communication and technology. By eliminating tariffs and other barriers to trade, the world's economy is becoming increasingly interlinked. This has advantages. Transaction costs and investment risks can be reduced and greater investment encouraged. The increased competition encouraged by regional integration fosters competition and innovation. Reduced costs in telecommunications and energy infrastructure are possible. For water, globalization enables economies of scale through access to bigger markets, facilitates improved cooperation over international waters, and allows a benefits-based approach towards regional water-resource systems and inter-country collaboration on water knowledge and skills

2. www.unep.org/
Documents.Multilingual/
Default.asp?DocumentID=
65&ArticleID=1291&l=en

BOX 1.4: MOBILE PHONES AND THE WATER SECTOR

Current available evidence suggests that promoting a widespread use of mobile phones (cell phones) may be a sensible way to encourage bottom-up development. Mobile phones help to raise long-term growth rates – an extra ten phones per 100 people in a typical lower-income country has been shown to raise GDP growth by 0.6 percentage points. Mobile phones help reduce transaction costs, widen trade networks and do away with the need for intensive travelling. In terms of the water sector, in low-income countries they are used by fishers and farmers to obtain the best prices for their produce, to help provide early warnings to communities of floods, to get information and help in treating water-related diseases and many more. The UN has set a target of access by 50 percent of the world population, although some three-quarters of that population already live within range of a mobile telephone network.

Source: Economist, March 2005.

transfer. Disadvantages, however, may include increased water scarcity and pollution, if water demand and pollution control are not carefully managed (World Development Report, 2005). Countries with poor or weakly enforced environmental regulations which allow pollution (air or water) to flow across international boundaries are of great concern (World Bank, 2005).

In some countries, there are moves to divert water from the production of low-value staple crops to higher-value cash crops such as vegetables, fruit and flowers. As exporting grows, there are concerns that the rules of the North American Free Trade Agreement (NAFTA) and the World Trade Organization (WTO) increase environmental risk due to their restrictions on the use of the precautionary principle.[3] Several critics of these organizations' rules argue that WTO, for example, has always put the interests of commerce before environmental protection. This may lead to countries bound by these trade rules to agree to bulk exports of water to other countries against their wishes (Figueres, 2003).

One of the consequences of globalization and the increase of market-based economics is that rights markets have been widely advocated as the way to manage natural resources. **Chapter 12** looks at economic valuation techniques for water.

Technological innovations and water
Technology can offer significant opportunities for the water sector. In water treatment, membrane technology is an example. Membranes are manufactured filtration systems that can separate a wide and growing range of substances – both organic and inorganic – which are present in water, from the water itself. They can be used for industrial and drinking water treatment, wastewater treatment, desalination of salt water and brackish water, and so on. The previous high costs are being reduced substantially and the technology is now increasingly available worldwide. The use of ultra-violet irradiation of waters for drinking water, in industrial water treatment, and for reducing the polluting burden of wastewater effluents, is spreading. New understanding is emerging of on-site wastewater treatment and recycling and small water and wastewater systems. This offers lower overall costs for water supply and sanitation systems and nutrient recovery options, as well as reducing the complexity of large centralized systems (Mathew and Ho, 2005). There is potential for using these technological applications and others in extending the provision of water supply and sanitation services to communities, as is covered in most chapters of the Report.

Much has been made of the digital divide, the uneven distribution worldwide of communications technology and access to, and use of, information. The UN has responded to this by creating the 'Digital Solidarity Fund' in March 2005, intended to enable people and countries presently excluded from the information society to gain access to it. Initiatives proposed include the construction and operation of regional telecentres where people have access to computers, the internet, telephones, and so on. Over and above this, the many applications of satellite surveillance and modelling can have substantial potential in water resources monitoring in lower-income countries, as described in **Chapters 4** and **13**. At the same time, focus needs to placed on basic knowledge and capacity-building, which can be more easily shared in this world of increased communications.

3. The Precautionary Principle, adopted by the UN Conference on the Environment and Development (1992), states that in order to protect the environment, a precautionary approach should be widely applied, meaning that where there are threats of serious or irreversible damage to the environment, lack of full scientific certainty should not be used as a reason for postponing cost-effective measures to prevent environmental degradation (from European Environmental Agency website glossary.eea.eu.int/ EEAGlossary/P/precautionary_ principle)

Part 2. Governing Water: A Shared Responsibility

In simple terms, the great challenge of this century is to find the means to develop human capital (socio-economically, culturally and equitably), while at the same time preserving and protecting natural capital. It is necessary to acknowledge that for far too long, the headlong pursuit of material prosperity for the few has *excluded* far too many poor people from well-being, health, food and environmental security; has *excluded* the interests of the natural environment; and has *excluded* adequate consideration of the interests of future generations. We have come to realize that adopting an *inclusive* approach is essential to securing the sustainability of all forms of life.

...water managers around the world agree that the only way forward is through an inclusive and integrated approach to water resources management...

WWDR1 (2003) concluded that governance issues form the central obstruction to sound and equitable water sharing and management worldwide. Sharing is at the heart of the governance issue and the title of this Report reflects this. Given the complexity, uncertainty and increasing vulnerability of both natural and human systems, water managers around the world agree that the only way forward is through an inclusive and integrated approach to water resources management (IWRM), which recognizes the need to ensure a holistic protection

system. **Chapter 2** on water governance leads off this discussion by looking in greater detail at obstacles connected to implementing an integrated approach to water resources management, thereafter taken up by most chapters of the Report.

Countries have to shift to a more inclusive set of values in the overall interests of the entire planet. Few have so far done so. Many of the richer countries have used redistributive taxation, education, equal opportunities and social welfare

Figure 1.1: The reiterative policy-making process

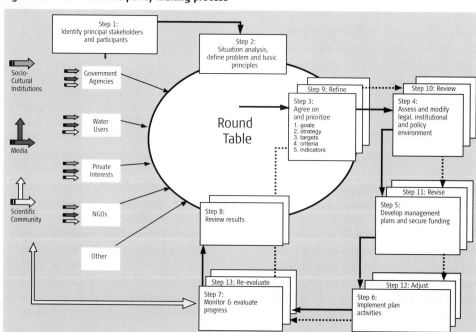

Source: Derived from Gutrich et. al., 2005.

BOX 1.5: THE EU AND SOUTH AFRICA: INCLUSIVE WATER MANAGEMENT

In the European Union (EU), redistributive taxation has transferred wealth from the richer northern countries to the poorer south and is now beginning to do so to the new accession states to the east, significantly raising their standards of living. The bigger market and increased competition are driving wealth generation. At the same time, one of the most comprehensive environmental protection regimes in the world is being put in place in the EU, which will greatly enhance environmental protection and improved water management.

At the other end of the world, driven by South Africa's reforming government, significant changes are occurring in the southern African region in both attitudes and techniques of water management. The biggest changes are in South Africa itself but their ideas are spreading to adjacent countries. There are some limited parallels with the EU in that change is being driven by a wealthier core nation with strong institutions and clearly articulated equitable values, and is rippling out from there. In both the EU and southern Africa, the process of change is underpinned by institutional values that emphasize inclusiveness of both the whole

population and the needs of the natural environment. The water laws and regulations of both of these regions, especially at the core of them, are characterized by commitment to equality of access to water for all and to environmental protection of a more sophisticated nature (relatively speaking in each case) than almost anywhere else on the planet. Given the present strong trend towards regional economic cooperation, reported later in this chapter, the experiences of these two regions are significant. To what extent their inclusive approach will spread remains to be seen.

---

programmes to release the wealth-generating, creative and upwardly mobile-potential of their citizens. They have put in place extensive environmental protection/rehabilitation measures. They have entrenched human and property rights and established the clear rule of law. Multinational companies based in these countries and aware of the precious nature of their corporate reputations, have shown an increased inclination to take environmental and employment concerns seriously when operating in lower-income countries. The wealthier countries have produced cadres of educated, competent people in relatively sound (though by no means perfect) institutions and organizations for water management, such as in the European Union (EU), but such an overall pattern of development is also observable in South Africa, for example (see **Box 1.5**).

2a. An integrated approach – IWRM

Reference has already been made to the need for an integrated and holistic approach to water resources management. Fundamentally, this is a response to the much-criticized, sector-by-sector approach to water management (irrigation, municipal, energy, etc.), highlighting instead the benefits that an integrated, overall approach to water management, on a catchment or basin basis, can deliver.

Integrated Water Resources Management (IWRM) promotes not only cross-sectoral cooperation, but the coordinated management and development of land, water (both surface water and groundwater) and other related resources, so as to maximize the resulting social and economic benefits in an equitable manner, without compromising ecosystem sustainability. It is not only the watershed or basin that must be considered in the IWRM approach, but any adjacent coastal and marine environments, as well as both upstream and downstream interests in the basin (see **Chapters 4, 5** and **11**).

The socio-economic dimension, with its focus on human concerns, is a crucial component of the approach, taking full account of:

- stakeholders having input in the planning and management of the resource, ensuring especially that the interests of women and the poor are fully represented

- the multiple uses of water and the range of people's needs

- integrating water plans and strategies into the national planning process and water concerns into all government policies and priorities, as well as considering the water resource implications of these actions

Girl collecting water from a community supply, Abidjan, Côte d'Ivoire

Replacing old water fittings with new, more efficient ones can produce good results in domestic and industrial water systems

■ the compatibility of water-related decisions taken at a local level with a country's national objectives

■ the water quantity and quality needs of essential ecosystems so that they are properly protected (GWP, 2004).

The 2002 World Summit on Sustainable Development (WSSD) sought to move the water sector worldwide towards more sustainable approaches to water management, building ecosystem considerations into overall IWRM management paradigms and calling all countries to produce IWRM and water-efficiency plans by 2005. **Chapter 2** reports that some progress on this is being made, but that many countries still have much to do. As the organizing principle of water management, IWRM is covered in most of the chapters of the Report. **Chapter 2** examines water governance, while **Chapters 8** and **9** examine how an integrated approach to water and industrial energy management respectively can provide big savings in all. **Chapter 10** stresses how disaster risk reduction has to be a key component of IWRM, while **Chapter 12** looks at the use of economic valuation techniques and pricing of water – important tools for IWRM.

2b. Demand management

Traditionally, the responses to pressures on water availability were solved by increasing supply: developing new sources and expanding and increasing abstractions from existing ones. As this is not sustainable, attention is switching rapidly towards more efficient and equitable approaches. The process of using water more efficiently and fairly, improving the balance between present supplies and demand, and reducing excessive use, is known collectively as demand management.

Consumer attitudes and behaviours (wrongful and wasteful use included) are a problem in which information campaigns and consumer education programmes can play important roles. **Chapter 13** discusses some of these. Economic incentives, in the form of water-use metering and the application of tariff systems that discourage wasteful use can be used to good effect, although allowances have to be made to ensure the poor are not disadvantaged (see **Chapter 12**). Replacing old water fittings with new, more efficient ones can produce good results in domestic and industrial water systems. Fixing the leaks in urban water distribution systems, where up to 60 percent or more of the water supplied can be lost through unrepaired leaks, offers much potential (see

---

BOX 1.6: THIRTEEN KEY IWRM CHANGE AREAS

The Global Water Partnership (GWP) has identified 13 key IWRM change areas within overall water governance, which together form the process of moving towards a more integrated water management approach. These key change areas are contained within a framework divided into the enabling environment, institutional roles and management instruments.

Enabling environment

1. Policies – setting goals for water use, protection and conservation
2. Legislative framework – defining the rules needed to achieve policies and goals
3. Financing and incentive structures – allocating financial resources to meet water needs.

Institutional structure

4. Creating an organizational framework – understanding resources and needs
5. Institutional capacity-building – developing human resources.

Management instruments

6. Water resources assessment – understanding resources and needs
7. Plans for IWRM – combining development options, resource use and human interaction
8. Demand management – using water more efficiently
9. Social change instruments – encouraging a water-oriented civil society
10. Conflict resolution – managing disputes and ensuring the sharing of water

11. Regulatory instruments – determining equitable allocations and water use limits
12. Economic instruments – valuing and pricing water for efficiency and equity
13. Information management and exchange – improving knowledge for better water management.

Source: GWP, 2004.

Chapter 3). In agriculture, changing cropping patterns, moving to more water-efficient crops, precision application of irrigation water (see **Chapter 7**), and improving the performance of water delivery and distribution systems can collectively produce improved water productivity. **Chapter 8** offers a detailed look at how industry has access to a growing range of cost-effective methods of optimizing water-use productivity and minimizing harmful industrial emissions. Combinations of all of these, as appropriate, can be very effective (GWP, 2004).

A big expansion in water harvesting, at domestic and community levels, is ongoing, particularly in Asia. Water recycling and reuse, already very prevalent in dry areas, is set to grow. Improvement in knowledge and understanding of treatment at different levels of sophistication is increasing, which will help to minimize risks to workers and consumers involved in the many different applications of wastewater reuse. The world has substantial deposits of brackish water, often in underground aquifers. As the cost of desalination is falling, because of technology improvements and lower energy costs, the prospects of desalinating brackish water – and also seawater in the case of coastal communities – are becoming more attractive.

Allocative efficiency – seeking to ensure that water is allocated to the highest value uses, while ensuring the interests of the poor and ecosystems are not neglected – may be sought through water rights, water markets and appropriate cost-benefit analyses (see **Chapters 2** and **12**). In low-income countries, it is essential that the role of water in poverty alleviation is fully factored in and that crucial environmental flows are maintained (GWP, 2004).

2c. Subsidiarity

There is an increasing trend towards delegation of water management responsibility to local authorities and water user groups, thereby promoting the principle of subsidiarity.

Devolution or decentralization of power from national governments and agencies to regional and/or local government authorities and organizations –including responsibility for water – is ongoing in many parts of the world. Examples of

this trend include new river basin management arrangements; transfer of responsibility for water supply and sanitation to municipal authorities, NGOs or community groups; and irrigation management transfer (IMT) to farmer/user groups (see **Chapter 7**). The potential benefits are good, as local management should better understand the needs, resources and demands of the situation. A degree of competition between local authorities can stimulate innovation while cooperation between stakeholders can be improved (World Bank, 2005).

The downside in practice is that many governments transfer water management responsibilities to a range of subnational entities that lack the capacity and resources to cope. Also there are larger-scale issues in water management that cannot readily be dealt with at the local level – allocations, pollution control, storage and others. Administrative areas may not coincide with river basins and watersheds. They may overlap adjacent basins, or several administrative units may share a basin. Some communities undergoing rapid socio-economic development involving significant upward and regional social mobility may find that membership of community management arrangements is unpopular, as people see no long-term advantage in participating (Moench et al, 2003). **Chapter 2** goes deeper into many of the problems associated with devolving responsibility for water management from central government to other entities and organizations and **Chapter 11** addresses the issue of moving water and resolving conflicts between countries, sectors, communities and other stakeholders.

2d. Gender mainstreaming

Of the 1.3 billion people living in abject poverty, the majority are women and children. They also happen to be the largest group systematically under-represented in water management arrangements. Water management is often gender specific at its different scales, reflecting the different ways men and women take responsibility. Generally, men take care of longer-term needs. Women, on the other hand, are mostly responsible for household hygiene, food and water. Often, this involves women and girls walking long distances to obtain water. Not only is the heavy burden of the water physically harmful, but the time lost means

A big expansion in water harvesting, at domestic and community levels, is ongoing, particularly in Asia. Water recycling and reuse, already very prevalent in dry areas, is set to grow

...women possess a lot of knowledge and experience of manage- ment and conservation of natural resources...

This community in Rajapur village, western Bangladesh, holds monthly meetings to discuss primary school attendance and other important issues

less available time for more productive purposes, such as livelihood activities and enhanced childcare. Ecosystems are frequently important food sources for poorer families and it is generally women who are involved in gathering food from them. Thus, ecosystem damage and species loss hit poor families and women particularly hard.

The pressing need of women for water supply and sanitation for their families gives them in a key role in community water service provision. Since many of the farming activities in poor communities are carried out by women, their needs for crop water are essential for family nutrition. Women, children and the elderly are also the most at risk from water-related hazards. Yet, all too often, women are excluded from important planning and decision-making in water management. This exclusion inevitably makes water service provision, in its many forms, less responsive to real need. Moreover, as Agenda 21 noted, women possess a lot of knowledge and experience of management and conservation of natural resources – including local water sources – as well as good water management skills. However, significant barriers to their participation in this role have arisen from a variety of causes – legal, constitutional, social, behavioural,

cultural and economic. In some societies, men have deep-seated insecurity about the idea of women owning property, including land or water rights for farming (Vyas, 2001).

Greater involvement of women in water matters enables better demand responsiveness to water provision and prevention of pollution. By ensuring their voices are fully heard in the water planning process, the effects on their subsistence and their development needs can be acknowledged and their interests protected. However, a number of issues have to be considered. The different status of men and women has to be acknowledged in its entirety; they have different needs and priorities and their life courses differ considerably. Equal treatment will not necessarily produce equal results and a gender equity approach is needed, requiring a good understanding of the frequently complicated relationships between domestic water use and its use in agriculture, industry and energy. Men and women often approach different decision-making differently. In addition, the institutional structures of general governance and water management determine the roles, rights and responsibilities of each sex in relation to control over and access to resources. Gender equity also requires that both sexes receive equitable benefits from any decentralized management structures and from new and improved infrastructure.

Gender issues in water management are well understood at the international level and, as a result, most guidelines produced by governments, designs for new projects, and programme policies now take account of gender issues. Gender mainstreaming (including gender equity matters in policies, programmes and procedures), gender budgeting (analysing all projects and policies to ensure equitable benefits for all), and affirmative action (to secure fully equitable participation in water planning and management), are all acknowledged as essential for greater gender equity and better water governance. However, much still remains to be done at the local level, a long-term task that will require persistence, capacity development and high-level political commitment. Despite all this, there is reason for cautious optimism, as progress is being made (Guerquin et al., 2003).

Part 3. Changing Natural Systems

The past twenty-five to thirty years have seen a substantial focus on the environmental impacts of water resources infrastructure development. Rather less attention has been paid to assessing the environmental impacts of water resource strategy. Recently, cooperation between ecologists and water managers has led to attempts to integrate an ecosystem approach into Integrated Water Resources Management (IWRM), although this is at an early stage. The task has been to conceptualize a catchment- or basin-based holistic approach, which recognizes the multiple roles of water both in ecosystems and human socio-economic systems. This involves consideration of terrestrial and aquatic ecosystems and the water links between them. Water managers are challenged with increasing their understanding of the biotic linkages between freshwater circulation and ecosystems. The process of photosynthesis (which consumes vast quantities of water resources) and the very significant changes in runoff from major land use changes, need to be better appreciated, as indicated in Chapters 4 and 5.

In a very short space of time, in planetary terms, we have sought to redesign and impose a new order on natural planetary systems built over aeons of time

3a. Human intervention

Humanity has embarked on a huge global ecological engineering project, with little or no preconception, or indeed full present knowledge, of the consequences. In a very short space of time, in planetary terms, we have sought to redesign and impose a new order on natural planetary systems built over aeons of time. In the water sector, securing reliable and secure water supplies for health and food, the needs of industrial and energy production processes, and the development of rights markets for both land and water have hugely changed the natural order of many rivers worldwide (see **Sections 3** and **4** of Report). There is a vast range of interactions between the biosphere and landscapes, as a result of which great variability results at a range of scales, and novel and unexpected properties of the system emerge. This variability is critical for the way ecosystems function, sustain themselves and evolve, and we never cease interfering with these natural systems. Land-use changes, urban development, dam construction and other river diversions all disrupt the natural pattern and rhythm of natural processes, without attention to the consequences and the negative effects on biodiversity.

Threatened environmental resilience

Left undisturbed, natural ecosystems have great resilience but a minimum composition of species must be maintained to ensure that the relations between the primary producers, consumers and decomposers can be sustained. Only thus can they continue the mediation of energy flow, the cycling of elements, and the spatial and temporal patterns of vegetation. Resilience is a buffer against disturbances and this buffer is best provided through maintaining biological diversity.

However, human impacts on the quantity and quality of available water seriously inhibit this resilience, leading to the risk of a retreat to a more vulnerable state. Pollution from agriculture, industry and domestic wastewater is making water resources, both surface water and groundwater, increasingly scarce and decreasingly poor in quality.

Loss of biodiversity is an important indicator of lowered resilience and the current deterioration in freshwater biodiversity (greater than either marine or terrestrial systems) is of great concern, as discussed in **Chapter 5**. Human reactions to environmental changes are less direct than for other species, because we are slow to become aware of changes before responding. Human resilience rests in the coping capacity of society and its institutions. As any resilience declines, whether social or ecological, it takes progressively smaller external changes to cause big problems. Reduced ecological resilience, from land degradation and drought, can increase social and environmental vulnerability, leading to the loss of livelihood and creating tension and conflict over freshwater and food. The key challenge facing water managers, now and in the future, is to try to optimize ecosystem resilience in response to human and

...land-use management and property ownership fragment the natural landscape in a totally different way from natural ecosystem processes

natural disturbances and protect this resilience with catchment-level life-support systems – in particular essential productivity functions (GWP, 2003).

3b. Climate variability and change

There is frequent confusion between climate change and climate variability. Climate change is associated with global warming and is a long-term change with its origins in natural factors and, as is now accepted, human activities. Climate variability, on the other hand, has always been part of the Earth's climate system, although it has so far received surprisingly little attention from the water sector. It affects water resources by way of floods, droughts, waterborne disease, and so on. It is not just the extremes of climate variability that are of concern to the water sector: the increasing and extreme variability in the hydrological cycle and climate systems, as is shown in **Chapter 4**, together with the dynamic processes that lie behind it, impact on countries' water resources and can make it difficult to meet the MDGs (Lenton, 2004).

Managing climate variability and the impacts of climate extremes is one of the challenges of sustainable development. In fact, the skills and knowledge obtained in dealing with these variability problems will be invaluable in confronting the longer-term challenges of climate change, as specifically discussed in **Chapters 10** and **13**. As ever, it is the poor who are at greatest risk. Thus climate effects have to be built into poverty reduction activities and included in national development plans and national water resources policies, using the IWRM approach. Both water managers and decision-makers have to be encouraged to engage in greater dialogue with climate and development specialists to better understand the climate-related challenges and how to deal with them. At the same time, but on a broader front, although there is a clear need to learn to adapt to the challenges of climate variability and change, all actions to mitigate the anthropological impacts of this must continue.[4]

On the broader issue of climate change several concerns are apparent. The yields of staple crop production (e.g. rice, corn and wheat) are sensitive to temperature increase, the most vulnerable period being during pollination, just before seed formation.

It had been thought that increased CO_2 levels would lead to higher grain yields, but the view is now that negative effects of temperature increase will outweigh this. **Chapter 7** examines key issues connected with the potential for climate effects to alter food production patterns. Warming over land will be greater than warming over the seas and this effect will be greater in the higher latitudes, affecting continental interiors more than coastal regions. This has major implications for grain-producing regions. The headwaters of many large Asian rivers originate in the Himalayas. The large amount of freshwater traditionally stored in the glaciers there is being reduced as glaciers shrink. This will likely alter seasonal runoff patterns, increasing flood extremes and affecting availability of critical irrigation waters (Lenton, 2003).

Chapter 8 points out the growing risk of what it calls 'Natech' disasters, where extreme climate events can severely damage industrial installations, reducing not only economic activity but also releasing gross pollution into the environment. Given the fact that much energy generation is located in fossil-fuel-powered electricity generating stations and that the very large greenhouse gas emissions from these are believed to have a big impact on climate, **Chapter 9** looks at the issues involved in connection with more sustainable energy provision as related to water.

3c. Ecological water management

As discussed, land-use management and property ownership fragment the natural landscape in a totally different way from natural ecosystem processes. Financial markets and global and national business cycles have their own patterns and cycles and the capitalistic approach needs market advantage and security of tenure, which is at variance with natural patterns and cycles. As a result, much policy is made within overly simple settings and fails to recognize the complexity of these different cycles and their interactions. The critical challenge is to recognize that the spatial and temporal scales of variability within ecosystems, society and the economy are strongly linked, but are not congruent.

Global development must be equitable and inclusive not only of the interests of humanity but also of

4. For more information on the Dialogue on Water and Climate see www.waterandclimate.org

BOX 1.7: THE PARTICULAR PROBLEMS OF TROPICAL COUNTRIES

Climate impacts are particularly severe in low-income countries in the tropics, which generally include those countries having the greatest difficulties working towards the MDGs. The problem is exacerbated by rainflow and streamflow, which are concentrated over a short period of a few months. Very significant seasonal and annual fluctuations around long-term historical averages are reflected in long dry periods and recurrent droughts and floods. Since these countries often rely substantially on natural resources, the impact of floods and droughts on development is amplified. Climate variability can also increase vector-borne disease outbreaks (e.g. malaria) and the incidence of diarrhoeal disease in the rainy season.

In addition, climate uncertainties lead to greater risk aversion by farmers in crop selection, planting and fertilization, as well as problems for reservoir managers responsible for irrigation and hydroelectrical production. Thus, the challenge is to greatly increase water storage to minimize the impact of climate variability, while trying to avoid the environmental and social disruption of large dams. Balancing the high demand for irrigation water against other uses adds to the challenge of obtaining greater water productivity from irrigation waters ('more crop per drop'). These storage and water productivity issues require further substantial research and development work (Lenton, 2004) See **Chapter 7** for more on the issue of storage.

Source: Lenton, 2004.

the natural planetary ecosystems that support us. The task of water managers is therefore far from easy. They must satisfy socio-economic needs, minimize the pollution burden and accept the consumptive use involved, including a better understanding of the limits of the self-cleansing capacities of ecosystems. Minimum ecological criteria have to be met.

Catchments have to be viewed as socio-eco-hydrological systems in which trade-offs are needed or will have to be made socially acceptable by appropriate institutions, regulations and finance. A key entry point is defining minimum criteria or 'bottom lines' for terrestrial ecosystems. In order to balance upstream and downstream interests, work has to start downstream, identifying bottom lines for the different components of the aquatic ecosystems, for example, uncommitted environmental flows and minimum water quality. Thereafter the process is carried on upstream, section by section, constantly seeking to identify resilience determinants to avoid ecosystem collapse. Agricultural water management and food production as major water users have to be very sensitive to ecosystem considerations to ensure sustainability.

Skills will have to be developed to achieve all of this – to negotiate trade-offs and define ecological bottom lines and sustainability, based on a fuller understanding of both ecosystem and societal resilience. Increased cooperation between ecologists, water managers and social scientists

is needed to make clear the water linkages connecting terrestrial ecosystems, human communities and resilience. Concepts of vulnerability and resilience, coupled with a better appreciation of the crucial and central role of both terrestrial and aquatic ecosystems to humanity, must be understood by all parties – technical specialists and other stakeholders alike, including the water consumer base. The ways in which these skills can be better developed are covered in **Chapter 13**. The subsequent attitude and behaviour changes that would follow such improved understanding would go a long way to furthering social, economic and environmental sustainability and enhance the effectiveness of IWRM (GWP, 2003). While IWRM is undoubtedly the essential approach to effective and optimum water management, implementing it can present its own challenges.

Global development must be equitable and inclusive...

Fishermen fixing a net, India

There are more than 1 million deaths each year from malaria, with between 300 and 500 million cases in total...

Part 4. Challenges for Well-being and Development

Balancing the increasing competition among the diverse and different water-using sectors – irrigation, municipalities, industry, environmental flows – including the demands of upstream and downstream users, is a challenge facing watersheds worldwide. Decisions on water allocations have to be made at different scales and a wide range of transboundary and other regulatory instruments and water and pollution management techniques need to be developed and shared: local scale in particular catchments, full-river basin scale where the geographical extent of the basin may well encompass several domestic political boundaries, national level to ensure that the potential of water to stimulate socio-economic development is realized, and at international level in the case of transboundary waters (Stockholm, 2002).

4a. Water and health: Reducing infectious diseases

In terms of threats to human security from premature death, infectious disease ranks at number one, being responsible for 26 percent of all premature deaths. The top five communicable diseases worldwide in 2002, in rank order, in terms of early mortality, were:

- Respiratory infections caused around 4 million deaths

- HIV/AIDs with some 2.8 million deaths

- Diarrhoea causing 1.8 million deaths

- Tuberculosis, causing 1.6 million deaths

- Malaria, accounting for 1.3 million deaths.

Although not all of these can be directly related to water issues, they are closely connected with water supply, sanitation and habitat challenges, which as noted earlier, the Commission on Sustainable Development wishes to be considered together in future. As discussed in **Chapter 6**, it is increasingly widely recognized that diarrhoea, the leading cause of deaths in children of developing countries, could be controlled by improving access to safe drinking water and sanitation. There are more than 1 million deaths each year from malaria, with between 300 and 500 million cases in total, affecting populations in tropical regions of Africa, Asia and the Americas. Approximately 40 percent of the total world population is at risk of infection,

particularly pregnant women, unborn babies and children under 5 (Concern/Guardian, 2005).

Every year, around 10.8 million children die before their fifth birthday and, of these, 4 million die before they reach 1 month old. Some 92 percent of all deaths of under-5 children occur in just forty-two lower-income countries. It is estimated that 63 percent of all deaths of under-5 children can be prevented using current knowledge and methods including oral rehydration for diarrhoea, antibiotics for pneumonia, mosquito nets and anti-malaria drugs for malaria, better water supply sanitation and domestic hygiene. The links between childhood sickness and death and inadequate water and sanitation availability, unsatisfactory hygiene, lack of better water management practices are clear.

4b. Water and food: Facing growing demand and competition

Over the last fifty years, agriculture has been facing the great challenge of providing food to a global population doubling in size. This has resulted in water withdrawals that largely exceed those of any other sector. However, 13 percent of the human population is still underfed, the majority living in rural areas of developing countries – the countries most likely to support the biggest share of demographic growth in the years to come. Increased competition for water and the need to integrate environmental issues threatens water for food and is an issue that cannot be tackled through a narrow sectoral approach. New forms of water management in agriculture, including irrigation management, must continue to be explored and implemented with focus on livelihoods as well as on productivity, as discussed in **Chapter 7**.

In agriculture, changing cropping patterns to lower water use crops, precision application of irrigation water at the critical point of the crop growing cycle, and better water delivery and distribution systems can collectively produce improved water productivity. Improved irrigation water application technologies enable a more precise and timely application of water at critical points in the plant life-cycle, improving irrigation water productivity and efficiency. These technologies are well known and widely applied by better resourced farmers. However, innovations in micro-irrigation techniques mean that these can now be made affordable to poor farmers. Very low-cost drip irrigation systems and treadle pumps allied to low-cost double walled plastic water storage tanks, which rest in easily dug earthen trenches, have now been developed. This combination costs one-fifth of the price of conventional ferro-cement tanks. These new methods, when applied to the increasingly smaller plots, enable them to produce a range of higher-value cash crops and significantly improve their income prospect (Polak, 2004). In the growing competition for increasingly scarce water resources, policy-makers are looking to the value generated by water use. As reforms to agriculture are forced to compete with industry and service sector developments, crop production has shifted from low-value staple crops to high-value horticultural crops.

4c. Water for industry and energy: Aiming for sustainability

Though not explicitly included in the MDGs, industry and energy are both water-related issues vital to socio-economic development. The World Summit on Sustainable Development held in Johannesburg in 2002 proposed a Plan of Implementation that makes a strong link between the related goals of industrial development, improved access to energy services, and poverty eradication and sustainable natural resource management. Industry is a significant engine of growth, accelerating particularly in highly indebted countries, and making up the bulk of the economy in East Asia and the Pacific. But to be sustainable, economic development also needs an adequate and steady supply of energy, for which water is a key resource – whether through hydropower, nuclear-based energy generation, coal slurry technology, small-scale hydels or other sources, as discussed in **Chapter 8**.

Both industrial growth and increased energy production are demanding an enlarged share of water resources. Currently, the total water withdrawal by industry is much greater than the water actually consumed. Industries have a dramatic effect on the state of the world's freshwater resources, both by the quantity of water they consume and their potential to pollute the water environment by their discharge. Industrial discharge returned without treatment has high organic content, leading to rapid growth of algae, bacteria and slime, oxygen-depleted water, and thermal pollution. Discharge can affect a relatively large volume of water and have numerous impacts on human health. Polluted water may affect fishing grounds, irrigated lands, municipalities located downstream and even bathing water. It is also recognized that water pollution can have significant transboundary effects.

Energy-intensive water delivery systems can also have dire impacts for areas with scarce water and energy resources. Some sources of water supply are more energy intensive than others, such as thermal desalination, which requires more energy than wastewater recycling. Pumping water is a major cost element everywhere, and consumes significant energy resources worldwide. Reducing the inefficiencies that occur in energy production (e.g. during electricity generation, transmissions, distribution and usage) will reduce electric power requirements, leading to higher water savings, as discussed in **Chapter 9**. In addition, changing environmental contexts demand that a greater investment be made in renewable energies.

Improving environmental governance is central to limiting industrial pollution and reducing the inefficiencies that occur within energy production and distribution systems. In industry, governance initiatives now exist at international and national levels, as well as at the level of industrial sectors and individual companies. The Basel Convention on the Control of Transboundary Movements of

Polluted water may affect fishing grounds, irrigated lands, municipalities located downstream and even bathing water

This self-propelled, centre-pivot irrigation machine drills for water 30 to 400 m below the surface in Ma'an, Jordan. A pivoting pipeline with sprinklers irrigates 78 hectares of land. Production of 1 ton of grain requires about 1,000 tons of water. At the current rate of use in Jordan, subterranean water reserves could dry up before 2010

BOX 1.8: INLAND WATER TRANSPORT: A TOOL FOR PROMOTING ECONOMIC AND SUSTAINABLE DEVELOPMENT

A well-functioning transport system is crucial to the development of a strong and vibrant economy, enabling ready access to both raw materials and markets. In the twentieth century, road transport has come to be regarded as a particularly effective mode of transport. Yet inland water transport (IWT), using rivers, canals and lakes, has been historically important in economic development worldwide and offers important environmental, economic and other practical benefits that make it one of the most advantageous modes of transport even today.

In Europe, where more than 35,000 kilometres (km) of waterways connect hundreds of cities and industrial regions, 125 billion ton-kilometres of freight were transported by inland waterways in 2003 alone. In the United States, where more than 25,000 km of inland, coastal and intracoastal waterways exist, water traffic represented 656 million tons in 2000. In fact, many well-known and prosperous cities, such as Paris, San Francisco, Rotterdam, Shanghai and London, developed as a result of their position as water transport hubs. Building on the natural advantages of geographic location and navigable waterways, developing countries may find IWT a cost-effective and sustainable way of developing transport infrastructure where constraints of land availability and cost inhibit the expansion of rail and road infrastructure.

The environmental benefits of water transport, when compared to other modes, are also apparent. Whereas road transport consumes large amounts of non-renewable energy and contributes significantly to air pollution, water transport is more energy efficient and environment-friendly. Energy consumption for water transport per ton-kilometre is half of that of rail and one-sixth that of road, while carbon dioxide emissions from IWT are approximately one-thirteenth of those of road freight transport. Additionally, while vehicular transport contributes to noise pollution and exacerbates land congestion and road accidents, water

transport can relieve pressure on overloaded road systems in densely populated regions, thus reducing both traffic accidents and noise levels.

The financial and environmental advantages of water transport make it a smart investment for many regions. The Asian and Pacific region, for example, has at least 280,000 km of navigable waterways, more than 340,000 large vessels and millions of traditional craft, carrying over 1 billion tons of cargo and a half billion passengers each year. In some countries, such as China, water transport is already well developed. China has more than 5,600 navigable rivers, with a total navigable length of 119,000 km, including the Yangtze River, which alone comprises 50 percent of the national total. The annual volume of IWT freight in China was about 690 million tons in 2000, the large majority of which moved along the Yangtze. Despite possessing vast IWT potential, other countries in the region have been slow in putting it to use. Such is the case of India, which has an extensive river system, including 14,500 km of navigable waterways, only 37 percent of which is currently utilized for motorized transport. In 2001/02, IWT accounted for a mere 0.1 percent of India's total domestic surface transport (compared to 68 percent for road and 30 percent for rail).

IWT is important in other regions as well. In Latin America, the Paraguay-Parana Waterway Project was proposed in the late 1980s as a means to promote the economic development and integration of countries within the La Plata Basin. The Paraguay and Paraná rivers are natural north-south transport corridors, extending through four countries (Argentina, Bolivia, Brazil and Paraguay) and accessible to a fifth (Uruguay), thus connecting the heart of South America to the Atlantic Ocean. Designed to expand navigation possibilities, the Project is meant to reduce the cost of transport within the region, improve links between commercial centres and provide an outlet to the sea for the

landlocked countries of Bolivia and Paraguay. Though the project is still in the planning stages, its legal framework was approved by all associated riparian countries in 1996.

Although IWT development is picking up speed on a global scale (particularly in Asia), vast lengths of navigable water remain underdeveloped. In some areas, the potential of IWT may be greatly limited by natural constraints. Where long and harsh dry seasons diminish water levels, IWT may be considered an unreliable mode of transport. In Bangladesh, for example, where an inland navigation network spans 24,000 km, the dry season (from December to May) limits access to the system to vessels of 100 deadweight tons or below. Throughout Africa, seasonal climate variation and unpredictable water depths limit the number of inland water bodies that are navigable. There are only three rivers classified as international waterways in Africa: the Congo, Nile and Zambezi rivers. Hydraulic work could, however, increase the number of potentially navigable rivers on the continent. Nigeria, for example, is estimated to have over 3,000 km of potentially navigable inland waterways were they to be developed.

Factors limiting the development of IWT systems include the poor recognition of IWT potential, lack of technology, limited financial resources, exclusion of IWT from Integrated Water Resources Management (IWRM) planning, insufficient institutional capacity, inadequate legal instruments, an absence of policies, limited information sharing and poor public awareness.

Despite the numerous advantages of water transport, it is not without negative environmental impacts. Hydraulic work undertaken to make rivers more navigable (e.g. constructing dikes, straightening canals, destroying rapids, dredging and sometimes even adding artificial waterways), can prove harmful to an ecosystem's balance and local biodiversity.

BOX 1.8: *continued*

For example, over the past 150 years, regulation work on the Danube River Basin has significantly damaged historical floodplains (see **Chapter 14**). Large dikes and disconnected meanders in this area were found to suppress the linkage between surface water and groundwater, reducing the recharge of aquifers important for local drinking water supplies. Concern has also been raised over the impact of navigation channel repairs on wetlands and biodiversity in the Niger Delta.

In order to promote efficient and sustainable waterway projects, it is essential that IWT be integrated into overall IWRM plans. Although inland transport may carry some environmental risks, overall IWT accounts for comparatively less environmental externalities than other modes of transport. In addition, Environmental Impact Assessments of IWT projects can greatly help to identify areas at risk so that mitigation measures can be undertaken. There is an urgent need for research as we need to find more environmentally friendly means to maintain the navigability of watercourses in addition to learning more about how the world's water resources may be impacted by new environmental challenges, such as climate change. Overall, IWT remains one of the most economically and environmentally sustainable modes of transportation, and when integrated with other transport development in the context of IWRM planning, it can help to re-establish a balance between the various modes of transport, making the transportation industry as a whole more responsive to broader societal goals.

Sources: EUROPA, 2005; US Army Corps of Engineers, 2004; ADB, 2003; River Bureau of the Ministry of Land, Infrastructure and Transport of Japan, 2003; UNECA, 2002; UNESCAP, 2003 and 2004.

Hazardous Wastes and their Disposal is a recent example of an international mechanism aimed at addressing issues of waste generation, movement, management and disposal. Yet, as discussed in **Chapter 8**, to be truly effective, efforts to curb industrial water pollution require that such agreements be translated into action through national policies and at the industry/sectoral level. Stepped water tariffs, subsidies for industries implementing innovative environmental technologies, and financial and advisory support for new research are just a few examples of measures that can be taken.

Environmental concerns, particularly over climate change and nuclear waste disposal, as well as safety and security of supply, are prompting governments to introduce policies aimed at accelerating the use of renewable energy and Combined Heat and Power (CHP). Total worldwide investment in renewable energy rose from $6 billion in 1995 to approximately $22 billion in 2003, and is increasing rapidly. This trend of investment in renewable energies can not only help to increase the production of more efficient energy, but is critical to our ability to face future challenges posed by environmental uncertainties.

Part 5. Management Responses and Stewardship

Over the last century, there has been a significant rise in water-related disasters, affecting an increasing number of people, particularly those living in developing countries. Resulting damages to property and losses of life and livelihoods compromise the gains of development.

5a. Managing risks: Dealing with increasing frequency

It is quite possible that climate change will not only cause rises in global temperature, but also lead to changes in the frequency of floods, droughts, storms, fires – bringing about more and greater unexpected effects. One example is the headwaters of many large Asian rivers originating in the Himalayas. The large amount of freshwater traditionally stored in the glaciers there is being reduced as glaciers shrink. This will likely alter seasonal runoff patterns, increasing flood extremes and affecting availability of critical irrigation waters (Lenton, 2003).

Over the past decade, progress has been made in risk management, thanks to scientific advancements and the recognition of the various dimensions of risk, including political, social and cultural issues. As discussed in **Chapter 10**, however, technical and organizational constraints remain high and slow down the design and implementation of efficient risk reduction.

5b. Sharing water: Facing increased competition

Over 260 of the international or transboundary basins, with over 50 percent of Earth's surface and 40 percent of the global population, are shared by one or more countries worldwide. Opinions vary as to the likelihood of cooperation rather than conflict, but experience has shown that cooperation is more likely, despite that fact that wars over natural resources – oil, minerals, metals, diamonds, timber, water, and so on – have been a feature of almost a quarter of recent wars. Given the importance of international or transboundary water resources and their potential for cooperation in development, plus the need to avoid conflict, **Chapter 11** explores in more detail the cooperation possibilities, which can produce benefits way beyond the water sector as covered in **Chapter 2**.

Cooperation may arise spontaneously from perceived threats or opportunities by riparian states, or from without by concerned intermediaries, for example, multilateral agencies and respected statesmen. Threats may be associated with increasing extremes of climate variability, while opportunities could relate to the socio-economic development potential of a cooperative approach. Once the principle of cooperation is established, trust-building measures such as cooperative research, joint data collection, knowledge and information sharing are important for building the basis for collaborative planning and management.

There are compelling arguments that riparian countries collect compatible data, which is analysed and shared to facilitate efficient use of shared water resources (Moench, 2003). Experience has shown that, with shared water resources, the greater the capacity to collect, process, interpret and accept data, the greater the range of policy options that can be generated – and the less likelihood there is of disagreement and conflict between riparian users. Data gathering may be best delegated to a trusted neutral organization, as suggested above. Alternatively, riparian states might set up data collection arrangements under their joint control. It is now widely accepted that jointly controlled data generation and analysis is an essential early step in building long-term riparian cooperation over shared water resources.

Perceptions of fairness for all parties are essential and it is important to recognize different riparian views of the benefits on offer. Trust is a *sine qua non* – riparian states need to move from their pre-cooperation positions over the water itself to fostering an interest in the benefits to be gained from cooperation. Unequal capacity between riparian states complicates negotiations. Self-financing institutions for basin cooperation are vital to pave the way for needed investment. The trade-offs between environmental, political and economic challenges have to be balanced. This is no easy task, and the key to resolving it lies in the choice of process and subsequent commitment to it. A small but growing body of needed expertise is now becoming available and the prospects of furthering the benefits-based approach set out above are improving (Grey, 2002).

5c. Enhancing knowledge and capacity: At all levels
Deficiencies in some countries in data collection and information sharing pose several challenges to water resources management. There is a serious dearth of detailed hydrometeorological data. Many

hydrometeorological measuring stations are degraded as a result of both lack of maintenance and skilled operating staff. Many of the instruments are out of date and poorly calibrated. Network characteristics and measured variables vary from country to country. Systems for storing, processing and managing data for water resources are often rudimentary. To try to cope with these deficiencies, a growing range of satellite-based remote sensing systems are under development, as explored in **Chapters 4** and **13,** although without a sound and accessible knowledge base and capacity-building efforts, these systems will not benefit the areas most in need of attention. Indeed, there is the matter of who collects the data, how it is to be interpreted, and who has access to it, other than the collection agency, and how people in lower-income countries can access the vast and growing literature and knowledge on solving problems that is accumulating in agencies, archives and organizations around the world.

Hydrological conditions are highly variable from season to season and year to year. Since much hydrometeorological data is in such short supply, the combination of these two factors means that the nature of many emerging water problems and the possible responses are often uncertain. Much hydrometeorological information in some parts of the world is held by hierarchically structured government departments functioning in specific water-using sectors, for example irrigation or water supply. The way information is generated, analysed, controlled and disseminated sets up the context in which perspectives are formed and solutions generated. Different organizations collect and analyse the information that supports their paradigm relating to the part of the water sector in which they operate. Thus, there are arguments to ensure that primary baseline information should be produced by organizations that are institutionally separated from executive functions in water.

As discussed earlier, there is an increasing trend towards delegation of water management responsibility to local authorities and water user groups. They must, therefore, have access to baseline data on water needs and availability. But data itself is useless without the capacity to interpret and analyse it meaningfully. A vast amount

There is an increasing trend towards delegation of water management responsibility to local authorities and water user groups

BOX 1.9: HURRICANE KATRINA

In August 2005, Katrina, a category-5 hurricane, ravaged the coastal regions of Florida, Alabama, Mississippi and Louisiana (especially in greater New Orleans). At least 1,336 lives were lost and over 4,000 people are still missing. Economic losses and infrastructure damages will exceed US$75 billion, 200,000 homes and businesses were severely damaged, and over 400,000 people were displaced. It was the costliest and one of the five deadliest hurricanes ever to hit the US.

The devastating effects of Katrina resulted from several unfortunate events: the inability or failure of many people to evacuate; the twenty-eight breaches of the New Orleans levees, caused by under-engineering and insufficient maintenance; and the slow and inadequate emergency response.

The effects of Hurricane Katrina were not unexpected. Since before 2000, hurricane and natural disaster management specialists, the local and national press, and the Federal Emergency Management Agency (FEMA) had all voiced concern. In July 2003, FEMA staged a five-day exercise called Hurricane Pam that simulated 50 cm of rain and winds of over 190 km per hour in the New Orleans area. In 2003, a study conducted by Dr. van Heeden of Louisiana State University's Center for the Study of Public Health Impacts of Hurricanes showed that in the event of a major hurricane, about 69 percent of New Orleans residents would evacuate the area; 10 percent would leave their homes but not the

area; and 21 percent of New Orleans residents would stay in their homes (57,000 families did not own a motor vehicle). This is likely linked to the fact that as of 1999, about 28 percent of residents were living below the poverty line. Furthermore in Louisiana, the majority of all hurricane fatalities were over the age of 60.

Despite these concerns and efforts, the US was not prepared for such a disaster, particularly with regard to the local population, and especially the poor. The Director of the Institute for Crisis, Disaster, and Risk Management, Dr. John Harrald, testified in September 2005 to the congressional Committee on Government Reform that the US Government had 'confused preparing the government with preparing the society at large'. Many people were unable to evacuate, because they had nowhere to go and no means to get there.

There was also a breakdown in links between federal, state and local response efforts. The Department of Homeland Security's 2004 National Response Plan (NRP) stated in its Catastrophic Incident Annex that federal response activities should begin even before a detailed situation and needs assessment is available, which 'may require mobilizing and deploying assets before they are requested via normal NRP protocols'.

However, on 29 August, the day Katrina hit New Orleans and after a state of emergency had

been declared, the director of FEMA 'urged all fire and emergency services departments not to respond to counties and states affected by Hurricane Katrina without being requested and lawfully dispatched by state and local authorities'. A lack of coordination and communication between different levels of government led to the exacerbation of an already serious situation. According to Dr. Harrald's testimony, there was 'a failure to act creatively and quickly'. As a result, those stranded in New Orleans were without food, water, sanitation and police protection for several days.

A more integrated approach to disaster prevention and management, in which federal, state and local levels of government are better coordinated, is needed. According to Dr. van Heeden, several medium- and long-term changes could help prevent or mitigate the effects of disasters like Katrina: in the short term, the revision of the levees' structural design and implementation, as well as the installation of floodgates to stop the water surge from moving up the canals to Lake Pontchatrain; and in the long term, the protection of wetlands, which act as a buffer against tropical storms and hurricanes, protecting coastal settlements.

The openness of the American media has allowed much information to be gleaned. The technological and scientific task of prediction was successful: computer models of the hurricane were able to predict the hurricane's path fairly accurately. However, the social aspect of disaster management needs to be better addressed. This would entail a realistic approach to warning and evacuation, taking into consideration the inability of many people to evacuate because of poverty, age or physical handicaps.

A house totally destroyed by Hurricane Katrina on Lake Pontchatrain near New Orleans

Sources: AGI, 2005; FEMA, 2006, 2005, 2004; Department of Homeland Security, 2004; Harrald, 2005; Knabb et al., 2005; State of Louisiana, 2005; Times Picayune, 2002; US Census Bureau, 2005; US Senate Committee on Homeland Security and Government Affairs, 2005; van Heeden, 2004, 2004/5; White House, 2005.

BOX 1.10: THE TSUNAMI DISASTER AND FUTURE PREPAREDNESS

The earthquake and tsunami of 26 December 2004 that swept through the Indian Ocean coastal states killed more than 283,100 people, according to the United States Geological Survey, making it one of the deadliest disasters in modern history. Beyond the loss of human lives, the tsunami also destroyed livelihoods, traumatized whole populations, and severely damaged habitats and freshwater resources.

UNESCO is actively participating in the international effort to assess the tsunami impact and identify priority needs in the recovery and reconstruction process. Particular focus is being made on environmental, cultural as well as educational damage assessment and rehabilitation. Furthermore, through its Intergovernmental Oceanographic Commission (IOC), UNESCO received a mandate to help Member States of the Indian Ocean rim establish a Tsunami Early Warning System. UNESCO's immediate response included an interim tsunami advisory information system, established as of 1 April 2005 under the aegis of UNESCO/IOC in cooperation with the Pacific Tsunami Warning Center (PTWC) in the US and the Japan Meteorological Agency (JMA).

Between May and September 2005, national assessments of sixteen countries in the Indian Ocean were conducted to identify capacity-building needs and support requirements for developing an Indian Ocean Tsunami Warning System (IOTWS). The overall regional summary indicates that:

- Most countries have established or strengthened their disaster management laws, national platforms and national and local coordination mechanisms to guide all-hazard disaster risk reduction and to establish clearer responsibilities for end-to-end early warning systems. Not all have specifically addressed the tsunami coordination aspect.

- All participating countries received international tsunami warnings from the Pacific Tsunami Warning Center (PTWC) and the Japan Meteorological Agency (JMA) except Somalia, and most countries received these warnings at facilities with back-up systems for receiving warning messages that operate around the clock. Few countries operate a national tsunami warning centre or have the capacity to receive or provide real-time seismic or sea-level data.

- Few participating countries have developed tsunami emergency and evacuation plans and signage or tested response procedures for tsunamis or earthquakes. Much of the information and data needed to develop these plans, such as post-event surveys, inundation modelling, and tsunami hazard and vulnerability assessment, has yet to be collected.

- Many participating countries have assessed local government capacity for disaster preparedness and emergency response but not community preparedness. Community education and outreach programmes are being developed but have not been implemented in most participating countries.

- Most countries have made progress in developing policies, assessing technological needs, and establishing coordination mechanisms at a national level for tsunami warning and mitigation. Local planning and preparedness activities are being carried out first in selected target areas, or cities and towns, rather than as comprehensive national programmes.

A recent evaluation by the United Nations International Strategy for Disaster Reduction (ISDR) and the IOC regarding the strengthening of early warning systems in countries affected by the 26 December 2004 tsunami evaluated that excellent progress has been made to establish the core technical elements of a regional tsunami early warning system (this system is on track for initial completion by July 2006) and that significant progress in awareness raising and capacity-building has been achieved, but much work remains to build the long-term capacities of countries for effective early warning and risk management.

IOC is now coordinating the creation of regional early warning systems for tsunamis and other coastal hazards in all regions of the world – the Indian Ocean, the Pacific Ocean, the North East Atlantic, the Mediterranean and connected seas, and the Caribbean Sea and adjacent region. Intergovernmental Coordination Groups are working in the Indian Ocean (ICG/IOTWS), the Caribbean and Adjacent Regions (ICG/CARTWS), the North Eastern Atlantic, and the Mediterranean and connected seas (ICG/NEAMTWS). These are joining the Pacific system (ICG/PTWS, formerly ITSU) established by IOC in 1965.

Source: A contribution of Intergovernmental Oceanographic Commission of UNESCO.

Mullaitivu, a town in Northeastern Sri Lanka ravaged by the tsunami of 26 December 2004

Charging is but one policy option, full-cost subsidy is another

of information is now available on solutions to many of the world's water problems. It seems increasingly that, someone, somewhere, has an answer with a potentially wide application. However, this information is not well analysed, collated and disseminated. It has been alleged that twenty years is the average period of time for a new idea to enter mainstream consciousness and understanding – this certainly seems true of many techniques developed for different parts of the water sector. Much more could usefully be done to collect and collate good international practice on key aspects of water use and management and disseminate it more widely, especially in lower-income countries. **Chapter 13** explores many of these issues and challenges. As the problem of data scarcity is widespread across all sectors of water, all the chapters point to the additional challenges that this poses, not only to water managers but also to international, national and local monitoring and policy-making.

5d. Valuing and charging for water: Market vs. non-market values

The water sector requires reform if the MDGs are to be met. Policy-makers must reform institutions, reformulate water policies, and initiate new ways of organizing water supply and sanitation. To select among the different policy options and programme initiatives on the table, policy-makers must have a means of determining which are likely to bring results that best meet society's goals, recognizing the many values of water.

The common method used for public policy analysis, that is, to differentiate the various options, is *benefit-cost analysis* (BCA). This method totals up the pluses and minuses of each option so that the net benefits of the different options may be considered, and the trade-offs (the different level of advantages and disadvantages associated with various sub-elements of the programmes/projects) between the different options may be clearly seen.

In preparing a BCA one needs to make an *economic valuation* (calculate in monetary terms) of the benefits and costs associated with each option. A problem emerges when the benefits or costs associated with a particular activity are not subject to any market transaction, which in turn means that

there is no market price by which to measure it. In such cases, economists have developed a variety of *non-market valuation techniques* that can be used to estimate surrogate (or shadow) prices that may be used to value the goods and services in question.

Many of the goods or services that must be evaluated in assessing alternative water policies are secondary effects or externalities of the policy or project in question and include social and environmental impacts. In certain cases, some members of the community may be resistant to (even offended by) the attempts to put a monetary value on certain social or environmental effects. In such cases, BCA must be complemented by public discourse open to all stakeholders and political negotiations in order to reach consensus on the most suitable policy/programme.

Charging for water is but one policy option (full-cost subsidy is another). It serves several objectives, including cost recovery, revenue raising, and demand management, all of which contribute to the ultimate objective of sustainable utilization of water resources respecting the societal principles of social equity, environmental preservation and economic efficiency. Determining the tariff structure – form and level – is essentially a political decision, which may also draw on the techniques of economic valuation and BCA to elucidate the net result of various options. Benefit capture analysis, that is, more focused attention on *who benefits* and *who bears the costs* is increasingly applied in order to understand more clearly the distributional aspects as well as the financial implications of various policy options.

Private-public partnerships, payment for environmental services and trade policies revised to reflect the concept 'virtual water' are all policy responses that recognize directly or indirectly the increasing value of water, and in which the valuation techniques described above would have been employed to determine their suitability to a particular situation. **Chapter 12** examines all the issues related to valuing and charging for water in greater detail.

Metering and charging for water consumption contribute to the sustainable utilization of water resources

BOX 1.11: POLICY-MAKING AND SCIENTIFIC INTEGRITY

Responsible water resources management in particular and good governance in general rely on sound policy decisions, which require the objective collection and analysis of data and information. As scientific knowledge becomes simultaneously wider in breadth and more specialized in depth, governments, and the people these governments represent, rely increasingly on the expert knowledge of scientists.

Today, however, the world is seeing a strong politicization of natural and social science, which is detrimental to both good governance and scientific inquiry. Political interference manifests itself through corruption; conflicts of interest; cronyism; the political vetting of scientific appointments; and governmental censorship, suppression and distortion of scientific findings.

These problems manifest themselves to varying degrees in many countries (both developed and developing) and can be found across the spectrum of scientific endeavour, including climate change, AIDS prevention, agricultural science, reproductive health, environmental protection, military intelligence, etc.

Due to the rapid pace of scientific advancement, a layman's education is no longer sufficient to evaluate many aspects of public policy, which means that the general public is increasingly dependent on the integrity of scientists and the institutions for which these scientists work.

Transparency and scientific integrity are critical for good, democratic governance. Policy-makers and the electorate need accessible information in order to make informed decisions about public

policy. Scientists, both in the private and public sectors, must be independent and not subject to political, financial or physical retribution in response to scientific findings that do not fall in line with a government's or corporation's policies and ideology. And in the case of publicly funded research, findings should be transparent and not subject to political manipulation or suppression. Scientists must maintain their integrity by resisting pressure from corporations, governments or other interested parties to compromise their research.

For more information, see UNESCO's Ethics of Science and Technology Programme (www.unesco.org/shs/est); the Union of Concerned Scientists (http://www.ucsusa.org/) and the International Council for Science (www.icsu.org).

Part 6. Water and Global Targets: Where Do We Stand?

Within the UN system, a clear wish has emerged to take a lead role in finding ways to share the world's available resources more equitably. Here we review some of the global targets involved in the UN's aspirations for poverty relief as they relate to water and WWAP's mandate, as the flagship programme of UN-Water – to undertake and report on assessment processes and refine the thinking behind better-adapted monitoring tools and indicator rationale for the water sector.

6a. The Millennium Development Goals (MDGs)

Long experience has shown that setting targets is vitally important for focusing attention and providing incentives to mobilize action on key issues of development. Recognizing the need to speed up poverty alleviation and socio-economic development, the 2000 UN General Assembly Millennium Meeting established eight Millennium Development Goals (MDGs), with targets, to be achieved by 2015, from a baseline of 1990 and with a major review in 2005. The role played by water in achieving the goals is summarized in **Box 1.12**.

The major UN conferences and other international water meetings (e.g. the World Water Forums, see WWDR1, 2003) have a history of global target setting.[5] However, all too frequently, these have not included detailed enough implementation plans or the necessary financial resources. As a result, although good progress was made on some early targets, they were rarely met in full. Global targets are just that—global ambitions that can only be met via the aggregate of local actions in communities worldwide. Without local commitment and the needed resources, targets will never be met in full.

5. http://www.unesco.org/
water/wwap/index.shtml

BOX 1.12 WATER AND THE MILLENNIUM DEVELOPMENT GOALS

GOAL 1. ERADICATE EXTREME POVERTY AND HUNGER*

Water is a factor of production in virtually all enterprise, including agriculture, industry and the services sector. Improved nutrition and food security reduces susceptibility to diseases, including HIV/AIDS, malaria among others. Access to electricity is key to improving quality of life in the modern age. Competition between the various sectors must be balanced by policies that recognize the ability and responsibility of all sectors to address the issues of poverty and hunger.

Targets:
- Halve, between 1990 and 2015, the proportion of people whose income is less than $1 a day
- Halve, between 1990 and 2015, the proportion of people who suffer from hunger

WWDR2 Water-related Indicators:
- Percentage of undernourished people
- Percentage of poor people living in rural areas
- Relative importance of agriculture
- Irrigated land as percentage of cultivated land
- Relative importance of agriculture water withdrawals in water balance
- Extent of land salinized by irrigation
- Importance of groundwater in irrigation
- Dietary Energy Supply (DES)

See Chapter 7: *Water for Food, Agriculture and Rural Livelihoods*

- Trends in industrial water use
- Water use by sector
- Organic pollution emissions by industrial sector
- Industrial water productivity
- Trends in ISO 14001 certification, 1997-2002
- Access to electricity and domestic use
- Electricity generation by fuel, 1971-2001
- Capability for hydropower generation, 2002
- Total primary energy supply by fuel
- Carbon intensity of electricity production, 2002
- Volume of desalinated water produced

See Chapter 8: *Water and Industry* and Chapter 9: *Water and Energy*

GOAL 2. ACHIEVE UNIVERSAL PRIMARY EDUCATION

Promotion of a healthy school environment is an essential element of ensuring universal access to education, and school enrolment, attendance, retention and performance are improved; teacher placement is improved. In this respect access to adequate drinking water and sanitation is key.

Target:
- Ensure that, by 2015, children everywhere, boys and girls alike, will be able to complete a full course of primary schooling

WWDR2 Water-related Indicator:
- Knowledge Index

See Chapter 13: *Enhancing Knowledge and Capacity*

GOAL 3. PROMOTE GENDER EQUALITY AND EMPOWER WOMEN

Educating women and girls will permit them to fulfil their potential as full partners in the development effort.

Target:
- Eliminate gender disparity in primary and secondary education, preferably by 2015 and in all levels of education no later than 2015

WWDR2 Water-related Indicator:
- Access to information, participation and justice in water decisions

See Chapter 2: *Challenges of Governance*

GOAL 4. REDUCE CHILD MORTALITY

Improvements in access to safe drinking water and adequate sanitation will help prevent diarrhoea, and lay a foundation for the control of soil-transmitted helminths and schistosomiasis among other pathogens.

Target:
- Reduce by two-thirds, between 1990 and 2015, the under-five mortality rate

WWDR2 Water-related Indicators:
- Mortality in children <5 yrs
- Prevalence of underweight children <5 yrs
- Prevalence of stunting among children <5 yrs

See Chapter 6: *Protecting and Promoting Human Health*

GOAL 5. IMPROVE MATERNAL HEALTH

Improved health and nutrition reduce susceptibility to anaemia and other conditions that affect maternal mortality. Sufficient quantities of clean water for washing pre-and-post birth cut down on life-threatening infection.

Target:
- Reduce by three-quarters, between 1990 and 2015, the maternal mortality rate

WWDR2 Water-related Indicator:
- DALY (Disability Adjusted Life Year)

See Chapter 6: *Protecting and Promoting Human Health*

GOAL 6. COMBAT HIV, AIDS, MALARIA AND OTHER DISEASES

Improved water supply and sanitation reduces susceptibility to/severity of HIV/AIDS and other major diseases.

Targets:
- Have halted by 2015 and begun to reverse the spread of HIV/AIDS
- Have halted by 2015 and begun to reverse the incidence of malaria and other major diseases

WWDR2 Water-related Indicator:
- DALY (Disability Adjusted Life Year)

See Chapter 6: *Protecting and Promoting Human Health*

BOX 1.12: *continued*

GOAL 7. ENSURE ENVIRONMENTAL SUSTAINABILITY

Healthy ecosystems are essential for the maintenance of biodiversity and human well-being. We depend upon them for our drinking water, food security and a wide range of environmental goods and services.

Target:
- Integrate the principles of sustainable development into country policies and programmes and reverse the loss of environmental resources

WWDR2 Water-related Indicators:
- Water Stress Index
- Groundwater development
- Precipitation annually
- TARWR volume (total annual renewable water resources)
- TARWR per capita
- Surface water (SW) as a % TARWR
- Groundwater (GW) as a % of TARWR
- Overlap % TARWR
- Inflow % TARWR
- Outflow % TARWR
- Total Use as % TARWR

See Chapter 4: *The State of the Resource*

- Fragmentation and flow regulation of rivers
- Dissolved nitrogen ($NO_3 + NO_2$)
- Trends in freshwater habitat protection
- Trends in freshwater species
- Biological Oxygen Demand (BOD)

See Chapter 5: *Coastal and Freshwater Ecosystems*

Targets:
- Halve by 2015 the proportion of people without sustainable access to safe drinking water and basic sanitation
- By 2020, to have achieved a significant improvement in the lives of at least 100 million slum dwellers

WWDR2 Water-related Indicators:
- Urban Water and Sanitation Governance Index
- Index of Performance of Water Utilities

See Chapter 3: *Water and Human Settlements in an Urbanizing World*

- Access to safe drinking water
- Access to basic sanitation

See Chapter 6: *Protecting and Promoting Human Health*

GOAL 8. DEVELOP A GLOBAL PARTNERSHIP FOR DEVELOPMENT*

Water has a range of values that must be recognized in selecting governance strategies. Valuation techniques inform decision-making for water allocation, which promote sustainable social, environmental and economic development as well as transparency and accountability in governance. Development agendas and partnerships should recognize the fundamental role that safe drinking water and basic sanitation play in economic and social development.

Targets:
- Develop further an open trading and financial system that is rule-based, predictable and non-discriminatory, includes a commitment to good governance, development and poverty reduction – nationally and internationally
- Address the special needs of landlocked and small island developing states

WWDR2 Water-related Indicators:
- Water sector share in total public spending
- Ratio of actual to desired level of public investment in water supply
- Rate of cost recovery
- Water charges as a percent of household income

See Chapter 12: *Valuing and Charging for Water*

- Water interdependency indicator
- Cooperation indicator
- Vulnerability indicator
- Fragility indicator
- Development indicator

See Chapter 11: *Sharing Water*

- Disaster Risk Index
- Risk and Policy Assessment Index
- Climate Vulnerability Index

See Chapter 10: *Managing Risks*

- Progress toward implementing IWRM

See Chapter 2: *Challenges of Governance*

* Only the most relevant targets have been listed for this goal.

Taking precautions at household level to prevent water-related disease through sound domestic hygiene practices is a challenge to the individual family

Extending the provision of water supply and sanitation facilities is a challenge to individual communities. Making more effective use of modern communications technology to speed up warnings to communities at risk of flooding and other water-related hazards must be tackled by individual governments and their agencies within particular river basins. Expanding water use productivity in both rainfed and irrigated croplands has to be done by individual farmers, irrigation managers and water user associations. Taking precautions at household level to prevent water-related disease through sound domestic hygiene practices is a challenge to the individual family. Yet it is the aggregate of all these many millions of actions which help to realize the MDGs. Governments and NGOs must help through the provision of needed resources, including education and training, to improve knowledge and skill and thus foster the development of the self-sufficiency and resilience needed to meet the MDGs.

Recognizing these problems, the UN has established a number of initiatives to help meet the MDG targets, as well as others. One such initiative is the Millennium Research Project – an independent advisory body set up by the UN Secretary General to propose the best strategies to meet the MDGs. In turn, the Project established ten Millennium Task Forces, charged with identifying what was needed to reach the MDG targets. One of these, Millennium Task Force 7, covers water supply and sanitation, and highlights from its latest findings can be found in **Chapter 6**. **Chapter 10** reviews the linkages between the Millennium Development Goals and disaster risk reduction. The UN has instigated Millennium Campaigns across the world, encouraging industrialized countries to increase support through aid, trade and debt relief. For lower-income countries, the Campaigns focus on engaging support for urgent action on the MDGs.

Progress on global goals and targets
Progress on the MDGs is monitored via the Millennium Development Goal Reports, which build on the national Human Development Reports,[6] and the UN Secretary General's reports to the General Assembly. The latest of these shows that, with only ten years to go on the MDGs, progress has been patchy and slow.

The 13th Commission on Sustainable Development (CSD) sessions in April 2005 set out a number of priority policy options designed to accelerate progress towards achieving the MDGs on water and sanitation and human settlements. It confirmed that, in future, water, sanitation and human settlements should be dealt with in an integrated manner. This is logical since the challenge for the poor is essentially one of securing acceptable habitation of which water, sanitation and shelter are key components.

It recommended also that countries should identify or establish an institutional base for sanitation and prioritize investment for it where needs are greatest and the impacts likely to be most substantial – in health centres, schools and workplaces. Both financial and human resources are required, as discussed in **Chapters 12** and **13**. Greater resources are needed for sanitation, together with more community involvement and an emphasis on low-cost technology options. Issues of strengthening national and local authority capacity to deliver and maintain water supply and sanitation systems, the contentious subject of cost recovery, approaches to the provision of wastewater systems, use of debt relief to mobilize resources for water and sanitation, and the greater use of grant aid, were also covered. A notable feature of the April meeting of the Commission was the announcement by Mikhail Gorbachev of Green Cross International, that he and others are promoting an initiative for a global convention on the right to water (Water 21, 2005).

The September 2005 meeting of the UN General Assembly was set to review progress and to agree an agenda for the next stage. This last point is crucial since it is widely agreed that 2005 is regarded as a make or break year for getting the MDG project on course to deliver the targets. Professor Jeffrey Sachs, director of the Millennium Project, has submitted to the Secretary General a plan showing how the MDGs can still be met. The plan sets out a coordinated programme of proposed investments in infrastructure, health and education in low-income countries, plus additional overseas development assistance (ODA) from the richer countries and progress in the Doha negotiations of the World Trade Organization (Sachs, 2005). Importantly, the September meeting of the UN General Assembly was able to reach agreement on

6. hdr.undp.org/

a commitment by countries to prepare, by 2006, comprehensive national development strategies designed to reach internationally agreed development goals and objectives, which include among them the MDGs.

Elsewhere, the UN Millennium Task Force 7 on Water and Sanitation has indicated that the targets for water and sanitation will not be met worldwide at present rates of progress. Good progress is being made with water supply but sanitation is lagging substantially behind. The Task Force has set out five guiding principles and ten critical actions (see **Box 1.13** below). These are vital to achieving the water and sanitation MDGs and ensuring that sound management and development of water resources are a fundamental component of the whole MDG programme. The slow progress on sanitation is of particular concern, since poor sanitation is implicated in much-water related disease, as is shown in **Chapter 6.**

6b. Indicator development and the World Water Assessment Programme (WWAP)

In order to check on progress towards meeting goals and targets, regular and reliable monitoring is required. In recognition of this and the critical role of water in poverty alleviation and socio-economic development, WWAP was established in 2000 under the auspices of the UN and charged with the responsibility to monitor and report on water around the world – its availability, condition and use, and the world's progress towards water-related targets and goals. The identification of the most crucial issues to monitor is a key part of the process as is the development of indicators relevant to the data and trends, goals and targets being monitored in areas where freshwater plays a key role.[7]

Experience in many sectors has shown that, done correctly, development and testing of indicators is a lengthy process. Indicators have to meet well-defined criteria and be selected through a carefully planned and implemented process, including stakeholder involvement. Understanding causal relationships in complex, dynamic systems requires information that is not always readily available. Indicator production can involve time-consuming collection, collation and systematization of large amounts of data. Because the same indicator may need to satisfy often conflicting but equally important social, political, financial and scientific goals and objectives, deriving indicators becomes an objective-maximization exercise constrained by available time, resources and partnership arrangements.

One critical challenge is to identify or develop indicators applicable to as many situations as possible so that cross-country and inter-regional comparisons can be made. With data gathered according to commonly agreed and standardized norms, it could be possible to derive 'lessons' that are relevant across many locations. Data plotted over time can reveal developing trends, while country-specific data collected in a common format facilitate inter-area comparison. Inter-country and inter-regional analyses illuminate success as well as stagnation and enable decision-makers to discern areas in need of attention. Selected to address the key concerns of decision-makers, indicators provide critical data for policy analysis, programme design and fiscal planning. Thus, prepared from carefully selected data, distilled into authoritative information and presented clearly and concisely in a user-friendly format, indicators play a key role in documenting global trends that are crucial to sustainable development. A balance has to be struck between the 'ideal' – indicators that are consistent with theoretical definitions – and the 'practical' or feasibly measurable variables that provide acceptable approximations to the ideal. Striking a balance is critically important in determining cost-efficient and cost-effective data collection.

Indicator development within WWAP focuses on utilizing and adapting existing knowledge, datasets and indicators to formulate and develop easy-to-use, easy-to-understand, yet robust and reliable indicators. These promote better water resource management by:

- providing a clear assessment of the state of water resources

- identifying the emergence of critical water resources issues

- monitoring progress towards achieving water policy objectives.

Data plotted over time can reveal developing trends, while country-specific data collected in a common format facilitates inter-area comparison

7. The first edition of the WWDR (WWDR1, 2003) provided an assessment of the world's progress in meeting critical water needs since the UNCED in Rio de Janeiro in 1992. WWDR1 identified challenges in eleven areas: meeting basic needs; securing the food supply; protecting ecosystems; sharing water resources; managing risks; valuing water, water governance; water use in industry and energy production; providing water for cities; and ensuring the water knowledge base.

BOX 1.13: RECOMMENDATIONS OF THE MILLENNIUM TASK FORCE ON WATER AND SANITATION

At the end of a three-year project, the UN Millennium Project Task Force on Water and Sanitation identified five guiding principles and ten critical actions essential to reaching the water and sanitation MDGs.

Five Guiding Principles
These are as follows:

1. There must be a *deliberate commitment* by donors to increase and refocus their development assistance and to target aid to the poorest countries.

2. There has to be *deliberate commitment* by middle-income country governments that do not depend on aid, to reallocate their resources so as to focus funding on their un-served poor.

3. There have to be *deliberate activities* to create support and ownership for water supply and sanitation initiatives among both women and men in poor communities.

4. There must be *deliberate recognition* that basic sanitation in particular requires community mobilization and actions that support and encourage such mobilization.

5. There must be *deliberate planning and investment* in sound water resources management and infrastructure.

Ten Critical Actions
These actions are needed not only to meet the water and sanitation targets but also to facilitate the sound management of water resources for all MDGs:

1. Governments and other stakeholders must move the sanitation crisis to the top of the agenda.

2. Countries must ensure that both policies and institutions for water supply and sanitation, and for water resources management and development, respond equally to the different roles, needs and priorities of women and men.

3. Governments and donor agencies must act together to reform investments for improved water supply, sanitation and water management.

4. Actions to meet the water and sanitation targets must focus on sustainable service delivery and not just on the construction of facilities.

5. Governments and donor agencies must ensure that local authorities and communities have the authority, resources and professional capacity required to manage water supply and sanitation service delivery.

6. Governments and utilities must ensure that those who can pay for services do pay, so that the revenues to fund operation, maintenance and service expansion are available, while at the same time ensuring that the needs of the poorest households are met.

7. Within the context of national poverty reduction strategies based on the MDGs, countries must produce coherent water resources development and management plans that will support the achievement of the MDGs.

8. Governments and their civil society and private sector partners must support a wide range of water and sanitation technologies and service levels technically, socially, environmentally and financially appropriate.

9. Institutional, financial and technological innovation must be provided in strategic areas.

10. UN agencies and Member States must ensure that the UN system and its international partners provide strong and effective support for the achievement of the water supply and sanitation targets and for water resources development and management.

In addition to these guiding principles and critical actions, the Task Force report sets out an Operational Plan specifying the steps that the key players – national and sub-national governments, donors, civil and community organizations and research institutions – need to follow in support of achieving the MDGs

Source: 'Health, Dignity and Development: What Will it Take?' – a summary of the key recommendations of the UN Millennium Task Force on Water and Sanitation's Final Report, prepared with the aid of the Stockholm International Water Institute (SIWI) www.siwi.org

On-going research in a water-related research centre, Delft, Netherlands

Indicators can be used to monitor performance and track changes, not only in the natural environment, such as the hydrological cycle, the aquatic environment, water quality, water availability and use, but also in the socio-economic and political environment of the water world – in governance, in sharing water and in water pricing and valuation, among others. Indices attempt to pull together variables or indicators from a variety of diverse elements to provide a more comprehensive assessment of a particular issue or challenge area.

WWAP seeks to make use of as many relevant existing indicators and on-going indicator development initiatives as possible. The indicator development process involves close cooperation with the members of UN-Water, interested UN member countries, NGOs and universities. This process follows the five-element DPSIR analytical framework – Driving forces, Pressures, State, Impacts and Responses – originally developed by the European Environmental Agency (see **Figure 1.2**). Relevant indicators have been selected,

carefully examined according to jointly agreed criteria and, where suitable, given support for further refinement and development. The original list of 176 indicators in the first WWDR (2003) has been refined by UN agencies and specialists from universities and NGOs and reduced to the present 63, although this number is by no means final. The process has involved WWAP working with participating countries to evaluate the original set of indicators through testing and adaptation (see **Chapter 14**) – a crucial part of the indicator development process. The goal is to develop a set of indicators that are accepted across the entire UN system.

As a result of the work done to date, WWAP has designated four categories of indicators:

- *Basic indicators*, which provide fundamental information not directly linked to policy goals (e.g. water resources, GNP and population), are well established, widely used and data is generally widely available around the world.

The goal is to develop a set of indicators that are accepted across the entire UN system

Figure 1.2: The DPSIR framework of analysis

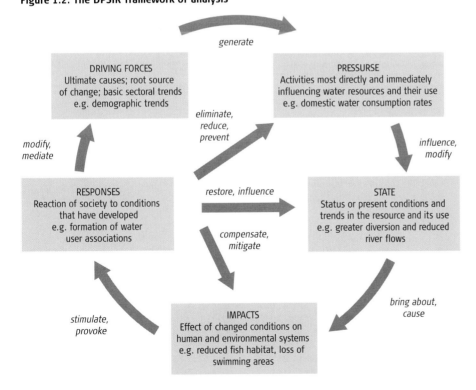

Source: Costantino et al., 2003.

...political relevance and quantifiable data are key features in indicator selection...

Table 1.1: Proposed WWDR2 indicators by challenge area

Challenge Area	Indicators[1]	DPSIR aspect[2]	Status[3]
Global	Index of non-sustainable water use	R	K
	Urban and rural population	D	B
	Relative Water Stress Index	S/P	K
	Domestic and Industrial Water Use	D	B
	Water Pollution Index	P	K
	Sediment Trapping Efficiency Index	P	K
	Climate Moisture Index (CMI)	D	K
	Water Reuse Index (WRI)	P	K
Governance	Access to information, participation and justice	R	D
	Progress toward implementing IWRM	R	K
Settlements	Index of Performance of Water Utilities	S	D
	Urban Water and Sanitation governance index	S	D
	Slum Profile in Human Settlements	P	D
Resources	Precipitation annually	D	B
	TARWR volume (total actual renewable water resources)	S	K
	TARWR per capita	S	D
	Surface water (SW) as a % of TARWR	S	D
	Groundwater development (GW % of TARWR)	S	K
	Overlap %TARWR	S	D
	Inflow as % TARWR	S	D
	Outflow as % TARWR	S	D
	Total use as % TARWR	S	D
Ecosystems	Fragmentation and flow regulation of rivers	S/I	K
	Dissolved nitrogen (NO_3+NO_2)	S	K
	Trends in freshwater habitat protection	S/R	K
	Trends in freshwater species populations	S	K
	Biological Oxygen Demand (BOD)	S	K
Health	DALY (Disability Adjusted Life Year)	I	K
	Prevalence of underweight children < 5 years old	I	D
	Prevalence of stunting in children < 5 years old	I	D
	Mortality in children < 5 years old	I	D
	Access to safe drinking water	S	K
	Access to basic sanitation	S	K
Agriculture	Percentage of undernourished people	S	K
	Percentage of poor people living in rural areas	S	K
	Relative importance of agriculture in the economy	S	K
	Irrigated land as a percentage of cultivated land	S/P	K
	Relative importance of agriculture water withdrawals in water balance	P	K
	Extent of land salinized by irrigation	S	K
	Importance of groundwater in irrigation	S/P	K
Industry	Trends in industrial water use	P	K
	Water use by sector	S	K
	Organic pollution emissions by industrial sector	I	K
	Industrial water productivity	R	K
	Trends in ISO 14001 certification, 1997-2002	R	K
Energy	Capability for hydropower generation, 2002	S	K
	Access to electricity and water for domestic use	S	K
	Electricity generation by fuel, 1971-2001	S	K
	Total primary energy supply by fuel, 2001	S	K
	Carbon intensity of electricity production, 2002	S	K
	Volume of desalinated water produced	R	K
Risk	Disaster Risk Index	S	K
	Climate Vulnerability Index (CVI)	P	K
	Risk and Policy Assessment Index	R	K

Table 1.1: *continued*

Challenge Area	Indicators[1]	DPSIR aspect[2]	Status[3]
Sharing[4]	Water interdependency indicator	S	C
	Cooperation indicator	S	C
	Vulnerability indicator	S	C
	Fragility indicator	S	C
	Development indicator	S	C
Valuing	Water sector share in total public spending	S	D
	Ratio of actual to desired level of public investment in water supply	P	D
	Rate of cost recovery	S	D
	Water charges as percent of household income	P	D
Knowledge	Knowledge Index	S	D

Notes:

1. Indicator Profile Sheet provides detailed definition and information on computation of indicator, and can be found on the accompanying CD.
2. DPSIR is the UNECE analytical framework employed in the assessment of the challenge areas, specifically **D**riving force, **P**ressure, **S**tate, **I**mpact and **R**esponse.
3. Level of development, highest to lowest: B = basic indicator; K = key indicator, for which there is an Indicator Profile Sheet and statistical data ; D = developing indicators for which there is an Indicator Profile Sheet* but not yet statistical presentation, and C = conceptual indicator for which there is conceptual discussion only.
4. A total of 25 potential indicators were proposed as the basis for discussion at the 'Indicators of Water Conflict and Cooperation Workshop' convened in Paris, November 2004, by UNESCO; here we present the few central indicators proposed for testing.

...well-designed indicators will enable complex information to be presented in a meaningful and under-standable way both for decision-makers and the public

- *Key indicators,* well defined and validated, have global coverage, and are linked directly to policy goals and convey important messages of the present Report

- *Developing indicators* are in a formative stage and may evolve into key indicators, following refinement of methodological issues or data development and testing

- *Conceptual indicators* require considerable methodological advancement, resolution of data issues and fieldwork before evolving into key indicators.

Bringing all of this together, WWAP has produced a catalogue of the indicator work done to date (see **Table 1.1** and the **CD-ROM**).

Each chapter of the present Report focuses only on the policy-relevant indicators most directly associated with their challenge area. Depending on the particular nature of the challenge areas, some chapters have more indicators than others. Since political relevance and quantifiable data are key features in indicator selection, it is easier to identify quantifiable variables in health, agriculture, industry and energy. Where institutional change is more relevant than infrastructure development – for example, in water governance, risk management, sharing and valuing water – identifying qualitative indicators that provide a meaningful measure of progress towards sectoral goals and objectives is more difficult. However, since qualitative aspects can be the deciding factor as to whether goals are achieved – despite being more challenging and costly to monitor – development work on them needs to continue. Furthermore, since indicator development is an ongoing process, not all indicators presented in the following chapters are elaborated to the same degree.

Experience has shown that the WWAP indicator development process has not only facilitated countries' utilization and testing of existing indicators but has also helped them to develop their own indicators. It is clear that both the product (the information produced) and the process (developing indicators and analysing the resulting data) contribute to building needed individual as well as organizational capacity.

As water management grows increasingly complex, well-designed indicators will enable complex information to be presented in a meaningful and understandable way both for decision-makers and the public. Indicators have an especially important role to play in IWRM, which requires information not only on water resources but also a variety of socio-economic factors and their impact on water systems. Appropriate indicators, by simplifying complex information, can provide better communication and cooperation between stakeholders. The organization of this second WWDR is such that its core chapters address the challenge areas recognized to be critical for effective IWRM.

6c. The crucial role of case studies

One of WWAP's significant features is a range of seventeen case studies in forty-one countries. Collectively, these illustrate the many different types of problems and challenges faced by policy-makers and water managers. **Chapter 14** introduces the case studies (and much more detail can be found on the WWAP website) and highlights their key features. Virtually all of the many factors influencing water resources management raised in this introductory chapter, can be observed, in one form or another, in the various case studies.

The case studies include profiles of highly developed countries, such as Japan, and some of the poorest countries in the world, such as Ethiopia. The studies also reflect the challenges in major transboundary river basins such as the Danube River Basin, the second largest in Europe, and the La Plata River Basin, the fifth largest basin in the world. Almost all the case study partners from WWDR1 have continued developing their pilot projects and have contributed to WWDR2 with national-scale case studies. In addition, five case studies have been developed in Africa to highlight the range of water challenges confronting the continent. The Mongolian case study has helped to provide a more complete picture of water problems in Asia. Efforts towards attaining global coverage will continue in subsequent editions of the WWDR as additional case study partners are sought.

The WWAP case studies clearly show that the approach towards sustainable utilization of water resources is evolving globally in the direction of IWRM. Integrating surface water and groundwater

resources within a basin and balancing competing sectoral interests with the needs of ecosystems within the integrity of the hydrological unit are becoming mainstream values that are increasingly accepted around the globe. However, political boundaries, which do not necessarily coincide with the natural borders of basins, make cooperation a necessity – cooperation not only in the international context, but also at national, sub-national and local levels. The WWAP case studies are in fact an instrument for promoting and enhancing cooperation among all relevant stakeholders, including NGOs, IGOs, research institutions, universities and water users themselves. Case study-related national meetings often bring all relevant agencies together, breaking the standard approach whereby each organization works in isolation. This process has helped WWAP's case study partners identify problems and reach consensus on the challenges to be addressed in the water sector.

The WWAP case studies also serve to benchmark the current situation and thus provide a basis on which to analyse change (positive or negative) in the water sector. They have facilitated the testing of the indicators suggested in WWDR1, which are critical to monitoring the impact of policy and programmes.

The case studies clearly highlight the diversity of circumstances and various challenges and priorities facing different regions. For example, in the Danube River Basin, homogeneous implementation of the Water Framework Directive among EU Members and non-EU Member countries is a priority, whereas in the La Plata Basin, poverty alleviation and curbing the health burden of increasing environmental pollution is high on the agenda. In Japan, thanks to the adoption of proper waste management techniques, water-borne diseases are no longer considered a major threat, whereas in Africa, water-related illnesses are common and still claim a great number of lives every year. In South Africa, due to the limited availability of surface water, hydro-power is minimal, and coal is the country's major source of energy production, whereas in the La Plata River Basin, the production of hydropower is a regional priority (over 90 percent of all energy used by Brazil comes from hydropower).

Perhaps one of the most important aspects of case studies is how they illustrate the importance of vertical integration. In other words, how the policies developed at national and sub-national levels are translated into action at the community level, and how decisions taken at the local level affect the decisions of higher management. For example, in Mongolia, lack of public involvement at the local level has limited the effectiveness of many policies and programmes. Facilitating the involvement of water users and stakeholders in managing water resources remains a challenge in many developing countries. Meanwhile, in countries such as Estonia, where the Water Framework Directive is being implemented, it is widely recognized that water management must respond to local actions and needs. A strong public information and consultation component is therefore a prerequisite for the preparation of river basin management plans.

The attainment of the MDGs remains high on the global agenda. Although global progress is being made, at present not all countries are on track. In countries such as South Africa, where the water and sanitation-related MDGs have already been attained, the governments are trying to further improve the livelihoods of their people. In other areas, for example in the Lake Titicaca Basin, many people are struggling with poverty and lack of access to safe water and sanitation. As with WWDR1, the WWDR2 continues to be an important advocacy tool for water supply and sanitation concerns. The WWAP case studies demonstrate the close link between inadequacies in the provision of water and sanitation facilities and a lack of financial and human resources. Low capacity in the water sector can be identified as the main reason behind the failure of countries to utilize water resources to contribute significantly to socio-economic development.

6d. Looking ahead: WWAP in the medium term

Harmonizing indicators at a global level requires considerable effort. Indicators developed for one location may not be applicable worldwide or suitable for scaling-up to a regional or global level. High-quality data may not be available for a theoretically relevant indicator. In fact data availability is a serious limitation for some indicators and some regions. For reasons reported in WWDR1 and re-emphasized here in **Chapter 13**, there is an ongoing deterioration in the systems of collecting

hydrometeorological data. Thus WWAP is tasked to develop simple objective indicators that can be supported by available data or data that is relatively easy to collect. WWAP will also refrain from the trend elsewhere of merging variables into ostensibly more comprehensive, yet, by their nature, more subjective and complex indices. Developing good, usable indicators is a slow, painstaking process.

The following specific areas are those on which WWAP intends to concentrate in the next few years:

Cooperation with participating countries in testing and evaluating indicators, improving data sets and developing monitoring programmes with indicators. Government officials, and all users and stakeholders need to be represented in the development process so that indicators accurately reflect experience on the ground. Indicator sets must be linked to national- and local-level strategies for water resource management and the targets and objectives from which these emanate. WWAP's intention is to work to improve the involvement of UN member countries in supplying data to the UN agencies and in working with WWAP to help the world improve the management of water resources.

Development of methods to enhance stakeholder participation at all levels in indicator development, assessment and monitoring. Emphasizing the need for stakeholder involvement in indicator development, WWAP will seek to tackle the problems of commitment to information production, reporting and application to decision-making. Encouraging countries to view indicator development within the wider context of planning and management, WWAP will seek to demonstrate how indicators are an important management tool to identify and minimize damage from environmental hazards.

Working with scientists to define and develop indicators proposed by our partner agencies and cooperating countries and identifying research needed to clarify linkages and provide the information needed to refine computer models. WWAP is aware of initiatives by a number of organizations – NGOs, institutes and universities – whose work it will endeavour to incorporate as applicable within the indicator development process.

The WWAP case studies demonstrate the close link between inadequacies in the provision of water and sanitation facilities and a lack of financial and human resources

Section 1: CHANGING CONTEXTS

Moving forward with developing geo-referenced data and mapping capability for the analysis of water-related challenges among member countries. The advent of spatially discrete, high-resolution earth system data sets is poised to enable a truly global picture of progressive changes to inland water systems to be produced, monitoring of water availability worldwide to be facilitated, and a consistent, 'political boundary-free' view of the main elements of the terrestrial water cycle to be produced. WWAP has responded by commencing assessment of the relevance of potentially useful data sets that these new systems will generate. Integrating the newly available information into its indicator development programme will provide a central challenge for WWAP, one which will require extensive investment in GIS technology and training.

Working with UN partner agencies to develop a corporate database and reconciling inconsistencies and incompatibilities of current data sets. UN-Water has identified the need to develop a user-friendly, uniform and consistent UN corporate database containing the key water indicators – a process which has begun.

It has become increasingly clear throughout the preparation of this Report that water resource issues are extremely complex and transcend the water sector. With the targets of the Millennium Development Goals facing today's water managers, it is urgent that we extend the horizon of concern to embrace the major social, cultural and economic issues that are fundamental to the forces driving the fast-paced change characteristic of our world today. Given the magnitude of the challenges we face if we will only meet the MDGs we recognize that managing water is a shared responsibility. Thus, we stress the importance of bringing together all parties to address key governance issues raised in this Report so that all may secure a better quality of life not only in the short to medium term but through sustainable development of water resources over the long term.

Wastewater treatment at a bottling factory, Indonesia

References and Websites

AGI (American Geological Institute). 2005. Summary of hearings on Hurricane Katrina. www.agiweb.org/gap/legis109/katrina_hearings.html#sep14.

Asian Development Bank (ADB). 2003. Inland Water Transport Development In India –the Role of the ADB. www.adb.org/Documents/Speeches/2003/sp2003008.pdf.

Braga, B. P. F. 2003. The role of regulatory agencies in multiple water use. *Water Science Technology*, Vol. 47, No. 6, London, IWA Publishing.

Braga, B. and Granit, J. 2003. Criteria for priorities between competing water interests in a catchment. *Water Science and Technology*, Vol. 47, No. 6, London, IWA Publishing.

Browne, L. 2005. *Outgrowing the Earth: the Food Security Challenge in an Era of Falling Water Tables and Rising Temperatures.* London, Earthscan.

Cincotta, R. P., Engelman, R. and Anastasion, D., 2003. *The Security Demographic: Population and Civil Conflict after the Cold War.* Washington DC, Population action International.

Concern Worldwide. 2005. Concern Worldwide and hunger: a briefing paper prepared for the UN World Summit. September. London, Concern Worldwide.

——. 2005. Looking into the future: a review of progress on the MDGs, prepared for the September 2005 UN World Summit. *The Guardian*, Manchester.

Costantino, C., Falcitelli, F., Femia, A. and Tudini, A. 2003. Integrated environmental and economic accounting in Italy. Paper. Workshop on Accounting Frameworks to Measure Sustainable Development, 14–16 May 2003. Paris, OECD.

Department of Homeland Security. 2004. Catastrophic Incident Annex. *National Response Plan.* Washington DC.

Economist. 2005. The real digital divide. *The Economist*, 12 March 2005. London, 2005.

——. 2004. World in 2005: a survey of key issues and likely trends worldwide in 2005. *The Economist*, London.

EUROPA. 2005. Inland Water Transport. European Commission. europa.eu.int/comm/transport/iw/index_en.htm.

European Commission. 2005. *Opening the Door to Development: Developing Country Access to EU Markets 1999-2003.* Brussels, European Commission.

FEMA (Federal Emergency Management Agency). 2006. By the numbers: FEMA recovery update in Louisiana. Press release, 11 January. Washington DC.

——. 2005. First responders urged not to respond to hurricane impact areas unless dispatched by state, local authorities. Press release, 29 August 2005. Washington DC.

——. 2004. Hurricane Pam exercise concludes. Press release, 23 July 2004. Washington DC.

Figueres, C., Tortajada, C. and Rockstrom, J. 2003. *Rethinking Water Management: Innovative Approaches to Contemporary Issues.* London, Earthscan.

Grey, D. and Sadoff, C. 2003. Beyond the River: the Benefits of Cooperation on International Rivers. *Water Science and Technology*, Vol. 47, No. 6. London, IWA Publishing.

——. 2002. Water Resources and Poverty in Africa: Essential Economic and Political Responses. Working paper prepared by the World Bank for the African Regional Ministerial Conference on Water (ARMCOW). Washington DC, World Bank.

Guerquin, F., Ahmed, T., Mi Hua Ikeda, T., Ozbilen, V. and Schuttelaar, M. 2003. *World Water Actions: Making Water Flow for All*, World Water Council, Water Actions Unit. London, Earthscan.

Gutrich J., Donovan D., Finucane M., Focht W., Hitzhusen F., Manopimoke S., McCauley D., Norton B., Sabatier P., Salzman J., Sasmitawidjaja V. 2005. Science in the public process of ecosystem management: lessons from Hawaii, Southeast Asia, Africa and the US Mainland. *Journal of Environmental Management* Vol. 76, No.3, pp.197–209.

GWP (Global Water Resources). 2004. *Catalyzing Change: A Handbook for Developing Integrated Water Resources Management (IWRM) and Water Efficiency Strategies.* Stockholm, GWP Technical Committee.

——. 2003. *Water Management and Ecosystems: Living with Change.* Draft document, Stockholm, GWP.

Harrald, J. R. 2005. Back to the drawing board: A first look at lessons learned from Katrina. Testimony for the House Committee on Government Reform Hearings, September 15, 2005. Washington DC. reform.house.gov/GovReform/Hearings/EventSingle.aspx?EventID=33985

Harris, G. 2002. Ensuring Sustainability: Paradigm Shifts and Big Hairy Goals, opening speech for the Enviro 2002 joint conference of the International Water Association and Australian Water and Wastewater Association, Melbourne.

Hawken, P., Lovins, A. B., and Lovins, L. H. 1999. *Natural Capitalism: the Next Industrial Revolution.* London, Earthscan.

Henderson, M. 2005. Rice genome is key to ending hunger. *The Times*, London, 11 August 2005.

Knabb, R. D., Rhome, R. J. and Brown, D. P. 2005. Tropical Cyclone Report: Hurricane Katrina – 23-30 August 2005. National Hurricane Center. www.nhc.noaa.gov/pdf/TCR-AL122005_Katrina.pdf

Lenton, R. 2004. Water and climate variability: development impacts and coping strategies. *Water Science and Technology*, Vol. 49, No. 7. London, IWA Publishing.

Mathew, K. and Ho, G. (eds). 2005. Onsite wastewater treatment, recycling and small water and wastewater systems. *Water Science and Technology*, Vol. 51, No. 8, London, IWA Publishing.

Mbeki, M. 2005. Eye Witness; *Sunday Times*, 3 July 2005, London.

Moench, M., Dixit, A., Janakarajan, S., Rathotre, M. S. and Mudrakarthe, S. 2003. *The Fluid Mosaic: Water Governance in the Context of Variability, Uncertainty and Change.* A Synthesis Paper, Institute of Development Studies (IDS), Institute for Social And Development Transition (ISET), Madras Institute of Development Studies (MIDS), Nepal Water Conservation Foundation (NWCF), Vikram Sarabhai Centre for Development Interaction (VIKSAT); NWCF, Kathmandu, Nepal and ISCT, Boulder, Colorado, USA.

Polak, P. Water and the other three revolutions needed to end rural poverty. *Water Science and Technology*, Vol. 51, No. 8, London, IWA Publishing.

River Bureau of the Ministry of Land, Infrastructure and Transport of Japan. 2003. Water and Transport Theme, Statement, Third World Water Forum. www.rfc.or.jp/IWT/PDF/Statement%20_adE_.pdf.

Sachs, J. 2005. *The End of Poverty: How We Can Make it Happen in Our Lifetime.* London, Penguin Books.

——. 2005. The African challenge: the mission. *Sunday Times*, London, 3 July 2005.

Smith, D. 2005. Can the politicians do it? *Sunday Times*, London, 3 July 2005.

State of Louisiana. 2005. State of Emergency – Hurricane Katrina. Proclamation No. 48 KBB 2005. Baton Rouge, State of Louisiana Executive Department.

Takahashi, K. 2004. Keynote address for the Stockholm Water Symposium. *Water Science and Technology* Vol. 51, No. 8, London, IWA Publishing.

——. 2001. Globalization and its challenges for water management in the developing world. *Water Science and Technology*, Vol. 45, No. 8, London, IWA Publishing.

Times-Picayune, The. 2002. Washing away. Five part series, 23-27 July 2002.

United Nations (UN) & World Water Assessment Programme. 2003. *UN World Water Development Report: Water for People, Water for Life.* Paris, New York and Oxford, UNESCO and Berghahn Books.

United Nations Economic and Social Commission for Asia and the Pacific (UNESCAP). 2004. Manual on Modernization of Inland Water Transport for Integration within a Multimodal Transport System. United Nations Publication, Bangkok. Available Online at: www.unescap.org/ttdw/Publications/TFS_pubs/Pub_2285/pub_2285_Ch5.pdf.

——. 2003. Review of Developments in Transport in the ESCAP Region. United Nations Publications, New York. Available Online at: www.unescap.org/ttdw/Publications/TPTS_pubs/pub_2307/pub_2307_ch11.pdf.

United Nations Economic Commission for Africa (UNECA). 2002. The Way Forward. www.uneca.org/eca_programmes/trade_and_regional_integration/THE%2520WAY%2520FORWARD-FINAL.doc

United States Census Bureau. 2005. Income, Poverty, and Health Insurance Coverage in the United States: 2004. Washington DC, United States Department of Commerce, Economics and Statistics Administration.

United States Senate Committee on Homeland Security and Government Affairs. 2005. Hurricane Katrina: Why did the levees fail? United States Senate Hearing, Washington DC.

US Army Corps of Engineers. 2004. Inland Waterway Navigation Value to the Nation. www.mvr.usace.army.mil/Brochures/InlandWaterwayNavigation.asp.

van Heeden, I. 2004/5. Storm that drowned a city. Interviews with Nova Science programming on National Public Radio. 19 October 2004, 10 September and 5 October 2005. www.pbs.org/wgbh/nova/orleans/vanheerden.html

——. 2004. Coastal land loss: Hurricanes and New Orleans. Baton Rouge, Center for the Study of Public Health Impacts of Hurricanes Louisiana State University Hurricane Center.

Vyas, JN. 2001. Dams environment and regional development: harnessing the elixir of life: water. *Water Science and Technology* Vol. 45, No. 8, London, IWA Publishing.

Water 21, 2005 – taken from articles in Water21, the magazine of the International Water Association, issue June 2005; publ. IWA Publishing, London, 2005.

White House. 2005. Statement on federal emergency assistance for Louisiana. White House press release, 27 August 2005. Washington DC, Office of the Press Secretary.

World Bank. 2004. *World Development Report 2005: A Better Investment Climate for Everyone.* Washington DC, World Bank.

Worldwatch Institute. 2005. *The State of the World 2005: Global Security.* London, Earthscan.

If the misery of our poor be caused not by the laws of nature, but by our institutions, great is our sin.

Charles Darwin

CHAPTER 2

The Challenges of Water Governance

By

UNDP
(United Nations Development Programme)

With

IFAD
(International Fund for Agricultural Development)

Line of buckets waiting to be filled by a slow tap at a water distribution point in Kansay, near Ngorongoro, Tanzania. People will carry these buckets of water up to 3 miles each way to and from their homes

Key messages:

In many countries water governance is in a state of confusion: in some countries there is a total lack of water institutions, and others display fragmented institutional structures or conflicting decision-making structures. In many places conflicting upstream and downstream interests regarding riparian rights and access to water resources are pressing issues that need immediate attention; in many other cases there are strong tendencies to divert public resources for personal gain, or unpredictability in the use of laws and regulations and licensing practices impede markets and voluntary action and encourage corruption.

■ Good water governance is a complex process, influenced by a given country's overall standard of governance, its customs, mores, and politics and conditions, events within and around it (e.g. conflict) and by developments in the global economy. There is no blueprint for good water governance.

■ Reforms of water governance are being driven by internal pressures on water resources and environmental threats, growing population and the focus of the international community on poverty alleviation and socio-economic development (e.g. Millennium Development Goals). However the rate of reform is patchy and slow.

■ There are significant and serious gaps in developing countries between land and water use policies and governance *and* between policy-making and its implementation, often due to institutional resistance to change, corruption, etc.

■ In the water sector, as worldwide, corruption is pervasive, though shortage of information about its extent in the water sector prevents a full picture from being obtained. It has had little attention to date in the water sector and much remains to be done.

■ Increasing recognition is accorded to the right to water, in terms of a human right to a supply of safe water, the role of water rights in helping to deal with local competition for water and in dealing with social, economic and environmental problems.

■ The privatization of water services displays uneven results. Many multinational water companies are currently decreasing their activities in developing countries. The potential of local small-scale companies and civil society organizations to help improve water services has largely been overlooked by governments and donors.

■ Many governments recognize the need to localize water management but fail to delegate adequate powers and resources to make it work. Local groups and individuals are often without access to information, are excluded from water decision-making, and thus lack a capacity to act.

Top: Hydraulic drilling stations, equipped with manual pumps, are gradually replacing the less sanitary, traditional village wells, as seen here in northern Côte d'Ivoire.

Above: Pipeline in the outskirts of Gangtok, Sikkim, India

Below: Fishermen preparing their nets for fishing on Surma River, Bangladesh

Part 1. Water Governance Today

A basic insight – which has not yet garnered enough attention – is that the insufficiency of water (particularly for drinking water supply and sanitation), is primarily driven by an inefficient supply of services rather than by water shortages. Lack of basic services is often due to mismanagement, corruption, lack of appropriate institutions, bureaucratic inertia and a shortage of new investments in building human capacity, as well as physical infrastructure. Water supply and sanitation have recently received more international attention than water for food production, despite the fact that in most developing countries agriculture accounts for 80 percent of total water use. It is increasingly agreed in development circles that water shortages and increasing pollution are to a large extent socially and politically induced challenges, which means that there are issues that can be addressed by changes in water demand and use and through increased awareness, education and water policy reforms. The water crisis is thus increasingly about how we, as individuals, and as part of a collective society, govern the access to and control over water resources and their benefits.

In many places of the world, a staggering 30 to 40 percent of water or more goes unaccounted for due to water leakages in pipes and canals and illegal tapping

Water governance is an overarching theme of the World Water Development Report. This chapter will present the state of and trends in key governance variables, such as ongoing water reform work and its implementation, the impacts of corruption on water development and water governance from below. Citizens and organized interests are demanding much more transparency and influence in water decision-making.

It will also illustrate that very complex and dynamic events and processes external to the water sector define how we relate to water. Changes in water use patterns are continuously redefined through such things as culture, macroeconomic and development trends, processes of democratization and social and political stability or unrest. This chapter will also look at how water governance is undertaken in practice and discuss and analyse various settings related to water and power politics. Examples are provided that point at complex urban and rural water use dynamics, the increasing need for integrated approaches to water, the range of international targets for local actions and the multitude of stakeholder interests. Finally, some overarching challenges are identified, which are taken up by the ensuing chapters and relate to specific water governance issues, challenges and potential solutions with respect to their fields.

1a. The water–poverty link

How societies choose to govern their water resources has profound impacts on people's lives and their ability to prosper, as well as on environmental sustainability. On the ground, this means that some groups or individuals will benefit while others will lose out when water allocation changes are made. Having a fair water provision can, for many people, be a matter of daily survival. How and for whom water is being governed has impacts on river flows, groundwater tables and pollution levels, affecting both upstream and tail-end water users. The capacity of countries to pursue poverty reduction strategies and Integrated Water Resources Management (IWRM) plans, meet new demands and manage conflicts and risks depends to a large extent on their ability to promote and put into place sound and effective governance systems.

Improved governance is essentially about improving people's livelihood opportunities, while providing the backbone for governments worldwide to alleviate poverty and increase the chances of sustainable development. **Box 2.1** provides an example of how governance, development and livelihood opportunities can be linked in practice.

One of the most striking features of the link between water and poverty is that each year, thousands of African and Asian children die from water- and sanitation-related diseases (see **Chapter 6**). In the poorest countries, one out of every five children fails to reach his or her fifth birthday, mainly due to infectious and environmental diseases that arise from poor water quality. Over the last two decades, the number and scale of water-related disasters - either too much water (floods) or too little (droughts) - have increased greatly because of changing climate patterns (see **Chapters 4** and **10**). Many countries in sub-Saharan Africa and the Indian and Pacific oceans, along with low-lying Small Island States, are the most vulnerable to climate change, because widespread poverty limits their capabilities to adapt to climate variability. Too

Over the last two decades, the number and scale of water-related disasters (either too much water or too little) has greatly increased due to changing climate patterns

often, those affected by such disasters are the poor, who do not have the means to escape poverty traps.

1b. The four dimensions of water governance

The conceptual development of water management has paved the way for an IWRM approach (see **Figure 2.1**), which is considered by many as an appropriate vehicle to resolve the world's water challenges. As defined by the Global Water Partnership (GWP), IWRM is 'a process which promotes the co-ordinated development and management of water, land and related resources in order to maximize the resultant economic and social welfare in an equitable manner without compromising the sustainability of vital ecosystems' (GWP, 2000). IWRM should be seen as a comprehensive approach to the development and management of water, addressing its management both as a resource and the framework for provision of water services (see also **Chapter 1**).

■ The social dimension points to the equitable use of water resources. Apart from being unevenly distributed in time and space, water is also unevenly distributed among various socio-economic strata of society in both rural and urban settlements. How water quality and quantity and related services are allocated and distributed have direct impacts on people's health as well as on their livelihood opportunities. It is estimated that daily water use per inhabitant totals 600 L in residential areas of North America and Japan and between 250 L and 350 L in Europe, while daily water use per inhabitant in sub-Saharan Africa averages just 10 L to 20 L. Currently, 1.1 billion people lack sufficient access to safe drinking water, and 2.6 billion people lack access to basic sanitation (see **Chapter 6**). People in slums have very limited access to safe water for household uses. A slum dweller may only have 5 L to 10 L per day at his or her disposal (see **Chapter 3**). A middle- or high-income household in the same city, however, may use some 50 L to 150 L per day, if not more. Similarly, water for food production often benefits large-scale farmers to the detriment of small-scale and landless farmers (see **Chapter 7**).

■ The economic dimension draws attention to the efficient use of water resources and the role of water in overall economic growth (see **Chapter 12**). Prospects for aggressive poverty reduction and economic growth remain highly dependent on water and other natural resources. Studies have illustrated that per capita incomes and the quality of governance

are strongly positively correlated across countries. Better governance exerts a powerful effect on per capita incomes. As recently as 200 years ago, per capita incomes were not very different across countries. Today's wide income gaps across countries reflect the fact that countries that are currently rich have grown rapidly over the past two centuries, while those that are poor have not. It has been suggested that a substantial fraction of these vast income gaps is due to 'deep historical differences in institutional quality' (Kaufmann and Kraay, 2003). Water use efficiency in developing countries is very low in both urban and rural areas, and there is great room for improving the water situation through improved water distribution and management.

■ The political empowerment dimension points at granting water stakeholders and citizens at large equal democratic opportunities to influence and monitor political processes and outcomes. At both national and international levels, marginalized citizens, such as indigenous people, women, slum dwellers, etc., are rarely recognized as legitimate stakeholders in water-related decision-making, and typically lack voices, institutions and capacities for promoting their water interests to the outside world (see **Chapter 13**). Empowering women, as well as other socially, economically and politically weak groups, is critical to achieving more focused and effective water management and actions to ensure greater equity.

■ The environmental sustainability dimension shows that improved governance allows for enhanced sustainable use of water resources and ecosystem integrity (see

Figure 2.1: Dimensions of water governance

Source: Tropp, 2005.

BOX 2.1: GENDER, POVERTY, IMPROVED GOVERNANCE AND WATER ACCESS IN PUNJAB, PAKISTAN

In Punjab, women and children are often the most affected by the lack of access to water. The Government of Pakistan has implemented the Punjab Rural Water Supply and Sanitation Sector Project. The project has provided safe drinking water and drainage facilities to about 800,000 people by using a community-based, demand-driven approach, wherein the local people participated from planning through construction and eventually became fully responsible for operation and maintenance work. The project also implied strict implementation of water fees. Both men and women formed and were part of community-based organizations to implement water-related activities and promote other development and livelihood activities. The main impact of the project was to free women and children from the hard labour of carrying water. Other positive impacts included increased household income by an average of 24 percent. It was reported that 45 percent of the time saved from carrying water is spent on income-generating activities. In addition, there is a reported 90 percent decrease in water-related diseases and as much as an 80 percent increase in the enrolment of school children in some communities. The Punjab project demonstrates that it is possible to combine an efficient and large-scale extension of services with actions to improve governance and that it is critical for any development effort to involve both women and men.

Source: Soussan, 2003.

Chapter 5). The sufficient flow of quality water is critical to maintaining ecosystem functions and services and sustaining groundwater aquifers, wetlands, and other wildlife habitats. A worrisome sign is that water quality appears to have declined worldwide in most regions with intensive agriculture and large urban and industrial areas (see **Chapters 7** and **8**). With the reduction and pollution of natural habitats, the diversity of freshwater flora and fauna is becoming increasingly threatened. Poor people's livelihood opportunities, in particular, depend directly upon sustained access to natural resources, including water – especially since they tend to live in marginalized areas that are prone to pollution, droughts and floods. The essential role of water for maintaining a healthy environment is being increasingly emphasized in the change of attitudes towards wetlands, which is an encouraging sign.

Decisions about water are being made by the minute around the world within urban and rural households, neighbourhoods, small businesses, corporate boardrooms, and in the offices of local, state and national governments, as well as on the international scale. The particular settings vary, as do the people and groups involved.

Water decisions are anchored in governance systems across three levels: government, civil society and the private sector. Facilitating dynamic interactions among them is critical for developed and developing countries alike.[1] The water sector is a part of a broader society and its politics and is thus affected by decisions that lie outside of the water sector. The governance of water in particular can be said to be made up of the range of political, social, economic and administrative systems that are in place, which directly or indirectly affect the use, development and management of water resources and the delivery of water services at different levels of society. Governance systems determine who gets what water, when and how and decide who has the right to water and related services and their benefits. The representation of various interests in water decision-making and the role of politics are important components in addressing governance dynamics.

Water is power, and those who control the flow of water in time and space can exercise this power in various ways. It is often claimed that clean water tends to gravitate towards the rich and wastewater towards the poor. Sandra Postel has aptly noted that 'water grabs and power plays are legendary in the western United States'. The water tensions of the American west have been captured in popular movies such as *Chinatown*, where farmers were being 'sucked dry' by Los Angeles (Postel, 1999). As water demands and uses increase at exponential rates due to population growth, stakes rise in many parts of the world. As opportunities to expand water supplies decrease, competition over current supplies escalate, creating the need for improved governance (see **Chapter 11**).

The way in which societies govern their water resources has profound impacts on settlements, livelihoods and environmental sustainability, yet governance has traditionally not received the same attention as technical issues. Any water governance system must be able to allocate water to ensure food and security but also be

Water is power, and those who control the flow of water in time and space can exercise this power in various ways

1. The United Nations Development Programme (UNDP) defines governance as 'the exercise of economic, political and administrative authority to manage a country's affairs at all levels. It comprises the mechanisms, processes and institutions, through which citizens and groups articulate their interests, exercise their legal rights, meet their obligations and mediate their differences'.

able to assess for whom and what purposes water is provided. In practice, trade-offs have to be made and the allocation of benefits and costs clarified. In short, governance is about making choices, decisions and trade-offs. Governance addresses the relationship between organizations and social groups involved in water decision-making, both horizontally, across sectors and between urban and rural areas, and vertically, from local to international levels. Operating principles include downward and upward accountability, transparency, participation, equity, rule of law, ethics and responsiveness (see **Box 2.2**). Governance is therefore *not* limited to 'government' but includes the roles of the private sector and civil society. The character of relationships (and the formal and informal rules and regulations guiding such relationships) and the nature of information flow between different social actors and organizations are both key features of governance (Rogers and Hall, 2003; GWP, 2003).

Water governance is sometimes equated with the actual water policy in place, but governance is more; it is about the exercise of power in policy-making and whether or not to implement particular policies. Which actors were involved in influencing the policy in question? Was the policy developed in a participatory and transparent fashion? Can revenues and public and bureaucratic support be raised to implement the policy? These are just some of the important questions involved, but they indicate that governance is about the process of decision-making, its content and the likelihood of policies and decisions to be implemented. To be able to understand why water is allocated in different ways, it is necessary to look into the dynamics of policy and decision-making, informal and formal legislation, collective action, negotiation and consensus-building and how these interact with other institutions.

1c. Privatization, conflicts and democratization

The past decades have witnessed tremendous social, political and economic changes. The end of the cold war and the process of decolonization continue to shape current societal events. Globalization and the increasing speed of information exchange have had tremendous impacts on societies. Terrorism has also had a major impact on how countries interact with each other and on how governments interact with their citizens. Some commentators worry that we are heading towards a more closed 'barbed-wire' society in an effort to keep out

threats, while others feel that our new means of communication and economic growth make for more open societies (see **Chapter 1**).

The way we perceive and govern our water resources is also rooted in culture. But although water is considered by most cultures to be something critical for all life, with a prominent place in cultural and religious beliefs, it is something of a paradox that water is often taken for granted and is increasingly polluted, with many people having limited access to clean drinking water and water for productive activities.

The development of governance and management systems within the water sector is closely related to overall development trends in which the role of the state has shifted from the provider to the enabler of development and welfare (the 'rolling back of the state'). By 2000, national, provincial and local governments in ninety-three countries had begun to privatize drinking water and wastewater services. Between 1995 and 1999, governments around the world privatized an average of thirty-six water supply or wastewater treatment systems annually (WRI, 2003). Despite the push for increased privatization, the water-services sector remains one of the last public 'bastions'. Water still remains an area that is generally heavily dependent on public investment and regulations in developed and developing countries alike.

War and social and political unrest demolish people's lives and livelihood, as well as destroy important water resources, disrupting water services and impacting negatively on governance. Between 1990 and 2000, 118 armed conflicts worldwide claimed approximately 6 million lives. War will have long-term effects and will continue to affect people's livelihood opportunities and access to natural resources and public services many years after the actual conflict has ended. In 2001 it was estimated that some 12 million refugees and 5 million 'internally displaced persons' were forced to settle in resource-scarce areas, putting further pressure on people, water and the environment (WRI, 2003). Recent conflicts in Kosovo, Afghanistan and Iraq have led to the destruction of economically vital water infrastructures, and many people are deprived of safe drinking water and basic sanitation as well as sufficient water for productive uses (WRI, 2003; see also **Chapters 1**, **3** and **11**).

The resolution of conflict and social and political instability can sometimes yield unexpected opportunities

Annually, between 1995 and 1999, governments around the world privatized an average of thirty-six water supply or wastewater treatment systems

BOX 2.2: CRITERIA FOR EFFECTIVE GOVERNANCE

■ Participation: all citizens, both men and women, should have a voice, directly or through intermediary organizations representing their interests, throughout the processes of policy- and decision-making. Broad-based participation hinges upon national and local governments following an inclusive approach.

■ Transparency: information should flow freely in society. Processes, institutions and information must be directly accessible to those concerned.

■ Equity: all groups in society, both men and women, should have the opportunities to improve their well-being.

■ Effectiveness and efficiency: processes and institutions should produce results that meet needs while making the best use of resources.

■ Rule of law: legal frameworks should be fair and enforced impartially, especially laws on human rights.

■ Accountability: governments, the private sector and civil society organizations should be accountable to the public or the interests they are representing.

■ Coherency: taking into account the increasing complexity of water resources issues, appropriate policies and actions must be coherent, consistent and easily understood.

■ Responsiveness: institutions and processes should serve all stakeholders and respond properly to changes in demand and preferences, or other new circumstances.

■ Integration: water governance should enhance and promote integrated and holistic approaches.

■ Ethical considerations: water governance has to be based on the ethical principles of the societies in which it functions, for example, by respecting traditional water rights.

Source: UN, 2003.

for fundamental changes in society that can lead to improved policy-making, which in turn can benefit a nation's water prospects (see **Chapter 14**). The political changes in South Africa in the early 1990s and the emergence of a democratic system have allowed for reform of the water sector in the areas of policy, organizational structure, water rights and legislation. South African water reform is a very comprehensive and innovative approach to water management, allowing for more holistic, people-centred and ecological approaches to the governance of water. It also aims at redistributing water resources to the benefit of poor people.

Democratization, macroeconomic changes, population growth and other demographic changes, and social and political instability often have much greater impacts on water use and demands than any water policy itself. Global market conditions and trade regimes are factors that affect crop choices and thus also have serious implications for water use and demands in agriculture. Market liberalization can contribute to improving the water situation for many people but can also increase pressures to overexploit water and the environment. The importance for water professionals to increase their understanding of social, economic and political conditions external to the water sector that have both

direct and indirect impacts on how water is being used and governed is highlighted in Waterbury's study of cooperation among the Nile Basin countries (2002; see also **Chapter 11**).

Improved governance and water shortages: A double challenge

Increasing water demands will lead to a decline of per capita supply in the future. Currently, an estimated twenty-six countries with a combined population of more than 350 million people are located in regions with severe water scarcity where the available water resources seem to be sufficient to meet reasonable water needs for development activities, *but* only if these countries take appropriate water demand and supply management measures. In many countries, there will also be additional, sometimes severe, local water scarcities, even within countries that have sufficient aggregate water resources, such as within the US and India.

A comparison of water shortages and governance challenges shows that many countries, particularly in the Middle East and North Africa, are facing a double challenge. It is also evident that countries that have bountiful water resources are facing governance challenges to provide water and sanitation services *and* protect water resources. For example, countries in Central Africa, which have ample water resources, have not been able to provide their citizens with a sufficient supply of water; hence the point that water provision is often less a question of available water resources than of properly functioning institutions and proper infrastructure management. Despite limited democratic provisions in some countries, water can still be managed in more democratic ways at the local level, such as through water-user associations or other types of local organizations.

It has been suggested that partial democratization, without the appropriate checks and balances, can, at least in the short run, lead to an increase in the exploitation of water, land and forests through patronage politics. It has also been suggested that within well-established democratic polities, politicians can make environmental 'pay-offs' to groups that financially support the campaigns of a particular party or candidate. These 'pay-offs' can include, for example, circumventing certain environmental regulations and allowing the lax enforcement of pollution control (WRI, 2003). Despite the fact that democracy has flourished in Western Europe, more than half of European cities are currently exploiting groundwater at unsustainable rates. Chronic water shortages already affect 4.5 million people in Catalonia, where authorities are pressing for the construction of a pipeline to divert water from the Rhone in France to Barcelona. It is thus apparent that many water development 'principles' – IWRM, participation, transparency, community involvement and decentralization – require improved governance in order to be successfully implemented. It is unlikely that effective participation and transparency within the water sector will take place unless there are overall changes in how societies and political systems function.

Water policy reform, IWRM implementation and meeting the Millennium Development Goals (MDGs) all require that we address issues in the water and development interface, as well as issues that have traditionally been considered outside the scope of water. If we wish to increase stakeholder participation, make decentralization more effective and hold water agencies and utilities accountable, enhanced democratization is required. Yet fairly little is known about the local and practical links between water shortages and democratization. The general notion is that democratization is beneficial to improved water governance and would open up for more transparency, decentralization and participation. But in

...partial democratization, without the appropriate checks and balances can, potentially lead to an increase in the exploitation of water, land and forests through patronage politics

which ways? How big an impact would it make? And what type of democratization makes the biggest impact? The cases of India (low levels of water services) and southern Spain (dwindling groundwater) indicate that democracy itself is not sufficient. It also depends on how political rights and civil liberties are exercised, as well as on other factors, such as demographic development, economic growth, institutional effectiveness and how welfare is generated and distributed within and between societies. This does not mean that water managers should refrain from trying to make a difference, but rather underlines an urgent need to collaborate with new actors outside the water realm and establish more inclusive water development networks. Political change has to begin somewhere, and in some cases the promotion of improved water governance may even serve as an avant-garde for inducing broad-based reform. It has been pointed out that cooperative water development in the Netherlands in the earlier part of the twentieth century was an important part of nation-building for the modern Dutch welfare state (Delli Priscoli, 2004).

Improved governance and impacts on water resources management and related services are both complex and dynamic. If a country lacks essential freedoms, like the freedom of speech and the right to organize, promoting participatory approaches in water development programmes is compromised. If citizens cannot access basic information on water quantity and quality, it seriously curtails their chances of halting environmentally unsound water projects or to hold relevant government agencies accountable. In southern India, due to tensions over sharing water from the Cauvery River between the states of Karnataka and Tamil Nadu, information on basic hydrological data is frequently withheld by the responsible authorities (see **Chapter 11**).

1d. International responses to improve water governance

The concept of water governance has gained a lot of ground and evolved over the past decade. Within the international political arena the concept has evolved from a political taboo in North–South development cooperation dialogue to gain wider acceptance as a critical issue at international, national and local levels. The framing of water challenges in terms of governance has allowed a broadening of the water agenda to include the scrutiny of democratization processes, corruption, power imbalances between rich and poor countries and between rich and poor people. Governance and politics are

BOX 2.3: DEVELOPING INDICATORS FOR ASSESSING GOVERNANCE

During recent years, the international water community has centre-staged governance as the most important challenge to improve water management and services provision. A serious weakness is that very few, if any, robust indicators exist to monitor and assess trends for water governance. A key challenge for all development actors is to publish disaggregated data on water governance issues to assess if countries are on the right track in their reform efforts. Water governance indicators should be useful to national stakeholders as a tool for priority setting, and strengthening the responsiveness of institutions and processes to the water needs of water users.

There has been great progress in quantifying and standardizing governance indicators. The progress in research, measurement and indicator development has helped identify the many components of governance. Improved governance can result in higher economic growth, more productive investments, lower transaction costs and more effective implementation of policies and legislation (UN Millennium Project, 2005). Thus, if a country's governance is improved through increased transparency, strengthening of local democracy, improvements in the judiciary system etc., such changes will also spill over to the water sector. But due to limited research and indicator development there is currently little evidence that can help us understand to what extent and how this is occurring. It is thus critical that the water governance knowledge base be enhanced on, among other things: What type of governance is favourable to improved water resources management and water services provisions? Are some governance components more critical to address than others to improve water supply and sanitation coverage as well as the sustainable use of water resources?

The development and application of appropriate water governance indicators will make a major contribution to the type of water policy interventions that are required by governments and the whole development community. Here we highlight some attempts and definitions of good governance.

- **Country policy and institutional assessments:** The World Bank evaluates economic management (debt, macroeconomic and fiscal policies), structural policies (trade, financial, private sector strategies), policies for social inclusion and equity and public sector management and institutions (rule of law, financial management, efficiency of public administration, transparency, accountability, corruption).

- **Freedom House:** The Freedom in the World rankings measure political freedoms and civil liberties. Political freedoms are measured by the right to vote, compete for public office, and elect representatives who have a decisive vote on public policies. Civil liberties include the freedom to develop opinions, institutions, and personal autonomy without interference from the state.

- **International Country Risk Guide:** The International Country Risk Guide ranks political, economic, and financial risks. Political risks include government stability, socio-economic conditions, investment profile, corruption, conflict, quality of bureaucracy, democratic accountability, law and order, and the presence of religion and the military in government. Economic risk measurements include per capita gross domestic product (GDP), GDP growth, inflation, and fiscal policies. Financial risk measurements include foreign debt, trade balances, official reserves, and exchange rate stability.

- **Governance Matters:** These data sets, produced by the Global Governance group at the World Bank Institute, rank seven aspects of governance: voice and accountability, political stability, absence of violence, government effectiveness, regulatory quality, rule of law, and control of corruption.

- **Millennium Challenge Account:** The Millennium Challenge Account was announced by the US Government in 2002 as a new foreign aid programme to assist countries that are relatively well governed. Governance is measured based on three broad categories: ruling justly, investing in people, and encouraging economic freedoms. Ruling justly is measured by scores on civil liberties, political freedoms, voice and accountability, government effectiveness, rule of law, and control of corruption. Investing in people is measured by public spending devoted to health and education, primary completion rates, and immunization rates. Encouraging economic freedoms is measured by fiscal and trade balances and the investor climate.

- **Transparency International:** Transparency International ranks countries on the basis of a Corruption Perceptions Index, a composite index that measures the degree to which corruption is perceived to exist among public officials, politicians and the private sector.

Source: UN Millennium Project, 2005. For additional information on existing governance indicators see, UNDP, 2004c. This guide provides, among other things, an outline of existing governance-related indicators.

Section 1: CHANGING CONTEXTS

10 percent – or some US $300 million – of the total aid in the water sector is directed to support the development of water policy, planning and programmes

increasingly viewed as an integral part of water crises and thus as a part of resolving them (see **Box 2.3**).

An important part of the work of bilateral and multilateral organizations has been to support the enhancement of capacities to strengthen national and local water agendas and policies, investment priorities, while providing useful examples for scaling up activities. Despite these efforts, water is not considered a main priority in most countries. Investment in the water sector is still at a very low level in developing countries, and despite promises of action-oriented outcomes by the world's governments at the WSSD, much remains to be done about water governance issues in donor budgets.

According to statistics from the Organisation for Economic Co-operation and Development (OECD), total aid to the water sector during recent years has averaged approximately US $3 billion a year. An additional US $1 to 1.5 billion a year is allocated to the water sector in the form of non-concessional lending, mainly by the World Bank.

Over three-quarters of the aid to the water sector is allocated to water supply and sanitation. The bulk of the aid for water supply and sanitation is allocated to a handful of large projects undertaken in urban areas. While such support is, of course, much needed and desired, it is disheartening from a governance point of view that only about 10 percent – or some US $300 million – of the total aid in the water sector is directed to support the development of water policy, planning and programmes.

The statistics also show that many countries where a large portion of the population have insufficient access to safe water received very little of the aid. Only 12 percent of total aid to the water sector in 2000–01 went to countries where less than 60 percent of the population has access to an improved water source, which includes most of the least developed countries. On the positive side, aid allocated to various types of low-cost and small-scale technologies (for example, treadle pumps, gravity-fed systems, rainwater harvesting, sustainable small-scale sanitation, etc.) seems to be increasing (OECD, 2002).

In 1999, the World Bank and the International Monetary Fund (IMF) started to develop a new framework for giving low-interest loans and debt relief to forty-two of the poorest countries in the world. The poor countries that want to be a part of this must formulate and put in place what is called a Poverty Reduction Strategy Paper (PRSP). The strategy is supposed to indicate how a government will use the funds for targeted poverty reduction in their country. The process leading to a PRSP is also supposed to be based on broad multi-stakeholder and participatory processes for their design, implementation and monitoring. It is seen as critical that PRSPs are driven and owned by the forty-two countries in question. In essence, the PRSPs represent a means of securing resources for development priorities and serve as countries' long-term development strategy. Both multi- and bilateral donors are increasingly using PRSPs to coordinate their development cooperation and to achieve coherence in development objectives with recipient governments. Considering the fact that PRSPs represent long-term development strategies, it is worrisome that water resources issues and related services have so far received very low priority in their design. Two PRSP assessments show that the key initial planning and resource commitments needed to achieve water-related targets are not being met. Water targets are not linked to key strategies that prioritize and fund action (see **Box 2.4**).

Within the water sector, there is a widespread belief that we now have most of the needed principles in place in order to make a lasting improvement to the world's water resources situation, which will also make a major contribution to the overall work of alleviating poverty. What is lacking today are the concerted actions and the means for effective implementation of various water policies and development programmes. The implementation of countries' existing water policies would go a long way in meeting the MDGs and the water targets set in Johannesburg.

Life on the Mekong River, Viet Nam

BOX 2.4: WATER IN POVERTY REDUCTION STRATEGY PAPERS (PRSPs)

A recent study by the Overseas Development Institute (ODI) and WaterAid on the extent to which water supply and sanitation (WSS) was given priority in PRSPs in sub-Saharan Africa concluded: 'WSS had been inadequately reflected both in terms of the process of PRSP preparation and the content of emerging PRSPs'. In total, seventeen African PRSPs were examined, and of these, only Uganda showed a high level of priority to water supply and sanitation (see **Chapter 14**). This is surprising, given the international prominence given to these issues through the MDG on water and sanitation and a strong demand from rural and urban communities to urgently improve water-related services. There are several reasons that can explain this, including a limited understanding of the social and economic benefits of improved water and sanitation, weak poverty diagnosis and limited dialogue and interaction between central ministries, local

governments and local communities within the sector. In other words, water supply and sanitation issues are under-represented in PRSPs, partly because the water sector has failed to articulate the needs and potential impacts on poverty of investments in this sector and partly due to critical national decision-makers' limited understanding of the issues.

A water resources assessment of nine Asian PRSPs found similar results. In the Asian cases, water resource issues, such as floods and droughts, as well as water supply and sanitation and irrigation, were frequently present in the analysis of issues in the PRSPs but were more rarely reflected in the programmes for action or priorities for investment. The failure of key water advocates and decision-makers has again been cited as the main reason for this. However, it is important not to forget key economic decision-makers outside the water sector, as well as their

failure to fully appreciate the importance of improved water resources management and water supply and sanitation for social and economic development. A wider assessment of forty interim and full PRSPs by the World Bank confirms this. The assessment showed that natural resources management and environmental protection were only included in limited ways. There were some exceptions, however, like in Mozambique where the protection and management of environmental and natural resources was seen as being prioritized. The assessment also indicated that the result for full or final PRSPs was slightly better than for the interim version, suggesting that priority issues of natural resources and environment improved as consultations were wider and more thorough.

Sources: Frans and Soussan, 2003; Bojö and Reddy, 2002; Slaymaker and Newborne, 2004.

Training around a new water pump with an instructor during water and sanitation programme in Budari, Uganda

Part 2. Water Governance in Practice: Trends in Reform and Rights

Water policies and reforms have too often been driven by assumptions about the need to increase supplies through investing in physical infrastructure

Governance is one of the biggest challenges within the water sector: Why and how are certain decisions made? What stakeholders are involved? What principles, rules and regulations (formal and informal institutions) apply? Governance is process-oriented and thus intrinsically linked to politics and preoccupied with how various actors relate to each other. Because of the varying characteristics of water resources and the myriad socio-economic and political frameworks, governing mechanisms vary considerably across countries, including differences such as the reformed items, the pace at which countries are moving towards implementing water reforms, the level of the reform and the degree of targeting environmental and social objectives.

2a. National water policies in the making

Reform of the water sector is now taking place in many countries around the world. Reasons for reforms can vary according to the particular situation. In most cases, water reforms in a particular sector appear to be associated with a larger reform agenda. For example, pricing reforms are often complicated by financial constraints and cost recovery affects the fiscal budget. In Pakistan, the central government has to subsidize the budgets of the irrigation departments and in Morocco the public budget used to be the sole source of funding of water services provided mainly by irrigation districts. The Republic of Yemen, where macroeconomic measures accompanied water reforms, provides a good example of the importance of having a wide-ranging agenda for reform (Wambia, 2000; Diao and Roe, 2000; Ward, 2000). In the case of South Africa, water reform was made part and parcel of the overall political changes in the early 1990s (see **Chapter 14**).

Water policies have too often been driven by assumptions about the need to increase supplies through investing in physical infrastructure. Current water sector reform, particularly in developing countries, has increasingly tried to balance issues of infrastructure and technology with governance and management issues, such as multi-stakeholder participation, as well as measures for enhancing demand management, decentralization and various elements of integrated and basin management approaches. Another example includes water reform in Nicaragua, where current water legislation covers a wide range of issues from water rights, participation and economic incentive instruments to technology.

In recent water sector reform – in South Africa, Zimbabwe, Kazakhstan and the European Union (EU) Water Framework

Directive management, for example – the basin level constitutes an important component of improved integrated management of water. Kazakhstan, for example, has established eight river basin organizations that are responsible for water resources governance and use, water plan preparation, water allocation and permit provision (see **Chapter 14**). The ongoing water reform in Kazakhstan also includes provisions for public involvement and decentralization that can take the form of local self-governance (UNDP – Kazakhstan, 2003). In many cases, water management at the basin level also extends to the management of river basins and groundwater aquifers shared between sovereign states. The increased water cooperation between nations will require building mutual trust as well as long-term commitments from the parties involved. Normally, this necessitates a whole sequence of cooperative actions that can start off with agreeing on methodologies and standards for collecting hydrological data and joint monitoring plans. Other actions can include the harmonization of water policies and joint water management plans. It is important that transboundary water management not stop with cooperation at a technical level, but that such cooperation 'spill-over' to joint development plans for a river basin or even to broader issues of sharing the benefits of river basins and groundwater aquifers. There is currently a wide range of various sub-regional river basin commissions on all continents; however, their mandates and impacts on water resources use efficiency and sustainability vary to a great extent (see **Chapter 11**).

Despite the growing call for integrated approaches, there is, in reality, limited practical experience of how it can be implemented. In the overall context of IWRM, relevant challenges to and opportunities for an improved integration of land and water governance have, surprisingly, received little attention. It has proven difficult

BOX 2.5: INTEGRATING LAND AND WATER GOVERNANCE

In spite of its obvious importance, the knowledge base on land and water governance is weak. A recent study undertaken by the International Fund for Agricultural Development (IFAD, 2004) examined several case studies in which the links between land and water governance were identified as a key to successful development.

■ One typical, poor community in Zimbabwe wished to improve livelihoods by investing in small-scale irrigation. However, they lacked the financial capital to do this and were unable to borrow money because they had no collateral for the loan. Normally, ownership of the land is sufficient security for a loan but in Zimbabwe all land titles are vested in the president, so this option was not available to them. To overcome this investment problem, the community, with international support, sought technical assistance from a regional platform for advocacy, the Women's Land and Water Rights in Southern Africa (WLWRSA), which also helped to secure community water rights

for irrigation. In return, WLWRSA uses this field experience to support its advocacy and increase its legitimacy.

■ In Bangladesh, major reforms in the governance of inland public water bodies, supported by several external agencies, have significantly improved the livelihoods of poor landless fishermen. Inland fisheries are critically important for food security and livelihoods, but access to lakes by poor, landless fishermen is problematic, because they require access rights to the shores as well. The wealthy tend to dominate the annual leasing arrangements, leaving most fishermen to work as share catchers with minimal reward. The lack of secure tenure means there is no incentive for them to invest in the lakes, so they remain in a poor, unproductive state. Over the past fourteen years, the situation has improved significantly following a package of reforms that introduced long-term lake and shore lease arrangements for the poor, decentralized resource management to

fishing groups and limited group membership to those below a set poverty limit to protect the poor. All this encouraged investment in the lakes, resulting in the improvement of lake productivity, fish stocking levels, fishermen's incomes and infrastructure. Women also benefited by the introduction of further reforms on ponds, which gave them full access to inputs and benefits.

These and other cases indicate, among other things, that local participation and empowerment are critical for making changes and that external partners can play an important role in achieving the desired impact on awareness and resulting policy changes. They also indicate that enhanced local participation has made a positive difference for women, which has been beneficial for both equity and efficiency in the work of the communities.

Sources: IFAD, 2004, www.ifad.org/events/water; NRSP, 2004.

to integrate or coordinate land and water in a meaningful way, particularly for the rural and urban poor who have been socially and politically marginalized, and largely excluded from access to land, water resources management and related services. The benefits of integrating land and water in decision-making are illustrated in **Box 2.5**.

Similar challenges (not having proper tenure and access to water services) are found within urban slums where local authorities do not provide proper public services, such as water, sanitation, transportation and electricity. Major questions remain to be worked out to put effective IWRM into practice, again in ways explicitly linked to governance issues (Moench et al., 2003):

■ Who is in charge of integration? Who implements integration? What are the roles and responsibilities of governments, the private sector and civil society and the international community?

■ Who decides what interests should be reflected in IWRM plans and policies? How should policy processes be governed to ensure that relevant stakeholder interests are duly reflected?

■ How should conflicting interests and disputes be resolved? What are the appropriate formal and informal institutions and conflict resolution mechanisms for efficient and equitable water decisions?

■ Is there really a need to integrate all water issues? Some water management issues, such as waste disposal control from a sewage treatment plant, will not need the same level of integrated decision-making as a major water allocation decision to construct a large-scale dam or irrigation scheme along a transboundary river.

Meeting the Johannesburg target on IWRM plans

Putting into place strategic and well-planned water projects will help countries to set the right priorities and undertake actions required to meet the Johannesburg target. These plans can thus become critical instruments for achieving domestic political targets as well as targets agreed on in the international arena, like the MDGs or regional transboundary water cooperation agreements. If we take the status of the recent Johannesburg target to 'develop integrated water resource management and water efficiency plans by 2005, with support to developing countries' as a proxy for improved reform and governance in the water sector, it reveals that progress is taking place but that much remains to be done.

At the end of 2003, an 'informal stakeholder baseline survey' was conducted by the GWP on the status of water sector reform processes in various countries of the world. The survey was conducted in 108 countries[2] and provides a number of qualitative elements allowing an assessment of countries' readiness to meet the 2005 Johannesburg implementation plan target on IWRM Plan preparation. In this respect, the level of awareness, political support and the countries' capacity to build on past and ongoing processes relating to water-related reforms and rely on existing multi-stakeholder platforms were some of the components that were assessed.

The survey provides a snapshot of where countries stand in terms of adapting and reforming their water management systems towards more sustainable water management practices. The pilot results show that of the 108 countries surveyed to date, about 10 percent have made good progress towards more integrated approaches and 50 percent have taken some steps in this direction but need to increase their efforts, while the remaining 40 percent remain at the initial stages of the process (see **Table 2.1**).

Several countries have begun, or have already been through, the process of putting into place IWRM elements. South Africa, Uganda and Burkina Faso have, with international assistance, gone through multi-year IWRM planning processes resulting in new national policies, strategies and laws for their water resources development and management. Other countries in Africa have also been identified as having good opportunities to advance their water agenda. For example, water legislation is being prepared in Congo-Brazzaville and Malawi, where the opportunity can be seized to promote

integrated approaches towards water management. Similar opportunities exist in Asia, such as with China's water policy work, and the water reform processes in countries like Sri Lanka and Pakistan. Development has also been rapid in Central Asia, where, for example, Kazakhstan and Kyrgyzstan have made headway towards developing IWRM approaches. In Latin America, Brazil's wastewater reform is an example of IWRM processes. Many of these are now in or on the verge of the implementation stage. Other countries in Latin America have also made headway (see **Chapter 14**). There are, for example, favourable political and institutional conditions in Honduras where the multi-stakeholder Water Platform (Platforma del Agua de Honduras) provides momentum to advance IWRM approaches and other water-related issues.

This qualitative assessment does not, however, allow for regional or country comparisons, as exemplified by the cases of Viet Nam and Sierra Leone, which have both been classified as in the initial stages of developing IWRM approaches. Sierra Leone is a conflict-ridden country where the main focus is on building peace and stability and reconstructing basic services such as water supply and sanitation; it is thus far from engaging in developing IWRM approaches. Viet Nam, on the other hand, has showed progress in recent years. In 1988, it adopted a national water act and a National Water Resources Council, and three river basin organizations were established in 2000 and 2001. It is clear that water is fairly high on the political agenda and Viet Nam is in a good position to advance implementation as well as incorporation of IWRM approaches.

The assessment indicated that the countries that have made the most progress in adapting and reforming their water management systems towards more sustainable water management practices have often started by focusing on specific water challenges, such as coping with perennial droughts or finding ways to increase water for agriculture while still ensuring access to domestic water in burgeoning urban areas. South Africa, for example, developed comprehensive policies, legislation and strategies starting in 1994, focusing outward from drinking water (and later on sanitation) to give expression to the political, economic and social aspirations and values of the new democratic political paradigm.

Recently, there have also been other IWRM plan assessments initiated to measure how much progress

---

2. Forty-five in Africa, forty-one in Asia and the Pacific and twenty-two in Latin America. For more information on this assessment see: www.gwpforum.org

Table 2.1: Country readiness to meet the Johannesburg target on IWRM planning by 2005

Region	Number of countries surveyed	Good progress	Some steps	Initial stage
Africa				
Central Africa	7		3	4
Eastern Africa	5	1	2	2
Med (North Africa)	5	1	3	1
Southern Africa	12	2	5	5
West Africa	16	2	4	10
Total	*45*	*6*	*17*	*22*
Asia and Pacific				
Central Asia	8	2	4	2
China	1	1		
South Asia	6		4	2
Southeast Asia	8		4	4
Pacific	18	2	8	8
Total	*41*	*5*	*20*	*16*
Latin America and the Caribbean				
Caribbean	6		6	
Central America	7	2	3	2
South America	9	1	5	3
Total	*22*	*3*	*14*	*5*
Total	**108**	**14**	**51**	**43**

Source: GWP, 2003.

countries have made towards adopting and implementing IWRM. A 2005 study of the status of IWRM plans in the Arab States indicated that progress is very uneven in the region. Some places, such as Jordan, Egypt, the Palestinian Autonomous Territories, Yemen and Tunisia, have national water policies, plans or strategies in place that incorporate many elements of IWRM. Eleven out of the twenty-two countries included in the study need major water policy enhancements to put IWRM plans in place. For most of these eleven countries, the study identified ambition and ongoing efforts to further progress of developing IWRM plans. For six of the countries included in the study the situation seems to be less progressive, with some of the countries even lacking ongoing efforts to develop IWRM plans (Arab Water Council, 2005).

It is important to stress that, even though many countries lack IWRM elements in their water reform attempts and aspirations, this should not refrain them from acting. It is more realistic to implement reforms incrementally than to await the 'perfect' policy document that may never get past the drawing board. **Box 2.6** highlights the fact that making and implementing water policy can take very different paths.

Water policies, politics and resistance
No reform is stronger than its weakest link, which is to say, implementation. Recent years have seen the development of sophisticated water policies and plans in many parts of the world, such as in South Africa, in Europe with the EU Water Framework Directive and in Chile with water privatization. Some of the reforms in developing countries have been assisted by the international community and have frequently been motivated, at least in part, by the active international debate on these issues. These achievements also need to be balanced, however, by a recognition that policy changes at the national level have often only been imperfectly followed through to effective implementation. A recent example is Zimbabwe, where the actual content of water reform is considered progressive, but where reform has stalled due to recent political instability and weak implementation capacity.

There is a tendency to separate policy-making processes from implementation. The notion is that policy-making is ascribed to decision-makers, while implementation is linked to administrative capacity. This kind of thinking is too rigid and fails to acknowledge that policies are often modified as they move through public administrations and

It is more realistic to implement reforms incrementally than to await the 'perfect' policy document that may never get past the drawing board

BOX 2.6: THE POLICY PROCESS: DECISION-MAKING VERSUS IMPLEMENTATION

The making of water policy involves a multitude of decisions, actors and processes. Two different models of the policy-making process are delineated: the first model shows the linear and idealized input-output version of policy-making. Typical stages for policy are inputs that constitute the basis for formulation of policy, the content of the policy, implementation and a feedback loop to input. While these stages can be identified in the policy process, the second model displays a more realistic version of how policy-making is done. Policy-making is not a straightforward linear process, but rather a 'messy' business, in which various actors with different interests, stakes and powers are trying to influence the policy outcome while different policy stages are interlinked and sometimes done in a simultaneous fashion. What the policy process looks like, what actors are involved and other concerns differ among various development contexts and depend on what water challenges the policy is intended to address. The non-linear model shows some of the critical factors that are shaping policy formulation and implementation.

In practice, decision-making for water is done by and through various kinds of organizations and formal and informal regulations, such as water licensing or customary allocation decisions. Many organizations are formally constituted and have legal rights and responsibilities, while many are informal and less visible to the observer. Local

levels are critical, since implementation must ultimately take place in local urban and rural settings. How these organizations function and their relationship dynamics both have an impact on water governance and the possibilities of more effective governance.

At the local level, many local entities are involved in water decision-making: irrigation, environmental and health departments and agencies, urban development agencies, rural and urban planning agencies, regulatory agencies, public water utilities, water-user associations, consumer groups and other types of NGOs, religious groups, farmer organizations and unions, municipalities, community leaders and local entrepreneurs, etc.

At the national level, parliaments, governments and their ministries, consumer groups, research institutes, NGOs and other interest groups, trade unions, private businesses, etc., play critical roles.

The international level and external pressures can many times have a critical influence on water reform in developing countries. Such influence may take the form of pressure to comply with a structure imposed by an international development agency as part of a large investment project. It can also take the form of loan conditionality. These are common features in structural adjustment projects that enhance price reforms in various sectors. Other types of

conditionality can be found in big national water resource projects that include large components of institutional or pricing reforms, as was the case in Pakistan and Mexico. Although not yet common or widely used in the water sector, several trade agreements that affect the agriculture sector may impose the restructuring of a price system in one country as part of a condition for that country to join the regional agreement. An example of such regional cooperation is the recent European Water Framework Directive (see **Chapter 14**). This is a legislative piece that will guide European water policies in the coming decades. In the specific case of the European Union, water-related economic dependencies (agriculture) and likely water-related environmental externalities are a driving force behind regional agreements (DFID, 2002). Another set of organizations that can affect domestic policies is regional river basin commissions (The Mekong River Commission and The Nile Basin Initiative are two examples). Various kinds of international NGOs are also important actors for advocacy and promoting cooperation and research. Also multinational water companies are now playing an important role in developing countries' policy development and implementation. There is a plethora of water organizations from the very local setting up to the global scale. The effectiveness of these organizations and how they relate to each other and other organizational entities determine governance outcomes.

The linear policy model

The Policy Process according to the Stages Model

Input	Policy	Output
Information	Regulation	Application
Perception/ Identification	Distribution	Enforcement
Organisation	Redistribution	Interpretation
Institutions	Capitalization	Evaluation
Demand	Ethical ruling	Legitimation
Support	Constitutional	Modification/ adjustment
Apathy	Symbolic	Retreat/ disavowal

The non-linear policy model

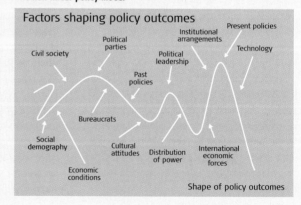

Factors shaping policy outcomes

Civil society · Political parties · Institutional arrangements · Present policies · Political leadership · Technology · Past policies · Bureaucrats · Social demography · Cultural attitudes · Distribution of power · International economic forces · Economic conditions

Shape of policy outcomes

Source: Gooch and Huitema, 2004.

national bureaucracies to local levels for ultimate enforcement. Policy implementers can respond to policy change in different ways: they can implement reforms fully or partly, or if internal reform resistance is high in combination with lax monitoring from policy-making levels, they can ignore new policies altogether. It is clear, then, that policy-makers should not escape the responsibility of implementation and making sure that adequate capacities and financing is available for effective implementation.

Because reforms change the status quo, one can expect both support for and opposition to reform agendas by affected groups. Institutional reforms generate the active involvement of interest groups that may be affected directly or indirectly. In some cases, the implementing agency may not have a reform agenda that coincides with that of the government initiating the reform. For example, a stakeholder analysis of the parties involved in a proposed pricing reform for the capital of Honduras, Tegucigalpa, shows that the public agency in charge of supplying water to the city was a major opponent to the reform. A main aim of the agency was to seek continued power over water allocation and administration. Interestingly, the case of Tegucigalpa also showed that the main support for reform was provided and driven by external international development agencies, but in this case, this was not sufficient for reform to take effect. It was noted that for the reform to go forward, it would require the support from critical national power centres, such as the president and key government ministries. (Strand, 2000; DFID, 2002). A similar situation of opposing reform is described by Wambia (2000) in the case of Pakistan, where certain government agencies along with ministries opposed reform because it was felt that it would affect them negatively. Part of the reform was to transfer power and financial resources from the irrigation ministry and its regional offices to what were called Area Boards.

In governance systems that foster a non-implementation policy climate, decision-making tends to be discretionary, unpredictable and largely non-conducive to influence by citizens. The implementation and allocation of resources tend to benefit the ruling elite or certain groups or individuals tightly connected to the ruling elite. Many civil society actors in developing countries are well aware of limited government commitment and capacities to follow through on policy development and external pressures. As a result, many NGOs and other civil society actors,

particularly those lacking political and financial clout, do not find it very worthwhile to engage in policy processes. Many NGOs tend to argue that the value of influencing a policy that will not be implemented is not worth the effort. Since limited implementation also opens up for discretionary decision-making, many NGOs instead attempt to manoeuvre in a highly informal local and national political setting, often through face-to-face relationships, through which they can influence decision-making in favour of their cause. Consequently, due to more discretionary and ad hoc decision-making, many NGOs in developing countries feel that it is better to influence the implementation of a policy through informal means rather than through policy content. There can be many other barriers, both overt and hidden, to the effective implementation of policy: the lack of capabilities in or resources available to government departments; resistance by sceptical officials; pressure by interest groups, such as industrialists or farmers; problems with other aspects of law and policy that can block effective implementation, such as changes of land management practices that also affect water resources. In the various international attempts to promote policy development, such developing country policy realities have yet to be fully acknowledged.

Water policy reform ahead: Where to begin?

So while there has been progress in water policy development during the past decade, this progress is uneven and considerable challenges remain. Many government reforms fail because once implemented, they yield unsatisfactory outcomes because they never get past implementation. How can the prospects for success be improved? It has been noted that 'a reform programme will be successful if there is economic rationality in its design, political sensitivity in its implementation and close and constant attention to political-economic interactions and socio-institutional factors, so as to determine in each case the dynamics to follow' (Cordova, 1994). In particular, there needs to be a more complete understanding of the forces that lead to policy development in the first place and, critically, a concerted drive to make sure that policies are followed through to implementation. There also needs to be effective feedback and assessment mechanisms, so that the consequences of policy implementation can inform future policy development. There are many components that are critical for successful policy reform. For example, some studies in policy implementation suggest that in order to maximize the likelihood of success, or minimize

...policy-makers should not escape the responsibility of implementation and making sure that adequate capacities and financing is available for effective implementation

The limited funding opportunities in low-income countries expose domestic decision-makers and policy development to pressure from international lending institutes and donors

failure, it is critical to address three key strategic issues: keeping the scope of change narrow, limiting the role of aid donors and giving reform firm leadership while simultaneously allowing for line management discretion (Polidano, 2001).

Some critical issues for overcoming policy obstacles are outlined below:

■ **Acknowledge the role of politics and develop strategies accordingly:** Even though most reforms require technical input the process itself is essentially political and thus involves political compromises, bargaining and negotiated outcomes. In most cases, the proper packaging, sequencing, alliance building and communication of reform can lead to more tolerable reform content that can be more easily implemented. The power balance between critical political, social and economic actors will have a significant impact on reform outcomes.

■ **Secure high-level political support and commitment:** Without high-level political commitment to undertake reform, it will be very difficult to go from policy formulation to implementation. The whole government needs to be involved to ensure that sufficient resources and capacities are provided to achieve the reform objectives.

■ **Focus on process and seize the moment:** Management of the policy process, which has so far received very little attention, is just as important. Some policy studies suggest that the process is even more important than the actual policy content. The timing of a reform is important. According to the 'crisis hypothesis', a perceived or real crisis due to floods and droughts is needed to create conditions under which it is politically possible to undertake the reform. The 'honeymoon hypothesis' suggests that it is easier to implement a reform immediately after a government takes office (Williamson, 1994).

■ **Participation and inclusiveness:** Effective policy formulation and implementation requires transparency and inclusiveness. For example, does media and civil society advocacy representing the needs and interests of vulnerable groups, such as indigenous people, women and children and threatened ecosystems, influence the timing and content of decisions on policy changes? Inclusiveness and active engagement do not

only refer to civil society but also to different government agencies at various levels.

■ **National ownership of policy process and content:** The limited funding opportunities in low-income countries expose domestic decision-makers and policy development to pressure from international lending institutes and donors. If a country does not develop a certain 'internationally' required policy, it may face difficulties attracting international loans and development project funds. It is not uncommon in many low-income countries that due to external pressure they are required to put into place policies, plans and development programmes that lack 'national ownership' and have little chance of actually being implemented.

■ **Allow for incremental change and proper time for successful reform:** Reforms should be well prepared, because once they are in place, they are often difficult to modify. If possible, keep the reform as simple as possible and avoid addressing many reform objectives at the same time. Reform is an incremental process, which sometimes can be painstakingly slow, and managing policy processes is laborious and time-consuming and should not be underestimated. Although policy reform is an ongoing process and modification occurs over time, it is important that the main thrust of the policy can be sustained over time.

■ **Compensate policy reform losers:** Adequate compensation mechanisms, negotiated with stakeholders, are an important part of a reform. Those who are losing out considerably in a reform should be adequately compensated: paying a fair amount of compensation is important for building support and avoiding social and political clashes that can jeopardize or slow down the reform. In the case of water pricing reforms, several groups or issues, such as the poor, or the environment, may need to be specifically addressed. For example, addressing the needs of poor people may mean including a differentiated tariff structure.

■ **Improve coordination:** Uncoordinated donor activities increase the risk of overloading the capacity of governments and slow down reform work. Donors should also allow greater flexibility in the design and implementation of reforms and allow for more experimentation by governments. Different

forms of tension and competition among various government agencies are common. It is thus critical that the political leadership of reform is intact to allow for an effective coordination and a broad buy-in from central government agencies down to local regional and local administrations.

- **Monitor implementation and impacts:** The monitoring of policy reform and implementation is an area often neglected by governments. Some attempts have been made towards more systematic monitoring of the progress and impact of water reform, but there still remains much work to be done in this area in order to actually examine if claimed progress in water reform also impacts positively on sustainable water resource use and improved water services. Effective monitoring will also imply that policies can be fine-tuned, allowing for financial reallocation between reform priorities.

2b. Water rights

Ownership or the right to use a resource means power and control. While it may seem simple, water rights and ownership often have a complex relationship with water governance. How property rights are defined, who benefits from these rights and how they are enforced are all central issues that need further clarification in current water policies and legislation. Insecurity of water rights, mismatches between formal legislation and informal customary water rights, and an unequal distribution of water rights are frequent sources of conflict that can lead to poor decisions on efficient water resource use and equitable allocation. Also, the problem of managing dwindling groundwater supplies or fish stocks – which many times lack clear user rights – is a problematic water governance challenge.

Water rights can be defined as 'authorized demands to use (part of) a flow of surface water and groundwater. Including certain privileges, restrictions, obligations and sanctions accompanying this authorization, among which a key element is the power to take part in collective decision-making about system management and direction' (Beccar et al., 2002). Water rights are inextricably linked to property.

Well-defined and coherent water rights are fundamental to dealing with situations of increased competition between water users, an important issue, which is addressed in more detail in **Chapter 11**. Water management is a complex activity; it is a mobile resource

that is attached to many different and sometimes competing, economic, social and environmental values. While water users compete for the same resource and struggle for increasing control, they also need to cooperate if they want to make effective use of water and sustain the water's quantity and quality in the long run. This often occurs in 'pluralistic' legal contexts, where formal and informal normative systems sometimes clash. For example, in South Africa, water management moved from a pre-colonial collective activity to a publicly regulated resource under Roman-Dutch law. It was then transformed under Anglo-Saxon jurisprudence when it was captured as a private resource to the benefit of a small minority. A main objective of the current 1998 Water Act in South Africa is to redistribute water rights by granting water permits to sections of society that were previously discriminated against.[3] The minimization and resolution of water conflicts and disputes require clear and coherent water rights that contain management principles and strategies that can cooperate with fluctuations of water supply and demand.

It has been noted, for example, that water rights provide the backbone of water management strategies in small-scale agriculture and in many local contexts basing their water use on customary traditions. Water rights define who has access to water and in what ways the user can take part in local water decision-making. They also specify roles and responsibilities regarding operation, maintenance, monitoring and policing. In this sense, water rights manifest social relationships and local power structures of who is included or excluded from the benefits of water and what the various rights and responsibilities include. Water management practices in the Andes, for example, have shown that social and political inequalities can prevent successful collective action. However, this also showed that collective management of water can lead to more equitable water distribution, in addition to strengthening the bargaining position of weaker stakeholders (Boelens and Hoogendam, 2002). The critical importance of water rights is not unique to small-scale agriculture or indigenous systems but is equally relevant to society at large.

From a formal legal point of view, water is considered a property that belongs to the state in many countries. Many governments have largely ignored informal customary or traditional water rights. This oversight was initiated during colonialism and continued under state-led

While it may seem simple, water rights and ownership often have a complex relationship with water governance

3. See www.thewaterpage.com/leestemaker.htm

development and the 'green revolution'. For example, in many developing countries, the state is in charge of large-scale irrigation networks and the distribution of water permits. But in a parallel track, water rights are still considered by many local framers and other water users a common property where communities manage water based on traditional rights. In some places, water rights are also being privatized, and water markets that include tradable rights are being set up. This multiplicity of water rights can lead to confusing and conflicting situations of entitlement roles and responsibilities among government agencies as well as unclear guidelines on operation and maintenance. There are far too many cases where water resources and related networks and infrastructure are, in practical terms, roughly equal to open access property (no one's property), which can result in a 'tragedy of the commons'. This can ultimately result in a management breakdown where no user or government feels responsible for sustaining surface- and groundwater resources (quantity and quality) and related infrastructure. Thus, within a single country, water rights can take multiple shapes and range from customary laws and local practices, government regulations and bureaucratic procedures to privatization and tradable water rights. Water reform should increasingly acknowledge this multiplicity of water rights, which would offer better and more realistic ways of improving current water distribution.

Despite the various views on the water rights continuum, there is a consensus that the establishment of well-defined and coherent water rights can lead to a number of social, economic and environmental benefits:

- It can promote equitable water use between existing user groups and facilitate improved access to water by groups that have been previously denied formal or informal water rights.

- It can improve the efficiency of existing water supply allocations. For example, those requiring additional water resources, such as growing cities, can increasingly meet their needs by acquiring the water rights of those who are using water for low value purposes.

- It can provide a basis for improving hydrological data and information to manage the resource more effectively.

...there is a consensus that the establishment of well-defined and coherent water rights can lead to a number of social, economic and environmental benefits...

- It increases the willingness to take economic risks to invest in improved water management and practices in both rural and urban contexts, thus impacting positively on productive livelihood opportunities. It can also reduce the pressure on water resources, as it is likely that those with water rights have incentives for sustaining water.

Custom and tradition in water rights

Local customs and traditions are important factors in defining community water management, allocation and conflict mediation. Customary rights are often based on community traditions and norms. In many cases, customary rights represent functioning water resources management decision-making systems that determine local water use and attached rights and obligations. For example, in the Andes in Latin America there is a multiplicity of local customary decision-making systems for community irrigation regarding who gets what water, when and how (Beccar, 2002). Customary rights can govern a number of water-related local social and economic activities, such as irrigation, household water, fisheries, livestock, plants and animals, funeral practices and the environmental services provided by watersheds. Generally, customary rights are not static but can evolve over time as a response to formal legislation and changing local hydrological and socio-economic conditions. Beccar (2002) has noted that often enough, customary rights are not taken into account in formal water rights legislation and development projects. A study conducted in the Pangani River Basin in Tanzania revealed that out of 2,265 water abstractions, only 171 were based on formal water rights (Hodgson, 2004). This can lead to clashes between formal and informal rights and rules and render legislation and development projects less effective. Enhancing local decision-making capacities and reflecting customary water management and rights in formal legislation in relation to irrigation practices and other water-related activities can create a more genuine way of recognizing local customary water rights and management systems (Beccar, 2002).

For example, customary water user rights are acknowledged in Japanese river law (see **Chapter 14**). The user rights in Japan are of two types, 'Permitted Water Rights' and 'Customary Water Rights'. Permitted Water Rights are granted by the River Administrator to the water user in pursuance of the River Law, while Customary Water Rights are awarded to river water users whose usage goes back to a time before the River Law was established (in 1896). Users with Customary Water Rights are obliged to notify to the River Administrator details of their water usage such as the purpose for which river water is used, the quantity used, the conditions of usage and the intake position on the river. The majority of Customary Water Rights are for agricultural irrigation water. At present, Customary Water Rights account for roughly 70 percent of the water rights for irrigated agriculture. Since 1960, there has been an increasing demand for domestic and industrial water and the conciliation of water rights in times of drought has become an important issue. Consequently, rules were established to try to protect the established Customary Water Rights and adjust water use under the New Water Law that came into effect in 1964. The Amended River Law of 1997 incorporated further measures to reconcile different water uses in order to avoid conflict in extreme droughts (Kataoka, 2005).

Importantly, there is a danger of romanticizing customary rights and rules by expecting them to automatically take into account the needs of the whole community. It has frequently been noted that customary ways of allocating and managing water resources can mirror unequal local power relations (Hodgson, 2004). In such cases, it should be an imperative of formal judiciary systems to make sure that powerless and economically weak groups are protected against local inequities in water distribution and management. More than anything, though, the lesson to be learned is that it is impossible to fully tap the potential benefits from water projects and management without taking into account customary water rights and local ways of managing water resources. Taking into account local water practices should form a part of any water reform to minimize the social and economic costs that can be associated with local opposition to water reforms and development projects.

2c. Water as a human right

It has been estimated that in order to ensure our basic needs, every individual needs 20 to 50 litres of water free from harmful contaminants each and every day (UN, 2003).

Over the last three decades, water has been addressed in a number of international conferences, which have recognized that water is a basic human need and some have gone as far as explicitly affirming the right to water. The right to water was only implicitly endorsed in the 1948 Universal Declaration of Human Rights (UNDHR). In 1977, the Mar del Plata Action Plan stated: 'all peoples, whatever their stage of development and their social and economic conditions, have the right to have access to drinking water in quantities and of a quality equal to their basic needs'.[4] The recognition of water as a right continued with the 1989 Convention of the Rights of the Child (CRC).[5] Article 24 of the CRC, paralleling Article 25 of the Universal Declaration of Human Rights, provides that a child has the right to enjoy the highest attainable standard of health. Measures taken to secure this right, include 'combat[ting] disease and malnutrition ... through, inter alia, ... the provision of adequate nutritious foods and clean drinking water'.

In 2002 the United Nations Committee on Economic, Social and Cultural Rights (ICESCR) adopted the General Comment on the right to water (see **Box 2.7**). The Committee emphasized the government's legal responsibility to fulfil the right and defined water as a social, cultural and economic good in addition to being an economic commodity. The right to water applies primarily to water of acceptable quality and quantity 'for personal and domestic uses' – in effect an emphasis on 'affordable' water supply and sanitation. The need for access to water for farming and other productive uses is

4. United Nations Water Conference held at Mar del Plata, Argentina, 7–18 March 1977.

5. www.unhchr.ch/html/menu2/ 6/crc/treaties/crc.htm

BOX 2.7: THE RIGHT TO WATER: GENERAL COMMENT 15

The United Nations affirmed the right to water on 26 November 2002, noting that such a right is 'indispensable for leading a life in human dignity' and 'a prerequisite for the realization of other human rights'. Through its General Comment 15, the Committee on Economic, Social and Cultural Rights of the United Nations Economic and Social Council stated: 'the human right to water entitles everyone to sufficient, safe, acceptable, physically accessible and affordable water for personal and domestic uses.' While the right to water has been implicit in the rights to health, housing, food, life, and dignity already enshrined in other international conventions, such as the International Bill of Human Rights and the Convention on the Rights of the Child, General Comment 15 is the first to focus explicitly on the right to water and the responsibilities that governments have in delivering clean water and adequate sanitation services to all.

Source: The full text of this General Comment is available at: www.unhchr.ch/tbs/doc.nsf/0/a5458d1d1bbd713 fc1256cc400389 e94?Opendocument

...a human right to water might have the unintentional effect of causing disputes between neighbouring countries that share water

acknowledged, but while 'water is required for a range of different purposes' (e.g., to secure economic production and livelihoods), 'priority in the allocation of water must be given to the right to water for personal and domestic uses'.

The General Comment provides for a 'progressive realization' of the right and acknowledges that there can be constraints due to water scarcities. It also refers to the role of donors and their responsibility to assist by providing financial and technical assistance. It is important not to confuse the right to water with water rights. Moreover, it would be wrong to interpret the General Comment as a right to free water. The Comment is clear on this and includes provisions regarding the 'economic accessibility' of water and water services defined as 'affordable'.

Only a few countries have made formal legal commitments to acknowledge a right to water, but even fewer have matched an explicit right to water in their constitutions with actual implementation (Centre on Housing Rights and Evictions, 2004). One such example is South Africa. Section 27 (1b) of the Bill of Rights of the Constitution of South Africa states: 'Everyone has the right to have access to sufficient food and water.'[6] Water policies and measures to implement this right in South Africa are now being developed (see **Chapter 14**).

There are current signs that the human right to water is gaining more national and international recognition. According to the Centre on Housing Rights and Evictions (2004), some of the more recent progress includes the following:

- In 2004, a Uruguayan referendum enacted the human right to water into the Constitution when more than 64 percent of the population voted in favour of the amendment. Kenya, in its 2004 draft constitution, is now considering the explicit inclusion of the right to water and sanitation in its legislation.

- Courts in India, Argentina, Brazil and South Africa have, in some cases, reversed decisions to disconnect water supply to poor people who cannot afford to pay.

- The Millennium Project Task Force on Water and Sanitation recommended that the international community explore ways to the right to water to influence national policy on water and sanitation.

Does it make a difference to recognize water as a human right?

The international community has for a long time explicitly acknowledged a human right to food. However, people continue to die of starvation and nearly a billion people remain undernourished. Does explicitly acknowledging a human right to water make a practical difference in people's lives?

Recognizing water as a human right can have a significant impact on national water law, policy, advocacy and development programmes. It can also be a way of promoting an enhanced effort by the international community and local governments to improve water resources management and to meet the MDGs on water supply and sanitation. It could further serve as a means to increase the pressure to translate the right to water into specific national and international legal obligations and responsibilities: 'To emphasize the human right of access to drinking water does more than emphasize its importance. It grounds the priority on the bedrock of social and economic rights, it emphasizes the obligations of states parties to ensure access, and it identifies the obligations of states parties to provide support internationally as well as nationally' (Jolly, 1998). Potentially, it can also support national and international legal frameworks that regulate access to water and contribute to water conflict resolution. Finally, explicitly acknowledging a human right to water can help to place water issues higher up on political agendas around the world.

But going beyond the legal sense of water as a human right, many critical practical questions remain for meaningful implementation: what aspects should such a right entail? What are the concrete economic and social benefits? What practical mechanisms would be required for its effective implementation? How can such a right be implemented if particular localities lack the hydrological necessities? Who should pay for it? Should the responsibility be placed on governments alone, or should it also be held by individuals, communities and private actors (Scanlon et al., 2004)? Some of the concerns that have been expressed include the fact that issues of water obligation and responsibility (e.g., the obligation not to pollute) are not sufficiently emphasized and minimum rights to water (for example 40 L/capita/day) can provide excuses for governments to 'lock' water provision at that level. As was previously noted, daily per capita water consumption in Europe and North America averages some 250 L to 600 L. It is thus clear that from

6. www.info.gov.za/documents/ constitution/1996/96cons2.htm

a development perspective more water or increased water use efficiency is required, apart from meeting basic needs. So far, the debate on water as a human right has revolved around safe drinking water, but very little on sanitation. Furthermore, basic water needs for direct economic activities, such as agricultural and industrial production, have not been a part of the water as a human right agenda.

Concern has also been raised that a human right to water might have the unintentional effect of causing disputes between neighbouring countries that share water. According to the 1997 UN Convention on the Law of the Non-Navigational Uses of International Watercourses, a country is not permitted to exploit a shared water resource in a manner that deprives individuals in a neighbouring country of access to their basic human needs (see **Chapter 11**). In practice, this kind of conflict seems unlikely to arise.

At the national level, governments are responsible for providing for adequate water supply and sanitation to their citizens and ensuring that citizens comply with existing water legislation. In practical terms, this means that relevant stakeholders participate in decision-making and that decisions are made transparent and information available, so that citizens can act on the information that is being provided to them. But to what extent do governments have an obligation to provide the right to water to their citizens? While the many international declarations and formal conference statements supporting a right to water do not directly require states to meet individual water requirements, Article 2(1) of the ICESCR binds governments to provide the institutional, economic and social environment necessary to help individuals to progressively realize those rights. It has been argued that

under certain circumstances, such as when individuals are unable to meet basic needs for reasons beyond their control, including natural disasters, discrimination, economic impoverishment or disability, governments should provide for basic water needs. Under such circumstances, the meeting of basic water needs may take precedence over other spending for economic development. It may also require new financial resources to be made available (Gleick, 1996).

According to a recent study, several cases (Centre on Housing Rights and Evictions, 2004) from both developed and developing countries demonstrate that there is a legally enforceable right to water. The Menores Comunidad Paynemil and Valentina Norte Colony cases from Argentina have required states to address the pollution of drinking water sources, as illustrated in **Boxes 2.8** and **2.9**. Cases from other parts of the world also suggest that taking the issue to court can be a successful way of addressing local water concerns. For example, court cases in India led the authorities to reconsider plans for digging wells on a set of islands. Local concerns raised the possibility that it might affect the water quality. In another case, local Indian authorities were ordered by the court to take immediate action to address the problem of inadequate sanitation. These cases show, among other things, that the effective implementation of water as a human right requires that the judiciary system function and that relevant water information be made accessible to the public.

2d. Is corruption draining the water sector?

Within public service institutions for water, corruption remains one of the least addressed challenges. Historically, bilateral and multilateral organizations and their clients have more or less tacitly accepted corruption

---

BOX 2.8: SAFEGUARDING WATER SUPPLY TO LOW-INCOME GROUPS

In the City of Cordoba, the water supply of nineteen low-income families was disconnected by a water service company on the grounds of non-payment. As a response, the families sued the water service company. They argued that the disconnection was illegal and that the company had failed to comply with its regulatory obligation to provide 50 L of water per day, which was to be supplied whether or not payment was made, They also claimed that even the minimum supply of 50 L was inadequate. The families requested the court to obligate the company to provide at least 200 L of water daily per family. The Judge rejected the argument that the decision by the company to cut or restrict the supply of water on the grounds of non-payment was illegal. However, the Judge recognized that the contractual obligation to provide a minimum of 50 L of water in all circumstances was clearly insufficient for a standard family and therefore required the company to provide a minimum of 200 L per household.

Source: Centre on Housing Rights and Evictions, 2004.

BOX 2.9: SAFEGUARDING THE RIGHT TO SAFE WATER SUPPLY

The Paynemil Mapuche Community in Neuquen, Argentina, 1995

People in the community discovered that the company's plant was polluting the local water source. The community, together with a university institute, filed complaints to six different local authorities about the potential heavy-metal pollution of the aquifers from which the community extracted its drinking water. As a part of the complaint, case studies were presented that showed that the water was non-potable. Studies ordered by local authorities showed that many children also had high levels of heavy metals (lead and mercury) in their bloodstream and urine. The public agents in the Provincial Ministry of Health brought the issue to a higher political level when they communicated their concern to the Health Minister. It was recognized that the water was unfit for human consumption and that traditional disinfection methods, such as boiling and filtering, were inappropriate. It was recommended that the

Minister intervene in order to provide water for the Community. In March 1997, the issue became a court case:

[T]he Children's Public Defender, filed an *accion de amparo* (a special expedited procedure) against the Government, arguing that the Province had neglected to fulfil its obligation to protect and guarantee the good state of health of the population. The court of first instance accepted the Public Defender's arguments and ordered the Provincial Executive Power to: (1) provide – within two days notice of the decision – 250 litres of drinking water per inhabitant per day; (2) ensure – within 45 days – the provision of drinking water to the affected people by any appropriate means; (3) set up – within 7 days – a procedure to determine whether the health of the population had been damaged by the existence of heavy metals, and in such a case, to provide the necessary treatment; and (4) take steps to protect the environment from pollution.

In May 1997, the Provincial Court of Appeals confirmed in all its terms the above-mentioned decision. Both courts based their decisions on the fact that the Government had not taken any reasonable measures to tackle the pollution problem that seriously affected the health of the Paynemil, even though it was well informed about the situation. The Court of Appeals stated: 'even though the Government has performed some activities as to the pollution situation, in fact there has been a failure in adopting timely measures in accordance with the gravity of the problem'. The Court of Appeals noted that, due to the serious consequences brought about by the pollution of water, any delay in providing resources and in adopting those steps necessary to reverse the present situation constituted an illegal omission violating the Paynemil community's constitutional rights to health and to a safe environment.

Source: Centre on Housing Rights and Evictions, 2004.

---

in public service delivery. Corruption has been seen as a 'necessary evil' and sometimes something that could 'grease the wheels' of development efforts. In recent years, there has been an important shift in thinking, and anti-corruption measures are now viewed as central to equitable and sustainable development. There is now also a growing body of research showing that corrupt practices are detrimental to economic efficiency and social equity and thus limit the scope for development opportunities.

Corruption is a symptom of poor governance in both private and public spheres.[7] In many countries, the legislative framework and judiciary systems are often inadequate and too weak. When this is combined with, for example, low wages, huge income disparities (both within and between countries) and accountability and transparency shortcomings, personal economic gains can be stronger than concern for the well-being of citizens, in terms of providing water-related services and the sustainable development of water resources. From an institutional perspective, corruption arises when public officials have wide authority, little accountability and

perverse incentives and when their accountability responds to informal patron–client linkages rather than adhering to existing rules, regulations and contracts. In countries where corruption is common and visible, there is often a high social acceptance of and tolerance towards corrupt practices that can lead to a deep-rooted 'culture of corruption'. It may even go as far as those engaging in it believing they have a right and entitlement to the benefits they reap and that working within the public sector is perceived as a fairly legitimate opportunity to enrich oneself and further one's personal or family interests.

Corruption undermines development efforts and makes it harder and much more costly to reach various national and international development targets, such as the MDGs. Fighting corruption is therefore not only a national priority but also a global challenge.

Corruption and its consequences for development and water service provision
More than US $1 trillion is paid in bribes each year worldwide in both rich and developing countries, according

7. The United Nations Development Programme (UNDP) defines corruption as 'The misuse of public power, office or authority for private benefit – through bribery, extortion, influence peddling, nepotism, fraud, speed money or embezzlement. Although corruption is often considered a sin of government and public servants, it also prevails in the private sector'. Corruption is not only about exchange of money and services, it also takes the form of cronyism, nepotism and various kinds of kickbacks. (TI, 2004, UNDP, 2004a.)

to estimates by the World Bank Institute (WBI, 2004). This is almost equal to the combined GDP of low-income countries. The $1 trillion figure, calculated using 2001–02 economic data, compares with an estimated size of the world economy at that time of just over US $30 trillion and does not include embezzlement of public funds or theft of public assets. It is very difficult to assess the extent of worldwide embezzlement of public funds, which is a very serious issue in many settings. For example, Transparency International (TI) estimates that former Indonesian leader Suharto embezzled anywhere between US $15 and 35 billion from his country, while Marcos in the Philippines, Mobutu in Zaire and Abacha in Nigeria may have embezzled up to US $5 billion each (TI, 2004).[8]

The estimation of global corruption costs does not take into account indirect costs in the form of alternative uses of funds for reducing poverty and economic inequalities and providing water, health care, education, etc. WBI research suggests that countries that tackle corruption and improve their rule of law can increase their national incomes by as much as four times in the long term, in addition to drastically improving service provision, such as water supply, sanitation and health.

Corruption costs the water sector millions of dollars every year. It siphons off scarce monetary resources and diminishes a country's prospects for providing water and sanitation for all. Corruption has several negative water development consequences (UNDP, 2004a):

- It reduces economic growth and discourages investments within the water sector. It undermines performance and effectiveness of both public and private sectors. The undermining of public and private institutions leads to inefficient and unequal allocation and distribution of water resources and related services.

- It decreases and diverts government revenues that could be used to strengthen budgets and improve water and other services, especially for poor people.

- It renders rules and regulations ineffective, thereby contributing to increased water pollution and over-abstraction of ground and surface water. It also breeds impunity and dilutes public integrity. Discretionary powers and uncertainties in policy and law enforcement create unpredictability and inequalities and can also lead to bypassing the rule of law and justice system.

Corruption is thus a serious problem within the water sector, but empirical evidence is still insufficient to make generalizations of the magnitude of the problem and by how much exactly it contributes to unsustainable uses of water resources, water pollution and how much it drains water development efforts. Corruption takes place in all countries, but in some countries, it occurs on a more systematic basis and is often seen as a part of how business is done between public agencies and citizens, and between public agencies and the private sector, as well as within the public sector itself. Typical features of systems prone to corruption include the following (UNDP, 2004a):

- concentration of political powers in the executive branch and private and public monopolies, in combination with weak or non-existing checks and

8. The global NGO, Transparency International (TI), was formed in 1993. TI has played an important role in putting the corruption issue on international as well as national agendas. It publicizes annual reports and a corruption index on the status of corruption in the world. The World Bank Institute (WBI) has also developed corruption indices. Another anti-corruption instrument in place is the OECD Convention on Combating Bribery of Foreign Public Officials in International Business Transactions.

BOX 2.10: **CORRUPTION IN THE IRRIGATION SECTOR IN PAKISTAN**

Agriculture is the largest sector in Pakistan's economy. It contributes one quarter of the country's gross domestic product (GDP) and employs almost half of the labour force. Agricultural production is highly dependent on irrigation, and the Indus Basin Irrigation System and groundwater resources irrigate 80 percent of Pakistan's cropland.

One important factor that hampers productivity in the rural sector is the difficulty of acquiring access to irrigation water, especially for tail-enders and small farmers in general. It has been observed that land inequality in rural Pakistan also reinforces inequities in access to critical resources like canal irrigation. Although canal irrigation substantially increases productivity, the pricing regime and delivery mechanism for canal water clearly benefits those who have large holdings. In addition, the flat-rate pricing mechanism for water often leads to wasteful and inefficient water use in a situation where water for irrigation is a highly valuable and scarce resource. Moreover, the ability to influence officials of the irrigation department to divert water to the highest bidder allows those with larger land holdings to skew water distribution in their favour. As a result, this can impose a threefold cost on the poor: they must pay water charges whether or not they get water, pay bribes to get the water which is their right, and suffer lower productivity due to uncertain and erratic water supplies.

Sources: World Bank, 2004 and *Daily Times*, Pakistan, 12 July 2004.

BOX 2.11: CORRUPTION IN THE WATER SUPPLY AND SANITATION SECTORS IN INDIA

A more systematic effort to map petty corruption and its *modus operandi* in India's water sector has recently been made. Results show the following:

- 41 percent of the customer respondents had made more than one small payment (median payment US $0.45) in the past six months to falsify metre reading to lower bills.
- 30 percent of the customer respondents had made more than one small payment (median payment US $1.90) in the past six months to expedite repair work.
- 12 percent of the customer respondents had made payment (median payment US $22) to expedite new water and sanitation connections.

The revenues lost due to falsifying water metres accumulates to large sums over time. This is money that alternatively could be spent on improved operation and maintenance, new investments to improve water and sanitation systems for economically weak groups, etc. Such alternative costs are rarely taken into account in corruption equations.

The study also indicates the frequency of side payments from contractors to public officials within the water and sanitation sector. According to public official respondents, side payments occur on a frequent basis:

- 17 percent said that it takes place every time
- 33 percent claimed it was quite common
- 8 percent said that it takes place about half the time
- 17 percent said that it occurs occasionally
- 25 percent said that it occurs infrequently/never.

The value of the kickbacks to public officials normally ranged from 6 percent to 11 percent of the contract value. The study also suggests that side payments for transfers of staff occur on a frequent basis. Interestingly, side payments for promotions were less common.

Source: Davis, 2004.

---

balances, poor transparency of decision-making and restricted access to information

- discretionary decision-making within both public and private sectors

- lack of accountability and weak systems of oversight and enforcement

- soft social control systems that provide a breeding ground for acceptance and tolerance for corrupt activities.

Two examples of corruption in the water and sanitation and irrigation sectors are provided in **Boxes 2.10** and **2.11**. While these examples are not intended to suggest that any one country or water sub-sector is worse than the rest, it is important to note that many countries that face severe water challenges are also ranked as corrupt countries.[9]

In places where corruption is endemic, the consequences are disproportionately borne by the poor, who have no resources to compete with those able and willing to pay bribes. In the end, corruption 'tightens the shackles of poverty' on countries or groups that can least afford it (UNDP, 2004a). Although in theory citizens have an option to stay out of corruption, it is often difficult to do so, since the choice can be between having access to

drinking water and going thirsty or having sufficient irrigation water for agricultural crop production or losing crops and farming income. Buying water from private water vendors is in general more costly than the municipal water supply. Water is something everybody needs, and there is no substitute for it. Thus, in reality, citizens often have few alternatives to bribing officials for receiving water and sanitation services.

Abating corruption

The global response to abate corruption has recently picked up speed. Many bilateral and multilateral organizations, governments, civil-society organizations and private businesses are currently developing internal and external governance guidelines, codes of conduct and sponsor anti-corruption/improved governance research and development programmes. In 2003, the United Nations Convention against Corruption was adopted by the General Assembly by resolution 58/4 of 31 October 2003. A major breakthrough of the Convention is that countries agreed on asset-recovery, which is stated explicitly as 'a fundamental principle of the Convention'. This is a particularly important issue for many developing countries where high-level corruption has plundered national wealth, and where resources are badly needed for reconstruction and the rehabilitation of societies under new governments. As of 11 October 2004, the Convention has 140 signatories and has entered into force. As a spin-off of the Convention work, the Global Compact adopted in 2004 a tenth principle on

9. For the most current corruption index, see: www.transparency.org/

corruption: 'Business should work against corruption in all its forms, including extortion and bribery.'

The battle against corruption is multi-faceted and requires actions at many levels in the form of public sector reform, increased salaries among public officials, strict enforcement of existing rules and regulations, improved accountability and transparency, multilateral cooperation and coordination to track financial flows and monitor international contracts, etc.[10]

Progress has been made in fighting corruption in some areas, but much still needs to be done. The main challenges lie ahead, and will require committed leadership from governments, the private sector and civil society backed by citizens' support to push ahead for effective anti-corruption measures and reform. Even though corruption in the water sector is facilitated by general inadequacies or a break-down of governance systems, private, public, local, national and international water decision-makers should all take responsibility for initiating and implementing adequate anti-corruption measures in the water sector. It is believed that changes within one sector can 'spill-over' and make a contribution to wider reforms of governance systems.

2e. Privatization of water services

The opening up of the water services sector to private competition is a part of water sector reform of many developing countries. This had not been possible without overall economic reforms of liberalizing and decentralizing economies that was initiated in the late 1980s and early 1990s. The ongoing water reforms in Ghana, India, Kenya, Niger, South Africa, Tanzania and Uganda, for example, are part and parcel of current economic and decentralization reforms where management and financial responsibilities are transferred to local governments, states or provinces.

The mobilization and effective use of existing financial resources necessary to manage water resources and water supply and sanitation services are major challenges facing public and private water decision-makers and managers around the world. Ultimately, the responsibility for financing the water resource infrastructure and water management needs rests with local or national governments. However, this does not automatically imply that funds necessarily have to come from governments. Investment from private and other external sources, such as donors and international lending institutions, is, and will continue to be, frequently required.

The current situation in many developing countries is problematic where even basic water functions, such as operation and maintenance of drinking water supply and sanitation utilities, are not adequately funded. Adequate investments for improved water resources management or ecosystem maintenance is typically even harder to obtain. In many developing countries, investment in water management and services is funded through insufficient and insecure sources by central governments with very limited funds. Water charges collected on the basis of resource use are often paid into a 'general exchequer', which can lead to lost opportunities to redefine financial roles and responsibilities between different water users and government agencies (Rogers and Hall, 2003).

Considerable discussion remains over public and private water services, institutional arrangements and the application of economic instruments to make water services delivery more efficient and equitable. Increased privatization, which in many cases has implied rising water prices, is in many social settings a heavily politicized issue that is creating social and political discontent, and sometimes outright violence, the most cited case being that in Cochabamba, Bolivia. There are also other cases where private operators have faced social protests against increasing user fees or private firms' performance. This has led to a situation where operations are handed back to public authorities. For example, Trinidad and Tobago is reforming its water sector, now with a strong reluctance to re-introduce private operators. There are, however, also examples, as in Port Vila, Vanuatu, where privatization can improve low-income households' access to water services. Experience demonstrates, among other things, the need for a well-planned concession contract, enforcement of regulatory powers and strong commitment by political leaders and participation by communities (see **Box 2.12**).

Private enterprises within the water supply sector skyrocketed during the 1990s. It went from almost none in the early 1990s to the current more than 2,350 private enterprises (UNDP, 2003). Until recently there have been expectations that the private sector will continue to expand its investments in developing countries. Estimation suggests that the private sector spent US $25 billion on water supply and sanitation in developing countries between 1990 and 1997, compared with US $297 million in 1984-90. Most of this investment was in Latin America and Asia, whereas Africa received less than one percent

The opening up of the water services sector to private competition is a part of water sector reform of many developing countries

10. See WBI recommendations at www.worldbank.org/wbi/governance

of the total investments made (UNEP, 2004). It now seems like this trend of increased privatization is reversing. Interestingly, many of the big multinational water companies are questioning their own role in increasing their investments in developing countries and playing an active role in resolving the water supply and sanitation crisis of many developing countries. Due to the political and economic high-risk operations, shrinking profit margins (in part due to currency instability), and increasing criticism affecting firms' business image, many of the multinational private water companies have started to retreat from water services contracts and

investments in developing countries. Saur, for example, has pulled out from some African countries like Mozambique and Zimbabwe. Another example is in 2003 when Suez started to downsize its water investments in developing countries with one-third of current investment levels. Likewise, other companies like Veolia and Thames Water are reconsidering their commitment in developing countries. In 2004, Thames Water pulled out from its water operations in Shanghai and is also facing difficulties in Jakarta, where it was involved in Jakarta's water privatization, which began in 1997, but where the company has not been able to

---

BOX 2.12: EXPERIENCES IN PRIVATE SECTOR INVOLVEMENT IN WATER SUPPLY AND SANITATION

Experiences in Côte d'Ivoire, Guinea and Senegal

In 2001 there was an assessment made of the results of water privatization in three African countries: Côte d'Ivoire, Guinea and Senegal. The contractual arrangements that were applied varied from medium-term leases to long-term concession contracts. All three countries displayed similar results from the privatization process: the connection rates had increased, and there were also tangible improvements for billing and collection of revenue. However, increased tariffs had made water supplies unaffordable for many of the poorest sections of society, which led to people getting disconnected from water supply due to inability to pay. It was also unclear to what extent poor people had benefited from water network connection expansions. Experiences confirmed that very poor sections normally tend to be excluded from being a part of privatized service extension. To provide the poorest section of society with adequate water services is typically viewed as a high-risk enterprise that largely lacks opportunities for economic return. Similar experiences have also been found in other places, such as in greater Buenos Aires, Argentina and in Bolivia.

Port Vila, Vanuatu: Water concession with pro-poor elements

In Port Vila, the poorest of the urban poor reside in overcrowded, informal, impermanent

housing on the city periphery. In many cases, conflicts over land tenure beyond the municipal boundary put a strain on the extension of services. Water supplies in many areas are comprised of hand-dug wells and shallow bores. In 1994, Union Electrique du Vanuatu (UNELCO, a subsidiary of Ondeo Services) signed a forty-year concession contract to supply water and electricity to Port Vila. A total investment of US $11.6 million was anticipated during the contract period, with US $580,000 per year for the first five years.

The concession aims at providing a 'self-regulatory system', where the government monitors the concession and facilitates the extension of services to new consumers. A structured fixed-tariff system was established to ensure affordable water rates for poorer sections of society as well. For low-income households, the US $0.20 per cubic metre (m^3) paid goes into a special fund to finance free connections. But households will still have to pay the quarterly consumption bill.

The company is regularly monitored by the government and provides regular financial reports and investment plans for five-year periods to the government for approval. In this particular case, UNELCO and the government has improved the water supply network and extended uninterrupted, affordable water services to many new households. Prior to the

contract, intermittent water supply was often common, but in most cases, water is now available 24 hours per day, and unaccounted-for water has decreased dramatically from 50 to 23 percent. The water tariff for the first 50 m^3 per month was reduced to US $0.58 per m^3 from US $0.75 per m^3. It has been reported that annual economic losses of up to US $440,000 in 1991 have been turned into a surplus of US $12,000 in 2000. Private sector involvement seems to work well in this small urban center. Groundwater not requiring treatment is inexpensive; tariffs are relatively higher than in other Asian cities; and the relatively high cost-recovery facilitates free connections for the poor. Some of the components that have made this enterprise work for poor people include a pro-poor concessional approach, with a clear investment plan and targets of network expansion; multiple service levels; and cross-subsidized connection fees. The government has shown a strong commitment and enforced its regulatory powers. In addition, the study also indicates the importance of strong community leadership, active participation by recipient communities and a high degree of awareness among water services consumers in general.

Sources: ADB, 2002; Mehta and Mirosa-Canal, 2004.
For more accounts of water services privatization in urban areas, see also UN-HABITAT, 2004.

make profit. Resistance from consumers and political unease have delayed planned rises of water rates, and the company also brought on bad publicity due to allegations of inadequate service performance. Many of the multinational water companies are consequently focusing even more on the less risky markets of Europe and North America.[11]

The private sector has been more effective at handling issues of efficiency in water distribution systems, like decreasing unaccounted-for water, improved billing and increased revenue collection than at meeting issues of equity of water distribution through the extension of water networks. Rapidly increased water charges have in some cases implied that poor people are being disconnected from services. Those who have benefited from private water services in developing countries are predominantly those living in relatively affluent urban pockets. Cost recovery for water services related to irrigation and water supply and sanitation in rural, peri-urban and slum areas has proven to be a much more complicated task and is often considered to be less economically viable. Often, the extension of services to poorer sections of society is not a part of public–private contracts. It is evident that the private sector is facing similar difficulties as many developing country governments are unable to reach poor people with improved water services. The record of private sector involvement has been mixed, and there are often wide gaps between the effectiveness of regulatory powers and private sector operations.

Weak regulatory powers and poor governance

The increasing privatization has implied a new division of roles: governments are increasingly becoming the service regulator, while the private sector becomes the service provider. Many governments in developing countries operate with a low regulatory capacity for enforcement of contracts and the distribution and monitoring of water licensing and permits. With regulatory functions often vested with government agencies, also responsible for water services provision, conflicts arise regarding service quality and a government agency's independence may be called into question. Successful water services privatization will require a clear set of rules that promotes both equity and efficiency in water distribution, effectively enforced by an independent government regulator adequately equipped with authority, finances and human capacities.

This development has taken place in both middle-income countries, like South Africa, and in some of the world's poorest countries, such as Niger. Obviously, these countries have different regulatory capacities. The regulatory authority is absent in Niger, while it is relatively effective in South Africa (Mehta and Mirosa-Canal, 2004). Adequate and enforced regulatory frameworks are critical for successfully increasing private sector involvement in water service delivery. But despite this insight, in many countries, privatization is taking place without a balance between financially strong private companies and financially strained government agencies with low institutional capacities. Some of the multinational companies that government authorities are going in to contract with and are supposed to regulate can sometimes have bigger annual turnover than the GDP of the country. For example, in 2002 the water-related revenue of Vivendi Environment exceeded the GDP of countries like Côte d'Ivoire and Kenya (Mehta and Mirosa-Canal, 2004). Such economic asymmetry can have an effect on the negotiation powers of going in to contracts and how contracts are interpreted once they are implemented. Governments, including donors, should increasingly support the establishment of independent and strong regulatory authorities in order to facilitate the enforcement of concessions or other types of arrangements between the public and private sectors.

Sound governance practices create enabling environments, which encourage equitable and efficient public and private sector investment. Beyond providing general support to achieve good governance conditions, specific attention should be directed to efforts that reduce risk and facilitate healthy capital markets, especially domestically. Some additional points on using improved governance to mobilize financial resources (Rogers and Hall, 2003):

- Water should be recognized as an economic, social, and environmental good; the full costs of water management and water services must be acknowledged, and the costs should be transparent and affordable (through tariffs, cross-subsidies, taxes, etc., see **Chapter 12**). All costs and benefits of improved water management and services should be taken into account.

- Capital should be made available at all levels, such as micro-credit, revolving loan programmes and the issuing of local bonds.

Sound governance practices create enabling environments, which encourage equitable and efficient public and private sector investment

11. *The Economist*, The flood dries up: International water companies, August 28 2004; Mehta and Canal, 2004.

■ It is essential to have institutional clarity regarding water access and allocation: water rights and permits, regulatory frameworks and management responsibilities.

■ Decision-making systems that are transparent, inclusive and that can be held accountable are necessary.

The importance of improved governance can be illustrated with Porto Alegre, Brazil, where the opening up of decision-making processes, such as participatory budgeting, led to improved governance systems (see **Box 2.13**). Controversies that have surrounded some of the public–private water service contracts could have been reduced if the contracting process had been open to public scrutiny.

Untapped potentials – local small-scale water companies

The bulk of discussions on the need for improved financing and increased privatization of water services has focused on the role of multinational companies. However, privately operated water utilities (excluding small-scale local operators) only supply approximately 5 to 10 percent of the world's population with drinking water and even less with sanitation (McGranahan, 2004a). Very little attention has been paid to increasing capacities and incentives of domestic water operators and local entrepreneurs or to exploring the role of local communities and various kinds of water user associations and community-based organizations. Multinational companies' comparative advantage is to construct, operate and manage large-scale infrastructure for drinking water and sanitation, hydropower and irrigation. In many instances, such large-scale infrastructure development

remains economically unviable. It is becoming increasingly acknowledged that many water multinationals, as well as governments, so far have not been able to provide the type of technology that developing countries need to meet the MDGs on water and sanitation or other international and national water targets. Considering the billions of people who are underserved with sufficient water supply and sanitation one can conclude that there exists huge demand for small-scale improvements as well as application of large-scale solutions where appropriate. Local entrepreneurs, communities and local organizations should increasingly be viewed as important stakeholders who can contribute to meeting the MDGs and other international water targets.

There is currently an increasing number of examples of private companies in developing countries that are involved in the water sector to provide more appropriate small-scale technology and where consumers may have greater opportunities to decide on technologies applied. It has been noted that small-scale independent water networks are best documented and perhaps also most common in Latin America. For example, in Paraguay it has been estimated that small-scale operators serve about 9 percent of the population. In other countries the coverage tends to be considerably lower but significant enough to compete with larger water providers (Solo, 2003). The small-scale water networks have mainly emerged due to failing public utilities. A study on the role of small-scale enterprises and water user associations in providing water supply to rural areas in Morocco showed that there are big potentials for both local entrepreneurs and stakeholders to be involved in a meaningful way. The potential of local entrepreneurs has also been noted in other contexts, such as in Uganda. In 1991, water services management

Recent trends indicate that water multinationals are scaling down their investment plans in developing countries

---

BOX 2.13: IMPROVED GOVERNANCE FOR BETTER WATER SERVICES IN BRAZIL

In Porto Alegre, Brazil, in the past decade, US $140 million was invested in water and sanitation systems, 80 percent of which was generated through tariffs. The tariff system includes cross-subsidies and a minimum water supply requirement of 10 m³ per month for low income households, but they only have to pay the cost of 4 m³. There are different tariff levels depending on income. Large consumers, like factories and shops, pay higher rates. It is

reported that 99.5 percent of the city is supplied with water of a good quality, and 84 percent of the city's sewage is being treated. Interestingly, this part of Brazil is known for being wealthier and having less social inequality than many other parts of Brazil; furthermore, there is a vibrant associational life. Hence, it seems like the governance system coupled with existing levels of wealth and civil society vibrancy has made it possible to raise required

investments to improve water supply and sanitation. Hence, it seems to be less of a question of whether or not a service should be public or private, and more a question of the strength of existing institutions and how well they can carry out public service tasks that benefit the whole society.

Source: Mehta and Mirosa-Canal, 2004.

BOX 2.14: THE ACTIVE ROLE OF WATER USER ASSOCIATIONS IN MOROCCO

Involving private companies in the management of urban water supply is considered one of the options for improving services in developing countries. But what about the management of rural water supply? How can small-scale entrepreneurs and other local actors be involved in a meaningful way? An example is provided from Morocco.

Water supply in rural areas of Morocco has developed much more slowly than in urban areas. Coverage in rural areas has been estimated at 40 percent. The central water authority Office National de l'Eau Potable (ONEP) faces challenges of increasing water supply coverage and sustaining such efforts. ONEP is now exploring measures of decentralizing, delegating and sub-contracting

rural water supply. A pilot study reveals that small-scale entrepreneurs (one to five employees) and water user organizations have a big potential role in improving rural water supply. Small-scale entrepreneurs have a significant role to play in pumping station operations, maintenance and caretaking, leakage control, repairs and maintenance of major pipes, maintenance of local piped network, standpipes and household connections, water metre reading, and water quality surveillance. Through the survey it became apparent that small-scale entrepreneurs have some technical experience but lack experience with commercial management. Communities preferred local entrepreneurs that they know. Interestingly, the entrepreneurs had a strong preference for service contracts rather than management

contracts. The service contract has the advantage of offering the operator a relatively 'safe' income, as it does not rely on the amount of water sold and its price. Management contracts normally insist that incomes are based on tariff and quantity of water sold and thus provide a more uncertain situation of future incomes. It was also clear from the study that water user associations have an important role to play in maintaining local water networks as well as mobilizing communities. Interestingly, water user associations were, along with local authorities, considered to play a significant role in commercial management of water, such as consumer administration, metre reading, and control of free-riders.

Source: Brikké, 2004.

---

contracts were awarded to local companies in Uganda for the management of water supplies in nine small towns. Prior to this, the World Bank had invested in infrastructure improvements. By 2003, the contracting out of water supplies management had extended to twenty-four towns. The basis for these management contracts can be traced back to the legal and institutional reforms undertaken in Uganda during the 1990s. There are indications that these management contracts have reached some of their objectives. Since a large share of the urban poor live in small cities and towns this example can illustrate that there is a potential for local small-scale companies to reach underserved, low-income households in urban areas (McGranahan, 2004b).

Local private enterprises often have several advantages: (1) knowledge of local conditions; (2) local procurement, production and employment; (3) medium and small-scale operation that can have a positive impact on services in secondary cities and towns as well as rural villages; (4)

niche markets and a multiplicity of appropriate technologies that can be explored in a more effective way. However, a greater involvement of local enterprises in operating water and sanitation utilities does not automatically mean that services to poor people will improve. There is not yet any conclusive evidence that local small-scale water network operators would be more effective than other types of set-ups to reach poor people and informal settlements (McGranahan, 2004b). The effectiveness of reaching poor people will continue to depend on the nature of contracts, sound governance systems, increased investments and enhanced local water management capacities. A disadvantage is that local entrepreneurs in many developing countries are inexperienced in managing and operating utilities (McGranahan, 2004b). As seen in the case of rural water supply in Morocco (see **Box 2.14**), there may also in other contexts be preferences and capacities that favour service contracts with the public sector rather than managing and operating utilities.

Part 3. Decentralization of Water Control and Decision-making

Principle 10 of Agenda 21 (UN, 1992) states:

> **Environmental issues are best handled with the participation of all concerned citizens, at the relevant level. At the national level, each individual shall have appropriate access to information concerning the environment that is held by public authorities, including information on hazardous materials and activities in their communities, and the opportunity to participate in decision-making processes. States shall facilitate and encourage public awareness and participation by making information widely available. Effective access to judicial and administrative proceedings, including redress and remedy, shall be provided.**

Yet the implementation of these rights has not kept pace with the expectations generated by this international declaration. The absence of information or mechanisms for participation and redress result in decisions that adversely impact, exclude, and are consequently opposed by, affected communities. Such decisions are rarely effective, frequently illegitimate and unjust, and undermine the ability to integrate environmental concerns into development processes. As the UN Economic Commission for Europe (UNECE) Convention on Access to Information, Public Participation in Decision-making, and Access to Justice in Environmental Matters in Member States (the Aarhus Convention) indicates, no country has fully developed policies or the organizational capacity to implement all pillars of Principle 10. This is even more true in many developing countries where transparency and inclusiveness in decision-making have a shorter history and often face stronger resistance by special interests inside and outside governments.

In the last decade or so, there has been a recognition that if efforts to improve the quantity or quality of water supply are to be successful, not only must they be technically sound and economically feasible, but they must also deal directly with poverty alleviation, local empowerment and environmental sustainability. In this regard, it has been recognized that local organizations and communities that are direct water users have strong local hydrological and socio-economic knowledge and also have the most at stake in water decisions.

The importance of participation and bottom-up approaches and the critical role of local community initiatives for resolving water challenges have been demonstrated by the Arwari River Parliament in

Rajasthan, India. The river parliament, formed in 1998, is based on a community-centred river basin approach and consists of some seventy villages and forty-six micro watersheds. The parliament meets twice a year and aims at a wide range of objectives, such as sustainable management of water resources, managing soil fertility and land erosion, stopping illegal mining activities, generating self-employment and alternative livelihood options, promoting women's groups and increasing agricultural productivity through application of local seeds and manure. It is reported that there have been positive social, economic and environmental impacts. Among other things, patterns of resource use are regulated, and there is also a platform to resolve disputes related to the allocation and management of water, forest and land (Moench et al., 2003).

Several alternative methods of widening and deepening people's participation are emerging in different countries to address water-related and other development challenges. Reforms regarding increasing inclusiveness in water decision-making are applied to the water sector in various degrees and forms.

3a. Benefits of decentralization

The concept of decentralization has evolved over time and acquired different interpretations. In a simplistic way, decentralization can be said to be a means of dispersing decision-making closer to the point of where the practical work is done. It has, among other things, been defined as the primary strategy for transferring responsibility from the central government to sub-national levels of government and representing a fundamental change in the institutional framework in which political, social and economic decisions are made (Rondinelli et al., 1984).

The term has now come to be used in senses that deviate in many ways from this characterization, since decentralization currently also refers to the transfer of responsibilities to civil society and the private sector. Decentralization should thus generally be perceived as a process in which the government relinquishes some of its decision-making powers and management responsibilities to lower levels of government, private sector or community and civil society organizations.

When it works well, decentralization has many benefits: it can allow for a democratization of decision-making through improved stakeholder inclusiveness, transparency and accountability. Appropriately implemented, it can empower people, particularly those lacking the social and political clout and financial means to have a voice and take part in decisions that define their livelihood opportunities. It can also encourage the integration of traditional knowledge and practices with innovative technologies and science to promote fair and efficient management of water resources and services.

Decentralized systems can make governments more responsive. It is more likely that local public sector officials will be held accountable by the local consumers of services than by central government officials. It can thus promote government agency responsiveness to address water services and administrative shortcomings. The existence of autonomous and flexible local governments can promote the production of appropriate and innovative practices. Similarly, local governments can be better positioned to work with local NGOs and other volunteer, community-based organizations in identifying problems and defining solutions. Decentralizing decisions can also lead to easier public access to water information, which also entails lower costs to citizens for obtaining information about government performance for local government services than for central government.

As pointed out, decentralization done appropriately can lead to increased democratization and bring decisions closer to water users. But there is also an economic efficiency rationale behind decentralization that often goes hand-in-hand with aims of democratizing decision-making. It is assumed that bringing decisions closer to water users can lead to better water resources allocation as well as services provision. In a highly centralized system of public goods and services, there is often a lack of reliable information about the actual costs of those services and to what extent services are being subsidized.

Moreover, there is limited information on performance and quality of services of public sector agencies. In contrast, the allocation of resources to public sector services is likely to be more efficient under a decentralized regime, with the assumption that local public institutions are more likely to have better information on citizens' preferences and needs, such as water demands among various user groups and the willingness and capability to pay for improved services. Another way of decentralizing is to make the direct water users more responsible for water management. For example, benefits of local community participation and the use of traditional water managers to mediate water conflicts, allocate water in efficient and equitable ways and maintain irrigation infrastructure have been demonstrated in the case of irrigation management in India. Such experiences can also serve as an entry point for decentralizing of irrigation-management responsibilities. The local modalities and premises for water allocation, management practices and mechanisms for dispute resolution can differ.

3b. Decentralization in practice

Despite ongoing reform work, many countries still lack legal provisions for local government water responsibility in water resources management/development. A survey of forty-two countries in Africa, Asia, Europe and Latin America indicates that twenty-four countries are lacking legal provisions for local government responsibility. However, in some countries, the legal provisions for local water resources management are fairly strong. For example, in Mongolia, regional and local governors have responsibilities for water management and plans for water collection, restoration and use. The water laws in Viet Nam call for more integrated approaches and have decentralized irrigation management to local water committees.[12]

What does the practice of decentralization tell us? Decentralization of the power to manage water (not just read metres and fix leaks) does not come easily. There are at times very strong social and political forces, both inside and outside government, which benefit from preserving status quo. A number of studies confirm the general hypothesis emerging from research that the theory and rhetoric of decentralization frequently fails to match the willingness of central governments and their attached agencies to relinquish power (Olowu and Wunsch, 2003).

Years of centralized authority may have undermined old traditional resource management systems, as in the case

Decentralization can empower people, particularly those lacking the social and political clout and financial means to have a voice and take part in decisions that define their livelihood opportunities

12. These figures are based on Water Law and Standards. See: www.waterlawandstandards.org

BOX 2.15: TRADITIONAL SYSTEMS UNDER THREAT IN INDIA

In the Indian States of Karnataka, Andhra Pradesh and Tamil Nadu, traditional water managers are called *Neerkatti*. They manage water tanks for irrigation. Their knowledge of the terrain, drainage and irrigation needs is much beyond that of present-day water engineers. Their role begins before the monsoon. The tank, which is common property, requires collective action to be maintained. The *Neerkattis* decide upon the date on which the community assists to desilt the tank and clean the catchment. They size up the work required and divide the labour among tank beneficiaries. With the first monsoon shower, they take stock of water availability and decide upon per capita allocation and what type of crops to grow. The *Neerkattis* ensure supply to every field on a rotational basis. How this is done can vary. In one village the tail-end users receive water first. The lands located closer to the tank will benefit from seepage water. The duration of irrigation depends on the crop being grown. If a crop begins to dry, *Neerkattis* have the authority to divert water to drying fields, even by closing the diversion to all other fields. They can also have other functions related to crop management. To minimize the risk of partisanship, the *Neerkattis* do not enjoy any formal political power or own land linked to the water-tank system.

Similar systems with traditional water managers are also found in other parts of India, such as in Ladakh, Uttaranchal and Maharashtra. For example, in Maharashtra's cooperative irrigation system, the Cultivator Committees appoint water managers. The post is often hereditary to ensure loyalty to local practice. To ensure neutrality in irrigation decision-making, landless village residents are usually chosen. Their task is to look after, regulate and maintain water flows in the main, distributary and field canals. The water managers play a critical role in resolving water disputes thanks to their knowledge of water flows in different canal systems. In some villages in Ladakh, a contract is written between the traditional water managers and villages. The contract stipulates obligations and responsibilities and the impartiality of the water managers. Water managers also supervise maintenance and mediate conflicts. If there are any serious disputes, the matter can be referred to other village institutions. How water is allocated differs. In some places, there has been a kind of lottery system developed to ensure impartial allocation. In other places, irrigation water is prioritized to the farmer who tills the land first, irrespective of location.

The role of traditional water managers has been greatly reduced over the past fifty years, mainly due to water becoming a state property. Irrigation management is consequently run by government departments and has shifted to a government induced trend from surface-water to groundwater irrigation. Consequently, traditional tank irrigation infrastructure has degraded, and canals have clogged up. Traditional water management often clashes with formal water licensing practices and regulations, hence creating uncertainties in management responsibilities. The governments of Andhra Pradesh and Tamil Nadu have recently enacted legislation to enhance participatory irrigation management, stemming from growing water disputes and the need to revive water tanks. However, the legislation has largely implied the creation of thousands of water user associations (70,000 in Andhra Pradesh alone) without reviving the role of traditional water managers. These water user associations have had very mixed results. Traditional water management knowledge is increasingly being lost.

Source: Centre for Science and Environment, 2003.

of years of colonial rule followed by centralized authority (**Box 2.15**). Past centralization of water decision-making in India has led to local ways of managing water resources being increasingly lost. In such cases, civil society of local water management institutions must be reinvented and reformulated.

Observed trends show many governments to be unwilling to decentralize adequate powers and resources to local bodies, which can severely stifle local bodies' options to manage water resources and deliver services. It has also been observed that governments may take back powers and financial resources from local bodies, often in response to pressure from bureaucrats and legislators unhappy with the loss of power (Manor, 2003). For example, in the Middle East

and North Africa, many cases of participatory irrigation management (PIM) indicate that governments give PIM rhetorical support, but in practice, they do not provide appropriate incentive structures, institutional mechanisms or regulations to allow for effective local irrigation management. Despite the evidence that PIM works, some governments are less enthusiastic about local water management than they claim to be. Several case studies demonstrate that with the application of PIM, water-use efficiency can increase by up to 30 to 50 percent and that the energy use for pumping was cut in half (Attia, 2004). Other, less well-documented benefits can include a reduction in local water conflict and a sense of community and individual empowerment that is said to improve family health and well-being (Brooks, 2002).

Other studies have looked at the potential of decentralization to enhance democratic participation and the empowerment of marginalized groups, including women. They demonstrate that democratic decentralization has a very mixed record as a means of reducing poverty. It has been noted that it may help to reduce poverty that arises from inequalities between regions but often does less to reduce poverty that arises from inequalities within regions. Recent evidence suggests that the impact of decentralization may be more positive than previously thought, particularly for women. There is clear evidence that the presence of significant numbers of women in local decision-making bodies sometimes increases the quantity and quality of crucial services, such as ante- and postnatal care (Manor, 2003).

3c. Information, development and access

People's participation in and access to relevant water information are essential preconditions for successful decentralization. One of the reasons why decentralization is claimed to be conducive to efficiency and equity in water allocation and distribution is because it enables local-level services to be tailored according to local needs and demands. What are the mechanisms through which local needs, demands and knowledge can be made known to decision-makers? The only feasible way is to have an inclusive and transparent process of local and central decision-making through which various stakeholders can voice their

rights and preferences. This point has been underlined in a comparison between successful and unsuccessful cases of community-based resource management in Mexico. Two different features of successful management were identified: the first emphasized the important role of vigorous, regular and well-attended community assemblies and the second stressed the importance of accounting and reporting practices that provide community members with a sufficient flow of information (Klooster, 2000). The difference in water services that such vigorous and transparent local community assemblies can make has also been reported from other parts of Latin America. A case study (Rosensweig, 2001) of the small town of Itagua, Paraguay showed that community-based water boards were successful in improving water services.

3d. Degree of public participation in water decision-making

The weak levels of participation and access to information have also been confirmed in a number of country surveys carried out by The Access Initiative (TAI) in 2001–03. **Table 2.2** shows that legal provisions for participation are weak.

In this first round of national assessments undertaken by TAI, access to information on water quality monitoring and the degree of public participation in water decision-making processes emerged as key areas for improvement in most countries.[13]

UNICEF school project in a camp for internally displaced persons in Sudan

Table 2.2: Public participation rights in constitutional and legal frameworks

Indicators	Weak	Intermediate	Strong
Constitutional guarantees to public participation	Constitution does not explicitly guarantee right to public participation in decision-making: **Chile, Hungary, India, Indonesia, Mexico, South Africa, United States**	No value offered: only two indicator choices were 'strong' and 'weak'	Constitution guarantees the right to public participation in decision-making: **Thailand, Uganda**
Comprehensiveness of notice and comment in different types of decision-making processes	Types of policy- and project-level decisions requiring public notice and comment are not specified: **Indonesia, Thailand**	Types of project-level decisions requiring public notice and comment are specified, but types of policy-level decisions are not: **Chile, Hungary, India, Uganda**	Types of both policy- and project-level decisions requiring public notice and comment are specified: **Mexico, South Africa, United States**
Public notice and comment requirements for environmental impact assessments	No requirement for public notice and comment for Environmental Impact Assessments (EIAs): **Thailand**	EIAs require public notice and comment at final stage: **Hungary, India, Indonesia, Mexico, Uganda**	EIAs require public notice and comment at various stages: **Chile, South Africa, United States**

Source: The Access Initiative, National Team Reports, see WRI, 2002.

13. In 2001–03, TAI conducted pilot assessments in nine countries to test its methodology and identify needs for improved access to information and participation. For example, a number of sub-indicators were used to evaluate a number of characteristics of public participation in specific decision-making cases. Some of these indicators examine the law and regulations governing public participation. Others focus on practice illustrated by selected decision-making processes. The results described here are based on the assessments found at www.accessinitiative.org

The TAI pilot assessments indicated the following:

- Public participation rights are insufficiently articulated in most pilot country legal and constitutional frameworks.

- National-level environmental policy-making cases recorded the strongest rankings for quality and accessibility.

- Regional or local planning processes demonstrated intermediate or strong levels of accessibility but considerable variability in the quality of effort made by public authorities.

- In productive, extractive, infrastructure and other sectors at the national level in the pilot countries, decision-making is generally less accessible to the public.

- Project-level decisions recorded highly variable public participation, in terms of both quality and accessibility.

- Participation tends to be weak at the earliest stages of decision-making and in the monitoring of implementation or review of performance.

- Decision-making processes usually place the onus of initiating participation on the public or affected communities.

- Meaningful public participation improves decisions. In the cases where the pilot-country government invested in supporting meaningful participation and actively solicited input – or where civil society organized, initiated a dialogue or provided input to which the government responded – the decisions adopted incorporated environmental and social aspects.

In 2004–05, over twenty countries carried out TAI assessments, and thus more information will be available on information disclosure and participation. Some initial results from this next round of assessments have shown specific gaps, particularly in information dissemination among minority groups as well as public involvement in planning processes. As countries develop IWRM plans, it is important to examine the degree to which the public is not simply consulted but rather directly involved in the formulation of the plans. **Box 2.16** highlights two cases regarding access to water information and participation in Ukraine and Estonia. In addition, TAI is beginning the process of

...the public rarely has easy access to useful information about the quality of drinking and surface water

developing water governance indicators to be pilot tested in 2005–06.

3e. Access to water quality information

Information about water may also mobilize public opinion to urge polluters and governments to reduce pollution and improve water quality. It was found in the TAI pilot assessments that the public rarely has easy access to useful information about the quality of drinking and surface water. **Table 2.3** indicates that even if the quality of information is good, the access to it is often very limited. As a result, individuals and communities cannot protect themselves from contaminated water or monitor the improvement of its quality. The TAI pilot assessments on access to water quality information also indicated the following results:

- The water quality monitoring systems assessed had operated for more than three years. All monitoring systems, except those in India, have provided regular reports at least over this period. Some of the monitoring systems have operated for decades. For example, RandWater in South Africa began operations in 1927.

- In terms of the breadth of parameters monitored, monitoring systems are categorized as either comprehensive or basic. In Hungary, India, Mexico, South Africa and the United States, it was found that the chosen systems monitor a comprehensive set of physical, bacterial, chemical and viral parameters in water. Systems in Thailand and Uganda monitor a more basic set.

- The quality of the system for providing water quality information also depends on how the monitoring networks are coordinated. Monitoring systems can cover a single urban area (as in Mexico or Indonesia), entire countries (as in Hungary, India, Chile and Thailand), or selected regions within a country (such as the State of California in the United States).

- In Thailand, the monitoring of drinking water from the tap, by contrast, is divided among several bodies – the Metropolitan Waterworks, Provincial Waterworks Authorities, and the private Universal Utilities Company. The monitoring of bottled water comes under another body – the Food and Drug Administration. In Thailand it was also found that, while some analytical data on the quality of tap water

BOX 2.16: ACCESS TO INFORMATION AND PUBLIC PARTICIPATION IN THE WATER SECTOR

Water-related emergencies in Ukraine

The case of an emergency in 2000 involving the pollution of groundwater in five settlements of the Pervomaysk district of the Mykolayiv oblast, caused by a number of extremely toxic chemical agents (the so called 'Accident in Boleslavchik'), was assessed. Different indicators were used to estimate access to information for environmental emergencies. Among the indicators, the highest score was given to the presence of information about environmental emergencies on the Internet. This type of information is freely located on the website of the Ministry on Emergencies of Ukraine. However, the public had no access to online information about the impact of this environmental emergency on people's health and environment, especially about its effect on the quality of drinking water. It was rather difficult to find empirical material related to public participation for the chemical sector. The research team of Ecopravo-Kyiv focused on the role of the public in the development of the National Environment Health Action Plan (NEHAP) and the 2002 Ukrainian law, 'On drinking water and water supply system'. Assessment of public participation in decision-

making related to the implementation of policy, strategy, plans, programmes and legislation was made using different indicators. It was found that there was no public participation in developing the above documents for the chemical sector. However, all of the documents were available and accessible to the public.

Drinking water monitoring and regional water plans in Estonia

The access to drinking water monitoring data in a small town in southeast Estonia was assessed, where drinking water problems are known to exist, and these problems are representative of the whole region. The town is also representative in terms of administrative capacity. When compared to air monitoring data, data for drinking water dissemination is not very developed. One of the reasons is perhaps that air monitoring belongs to the jurisdiction of the Ministry of the Environment (where several trainings and projects have been carried out to implement the Aarhus convention), but drinking water falls under the control of the Ministry of Social Affairs. There is a strong legal mandate for drinking water monitoring, but there is a

problem with making this data public. However, the Ministry of Social Affairs has initiated a project for disseminating drinking water monitoring data to the public via the Internet. There are eight sub-basins in Estonia for which the government has started drafting water management plans. The Pandivere basin is located in the central Estonia and is a nitrate-sensitive region. Overall, the public was invited to participate in drafting the water management plan and was given reasonable time for commenting. Public input was incorporated into the final decision. However, from the perspective of the engagement of minorities, the opportunity for public participation was negative. No special efforts were made to invite Russian-speaking people (approximately 5 percent or less of the population in the basin) to participate in the plan's development. Although the law prescribes conditions for participation (e.g., time for commenting, number of meetings, etc.), the law does not say anything about taking special efforts to involve minorities.

Source: The Access Initiative, National Team Reports, EcoPravo, Kiev, Ukraine and Stockholm Environment Institute, Tallinn, Estonia, 2004.

could be obtained from water authorities, the country's Food and Drug Administration (FDA) provides no analytical monitoring on contaminants in bottled water. The FDA only notes the conclusion of its analysis: whether the quality of the drinking water of the selected brand is 'safe'. This policy means that consumers cannot check for the presence of specific contaminants. This lack of detail can be particularly relevant to vulnerable populations such as children, pregnant women, and older people.

- In Hungary, different aspects of drinking and surface water monitoring, however, are managed by different agencies, and neither the system nor the data are coordinated or integrated. The lesson to be drawn from both the Thai and the Hungarian examples is straightforward: unified and integrated systems provide a more coherent picture of water quality and present less of a challenge in obtaining information.

- Information technologies facilitate public access to information. Websites increasingly provide an opportunity for the public to learn more about water quality monitoring issues. In Hungary, the National Health Action Program website provides widespread coverage of environmental and health issues. In California, a website for the Environmental Justice Coalition for Water encourages citizens to become involved in monitoring the water quality in their communities.

- Two countries provide examples of how water data can be disseminated. In South Africa, RandWater has created a website to provide users with updates on water issues. For example, a map highlights areas where water should not be used for drinking without treatment and where contact should be avoided because of microbiological health effects. In the United States, water suppliers disseminate annual reports to customers about their drinking water.

Table 2.3: Quality and accessibility of water data, selected countries

Country	Quality of information[1]	Accessibility
Hungary[2]	Intermediate	Weak
India	Intermediate	Weak
Indonesia[3]	Weak	Weak
Mexico[4]	Strong	Weak
South Africa	Intermediate	Intermediate
Thailand	Weak	Weak
Uganda	Weak	Weak
United States: California	Intermediate	Strong

1. Systems score weak when only a few parameters on quality of water are collected.
2. Data from almost all 12 inspectorates and from 7 of 19 public health offices in four weeks; 7 of the 19 offices responded on drinking water.
3. Indonesia submitted a single value for both air and water quality information.
4. Mexico disseminates drinking water information at the state level but not by individual water supply.

Source: The Access Initiative, National Team Reports, 2004.

■ Detailed information on drinking and surface water quality, on the other hand, is difficult to obtain in all but two of the pilot countries: the United States and South Africa. Under the 1996 amendments to the Safe Drinking Water Act, the United States requires water suppliers to provide customers with annual reports. These reports are usually mailed with bills; many are also posted on the Internet. Teams in five countries (Hungary, India, Mexico, Thailand and Uganda) found no active dissemination of data on drinking water quality for the public on the Internet or in the press. In Mexico and Uganda, teams could not obtain the data at all; in India, data could be obtained only through a personal contact.

In short, there are considerable differences in the performance of government agencies in providing information to the public about drinking or surface water quality. Collectively, performance scores in providing water quality information are weak. This should be contrasted to the assessment of information disclosure and public participation with regard to air quality that the assessment found to be strong.

Men fishing at sunset with square nets in Dhaka, Bangladesh

Part 4. Water Governance Ahead

In the past decade, water and its governance have featured prominently on the international political agenda and will continue to be an international priority through 2015 within the Decade on Water for Life. International efforts to foster water institutional reform have included recommendations on good water management practice, and the setting of goals and targets for improved water service provision to the poor and for greater environmental sustainability via the MDG structure. High hopes and expectations are now vested in recent international water targets from the Millennium Summit and the WSSD to improve the water situation for billions of people. It is a paradox that while various international fora have intensified their work towards improving the world's water situation and implementing time-bound water targets, the actual funding to the water sector in developing countries is diminishing, or stagnating at best. Funding from donors remains stagnant, and additional investments from multinational water companies to improve water governance and access to water are currently decreasing. There is very little evidence that governments in developing countries are strengthening their water budgets.

It has been demonstrated that water governance is nested in the setting of overall national governance and is correspondingly influenced, for better or for worse, by that, by the national culture, and by events local to the country and its surrounds (e.g., conflict) and developments in the global economy. Some of the trends of water governance include the following:

- As a response to internal pressures and to pressures from the international community and regional organizations such as the EU, a widespread process of reform of water governance is now underway. Progress is patchy but generally slow, as evidenced by the limited achievements in the production of national IWRM plans and the weak coverage of water in PRSPs. In developing countries there are often significant and serious gaps between policy-making and its implementation, not least because of institutional resistance by public sector water organizations.

- Progress is being made in water rights – in recognizing their importance in dealing with social and economic problems, in recognizing the importance of local traditions and customs, in facilitating the management of local competition for water and in recognition of human rights to safe water.

- Corruption is a major issue in the water sector, as in many other sectors, but the impact of it is not well quantified because of a lack of detailed information. It is one of the least addressed challenges in the sector and much remains to be done.

- Privatization of water services in developing countries has not been able to meet the high expectations on improved and extended water supply and sanitation services. Much of the privatization debate has had a biased focus on multinational water companies. Local and small-scale water companies are mushrooming in both urban and rural settings and their potential to improve water supply and sanitation largely remains unexplored. There is thus a need to refocus privatization to more systematically explore how local water enterprises, including both water companies and civil society organizations, can contribute to improved water services. It is also high time to bring the government back in and re-emphasize its importance in raising and stimulating adequate investment funds, as well as its critical role in regulatory and other governance functions.

- Recent moves by governments in lower-income countries to delegate responsibility for water management to lower levels of administration have had limited success. Progress is slow, governments are not delegating the needed powers and resources and have in some cases taken back the delegated responsibility. Often local governments do not have the capacity to do what is required. Local groups and individuals are hampered by lack of access to key information and frequently by exclusion from participating in water decision-making. There should also be a more sober view on decentralization itself and what types of decentralization are useful for improved water resources management and services provision.

In developing countries there are often significant and serious gaps between policy-making and its implementation, not least because of institutional resistance by public sector water organizations

Decentralization without the right checks and balances may lead to local elites strengthening their positions at the expense of politically and economically marginalized groups.

At present, more effective water institutions are yet to evolve in many countries. Much of the conceptual development and division of roles and responsibilities among government agencies, private businesses and civil society tend to become mired in politics and do not reflect true on-the-ground needs. Governments and donors should increasingly support the establishment of independent and strong regulatory authorities to facilitate enforcement of concessions or other types of arrangements between the public and private sectors. There should be increased efforts to strengthen regulatory capacities as well as to make them independent.

Experience has shown that development can be more deeply rooted in systems where governments, private firms and civil society can work together in various constellations. There needs to be an improved water trialogue between governments, civil society and the private sector. The enhancement of governance, coupled with an integrated management approach, can be a vehicle for increased transparency, participation and a climate of trialogue and trust-building, aimed at increasing negotiation and minimizing differences within the water sector. It is perhaps naive to think that all disputes and differences can be bridged, but a society that claims to attack water problems must make serious efforts to address differences and be able to come up with legitimate institutions and processes that can mediate disputes (by the judiciary system, informal conflict resolution mechanisms and elections), or at least minimize their impacts (compensation to disfavoured groups, etc.).

It has been demonstrated that governance matters for the equitable, efficient and sustainable management of water resources and related services and contributes to achieving international water development targets. Governance systems are intrinsically linked to political processes and power. Therefore, the road to improved governance cannot avoid politics and manoeuvring in highly politicized contexts. Addressing improved water governance is challenging, since it needs to take place within a larger context of reform. Still, stakeholders within the water sector can do their part by striving towards integrated policies that also allow for multi-

There is very little evidence that governments in developing countries are strengthening their water budgets

stakeholder participation and subsidiarity. Water stakeholders at all levels should not refrain from attempting to play a role in shaping policy outcomes and influencing political will. They must be increasingly prepared to manoeuvre in different social and political contexts. This is not to suggest that water managers should choose political sides, but rather that they should be aware that policy-making and implementation involve politics. By knowing the political game and rules, they can make more strategic inputs into policy-making processes and other decision-making processes. In continuing the work to improve water governance it is critical to have the following items in mind:

■ Water sector reform goes hand-in-hand with overall governance reform. It is highly unlikely that more effective participation, transparency, etc. will take a firm root in the water sector, unless the country's overall governance system allows it to do so. As a part of broadening the water agenda, there is an increasing need to harmonize and coordinate international water targets and principles with other international regimes, such as with global or regional trading regimes. Unless water concerns are made part of broader national and international trade processes, stability and democratization, the chances of achieving the international water targets remain difficult. There is thus a need to collaborate with new actors outside the water realm and to form more inclusive water development networks.

■ Water reform and implementation is progressing, although sometimes at a painstakingly slow pace. In many developing country settings, the water sector and its institutions are plagued by fragmentation, marginalization and low capacities. Further, the marginalization of water departments and ministries in a country's overall political affairs is the rule rather than the exception. International water targets and national IWRM plans and policies mean little unless they are translated into legislation, institutional reform, participation implementation, sufficient funding, etc. Actions must be intensified towards the implementation of water policies and plans. Increased political commitment to implement existing water policies and legislation would go a long way towards achieving the international water targets.

■ The representation and participation of various interests in water decision-making is an important component in addressing urban and rural water and food security and governance dynamics. Issues of power and representation should be made explicit, for example, while negotiating fair compensation to farmers for water transfers to cities.

4a. No blueprint for progress

There is no blueprint for improved governance. With social, political and economic preconditions as a base any society must find its own way of improving governance. Despite the variety in design and implementation of improved governance across the world, there are also certain characteristics of the water sector that must be taken into account:

- Water is a renewable resource, as it reproduces itself through the hydrological cycle. While there is plenty of freshwater at the global scale, there is a challenge of having sufficient water, of the right quality, at the right place and at the right time.

- The nature of water resources is multi-purpose and hydrologically interconnected.

- Water is mainly considered a public good, but due to its multi-purpose nature, it is also at times an economic good. Importantly, water has economic, social and environmental value, which, among other things, calls for dialogue between water users and enhancement of participation and multi-stakeholder processes.

- The provision of water-related services is often characterized as being close to monopoly situations, particularly for water supply and sanitation. This can limit the effectiveness of water markets and can also require regulated price ceilings to protect consumers from monopolistic power abuses.

- The capital-intensive nature of water-related infrastructure is often combined with low-cost recovery and heavy subsidization. Water infrastructure investment is also considered a 'sunk cost', meaning that investments made to provide water services cannot be transferred or redeployed for other purposes, hence increasing capital investment risks.

It is important to develop institutions and governance systems that can effectively respond to situations characterized by variability, risk, uncertainties and change. Conventional water planning remains rigid, and the challenge remains to develop governance frameworks and institutions that are flexible and adaptive. More attention needs to be given to resilient institutions and approaches that can govern or guide the complex, often surprise-laden, process of water governance central to long-term management at a regional, basin, aquifer or even local level (Moench et al., 2003). This suggests that specific solutions – the ideal solution – may be less relevant and emphasizes the importance of enabling processes and frameworks that can be applied to resolve certain issues in situations of economic or other constraints and in contexts of change, that is, 'second or third best' solutions.

Many countries are currently at a crossroads about whether to provide the required political and financial capital to enhance efforts to improve water governance. 'Business as usual' is no longer a viable option. If investment levels and reform speed are not stepped up, countries' abilities to provide water and sanitation for all, increase food production, while maintaining the environment will be seriously compromised. This will, in the long- and short-run, curtail societies' prospects for development.

...there is a challenge of having sufficient water, of the right quality, at the right place and at the right time

References and Websites

ADB (Asian Development Bank). 2002. *Beyond Boundaries Extending Services to the Urban Poor.* Manila, ADB.

Arab Water Council, United Nations Development Programme and Centre for Environment and Development for the Arab Region and Europe. 2005. *Status of Integrated Water Resources Management (IWRM) Plans in the Arab Region.* Cairo, Egypt.

Attia, B. 2004. Comparative Analysis: Case Studies of Tunisia, Turkey, Yemen and Egypt. IDRC. *Water Demand Management Forum – Middle East and North Africa: Advocating Alternatives to Supply Management of Water Resources.* CD-ROM.

Beccar, L. et al. 2002. Water rights and collective action in community irrigation, R. Boelens and P. Hoogendam (eds), *Water Rights and Empowerment.* Assen, The Netherlands, Gorcum Publishers.

Boelens, R. and Hoogendam, P. (eds.). 2002. *Water Rights and Empowerment.* Gorcum Publishers, Assen, The Netherlands.

Bojö and Reddy. 2002. Poverty reduction strategies and the environment, World Bank, *Environmental Economics Series Paper,* No. 86.

Brikké, F. 2004. Act Locally! How to involve small-scale enterprises in providing water supply to rural areas. *Industry and Environment,* Vol. 27, No.1, UNEP.

Briscoe, J. 1997. Managing water as an economic good: Rules for reformers. *Water Supply,* Vol. 15, No. 4.

Brooks, David B. 2002. *Local Water Management.* Ottawa, Canada, IDRC Books.

Bruns B. R. and Meinzen-Dick, R. S. (eds). 2000. *Negotiating Water Rights.* New Delhi, IFPRI, Vistaar Publications.

Centre for Science and Environment. 2003. Apt arrangers, *Down To Earth.* 31 October. www.downtoearth.org.in/

Centre on Housing Rights and Evictions. 2004. Legal resources for the right to water: International and national standards. Amsterdam, The Netherlands. www.cohre.org/water/

Cordova, J. 1994. Mexico. J. Williamson (ed.), *The Political Economy of Policy Reform.* Institute for International Economics, Washington DC.

Crook R. C. and Manor, J. 1999. *Democracy and Decentralization in South Asia and West Africa Participation, Accountability and Performance.* Cambridge, UK, Cambridge University Press.

Davis, J. 2004. Corruption in public service delivery: Experience from South Asia's water and sanitation sector. *World Development Report,* Vol. 32, No. 1, pp. 53–71.

Delli Priscoli, J. 2004. What is public participation in water management and why is it important? *Water International,* Vol. 29, No. 2, pp. 221–27.

Department for International Development (DFID). 2002. The political economy of water reform. Unpublished working paper.

Diao, X. and Roe, T. 2000. The win-win effect of joint water market and trade reform on interest groups in irrigated

agriculture in Morocco. A. Dinar (ed.), *The Political Economy of Water Pricing Reforms.* New York, Oxford University Press.

Dinar, A. (ed.). 2000. *The Political Economy of Water Pricing Reforms.* New York, Oxford University Press.

Frans and Soussan. 2003. *Water in Asian PRSPs.* Asian Development Bank Water and Poverty Initiative. Manila, ADB.

Gleick, P. H. 2003. Global freshwater resources: Soft-path solutions for the 21st century. *Science.* Vol. 302, pp. 1524–28.

——. 2000. The human right to water. P. Gleick (ed.), *The World's Water: 2000–2001,* Washington, DC, Island Press.

——. 1996. Basic water requirements for human activities: Meeting basic needs. *Water International,* Vol. 21, No. 2, pp. 83–92.

Global Water Partnership (GWP). 2003. Effective water governance: Learning from the dialogues. www.gwpforum.org/gwp/library/ Effective%20Water%20Governance.pdf

——. 2000. *Towards Water Security: A Framework for Action.* Stockholm, GWP.

——. 2000. Integrated Water Resources Management. TEC Background Paper, No. 4, Stockholm.

Gooch, G. D. and Huitema, D. 2004. Improving governance through deliberative democracy: Initiating informed public participation in water governance policy processes. Paper presented at the Stockholm Water Symposium, 18 August 2004.

Hodgson, S. 2004. *Land and Water – The Rights Interface,* FAO Legal Papers No. 36. FAO, Rome.

Holmes, Paul R. 2003. On risky ground: The water professional in politics. Paper presented at Stockholm Water Symposium, 2003.

IFAD (International Fund for Agricultural Development). 2004. Linking land and water governance – IFAD experience. Proceedings of workshop, Farmers' views first: Land and water governance. World Water Week, Stockholm 2004. www.ifad.org/events/water

Johnson, R. W. and Rondinelli, D. 1995. *Decentralization Strategy Design: Complementary Perspectives on a Common Theme.* North Carolina, US, Research Triangle Institute.

Johnson, R. W. and Minis, H. P. Jr. 1996. *Towards Democratic Decentralization: Approaches to Promoting Good Governance.* North Carolina, US, Research Triangle Institute.

Jolly, R. 1998. Water and human rights: challenges for the twenty-first century. Address at the Conference of the Belgian Royal Academy of Overseas Sciences, 23 March, Brussels.

Kataoka N. 2005. Conservation of the waterfront environment along Japan's rivers: Institutions and their reforms of river basin management. J. L. Turner and K. Otsuka (eds) *Promoting Sustainable River Basin Governance - Crafting*

Japan-U.S. Water Partnerships in China, Institute of Developing Economies, IDE, Chiba, Japan.

Kaufmann, D. and Kraay, A. 2003. *Causality which Way? Evidence for the World in Brief.* Washington, DC, World Bank Institute.

Kaufman, D., Kraay A. and Zoido-Lobaton. 1999. *Governance Matters.* Washington, DC, World Bank Institute.

Keohane, R. and Ostrom, E. (eds). 1995. *Local Commons and Global Interdependence: Heterogeneity and Cooperation in Two Domains.* Berkeley, California, Sage Press.

Klooster, D. 2000. Institutional choice, community, and struggle: A case study of forest co-management in Mexico. *World Development,* Vol. 28, No.1, pp. 1–20.

Manor, J. 2003. Local governance. Paper prepared for Sida, Stockholm.

McGranahan, G. 2004a. Getting the private sector to work for the urban poor: Revisiting the privatization debate. Issue Paper in United Nations Human Settlements Programme (UN-HABITAT). 2004. *Urban Service Dialogue: Getting the private sector to work for the poor.* Second World Urban Forum, September 2004, Barcelona.

——. 2004b. Getting Local Water and Sanitation Companies to Improve Water and Sanitation Provision for the Urban Poor. Issue Paper in United Nations Human Settlements Programme (UN-HABITAT). 2004. *Urban Service Dialogue: Getting the Private Sector to Work for the Poor.* Second World Urban Forum, September 2004, Barcelona.

Mehta, L. and Mirosa-Canal, O. 2004. Financing water for all: Behind the border policy convergence in water management. IDS Working paper, No. 233.

Meinzen-Dick R. S. and Bakker M. 2001. Water rights and multiple water uses: Issues and examples from Kirindi Oya irrigation system, Sri Lanka. *Irrigation & Drainage Systems,* Vol. 15, pp. 129–48.

Moench, M., Dixit, A., Janakarajan, S., Rathore, M. S. and Mudrakartha, S. 2003. *The Fluid Mosaic: Water Governance in the Context of Variability, Uncertainty and Change,* Ottawa International Development Research Centre (IDRC).

Natural Resources Systems Programme (NRSP). 2004. Research Highlights 2002–03. Hemel Hempstead, UK: NRSP.

OECD (Organization for Economic Co-operation and Development). 2002. Supporting the development of water and sanitation services in developing countries, *Development Co-operation Report,* Paris.

Olowu, D. and Wunsch, J. S., 2003. *African Decentralization and Local Governance.* Boulder, Colorado.

Pierre, J. (ed.) 2000. *Debating Governance.* Oxford, Oxford University Press.

Polidano, C. 2001. Why civil service reforms fail: Public policy and management, Working Paper No. 16, Institute for Development Policy and Management, Manchester.

Postel, S. 1999. *Pillar of Sand: Can the Irrigation Miracle Last?* New York, Norton.

Rogers P. 2002. Water Governance. Paper prepared for IADB annual meeting, March 2002.

Rogers, P. and Hall, A. 2003. Effective water governance. *TEC Report No. 7*, Global Water Partnership, Stockholm.

Rondinelli, D., Nethis, J. and Chemma, S. 1984. *Decentralization in Developing Countries: A Review of Current Experience.* Washington, DC, World Bank.

Rosensweig, F. 2001. *Case Studies on Decentralization of Water Supply and Sanitation Services in Latin America.* Strategic Paper No. 1, Environmental Health Project, USAID, Washington.

Scanlon, J., Cassar, A. and Nemes, N. 2004. Water as a Human Right? *IUCN Environmental Policy and Law Paper No. 51*, Gland, Switzerland.

Shivakoti, G. P. and Ostrom, E. (eds). 2001. *Improving Irrigation Governance and Management in Nepal.* ICS Press.

Slaymaker and Newborne. 2004. Implementation of water supply and sanitation programmes under PRSPs. London, Overseas Development Institute.

Solo, T. M. 2003. *Independent Water Entrepreneurs in Latin America: The Other Private Sector in Water Services.* World Bank, Washington D.C.

Soussan, J. 2003. *Water and Poverty: Fighting Poverty through Water Management.* Manila, ADB.

Strand, J. 2000. A political economy analysis of water pricing in Honduras's capital, Tegucigalpa. A. Dinar (ed.), *The Political Economy of Water Pricing Reforms.* New York, Oxford University Press.

Transparency International (TI). 2004. *Global Corruption Report 2004.* London, Sterling, VA, Transparency International and Pluto Press.

Tropp, H. 2005. Building New Capacities for Improved Water Governance. Paper presented at the International Symposium on Ecosystem Governance, 2005, South Africa, organized by CSIR (Council for Scientific and Industrial Research).

UN (United Nations). 1992. Agenda 21. Official outcome of the United Nations Conference on Environment and Development (UNCED), 3–14 June 1992, Rio de Janeiro.

UN-WWAP (World Water Assessment Programme). 2003. *World Water Development Report: Water for People Water for Life.* Paris, UNESCO and New York, Berghahn Books.

UNDP (United Nations Development Programme). 2004a. *Anti-corruption.* New York, Practice Note.

——. 2004b. *Decentralised Governance for Development: A Combined Practice Note on Decentralization, Local Governance and Urban/Rural Development.* New York, Practice Note.

——. 2004c. *Governance Indicators: A User's Guide.* Oslo Governance Centre.

——. 2003a. *Human Development Report: Millennium Development Goals – A Compact Among Nations to End Human Poverty.* Oxford and New York, Oxford University Press.

——. 2003b. *Kazakhstan National Human Development Report 2003: Water as a Key Human Development Factor.* Almaty, Kazakhstan.

——. 2002. *Public Administration Reform.* New York, Practice Note.

UNEP (United Nations Environment Programme). 2004. Water and development: Industry's contribution. *Industry and Environment*, Vol. 26, No. 1.

United Nations Human Settlements Programme (UN-HABITAT). 2004. Urban service dialogue: getting the private sector to work for the poor. Paper presented at the Second World Urban Forum, September 2004, Barcelona.

UN Millennium Project. 2005. *Investing in Development: A Practical Plan to Achieve the Millennium Development Goals.* New York. www.unmillenniumproject.org/reports/fullreport.htm

Wambia, J.M.. 2000. The political economy of water resources: Institutional reform in Pakistan. A. Dinar (ed.), *The Political Economy of Water Pricing Reforms.* New York, Oxford University Press.

Ward, C. 2000. The political economy of irrigation water pricing in Yemen. A. Dinar (ed.), *The Political Economy of Water Pricing Reforms.* New York, Oxford University Press.

Waterbury, J. 2002. *The Nile Basin: National Determinants of Collective Action.* New Haven and London, Yale University Press.

WBI (World Bank Institute). 2004. The costs of corruption. *News & Broadcast.* web.worldbank.org/WBSITE/EXTERNAL/NEWS/0,,contentMDK:20190187%7EmenuPK:34457%7EpagePK:34370%7EpiPK:34424%7EtheSitePK:4607,00.html

Williamson, J. (ed.). 1994. *The Political Economy of Policy Reform.* Washington, DC, Institute for International Economics.

World Bank Report. 2004. *Pakistan Poverty Assessment.* Washington DC, World Bank.

WRI (World Resources Institute). 2003. *World Resources 2002-2004: Decisions for the Earth – Balance, Voice and Power.* Washington, DC, World Resources Institute.

——. 2002. *Closing the Gap: Information, Participation and Justice in Decision-making for Environment*, Washington DC, World Resources Institute.

The Access Initiative: www.accessinitiative.org

The African Water Page: www.thewaterpage.com

Centre on Housing Rights and Evictions (COHRE): www.cohre.org/water/

Freedom House: www.freedomhouse.org

Global Water Partnership: www.gwpforum.org

Transparency International: www.transparency.org/

Utstein Anti-Corruption Resource Centre: www.u4.no/

UNDP Governance Centre, Sources for Democratic Governance Indicators, 2003. www.undp.org/oslocentre

UNDP Water Governance Facility at Siwi: www.watergovernance.org

UNDP: www.undp.org/water

Water Law Standards, see: www.waterandstandards.org

The World Bank Institute: www.worldbank.org/wbi/governance/

World Resources Institute: www.wri.org

The sewer is the conscience of the city

Victor Hugo, *Les Misérables*

CHAPTER 3

Water and Human Settlements in an Urbanizing World

By

UN-HABITAT
(United Nations Human Settlements Programme)

Low-income neighbourhood built on a hill in Mexico City

Key messages:

Urban populations have exploded worldwide in the last fifty years creating unprecedented challenges, among which provision for water and sanitation have been the most pressing and painfully felt when lacking. Those who suffer the most are the poor, often living in slum areas that are left out of water development schemes, due often to failures in governance at many levels. Yet new partnerships emerge, where local communities are empowered to build innovative and efficient models that integrate socio-economic realities and improve water and sanitation provision.

Above: Downtown skyscrapers in Los Angeles, United States

Below: A mother washing her child in a slum, India

Bottom: Cairo, Egypt

■ It is within human settlements that virtually all non-agricultural water use is concentrated, and also where most water-related diseases are contracted. The role of planning for the water needs of human settlements becomes increasingly pressing in an increasingly urbanized world where water provision for urban centres (both for production and for human use) represents an increasing proportion of total freshwater use.

■ Human settlements provide a concrete context for action. The struggle to achieve the Millennium Development Goals (MDGs) for water and sanitation will have to be achieved in human settlements – in our cities, towns and villages. Here is where the actions have to be coordinated and managed. It is at this level that policy initiatives become an operational reality and need both political and administrative support: conflicts have to be resolved and consensus found among competing interests and parties.

■ The water and sanitation MDG targets will not be met without better urban governance but this has to be embedded within regional water governance arrangements that can manage water stress. Human settlements are the loci of water service provision; many are also the major polluters of water resources. Metropolitan, city and municipal governments have critical roles in water governance to ensure adequate provision for water and sanitation within their boundaries.

■ International agencies need to recognize the key role of local processes and local institutions in meeting the water- and sanitation-related MDGs and in better water management – and the involvement within this of local governments and civil society. The institutional framework through which national governments and international agencies support these local processes has to change in most low- and middle-income nations.

■ More attention is needed to generate the information base to support the improvement and extension of provision for water and sanitation to those who are unserved or inadequately served. This information base is often weakest in the areas where provision is worst – in large rural settlements and in the informal settlements where much of the unserved urban population live. Details are needed for each household and housing unit of the quality and extent of provision (if any) combined with maps of each settlement which show each housing unit, existing water pipes, sewers and drains, and paths and roads.

Part 1. The Changing Face of Human Settlements

During the twentieth century, the world's urban population increased more than tenfold, while rural population increased but twofold.[1] Today, half the world's population lives in urban centres, compared to less than 15 percent in 1900.[2] In 1900, 'million cities' (cities with more than one million inhabitants) were unusual and cities with over 10 million unknown; by 2000, there were 387 million cities and 18 with more than 10 million inhabitants. This trend of urbanization, in addition to the increasing importance of large cities, was underpinned by the transformations of national political structures (especially the virtual disappearance of colonial empires) and of national economies and employment patterns (most economic growth was in urban-based industry and services). Virtually all nations experienced an employment shift away from agriculture, and most of the growth in those working in industry and services were in urban areas.

By 2000, Asia alone had nearly half the world's urban population and more than half its million cities

In the second half of the twentieth century, most of the growth in the world's urban population was in low- and middle-income nations. By 2000, Asia alone had nearly half the world's urban population and more than half its million cities. **Tables 3.1 and 3.2** show the growth in urban populations and urbanization levels between 1950 and 2000. **Figure 3.1** highlights the regions where most of this growth in urban population took place. On a global scale, this increasing concentration of population and economic activities in urban areas is likely to continue, with most of the increase over the next twenty to thirty years likely to be in urban areas in Africa, Asia

and Latin America (UN, 2004). This chapter thus focuses on improving water and sanitation provision in urban areas.

1a. Trends in an urbanizing world

The trend towards more urbanized societies and the growing number of people living in large cities have very large implications for freshwater use and wastewater management. Although within virtually all national economies, agriculture remains the largest user of freshwater resources, the water demands from city enterprises and consumers have become increasingly important, and many major cities have had to draw

Figure 3.1: Comparative distribution of the world's urban population, 1950–2000

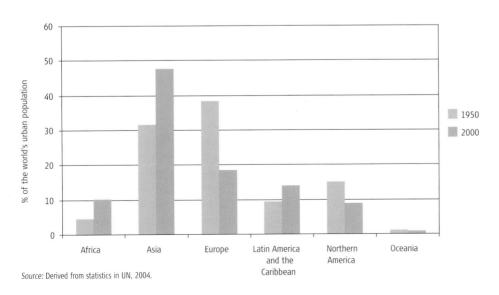

Source: Derived from statistics in UN, 2004.

1. Unless otherwise stated, this chapter draws its population statistics from United Nations, 2004.

2. United Nations estimates suggest that 48.3 percent of the world's population lived in urban areas in 2003 and that this figure will reach 50 percent around 2007. Figures for rural and urban populations for 1900 are drawn from Graumann, 1977.

Table 3.1: The distribution of the world's urban population by region, 1950–2010

Region	1950	1970	Year 1990	2000[1]	Projection for 2010
Urban population (millions of inhabitants)					
World	733	1,330	2,273	2,857	3,505
Africa	33	83	199	295	417
Asia	232	486	1,012	1,367	1,770
Europe	280	413	516	529	534
Latin America and the Caribbean	70	163	314	393	472
Northern America	110	171	214	250	286
Oceania	8	14	19	23	26
Population living in urban areas (%)					
World	29.1	36.0	43.2	47.1	51.3
Africa	14.9	23.2	31.9	37.1	42.4
Asia	16.6	22.7	31.9	37.1	42.7
Europe	51.2	62.9	71.5	72.7	74.2
Latin America and the Caribbean	41.9	57.4	71.1	75.5	79.4
Northern America	63.9	73.8	75.4	79.1	82.3
Oceania	60.6	70.6	70.1	72.7	73.7
World's urban population (%)					
Africa	4.5	6.2	8.7	10.3	11.9
Asia	31.7	36.5	44.5	47.8	50.5
Europe	38.3	31.0	22.7	18.5	15.2
Latin America and the Caribbean	9.50	12.3	13.8	13.8	13.5
Northern America	15.0	12.9	9.4	8.8	8.2
Oceania	1.1	1.0	0.8	0.8	0.7

1. The statistics for 2000 in this table are an aggregation of national statistics, many of which draw on national censuses held in 1999, 2000 or 2001 – but some are based on estimates or projections from statistics drawn from censuses held around 1990. There is also a group of countries (mostly in Africa) for which there are no census data since the 1970s or early 1980s so all figures for their urban (and rural) populations are based on estimates and projections.

Source: Derived from UN, 2004.

freshwater from increasingly distant watersheds, as local surface and groundwater sources no longer meet the demand for water, or as they become depleted or polluted. There has also been a rapid growth in the contribution of city-based enterprises and consumers to water pollution – although in many high-income and some middle-income nations, this has been moderated and on occasion reduced by more effective wastewater treatment, pollution control and economic shifts away from industry. As this chapter will describe in more detail, provision for water and sanitation in urban areas has expanded much more slowly than population growth in most low-income and many middle-income nations.

Another important human settlement trend is change in the employment structure of many rural settlements and small urban centres. Although in most nations, available data do not allow any precise documentation of this phenomenon, a large part of the rural population lives in what could be called 'large villages' or 'small towns', with populations

ranging from a few hundred to several thousand inhabitants. For water and sanitation and wastewater management, these 'large villages' or 'small towns' have great importance, given the fact that in virtually all nations, a large proportion of the national population live in them, and that they have key 'urban' characteristics that have relevance for how best to improve provision for water and sanitation, even if they are classified as 'rural'.

1b. Challenges of an urbanizing world: Inadequate provision, increased inequalities
The main water-related urban challenge in low- and middle-income nations remains ensuring adequate provision for water and sanitation and sustainable wastewater management. As data about deficiencies in provision in urban areas have improved, so has the extent of these deficiencies become more evident. According to the (WHO/UNICEF) Joint Monitoring Programme (JMP), if the Millennium Development Goal (MDG) of halving the proportion of people without sustainable access to safe

Table 3.2: The distribution of the world's largest cities by region, 1800–2000

Region	Year 1800	1900	1950	2000[1]
Number of million cities				
World	2	17	86	387
Africa	0	0	2	35
Asia	1	4	31	194
Europe	1	9	30	62
Latin America and the Caribbean	0	0	7	49
Northern America	0	4	14	41
Oceania	0	0	2	6
Regional distribution of the world's largest 100 cities (%)				
Africa	4	2	3	8
Asia	65	22	37	44
Europe	28	53	34	15
Latin America and the Caribbean	3	5	8	16
Northern America	0	16	16	15
Oceania	0	2	2	2
Average size of the world's 100 largest cities (thousands of inhabitants)				
World	187	725	2 200	6 300

1. Some figures for city populations for 2000 are based on estimates or projections from statistics drawn from censuses held around 1990. There is also a group of countries (mostly in Africa) for which there is no census data since the 1970s or early 1980s so all figures for their city populations are based on estimates and projections. The regional distribution of cities in 1950 and 2000 is in part influenced by the way that cities/urban agglomerations are defined within nations.

Source: Satterthwaite, 2005.

The main water-related urban challenge in low- and middle-income nations remains ensuring adequate provision for water and sanitation and sustainable wastewater management

drinking water supply and basic sanitation is to be met by 2015, 961 million urban dwellers must gain access to improved water supply and over 1 billion must gain access to improved sanitation (WHO and UNICEF JMP, 2004).

In 2000, more than 900 million urban dwellers lived in slums, most under life- and health-threatening circumstances in low- and middle-income nations (UN-HABITAT, 2003a).[3] This represents nearly a third of all urban dwellers worldwide. In most urban areas in low- and middle-income countries, between one-quarter and one-half of the population lacks provision for water and sanitation of a quality that greatly reduces the risk of human contamination with faecal-oral pathogens (UN-HABITAT 2003b). In most smaller urban centres, the proportion of people lacking good provision for sanitation is even higher. Most urban centres in these nations have no sewers at all and little or no other public infrastructure for good quality sanitation (Hardoy et al., 2001., UN-HABITAT, 2003b). Not surprisingly, at any given time, close to half the population in low- and middle-income nations is suffering from one or more of the main diseases associated with inadequate water and sanitation and, as **Chapter 6**, describes, virtually all these diseases and health burdens could easily be prevented (WHO, 1999; Millennium Project, 2005a).

The impact of these inadequacies in provision are difficult to convey using conventional indicators and quantitative data. The indicators of inadequate provision rarely identify on whom the impacts fall most heavily in terms of health and the burden of managing provision for water and sanitation within the home. **Box 3.1** provides some insights into these impacts and shows how they tend to fall most heavily on women and girls.

All too often, the advantages of urban areas (e.g. economies of scale and agglomeration, particularly for piped systems, greater potential for wastewater reuse) are not fully exploited, while their potential disadvantages (e.g. the greater risk of infectious disease outbreaks when the water and sanitation systems fail) pose great risks. In effect, urban populations have grown more rapidly than the capacity of governments to manage them, including putting into place institutional structures to ensure good provision for water, sanitation and wastewater management within each locality. This is part of a larger failure to support the development of competent, accountable city and municipal governments. Thus, the key need is for what might be termed 'good local governance', part of which is 'good water governance' (see **Chapter 2**).

*In principle,
sound water
governance
should be open
and transparent,
inclusive and
communicative,
coherent and
integrative, and
equitable and
ethical*

*Opposite: A favela (slum
housing) in Rio de
Janeiro, Brazil*

Particularly challenging issues include:

■ meeting water, sanitation and wastewater management
needs in the largest and fastest growing cities,
especially on their peripheries

■ changing water management systems to cope with the
more decentralized patterns of urban development evident
in most high-income nations and many middle-income
nations and low-density sprawl around urban centres

■ improving provision in large villages and small towns,
especially to the high proportion of the population
with very limited capacity to pay

■ recognizing the importance of regular and convenient
water supplies for the livelihoods of low-income
households, as well as for health, including urban
agriculture, for instance, which makes up an important part
of the livelihoods of tens of millions of urban households
(Smit et al., 1996) and for household enterprises.

Few valid generalizations can be made as to what
approach should be taken, because the most effective
means of addressing these provision deficiencies varies
so much from urban centre to urban centre. However,
in most instances, the following is true:

■ Provision deficiencies are not a problem that either
the private sector or the public sector can solve alone.

■ In many nations, at least in the next five to ten years, it
will not be possible for the provision deficiencies in most

urban areas to be addressed by the conventional model
of a (public or private) water utility extending piped
water supplies and sewers to individual households.

■ It will be impossible to meet the MDG targets in urban
areas, unless there are policies for improving water
and sanitation provision for low-income households
and community organizations, including brokering
agreements for those living in illegal settlements.

■ What is normally considered part of water and
sanitation provision must be expanded to include
slum and squatter upgrading programmes and
provisions for housing finance, as these play
important complementary roles.

■ Whether formal provision for water and sanitation is
undertaken by public or private utilities, city and
municipal governments have a critical role to play in water
governance, both in terms of ensuring provision for water,
sanitation and wastewater removal and in improving
sustainable water management within their boundaries.

■ The MDG targets for water and sanitation will not be met
without better urban governance. These also need to be
embedded within regional water governance arrangements
that often require agreements developed with freshwater
users upstream of the city and more attention to reducing
the impact of water pollution and urban runoff on water
quality for users downstream of the city (see, for instance,
Guadalajara in von Bertraub, 2003).

BOX 3.1: WATER SUPPLY INADEQUACIES WEIGH HEAVIEST ON WOMEN

There is no water to wash our hands when we use the
nearby bushes, plastic bags or the only public toilet
available some distance from our homes. There is always
fighting on who will be next, although there is a queue.
Everyone watches. There are no doors for privacy. How
long are we going to live this way? It is affecting our
pride and dignity. ... Sometimes we have to go to the
back of our house to defecate in a plastic bag and
throw it in nearby bushes or in the gully – this is called
'kitting'. The problem gets worse during menstruation
both for us and our daughters – they too can't attend
school as there is nowhere at school for them to clean
themselves, and we the mothers don't have enough
water to wash our bodies and to feel clean.
　　　　　　– Charlene, living in a slum in the Caribbean

We have been in this settlement (in Kothrud, in the
western part of Pune) for more than twelve years,
since we worked as labourers on the construction of
these apartment blocks that you see all around here.
Nearly 700 families live here now. When the
construction work was in progress, we got water at our
work sites. But now we face acute shortage of water.
We have public standposts in the settlement, but the
water is available for only two to three hours a day.
In such a short period of time, it is not possible for
all of us to fill water. There is always a long queue and
frequent fights. Women come to blows because some
try to fill many *handaas* (small water containers) or
jump the queue. Those who do not get their turn
before the water is turned off have to walk 20 to 30

minutes to fetch water. Some pay up to five rupees for
one *handaa* of water. Some collect the water that keeps
percolating in a small ditch by the side of the path near
the water taps. As you can see the water is turbid.
We cannot drink it, but we can use it for washing. For
a few weeks before the municipal elections, one of the
candidates who lives just on the other side of this hill
used to supply water to us in long hosepipes from
taps in his house. After the elections, the hosepipes
disappeared and our water supply stopped. Now if we
go to him to ask for water he drives us away as if we
are beggars. It is so humiliating!
　　　– from interviews with women in Laxminagar, Pune, India.

Sources: UN-HABITAT, 2004, as quoted in Millennium Project
2005a (for the Caribbean); Bapat and Agarwal, 2003 (for India).

Part 2. Developing Pro-poor Urban Water and Sanitation Governance

The groups most affected by the adverse consequences of expanding urban water systems are usually those lacking economic and/ or political influence...

Although there has been much debate about the relative merits of public versus private provision for water and sanitation in urban areas, this is actually of secondary importance to improving local governance – which would then allow more locally appropriate choices to be made about the roles and responsibilities of various agencies. Government agencies are always going to have responsibility for ensuring and overseeing provision, whoever the actual service providers are. And even in public water and sanitation agencies, it is common for much of the actual work to be contracted out to private enterprises. In addition, it is clear that the potential for international companies and corporations to contribute to improving and extending provision was greatly overstated during the 1990s – which in part explains these companies' declining interest in this sector. It is also clear that too little attention has been given by international aid agencies and development banks to improving the capacity and competence of national and local public utilities and private companies and to potential partnerships between them (Budds and McGranahan, 2003; UN-HABITAT, 2003b).

Thus, the focus of this section is on the changes that support what might be termed 'pro-poor governance' for water and sanitation. These fall into four main categories: those that increase the power and voice of the urban poor to make demands; those that make the government more responsive to their demands; those that make the (public, private, non-governmental organization or community) providers more responsive to their demands; and those that prevent corrupt and clientelistic practices from undermining the relationships between poor communities and their governments and water and sanitation providers. The latter issue of corruption is considered in Chapter 2.

This section also discusses the role of public utilities in smaller urban centres, and as these are not attractive to private investments, how public utilities can be made more responsive to providing water and sanitation to the urban poor. The issue of privatization is covered in detail in **Chapter 2** and **12**.

2a. Facing the challenges of the urban poor in terms of water and sanitation delivery

Better water and sanitation could improve the lives of hundreds of millions of urban dwellers who are currently unserved or inadequately served by formal utilities and lack the financial and organizational resources to develop adequate, safe alternatives. Most of these urban dwellers also suffer from other poverty-related deprivations. The key issue is what needs to change. There is a growing consensus that a central element to improving their water and sanitary conditions is to ensure that water and sanitation providers and those who work with them (and that oversee them) are made more accountable to urban poor groups (World Bank, 2003). In other words, by changing the governance framework in this way, it redresses current inequalities both in provision and in influence over policies and priorities.

Increasing the power and voice of the urban poor
The urban poor usually lack influence within government agencies or water providers. Influencing the state typically involves different actions from influencing water providers – voting or lobbying rather than paying, for example. Nevertheless, many of the changes that help people rise out of poverty, from receiving a good education to gaining income-earning opportunities, can simultaneously help them to influence governments and strengthen demands on water providers, be they private or public. Four particularly relevant changes are as follows:

- higher incomes, which allow people to pay more for water services, and to live in better-served locations, as well as often contributing to their political influence

- greater housing legality and security, which can confer political legitimacy (and a legal address may be needed to be able to vote) and increase resident capacity to negotiate with water providers and influence their willingness to invest their own time and resources in water-related infrastructure (this applies to renters as well as actual or aspiring owner-occupiers)

- better information, which can provide residents with an improved basis for setting their own water and sanitation objectives and for negotiating with others in pursuing these

- better-organized communities, who are in a stronger position to negotiate with both government and water providers (and, in some cases, are in a better position to make local investments in water infrastructure).

Increasing the responsiveness of the state

The capacity of urban poor groups to influence water policies and water providers also depends on the responsiveness of the government and water providers. Politicians often promise better water services. Democracy should help to increase the accountability of politicians and make governments more responsive to the water demands of their underprivileged citizens. Democratization and decentralization should make governments more responsive to water demands: in Latin America, for example, this combination helps to explain why public water and sanitation services improved in many urban centres even while their economies were not growing during the 1980s and 1990s. But democratization and decentralization are no guarantee that the government will respond to the demands of urban poor groups. Also, even with non-democratic regimes, states can be more or less responsive to the demands of urban poor groups (see **Chapter 2**).

Many city governments have recognized a need to give lower-income groups and other groups with ill-served needs a greater say in 'governance' by the means of conventional representative democratic structures. Participatory budgeting is one of the most significant innovations in this respect, and it is being applied in about 250 cities (Cabannes, 2004; Menegat, 2002), many in Brazil, as well as in other Latin American nations and some European nations. Participatory budgeting means more scope for citizen groups and community-based representatives in setting priorities for local government expenditures; it also implies a local government budgeting system that is more transparent and available to public scrutiny. While this process does not lie outside representative democratic systems since the municipal council is still responsible for approving the budget, at least more scope is given to civil society groups to influence it (Cabannes, 2004; Menegat, 2002; Souza, 2001). In Porto Alegre, the politicians who introduced participatory budgeting were surprised when participants from low-income communities prioritized

sanitary improvements, but they supported it once these priorities had been made clear.

Increasing the responsiveness of providers

Similarly, the capacity of urban poor groups to influence water providers directly depends on how responsive these providers are and to what they are responsive. This applies to both privately- and publicly-operated utilities, and in many circumstances the distinction between negotiating with large utilities as opposed to small enterprises is more significant, especially since large private utility operators are almost always working under contract. If the company's contract gives them a strong incentive to do so, they will be responsive to the demands of the urban poor. If the contract does not give such incentives, they will be less responsive. Market conditions matter, but are mediated by the state. Good regulation is important whether for public or private utilities, but this needs sufficiently qualified and motivated staff and resources as well as protection from political interference. A small-scale water vendor who earns all revenue from sales has different motivations for responding to demands. Here, much will depend on the level of competition in the market (rather than for the market, as is the case with competition for large concessions), and on other factors that determine whether the water vendor needs to be concerned about losing sales.

One particular difficulty is making water and sanitation providers responsive to the needs of those with very low incomes and very limited capacities to pay, especially if this is also accompanied by a commitment to charge customers prices that reflect their costs. One way to ensure provision for all within an overall system that seeks to recover costs wherever possible is through systems that provide a minimum amount of water at very low cost or no cost, with rising unit costs as more is used. This can be done through differential services (for instance public standpipes where water is free and household connections where water has to be paid for) or differential prices (for instance, 'lifeline' tariffs for a given minimum quantity of water with very low or no unit costs and higher unit costs for consumption above this). Both present problems in actual implementation. Water utilities that seek to maximize their revenues dislike both of these, unless compensated by the government – and governments are often reluctant to pay for this (see Connors, 2005). Lifeline tariffs can only work where low-income households have a water connection and the terms set can be so restrictive or inappropriate that they

The capacity of urban poor groups to influence water policies and water providers also depends on the responsiveness of the government and water providers

bring little benefit to poorer households (Sohail and Cotton, 2003; see also **Chapter 12**).

2b. Getting the best out of public utilities

Much of the urban population in low- and middle-income nations live in settlements that are of little interest to the private sector. Urban water and sanitation utilities were developed in response to public health threats, and in the belief that, if left to the market, piped water networks and sanitation systems would remain underdeveloped. The need for water supplies to fight fires also helped encourage public provision. Unfortunately, while markets do not give private water and sanitation enterprises sufficient incentives to provide adequate services, public water and sanitation utilities are also difficult to manage effectively. The notion that utilities can be effectively centrally organized is probably just as misleading as the notion that they can be left to the market. To be effective, both publicly- and privately-operated utilities need to be regulated and to engage in effective negotiations with government agencies, private enterprises and civil society groups, as well as with their actual and potential customers.

A number of the problems commonly associated with under-performing public utilities are largely beyond the utility's control: government agencies failing to pay their water bills; price controls imposed on the utilities that are not matched by the requisite financial subsidies; and the lack of a mandate to serve most informal settlements (or even regulations preventing them from doing so), even when 10 to 50 percent of a city's population live in such settlements. Other difficulties are not specific to water governance and include corruption, aid-dependence and development agendas, clientelism, political instability and authoritarianism. Given the range of political, economic and regulatory environments within which public utilities operate, it is impossible to generalize about what governance changes are likely to yield the greatest benefits in the form of improved services. There are, however, a number of areas where the quality of governance is likely to be particularly critical to the urban poor, and where increasing the transparency, accountability and equity of utility operations is likely to be especially beneficial.

Expansion plans and support for non-networked systems

For those currently without access to the utility's water and sanitation services, the expansion plans can be of critical importance, along with the mechanisms through which these plans are to be realized. Particularly where a large share of the population remains unconnected to networked water and sanitation systems, the extent to which the government supports alternative, non-networked, water and sanitation systems is also likely to be a critical concern for the urban or peri-urban poor. Often, the plans are not open to public scrutiny or influence, and people are often not even aware of the legal status of non-official water and sanitation providers. These issues only enter public debate when things go seriously wrong. In many circumstances, better governance requires more transparent planning procedures, and more accountability to low-income groups as well as to the plans themselves.

Connection and disconnection procedures

People living in conditions of poverty constitute a large share of the unconnected, and often face special difficulties making the 'lump' payments typically required to connect to the piped water network or sewer, or to pay for an alternative in-house sanitation facility. Some utilities provide more options, including for example the use of pre-payment meters, as used in parts of South Africa. Again, better governance is likely to require more open negotiation with urban poor groups and their representatives, both in the design and in the procedures themselves. In some urban areas, this may require special measures to enable renters to connect, as in Bangalore, India, where the utility recently made special provisions for renters, enabling them to negotiate directly with the utility (Connors, 2005). Also, the urban poor are particularly vulnerable to being disconnected (although in many countries it is public organizations rather than low-income residents that are most delinquent in paying their water bills). Those disconnected may be unaware of their rights, including their rights of appeal. Again, better governance is likely to require more effective negotiation, transparency and accountability (see **Chapter 2**).

Price controls and subsidies

The effects of price regulations and subsidies also depend heavily on the quality of local governance. Proponents will argue that price controls and subsidies are necessary to ensure that the poor can afford the services. Opponents will point out that price controls and subsidies favour the affluent, since they are the ones who already have access to water and sanitation services (a typical recommendation being that connection costs be cross-subsidized through the water tariff). Unfortunately, when governance problems prevail, both positions may be

correct. If, for example, the subsidy provided is sufficient to provide existing customers with water at the controlled rates, but not sufficient to finance expansion, the urban poor may actually lose out from price controls. In this example, however, the problem does not lie with the subsidies but with governance processes that do not allow for transparent and effective negotiation over the tariffs, subsidies and taxes (see **Chapter 12**).

Standards and their enforcement

As with price controls and subsidies, governance problems can undermine the operation of standards. Standards that are too low may leave the urban poor at risk, while standards that are too high may exclude them, particularly if they are not backed up by appropriate financial mechanisms. High sanitary standards that low-income households cannot meet do not stimulate sanitary improvements, and often contribute to housing insecurity both for renters and owner-occupiers. When standards are imposed on water and sanitation utilities, even if the water users do not pay directly, they often pay indirectly, for instance through higher rates or the utility being unable to expand according to plan. Again, good governance is likely to require negotiation and greater accountability, so that standards are not agreed upon if the basis for achieving them is not going to be put in place.

Part 3. Expanding Provision in Slums and Squatter Settlements in Low- and Middle-income Nations

A girl balances on a flimsy bridge over a polluted and flooded alley in a Phnom Penh slum, Cambodia

Improving and extending provision for water and sanitation in slums and squatter settlements presents any formal service provider with difficulties. These include the uncertainty as to who within each house, apartment or shack has responsibility for ensuring the payments (there are often multiple families and many are tenants or sub-tenants). For most informal settlements, there are the uncertainties regarding who owns the land and a lack of an official map showing plot boundaries, roads and paths (without which it is impossible to design and lay piped systems). There is usually no register of households and no official addresses assigned to dwellings. Furthermore, many informal settlements have sites and site lay-outs for which it is difficult to provide piped services – difficult terrain (steep slopes, waterlogged sites) and lack of public roads and footpaths alongside or under which a piped system could be installed.

In addition, many of the ways in which provision for water and sanitation can be improved and extended depend on the actions and investments of groups other than official water and sanitation service providers. These include individuals and households as they entail the following:

- Investing in provision in their existing homes: this may involve official service providers – for instance as the household pays to connect to official systems and takes responsibility for internal plumbing to support these connections – or may be independent of such service providers – for instance tapping groundwater, building or improving personal on-site sanitation provision.

- Investing in improved provision in their neighbourhood – for instance as they join together with other households in their street or neighbourhood to build sewers. This too may or may not involve official service providers. The model developed by the Orangi Pilot Project in Pakistan is particularly significant in this respect, not only in terms of the very large number of households reached with good sanitation (in Orangi, in other parts of Karachi, and in many other urban centres in Pakistan) but also in its demonstration of the potential effectiveness of government–community partnerships in helping to develop the wider piped system. (This is described in more detail further on in the chapter.)

- Developing new homes on land they purchase or occupy and through which they get better quality provision for water and sanitation.

For most informal settlements, there are the uncertainties regarding who owns the land and a lack of an official map showing plot boundaries

Most inter-national aid agencies and development banks see 'housing' as somehow distinct from improving provision for water and sanitation...

Map 3.1: Urban slum incidence in developing countries, 2001

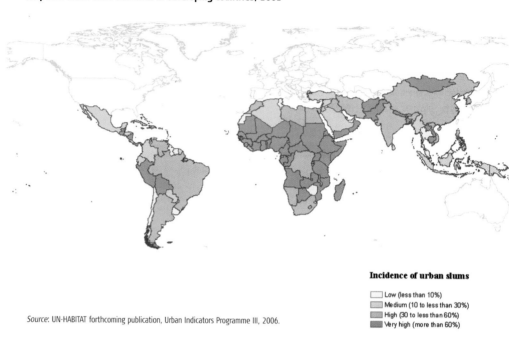

Incidence of urban slums

- ☐ Low (less than 10%)
- ☐ Medium (10 to less than 30%)
- ▨ High (30 to less than 60%)
- ■ Very high (more than 60%)

Source: UN-HABITAT forthcoming publication, Urban Indicators Programme III, 2006.

As the discussion below describes, there has been considerable innovation in many nations in helping low-income individuals and households to get better water and sanitation provision either through upgrading or through support for incremental home development – and these have great importance for achieving the water and sanitation MDGs, even if none are officially classified as 'water and sanitation' interventions, because they are considered as housing interventions.

The conventional model of urban provision remains piped water to each dwelling (house or apartment) and a sewer connection for wastewater and toilets provided by a public or a private company. Virtually all urban households are served by this in high-income nations; much of the urban population in many middle-income nations also enjoy this level of provision. Where the water supply is regular and of good quality, this model has proved popular because of its convenience for households and generally the low proportion of household income that it costs; it has also proved very effective in terms of improving public health.

Issues have been raised regarding the ecological sustainability of this model, both at the input end (because of the large volumes of freshwater needed, especially for larger cities) and at the output end

(through the high cost of treating the sewage/wastewater or the ecological damage to water bodies from inadequately treated wastewater flows).

The success of this conventional model for about half the world's urban population has provided the assumption that this is the model that should be extended to all urban dwellers, even though it has not covered hundreds of millions of urban households, despite its thirty years of advocacy and promotion by international agencies.

Over the last two decades, major changes in how water and sanitation provision is to be financed have also been promoted by many international agencies, but not in the form of provision. Conventional water and sanitation systems were to be funded through the shift from public-sector provision to privatization and private-public sector partnerships – but as described in **Chapter 2**, this did not provide the additional capital investment flows that had been anticipated and did not provide the hoped-for expansion in provision. The inadequacies in provision could not be solved by the efficiency gains that private-sector provision could bring. Many other factors contributed to this failure, such as the weakness and ineffectiveness of local governments, but at the problem's root is the fact that there is not enough capital to finance the high costs of expanding and extending

Map 3.2: Change in slum population in developing countries, 1990–2001

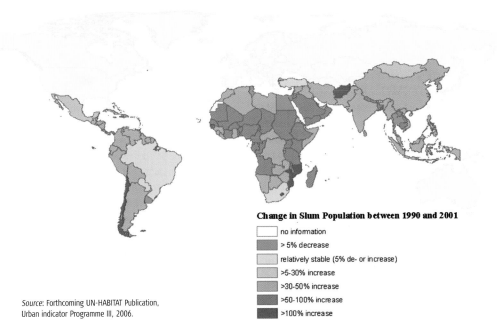

Change in Slum Population between 1990 and 2001

	no information
	> 5% decrease
	relatively stable (5% de- or increase)
	>5-30% increase
	>30-50% increase
	>50-100% increase
	>100% increase

Source: Forthcoming UN-HABITAT Publication,
Urban indicator Programme III, 2006.

*Open sewers in slum area
of Addis Ababa, Ethiopia*

provision of household water and sewer connections
and of building the institutional capacity to undertake
this – and manage the systems once they are constructed
– whether publicly or privately. In most cities, in addition
to the difficulties of extending provision to the large
sections of the population who live in illegal settlements,
this lack of finance (and often professional capacity) is
due, as noted earlier, to the lack of maps and addresses
for each household, the opposition from landowners
and government agencies to providing services to illegal
settlers and the difficulties of actually installing the pipes
in irregular layouts and on difficult sites.

3a. Lowering the cost of conventional household systems

Many innovations in urban areas of low- and middle-
income nations show how it is possible to overcome
these difficulties. The fact that these examples are drawn
from a wide range of cities, including some in very poor
nations, and that many are funded by local resources,
suggests that it is possible to improve and extend
provision to meet the MDGs in urban areas.

Perhaps the best known example of dramatically lowering
the cost of conventional household systems is the work
of the Pakistani NGO, the Orangi Pilot Project (OPP). This
project supports community-managed sewer construction

in Orangi, the largest informal settlement in Karachi
(Hasan, 1997; Zaidi, 2000), and was able to reduce the
cost of sewer provision per household to a fifth of what
municipal authorities charged, thereby making it
affordable to low-income households. In so doing, sewer
provision has reached hundreds of thousands of people,
and most of the costs are covered by what low-income
households can pay. This also developed the concept of
what OPP termed 'component sharing', where the
inhabitants of streets and neighbourhoods take
responsibility for the pipes, sewers and drains in their
neighbourhood and official service providers install the
water mains and sewer and drainage trunks to which
neighbourhood systems can connect.

There are three other important aspects to OPP's work
that are perhaps less well known. The first is the extent
to which their model has been used in many other areas
of Karachi and in many other urban centres in Pakistan,
including its widespread adoption by official (national and
local) government agencies (Rahman, 2004a). The second
is OPP's demonstration of how to develop detailed maps
of informal settlements to provide the information base
that allows official agencies to install water and sanitation
– which will be described in more detail later. The third is
to demonstrate a much cheaper and more effective means
of improving the city-wide system of sewers and drains –

*...there is not
enough capital
to finance the
high costs of
expanding and
extending
provision of
household water
and sewer
connections*

into which the community systems can integrate (Orangi Pilot Project Research and Training Institute, 1998).

OPP-supported local and city-wide sanitation schemes have achieved what is often said to be impossible by private or public water and sanitation utilities: provision of good quality sewers to each household with cost recovery and a city-wide system into which these can feed that does not require large external funding. Official water and sanitation agencies usually refuse to consider extending sewers to low-income settlements, because it is thought to be too expensive or because they do not believe that residents will pay for services. Were the OPP 'component sharing' model to be adopted in other cities in Asia and Africa, it would be possible to develop sewers (and the larger sewer system into which these integrate) capable of reaching hundreds of millions of low-income groups with good quality sanitation for a reasonable total cost, much of which could be covered by household payments. This model illustrates how the achievement of ambitious targets for improved water and sanitation is as much about the development of competent, capable, accountable, local agencies or utilities that can work with community organizations as it is about external finance.

In Brazil, PROSANEAR (Water and Sanitation Program for the Low-income Urban Population) combined community participation with more cost-effective technologies, which reduced unit costs and made the investments work better, consequently providing close to a million low-income people with water piped to their homes and connection to sewers (see **Chapter 14**).

3b. Community provision

Although most governments and international agencies have supported public or communal facilities for water, they generally avoid provision of communal facilities for sanitation. Three recent large-scale programmes show the possibilities for community provision. The first is in the community-designed, built and managed toilets in Indian cities. In the early 1990s, the National Slum Dwellers Federation and *Mahila Milan* (savings cooperatives formed by women slum and pavement dwellers) began to experiment with designing, building and managing community-toilet blocks in slums where there was insufficient room or funding for household provision. Large-scale community toilet block construction programmes then developed in Pune and Mumbai, when local government staff saw how much better the community-designed, built and managed toilets worked

than the contractor-built public toilets they had previously had made. Over 500 community designed and managed toilet blocks were built, serving hundreds of thousands of households, and comparable toilet programmes are also developing in other cities. There are plans to promote this approach in smaller towns (Burra et al., 2003: see also **Chapter 6**).

The second example is from Dhaka and Chittagong, Bangladesh's two largest cities. Here, the UK charity, WaterAid, supported community provision through seven local NGOs. This provided for water points supplied through legal connections to the metropolitan water authority lines or tube wells (where such connections were not possible), sanitation blocks (with water points, bathing stalls and hygienic latrines), community/cluster toilets with septic tanks, household water-seal pit latrines, the construction of footpaths, drainage improvements, solid waste management and hygiene education. Most facilities are provided on a full cost-recovery basis with recovered funding supporting additional slum projects (Hanchett et al., 2003).

The third example is from Luanda, Angola, where a local NGO (Development Workshop Angola) has supported the construction and management of 200 standpipes, each serving about 100 families. This programme has supported the development of locally elected water committees to manage these standpipes, working in collaboration with the water utility and the local authorities. This is another example of community organizations and local NGOs doing the 'retail' part of water provision. Where local water agencies are too weak to be able to extend provision to unserved low-income communities, this kind of NGO–Community partnership can have particular importance (Cain et al., 2002).

One possible reaction to these examples from India, Bangladesh and Angola is that the inhabitants of slums and squatter settlements deserve better facilities – for instance good provision for water and sanitation within their homes, not in public facilities. And why should there be such a focus on keeping down costs and generating revenues when serving among the poorest groups? However, there are two good reasons why these approaches were adopted: the cheaper the solution (and the more it can generate cost recovery), the greater the potential scale; and if a relatively low-cost solution can be developed locally by local institutions with little or no reliance on external funding, the scaling up may require no external funding (see also **Chapter 12**).

3c. The role of slum and squatter upgrading

Slum and squatter upgrading programmes are the principal means by which water and sanitation provision is improved for lower-income groups in most cities in low- and middle-income nations. They became common in the early 1970s, in part because governments recognized that these were among the cheapest ways of improving conditions for poorer groups, and because international funding was available to support them. However, many were not effective because the quality and extent of the upgrading was so limited (for instance, for water and sanitation, standpipes were shared by dozens or hundreds of households without any sanitation provision) and provision for maintenance was inadequate (for instance, nothing was set up to ensure that the utility responsible for water and sanitation would maintain the newly installed infrastructure). Most also failed to solve more fundamental problems, especially providing the inhabitants of illegal settlements with secure tenure, so the inhabitants still felt too insecure to invest in improving their homes. Most upgrading programmes also had little scope for participation.

However, there are upgrading schemes that have learnt from these limitations and are improving and extending provision for water and sanitation in ways that are more sustainable and on a larger scale. One of the largest is the national Baan Mankong (secure housing) Programme in Thailand that seeks to reach 300,000 households in 2,000 poor communities in 200 Thai cities between 2003 and 2007 (Boonyabancha, 2003 and 2005). The programme channels infrastructure subsidies and housing loans through the national government agency, the Community Organizations Development Institute, directly to low-income communities, that plan and manage the implementation of improvements of their housing and basic services. This not only improves provision for basic infrastructure (including water and sanitation provision) but also ensures a stable, legal relationship between households and water utilities by providing secure tenure for residents. As housing tenure becomes secure and as housing improves, so too does the management of water and wastewater within homes. This programme also has loan programmes to support low-income households to improve and extend their homes (including improving provision for water and sanitation within the home). It also encourages organizations of the urban poor within each urban centre to join together with the local government, other government bodies and other stakeholders in order to develop city-wide plans.

This combination of grant-funded upgrading and micro-finance loans for individual households is also a characteristic of several housing and local development programmes in Central America to which the Swedish International Development Cooperation Agency (Sida) contributed funding. From 1998 to 2003, these programmes improved conditions for around 80,000 low-income households in five nations. Although their form differed from nation to nation, each programme was based on lending to low-income families to improve or expand their existing housing or to build new houses. Household savings and self-help efforts contributed around 20 percent of all funding provided by external agencies. In some programmes, family efforts were also supported by subsidies. Some programmes also provided municipal governments with credit or matching funds in order to provide basic services in new and existing low-income settlements. These programmes are unusual on at least two counts: first, from the outset, the international donor sought to build institutions within each nation for implementation; and second, it built the capacity to provide loans to low-income households and ensure good cost recovery performance with the funds recovered from loan repayments going to support other loans (Stein and Castillo, 2005).

Many other innovative upgrading programmes have demonstrated how water and sanitation provision can be greatly improved through a combination of cheaper technologies, strong community participation and loan financing, as is the case in the slum networking approach developed in Ahmedabad, India. The municipal authorities in Sao Paulo, Brazil implemented a major upgrading programme between 2000 and 2004, and, perhaps more importantly, recognized that working on a large scale necessitates a strong legislative, administrative and financial base, as well as land tenure legalization and the legislative changes that this requires (Budds et al., 2005).

Micro-finance systems that support households to improve and extend their housing can also play an important role in improving water and sanitation provision, as shown by the above examples. Loan financing for housing can also support land tenure regularization, which then allows or encourages better provision from official water and sanitation utilities. For instance, in Bolivia, many of the housing loans provided by the NGO, PROA (Centro de Servicios Integrados para el Desarrollo Urbano), are for housing improvements (including improved water and sanitation provision), regularization and new construction (Ferguson, 1999).

Slum and squatter upgrading programmes are the principal means by which water and sanitation provision is improved for lower-income groups in most cities in low- and middle-income nations

...the South African Government has provided hundreds of thousands of low-income households with one-off grants

3d. Providing alternatives to slums for low-income households

One important way to extend provision for water and sanitation to low-income households is to increase their capacity to buy, rent or build new accommodations with better provision. The Millennium Project Task Force on improving the lives of slum dwellers emphasized the importance of providing low-income households with alternatives to slums, i.e. more possibilities of acquiring new housing with infrastructure that includes water and sanitation provision (Millennium Project, 2005b). Given the speed with which urban populations are growing in most low- and middle-income nations, meeting the water and sanitation MDGs will mean not only improving provision in existing slums and squatter settlements, but also ensuring that expanding urban populations do not create new slums and squatter settlements with inadequate water and sanitation provision. Since purchasing new complete accommodation is beyond the means of most urban households, this will largely be through the capacity to legally purchase land for housing and the establishment of finance systems that help residents to purchase the land and build incrementally. Although pro-poor land management and housing finance systems are not normally considered key mechanisms for improving water and sanitation provision, they are often the means by which poorer families can acquire better provision.

Governments have moved away from direct provision of 'alternatives to slums'. Twenty to thirty years ago, large public housing programmes were common – and seen as the means by which low-income households could get better quality accommodation, including access to piped water and sanitation. Large site and service and core housing programmes were also based on the same principle, although residents were responsible for developing the house, as this kept down unit costs.

Most governments have stopped or greatly scaled down these programmes, in part because they often proved ineffective (many serviced site schemes were in locations too far from income-earning opportunities) but also because the units were assigned to (or soon purchased by) non-poor households. However, some governments have sought to work with the market in new housing provision rather than provide an alternative to the market through providing one-off grants to low-income households in order to help them buy a house or land with infrastructure on which they can build. For instance, the South African Government has provided hundreds of thousands of low-income households with one-off grants to support them in obtaining housing, which usually also means improved provision for water and sanitation (see **Chapter 14**). The government of Ecuador set up a housing incentive system that combines a grant from the government, savings from recipient households and loans from private institutions – all of which can support housing improvements or new housing. Although the experiences with these one-off housing subsidies are mixed – for instance, where these are channelled through housing contractors, the units constructed are often poor quality and in inappropriate locations – they have worked well in some instances and illustrate new ways by which governments and international agencies can support improved provision for water and sanitation.

In a similar approach, the Salvadoran Integral Assistance Foundation (FUSAI) in El Salvador supports new housing schemes for low-income households as well as upgrading. For the new housing schemes, FUSAI develops new land sites with infrastructure and services and aids low-income households in developing their homes on these sites. Households receive loans to support this development and then receive a land title when they have repaid the loan; the amount financed by the loan equals the price of the house, including road and infrastructure development, minus the subsidy received from the state and the value of the family's contribution (Stein and Castillo, 2005).

Thus, there is a multiplicity of means and methods by which provision for water and sanitation can be improved and extended and this usually requires a mix of household and community action and investment in addition to support from formal (private and public) agencies. Examples given in this section suggest that the potential role of community-based organizations and their federations working in partnership with city governments and other formal agencies has been underestimated.

A shanty township near Johannesburg, South Africa

Part 4. Addressing Water and Management Needs in Different Size Settlements

The size of a settlement and the nature of its economic base obviously influence the most appropriate means by which to provide or improve provision for water and sanitation and manage wastewater. Rural and small urban settlements in low- and middle-income countries need particular attention in this respect, for this is where much of the population with the worst water and sanitation conditions live. Another important theme is the water resource problems that often afflict large cities or urban regions, including those in high-income countries.

The percentage of households with piped or well water on the premises or with flush toilets generally decline with city size

4a. Large villages versus small urban centres

The population living in large villages (from a few hundred to several thousand inhabitants) and small urban centres have not been given sufficient attention: the proportion of the inadequately and unserved population remains high, and the interventions most appropriate for improving provision have yet to be applied, despite the fact that the concentrations of people and enterprises lower the unit cost of providing treated piped water supplies and most forms of sanitation and drainage.

Virtually all governments accept that settlements with more than 20,000 inhabitants are urban centres but disagree about where to draw the line between urban and rural for settlements with fewer than 20,000 inhabitants. Some classify all settlements with only a few hundred inhabitants as 'urban' while others consider most or all settlements with up to 20,000 inhabitants as 'rural'. This has significance for two reasons: a very high proportion of people live in settlements with between 500 and 20,000 inhabitants; and their designation as urban populations generally means more government structures of capacities gives a larger scope for supporting gives improved provision for water and sanitation.

The key characteristics relevant to improving water and sanitation in these urban centres include the following:

- spatial concentrations of households that reduce the unit costs of piped water, sanitation and drainage systems

- significant proportions of the economically active population working (part time or full time) in non-agricultural activities

- many households having members who have migrated temporarily to earn income and who send back remittances (although this is most apparent in low- and middle-income nations)

- a range of non-agricultural economic activities that require regular water supplies and wastewater disposal.

The issue of which settlements are to be defined as urban centres is politically charged in so far as both governments and international agencies make decisions about resource allocations between rural and urban areas depending on the proportion of the population that lives in them. The contestation between rural and urban proponents on what should receive priority in development investment and poverty reduction programmes has been one of the dominant development debates of the last thirty years. If India, Pakistan or Egypt were reclassified as predominantly urban nations – if their large villages were redefined as small urban centres, a conclusion that is possible to substantiate from a demographic and an economic perspective – it would change the perceptions and programmes of most international agencies working there.

Initiatives to improve urban water and sanitation provision tend to forget smaller urban centres. For more urbanized nations, 20 to 40 percent of their total population lives in urban centres with fewer than 200,000 inhabitants; for less urbanized nations, the majority of the population often lives in urban centres with fewer than 200,000 inhabitants. An analysis of provision for water and sanitation in urban areas of different sizes in forty-three low- and middle-income nations showed that in almost every case, the smaller the size-class for urban centres, the worse the provision. It was found that the percentage of households with piped or well water on the premises or with flush toilets generally declined with city size, and that generally the worst served urban populations were those in urban centres with fewer than 100,000 inhabitants (Montgomery et al., 2003).

...an important part of the demand for water comes from enterprises, and there may be important synergies between this demand and the possibilities for investing in improved provision...

As mentioned earlier, these interventions operate through economies of proximity (relating mostly to population density, which means less pipe and drain and ditch digging per household reached) and economies of scale (relating mostly to population size thresholds, which means lower unit costs per household served for water-treatment plants and maintenance and billing departments). With respect to the first, for most forms of infrastructure (including piped water, sewers and drains), the costs per person served in a settlement of 500 people per hectare are usually half of those for a settlement with 150 people per hectare (Banes et al., 1996); obviously more scattered lower density populations are much more expensive to reach. With respect to economies of scale for water and sanitation, what little work has been done suggests that most come into play at a relatively low population threshold, i.e. they work for most small urban centres, and some also work for large villages. For some infrastructure and services, the opposite may be true as cities get larger; although this may be related more to poor governance or outweighed by higher productivity (Linn, 1982; Hardoy et al., 2001). Many small urban centres and large villages also have higher average cash incomes per person than more dispersed rural populations, which means a higher capacity to pay for water piped to the home and good sanitation.

The potential economies of scale and proximity for large villages and small urban centres often go unrecognized by governments and international agencies. There are at least three reasons why this is so:

- the failure to recognize the proportion of national (and urban and rural) populations living in them

- the tendency for urban provision to concentrate in larger urban centres

- the failure of agencies responsible for water and sanitation provision in rural areas to exploit the economies of scale and proximity in large villages.

It has to be stressed that a significant proportion of the population and of economic activities in all nations are in small towns and large villages, which have unmet water, sanitation and water waste management needs. This is completely independent of whether the settlement is classified as a village, town or urban centre, even if this classification does influence the scope of its local government and often the potential for funding. There

are economies of scale and proximity in most of these settlements, which can lower unit costs for better water and sanitation provision.

In many such settlements, an important part of the demand for water comes from enterprises, and there may be important synergies between this demand and the possibilities for investing in improved provision, which also benefits households. This link between economic activities and domestic needs may also span rural-urban definitions, as demand for water for livestock and crops can help fund improved provision for water that also serves domestic needs. In many such settlements, there may also be sufficient demand for electricity and economies of scale and proximity, which make water and sanitation provision economically feasible and thus brings obvious advantages with regard to power for water pumping.

Improving provision for water and sanitation and integrating water use and wastewater management within broader regional water management systems are particularly challenging for these settlements, as they generally have the weakest local government systems. The institutions charged with water governance in these settlements often face specific challenges. For example, small urban centres may be large enough to justify a water and/or sewerage network but too small to support a locally managed utility. The challenges evident in these kinds of settlements are highlighted in a study of small urban centres in Kenya, Uganda and Tanzania that are clustered around Lake Victoria (see **Box 3.2; Chapters 5 and 14**). Rehabilitating existing infrastructure in these urban centres and building capacity to ensure the efficient operation of the utilities and provide the revenue base to operate and maintain the systems do not require very large investments. They do, however, require long-term support for building this capacity and some immediate investments to address the most serious problems. These systems would also help to address the need to reduce the severe impact that rapid urbanization is having on the environment (see also **Chapter 5**).

4b. Land and water resources management in cities and city regions

Urbanization and growing urban water consumption and wastewater generation in particular pose a number of challenges for watershed and river basin management. The conventional approach to meeting growing urban water demands has been to intensify withdrawals from local water aquifers or to divert more distant surface

BOX 3.2: LACK OF PROVISION IN SECONDARY URBAN CENTRES AROUND LAKE VICTORIA (KENYA, TANZANIA AND UGANDA)

Bukoba is a regional and district headquarters in Tanzania with 81,221 inhabitants in 2002, of which about 63 percent receive water services from the Bukoba Water and Sewerage Authority. There are no sewers in the town; residents rely on pit latrines or septic tanks. As a result, sullage and septic tank effluent is discharged into storm water drains, contaminating Lake Victoria, which is also the town's main source of water. The water distribution system was constructed in the 1940s, and small sections of the distribution system were last rehabilitated in 1986, but 60 percent of the water remains unaccounted for: two-thirds of which is a result of leaky pipes is and a third attributable to administrative losses, including illegal connections. The town has only one (very old and run-down) vehicle for collecting solid wastes; collection is irregular and services only the central business district. In the lower-income areas, it is common to find overflowing waste piles spilling onto streets and into adjacent storm water drains. The storm drainage system is very limited, and many of the drains are blocked. Homa Bay is a trading town, fishing centre and district headquarters in southwest Kenya of about 32,600 inhabitants. The water supply system there was constructed in 1958 and last rehabilitated in 2001. There is a full treatment plant but water quality is often poor and water volume is far below

demand, as there are regular electro-mechanical breakdowns, and the filtration system is wearing down. The plant currently serves about 15,000 residents through 1,672 legal connections. Water supply is not continuous, and the system suffers from low pressure, vandalism, illegal connections, leakages, old age and blockages (40 percent of its water goes unaccounted for). The town has several unplanned informal settlements, and most of their inhabitants get their water directly from the lake. Preliminary investigations indicated that only 22 percent of the population is connected to sewers; most people use pit latrines or toilets connected to septic tanks or the bush. Overflowing toilets and sewers are common during rainy seasons. There is

no exhauster tanker to empty pit latrines and septic tanks. Storm drains are not available for most of the town, and provision for the collection of solid wastes is highly inadequate, so it is common for drainage networks to be blocked.

Kyotera Town is the busiest centre in its district in Uganda, because it is on one of the main roads leading to Tanzania. About 10,000 inhabitants live there, but the daily influx of people increases the population to 25,000; overall the population is growing rapidly. There is no public water supply and the town relies mainly on piped water supplied by a local church project, although the water is of poor quality and not adequately treated and supply is frequently interrupted due to power failures. Other water sources are boreholes, wells and rainwater, even though most households cannot afford to collect and store rainwater. A groundwater supply system is being developed, but there are no sewers. About two-thirds of the population have pit latrines, and about 20 percent use four public toilets. There is no domestic solid waste collection service, and the storm drain system is very inadequate. The few drains that do exist are clogged with solid waste.

Source: UN-HABITAT, 2004b.

---

water flows. While municipal water demands are still several times smaller than the demands of irrigated agriculture, urban demands are spatially concentrated, and when conflicts are not well managed, the results can be inequitable and politically contentious.[3] More generally, as water infrastructure expands, local supply-demand imbalances are transferred to the regional scale, increasing the need for integrated management and adding to the pressure on ecological water requirements (the water needed to maintain ecosystem function and local hydrological cycles), even in areas distant from the main demand centres. While this process is more evident in affluent countries and regions, it is also occurring to some degree in most parts of the world. Research on the changing urban water systems in Africa, where insufficient infrastructure is a major problem, indicates that while in the early 1970s many major cities still used groundwater supplies as their primary water source, by

the 1990s primary sources were more likely to be rivers, and increasingly these river sources were more than 25 kilometres away (Showers, 2002).

The conventional approach to urban water supply planning started with projecting populations and per capita water consumption levels and multiplying them together to estimate future water requirements. The demand projections identified the quantities of water that had to be made available by increasing the exploitation of existing supplies or identifying and securing new sources. But securing new sources often became increasingly expensive, both in financial and in ecological terms. In the United States, for example, during the last decades of the twentieth century, many water utilities encountered problems in applying this demand-driven approach. Already in the 1970s, the US National Water Commission began to examine the

Favelas in Curitiba, Brazil

---

3. See, for instance, the Owens Valley-Los Angeles water controversy in Kahrl, 1982.

BOX 3.3: DEMAND-SIDE WATER MANAGEMENT: BEIJING'S WATER SCARCITY

Beijing has been facing growing water scarcity, with falling groundwater tables and problems securing sufficient surface water of adequate quality. There are plans to divert large quantities of water from the south to the north, a practice very much in the tradition of demand-driven water planning. The financial, environmental and social costs of the project are very high, and some have argued that it would be more beneficial to invest in demand-side management and improvements to upstream water systems rather than investing tens of billions of dollars to divert billions of cubic metres of water annually over distances of more than 1,000 km. Partly because of the focus on

infrastructure, however, the options and costs of upstream investments are not well understood or documented. While the scale of the water infrastructure projects planned and being implemented for Beijing is exceptional, the tendency to ignore the possibility of investing in improving water and land use upstream is not. Few cities consider negotiating for upstream land and water use practices that could help them secure better water supplies within their own watershed, partly because the institutional basis is lacking. New York was exceptional when it invested heavily in more ecologically suitable upstream land use practices to secure better quality water supplies.

Source: Platt, 2004.

...water resources management requires land-use management into which water resources management goals are integrated...

potential for urban water conservation, and by the end of the century, urban water demand management was being advocated for the following reasons:

- untapped sources of water were becoming rarer, and the depletion and contamination of groundwater sources had further limited supplies (see **Chapter 4**)

- the increased frequency of droughts had increased competition for water between urban and agricultural interests (see **Chapters 8** and **10**)

- environmental concerns about increased water use had intensified to the point where the development of new supplies was politically unfeasible, and the prospects for financing major construction programme were discouraging for many water agencies (Baumann and Boland, 1998) (see **Chapter 5**).

Urban development in California, and in particular in the Los Angeles region, already has a long history of water conflict (Kahrl, 1982; Reisner, 2001). More recently, there have also been a number of concerted efforts to improve the efficiency of water use and promote reuse (Gleick, 2000). Moreover, industrial water demands have been declining as a result of the changing economic structure. Water resources management remains a major challenge. At least for the foreseeable future, however, water scarcity is unlikely to become a major public health issue.

Many large metropolitan areas in low- and middle-income countries also face serious water problems that

infrastructure solutions cannot easily address, and often low-income groups within these areas also face problems getting adequate water and sanitation. Many large cities, one example of which can be seen in **Box 3.3**, are relying on increasingly long and costly water diversions *(Lundqvist et al., 2004; Tortajada and Castelan, 2003).*

Urban development also affects water users downstream, and the development of large metropolitan areas and urban regions typically has profound impacts on the water and land-use of the surrounding region. It is almost axiomatic that being upstream of a major city is preferable to living downstream. Around Hanoi, Viet Nam, for example, water-related problems downstream are felt by farmers and developers alike, and upstream locations are environmentally and economically more attractive (van den Berg et al., 2003, Showers, 2002).

Finally, it is obvious that sound water resources management goes hand-in-hand with holistic and integrated land-use management (as highlighted in **Chapter 2**), which in turn requires urban planning measures that define, restrict or control land-use changes. In most urban centres in low- and middle-income nations, either there is no coherent land-use management plan or it is not adhered to, whether by new enterprises, developers or squatters, and urban centre expansion is largely defined by where new investments are made. It is also common for polluting industries to locate or relocate on urban peripheries to escape pollution controls. So urban regions expand with a patchwork of developments to which it is expensive to extend infrastructure, including that related to water and wastewater. Many squatter settlements develop on land ill-suited to development, including steep slopes and flood plains. Protected watersheds may be particularly attractive to illegal land developers and squatters. And even in high-income nations, the power of land developers and their customers can subvert land-use management plans. There is intense competition for the best sites within and around urban centres, and the profits that can be made from land development (land value multiplies many times as it changes from agricultural to industrial, commercial or residential use) make it difficult to develop an effective governance framework, whether for ensuring land use management that contributes to water management or for other key tasks, such as ensuring sufficient, appropriately located land for housing for low-income groups. There have been important advances in this, as described in **Chapter 2**. Some cities also have programmes to restrict future developments in watersheds and address the water and sanitation needs of those who live there, while reducing their wastewater impacts on the watershed (van Horen, 2001; Jacobi, 2004).

4c. Access to clean water in settlements for refugees and internally displaced persons

There are currently about 10 million refugees and 25 million internally displaced persons (IDPs) throughout the world, and clean water takes on particular importance for these people (UNHCR, 2004). Forced to flee their homes and seek refuge either in a foreign land, usually in a neighbouring country, or in a different region of their own country, refugees and IDPs are cut off from their normal livelihoods, deprived of access to basic services, relegated to inhospitable environments and often live in crowded camps ranging from scattered spontaneous settlements to more organized camps. Populations in refugee settlements vary from a few thousand to over 100,000 in some instances. In a camp-like situation, however, manageable size cannot exceed 20,000 people. Providing clean water to refugees poses a range of challenges and deserves special considerations, as they have traditionally faced difficulties in fully exercising their rights and are very prone to exploitation (Shrestha and Cronin, 2006).

In a refugee emergency, especially when there is a large influx in a short period, water is often not available in adequate quantity and quality, creating major public health hazards in refugee settlements, with young children being primary victims. In 1994, when one million Rwandans fled the country after the genocide to neighbouring Democratic Republic of Congo, as many as 60,000 children died from a vicious cycle of water shortage and water-borne diseases, including cholera (see **Chapter 6**). In such a situation, the challenge is to

maintain a strong resource preparedness (equipment and personnel) while having a fail-safe mechanism to deliver water service provision within a matter of days.

Beyond the emergency phase, the careful design and management of water supply and distribution is essential to prevent the protection concerns of refugees from being aggravated and to improve their health and well-being. Even those individuals who may have traditionally lived on less than the recommended amount of water will require more water when living in a refugee or IDP camp, because of crowding and environmental factors. If the available water is limited or too far away (the UNHCR standard in a camp-like situation is to have a water point within 200 metres from the farthest dwelling), refugees and IDPs, in particular women and children, may be exposed to exploitation or attacks (as is currently happening in Darfur, Sudan), besides spending an inordinate amount of time and energy collecting water (UNHCR, 2000).

Another dimension of the challenge is the environmental impact. The presence of a large number of refugees or IDPs in an area that was previously sparsely populated, exerts pressure on the often marginal and fragile ecosystems. If sustainable systems are not put in place, water sources may be depleted and/or contaminated, which eventually could be a source of serious friction with local host communities.

The principles of water supply in refugee and IDP situations are based on core values that help protect their safety, rights and dignity (see **Box. 3.4**). These include the following:

- an equitable distribution of at least 20 L per person per day of safe water, so that it does not become a source of power that can be abused for various forms of exploitation

- secure access to water points so that the potential for sexual and gender-based violence is mitigated

- an adequate number of water distribution points in close proximity to the dwellings, so that physical burden (time and energy) on women and children is lessened

- participatory planning in place with the refugee community, so that development and operation management of the water supply system and sanitation and hygiene promotion activities are in accordance with their particular needs and cultural practices.

Providing clean water to refugees poses a range of challenges and deserves special considerations, as they have traditionally faced difficulties in fully exercising their rights and are very prone to exploitation

Although the number of refugees decreased by approximately 24 percent between 2000 and 2004, the number of IDPs increased by 43 percent in a single year (2003–04), many of whom are in Sudan and Colombia. This can be catastrophic, both in terms of human suffering and geopolitical conflict, thus it is important that the safety, rights and dignity of refugees and IDPs be safeguarded while it is still unsafe for them to return home (UNHCR, 2005).

BOX 3.4: PROVIDING FUNDAMENTAL RIGHTS TO REFUGEES: EXCERPTS FROM AN ALL-STAFF MESSAGE BY UNHCR'S ACTING HIGH COMMISSIONER ON WORLD WATER DAY 2005

'Our protection goals are to ensure that refugees' fundamental rights are respected, including their access to water. Fulfilling that basic right is essential for the life, health and dignity of the people of concern to UNHCR, as well as a benchmark for every relief operation. A clean water supply is just as indispensable to a refugee's survival in stable camp situations as it is in emergencies; studies show that between one-third and half of all illnesses in refugee camps are caused by poor water supply, inadequate sanitation services and deficient hygiene practices. A refugee's access to water depends not just on what we can provide but how we provide it. In one country, because of an inadequate water supply, over 40 percent of refugee school children regularly skip classes to help their mothers collect water. Refugees in another wait six hours on average every day to collect water. As a result, they collect unsafe water at unguarded locations in an effort to save time, exposing women to sexual assaults and multiplying the prevalence of diarrhoeal diseases. These terrible social costs can in no way be accounted for in a simple cost-benefit analysis. UNHCR must provide refugees with adequate safe water – and we must do so without putting women and children at risk. Protection concerns must be fully incorporated in the planning and operation of water-related facilities.'

Source: www.un.org/waterforlifedecade/statements.html

Part 5. Local Information for Global Goals

It is important that efforts to develop internationally comparable indicators do not detract from also developing indicators needed to ensure progress at the local level.

5a. The importance of community-driven assessments

The MDGs have encouraged greater attention to monitoring progress towards the stated targets and have stimulated efforts to identify indicators that can provide the basis for this monitoring. Although it might be assumed that these indicators will provide a stronger information base for action, this is not necessarily the case. This section suggests an approach that relies, to a limited extent, on standardized and representative surveys, and emphasizes indicators designed primarily to inform and support local action and only secondarily to corroborate or contribute to national and international indicators.

Data from household surveys based on a representative sample of national populations have become increasingly important information sources for showing the quality and extent of provision for water and sanitation. This is largely because international donors wanted more regular information regarding outcomes they considered important for development. National censuses are too infrequent (usually held once every ten years) and there are often long delays between when the census takes place and when the data generated by the census become available. There is also a problem due to a lack of censuses; many of the nations with the largest deficiencies in water and sanitation provision have not had a census recently. In nations where much of the population has provision from official service providers, the most detailed data on provision can come from these providers. This is not much use, however, in low-income nations and in most middle-income nations, because a high proportion of the population is not served by such providers.

While household surveys can show the extent of deficiencies in provision for water and sanitation for nations and the sample size can be made sufficient to provide accurate statistics for the largest city and for rural and urban areas, they often do not provide the data most needed for addressing the provision deficiencies:

the identification of the actual households, neighbourhoods and districts where provision is inadequate or non-existent. It does not help a water utility much to know the percentage of the population in the capital city that lacks water piped to their home if it does not know which households lack provision and which neighbourhoods have the largest deficiencies. In theory, censuses should provide this, but it is rare for the data collected on water and sanitation provision to be made available to urban governments and water and sanitation agencies that can help improve and extend provision. Or, if this data is provided, it is available in a form that is too aggregated to allow the identification of where provision is inadequate (Navarro, 2001). Household surveys and censuses also include only a limited range of questions about the quality and extent of water and sanitation provision. This may be enough to provide national governments and international agencies with a broad picture, but it does not show who has adequate provision for water and sanitation, let alone why (UN-HABITAT 2003b).

To improve or extend provision for water and sanitation to those who are unserved or inadequately served, details are needed for each household and housing unit on the quality and extent of provision (if any) combined with maps of each settlement that show each housing unit, existing water pipes, sewers and drains, and paths and roads. This information base is often weakest in the areas where provision is worst: small urban centres in informal settlements where much of the unserved urban population live. There are usually no official maps of these areas, or if they exist, they are inaccurate or lacking in detail. Officials – whether from water and sanitation agencies or municipal officials or those charged with undertaking surveys – may be afraid to enter squatter settlements.

Through detailed enumerations of all households in informal settlements and detailed maps showing each housing unit and plot boundaries and existing infra-structure, these problems may be solved. One example of this is the neighbourhood mapping programme supported by the Orangi Pilot Project's Research and Training Institute (OPP-RTI) which covers most of Karachi, which is an extension of the Institute's support for household- and neighbourhood-managed sewers and drains described earlier. The OPP-RTI noted the large investments being made in water and sanitation on a self-help basis by communities outside Orangi and recognized the need to document this work for the following reasons:

- to understand the extent of community initiatives being undertaken

- to avoid duplication of work being done by the government

- to enable people to realize the extent of their work and strengthen their capacity

- to inform the government of the scale and nature of this informal investment in water and sanitation so their policies could support existing work.

Surveys of 334 informal settlements have been prepared, encompassing 224,299 houses in 19,463 lanes. These were done by youths who received training from OPP-RTI. The neighbourhood surveys serve two functions: first, to demonstrate to the government the scale and breadth of household and community investment; and second, to provide detailed maps of each neighbourhood that allow government investments to complement household and community investments in the component sharing model described earlier (Rahman, 2004b; Orangi Pilot Project Research and Training Institute, 2002).

The urban poor federations whose work was mentioned earlier have also shown how it is possible to organize very detailed slum enumerations and surveys that draw information from each household, while developing the kinds of detailed maps that improved provision for water and sanitation. Those who undertake these surveys visit each household, so everyone is informed about why the enumeration is being done. The information collected is then returned to community organizations for verification. It then provides the base for detailed improvement plans. These enumerations cost much less than professionally managed surveys, yet are more detailed, more relevant to local action and less prone to serious errors and misrepresentation. These slum surveys also provide the organizational base from which to plan upgrading and new-house development, as illustrated by the findings from the Huruma enumeration in Nairobi, Kenya, which provided the basis for a settlement-wide upgrading programme (Weru, 2004).

These self-surveys and enumerations also provide urban poor organizations with a powerful tool with which to negotiate with local governments and water and sanitation providers. They no longer make arguments for improved provision for water and sanitation based on

...it is rare for the data collected on water and sanitation provision to be made available to urban governments and water and sanitation agencies that can help improve and extend provision

their poverty, but rather on detailed facts and maps. This community-driven production of detailed data also contributes to a more equal relationship with external agencies as it is produced and owned by the communities, not by external agencies. The surveys also give each person and household an official identity, as their occupation of land and housing is recorded – often for the first time. Some of the urban poor federations and their support NGOs have also undertaken city-wide slum surveys that provide documentation of all slums, informal settlements or pavement dwellings (Patel, 2004; ACHR, 2004; Boonyabancha, 2005).

These slum surveys also provide the organizational base from which to plan upgrading and new-house development (Patel, 2004).

5b. The information base that drives good governance

Community-based data collection of detailed information on households in poor settlements is built on the assumption that local government authorities are either incapable or unwilling to collect similar data. However, where there are legitimately elected local authorities and professional local government staff, an effort must be made to build their capacity to address the problems of water and sanitation in the poor communities. This includes the ability to collect and compile detailed information on poor communities.[4] This is a critical aspect of good governance. Fortunately, advances in technology make this easier to do than before.

Whereas it was once difficult to map a community, it is now relatively easy with the use of satellite imagery and geographic information systems (GIS; see **Chapters 4** and **13**). In today's market, satellite images are no longer expensive. This development, coupled with a UN-HABITAT programme to provide GIS capability in up to 1,000 cities globally, makes it feasible to bring this technology to local authorities. In high-income countries, the use of GIS by local authorities is well advanced. According to a survey conducted by the Public Technology Institute in the US, in 2003, 97 percent of US cities with 100,000 or more inhabitants used GIS; 88 percent of those between 50,000 and 100,000 used it; and 56 percent of those below 50,000. Hence, with the objective of bringing these advances to local authorities in low- and middle-income countries, UN-HABITAT is supporting local government authorities in their ability to develop sound pro-poor policy analysis as part of the Lake Victoria Water and Sanitation

Initiative. This initiative is to provide satellite images for seventeen towns in the Lake Victoria region. These images form the basis for a GIS database that is to be supplemented with field observations, household survey data, existing land-use and infrastructure information.[5] The effort is intended to bring the MDG indicators down to the local level, so that water and sanitation interventions can benefit the poorest communities.

5c. The focus of indicator development

This chapter has made clear the difference between the data commonly collected to monitor the quality and extent of water and sanitation provision on a national level and the data needed to drive improvements in the quality and extent of provision in each locality. Of course, national governments and international agencies need to monitor conditions and trends through household surveys based on representative samples, and these should influence priorities and resource allocations within central and state/provincial governments. International agencies also rely on these surveys for providing the information base for establishing priorities as well as for monitoring progress. But these surveys rarely provide the information base on the quality and extent of provision for water and sanitation with sufficient detail to assess the adequacy of provision (UN-HABITAT, 2003b). They have not yet been able to provide the information base needed by local water and sanitation service providers and local governments to indicate where the deficiencies in provision exist (both spatially and in terms of which households have deficient provision). It is important to balance the attention given to improving national monitoring of provision to serve the requirements of national governments and international agencies with that given to developing the local data needed to support improved provision.

As we have seen, the use of GIS by local authorities, commonplace in high-income countries, provides a basis for collecting the kinds of spatial information needed for pro-poor governance. Improved indicators are most needed to serve and inform the institutions with the capacity or potential to contribute to improving and extending provision and inform them as to where and with whom their action should concentrate. For water and sanitation, these are overwhelmingly local groups – water companies and utilities, local governments, NGOs and community organizations. In regard to indicators, the priority should be developing the capacity of these local bodies to identify relevant indicators and collect relevant data to monitor the state of water and sanitation in their communities.

4. In many countries, the collection of information about individuals and businesses is governed by an official statistics acts. These acts are intended to ensure the confidentiality of information that has been collected about a person or a business. Some of the information collected is of a sensitive nature, such as income, health issues or political activism. The individual or business needs to understand that information collected about them remains confidential.

5. UN-HABITAT recently completed a Cities Without Slums (CWS) initiative in Kisumu, Kenya, whose objective was to improve the livelihoods of people living and working in informal settlements in Kisumu. The programme was led by the Municipal Council of Kisumu (MCK), which is the key planning institution for urban development within the Municipality. MCK is to establish a City Development Strategy/CWS/GIS secretariat within their planning department. The secretariat will be in charge of the creation and maintenance of a digital map of Kisumu using a high-resolution satellite image as its basis. Other responsibilities will be to collect, edit, analyse and manage information on informal settlements, population, housing, land-use and urban transport mobility. The secretariat will become a centre of excellence for the use of GIS in local planning issues.

As emphasized throughout, much of the innovation in local governance for improved water and sanitation is rooted in partnerships between local governments and community-based organizations formed by households who generally have the worst provision for water and sanitation (and often local support NGOs). Discussions of indicators that are to have policy relevance and scientific rigour need to reflect this. At this stage, perhaps the most important issue is to support the development of methodologies that can be applied by local bodies that serve to provide the local information base needed to improve and extend water and sanitation provision. The continued development and reduced pricing of satellite imagery and GIS are important factors in enhancing the ability of local government to provide pro-poor solutions.

Part 6. Local Actions for Local Goals

Governments in low- and middle-income nations and international agencies need to recognize the long-term trend towards increasingly urbanized societies and economies and to support capacities within each urban centre for improving water and sanitation provision and wastewater management. Most governments and international agencies underestimate the scale of the deficiencies in provision in urban areas. Good local governance within each of the tens of thousands of urban centres in low- and middle-income nations is important for addressing these deficiencies, both in terms of governments that are more efficient and competent and of governments that work with and are accountable to their populations, especially those with the lowest incomes.

A significant proportion of the rural population lacking adequate provision for water and sanitation live in large villages with urban characteristics in their population size, density and concentration of non-agricultural enterprises. Many such villages should be reclassified as urban centres with attention given to supporting more competent, accountable local governments based there.

New approaches are needed, if the MDGs for water and sanitation are to be met, especially in smaller urban centres, large villages and poorer municipalities within metropolitan areas. In most such settlements, the deficiencies in provision cannot be addressed by the conventional model of a (public or private) water utility extending piped water supplies and sewers to individual households. What is needed is locally developed responses that make best use of local knowledge, resources and capacities. Some of the most cost-effective solutions have been developed by community organizations and federations formed by slum and shack dwellers and the local NGOs that work with them – and as this chapter describes, these have great potential to increase their scale and scope, where local governments and private water utilities work with them.

Many non-water-related policies and programmes are important for improving water and sanitation provision – for instance, housing finance systems, land-use management policies and household and community investments in better housing and neighbourhood infrastructure. Official support for upgrading slums and squatter settlements and for land subdivision for new houses can also contribute much to improved provision for water and sanitation. Housing finance programmes can also make an important contribution, as they support more households to buy, build or improve their homes, including improved provision for water and sanitation. In addition, better water governance usually depends on better local governments – strengthened and supported by appropriate decentralization and democratization.

Whether formal provision for water and sanitation is undertaken by public utilities or international, national or local private utilities, city and municipal governments have a critical role in providing the planning and governance framework. They also have the central role in embedding this within regional water governance arrangements that often require agreements developed with freshwater users upstream of the city and more attention to reducing the impact of water pollution and urban runoff on water quality for users downstream of the city.

It is difficult for very centralized international development agencies to provide the kinds of decentralized support that improving provision for water and sanitation requires. How can they know which factors are most important in each particular locality and how to support them? How can they help ensure the development of more competent, effective

Top to bottom:
Shibuya station, Japan

A class of children from Singapore drew their vision of their environment for the 'Scroll around the world' project

Cans for water provision in Santiago, Cape Verde

local water-and sanitation-providing organizations in which the unserved and ill-served have influence? This is an issue especially where smaller urban centres have governments that lack resources and the right to raise resources and suffer from very weak technical capacity. Furthermore, local social and political structures in these settlements often marginalize or exclude most of the unserved and ill-served. It is correct to say that good local governance is the solution, but saying this does not mean that these agencies know how to achieve it.

The information base needed to support good local governance for water and sanitation differs from the information base being developed by higher levels of government and international agencies to monitor trends in provision for water and sanitation to evaluate progress towards meeting the MDGs. More attention should be given to supporting local data collection and local monitoring to serve local action and evaluation.

References and Websites

Bairoch, P. 1988. *Cities and Economic Development: From the Dawn of History to the Present*. London, Mansell.

Banes, C., Kalbermatten, J. and Nankman, P. 1996. Infrastructure provision for the urban poor. Unpublished World Bank study, Washington DC.

Bapat, M. and Agarwal, I. 2003. Our needs, our priorities; women and men from the 'slums' in Mumbai and Pune talk about their needs for water and sanitation. *Environment and Urbanization*, Vol. 15, No. 2, pp. 71-86.

Baumann, D. D., Boland, J. J. and Hanemann, W. M. 1998. *Urban Water Demand Management and Planning*. New York, McGraw-Hill.

Bond, P. 2000. *Cities of Gold, Townships of Coal*. Trenton, NJ, Africa World Press.

Boonyabancha, S. 2005. Baan Mankong; going to scale with slum upgrading in Thailand. *Environment and Urbanization*, Vol. 17, No. 1.

——. 2003. A decade of change: From the Urban Community Development Office (UCDO) to the Community Organizations Development Institute (CODI) in Thailand. Poverty Reduction in Urban Areas Working Paper 12, London, IIED.

Budds, J. and McGranaham, G. (2003) Are the debates on water privatization missing the point? Experiences from Africa, Asia & Latin America, Environment & Urbanization.

Budds, J., Teixeira, P. and SEHAB. 2005. Building houses, building citizenship: integrated housing, urban development and land tenure legalization for low-income groups in São Paulo, Brazil. *Environment and Urbanization*, Vol. 17, No. 1.

Burra, Sundar. 2005. Towards a pro-poor slum-upgrading framework for Mumbai, India. *Environment and Urbanization*, Vol. 17, No. 1.

Burra, S., Patel, S. and Kerr, T. 2003. Community-designed, built and managed toilet blocks in Indian cities. *Environment and Urbanization*, Vol. 15, No. 2, pp. 11-32.

Cabannes, Yves. 2004. Participatory budgeting: a significant contribution to participatory democracy. *Environment and Urbanization*. Vol. 16, No. 1, pp. 27-46.

Cain, A., Daly, M. and Robson, P. 2002. *Basic Service Provision for the Urban Poor; The Experience of Development Workshop in Angola*, IIED Working Paper 8 on Poverty Reduction in Urban Areas.

Connors, G. 2005. Pro-poor water governance in Bangalore: A city in transition. *Environment and Urbanization*, Vol. 17, No. 1.

Ferguson, B. 1999. Micro-finance of housing: a key to building emerging country cities? *Environment and Urbanization*, Vol. 11, No.1, pp. 185-99.

Gleick, P. H.. 2000. The changing water paradigm – A look at twenty-first century water resources development. *Water International*, Vol. 25, No. 1, pp. 127-38.

——. 2003. Water use. *Annual Review of Environment and Resources*, Vol. 28, pp. 275-314.

Graumann, J. V. 1977. Orders of magnitude of the world's urban and rural population in history. *United Nations Population Bulletin 8*, New York, United Nations, pp. 16-33.

Hanchett, S., Akhter, S. and Khan, M. H. Summarized by Mezulianik, S. and Blagbrough, V. 2003. Water, sanitation and hygiene in Bangladesh slums; a summary of WaterAid's Bangladesh Urban Programme Evaluation. *Environment and Urbanization*, Vol. 15, No. 2, pp. 43-56.

Hardoy, J. E., Mitlin, D. and Satterthwaite, D. 2001. *Environmental Problems in an Urbanizing World*. London, Earthscan Publications.

Hasan, A. 1997. *Working with Government: The Story of the Orangi Pilot Project's Collaboration with State Agencies for Replicating its Low Cost Sanitation Programme*. Karachi, City Press.

Jacobi, P. 2004. The challenges of multi-stakeholder management in the watersheds of São Paulo. *Environment and Urbanization*, Vol. 16, No. 2, pp. 199-212.

Kahrl, W. L. 1982. *Water and Power: The Conflict over Los Angeles' Water Supply in the Owens Valley*. Berkeley, University of California Press.

Linn, J. F. 1982. The costs of urbanization in developing countries. *Economic Development and Cultural Change*, Vol. 30, No. 3.

Lundqvist, J., Biswas, A., Tortajada, C. and Varis, O. 2004. *Water Management in Megacities*, Stockholm, World Water Week.

Menegat, R. 2002. Participatory democracy and sustainable development: integrated urban environmental management in Porto Alegre, Brazil. *Environment and Urbanization*, Vol. 14, No. 2, pp. 181-206. www.ingentaselect.com/09562478/v14n2/.

Mitlin, D. and Mueller, A. 2004. Windhoek, Namibia: towards progressive urban land policies in Southern Africa. *International Development Planning Review*, Vol. 26, No. 2.

Montgomery, M.R., Stren, R., Cohen, B. and Reed, H.E. (eds.) 2003. *Cities Transformed; Demographic Change and its Implications in the Developing World*. Washington DC, The National Academy Press, London, Earthscan.

Navarro, L. 2001. Exploring the environmental and political dimensions of poverty: the cases of the cities of Mar del Plata and Necochea-Quequén. *Environment and Urbanization*, Vol. 13, No.1, pp. 185-99.

Orangi Pilot Project – Research and Training Institute. 2002., *Katchi Abadis of Karachi: Documentation of Sewerage, Water Supply Lines, Clinics, Schools and Thallas – Volume One: The First Hundred Katchi Abadis Surveyed*. Karachi, Orangi Pilot Project.

——. 1998. *Proposal for a Sewage Disposal System for Karachi*. Karachi, City Press.

Patel, S. 2004. Tools and methods for empowerment developed by slum dwellers federations in India. *Participatory Learning and Action 50*, London, IIED.

Platt, R. H. 2004. *Land Use and Society: Geography, Law, and Public Policy*. Washington DC, Island Press.

Rahman, P. 2004a. *Update on OPP-RTI's Work*. Karachi, Orangi Pilot Project – Research and Training Institute.

——. 2004b. *Katchi Abadis of Karachi; a Survey of 334 Katchi Abadis*. Karachi, Orangi Pilot Project – Research and Training Institute.

Reisner, M. 2001. *Cadillac Desert: The American West and its Disappearing Water*. London, Pimlico.

Satterthwaite, D. 2005. The scale of urban change worldwide 1950-2000 and its underpinnings. IIED Working Paper. www.iied.org/urban/index.html.

——. 2002. Coping with rapid urban growth. RICS International Paper Series, London, Royal Institution of Chartered Surveyors.

Showers, K. B. 2002. Water scarcity and urban Africa: an overview of urban-rural water linkages. *World Development*, Vol. 30, No. 4, pp. 621–48.

Shrestha, D. and Cronin, A. A. 2006. The right to water and protecting refugees. *Waterlines*, Vol. 24, pp. 12-140. *World Development*, Vol. 30, No. 4, pp. 621–48.

Smit, J., Ratta, A. and Nasr, J. 1996. *Urban Agriculture: Food, Jobs and Sustainable Cities.* Publication Series for Habitat II, Vol. 1, New York, UNDP.

Sohail Khan, M. and Cotton, A. 2003. Public private partnerships and the poor in water supply projects. WELL Factsheet, Loughborough, WEDC, Loughborough University. www.lboro.ac.uk/well/resources/fact-sheets/fact-sheets-htm/PPP.htm.

Souza, C. 2001. Participatory budgeting in Brazilian cities: Limits and possibilities in building democratic institutions. *Environment and Urbanization*, Vol. 13, No 1, pp. 159–84. www.ingentaselect.com/09562478/v13n1/.

Stein, A. and Castillo, L. 2005. Innovative financing for low-income housing improvement: Lessons from programmes in Central America. *Environment and Urbanization*, Vol. 17, No. 1.

Tortajada, C. and Castelan. E. 2003. Water management for a megacity: Mexico City Metropolitan Area. *Ambio*, Vol. 32, No. 2, pp. 124–29.

UN (United Nations). 2004. *World Urbanization Prospects: The 2003 Revision.* New York, United Nations Population Division, Department of Economic and Social Affairs, ST/ESA/SER.A/237.

——. *2004 Global Refugee Trends: Overview of Refugee Populations, New Arrivals Durable Solutions, Asylum-seekers, Stateless and Other Persons of Concern to UNHCR.* Geneva.

UN (United Nations) Millennium Project. 2005a. *Health, Dignity and Development; What Will it Take?* Task Force on Water and Sanitation, London and Sterling VA, Earthscan.

UN-HABITAT/(United Nations Human Settlement Programme). 2005b. *A Home in the City.* TaskForce on Improving the Lives of Slum Dwellers, draft.

——. 2004a. *Unheard Voices of Women in Water and Sanitation.* Manila and Nairobi, Water for Asian Cities Programme.

——. *Lake Victoria Region Water and Sanitation Initiative; Supporting Secondary Urban Centres in the Lake Victoria Region to Achieve the Millennium Development Goals.* Nairobi, Water for African Cities Programme, UN-HABITAT.

——. *The Challenge of Slums: Global Report on Human Settlements 2003.* London, Earthscan.

——. *Water and Sanitation in the World's Cities: Local Action for Global Goals.* London, Earthscan.

——. *Handbook for Emergencies.* United Nations High Commissioners for Refugees. Geneva.

UNHCR. 2004. *The World's Stateless People.* United Nations Questions and Answers. Geneva.

van den Berg, L. M., van Wijk, M.S. and Van Hoi. P. 2003. The transformation of agriculture and rural life downstream of Hanoi. *Environment and Urbanization*, Vol. 15, No. 1, pp. 35-52.

van Horen, B. 2001. Developing community-based watershed management in Greater São Paulo: The case of Santo André. *Environment and Urbanization*, Vol. 13, No. 1, pp. 209–22.

von Bertrab, E. 2003. Guadalajara's water crisis and the fate of Lake Chapala: A reflection of poor water management in Mexico. *Environment and Urbanization*, Vol. 15, No. 2, pp. 127–40.

Weru, J. 2004. Community federations and city upgrading: The work of Pamoja Trust and Muungano in Kenya. *Environment and Urbanization*, Vol. 16, No. 1, pp. 47–62.

WHO (World Health Organization). 1999. Creating healthy cities in the 21st Century. D. Satterthwaite, (ed.), *The Earthscan Reader on Sustainable Cities.* London, Earthscan Publications, pp. 137–72.

WHO (World Health Organization) and UNICEF (United Nations Children's Fund). 2000. *Global Water Supply and Sanitation Assessment 2000 Report.* New York/Geneva, WHO/ UNICEF.

WHO (World Health Organization) and UNICEF (United Nations Children's Fund) JMP (Joint Monitoring Programme). 2004. *Meeting the MDG Drinking Water and Sanitation Target; A Mid-term Assessment.* Joint Monitoring Programme for Water Supply and Sanitation, Geneva.

World Bank. 2003. *World Development Report 2004: Making Services Work for Poor People.* Washington DC, The World Bank and Oxford, Oxford University Press.

Zaidi, S.A. 2000. *Transforming Urban Settlements: The Orangi Pilot Project's low-cost sanitation model.* Karachi, City Press.

www.unhabitat.org
www.unwac.org
www.un-urbanwater.net
www.iied.org
www.measuredhs.com
www.unhabitat.org
www.adb.org/water/default.asp
www.un.org/waterforlifrdecade/statements.html
www.unher.org

SECTION 2

Changing Natural Systems

Both naturally occurring conditions and human impacts are asserting strong pressure on our water resources today, in the form of warming temperatures, rising sea levels, ecosystems damage and increased climatic variability, among others. Human influence is arguably becoming more important than natural factors. The construction of dams and diversions continue to affect river regimes, fragmenting and modifying aquatic habitats, altering the flow of matter and energy, and establishing barriers to the movement of migratory species. Deforestation, increasing areas of farmland, urbanization, pollutants in both surface and sub-surface water bodies and so on, all influence the timing and quantities of flows and are having a huge impact on the quality and quantity of freshwater.

It is against this background that we must assess the state of the water resources. Assessment is a critical and necessary first step to ensuring that the dual goals of water for environmental and human needs are met. This section presents an overview of the state of water resources and ecosystems and explores current assessment techniques and approaches to Integrated Water Resources Management (IWRM).

Global Map 3: *Relative Water Stress Index*
Global Map 4: *Sources of Contemporary Nitrogen Loading*

Chapter 4 – **The State of the Resource** (UNESCO & WMO, with IAEA)

This chapter reviews the main components of the water cycle and provides an overview of the geographical distribution of the world's total water resources, their variability, the impacts of climate change and the challenges associated with assessing the resource.

Chapter 5 – **Coastal and Freshwater Ecosystems** (UNEP)

Natural ecosystems, rich in biodiversity, play a critical role in the water cycle and must be preserved. In many areas, a variety of pressures on freshwater ecosystems are leading to their rapid deterioration, affecting livelihoods, human well-being and development. To reverse this trend, protecting ecosystems and biodiversity must become a fundamental component of Integrated Water Resources Management (IWRM).

Section 2: CHANGING NATURAL SYSTEMS

Relative Water Stress Index

Water stress is commonly evaluated by comparing the volume of renewable water resources per capita at a national level. New mapping capabilities allow the geography of water stress to be better defined. High resolution water stress indices can be computed based on the ratio of total water use (sum of domestic, industrial and agricultural demand or DIA) to renewable water supply (Q), which is the available local runoff (precipitation less evaporation) as delivered through streams, rivers and shallow groundwater. Developed from actual statistics, the Relative Water Stress Index (RWSI),

also known as Relative Water Demand, is useful because it is a dimensionless quantity, which can be applied at different scales. The map below shows populations living in water stressed (RWSI ≥ 0.4) and relatively unstressed (RWSI < 0.4) conditions highlighting substantial within-country differences that national estimates often obscure. The map shown below (at approximately 50 km resolution globally) nearly tripled earlier nation-wide estimates of those people living under severe water stress, with obvious impacts on the degree to which water problems can be appropriately identified and managed.

Population (in thousands) above (reds) and below (blues) water stress threshold (RWSI=0.4)

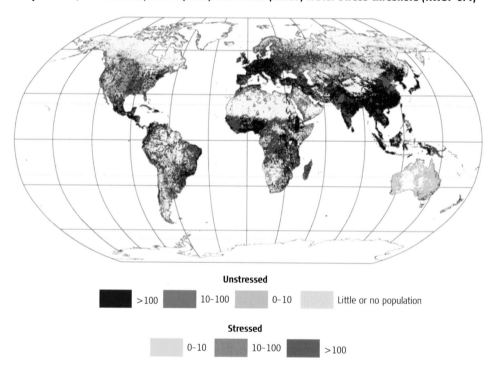

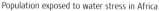

Unstressed

| >100 | 10-100 | 0-10 | Little or no population |

Stressed

| 0-10 | 10-100 | >100 |

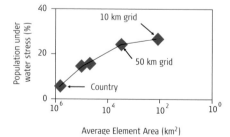

Population exposed to water stress in Africa

The scale of the analysis is critically important when assessing the level of water stress. For instance, the graph on the left shows that when water stress was computed at the country scale, about 4% of the population in Africa was identified as suffering under severe water stress. However, when evaluated using geospatial data at a 10-km resolution, the percentage of the population experiencing stress increased to 26% (Douglas et al., 2006; Vörösmarty et al., 2005a).

Source: Water Systems Analysis Group, University of New Hampshire. Datasets available for download at http://wwdrii.sr.unh.edu/

Sources of Contemporary Nitrogen Loading

Nitrogen actively cycles through the atmosphere, the continental land mass and the world's oceans, and represents a critical nutrient upon which plant, microbial, and animal life depend. Nitrogen, the most abundant gas in the atmosphere, is delivered to watersheds through natural processes including chemical transformation and washout from precipitation as well as biological fixation. The pathways that nitrogen follows as it travels through the environment are complex. Contemporary human activities have greatly accelerated the transport of reactive nitrogen through river basins that ultimately deliver this nutrient into coastal receiving waters (Galloway et al., 2004). Globally there has been a two-fold increase in the delivery of this nutrient to the oceans, with more than ten-fold increases in some rivers draining industrialized regions (Green et al., 2004). These increases arise from the widespread application of fertilizer, animal husbandry and point source sewage inputs.

These human induced changes to the nitrogen cycle have far reaching impacts on water quality and public health, protein supply for humans, and even the planetary heat balance through the emission on nitrogen-based greenhouse gases. The map below shows the predominant source of nitrogen within each grid cell. Fixation is the primary source throughout South America, Africa, Australia, and the northernmost reaches of Asia and North America. Atmospheric pollution and subsequent nitrogen deposition plays a dominant role throughout the industrialized northern temperate zones of Europe, Asia and North America. Fertilizers are the predominant source across major food producing regions. Livestock constitutes the most important source in Eastern Europe and India. Urban sewage loads create localized 'hotspots' for pollution. Understanding the patterns of such loadings is critical to the design of management interventions to protect society and well-functioning ecosystems.

Predominant Sources of Contemporary Nitrogen Loading

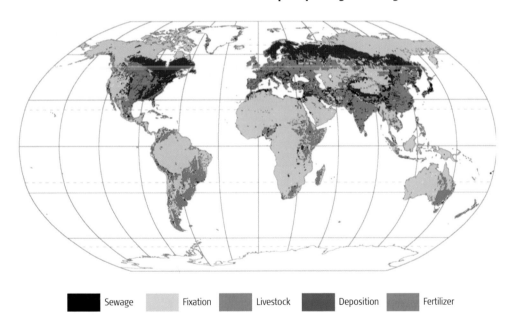

Sewage   Fixation   Livestock   Deposition   Fertilizer

Source: Water Systems Analysis Group, University of New Hampshire. Datasets available for download at http://wwdrii.sr.unh.edu/

I would feel more optimistic about a bright future for man if he spent less time proving he can outwit Nature and more time tasting her sweetness and respecting her seniority.

CHAPTER 4

The State of the Resource

By

UNESCO
(United Nations Educational, Scientific and Cultural Organization)

WMO
(World Meteorological Organization)

IAEA
(International Atomic Energy Agency)

Section 2: CHANGING NATURAL SYSTEMS

Top to bottom:
Perito Moreno Glacier,
Argentina

Man collecting stagnant
water for drinking,
Uganda

Bus driving across flooded
plateau in the Andes,
Bolivia

Key messages:

Our water resources, irregularly distributed in space and time, are under pressure due to major population change and increased demand. Access to reliable data on the availability, quality and quantity of water, and its variability, form the necessary foundation for sound management of water resources. The different options for augmentation expand the boundaries of the water resource in a conventional sense, helping to match demand and supply. All components of the hydrological cycle, and the influence of human activities on it, need to be understood and quantified to efficiently and sustainably develop and protect our water resources.

■ Climate change is having a significant impact on weather patterns, precipitation and the hydrological cycle, affecting surface water availability, as well as soil moisture and groundwater recharge.

■ The growing uncertainty of surface water availability and increasing levels of water pollution and water diversions threaten to disrupt social and economic development in many areas as well as the health of ecosystems.

■ Groundwater resources can, in many instances, supplement surface water, particularly as a source of drinking water. However, in many cases, these aquifers are being tapped at an unsustainable rate or affected by pollution. More attention should be paid to sustainable management of non-renewable groundwater.

■ Many traditional practices are being refined (e.g. rainwater harvesting), while more recent advances (e.g. artificial recharge, desalination and water reuse) are being developed further. More support needs to be given to policy options, such as demand management, which stress more efficient use of water resources, as well as to technical solutions on the supply side.

■ The projected increased variability in the availability and distribution of freshwater resources demands political commitment to supporting and advancing technology for the collection and analysis of hydrological data. More up-to-date information will enable policy-makers to make better informed decisions regarding water resources management.

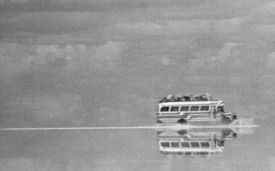

Part 1. Global Hydrology and Water Resources

The need to develop more sustainable practices for the management and efficient use of water resources, as well as the need to protect the environmental ecosystems where these resources are located, has led to fundamental shifts in awareness and public concern over the past decade. However, despite increased awareness of the issues at stake, economic criteria and politically charged reasoning are still driving water resource development decisions at most local, regional, national and international levels. Though the long-term benefits of an integrated approach to achieving sustainable water resources development have been cited in many of the global water conferences over the past decade, considerable time and change in policy will be required to implement such an approach. At present, best available practice and scientific knowledge are rarely adequately factored into decision-making or well represented when establishing water resource policy or implementing management practices. In the meantime, the pressures on our water resources are increasing.

At present, best available practice and scientific knowledge are rarely adequately factored into decision-making

1a. The driving forces and pressures on our water resources

The combination of both naturally occurring conditions and humanity's actions creates pressure on our water resources. Climate change and natural variability in the distribution and occurrence of water are the natural driving forces that complicate the sustainable development of our water resources. Some of the main driving forces affecting water resources include:

■ population growth, particularly in water-short regions

■ major demographic changes as people move from rural to urban environments

■ higher demands for food security and socio-economic well-being

■ increased competition between users and usages

■ pollution from industrial, municipal and agricultural sources.

While many issues remain on how to deal with and alleviate the pressures on our water resources, the progress being made in some sectors is worth noting. Natural units, such as river basins and aquifer systems, are becoming institutionally recognized: one example is the EU Water Framework Directive. Basin-oriented water resources assessment is increasingly being adopted by national and regional programmes and due consideration is given to the need to identify the critical volume and quality of water needed to maintain ecosystem resilience (environmental flows; see **Chapter 5**).

We are also seeing the emergence of highly detailed analyses of the processes involved as well as results-based diagnoses from catchment agencies, basin commissions and watershed and aquifer management authorities. These activities are being carried out globally in a variety of different economic and cultural settings and at different sizes and scales. Most of these organizations were created relatively recently for

Figure 4.1: Global distribution of the world's water

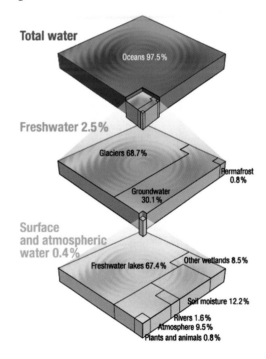

Source: Data from Shiklomanov and Rodda, 2003. Freshwater has a global volume of 35.2 million cubic kilometres (km³).

jurisdictions that correspond to physical hydrological limits rather than historically defined administrative boundaries (Blomquist et al., 2005; WWF, 2003). Moving away from historically administrative boundaries to a consideration of water resources management practice based on physical hydrological limits allows us to better respond to nature's variability.

To better combat flooding, the Associated Programme of Flood Management (APFM), a WMO and GWP joint effort as well as UNESCO's International Flood Initiative (IFI) outline new approaches that are being developed for a better understanding of the links between natural settings and the legal, environmental and social conditions inherent to flooding and the mitigation of its impacts. In this way, communities commonly faced with flooding can now develop more sustainable methods to reduce the socio-economic effects of such high-impact events (see **Chapter 10**).

Further illustrations of emerging progress on countering the pressures on water resources are highlighted in **Chapters 6** through **13** and in some of the case study examples cited in **Chapter 14**.

Ecohydrology stresses the important relationships and pathways shared among hydrological and ecological systems

1b. Water's global occurrence and distribution

The world's water exists naturally in different forms and locations: in the air, on the surface, below the ground and in the oceans (**Figure 4.1**).

Although a large volume of freshwater exists 'in storage', it is more important to evaluate the renewable annual water flows, taking into account where and how they move through/ the hydrological cycle (**Figure 4.2**). This schematic of the hydrological cycle illustrates how elements can be grouped as part of a conceptual model that has emerged from the new discipline of ecohydrology, which stresses the important relationships and pathways shared among hydrological and ecological systems (Zalewski et al., 1997). This conceptual model takes into consideration the detail of the fluxes of all waters and their pathways while differentiating between two components: 'blue water' and 'green water'. Blue waters are directly associated with aquatic ecosystems and flow in surface water bodies and aquifers. Green water is what supplies terrestrial ecosystems and rain-fed crops from the soil moisture zone, and it is green water that evaporates from plants and water surfaces into the

Figure 4.2: Schematic of the hydrologic cycle components in present-day setting

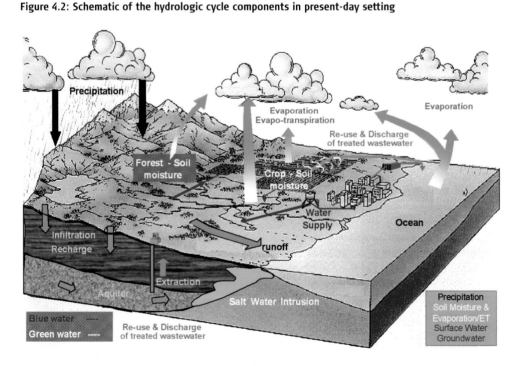

atmosphere as water vapour. This concept was developed by Falkenmark and Rockström (2004) who contend that the introduction of the concepts of 'green water' and 'blue water', to the extent that they simplify the discussion for non-technical policy-makers and planners, may help to focus attention and resources on the often neglected areas of rain-fed agriculture, grazing, grassland, forest and wetland areas of terrestrial ecosystems and landscape management.

Part 2. Nature, Variability and Availability

The Earth's hydrological cycle is the global mechanism that transfers water from the oceans to the surface and from the surface, or subsurface environments, and plants to the atmosphere that surrounds our planet. The principal natural component processes of the hydrological cycle are: precipitation, infiltration, runoff, evaporation and transpiration. Human activities (settlements, industry, and agricultural developments) can disturb the components of the natural cycle through land use diversions and the use, reuse and discharge of wastes into the natural surface water and groundwater pathways.

A young child plays in the monsoon rain in Thailand

2a. Precipitation

The Earth's atmosphere contains approximately 13,000 km^3 of water. This represents 10 percent of the world's freshwater resources not found in groundwater, icecaps or permafrost (**Figure 4.1**). This is similar to the volumes found in soil moisture and wetlands. However, of more importance is the fact that this vapour cycles in the atmosphere in a 'global dynamic envelope', which has a substantive annually recurring volume, estimated to be from 113,500 to 120,000 km^3 (Shiklomanov and Rodda, 2003; FAO-AQUASTAT, 2003). Precipitation occurs as rain, snow, sleet, hail, frost or dew. These large volumes illustrate precipitation's key role in renewing our natural water resources, particularly those used to supply natural ecosystems and rainfed crops. About 40 percent of the precipitation that falls on land comes from ocean-derived vapour. The remaining 60 percent comes from land-based sources. It is particularly pertinent to recognize that snowfall can contribute a large percentage of a region's total precipitation in temperate and cold climate regions. For example, in the western US, Canada and Europe, 40 to 75 percent of regional precipitation can occur as snow.

The International Panel on Climate Change (IPCC) has published the international reference for each country's average annual precipitation, based on the period of record from 1961 to 1990 (New et al., 1999; Mitchell et al., 2002). Countries' precipitation ranges from 100 mm/yr in arid, desert-like climates to over 3,400 mm/yr in tropical and highly mountainous terrains. Together with temperature, they define the significant variables in global climatic and ecosystem biodiversity settings. This long-term record base determines averages and defines predictable variability both in time (monthly, annually, seasonally) and place (nations, monitoring locations). This record is significant as its 30-year standard is commonly compared with actual annual amounts to define the relative current variability, frequently tied to regional and global evaluations of drought and climate change.

It is essential to water resources development to understand the pathways of water as it arrives in the form of precipitation and migrates through the cycle components. **Table 4.1** illustrates how precipitation, in three relatively diverse climatic zones, generally either returns by evaporation or evapotranspiration back into the atmosphere, becomes surface water through runoff, or recharges groundwater.

Mapping precipitation's isotopic composition ($_3$H, $_{18}$O and $_2$H) can help trace water movement through the water cycle components. This is routinely done as part of the Global Network of Isotopes in Precipitation (GNIP),[1] operated jointly by IAEA and WMO at 153 stations in 53 nations. IAEA has initiated several projects to study and distinguish among moisture sources and to better understand the cycle transport patterns using applied isotope techniques. Particular case studies have been carried out in India (Bhattacharya et al., 2003), Southeast Asia (Aggarwal et al., 2004) and with twenty-one research groups participating globally to monitor many other major rivers (**Figure 4.3**). This approach is of further significance as it assists in the evaluation of the hydrological cycle's response to climatic fluctuations and

It is essential to understand the pathways of water as it arrives in the form of precipitation and migrates through the cycle components

---

1. See isohis.iaea.org for more information.

Table 4.1: Precipitation distribution into surface water and groundwater components (by climate region)

	Temperate climate		Semi-arid climate		Arid climate	
	%	mm	%	mm	%	mm
Total precipitation	100	500–1,500	100	200–500	100	0–200
Evaporation/ Evapotranspiration	~ 33	160–500	~ 50	100–250	~ 70	0–140
Groundwater recharge	~ 33	160–500	~ 20	40–100	~ 1	0–2
Surface runoff	~ 33	160–500	~ 30	60–150	~ 29	0–60

Source: Hydrogeology Center, University Neuchâtel, 2003.

can be used to calibrate and validate atmospheric circulation models used in climate change studies.

2b. Evapotranspiration and soil moisture

The processes of evaporation and transpiration (evapotranspiration) are closely linked to the water found in soil moisture; these processes act as driving forces on water transferred in the hydrological cycle. Movement through soil and vegetation is large and accounts for 62 percent of annual globally renewable freshwater. Evapotranspiration rates depend on many locally specific parameters and variables that are difficult to measure

and require demanding analyses in order to calculate an acceptable level of accuracy. Other hydrological, cycle-related and meteorological data are also considered in the estimation of the rates. Today, however, local water management in basins or sub-basins can better calculate transpiration rates.

Evaporation from surface water bodies such as lakes, rivers, wetlands and reservoirs is also an important component of the hydrological cycle and integral to basin development and regional water management. In the case of artificially-created reservoirs, it has been estimated by Rekacewicz (2002) that the global volumes evaporating since the end of the 1960s have exceeded the volume consumed to meet both domestic and industrial needs.

Figure 4.3: Oxygen-18 content of stream water along the main stem of large rivers

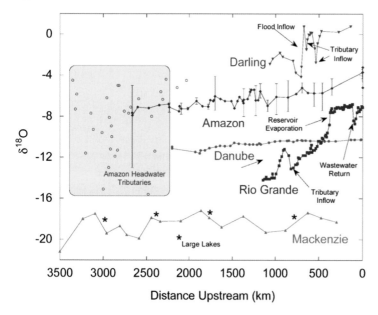

From the standpoint of food production and ecosystem maintenance, soil moisture is the most important parameter to net primary productivity (NPP) and to the structure, composition and density of vegetation patterns (WMO, 2004). Near-surface soil moisture content strongly influences whether precipitation and irrigation waters either run off to surface water bodies or infiltrate into the soil column. Regionally, mapping soil moisture deficit is becoming a widely used technique to link climatological and hydrological information in agriculture (e.g. Illinois, US) and to reflect drought conditions (US Drought Mitigation Center, 2004). Soil moisture distribution is now identified as a prerequisite for effective river-flow forecasting, irrigation system maintenance, and soil conservation (Haider et al., 2004). Its distribution in time and place are now viewed as essential to hydrological, ecological and climatic models – both at the regional and global level (US NRC, 2000).

Note: Surveys of oxygen-18 along the main stem of large rivers, such as Darling, Amazon, Danube, Rio Grande and Mackenzie show the contribution and mixing of runoff sources to rivers, such as tributaries, irrigation water and wastewater. Isotopes also reflect impacts of climate and land use pattern changes on the water balance such as an evaporative enrichment of river water in arid regions.

Source: IAEA, 2002.

The Global Soil Moisture Data Bank (Robock and Vinnikov, 2005; Robock et al, 2000) archives contain data sets of national soil moisture records but the data sets are incomplete in terms of global coverage.

Satellite data can provide broader coverage with current results that can be more closely representative when combined with ground validation. From 2002, NASA's climate-monitoring 'Aqua' satellite has daily records of 50 to 60 km resolution data, readily obtained from NOAA (Njoku, 2004; Njoku et al., 2004). From 2010, the 'Hydros' satellite will exclusively monitor daily soil moisture changes around the globe with an improved spatial resolution of 3 to 10 km (Entekhabi et al., 2004; Jackson, 2004). This will be an important upgrade for remotely-sensed soil moisture data, which are becoming increasingly relied upon by agricultural marketing and administrative boards, commodity brokers, large-scale farms, flood- and drought-monitoring and forecasting agencies, water resources planning and soil conservation authorities and hydroelectric utility companies.

2c. Snow and ice

About three-quarters of the world's entire natural freshwater is contained within ice sheets and glaciers. However, most (97 percent) is not considered as a water resource as it is inaccessible, located in the Antarctic, Arctic and Greenland ice sheets. However, land-based glaciers and permanent snow and ice – found on all continents except Australia – cover approximately 680,000 km[2] and are critical to many nations' water resources. Even in situations where ice covers only a small percent of a basin's upland mountainous terrain (e.g. in the Himalayas, Rockies, Urals, Alps, Andes), glaciers can supply water resources to distant lowland regions. Thus, glacial ice and snow represents a highly valuable natural water reservoir. Typically it affects stream-flow quantity in terms of time and volume since glaciers temporarily store water as snow and ice and release runoff on many different time scales (Jansson et al., 2003; Hock et al., 2005). Glacial runoff characteristically varies with daily flow cycles that are melt-induced and seasonal since concentrated annual runoff occurs in summer when the water stored as snow in winter is released as stream flow. The seasonal runoff benefits occur principally in nations in the mid- and high latitudes where there are otherwise only periods of low flow, but benefits also occur in many semi-arid regions. Glaciers can also affect long-term annual water availability since runoff either increases or decreases as their mass balance decreases or increases, respectively. Finally, glaciers tend to act as stream-flow regulators that can minimize year-to-year variability when catchment areas are moderately (10 to 40 percent) glaciated. Runoff variability rises as glaciated percentage both increases

and decreases. Glacier conditions are now monitored globally since climate change is affecting their size and mass balance.

2d. Surface waters

Surface waters include the lakes (as well as ponds), reservoirs, rivers and streams and wetlands our societies have depended upon and benefited from throughout history. The flow into and through these surface water bodies comes from rainfall, runoff from melting snow and ice and as base-flow from groundwater systems. While surface waters volumetrically hold only a small volume (0.3 percent) of the Earth's total freshwater resources, they represent about 80 percent of the annually renewable surface and groundwater. Ecosystem services from surface waters are widespread and diverse as well as being of critical importance. Reservoirs and large lakes effectively counteract high seasonal variability in runoff by providing longer-term storage. Other services supported by surface waters include shipping and transport, irrigation, recreation, fishing, drinking water and hydropower.

Lakes

Meybeck (1995), Shiklomanov and Rodda (2003) and most recently Lehner and Döll (2004) have provided extensive data characterizing the world's lakes on a global scale. Lakes store the largest volume of fresh surface waters (90,000 km[3]) – over forty times more than is found in rivers or streams and about seven times more than is found in wetland areas. Together with reservoirs, they are estimated to cover a total area of about 2.7 million km[2], which represents 2 percent of the land's surface (excluding polar regions) (Lehner and Döll, 2004). Most lakes are small. The world's 145 largest lakes are estimated to contain over 95 percent of all lake freshwater. Lake Baikal (Russia) is the world's largest, deepest and oldest lake and it alone contains 27 percent of the freshwater contained in all the world's lakes. Lake waters serve commerce, fishing, recreation, and transport and supply water for much of the world's population. However, detailed hydrological studies have been conducted on only 60 percent of the world's largest lakes (Shiklomanov and Rodda, 2003). LakeNet[2] is one example of an organization working with local and regional governments, NGOs and IGOs in over 100 countries in order to address this knowledge deficit, to tackle degrading conditions, and to develop lake basin management programmes that include important protection strategies. Recently, a global database of

Glacier conditions are now monitored globally since climate change is affecting their size and mass balance

Torre del Paine, Chile. Glacial ice and snow represent a highly valuable natural water reservoir

2. See www.worldlakes.org for more information.

...the world's total runoff is unevenly distributed throughout the year for most regions of the globe...

lakes, reservoirs and wetlands (GLWD) has been created and validated at the Center for Environmental Systems Research, University of Kassel (CESR, Germany) in cooperation with the World Wildlife Fund (WWF) (Lehner and Döll, 2004). The primarily digital map-based approach, complete with fully downloadable data, facilitates the linking of existing local and regional registers and remotely sensed data with the new inventory. As such, it is an important achievement related to global hydrological and climatological models.

Rivers and streams

An estimated 263 international river basins have drainage areas that cover about 45 percent (231 million km[2]) of the Earth's land surface (excluding polar regions) (Wolf et al., 1999, 2002). The world's twenty largest river basins have catchment areas ranging from 1 to 6 million km[2] and are found on all continents. The total volume of water stored in rivers and streams is estimated at about 2,120 km[3]. The Amazon carries 15 percent of all the water returning to the world's oceans, while the Congo-Zaire basin carries 33 percent of the river flow in Africa (Shiklomanov and Rodda, 2003).[3]

Variability in runoff is depicted by river/stream flow vis-à-vis time graphs (hydrographs). In terms of variability, **Figure 4.4** (Digout, 2002) illustrates the three low and three high runoff periods that were experienced in the twentieth century by documenting the natural fluctuations in river runoff in terms of both time and place. These types of periodic variations are not particularly predictable as they occur with irregular frequency and duration. In contrast, we are commonly able to predict runoff variability on an annual and seasonal basis from long-term measurement records in many river locations. River-flow graphs representative of the principal climatic regions are illustrated in **Figure 4.5** (Stahl and Hisdal, 2004). Shown together with monthly precipitation and evaporation, they portray the annual variability that is relatively predictable and similar according to principal climatic regions of the world. From this climatic zone perspective, tropical regions typically exhibit greater river runoff volumes while arid and semi-arid regions, which make up an estimated 40 percent of the world's land area, have only 2 percent of the total runoff volume (Gleick, 1993).

Monitoring networks for river flow and water levels in rivers, reservoirs and lakes, supplemented by estimates for regions where there is no extensive monitoring, help understand runoff and evaluate how to predict its

3. Statistics related to the world's river systems (length, basin area, discharge, principal tributaries and cities served) are currently updated online at www.rev.net/~aloe/river/, as part of an open source physical sciences information gateway (PSIGate).

Figure 4.4: Variations in continental river runoff through most of the twentieth century (deviations from average values)

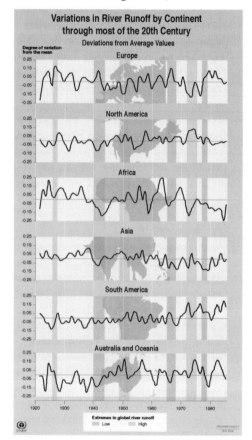

Source: Digout, 2002; UNEP/GRID Arendal; Shiklomanov, 1999.

variability. Measurement networks are relatively common in many developed populated areas. Most of the world's major contributing drainage areas have relatively adequate monitoring networks in place. The Global Runoff Data Center (GRDC, Koblenz, Germany), under WMO's auspices, routinely acquires, stores, freely distributes and reports on river discharge data from a network of 7,222 stations, about 4,750 of which have daily and 5,580 of which have monthly data (GRDC, 2005; **Map 4.1**). Other international programmes such as the European Water Archive (Rees and Demuth, 2000) and national data centres supplement this (data from private institutions are not included). The longer the flow record, the better we can predict variability in runoff – input that is especially important in the context of flood forecasting, hydropower generation and climate change studies. The quality and adequacy of data records for runoff vary tremendously. While some

Figure 4.5: Typical hydrographs in accordance with climatic settings

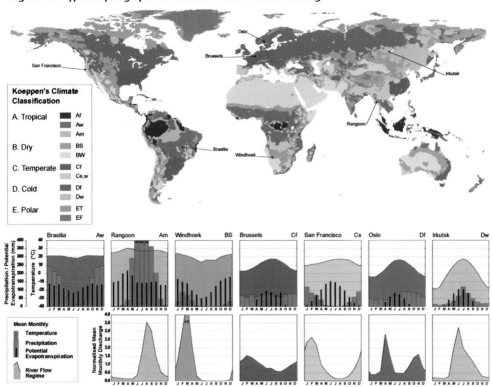

Note: For tropical climates close to the equator (Af), perennial rivers flow all year. Towards the north and south the tropical climates have a distinct rainy season and a dry season (Am and Aw). In dry climates (B) rivers are often ephemeral and only flow periodically after a storm. In the temperate Cf climate, there is no distinct dry or wet season, whereas the 'Mediterranean climate' (Cs) has a pronounced seasonal water deficit in the summer and a rainy winter reflected in the hydrograph. The cold climates (D) have a distinct snowmelt runoff peak and the Df climate has an additional peak in the autumn caused by rain.

Source: Stahl and Hisdal 2004.

records extend back 200 years in Europe and 100 to 150 years on other continents, in many developing nations the data record is generally of insufficient length and quality to carry out either reliable water resources assessments or cost-effective project designs. As a result, for these regions, data is rarely compiled or distributed effectively on a global scale (WMO, 2005).

2e. Wetlands

Wetlands are water-saturated environments and are commonly characterized as swamps, bogs, marshes, mires and lagoons. Wetlands cover an area about four times greater than the world's lakes. However, they contain only 10 percent of the water found in lakes and other surface waters. During the last century, an extensive number of wetlands were destroyed or converted to other forms of land-use. The role they play in terms of ecosystems and water services are more fully described in

Chapter 5. However, because they total about 6 percent of the Earth's land surface (OECD, 1996), they are critical areas to consider and protect in terms of surface water and, in some regions, groundwater resources. Currently, extensive work is being done through the 'Wise Use' campaigns sponsored principally by Ramsar, WWF and UNEP. These campaigns seek to maintain critical services in water and related livelihood and food production areas. An important new study on variability in the role of wetlands was carried out by Bullock and Acreman (2003), wherein they assess the differences in wetland water quantity functions based on 169 worldwide studies conducted from 1930 to 2002. They believe this new review 'provides the first step towards a more scientifically defensible functional assessment system (of wetlands)' and establishes 'a benchmark for the aggregated knowledge of wetland influences upon downstream river flows and aquifers'. They conclude that

Everglades National Park, United States, is 1 of the 1,558 wetland sites currently protected under the Ramsar Convention

Map 4.1: Distribution of GRDC water measurement stations, March 2005

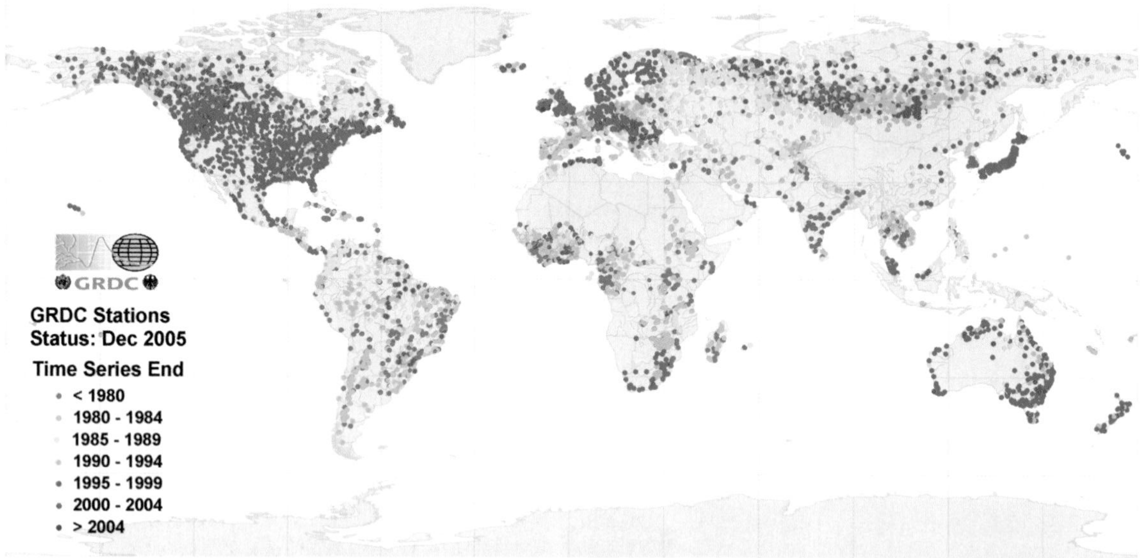

GRDC Stations
Status: Dec 2005

Time Series End
- < 1980
- 1980 - 1984
- 1985 - 1989
- 1990 - 1994
- 1995 - 1999
- 2000 - 2004
- > 2004

'there is only limited support to the generalized model of flood control, recharge promotion and flow maintenance portrayed throughout the 1990s as one component of the basis of wetland policy formulation', noting that support is confined largely to floodplain wetlands. They also note that: 'Less recognized are the many examples where wetlands increase floods, act as a barrier to recharge or reduce low flows' and that 'generalized and simplified statements of wetland function are discouraged because they demonstrably have little practical value'. Overall they conclude that wetlands cannot be considered to have the same role in every hydrological setting. They recommend that future water management actions for both basins and aquifers carefully evaluate each wetland's characteristics as they will exhibit different performance and functional roles according to their location in the watershed, their climate, and the extent of other development features.

2f. Groundwater

Global groundwater volume stored beneath the Earth's surface represents 96 percent of the Earth's unfrozen freshwater (Shiklomanov and Rodda, 2003). Groundwater provides useful functions and services to humans and the environment. It feeds springs and streams, supports wetlands, maintains land surface stability in areas of unstable ground, and acts as an overall critical water resource serving our water needs.

UNESCO and WMO support the International Groundwater Resources Assessment Centre (IGRAC, hosted in Utrecht, The Netherlands). IGRAC estimates that about 60 percent of withdrawn groundwater is used to support agriculture in arid and semi-arid climates. Morris et al. (2003) report that groundwater systems globally provide 25 to 40 percent of the world's drinking water. Today, half the world's megacities and hundreds of other major cities on all continents rely upon or make significant use of groundwater. Small towns and rural communities particularly rely on it for domestic supplies. Even where groundwater provides lower percentages of total water used, it still can serve local areas with relatively low-cost good-quality water where no other accessible supply exists. Finally, groundwater can bridge water supply gaps during long dry seasons and during droughts.

Occurrence and renewability

Recent, globally focused groundwater publications (Zekster and Everett, UNESCO Groundwater Series, 2004; UNEP, 2003), point out that large variations in groundwater exist in terms of occurrence, rate of renewal and volumes stored in different types of aquifers. Geological characteristics are also an important factor. While shallow basement aquifers contain limited storage, large volumes of groundwater are stored in thick sedimentary basins. Aquifers in folded mountain zones tend to be fragmented, while volcanic rock environments have unique hydraulic conditions. Shallow aquifer systems have near-surface water tables that are strongly linked to and interchange with surface water bodies. **Map 4.2** illustrates the thirty-six Global Groundwater Regions identified by IGRAC (2004), which compares predominant hydrogeological environments found around the world. The UNESCO-led World-wide Hydrogeological Mapping and Assessment Programme (WHYMAP) also contributes to mapping aquifer systems, collecting and disseminating information related to groundwater at a global scale (see **Chapter 13**).

Groundwater, as a potential resource, can be characterized by two main variables: its rate of renewal and its volume in storage. Much of groundwater is derived from recharge events that occurred during past climatic conditions and is referred to as 'non-renewable groundwater' (IAEA). The actual recharge of these aquifer systems is negligible. The world's largest non-renewable groundwater systems (**Table 4.2**) are located in arid locations of Northern Africa, the Arabian Peninsula and Australia, as well as under permafrost in Western Siberia. Their exploitation will result in a reduction in stored

volumes. A debate has arisen about how and when to use these groundwater resources as sustainable groundwater development is understood as 'exploitation under conditions of dynamic equilibrium leaving reserves undiminished'. However, nations may decide that the exploitation of such reserves is justified where undesired side-effects would not be produced (Abderrahman, 2003). UNESCO and the World Bank have jointly prepared the publication *Non-renewable groundwater resources, a guidebook on socially-sustainable management for policy makers* (forthcoming, 2006).

Transboundary groundwater

In terms of shared water resources, groundwater does not respect administrative boundaries. Most of the large non-renewable reserves in **Table 4.2** are shared. However, in addition to these aquifer systems, there are numerous smaller renewable transboundary aquifers located worldwide. Attention to shared groundwater resources management is increasing with strong support from

Groundwater, as a potential resource, can be characterized by two main variables: its rate of renewal and its volume in storage

Map 4.2: Global groundwater regions: predominant mode of groundwater occurrence and mean rate of renewal

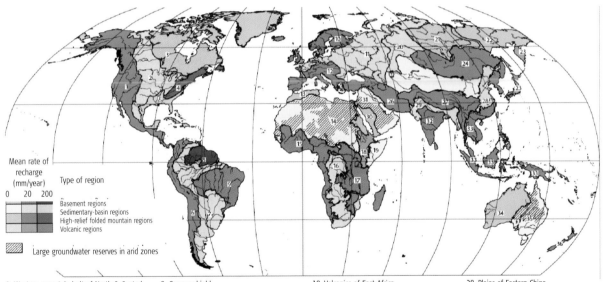

Mean rate of recharge (mm/year)   Type of region

0   20   200

Basement regions
Sedimentary-basin regions
High-relief folded mountain regions
Volcanic regions

Large groundwater reserves in arid zones

1 Western mountain belt of North & Central America	8 Guyana shield	18 Volcanics of East Africa	28 Plains of Eastern China
2 Central plains of North & Central America	9 Brazillian shield and associated basins	19 Horn of Africa basins	29 Indo-Gangelic-Brahmaputra Plain
3 Canadian shield	10 Baltic and Celtic shields	20 West Siberian platform	30 Nubian & Arabian shields
4 Appalachian highlands	11 Lowlands of Europe	21 Central Siberian plateau	31 Levant & Arabian platform
5 Caribbean islands and costal plains of North & Central America	12 Mountains of Central & Southern Europe	22 East Siberian highlands	32 Peninsular India & Sri Lanka
6 Andean belt	13 Atlas Mountains	23 Northwestern Pacific margin	33 Peninsulars & islands of South-East Asia
7 Lowlands of South America	14 Saharan basins	24 Mountain belt of Central & Eastern Asia	34 Western Australia
	15 West African basement	25 Basins of West & Central Asia	35 Eastern Australia
	16 Subsaharan basins	26 Mountain belt of West Asia	36 Islands of the Pacific
	17 East African basement & Madagascar	27 Himalayas & associated highlands	

Note: Small-scale world map showing 36 Global Groundwater Regions depicting predominant hydrogeological setting (Basement (red), Sedimentary Basin (yellow), High-Relief Folded Mountain (green) and Volcanic (blue). Higher groundwater renewal rates, as averaged over each of the mentioned Global Groundwater Regions, are reflected in the figure by higher colour intensities. The hatched zones depict areas of limited groundwater renewal that contain extensive (non-renewable) groundwater reserves which were created in the past.

Source: IGRAC, 2004.

Most renewable groundwater is of a high quality, is adequate for domestic use, irrigation and other uses, and does not require treatment

Table 4.2: Selected large aquifer systems with non-renewable groundwater resources

Countries	Aquifer system	Area (km²)	Estimated total volume (km³)	Estimated exploitable volume (km³)	Estimated annual recharge (km³)	Estimated annual abstraction (km³)
Egypt, Libya, Sudan, Chad	Nubian Sandstone Aquifer System	2,200,000	150,000 to 457,000	> 6,500	13	1.6
Algeria, Libya, Tunisia	NW Sahara Aquifer System	1,000,000	60,000	1,280	14	2.5
Algeria, Libya, Niger	Murzuk Basin	450,000	> 4,800	> 60 to 80	n.a.	1.75
Mali, Niger, Nigeria	Iullemeden Aquifer System	500,000	10,000 to 15,000	250 to 550	50 to 80	n.a.
Niger, Nigeria, Chad, Cameroon	Chad Basin Aquifer	600,000	n.a.	>170 to 350	n.a.	n.a.
S.Arabia, UAR, Bahrain, Qatar	Multilayer Aquifer Arabian Platform	250,000	n.a.	500?	30	13.9
Australia	Great Artesian Basin	1,700,000	20,000	170	50	0.6
Russia	West Siberian Artesian Basin	3,200,000	1,000,000	n.a.	55	n.a.

Source: Jean Margat, personal communication, 2004.
(Adapted from the UNESCO Working Group on Non-Renewable Groundwater Resources, 2004).

several international organizations that are addressing sustainable management strategies which would enable shared socio-economic development of such aquifers. At present, the UNESCO Internationally Shared Aquifer Resources Management (ISARM) project is compiling an inventory of transboundary aquifers.

Natural groundwater quality
Most renewable groundwater is of a high quality, is adequate for domestic use, irrigation and other uses, and does not require treatment. However, it should be noted that uncontrolled development of groundwater resources, without analysis of the chemical and biological content, is an unacceptable practice that can (as in the example of fluoride and arsenic problems in Southeast Asia) lead to serious health problems. Some waters have beneficial uses owing to naturally high temperatures and levels of minerals and gas. This is the case for thermal waters where these properties have been created by high geothermal gradients, volcanic settings or natural radioactive decay. In most cases, these groundwaters are highly developed and used for health and recreation (spa) and geothermal energy services.

Groundwater monitoring networks
Groundwater monitoring networks, as with surface water systems, operate differently at national, regional and local levels. Groundwater levels constitute the most

observed parameter, whereas widespread and continuous water quality and natural groundwater discharge and abstraction networks are operational in only a few countries (Jousma and Roelofson, 2003). Several large-scale efforts are underway to upgrade monitoring and networks, for example, in Europe (Proposal for new Directive on Groundwater Protection [EC 2003] and in India [World Bank, 2005]). However, groundwater assessment, monitoring and data management activities are for the most part minimal or ineffective in many developing countries and are being downsized and reduced in many developed counties (see **Chapter 13**). Lack of data and institutional capacity is endemic, making adequate groundwater development and management difficult. GEMS/Water (a UNEP programme) is currently adding national groundwater data to its international water quality database (described in Part 3). This will supplement the current global knowledge of groundwater quality information collected and displayed by IGRAC on its website, which includes special reports on both arsenic and fluorides in groundwater (IGRAC, 2005a, 2005b).

2g. Water availability
Efforts to characterize the volume of water available to a given nation have been ongoing for several decades. The primary input for many of these estimates is an information database (AQUASTAT) that has historically

been developed and maintained by FAO. It is based on data related to the quantity of water resources and uses a water-balance approach for each nation (FAO, 2003a). This database has become a common reference tool used to estimate each nation's renewable water resources. FAO has compiled an Index of Total Actual Renewable Water Resources (TARWR). The details of how the TARWR Index

and its national Per Capita Equivalent of 'Availability' (PCA) are determined and some of the considerations that should be taken into account when using the database index are explained in **Box 4.1**. The TARWR and PCA results for most nations from the latest 2005 update of the FAO AQUASTAT database are found in **Table 4.3**.

BOX 4.1: INDEX OF WATER RESOURCES AVAILABILITY – TOTAL ACTUAL RENEWABLE WATER RESOURCES (TARWR)

Total Actual Renewable Water Resources (TARWR) is an index that reflects the water resources theoretically available for development from all sources within a country. It is a calculated volume expressed in km³/year. Divided by the nation's population and adjusted to m³/yr, it is expressed as a per capita volume more readily allowing a relative evaluation of the resource available to its inhabitants. It estimates the total available water resources per person in each nation taking into account a number of individual component indicators by:

■ adding all internally generated surface water annual runoff and groundwater recharge derived from precipitation falling within the nation's boundaries,

■ adding external flow entering from other nations which contributes to both surface water and groundwater,

■ subtracting any potential resource volumes shared by the same water which comes from surface and groundwater system interactions, and

■ subtracting, where one or more treaty exists, any flow volume required by that treaty to leave the country.

It gives the maximum theoretical amount of water actually available for the country on a per

capita basis. Beginning in about 1989, TARWR has been used to make evaluations of water scarcity and water stress.

Considerations related to availability in the TARWR index
It is important to note that the FAO estimates are maximum theoretical volumes of water renewed annually as surface water runoff and groundwater recharge, taking into consideration what is shared in both the surface and groundwater settings. These volumes, however, do not factor in the socio-economic criteria that are potentially and differentially applied by societies, nations or regions to develop those resources. Costs can vary considerably when developing different water sources. Therefore, whatever the reported 'actual' renewable volume of water, it is a theoretical maximum, and the extent to which it can be developed will be less for a variety of economic and technical reasons. For example, Falkenmark and Rockstrom (2004) point out that, globally, approximately 27 percent of the world's surface water runoff occurs as floods. That is not considered a usable water resource even though it would be counted as part of the annual renewable surface water runoff component of TARWR. Therefore, the usable volumes available as resources to meet societal demands will be considerably less than the maximum number given as a nation's TARWR.

Four additional limitations are inherent in the TARWR information. First, seasonal variability in precipitation, runoff and recharge, which is important to regional and basin-level decision making and water storage strategies, is not well reflected in annualized quantities. Second, many large countries have several climatic settings as well as highly disparate population concentrations and the TARWR does not reflect the ranges that can occur within nations. The recently developed small-scale Relative Stress Index Map (Vörösmarty) could assist in overcoming this oversight. Third, there is no data in TARWR that identifies the volumes of 'green' water that sustain ecosystems – the volumes that provide water resources for direct rain-fed agriculture, grazing, grasslands and forests – nor does it account for the volumes of water that are potentially available from non-conventional sources (reuse, desalination, non-renewable groundwater). Finally, while the accounting-based method for a nation's TARWR adds all water that enters from upstream countries, it does not subtract any part of the water that leaves the nation in the TARWR number although estimates of those volumes are available for each country from the database.

Source: FAO, 2003a; FAO-AQUASTAT, 2005.

Table 4.3: Water availability information by country (AQUASTAT, FAO 2005)

Country	Population (1,000,000s)	Precip Rate[1] (mm/yr)	TARWR Volume 2005 (km²/yr)	TARWR Per Capita 2000 (m³/yr)	TARWR Per Capita 2005 (m³/yr)	Surface water % TARWR	Ground-water % TARWR	Overlap[2] % TARWR	Incoming Waters % TARWR	Outgoing[3] Waters % TARWR	Total Use % TARWR
1 Afghanistan	24,926	300	65	2,986	2,610				15%	77%	36%
2 Albania	3,194	1,000	42	13,306	13,060	55%	15%	6%	35%	0%	4%
3 Algeria	32,339	100	14	478	440	12%	92%	6%	3%	3%	42%
4 Angola	14,078	1,000	148	14,009	10,510	98%	39%	21%	0%	80%	0.2%
5 Antigua and Barbuda	73	2,400	0.1	800	710				0%	0%	
6 Argentina	38,871	600	814	21,981	20,940	34%	16%	16%	66%	14%	4%
7 Armenia	3,052	600	10	2780	3,450	60%	40%	13%	14%	31%	28%
8 Aruba	101										
9 Australia	19,913	500	492	25,708	24,710	89%	15%	4%	0%	0%	5%
10 Austria	8,120	1,100	78	9,616	9,570	71%	8%	8%	29%	100%	3%
11 Azerbaijan	8,447	400	30	3,765	3,580	20%	22%	14%	73%		57%
12 Bahamas	317	1,300	0.02	66	63	nd	nd	nd	0%	0%	
13 Bahrain	739	100	0.1	181	157	3%	0%	0%	97%	0%	258%
14 Bangladesh	149,664	2,700	1,211	8,809	8,090	7%	2%	0%	91%	0%	7%
15 Barbados	271	2,100	0.1	307	296	10%	92%	2%	0%	0%	105%
16 Belarus	9,852	600	58	5,694	5,890	64%	31%	31%	36%	96%	5%
17 Belgium	10,340	800	18	1,786	1,770	66%	5%	5%	34%	60%	
18 Belize	261	2,200	19	82,102	71,090				14%	0%	1%
19 Benin	6,918	1,000	26	3,954	3,820	38%	7%	6%	61%	22%	1%
20 Bermuda	82	1,500									
21 Bhutan	2,325	1,700	95	45,564	40,860	100%	0%	95%	0.4%		
22 Bolivia	8,973	1,100	623	74,743	69,380	45%	21%	17%	51%	93%	0.2%
23 Bosnia and Herzegovina	4,186	1,000	38	9,429	8,960					100%	
24 Botswana	1,795	400	12	9,345	6,820	7%	14%	1%	80%	5%	1%
25 Brazil	180,654	1,800	8,233	48,314	45,570	66%	23%	23%	34%	6%	1%
26 Brunei Darussalam	366	2,700	9	25,915	23,220	100%	1%	1%	0%	0%	
27 Bulgaria	7,829	600	21	2,680	2,720	94%	30%	26%	1%	92%	49%
28 Burkina Faso	13,393	700	13	1,084	930	64%	76%	40%	0%	100%	6%
29 Burundi	7,068	1,200	15	566	2,190	65%	48%	48%	35%	14%	2%
30 Cambodia	14,482	1,900	476	36,333	32,880	24%	4%	3%	75%	99%	1%
31 Cameroon	16,296	1,600	286	19,192	17,520	94%	35%	33%	4%	14%	0.3%
32 Canada	31,744	500	2,902	94,353	91,420	98%	13%	12%	2%	5%	2%
33 Cape Verde	473	400	0.3	703	630	60%	40%	0%	0%	0%	9%
34 Central African Rep.	3,912	1,300	144	38,849	36,910	98%	39%	39%	2%	98%	0.02%
35 Chad	8,854	300	43	5,453	4,860	31%	27%	23%	65%	9%	0.5%
36 Chile	15,996	700	922	60,614	57,640	96%	15%	15%	4%	0%	1.4%
37 China	1,320,892	600	2,830	2,259	2,140	96%	29%	26%	1%	25%	
38 China, Taiwan Prov.	22,894	2,400	67		2,930	94%	6%	0%	0%		
39 Colombia	44,914	2,600	2,132	50,635	47,470	99%	24%	24%	1%	50%	1%
40 Comoros	790	1,800	1.2	1,700	1,520	17%	83%	0%	0%	0%	
41 Congo, Dem Rep.	54,417	1,500	1,283	25,183	23,580	70%	33%	33%	30%	0%	0.03%
42 Congo	3,818	1,600	832	275,679	217,920	27%	24%	24%	73%	23%	0.005%
43 Costa Rica	4,250	2,900	112	27,932	26,450	67%	33%	0%	0%	7%	2%
44 Côte d'Ivoire	16,897	1,300	81	5,058	4,790	91%	47%	43%	5%	15%	1%
45 Croatia	4,416	1,100	106	22,669	23,890	26%	10%	0%	64%	38%	
46 Cuba	11,328	1,300	38	3,404	3,370	83%	17%	0%	0%	0%	22%
47 Cyprus	808	500	0.8	995	970	72%	53%	24%	0%	0%	31%
48 Czech Rep	10,226	700	13	1,280	1,290	100%	11%	11%	0%	100%	20%

Table 4.3: *continued*

Country	Population (1,000,000s)	Precip Rate[1] (mm/yr)	TARWR Volume 2005 (km²/yr)	TARWR Per Capita 2000 (m³/yr)	TARWR Per Capita 2005 (m³/yr)	Surface water % TARWR	Ground-water % TARWR	Overlap[2] % TARWR	Incoming Waters % TARWR	Outgoing[3] Waters % TARWR	Total Use % TARWR
49 Denmark	5,375	700	6	1,128	1120	62%	72%	33%	0%	0%	21%
50 Djibouti	712	200	0.3	475	420	100%	5%	5%	0%	0%	3%
51 Dominica	79	3,400									
52 Dominican Republic	8,872	1,400	21	2,507	2,370	100%	56%	56%	0%	5%	16%
53 Ecuador	13,192	2,100	424	34,161	32,170	102%	32%	32%	0%	36%	4%
54 Egypt	73,390	100	58	859	790	1%	2%	0%	97%	0%	118%
55 El Salvador	6,614	1,700	25	4,024	3,810	70%	24%	24%	30%	0%	5%
56 Equatorial Guinea	507	2,200	26	56,893	51,280	96%	38%	35%	0%	0%	0.4%
57 Eritrea	4,297	400	6	1,722	1,470				56%	35%	5%
58 Estonia	1,308	600	13	9,195	9,790	91%	31%	23%	1%	3%	1%
59 Ethiopia	72,420	800	122	1,749	1,680	16%	100%	16%	0%	80%	2%
60 Fiji	847	2,600	29	35,074	33,710				0%	0%	0.2%
61 Finland	5,215	500	110	21,268	21,090	97%	2%	2%	3%	25%	2%
62 France	60,434	900	204	3439	3,370	87%	49%	48%	12%	7%	20%
63 French Guiana	182	2,900	134	812,121	736,260				0%	0%	
64 French Polynesia	248										
65 Gabon	1,351	1,800	164	133,333	121,390	99%	38%	37%	0%	0%	0.1%
66 Gambia	1,462	800	10	6,140	5,470	38%	6%	6%	63%	0%	0.4%
67 Palestinian Territories	1,376	300	0	52	41	0%	82%	0%	18%	0%	
68 Georgia	5,074	1,000	63	12,035	12,480	90%	27%	25%	8%	19%	6%
69 Germany	82,526	700	154	1,878	1,870	69%	30%	29%	31%	59%	31%
70 Ghana	21,377	1,200	50	2,756	2,490	55%	49%	47%	43%	0%	1%
71 Greece	10,977	700	74	6,998	6,760	75%	14%	11%	22%	2%	10%
72 Greenland	57	600	603	10,767,857	10,578,950				0%	0%	
73 Grenada	80	1,500									
74 Guadeloupe	443	200									
75 Guatemala	12,661	2,700	111	9,773	8,790	91%	30%	23%	2%	47%	2%
76 Guinea	8,620	1,700	226	27,716	26,220	100%	17%	17%	0%	45%	1%
77 Guinea-Bissau	1,538	1,600	31	25,855	20,160	39%	45%	32%	48%	0%	0.4%
78 Guyana	767	2,400	241	316,689	314,210	100%	43%	43%	0%	0%	1%
79 Haiti	8,437	1,400	14	1,723	1,660	77%	15%	0%	7%	0%	7%
80 Honduras	7,099	2,000	96	14,949	13,510	91%	41%	31%	0%	0%	1%
81 Hungary	9,831	600	104	10,433	10,580	6%	6%	6%	94%	100%	7%
82 Iceland	292	1,000	170	609,319	582,190	98%	14%	12%	0%	0%	0.1%
83 India	1,081,229	1,100	1,897	1,880	1,750	64%	22%	20%	34%	68%	34%
84 Indonesia	222,611	2,700	2,838	13,381	12,750	98%	16%	14%	0%	0%	3%
85 Iran, Islamic Rep.	69,788	200	138	1,955	1,970	71%	36%	13%	7%	7%	53%
86 Iraq	25,856	200	75	3,287	2,920	45%	2%	0%	53%		57%
87 Ireland	3,999	1,100	52	13,673	13,000	93%	21%	19%	6%	0%	2%
88 Israel	6,560	400	2	276	250	15%	30%	0%	55%		122%
89 Italy	57,346	800	191	3,325	3,340	89%	22%	16%	5%	0%	23%
90 Jamaica	2,676	2,100	10	3,651	3,510	59%	41%	0%	0%	0%	4%
91 Japan	127,800	1,700	430	3,383	3,360	98%	6%	4%	0%	0%	21%
92 Jordan	5,614	100	1	179	160	45%	57%	25%	23%		115%
93 Kazakhstan	15,403	200	110	6,778	7,120	63%	6%	0%	31%		32%
94 Kenya	32,420	700	30	985	930	57%	10%	0%	33%	30%	5%
95 Korea, Dem. People's Rep.	22,776	1,400	77	3,464	3,390	86%	17%	16%	13%	6%	12%
96 Korea, Rep.	47,951	1,100	70	1,491	1,450	89%	19%	15%	7%		27%

Table 4.3: *continued*

Country	Population (1,000,000s)	Precip Rate[1] (mm/yr)	TARWR Volume 2005 (km²/yr)	TARWR Per Capita 2000 (m³/yr)	TARWR Per Capita 2005 (m³/yr)	Surface water % TARWR	Ground-water % TARWR	Overlap[2] % TARWR	Incoming Waters % TARWR	Outgoing[3] Waters % TARWR	Total Use % TARWR
97 Kuwait	2,595	100	0.02	10	8	0%	0%	0%	100%	0%	2,227%
98 Kyrgyzstan	5,208	400	21	4,182	3,950	214%	66%	54%	0%	36%	49%
99 Lao Peoples Dem. Rep.	5,787	1,800	334	63,184	57,640	57%	11%	11%	43%	100%	1%
100 Latvia	2,286	600	35	14,642	15,510	47%	6%	6%	53%	2%	1%
101 Lebanon	3,708	700	4	1,261	1,190	93%	73%	57%	1%	11%	31%
102 Lesotho	1,800	800	3	1,485	1,680	173%	17%	17%	0%	57%	2%
103 Liberia	3,487	2,400	232	79,643	66,530	86%	26%	26%	14%	0%	0.05%
104 Libyan Arab Jamahiriya	5,659	100	1	113	106	33%	83%	17%	0%	117%	802%
105 Lithuania	3,422	700	25	6,737	7,280	62%	5%	4%	38%	20%	1%
106 Luxemburg	459	900	3	7,094	6,750	32%	3%	3%	68%	100%	
107 Macedonia, Fr Yugoslav Rep.	2,066	600	6	3,147	3,100	84%	0%	0	16%	100%	
108 Madagascar	17,901	1,500	337	21,102	18,830	99%	16%	15%	0%	0%	4%
109 Malawi	12,337	1,200	17	1,528	1,400	93%	8%	8%	7%	93%	6%
110 Malaysia	24,876	2,900	580	26,105	23,320	98%	11%	9%	0%	0%	2%
111 Maldives	328	2,000	0.03	103	91	0%	100%	0%	0%	0%	
112 Mali	13,409	300	100	8,810	7,460	50%	20%	10%	40%	52%	7%
113 Malta	396	400	0.1	129	130	1%	99%	0%	0%	0%	110%
114 Martinique	395	2,600	nd								
115 Mauritania	2,980	100	11	4,278	3,830	1%	3%	0%	96%	0%	15%
116 Mauritius	1,233	2,000	3	1,904	2,230	86%	32%	18%	0%	0%	22%
117 Mexico	104,931	800	457	4,624	4,360	79%	30%	20%	11%	0%	17%
118 Moldova, Rep.	4,263	600	12	2,712	2,730	9%	3%	3%	91%	85%	20%
119 Mongolia	2,630	200	35	13,739	13,230	94%	18%	11%	0%	76%	1%
120 Morocco	31,064	300	29	971	930	76%	34%	10%	0%	1%	44%
121 Mozambique	19,182	1,000	217	11,814	11,320	45%	8%	6%	54%	0%	0.3%
122 Myanmar	50,101	2,100	1,046	21,898	20,870	84%	15%	14%	16%	5%	3%
123 Namibia	2,011	300	18	10,211	8,810	23%	12%	0%	66%	72%	2%
124 Nepal	25,725	1,300	210	9,122	8,170	94%	10%	10%	6%	100%	5%
125 Netherlands	16,227	800	91	5,736	5,610	12%	5%	5%	88%	0%	9%
126 New Caledonia	233	1,500									
127 New Zealand	3,904	1,700	327	86,554	83,760	0%	0%	1%			
128 Nicaragua	5,597	2,400	197	38,787	35,140	94%	30%	28%	4%	0%	1%
129 Niger	12,415	200	34	3,107	2,710	3%	7%	0%	90%	96%	6%
130 Nigeria	127,117	1,200	286	2,514	2,250	75%	30%	28%	23%	0%	3%
131 Norway	4,552	1,100	382	85,478	83,920	98%	25%	24%	0%	3%	1%
132 Oman	2,935	100	1	388	340	94%	97%	91%	0%	0%	137%
133 Pakistan	157,315	300	223	2961	1,420	21%	25%	22%	76%	3%	76%
134 Panama	3,177	2,700	148	51,814	46,580	97%	14%	12%	0%	0%	1%
135 Papua New Guinea	5,836	3,100	801	166,563	137,250	100%			0%	0%	0.01%
136 Paraguay	6,018	1,100	336	61,135	55,830	28%	12%	12%	72%	99%	0.1%
137 Peru	27,567	1,500	1,913	745,46.0	69,390	84%	16%	16%	16%	94%	1%
138 Philippines	81,408	2,300	479	6,332	5,880	93%	38%	30%	0%	0%	6%
139 Poland	38,551	600	62	1,596	1,600	86%	20%	19%	13%	3%	26%
140 Portugal	10,072	900	69	6,859	6,820	55%	6%	6%	45%	0%	16%
141 Puerto Rico	3,898	2,100	7	1,814	1,820				0%	0%	
142 Qatar	619	100	0.1	94	86	2%	94%	0%	4%	0%	554%
143 Reunion	767	2,100	5	6,935	6,520	90%	56%	46%	0%	0%	
144 Romania	22,280	600	212	9,445	9,510	20%	4%	4%	80%	0%	11%

Table 4.3: *continued*

Country	Population (1,000,000s)	Precip Rate[1] (mm/yr)	TARWR Volume 2005 (km²/yr)	TARWR Per Capita 2000 (m³/yr)	TARWR Per Capita 2005 (m³/yr)	Surface water % TARWR	Ground-water % TARWR	Overlap[2] % TARWR	Incoming Waters % TARWR	Outgoing[3] Waters % TARWR	Total Use % TARWR
145 Russian Federation	142,397	500	4,507	30,980	3,1650	90%	17%	11%	4%	0%	2%
146 Rwanda	8,481	1,200	5	683	610	100%	69%	69%	0%	81%	1%
147 Saint Helena	5	800									
148 Saint Kitts and Nevis	42	2,100	0.0	621	560	15%	85%	0%	0%	0%	
149 Saint Lucia	150	2,300									
150 Saint Vincent and the Grenadines	121	1,600									
151 Samoa	180	3,000									
152 Sao Tome and Principe	165	2,200	2.2	15,797	13,210				0%	0%	
153 Saudi Arabia	24,919	100	2.4	118	96	92%	92%	83%	0%	6%	722%
154 Senegal	10,339	700	39	4,182	3,810	60%	19%	13%	33%	14%	4%
155 Serbia and Montenegro	10,519				19,820	20%	1%	1%	79%		
156 Seychelles	82	2,000									
157 Sierra Leone	5,168	2,500	160	36,322	30,960	94%	31%	25%	0%	0%	0.2%
158 Singapore	4,315	2,500	0.6	149	139				0%	0%	
159 Slovakia	5,407	800	50	9,279	9,270	25%	3%	3%	75%	27%	
160 Slovenia	1,982	1,200	32	16,031	16,080	58%	42%	42%	41%	60%	
161 Solomon Islands	491	3,000	45	100,000	91,040				0%	0%	
162 Somalia	10,312	300	14	1,538	1,380	40%	23%	21%	56%	0%	23%
163 South Africa	45,214	500	50	1,154	1,110	86%	10%	6%	10%	19%	31%
164 Spain	41,128	600	112	2,794	2,710	98%	27%	25%	0%	31%	32%
165 Sri Lanka	19,218	1,700	50	2,642	2,600	98%	16%	14%	0%	0%	25%
166 Sudan	34,333	400	65	2,074	1,880	43%	11%	8%	77%	30%	58%
167 Suriname	439	2,300	122	292,566	277,900	72%	66%	66%	28%	0%	1%
168 Swaziland	1,083	800	4.5	4,876	4,160				41%	100%	18%
169 Sweden	8,886	600	174	19,679	19,580	98%	11%	11%	2%	2%	2%
170 Switzerland	7,164	1,500	54	7,462	7,470	76%	5%	5%	24%	76%	5%
171 Syrian Arab Rep.	18,223	300	26	1,622	1,440	18%	16%	8%	80%	119%	76%
172 Tajikistan	6,298	500	16	2,625	2,540	396%	38%	19%	17%		75%
173 Tanzania	37,671	1,100	91	2,591	2,420	88%	33%	31%	10%	14%	2%
174 Thailand	63,465	1,600	410	6,527	6,460	48%	10%	7%	49%	79%	21%
175 Togo	5,017	1,200	15	3,247	2,930	73%	39%	34%	22%	54%	1%
176 Tonga	105	2,000								0%	
177 Trinidad and Tobago	1,307	1,800	3.8	2,968	2,940				0%	0%	8%
178 Tunisia	9,937	300	4.6	482	460	68%	32%	9%	9%	4%	60%
179 Turkey	72,320	600	214	3,439	2,950	87%	32%	13%	1%	29%	18%
180 Turkmenistan	4,940	200	25	5,218	5,000	4%	1%	0%	97%		100%
181 Uganda	26,699	1,200	66	2,833	2,470	59%	44%	44%	41%	56%	0%
182 Ukraine	48,151	600	140	2,815	2,900	36%	14%	12%	62%	22%	27%
183 United Arab Emirates	3,051	100	0.2	58	49	100%	80%	80%	0%	0%	1,538%
184 United Kingdom	59,648	1,200	147	2,465	2,460	98%	7%	6%	1%	0%	6%
185 United States of America	297,043	700	3,051	10,837	10,270				8%		16%
186 Uruguay	3,439	1,300	139	41,654	40,420	42%	17%	17%	58%	0%	2%
187 Uzbekistan	26,479	200	50	2,026	1,900	19%	17%	4%	77%		116%
188 Venezuela, Bolivarian Rep.	26,170	1,900	1,233	51,021	47,120	57%	18%	17%	41%	6%	1%

Table 4.3: *continued*

Country	Population (1,000,000s)	Precip Rate[1] (mm/yr)	TARWR Volume 2005 (km²/yr)	TARWR Per Capita 2000 (m³/yr)	TARWR Per Capita 2005 (m³/yr)	Surface water % TARWR	Ground-water % TARWR	Overlap[2] % TARWR	Incoming Waters % TARWR	Outgoing[3] Waters % TARWR	Total Use % TARWR
189 Viet Nam	82,481	1,800	891	11,406	10,810	40%	5%	4%	59%	4%	8%
190 Palestine Territories	2,386		0.8		320	10%	90%	0%	0%	28%	
191 Yemen	20,733	200	4	223	198	98%	37%	34%	0%	0%	162%
192 Zambia	10,924	1,000	105	10,095	9,630	76%	45%	45%	24%	100%	2%
193 Zimbabwe	12,932	700	20	1,584	1,550	66%	25%	20%	39%	71%	13%

Source: FAO-AQUASTAT, 2005.

Notes:
1. Average precipitation (1961–90 from IPCC (mm/year). As in the FAO-AQUASTAT Database, for some countries large discrepancies exist between national and IPCC data on rainfall average. In these cases, IPCC data were modified to ensure consistency with water resources data.
2. Overlap is the water that is shared by both the surface water and groundwater systems.
3. Outflow – Sep. 2004 for surface water and Aug. 2005 for groundwater.

Part 3. Human Impacts

A number of forces continue to seriously affect our natural water resources. Many of these are primarily the result of human actions and include ecosystem and landscape changes, sedimentation, pollution, over-abstraction and climate change.

...each type of landscape change will have its own specific impact, usually directly on ecosystems and directly or indirectly on water resources...

The removal, destruction or impairment of natural ecosystems are among the greatest causes of critical impacts on the sustainability of our natural water resources. This issue is dealt with more broadly in **Chapter 5**. However, it should be emphasized that the ecosystems with which we interact are directly linked to the well-being of our natural water resources. Although it is difficult to integrate the intricacies of ecosystems into traditional and more hydrologically-based water assessment and management processes, this approach is being strongly advocated in some sectors and scientific domains (e.g. Falkenmark and Rockström, 2004; Figueras et al., 2003; Bergkamp et al., 2003). The basis of this approach is the recognition that each type of landscape change will have its own specific impact, usually directly on ecosystems and directly or indirectly on water resources. The magnitude of the impacts will vary according to the setting's conditions with a wide range of possible landscape changes. Changes that can occur to landscapes include: forest clearance, crop- or grazing lands replacing grasslands or other natural terrestrial ecosystems, urbanization (leading to changes in infiltration and runoff patterns as well to pollution), wetlands removal or reduction, new roadwork for transportation, and mining in quarries or large-scale open pits.

3a. Sedimentation

Sediments occur in water bodies both naturally and as a result of various human actions. When they occur excessively, they can dramatically change our water resources. Sediments occur in water mainly as a direct response to land-use changes and agricultural practices, although sediment loads can occur naturally in poorly vegetated terrains and most commonly in arid and semi-arid climates following high intensity rainfall. **Table 4.4** summarizes the principal sources of excessive sediment loads and identifies the major impacts that this degree of sediment loading can have on aquatic systems and the services that water resources can provide. A recently documented and increasing source of high sediment loads is the construction of new roads in developing countries where little consideration is given to the impacts of such actions on aquatic systems and downstream water supplies. Globally, the effects of excessive sedimentation commonly extend beyond our freshwater systems and threaten coastal habitats, wetlands, fish and coral reefs in marine environments (see **Chapter 5**). The importance of sediment control should be an integral consideration in any water resources development and protection strategy. UNESCO's International Sediment Initiative (ISI) project will attempt to improve the understanding of sediment phenomena, and provide better protection of the aquatic and terrestrial environments.

3b. Pollution

Humans have long used air, land and water resources as 'sinks' into which we dispose of the wastes we generate. These disposal practices leave most wastes inadequately treated, thereby causing pollution. This in turn affects

precipitation (**Box 4.2**), surface waters (**Box 4.3**), and groundwater (**Box 4.4**), as well as degrading ecosystems (see **Chapter 5**). The sources of pollution that impact our water resources can develop at different scales (local, regional and global) but can generally be categorized (**Table 4.5**) according to nine types. Identification of source types and level of pollution is a prerequisite to assessing the risk of the pollution being created to both

Table 4.4: Major principal sources and impacts of sedimentation

Pertinence	Sector	Action or mechanism	Impacts
SOURCES			
Agriculture areas, downstream catchments	Agriculture	▪ poor farming with excessive soil loss	▪ increase soil erosion ▪ add toxic chemicals to the environment ▪ sediment and pollutants are added to streams ▪ irrigation systems maintenance cost increased
Forest and development access areas, downstream catchments	Forestry, Road Building, Construction, Construction, Mining	▪ extensive tree cutting ▪ lack of terrain reforestation ▪ lack of runoff control in steep terrain	▪ increase natural water runoff ▪ accelerated soil erosion creating more sediment
MAJOR IMPACTS			
Major rivers and navigable waterways	Navigation	▪ deposition in rivers or lakes ▪ dredging (streams, reservoirs, lakes or harbors)	▪ decreases water depth making navigation difficult or impossible. ▪ releases toxic chemicals into the aquatic or land environment.
Aquatic ecosystems	Fisheries / Aquatic habitat	▪ decreased light penetration ▪ higher suspended solids concentrations ▪ absorbed solar energy increases water temperature ▪ carrying toxic agricultural and industrial compounds ▪ settling and settled sediment	▪ affects fish feeding and schooling practices; can reduce fish survival ▪ irritate gills of fish, can cause death, destroy protective mucous covering on fish eyes and scales ▪ dislodge plants, invertebrates, and insects in stream beds affecting fish food sources resulting in smaller and fewer fish, increased infection and disease susceptibility ▪ stress to some fish species ▪ release to habitat causes fish abnormalities or death ▪ buries and suffocates eggs ▪ reduces reproduction
Lakes, rivers, reservoirs as water supplies	Water supply	▪ increased pump/turbine wear ▪ reduced water supply usability for certain purposes ▪ additional treatment for usability required	▪ affects water delivery, increases maintenance costs ▪ reduces water resource value and volume ▪ increased costs
Hydroelectric facilities	Hydropower	▪ dams trap sediment carried downstream ▪ increased pump/turbine wear	▪ diminished reservoir capacity ▪ shortened power generation lifecycle ▪ higher maintenance, capital costs.
All waterways and their ecosystems	Toxic chemicals	▪ become attached or adsorbed to sediment particles	▪ transported to and deposited in, other areas ▪ later release into the environment.

Source: Adapted from Environment Canada (2005a), www.atl.ec.gc.ca/udo/mem.html

Note: Water transforms landscapes and moves large amounts of soil and fine-grained materials in the form of *sediment*.
Sediment is: 1) eroded from the landscape, 2) transported by river systems and eventually 3) deposited in a riverbed, wetland, lake, reservoir or the ocean. Particles or fragments are eroded naturally by water, wind, glaciers, or plant and animal activities with geological (natural) erosion taking place slowly over centuries or millennia. Human activity may accelerate the erosion. Material dislodged is transported when exposed to fluvial erosion in streams and rivers. Deposition occurs as on flood plains, bars and islands in channels and deltas while considerable amounts end up in lakes, reservoirs and deep river beds.

BOX 4.2: ACID RAIN IMPACTS ON WATER RESOURCES

Atmospheric contamination from industrial plants and vehicle emissions leads to dry and wet deposition. This causes acidic conditions to develop in surface water and groundwater sources and at the same time leads to the destruction of ecosystems. Acid deposition impairs the water quality of lakes and streams by lowering pH levels (i.e. increasing acidity), decreasing acid-neutralizing capacity, and increasing aluminum concentrations. High concentrations of aluminium and increased acidity reduce species diversity and the abundance of aquatic life in many lakes and streams. While fish have received most attention to date, entire food webs are often negatively affected. Despite improvements, it still remains a critical situation that impacts water resources and ecosystems in some developed regions of Europe and in North America. The situation remains an

important issue in several developing countries (for example in China, India, Korea, Mexico, South Africa and Viet Nam) where there are typically lower emission controls and inadequate monitoring and evaluation (Bashkin and Radojevic, 2001). In recognition of this, UNEP and the Stockholm Environmental Institute are sponsoring programmes such as RAPIDC (Rapid Air Pollution in Developing Countries) with the aim of identifying sources and sensitive areas and measuring levels of acid rain. Extensive funding from ADB is now being used to source reductions in several Asian nations. The problem has broad transboundary implications as acid rain can get carried over long distances from polluting areas to other countries. For example, Japan is impacted by Korean and Chinese emissions, while Canada, in addition to its own sources, receives substantive emissions from the US.

As reported by Driscoll et al. (2001), there are still impacts to water quality in northeastern US and eastern Canada, even though improved conditions developed after the introduction of the Clean Air Act and its amendments (1992).

41 percent of lakes in the Adirondacks of New York and 15 percent of all lakes in New England exhibit signs of chronic and/or episodic acidification. Only modest improvements in acid-neutralizing capacity have occurred in New England with none in the Adirondacks or Catskills of New York. Elevated concentrations of aluminum have been measured in acid-impacted surface waters throughout the Northeast.

Figure 4.6: Acid rain and its deposition processes

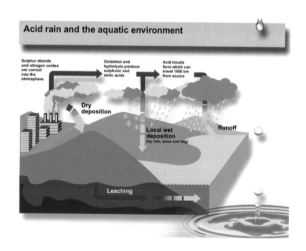

Figure 4.7: Five-year mean of the pH level in rainfall in the eastern regions of Canada and the US

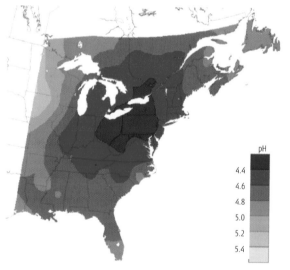

Source: Environment Canada, 2005c.

BOX 4.3: IMPACTS TO SURFACE WATER QUALITY FROM HUMAN ACTIVITY

The challenge of how to improve water quality by rehabilitation and protection of lakes, streams, reservoirs, wetlands and related surface water bodies is a growing global concern, typified by the recent European Commission Water Framework Directive (EC, 2000). However, surface water pollution risks, particularly in developing nations, remain relatively widespread. A valuable initial step in identifying the nature and extent of water quality impacts linked to pollution is to distinguish their point (PS) and non-point sources (NPS). PS pollution is commonly linked directly to end-of-pipe releases from industry and municipal wastes. Its control is more direct and quantifiable and in many developed countries its mitigation has been linked to treatment achieving lower contaminant concentrations before discharge. NPS pollution occurs when contaminants from diverse and widely spread sources are transported by runoff into rivers, lakes, wetlands, groundwater and coastal areas. This type of pollution is more difficult to address as there are a large number of sources, for example, varied agricultural areas all of which are using pesticides and nutrients. Today, however, NPS pollution is receiving more attention as its

impacts are becoming evident over large areas in lakes, streams and groundwater and can also be linked to the degradation of aquatic freshwater and marine ecosystems.

Further detail on pollution impacts are found in the chapters on human settlements (**Chapter 3**), agriculture (**Chapter 7**) and industry (**Chapter 8**).

Emerging Issues

Only a small percentage of chemicals are regulated locally, nationally or internationally (Daughton 2004). An emerging concern is contaminants in high population settings that are neither traditionally measured nor regulated, for example pharmaceuticals (Wiegel et al. 2004). Reynolds (2003) reports:

> Scientists are becoming increasingly concerned about the potential public health impact of environmental contaminants originating from industrial, agricultural, medical and common household practices, i.e., cosmetics, detergents and toiletries. A variety of pharmaceuticals including painkillers, tranquilizers, anti-depressants,

antibiotics, birth control pills, estrogen replacement therapies, chemotherapy agents, anti-seizure medications, etc., are finding their way into the environment via human and animal excreta from disposal into the sewage system and from landfill leachate that may impact groundwater supplies. Agricultural practices are a major source and 40 percent of antibiotics manufactured are fed to livestock as growth enhancers. Manure, containing traces of pharmaceuticals, is often spread on land as fertilizer from which it can leach into local streams and rivers.

Reynolds further notes that conventional wastewater treatment is not effective in eliminating the majority of pharmaceutical compounds. Since various contaminants do not always have coincident pollution patterns, single indicators for all contaminants are not effective. Reynolds (2003) suggests that 'pharmaceutical contamination in the environment will involve both advanced waste and water treatment technologies and source control at the point of entry into the environment ... all of which are issues of ongoing scientific research'.

---

the aquatic systems and, through that system, to humans and the environment. With the knowledge of the principal sources of the pollution, the appropriate mitigation strategy can be identified to reduce the impact on the water resources.

The potential impacts from the different pollution types based on the area (scale) affected, the time it takes to contaminate, the time needed to clean up (remediate) a contaminated area, and the links to the major controlling factors are illustrated in **Table 4.6** (Peters and Meybeck, 2000). With the exception of pathogenic contaminants, all other forms of pollution can extend to a regional scale. The fact that it takes considerably longer to remediate a contaminated area than to pollute it clearly highlights the need for adopting the precautionary principle and prioritizing protection strategies rather than costly ad-hoc restoration measures.

Developed countries have historically experienced a succession of water quality problems relating to pathogens, eutrophication, heavy metals, acidification, organic compounds and micro-pollutants and sediments from municipal, industrial and agricultural waste sources (Webb, 1999; Meybeck et al., 1989; Revenga and Mock, 2000). In rapidly developing countries – such as Brazil, China and India – similar sequences of water problems have emerged over the last few decades. In other developing countries, water pollution still remains problematic and is one of the single leading causes of poor livelihood and bad health (Lenton, 2004; and see **Chapter 6**).

Global water quality and pollution information

Assessing water quality enables the natural characteristics of the water to be documented and the extent of the pollution to be determined; however, today monitoring is

BOX 4.4: IMPACTS TO GROUNDWATER QUALITY FROM HUMAN ACTIVITY

Protection of groundwater sources is becoming a more widespread global concern as typified by the recent European Commission directive which focuses on preventing rather than cleaning up pollution (EC 2003). Incidents of groundwater pollution arising from human actions, particularly in developing nations, remain relatively widespread and its impacts in terms of degraded water quality are summarized in Zektser and Everett (2004). Throughout the world, most countries' practices of urbanization, industrial development, agricultural activities and mining enterprises have caused groundwater contamination and its most typical sources are illustrated in **Figure 4.8**. A 2002 joint World Bank, GWP, WHO and UNESCO online guidance document (Foster et al. 2002) states *'There is growing evidence of increasing pollution threats to groundwater and some well documented cases of irreversible damage to important aquifers, following many years of widespread public policy neglect'*. This guide is supplemented by recommendations in a 2003 joint FAO, UNDESA, IAEA and UNESCO report directly addressing the universal changes needed

in groundwater management practice (FAO 2003b) to arrive at more sustainable water development and use.

Groundwater pollution contrasts markedly in terms of the activities and compounds that most commonly cause surface water pollution. In addition, there are completely different controls that govern the contaminant mobility and persistence in the two water systems' settings. Foster and Kemper (2004), UNEP (2003), FAO (2003b) and Burke and Moench (2000) point out that groundwater management commonly involves a wide range of instruments and measures (technical, process, incentive, legal and enforcement actions/sanctions and awareness raising) to deal with resources that are less visible than those in our surface water bodies.

Mapping groundwater vulnerability
Groundwater is less vulnerable to human impacts than surface water. However, once polluted, cleaning it up (remediation) takes a relatively long time (years), is more technically

demanding, and can be much more costly. While this has been recognized for several decades (Vrba 1985), this important message has not been adequately or consistently conveyed to the policy-makers or the public. To address this gap, groundwater vulnerability assessment methods are being developed. These emerging 'vulnerability maps' have historically been applied to other risks such as flooding and landslides and they can now be used as direct input to water resources and land planning (Vrba and Zaporozek 1994). Results of such studies are absolutely critical where aquifers are used for water supplies and have sensitive ecosystem dependencies. In conjunction with other environmental input, they have become effective instruments used to regulate, manage and take decisions related to impacts from existing and proposed changes in land use, ecosystems and sources of water supplies. Large-scale groundwater vulnerability maps (e.g. France, Germany, Spain, Italy, The Czech Republic, Poland, Russia and Australia) serve as guidelines for land use zoning at national or regional levels.

Figure 4.8: Primary sources of groundwater pollution

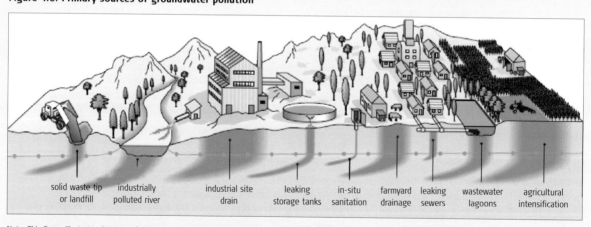

| solid waste tip or landfill | industrially polluted river | industrial site drain | leaking storage tanks | in-situ sanitation | farmyard drainage | leaking sewers | wastewater lagoons | agricultural intensification |

Note: This figure illustrates the type of sources that should be inventoried for cataloging potential sources of groundwater contamination.

Source: Foster et al., 2002.

Table 4.5: Freshwater pollution sources, effects and constituents of concern

Pollution type	Primary sources	Effects[1]	Constituents of concern[2]
1 Organic matter	Industrial wastewater and domestic sewage.	Depletion of oxygen from the water column as it decomposes, stress or suffocating aquatic life.	Biological Oxygen Demand (BOD), Dissolved Organic Carbon (DOC), Dissolved Oxygen (DO)
2 Pathogens and microbial contaminants	Domestic sewage, cattle and other livestock, natural sources.	Spreads infectious diseases through contaminated drinking water supplies leading to diarrhoeal disease and intestinal parasites, increased childhood mortality in developing countries.	Shigella, Salmonella, Cryptosporidium, Fecal coliform (Coliform), Escherichia coli (mammal faeces – E. Coli)
3 Nutrients	Principally runoff from agricultural lands and urban areas but also from some industrial discharge.	Over-stimulates growth of algae (eutrophication) which then decomposes, robbing water of oxygen and harming aquatic life. High levels of nitrate in drinking water lead to illness in humans.	Total N (organic + inorganic), total P (organic + inorganic) For eutrophication: (Dissolved Oxygen, Individual N species (NH4, NO2, NO3, Organic N), Orthophosphate)
4 Salinization	Leached from alkaline soils by over irrigation or by over-pumping coastal aquifers resulting in saltwater intrusion.	Salt build-up in soils which kills crops or reduces yields. Renders freshwater supplies undrinkable.	Electrical conductivity, Chloride (followed, post characterization by full suite of major cations (Ca, Mg), anions
5 Acidification (precipitation or runoff)	Sulphur, Nitrogen oxides and particulates from electric power generation, industrial stack and auto/truck emissions (wet and dry deposition). Acid mine drainage from tailings as well as mines.	Acidifies lakes and streams which negatively impacts aquatic organisms and leaches heavy metals such as aluminium from soils into water bodies.	pH
6 Heavy metals	Industries and mining sites.	Persists in freshwater environments such as river sediments and wetlands for long periods. Accumulates in the tissues of fish and shellfish. Can be toxic to both aquatic organisms and humans who consume them.	Pb, Cd, Zn, Cu, Ni, Cr, Hg, As (particularly groundwater)
7 Toxic organic compounds and micro-organic pollutants.[3]	Wide variety of sources from industrial sites, automobiles, farmers, home gardeners, municipal wastewaters.	A range of toxic effects in aquatic fauna and humans from mild immune suppression to acute poisoning or reproductive failure.	PAHs, PCBs, pesticides (lindane, DDT, PCP, Aldrin, Dieldrin, Endrin, Isodrin, hexachlorobenzene)
8 Thermal	Fragmentation of rivers by dams and reservoirs slowing water and allowing it to warm. Industry from cooling towers and other end-of-pipe above-ambient temperature discharges.	Changes in oxygen levels and decomposition rate of organic matter in the water column. May shift the species composition of the receiving water body.	Temperature
9 Silt and suspended particles	Natural soil erosion, agriculture, road building, deforestation, construction and other land use changes.	Reduces water quality for drinking and recreation and degrades aquatic habitats by smothering them with silt, disrupting spawning and interfering with feeding.	Total suspended solids, turbidity

Other pollutants include Radioactivity, Fluoride, Selenium.

Sources and notes:

1 Principally from Revenga and Mock, 2000. Their compilation from Taylor and Smith, 1997; Shiklomanov, 1997; UNEP/GEMS, 1995.

2 From R. Peters, W. Beck, personal communication, 2004.

3 Micro-organic pollutant list now includes a suite of endocrin disrupters, antioxidants, plasticizers, fire retardants, insect repellents, solvents, insecticides, herbicides, fragrances, food additives, prescription drugs and pharmaceuticals (e.g., birth control, antibiotics, etc.), non-prescription drugs (e.g., caffeine, nicotine and derivatives, stimulants).

Table 4.6: Spatial and time scales within which pollution occurs and can be remediated

Major Causes / Issues	Major Related Issues [1]	Scale [2] Local	Region	Global	Time to Pollute [3] <1	1 to 10	10 to 100	Time to Remediate [4] <1	1 to 10	10 to =>100	Major Controlling Factors Biophysical	Human
Population	Pathogens											Density & Treatment
	Eutrophication (*)											Treatment
	Micro-pollutants											Various
Water Management [4]	Eutrophication (*)										Hydrodynamics	Flow
	Salinization											Water Balance
	Parasites									> 100		Hydrology
Land Management	Pesticides											Agrochemicals
	Nutrients											Fertilizer
	Suspended Solids (*)											Construction/clearing
	Physical Changes									> 100		Cultivation, Mining, Construction, Clearing
Atmospheric Transport	Acidification (*)											Cities, melting and fossil fuel emissions
	Micro-pollutants											Cities
	Radionuclides									>> 100		Industry
	Mega Cities Pathogens											Population & Treatment
	Mega Cities Micro-pollutants											
	Mines Salinization											Types of Mines
	Mines Metals											
	Nuclear-Radionuclides									> 100		
Global Climate Change	Salinization										Temperature & Precipitation	Fossil fuel emissions & Greenhouse gases
Natural Ecology	Parasites (*)				Permanent			Permanent			Climate, Hydrology	
Natural Geochemistry	Salts										Climate, Lithology	
	Fluoride (**)				Permanent			Permanent				
	Arsenic, Metals (**)										Lithology	

Notes:
The nutritional status of most regions of the world has improved in all developing regions. Sub-Saharan Africa and South Asia have also improved their nutritional level, but they lag behind and are host to the majority of the undernourished people in the world.
1 Relevant primarily to * surface water, ** groundwater,
2 Local < 10000 km², region->10⁴ to 10⁶ km² global->10⁶ to 10⁸ km².
3 Lag between cause and effect.
4 Longest time scale is for groundwater, followed by lakes, and shortest for rivers and streams.

Category Shading:
Scale – the colour intensity increases as impact dimension becomes greater.
Time to pollute and Time to remediate are highlighted in red for most critical, orange for moderately critical, and yellow for the least critical situations. Green is shown for the situation where remedial actions could be less than one year (pathogens).

Source: Modified from Peters and Meybeck, 2000.

Food remains in the Mekong River after the daily market activities, Viet Nam

4. See www.gemstat.org for more information

a more holistic process relating to health and other socio-economic issues. The international compilation of surface water and groundwater quality data sets at a global scale is still in its relative infancy as compared to precipitation or surface water runoff data. Although some facilities have existed for several decades to collect and disseminate this type of data, it has been historically difficult to collect. This is attributable to several reasons. National centres have not always been linked to institutional networks. Most nations are simply not used to providing this information to anyone other than their immediate institutions and users for either national or specific project purposes. In addition, data in many developing countries is not extensive and even where it has been collected, making it publicly available as a data set is frequently not a priority for the already overloaded and meagrely resourced national and subnational water resource institutions. However, progress has been made in the past three years in this area. The GEMS/Water international water quality database[4] went online in March 2005 and now has begun to work with a broad range of agencies, NGOs and data quality groups to harmonize the reporting of water data and information. They have established a QA/QC (quality assurance/quality control) programme that includes laboratory evaluations based on a freely available published set of methods that are used by most of the laboratories that report their data to GEMS/Water. GEMS/Water (2005) reports that data is now received from about 1,500 stations globally, including about 100 for lakes and groundwater.

Increased awareness of the need for water quality data to evaluate impacts and design improved water use and reuse strategies in order to meet quality and quantity demands is emerging at national and river-basin levels. Moreover, there is increasing use and future development of shared aquifers and river basins – many of which are being supported extensively by programmes of the GEF (Global Environment Facility) and UNESCO.

3c. Over-abstraction
The problems of over-abstraction in surface water bodies and groundwater, sometimes tied directly to upstream diversions, reservoirs and deforestation, are well documented. The problems commonly become exacerbated when combined with extended natural dry periods. Notable examples of substantive reductions in large major river flows can be found around the world. Some of the basins suffering from this reduction are: Niger, Nile, Rwizi, Zayandeh-Rud (Africa); Amu Darya, Ganges, Jordan, Lijiang, Syr Darya, Tigris and Euphrates, Yangtze and Yellow (Asia); Murray-Darling (Australia); and Columbia, Colorado, Rio Grande and San Pedro (North America). Examples of lakes and inland sea areas decreasing dramatically in size and volume include: Lakes Balkhash, Drigh, Hamoun, Manchar, and the Aral and Dead Seas (Asia); Lakes Chad, Nakivale and in the Eastern Rift Valley Area, e.g. Nakuru (Africa); Lake Chapala (North America); and Mono Lake and the Salton Sea (North America). Dramatically lowered water levels in aquifers are increasingly reported, for example in the Mexico City and the Floridian and Ogallala aquifers (North America), as well as in China, India, Iran, Pakistan and Yemen (Asia).

Despite years of clear over-use with evident changes in both water and related ecosystem conditions, many of the same causes persist. Among the most prominent are the highly inefficient water supply provisioning practices for agriculture and municipal use, deforestation, and the basic lack of control over exploitation of the actual surface and groundwater sources. Inappropriate development of reservoirs and diversions combined with inadequate considerations of alternatives in conservation and use minimization (demand management) have further complicated and increased the impacts on existing water

resources. While there are some hopeful signs of change emerging in selected local actions (see **Chapters 5** and **7**), these are few in comparison to the broad-based and fundamental modifications needed in national, regional and subnational practices to reverse and counteract these ongoing substantive impacts.

Groundwater over-abstraction represents a special situation as the visual evidence is typically less obvious and the effects are more difficult to recognize and react to. Increased pumping from aquifers has increased globally, particularly during the second half of the twentieth century. While this has produced a number of important benefits, some have been sustainable over only relatively short periods and have had significant negative side-effects (UNEP, 2003; FAO, 2003b; Burke and Moench, 2000). We see, for example, that an initially impressive benefit was experienced in India where shallow groundwater development allowed irrigated land area to be essentially doubled, thereby dramatically increasing food production. However, it also caused momentous changes to local water regimes that resulted in a variety of impacts, including lowered water tables and entirely depleted groundwater resources in some areas. Similar cases from all climatic regions of the world illustrate that over-abstracting groundwater is relatively common. The results of groundwater over-abstraction can be seen in: reduced spring yields; rivers drying up and having poorer water quality because of lowered base-flow contributions; intrusion of saline waters or other poor quality water into the freshwater zones of aquifers; lowered or abandoned productivity as water levels decline in wells; higher production costs from wells or the need to extend underground aqueducts (qanats) as inflow rates decrease; and diminished groundwater-dependent ecosystems, including wetlands, as they become stressed or lose resilience from inadequate water sources. Subsidence is another particularly widespread impact that occurs from excessive over-pumping, with some notable examples in a number of major cities in China, Japan, Mexico and the US. However, this type of impact can be stopped when the over-pumping of the aquifer is discontinued, although the effects are not usually reversible. Llamas and Custodio (2003) provide a recently updated compilation of papers that illustrate the wide-ranging impacts of intensive groundwater exploitation by identifying examples of criteria that have led to over-abstraction actions and by explaining how these criteria can be part of sustainable development strategies.

Tigris River, Iraq

Groundwater over-abstraction represents a special situation as the visual evidence is typically less obvious and the effects are more difficult to recognize and react to

Map 4.3: Groundwater abstraction rate as a percentage of mean recharge

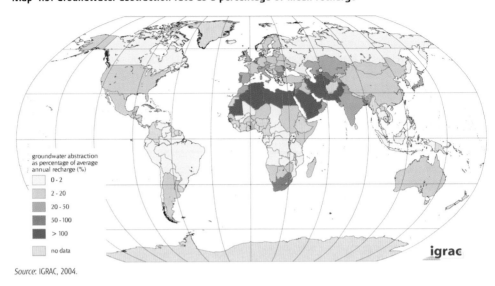

groundwater abstraction
as percentage of average
annual recharge (%)

- 0 - 2
- 2 - 20
- 20 - 50
- 50 - 100
- > 100
- no data

Note: Low percentages indicate underdeveloped groundwater resources, high percentages point to development stress or eventually overexploitation.

Source: IGRAC, 2004.

High levels of exploitation are currently taking place in many countries in the Middle East, Southern and Northern Africa, Asia, selected countries in Europe, and in Cuba

Map 4.3 introduces a groundwater development indicator that compares the degree of groundwater use in each nation to the volume of estimated recharge. Exploitation, for example of more than 50 percent of recharge, will likely result in particular stress on the aquifer sustainability of groundwater systems. High levels of exploitation are currently taking place in many countries in the Middle East, Southern and Northern Africa, Asia, selected countries in Europe, and in Cuba. In addition, as noted above, parts of China, India, Mexico, Pakistan and the US are also being overexploited in selected regions where there is high aridity and population density. Tracking groundwater use as compared to recharge volumes at national and subnational levels – and particularly for individual aquifers – should be practised and implemented to identify and take corrective action as needed to maintain groundwater development sustainability.

3d. Global warming and climate change
As noted above, there is empirical evidence of impacts on water resources from global warming. The IPCC, in cooperation with new partners, has begun to address this issue in addition to their more traditional focus on greenhouse gases and temperature changes. A recent IPCC expert meeting (IPCC, 2004, p. 27) identified two issues related to water and the impacts from global warming: one related to impacts and the other to knowledge gaps. These two issues, as taken from the IPCC report, are as follows:

■ 'The extreme event frequency and magnitude will increase even with a small increase in temperature and will become greater at higher temperatures. The impacts of such events are often large locally and could strongly affect specific sectors and regions. Increased extreme events can cause critical design values or natural thresholds to be exceeded, beyond which the impacts' magnitudes increase rapidly.'

■ Knowledge gaps related to the water sector were identified as:
  (1) Insufficient knowledge of impacts in different parts of the world (especially in developing countries),
  (2) Almost complete lack of information on impacts under different development pathways and under different amounts of mitigation,
  (3) No clear relationship between climate change and impacts on water resources,
  (4) Little analysis of the capacity and cost of adaptation, and
  (5) Lack of understanding of how changes in variability affect the water environment.

Arnell (2004) also assessed predicted impacts of both population and climate on water-stressed regions, based on population growth scenarios and climate change models. He concludes:

Climate change increases water resources stresses ... where runoff decreases, including around the

Mediterranean, in parts of Europe, central and southern America, and southern Africa. In other water-stressed parts of the world – particularly in southern and eastern Asia – climate change increases runoff, but this may not be very beneficial in practice because increases tend to come during the wet season and extra water may not be available during the dry season.

However, he further points out that model results differ by up to four times in terms of persons impacted according to different population and climate scenarios.

Shiklamanov and Rodda (2003) conclude that only general predictions and observations have been developed based on the assessments of global warming impacts on water resources to date. They agree with Arnell (2004) that assessments of future water resources can only be obtained by using estimates of possible regional (rather than global) changes in climate (primarily precipitation and temperature by seasons and months). They specify that the existing climate change estimates are extremely unreliable even for the largest regions and river basins. Furthermore, they suggest that the gap in knowledge related to the specific impacts of global warming on water resources is one of the largest scientific challenges in hydrology today.

BOX 4.5: ACCELERATING GLACIAL DECLINE

Land-based and mountain glaciers have generally experienced a worldwide retreat and thinning during the last century. Notably, glacier decline has considerably accelerated on a global basis during recent years (Arendt et al. 2002; Dyurgerov 2003). The mean mass balance decrease that took place during the period 1990–99 was three times greater than that of the previous decade (Frauenfelder et al. 2005). Data for this figure are based on measured changes in glacier mass balance made at thirty glaciers located in nine high mountain regions of Asia, Europe and North and South America.

As a specific country example we can look to China. In 2004, AFP (L'Agence France-Presse) cites renewed concerns of disappearing glaciers being broadcast in Asia, notably in China and Nepal. Yao Tangdong, China's foremost glaciologist, was quoted in state media as saying, 'An ecological catastrophe is developing in Tibet because of global warming and that most glaciers in the region could melt away by 2100'. His conclusion was based on the results of a forty-month study by a group of twenty Sino-American scientists which showed

separated ice islands that used to be connected with the glaciers at levels above 7,500 m. While Tibet's glaciers have been receding for the past four decades due to global warming, the rate of decline has increased dramatically since the early 1990s. It was initially thought that the water from the melting glaciers could provide additional water for China's arid north and west.

However, this hope has not been realized as much of the glacier runoff evaporates long before it reaches the country's drought-stricken farmers. 'The human cost could be immense' states AFP (2004), as 300 million Chinese live in the country's arid west and depend on the water flowing from the glaciers for their livelihoods.

Map 4.4: Principal regions of land-based and mountain glaciers

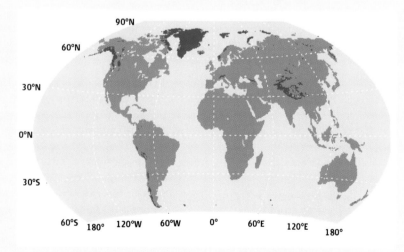

Source: GLIMS, 2005 (Global Land Ice Measurements from Space nsidc.org/data/glims/ lvOct05).

Section 2: CHANGING NATURAL SYSTEMS

Part 4. Matching Demands to Supply

Numerous responses have been put forward to meet the ever-increasing demand for water. In some cases, the response focuses on how to compensate for the natural variability in the hydrological cycle in order to provide a continuously available resource. In other circumstances, the response focuses on overcoming the reduced availability in water quantity or quality that results from human and development impacts, from a demand management perspective.

Most water-short regions of the world with dry climates have long-standing water conservation traditions. These are being maintained or supplemented with demand-management practices. To meet increased demands, water resource management practitioners are augmenting the limited natural water supply with desalination, water reuse, enhanced groundwater recharge and inter-basin transfers.

However, regions with abundant water (tropical and cold climates) are accustomed to water supply schemes and tend to adopt management practices that are particularly adapted to those specific settings. It is often taken for granted that resources will remain relatively abundant and could be readily treated or replaced if polluted; that any disruption in ecosystem balance could be remedied; and that adequate water could be diverted and stored to overcome the inconvenience of seasonal flow variations. However, in these regions, impacts from human development have been more severe than anticipated. Water resources have been diminished in quantity and quality, and ecosystem habitats have become endangered

to a point below their resilience levels. As a result, responses are emerging that include some of the same practices in demand management used in dry climates. In both water settings, it is increasingly recognized that maintaining and, where possible, restoring the state of the environment by keeping both aquatic and terrestrial aquatic ecosystems above resilience levels can provide substantial long-term benefits to a region's water resources.

4a. Environmental flows for preserving ecosystems and increasing water resources

The heightened awareness of the important role played by ecosystems in terms of water resources and sustainability is a result of the recent focus on 'environmental' or 'in-stream' flows. Dyson et al. (2003) define environmental flows as follows:

> the water regime provided within a river, wetland or coastal zone to maintain ecosystems and their benefits. They provide critical contributions to river health, economic development and poverty alleviation.

BOX 4.6: MANAGEMENT OF AQUIFER RECHARGE (MAR) – AN EXAMPLE FROM VIET NAM

The Binh Thuan province is located along the coastal plain in the lower part of central eastern Viet Nam; its principal city is Phan Tiet, 200 km East of Ho Chi Minh City. The area of the province is approximately 8,000 km², with a total population of 1 million.

Before 1975, the area was covered by a dense tropical forest, which was cleared to make room for rice fields and resulted in massive desertification. Due to an uneven rainfall distribution and a four-month period (from December to March) of very little precipitation, the area suffers from considerable water shortage during the dry season.

In order to combat desertification, improved practices in ecosystem rehabilitation as well as remediation techniques to restore aquifer systems and groundwater storage capacity are being developed. In particular, these techniques are being used in the Hong Phong sub-district (Bac Binh district), located about 25 km northeast of Phan Tiet, with an area of approximately 300 km² encompassing three villages.

The geo-hydrological assessment of the area, consisting of a semi-permeable bedrock and porous material (sand dunes) with a thickness of up to 150 m, allows for the use of SAR (storage and aquifer recovery) techniques by

redirecting rainfall during the rainy season and making use of the resource during the dry period (December–March).

The project's implementation by UNESCO is ongoing and the results achieved thus far have allowed for the selection of the site for the Aquifer Recharge Project in the morphological depression of Nuoc Noi, where the aquifer water table is very close to the ground level. The use of the bank filtration technique is already producing satisfactory results as water quality increases. Groundwater can be abstracted and used, after natural filtration, for different purposes (human and agricultural).

The means for maintaining and restoring these flows under multi-use and competing demand situations are increasingly being considered in detail in many nations and basins. In some regions, environmental flow considerations are being integrated into water policy, legislation and regulations, and water management practices. South Africa (1997), Australia (CSIRO, 2004) and several USA states (e.g. Connecticut, Texas), among others, already have broadly encompassing legislation and in-field practices that take into account environmental flows. More research is needed to understand the water volumes, levels and quality needed to keep ecosystems resilient during seasonal variations and periods of climatic stress. Furthermore, a recognized additional challenge is how to introduce and embed this concept in the predominantly engineering-driven water management agencies of many developing countries so that the resilience of their basin- and watershed ecosystem is less at risk (see **Chapter 5**).

4b. Combating natural variability

Dealing with variability in water runoff in particular has led to centuries-old practices of intercepting, diverting and storing water so that adequate volumes would be available to match the needs and demands of the users.

Rainwater harvesting

Rainwater management, also known as harvesting, is receiving renewed attention as an alternative to or a means of augmenting water sources. Intercepting and collecting rainwater where it falls is a practice that extends back to pre-biblical times (Pereira et al., 2002). It was used 4,000 years ago in Palestine and Greece; in South Asia over the last 8,000 years (Pandey et al., 2003); in ancient Roman residences where cisterns and paved courtyards captured rain that supplemented the city's supply from aqueducts; and as early as 3000 BC in Baluchistan where farming communities impounded rainwater for irrigation. Recently in India, it has been used extensively to directly recharge groundwater at rates exceeding natural recharge conditions (UNESCO, 2000; Mahnot et al., 2003). Reports from other international organizations focusing on this area[5] indicate that eleven recent projects across Delhi resulted in groundwater level increases of from 5 to 10 metres in just two years. In fact, the application of rainwater management in India is likely to be one of the most updated and modern in the world. The site www.rainwaterharvesting.org provides links to cases where rainwater management has been successfully applied in different nations in both urban and rural

settings. An advantage of the technique is that its costs are relatively modest and that individual or community programmes can locally develop and manage the required infrastructures (collection devices, basins, storage tanks, surface or below-ground recharge structures or wells). Larger rain harvesting schemes, which intercept runoff using low-height berms or spreading dikes to increase infiltration, have also been introduced in upstream catchments where deforestation has decreased water availability. The various methods of rainwater harvesting that have the potential to satisfy local community and crop demands are described in UNEP (2005).

Water diversions

Diverting surface waters into nearby spreading basins/infiltration lagoons, ditches, recharge pits or injection wells to recharge alluvial or other types of aquifers are techniques used to deal with natural variability in flow, reduce evaporative losses, and obtain better quality water. Water diversion programmes being established around the globe are referred to as ASR (artificial storage and recovery) or MAR (managed aquifer recharge) (see **Box 4.6**). This practice is being applied in arid and semi-arid locations throughout the Middle East and Mediterranean regions. Runoff in 'wadis' (dry riverbeds that only contain water during times of heavy rain) that otherwise would discharge into the sea or evaporate, is collected behind earthen berms following infrequent but heavy rainfall. The water infiltrates into the underlying alluvial gravel thereby remaining available for substantively longer periods without the excessively evaporative losses that would typically occur from surface storage. In wetter areas, diversions into alluvium are used as a means not only to store and maintain groundwater-dependent ecosystems, but also to reduce the treatment needed for the water supplies systems taken from the alluvium further downstream.

Professional associations such as the US National Ground Water Association (US NGWA) and the IAH (International Association of Hydrogeologists) Commission on Managing Aquifer Recharge (MAR)[6] in cooperation with UNESCO and other international donors, are actively supporting MAR with applied research, capacity-building and pilot projects. MAR programmes, some including injection of treated wastewaters, are being carried out in both developed and developing countries (e.g. in Australia, China, Germany, Hungary, India, Kenya, Mexico, Oman, Pakistan, the southern Africa region, Switzerland and the US).

Kakadu National Park, Australia. Most predictions indicate that precipitations will increase in Australia due to climate change

Intercepting and collecting rainwater where it falls is a practice that extends back to pre-biblical times...

5. See www.irha-h2o.org for more information.
6. www.iah.org/recharge/ MAR.html

...the major hydrological challenge will be to achieve more equilibrium between the stored volumes needed to meet users' demands and the incoming and outgoing flow...

Waste from a water desalination plant in the sea of Al-Doha, Kuwait. Several seawater desalination plants produce 75% of the country's water supply

Storing water in reservoirs

The construction of dams to create reservoirs has frequently been our response to growing demands for water to provide hydropower, irrigation, potable supplies, fishing and recreation, as well as to lower the impacts and risks to our well-being from high-intensity events such as floods and droughts. These facilities collect natural runoff, frequently quite variable in its location, duration and magnitude, and store it so that its availability is more constant and reliable. Good information on the number and capacity of dams is essential to assess impacts and responses at the local, national and regional levels in order to optimize water resources management, but it is also needed to address issues related to global climate and water availability scenarios (see **Chapter 5**).

Though the creation of reservoirs enables higher water availability when and where it is needed, the construction of these facilities has had a considerable impact, both positive and negative, on the Earth's ecosystems and landscapes and has resulted in modifications to the interactions among the components of the hydrological cycle. Despite increased benefits derived from the services reservoirs provide, there is ongoing debate about how to prevent and reduce the social and environmental consequences that come from building dams and creating reservoirs. Following considerable media attention and local actions some practices are changing. Large dam construction rates have slowed, at least temporarily, and there have been advances in the reconsideration of alternatives and design criteria. Some existing dams that no longer provide extensive services have been decommissioned. Lastly, existing reservoir operations and structures have been modified to allow releases. A balance between what enters and what is released is required to have a site's upstream and downstream hydrological settings and supporting ecosystems sustained. When such a balance is achieved, the results are substantial. There are both added benefits and potential further value to the role of reservoirs in development scenarios.

Transferring water among basins

The transfer of water from one river or aquifer basin to another basin has long been used as a way to meet water demands, particularly in arid and semi-arid regions. It occurs often when large populations or, more commonly, agricultural demands have outstripped existing water resources. Even in advanced national development stages, some basins can have surplus water resources while

others face shortages. Major long-distance schemes exist in many nations and new ones are in development. Linking the Ganga-Brahmaputra-Meghna system with other rivers in India is part of the solution being offered to counteract extensive recurring droughts and floods. For example, Shao et al. (2003) present the situation in China where there are seven existing major transfers and seven more planned or under consideration. They describe a large-scale south-to-north basin transfer involving the Yangtze and Yellow Rivers' basins which, when completed, would divert 450 km^3/yr. They also point out some of the impacts of such a large scheme. Multi-disciplinary approaches allow evaluation of the feasibility and sustainability of transfer schemes. Global experience has shown that although the transfer of water among basins has been identified as a hydraulically and technically feasible response, before proceeding with such potential changes, broad social and environmental considerations must be taken into account.

4c. Water reuse

Asano and Levine (2004) recently summarized the more important challenges associated with water reclamation and reuse. They noted that the technique of water reuse is being applied in many countries including the United States, Mexico, Germany, Mediterranean and Middle Eastern countries, South Africa, Australia, Japan, China and Singapore. Its increased application is being facilitated by modern wastewater treatment processes, which advanced substantially during the twentieth century. These processes can now effectively remove biodegradable material, nutrients and pathogens so the treated waters have a wide range of potential applications (**Table 4.7**). On a global scale, non-potable water reuse is currently the dominant means of supplementing supplies for irrigation, industrial cooling, river flows and other applications (Asano, 1998). The reuse of potable waters has been an accepted global practice for centuries. Settlements downstream produced their potable water from rivers and groundwater that had circulated upstream through multiple cycles of withdrawal, treatment and discharge (Steenvorden and Endreny, 2004; Asano and Cotruvo, 2004; GW MATE, 2003). San Diego gets 90 percent of its current municipal water supply from a wholesale water provider but in future that amount will decrease to 60 percent with the supplementary supply coming from reclaimed water and desalination (USGS, 2005). Similar programmes are emerging in many other large urban centres worldwide where there are limited or less readily available freshwater supplies. Similarly, riverbeds or percolation ponds have been used to artificially recharge underlying groundwater aquifers mainly with wastewater.

Recent documents from WHO (Aertgeerts and Angelakis, 2003) and the US EPA (2004) address the state-of-the-art aspects and future trends in water use, both of which predict increased development and use of the above-mentioned practice to augment water supply sources in order to meet demands. The WHO guidelines for wastewater reuse first published in 1995 are being updated with a planned release date of 2006 (WHO, 2005). According to water reuse surveys (Lazarova, 2001; Mantovani et al., 2001), the best water reuse projects in terms of economic viability and public acceptance are those that substitute reclaimed water in lieu of potable water for use in irrigation, environmental restoration, cleaning, toilet flushing and industrial uses.

The annual reclaimed water volumes total about 2.2 billion m^3, based on 2000 and 2001 figures from the World Bank. Recent projections indicate that Israel, Australia and Tunisia will use reclaimed water to satisfy 25 percent, 11 percent and 10 percent, respectively, of their total water demand within the next few years (Lazarova et al., 2001). In Jordan, reclaimed water volumes are predicted to increase more than four times by 2010 if demands are to be met. By 2012, Spain will need to increase its reclaimed water use by 150 percent and, by 2025, Egypt will need to increase its usage by more than ten times. A number of Middle Eastern countries are planning significant increases

in water reuse to meet an ultimate objective of 50 to 70 percent reuse of total wastewater volume. The growing trend of water reuse is not only occurring in water-deficient areas (Mediterranean region, Middle East and Latin America), but also in highly populated countries in temperate regions (Japan, Australia, Canada, north China, Belgium, England and Germany). This method of augmenting natural water sources is becoming an integral component to many water resources management plans and future use policies.

4d. Demand management

Conserving available water and reducing demand is a necessary measure in water-short regions, especially those in arid climates. Programmes of conservation and demand reduction are referred to as water demand management (WDM). This approach differs from the traditional supply-driven method, which makes all existing water available. WDM applies selective economic incentives to promote efficient and equitable water use. It also identifies water conservation measures that are aimed at raising society's awareness of the scarcity and finite nature of the resource.

Conservation measures have not been readily implemented, particularly where water was perceived as abundant. However, the benefits in the extended useful life of water supply and treatment plants and in the operating efficiency and duration of sewage disposal systems can be considerable in terms of higher economic return on investment. On the environmental front, conservation

At inland locations or where desalination is too costly, reclaimed water can now significantly contribute to the overall water supply used for irrigation or industry...

Table 4.7: Potential applications for reclaimed water

Application settings	Examples
Urban use	
Unrestricted	Landscape irrigation (parks, playgrounds, school yards), fire protection, construction, ornamental fountains, recreational impoundments, in-building uses (toilets, air conditioning)
Restricted-access irrigation	Irrigation of areas where public access is infrequent and controlled (golf courses, cemeteries, residential, greenbelts)
Agricultural irrigation	
Food crops	Crops grown for human consumption and consumed uncooked
Non-food crops, food crops consumed after processing	Fodder, fibre, seed crops, pastures, commercial nurseries, sod farms, commercial aquaculture
Recreational use	
Unrestricted	No limitations on body contact (lakes and ponds used for swimming, snowmaking)
Restricted	Fishing, boating, and other non-contact recreational activities
Environmental use	Artificial wetlands, enhanced natural wetlands, and sustained stream flows
Groundwater recharge	Groundwater replenishment, saltwater intrusion control, and subsidence control
Industrial reuse	Cooling system makeup water, process waters, boiler feed water, construction activities, and washdown waters
Potable reuse	Blending with municipal water supply (surface water or groundwater)

Source: Asano and Leavine, 2004.

One interesting emerging concept proposes combining desalinated water with aquifer storage and recovery...

allows for the diversion of the unused volumes to sustain ecosystems and also lowers the pollution loadings to lakes, rivers and groundwater. Such steps lead to improved protection of drinking water sources and overall ecological balance (Environment Canada, 2005b).

WDM advocates a wide range of measures that go beyond conservation to broader sustainable resource management. It applies to the protection of water quality sources; reduction of wastage both in infrastructure leakage and by users; improvement of water allocation among competing uses, and creation of appropriate pricing mechanisms. One example of a situation where conservation measures are needed is the case of 'undelivered water' – a commonly accepted result of utilities supplying water through piped distribution systems. The leakage from degraded pipes provides 'unaccounted for' water that results in both a physical shortage and reduced revenue. In terms of inefficiency of resources and operations, losses are routinely reported as 40 percent and as high as 60 to 70 percent in some major cities. Though it is an endemic problem for most water utilities, its impact on society in terms of wasted water resources is even more substantial.

Further water conservation can be achieved after delivery by improving use practices in households. Reductions in community water use after conservation measures have been applied are reported to be as high as 40 percent. These two situations illustrate to what extent the water that is currently supplied may not actually be needed. By reducing leakage and demand, substantial reductions in the source volumes could be achieved. This should be a clear message in development settings. WDM may obviate the need for some of the proposed large-scale physical or infrastructure investments and thereby provide real efficiency gains to society (GWP, 2005a).

4e. Desalination
Desalination is used mainly in water-scarce coastal arid and semi-arid areas that are located inland where the only available water source is saline or brackish groundwater. The technology has been well established since the mid-twentieth century and has evolved substantially to meet the increased demands of water-short areas. Awerbuch (2004) and Schiffler (2004) report on the global application of desalination capacity and the most recent advances and challenges. According to the latest statistics in 2002 from IDA (International Desalination Association),[7] about 50 percent of global desalination takes place in the

Middle East, followed by North America (16 percent), Europe (13 percent), Asia (11 percent) Africa (5 percent) and the Caribbean (3 percent). South America and Australia each account for about 1 percent of the global desalination volume. Globally, the contracted capacity of desalination plants is 34.2 million m^3/day converting principally seawater (59 percent) and brackish water (23 percent). In terms of the uses of desalinated water, municipalities are the largest users (63 percent), followed by substantial industry use (25 percent). The cost of producing desalinated water has fallen dramatically in the past two decades. Recently built large-scale plants produce fresh water for US$ 0.45/m^3 to US$ 0.50/m^3 using reverse osmosis (RO) systems and US$ 0.70/m^3 to US$ 1.0/m^3 using distillation systems. The energy consumed to drive the conversion is a significant part of the cost and ranges from 4 to 15kWh/m^3 depending on factors such as the technique used, the production rate of the facility, and the quality of the equipment (US NRC, 2004).

Much of the conversion is likely to continue to be heavily reliant on fossil fuels with its associated air pollution. The challenge of what to do with the brine waste by-product remains. Today it is disposed of by discharge into the ocean or surface waters, sewage treatment plants, deep-well injection, land application or further evaporation in ponds. Each of these methods has potentially adverse environmental impacts. The cost of concentrate disposal for inland locations often limits its applicability in these locations. Schiffer (2004) recommends the establishment of an internationally agreed-upon environmental assessment methodology for desalination plants to enable the impacts from different facilities to be consistently compared.

Future uses for desalination are emerging and IDA expects that, with increasing demand and the up-scaling of processes, it will continue to be applied for the development of economies in coastal areas to partially meet the demands of recreation and tourism, environmental protection, the military, and irrigated agriculture. One interesting emerging concept proposes combining desalinated water with aquifer storage and recovery (DASR) (Awerbuch, 2004; Pyne and Howard, 2004). This approach has the advantages of allowing storage and recovery of large volumes of water while minimizing facility throughput with lowered operating costs. Stored volumes could be used to meet daily or seasonal peaks in water demands while maintaining a steady desalination rate.

---

7. See www.idadesal.org for more information.

4f. Water Resources Assessment (WRA)

Water resources assessments (WRAs) are designed to be analyses of available water sources from the perspective of potential water use. Since Rio '92, and in particular the Dublin 2000 considerations, water resources have come to be more broadly considered within the dimensions of social equity, economics and ecosystem/ecohydrology. The modern WRA process can be adapted and updated to include these relationships (GWP, 2005b).

Hydrological data and information systems and networks provide the basic and critical input to WRA, whether the assessment is done within an IWRM perspective at the national or basin/sub-basin/aquifer level or otherwise. Factors that affect the accuracy of hydrological input to WRAs include: the number of gauging stations, station distribution within physiographic regions, duration and continuity of observations, quality of measurements, and data processing. The commonly measured parameters include precipitation, evaporation, soil moisture, river level and discharge, groundwater (well) depths, sediment and water quality data on a continuous, hourly, daily or monthly basis.

However, reliability and availability of data have declined sharply since the mid-1980s, particularly in Africa and in Eastern Europe (Rodda, 1998), and that situation has not changed substantively since the turn of the century. Investment in national networks has fallen drastically and is still decreasing. Hydrometric networks, while they are costly to maintain, provide basic WRA input that cannot be collected dependably by any other means (see **Chapter 13**).

The development of more decentralized and basin-type approaches for WRA is inherent in the internationally agreed upon IWRM principles. It is widely recognized that it will take several decades of institutional adjustment (Blomquist et al., 2005) to reorient water management practices on basins. However, such changes are beginning at the basin level and there are examples of decentralized approaches on most continents in terms of water management processes. An important element of the World Water Assessment Programme's mission is to assist partner case study countries in developing their own assessment capacity (see **Chapter 14**). Sovereignty issues and competition will always remain factors in managing the resource. However, the basic WRA scope which broadly defines the extent of available water quantity and quality, including aspects related to environment, pollution and water use, is the basis for effective management. This information can be collected and jointly developed by the nations sharing the resource (see **Chapter 11**). These will give forward-looking direction not only in water technology areas but also on how improving data, information and assessment practices for water resources will provide critical knowledge that will greatly benefit society, human livelihoods and the environment.

Many developing nations, where the demand for water is growing the fastest, have the worst capability for acquiring and managing water data

Part 5. The Challenge of Sustainable Development

Climate change and the hydrological variability of water's distribution and occurrence are natural driving forces that, when combined with the pressures from economic growth and major population change, make the sustainable development of our water resources a challenge.

5a. Driving forces and pressures

The combination of these factors commonly results in increased water use, competition and pollution in addition to highly inefficient water supply practices. These results can be traced back to the fact that most decisions in water resources management, at almost all levels, remain principally driven by short-term economic and political considerations that lack the long-term vision needed to implement sustainable development practices. Water management plans should consider the best existing practices and the most advanced scientific breakthroughs.

The scientific community has to convey more effectively its recommendations to decision-makers to enable the latter to develop and maintain multidisciplinary integrated approaches and solutions. Societies should realize that today's water-related challenges are no longer readily solved just by using last century's hydraulic schemes. Increased funding and resources need to be provided for the collection of detailed water data and information.

Section 2: CHANGING NATURAL SYSTEMS

Overall, there are reasons to be hopeful as new water programmes are emerging that finally emphasize the application of more sustainable practices to reduce impacts

5b. State of our natural water resources

The roles and interdependencies of the different hydrological cycle components are often not fully appreciated. As a result, it is difficult to set up adequate protection and prevention strategies.

All components of the hydrological cycle should be taken into account when developing water management plans. Each component has a specific role that must be better understood. For example, rain and snow directly supply terrestrial ecosystems and soil moisture is a unique water source for both agricultural development and terrestrial ecosystems. Furthermore, glacial melting has a strong influence on water availability in many nations and as a result more comprehensive global assessments are needed.

We can substantively predict annual variability in surface runoff and have created solutions to deal with it. However, overcoming the less predictable five- to ten-year global cycles of distinctly lower and higher runoff remains a challenge. Groundwater resources could provide a valuable contribution to overcoming climate variability and meeting demands during extended dry periods. A surplus of surface water runoff during wet periods can be used to replenish aquifer systems.

However, we do not have enough data on groundwater and aquifer systems, especially in developing countries where the lack of adequate surface water resources is most extreme. This is particularly true in both Asia and Africa where there has been a dramatic reduction in water monitoring programmes.

Water quality monitoring programmes are inadequate or lacking in most developing nations; thus safeguarding human health is difficult. Despite two decades of increased international scientific attention and concern, attempts to collect, compile and gain knowledge from consumption, pollution and abstraction data and information at a global scale are still piecemeal and in relatively early stages of applicability.

5c. Impacts

Poor quality water and unsustainable supplies limit national economic development and can lead to adverse health and livelihood conditions.

Landscape modifications further complicate our understanding of and ability to predict the impacts on water resources since these changes disrupt natural

hydrological and ecosystem functioning. This becomes more important when we seek to advance our understanding of the future impacts of climate change at local and regional scales. We know that detailed estimates of climate change impacts on water resources at regional or global scales are currently very problematic due to inadequate water data.

We have reached a reasonable level of knowledge towards recognizing impacts on water quality and quantity from pollution and excessive groundwater and surface water withdrawals. The focus must now be on reducing these impacts. In most developing countries, specific and well-targeted programmes should be funded to reduce impacts on water quality and quantity.

Overall, there are reasons to be hopeful as new water programmes are emerging that finally emphasize the application of more sustainable practices to reduce impacts.

5d. Responses

Prevention strategies and new technologies that augment existing natural water resources, reduce demand, and achieve higher efficiency are part of the response to meet today's increasing demands on our available water resources.

To meet current and future water demands, increased attention should be given to precautionary approaches such as innovative uses of natural supplies and new technologies. In the past we have responded by storing runoff in reservoirs, diverting flows from water-abundant to water-scarce regions, and extracting aquifer resources – methods that provided ample water where and when it was needed. These methods are likely to remain part of most water resources development strategies. Non-conventional water resources, such as water reuse and desalination, are being increasingly used and new technologies such as artificial recharge are also becoming more and more common. Capturing rain at the source through rainwater harvesting is yet another method used to increase the availability of natural water sources. In certain regions, an extreme response has been adopted. In some arid countries, where sufficient renewable water resources are not available, non-renewable groundwater reserves are being exploited to support development.

Demand management and conservation are methods that target efficiency. Conservation begins by reducing high losses from water supply distribution systems. Demand management has gone largely unaddressed since most water utilities still focus on infrastructure development rather than on conservation.

It is worth noting that industry's approach in recent years has been to reduce wastewater and minimize the quantity of processed water needed as this method has proven to be technically feasible and economically advantageous. The demand reduction and efficiency approach should be an integral part of modern water resources management. Its applicability should be promoted while recognizing that it requires a distinct change in the behavioural patterns of institutions, utilities and individuals – a change that will require education, awareness-raising and political commitment to achieve effective implementation.

Institutional responses at different levels are also needed. Some nations have implemented new laws and regulations that point the way forward toward protecting and restoring our water sources. Many nations are adapting emerging technical practices to secure and protect their existing natural water resources and use local knowledge as part of sustainable resource development.

5e. The benefits

There will be economic, social and environmental benefits from carrying out regular Water Resources Assessments (WRAs) in all basins and aquifers in individual nations as well as regionally, where transboundary shared water resources are present.

Modern approaches to WRA are rapidly emerging and now go well beyond the traditional hydraulic and supply-biased studies carried out during the last century. WRAs have been extended to take advantage of the recently recognized benefits that come from using an integrated approach (IWRM) and including ecosystems' services (ecosystem approach). WRAs continue to fundamentally require well-documented hydrological cycle component data – without this data the evaluation results are unreliable. To be comprehensive and assist in sustainable practices, WRAs should include well-documented user consumption and water quality requirements, accurate use data, estimates of the environmental flow volumes needed to maintain ecosystem resilience, characterization of both point and non-point sources of pollution and the quality of the receiving waters, and the extensive engagement of all water users and other pertinent stakeholders.

Providing incentives to improve demand management efficiencies has proven highly effective in augmenting natural water supplies. WRAs should consider new capacities to use non-conventional water supplies and new technologies to augment existing supplies. A comprehensive WRA must also include social and economic considerations as well as ecosystem needs and contributions.

If climate change follows the projected scenarios, we can expect more erratic weather in the future, including increased variability in precipitation, which will threaten crop yields in both developed and developing countries, while placing more than 2.8 billion people at risk of water shortage. Understanding all aspects of the hydrological cycle is critical if our society is to be able to cope with the many changes we observe.

Many nations are adapting emerging technical practices to secure and protect their existing natural water resources and use local knowledge as part of sustainable resource development

References and Websites

Abderrahman, W. A. 2003. Should intensive use of non-renewable groundwater resources always be rejected? R. Llamas and E. Custodio (eds) *Intensive Use of Groundwater: Challenges and Opportunities*. Lisse, The Netherlands, Balkema.

Aertgeerts, R. and Angelakis, A. 2003. *Health Risks in Aquifer Recharge using Reclaimed Wastewater: State of the Art Report*, SDE/WSH/03.08. (www.who.int/water_sanitation_health/wastewater/wsh0308/en/index.html lvOct05)

Aggarwal, P. K. and Kulkarni, K. M. 2003. *Isotope Techniques in Hydrology: Role of International Atomic Energy Agency*, Advances in Hydrology (Proc. Int. Conf. Water and Environment 2003, Bhopal, India), New Delhi, Allied Publishers Pvt. Ltd. pp. 361–69.

Aggarwal, P. K., Froehlich, K., Kulkarni, K. M. and Gourcy, L. L. 2004. *Stable Isotope Evidence for Moisture Sources in Asian Summer Monsoon under Present and Past Climate Regimes*. Geo. Res. Letters, Vol. 31.

AFP (L'Agence France-Presse). 2004. China warns of 'ecological catastrophe' from Tibet's melting glaciers, News article. *Terradaily*.

Arendt, A. A., Echelmeyer, K. A., Harrison, W. D., Lingle, C. S. and Valentine, V. B. 2002. Rapid wastage of Alaska glaciers and their contribution to rising sea level, *Science*, Vol. 297, No. 5580, pp. 382–86.

Arnell, N. W. 2004. Climate change and global water resources: SRES emissions and socio-economic scenarios, *Global Environmental Change*, Vol. 14, No. 1, pp. 31–52.

Asano, T. (ed.). 1998. Wastewater Reclamation and Reuse, *Water Quality Management Library Volume 10*. Boca Raton, Florida, CRC Press.

Asano, T. and Cotruvo, J. A. 2004. A Review: Groundwater Recharge with Reclaimed Municipal Wastewater: Health and Regulatory Considerations, *Water Research*, Vol. 38, pp. 1941–51. (www.med-reunet.com/docs/asano.pdf lvOct05)

Asano, T. and Leavine, A. D. 2004. Recovering sustainable water from wastewater. *Environmental Science and Technology*, June, pp. 201–08.

Awerbuch, L. 2004. Status of desalination in today's world. S. Nicklin (ed.) *Desalination and Water Re-use*. Leicester, UK, Wyndeham Press, pp. 9–12.

Bashkin, V and Radojevic, M. 2001. A Rain Check on Asia. *Chemistry in Britain*, No. 6. (Online at: www.chemsoc.org/chembytes/ezine/2001/bashkin_jun01.htm lvOct05)

Bergkamp, G., Orlando, B. and Burton, I. 2003. *Change: Adaptation of water resources management to climate change*. Gland, Switzerland, IUCN.

Bhattacharya, S. K., Froehlich, K., Aggarwal, P. K. and Kulkarni, K. M. 2003. Isotopic Variation in Indian Monsoon Precipitation. Records from Bombay and New Delhi, *Geophysical Research Letters*, Vol. 30, No. 24, p. 2285.

Blomquist, W., Dinar, A. and Kemper, K. 2005. Comparison of Institutional Arrangements for River Basin Management in Eight Basins. World Bank Policy Research Working Paper 3636, June 2005. (wdsbeta.worldbank.org/external/default/WDSContentServer/IW3P/IB/2005/06/14/000016406_20050614124517/Rendered/PDF/wps3636.pdf lvOct05).

Bullock, A. and Acreman, M. 2003. The role of wetlands in the hydrologic cycle. *Hydrology and Earth System Sciences*, Vol. 7, No. 3, pp. 358–89.

Burke, J. and Moench, M. 2000. *Groundwater and Society: Resources, Tensions and Opportunities*. New York, UNDESA (United Nations Department of Economic and Social Affairs) E.99.II.A.1.

CSIRO. 2004. Returning the lifeblood to rivers. A drought experiment - environmental flows resurrect irrigation country. How healthy river habitats suffer from altered flows. Clever planning and management approaches. Where wild things are dammed, *ECOS magazine*, Issue 122, pp. 11–19, CSIRO Publishing. (www.publish.csiro.au/ecos/index.cfm?sid=10&issue_id=4847 lvOct05)

Daughton, C. G. 2004. Non-regulated water contaminants: emerging research, *Environmental Impact Assessment Review*, Vol. 24, pp. 711–32.

Digout, D. 2002. *Variations in River Runoff by Continent through Most of the 20th Century – Deviations from Average Values*. (UNEP based on source material in Shiklomanov (1999) and UNESCO 1999 at www.unep.org/vitalwater/07.htm lvOct05).

Driscoll, C. T., Lawrence, G. B., Bulger, A. J., Butler, T. J., Cronan, C. S., Eagar, C., Lambert, K. F., Likens, G. E., Stoddard, J. L. and Weathers, K. C. 2001. *Acid Rain Revisited: Advances in scientific understanding since the passage of the 1970 and 1990 Clean Air Act Amendments*, Hubbard Brook Research Foundation. Science Links Publication, Vol. 1, No. 1. (www.hbrook.sr.unh.edu./hbfound/report.pdf lvOct05)

Dyson, M., Bergkamp, G. and Scanlon, J. (eds) 2003. *Flow. The Essentials of Environmental Flows*. Gland, Switzerland and Cambridge, UK, IUCN.

Dyurgerov, M. 2003. Mountain and subpolar glaciers show an increase in sensitivity to climate warming and intensification of the water cycle, *Journal of Hydrology*, Vol. 282, pp. 164–76.

EC (European Commission). 2003. Proposal for a Directive of the European Parliament and of the Council on the 'Protection of Groundwater against Pollution'. (europa.eu.int/eur-lex/en/com/pdf/2003/com2003_0550en01.pdf lvOct05).

——. 2000. Directive 2000/60/EC of the European Parliament and of the Council Establishing a Framework for the Community Action in the Field of Water Policy, Brussels. (europa.eu.int/comm/environment/water/water-framework/index_en.html lvOct05).

Entekhabi, D., Njoku, E. G., Houser, P., Spencer, M., Doiron, T., Kim, Y., Smith, J., Girard, R., Belair, S., Crow, W., Jackson, T. J., Kerr, Y.H., Kimball, J. S., Koster, R., McDonald, K. C., O'Neill, P. E., Pultz, T., Running, S. W., Shi, J., Wood, E., van Zyl, J. 2004. The Hydrosphere State (Hydros) Satellite Mission: An Earth System Pathfinder for Global Mapping of Soil Moisture and Land Freeze/Thaw. IEEE Trans. *Geoscience And Remote Sensing*, Vol. 42, No. 10, pp. 2184–95. (hydros.gsfc.nasa.gov/pdf/TGARSHydros.pdf lvOct05)

Environment Canada. 2005a. *Water – The Transporter* (www.atl.ec.gc.ca/udo/mem.html lvOct05)

——. 2005b. *The Bottom Line – Water Conservation*. (www.ec.gc.ca/water/en/manage/effic/e_bottom.htm lvOct05)

——. 2005c. Acid Rain (www.ec.gc.ca/acidrain/ lvOct05)

Falkenmark, M. and Rockstrom, J. 2004. *Balancing Water for Humans and Nature: The New Approach in Ecohydrology*, Earthscan, UK.

FAO. 2002. *World Agriculture: Towards 2015/2030. Summary Report*, FAO, Rome. (available at: www.fao.org/documents/show_cdr.asp?url_file=/docrep/004/y3557e/y3557e11.htm lvOct05)

——. 2005. Geo-referenced database on African dams. (www.fao.org/ag/agl/aglw/aquastat/damsafrica/index.stm lvOct05).

——. 2004. Personal communication from FAO containing National Downstream Volumes datafile, 6 Sept 2004, FAO AQUASTAT staff, Rome. (Supplemented 18 Aug 2005 with limited groundwater data available).

——. 2003a. *Review of World Water Resources by Country*. Water Report 23. (ftp://ftp.fao.org/agl/aglw/docs/wr23e.pdf lvOct05).

——. 2003b. *Groundwater Management: The Search for Practical Approaches*, FAO Water report 25 (ftp://ftp.fao.org/agl/aglw/docs/wr25e.pdf lvOct05)

FAO-Aquastat. 2005. (www.fao.org/ag/agl/aglw/aquastat/main/ lvOct05). Groundwater to surface water renewal ratio calculated from total annual internally generated groundwater and surface water volumes in the Aquastat database.

Figueras, C., Tortajada, C. and Rockstrom, J. 2003. *Rethinking Water Management*. UK, Earthscan.

Foster, S., Hirata, R., Gomes, D., D'Elia, M. and Paris, M. 2002. *Groundwater Quality Protection - a guide for water utilities, municipal authorities, and environment agencies*. GWMATE in association with GWP, 112 p. (DOI: 10.1596/0-8213-4951-1) (www-wds.worldbank.org/servlet/WDSContentServer/WDSP/IB/2002/12/14/000094946_02112704014826/Rendered/PDF/multi0page.pdf lvOct05).

Foster, S. and Kemper, K. 2004. *Sustainable Groundwater Management: Concepts and Tools*, World Bank GW MATE Briefing Note Series Profile (list of all fifteen briefing notes). (siteresources.worldbank.org/INTWRD/903930-1112347717990/20424234/BN_series_profileMay04.pdf lvOct05).

Fountain, A. and Walder, J. 1998. Water Flow through Temperate Glaciers, *Review of Geophysics*, Vol. 36, No. 3, pp. 299–328.

Frauenfelder, R., Zemp, M., Haeberli, W. and Hoelzle, M. 2005. World-Wide Glacier Mass Balance Measurements: Trends and First Results of an Extraordinary Year in Central Europe, *Ice and Climate News*, No. 6. pp. 9–10. (clic.npolar.no/newsletters/archive/ice_climate_2005_08 _no_06.pdf lvOct05)

GEMS/WATER. 2005. *2004 State of the UNEP GEMS/Water Global Network and Annual Report.* (www.gemswater.org/ common/pdfs/gems_ar_2004.pdf lvOct05)

Gleick, P. H. (ed.). 1993. *Water in Crisis: A Guide to the World's Freshwater Resources.* New York, Oxford University Press.

GLIMS (Global Land Ice Measurements from Space). 2005. *Project Description: Global Land Ice Measurements from Space* (nsidc.org/data/glims/ lvOct05).

Govt. South Africa. 1997. White Paper on Water Policy (Section B: New National Water Policy) CH 5. Water Resource Policy, Subchapter 5.2 Priorities – The Basic Needs and Environmental Reserve and International Obligations. (www.polity.org.za/html/govdocs/ white_papers/water.html#Contents lvOct05).

Govt. Western Australia. 2005. (portal.environment.wa.gov.au/ portal/page?_pageid=55,34436&_dad=portal&_schema= PORTAL lvOct05).

GRDC (Global Runoff Data Center). 2005. (grdc.bafg.de/ servlet/is/1660/, grdc.bafg.de/servlet/is/943/ lvOct05).

Greenhalgh, S. and Sauer, A. 2003. *Awakening the Dead Zone: An Investment for Agriculture, Water Quality, and Climate Change*, WRI Issue Brief. (pdf.wri.org/hypoxia.pdf lvOct05).

GWP. 2005a. *Efficiency in Water Use – Managing Demand and Supply.* (gwpforum.netmasters05.netmasters.nl/en/ content/toolcategory_453AAC8B-A128-11D5-8F08- 0002A508D0B7.html lvOct05)

——. 2005b. *Water Resources Assessment – Understanding Water Resources and Needs.* (gwpforum.netmasters05. netmasters.nl/en/content/toolcategory_5E1CD3DC-3B4A- 4D82-B476-82DEF0EEE0186.html lvOct05).

GW MATE. 2003. Urban Wastewater as Groundwater Recharge: evaluating and managing the risks and benefits, Sustainable Groundwater Management: Concepts and Tools, World Bank GW Mate Briefing Series Note 12. (siteresources.worldbank.org/INTWRD/903930- 1112347717990/20424258/BriefingNote_12.pdf lvOct05)

Haider, S. S., Said, S., Kothyari, U. C. and Arora, M. K. 2004. Soil Moisture Estimation Using Ers 2 Sar Data: A Case Study in the Solani River Catchment. *Journal of Hydrological Science*, pp. 323–34. (www.extenza- eps.com/extenza/ loadPDF?objectIDValue=34832 lvOct05).

Hock, R., Jansson, P. and Braun, L. 2005. Modelling the Response of Mountain Glacier Discharge to Climate Warming. U. M. Huber, H. K. M. Bugmann and M. A. Reasoner (eds), *Global Change and Mountain Regions –*

An Overview of Current Knowledge. Series: Advances in Global Change Research. Vol. 23, Springer.

IAEA. 2002. Isotope studies in large river basins: A new global research focus. *EDS* 83, pp. 613–17.

IGRAC. 2005. Global Groundwater Information System Database. (igrac.nitg.tno.nl/ggis_map/start.html lvOct05).

——. 2005a. Arsenic in Groundwater Worldwide (igrac.nitg.tno.nl/arsmain.html lvOct05).

——. 2005b. Fluoride in Groundwater Worldwide (igrac.nitg.tno.nl/flumain.html lvOct05).

——. 2004. Global Groundwater Regions. (igrac.nitg.tno.nl/pics/region.pdf lvOct05).

IPCC. 2004. *Expert Meeting on the Science to Address UNFCCC Article 2 including Key Vulnerabilities*, Buenos Aires, Argentina 18-20 May 2004, Short Report. (www.ipcc.ch/wg2sr.pdf lvOct05).

——. (Intergovernmental Panel on Climate Change). 2001. *Third Assessment Report: Climate Change 2001.* (www.ipcc.ch/pub/reports.htm lvOct05).

IRHA (International Rainwater Harvesting Association). 2004. *How RHW benefits water resources management (unpublished)* (www.irha-h2o.org – lvOct05).

Jackson, T. 2004. How Wet's Our Planet? Scientists want to be able to measure soil moisture everywhere, every day! Agric. Res. Vol. 52, No. 3, pp. 20–22. (www.ars.usda.gov/is/ AR/archive/mar04/planet0304.htm?pf=1 lvOct05).

Jansson, P., Hock, R. and Schneider, T. 2003. The Concept of Glacier Storage – A Review. *Journal of Hydrology*, Vol. 282, Nos. 1–4, pp. 116–29.

Jousma, G. and Roelofsen, F. J. 2003. *Inventory of existing guidelines and protocols for groundwater assessment and monitoring*, IGRAC. (igrac.nitg.tno.nl/pics/ inv_report1.pdf lvOct05).

Lakenet. 2005. (www.worldlakes.org lvOct05).

Lazarova, L. 2001. *Recycled Water: Technical-Economic Challenges for its Integration as a Sustainable Alternative Resource.* Proc. UNESCO Int'l. Symp. *Les frontiéres de la gestion de l'eau urbaine: impasse ou espoir?* Marseilles, 18–20 June 2001.

Lazarova, V., Levine, B., Sack, J., Cirelli, C., Jeffrey, P., Muntau, H., Salgot, M. and Brissaud, F. 2001. Role of water reuse for enhancing integrated water management in Europe and Mediterranean countries. *Water Science and Technology*, Vol. 43, No. 10, pp, 23–33.

Lehner, B. and P. Döll. 2004. Development and validation of a global database of lakes, reservoirs and wetlands. *Journal of Hydrology*, Vol. 296, Nos. 1–4, pp. 1–22.

Lenton, R. 2004. Water and climate variability: development impacts and coping strategies, *Water Science and Technology*, Vol. 49, No. 7, pp. 17–24.

Llamas, R. and Custodio, E. (eds). 2003. *Intensive use of groundwater, Challenges and Opportunities.* Balkema.

Mahnot, S. C., Sharma, D. C., Mishra, A., Singh, P. K. and Roy, K. K. 2003. *Water Harvesting Management*, Practical Guide Series 6, V. Kaul (ed.). SDC/Intercooperation Coordination Unit. Jaipur, India.

Mantovani, P., Asano, T., Chang, A. and Okun, D. A. 2001. Management Practices for Non-potable Water Reuse. Water Environment Research Foundation Report 97-IRM-6.

Meybeck, M. 1995. Global distribution of lakes. A. Lerman, D. M. Imboden and J. R. Gat (eds), *Physics and Chemistry of Lakes*, Springer, Berlin, pp. 1–36.

Meybeck, M., Chapman, D. and Helmer, R. (eds). 1989. *Global Freshwater Quality: A First Assessment.* Blackwell Ref. Oxford, UK.

Mitchell, T. D., Hulme, M. and New, M. 2002. *Climate Data for Political Areas.* Area, Vol. 34, pp. 109–12. (www.cru.uea.ac.uk/cru/data/papers/mitchell2002a.pdf lvOct05).

Morris, B. L., Lawrence, A. R. L., Chilton, P. J. C., Adams, B., Calow, R. C. and Klinck, B. A. 2003. Groundwater and its Susceptibility to Degradation. A Global Assessment of the Problem and Options for Management. Early Warning and Assessment Report Series, RS. 03-3. United Nations Environment Programme/DEWA, Nairobi, Kenya.

NSIDC (National Snow and Ice Data Center). 1999. update 2005. World glacier inventory. World Glacier Monitoring Service and National Snow and Ice Data Center/World Data Center for Glaciology. Boulder, CO. Digital media. (nsidc.org/data/docs/noaa/g01130_glacier_inventory/ lvOct05).

New, M., Hulme, M. and Jones, P. D. 1999. Representing Twentieth Century Space-Time Climate Variability. Part 1: Development of a 1961–1990 Mean Monthly Terrestrial Climatology. *Journal of Climate*, Vol. 12, pp. 829–56. (ams.allenpress.com/amsonline/?request=get-abstract &issn=1520-0442&volume=012&issue=03&page=0829 lvOct05).

Njoku, E. 2004. *AMSR-E/Aqua Daily L3 Surface Soil Moisture, Interpretive Parms, and QC EASE-Grids V001*, March to June 2004. Boulder, CO, US: National Snow and Ice Data Center. Digital media – updated daily.

Njoku, E., Chan, T., Crosson, W. and Limaye, A. 2004. Evaluation of the AMSR-E data calibration over land. *Italian Journal of Remote Sensing*, Vol. 29, No. 4, pp. 19–37. (nsidc.org/data/docs/daac/ae_land3_l3_soil_moisture.gd.html lvOct05).

OECD. 1996. Guidelines for aid agencies for improved conservation and sustainable use of tropical and sub- tropical wetland. OECD Development Assistance Committee: Guidelines on Aid and Environment. No. 9.

Pandey, D. N., Gupta, A. K. and Anderson, D. M. 2003. Rainwater harvesting as an adaptation to climate change, *Current Science*, Vol. 85, No. 1, pp. 46–59. (www.irha- h2o.org/doc/text/pandey00.pdf lvOct05).

Pereira, L., Cordery, I. and Lacovides, L. 2002. Coping with water scarcity, IHP-VI Tech. Documents in Hydrology No. 58, UNESCO.

Peters, N. E. and Meybeck, M. 2000. Water quality degradation effects on freshwater availability: Impacts of human activities, Int'l Water Res. Assoc., *Water International*, Vol. 25, No. 2, pp. 185–93.

Peters, N. E. and Webb, B. 2004. Personal communication – Water quality parameters to measure related to pollution.

Pyne, R. D. G. and Howard, J. B. 2004. Desalination/Aquifer Storage Recovery (DASR): a cost-effective combination for Corpus Christi, Texas, *Desalination*, Vol. 165, pp. 363-67. (www.desline.com/articoli/5744.pdf lvOct05).

Rees, G. and Demuth, S. 2000. The application of modern information system technology in the European FRIEND project. Moderne Hydrologische Informations Systeme. *Wasser und Boden*, Vol. 52, No. 13, pp. 9-13.

Rekacewicz, P. 2002. *Industrial and Domestic Consumption Compared with Evaporation from Reservoirs.* (UNEP based on Shiklomanov (1999) and UNESCO 1999 at www.unep.org/vitalwater/15.htm lvOct05).

Revenga, C. and Mock, G. 2000. *Dirty Water: Pollution Problems Persist.* World Resources Institute Program, Pilot Analysis of Global Ecosystems: Freshwater Systems. (earthtrends.wri.org/pdf_library/features/wat_fea_dirty.pdf lvOct05).

Reynolds, K. 2003. Pharmaceuticals in Drinking Water Supplies, *Water Conditioning and Purification Magazine*, Vol. 45(6). (www.wcp.net/column.cfm?T=T&ID=2199 lvOct05).

Robock, A. and Vinnikov, K. Y. 2005. Global Soil Moisture Data Bank (climate.envsci.rutgers.edu/soil_moisture/ lv Jul2004).

Robock, A., Vinnikov, K. Y., Srinivasan, G., Entin, J. K., Hollinger, S. E., Speranskaya, M. A., Liu, S. and Namkhai, A. 2000. The Global Soil Moisture Data Bank. *Bull. Amer. Meteorol. Soc.*, Vol. 81, pp. 1281-99.

Rodda, J.C. 1998. *Hydrological Networks Need Improving!* In: H. Zebedi (ed.), *Water: A Looming Crisis?* Proc. Int. Conf. on World Water Resources at the Beginning of the 21st century. Paris, UNESCO/IHP.

Schiffler, M. 2004. Perspectives and challenges for desalination in the 21st century. *Desalination*, Vol. 165, pp. 1-9.

Shao, X., Wang, H. and Wang, Z. 2003. Interbasin transfer projects and their implications: A China case study, *International Journal of River Basin Management*, Vol. 1, No. 1, pp. 5-14. (www.jrbm.net/pages/archives/JRBMn1/Shao.PDF lv Oct05)

Shiklomanov, I. A. 1999. World Freshwater Resources: World Water Resources and their Use. webworld.unesco.org/water/ihp/db/shiklomanov/index.shtml›

——. 1997. *Assessment of Water Resources and Availability in the World.* In Comprehensive Assessment of the Freshwater Resources of the World. Stockholm Environment Institute.

Shiklomanov, I. A. and Rodda, J. C. 2003. *World Water Resources at the Beginning of the 21st Century.* Cambridge, UK, Cambridge University Press.

Stahl, K and Hisdal, H. 2004. Hydroclimatology. Tallaksen, L. and van Lanen, H. (eds) *Hydrological Drought - Processes and Estimation Methods for Streamflow and Groundwater.* (Developments in Water Science, 48). p. 22. New York, Elsevier. Reprinted with permission from Elsevier.

Steenvorden, J. and Endreny, T. 2004. Wastewater Re-use and Groundwater Quality, IAHS Pub. 285.

Tallaksen, L. M. and Van Lanen H. A. J. (eds), *Developments in Water Science*, 48, Elsevier, The Netherlands.

Taylor, R. and Smith, I. 1997. *State of New Zealand's Environment 1997.* Wellington, New Zealand: The Ministry for the Environment. Revenga and Mock 2000.

UNEP. 2005. *Sourcebook of Alternative Technologies for Freshwater Augmentation in Africa.* ITEC. (www.unep.or.jp/ietc/Publications/TechPublications/TechPub-8a/index.asp lvOct05).

——. 2003. *Groundwater and its Susceptibility to Degradation. A global assessment of the problems and options for management,* UNEP/DEWA, Nairobi. (www.unep.org/DEWA/water/groundwater/groundwater_report.asp lvOct05).

UNEP/GEMS (United Nations Environment Program Global Environment Monitoring System/Water). 1995. *Water Quality of World River Basins.* Nairobi, Kenya: UNEP. Revenga and Mock 2000.

UNESCO. 2004. WHYMAP. *Groundwater Resources of the World.* Map 1:50 m. Special edn, August, BGR Hanover/UNESCO, Paris.

——. 2000. *Catch the water – where it drops. Rain water harvesting and artificial recharge to ground water. A guide to follow.* IHP program document.

UNESCO and World Bank. Forthcoming 2006. Non-renewable groundwater resources, a guidebook on socially sustainable management for policy matters.

US Drought Mitigation Center. 2005. *Drought Map – April 2004.* (www.drought.unl.edu/pubs/abtdrmon.pdf lvOct05).

US EPA (United States Environmental Protection Agency). 2004. *Guidelines for Water Reuse,* EPA 625/R-04/108. (www.epa.gov/ORD/NRMRL/pubs/625r04108/625r04108.htm lvOct05).

US GS (United States Geological Survey). 2005. *Reclaimed wastewater: Using treated wastewater for other purposes.* (ga.water.usgs.gov/edu/wwreclaimed.html lvOct05).

US NRC (United States National Research Council) 2004, *Review of the desalination and water purification technology roadmap.* (www.nap.edu/books/0309091578/html/R1.html lvOct05).

——. 2000. *Issues in the Integration of Research and Operational Satellite Systems for Climate Research: Part I. Science and Design, Part 6, Soil Moisture.* pp. 68–81. Commission on Physical Sciences, Mathematics, and Applications; Space Studies Board. National Academy Press, Washington DC. (print.nap.edu/pdf/0309069858/pdf_image/68.pdf lvOct05).

——. 1998. *Issues in Potable Reuse: The Viability of Augmenting Drinking Water Supplies with Reclaimed Water.* Washington DC, National Academy Press.

Vrba, J., 1985. Impact of domestic and industrial wastes and agricultural activities on groundwater quality. *Hydrogeology in the service of man*, Vol. 18, No. 1, pp. 91–117, IAH Memoirs of the 18th Congress, Cambridge, UK.

Vrba, J. and Zaporozec, A. (eds). 1994. *Guidebook on mapping groundwater vulnerability.* Vol. 16, 131. IHP-IAH International Contribution to Hydrogeology, Verlag H. Heise, Germany.

Webb, B.W. 1999. *Water quality and pollution.* Pacione, M. (ed.) *Applied Geography: Principles and Practice*, pp. 152-71. London and New York, Routledge.

Wiegel, S., Aulinger, A., Brockmeyer, R., Harms, H., Loeffler, J., Reincke, H., Schmidt, R., Stachel, B., von Tuempling, W. and Wanke, A. 2004. Pharmaceuticals in the River Elbe and its Tributaries, *Chemosphere*, Vol. 57, pp. 107-26.

WHO. 2005. Wastewater use. (www.who.int/water_sanitation_health/wastewater/en/ lvOct05).

Wolf, A. T., Natharius, J. A., Danielson, J. J., Ward, B. S. and Pender, J. K. 2002. International river basins of the world. International Journal of Water Resources Development, Vol. 15, No. 4, pp. 387-427. www.transboundarywaters.orst.edu/publications/register/tables/IRB_table_4.html

World Bank. 2005. *India's Hydrology Project Phase II.* (web.worldbank.org/external/projects/main?pagePK=104231&piPK=73230&theSitePK=40941&menuPK=228424&Projectid=P084632 lvOct05).

WMO. 2005. Analysis of data exchange problems in global atmospheric and hydrological networks, WMO/TD No. 1255, GCOS No. 96. (www.wmo.ch/web/gcos/Publications/gcos-96.pdf lvOct05).

——. 2004. Soil Moisture – Details of Recommended Variables. (www.wmo.ch/web/gcos/terre/variable/slmois.html lv-Oct05).

WWF. 2003. *Managing Rivers Wisely – Lessons Learned from WWF's Work for Integrated River Basin Management.* (www.panda.org/about_wwf/what_we_do/freshwater/our_solutions/rivers/irbm/cases.cfm lvOct05).

Zalewski, M., Janauer, G.A. and Jolankai, G. 1997. Ecohydrology. A New Paradigm for the Sustainable Use of Aquatic Resources. IHP-V. Technical Documents in Hydrology. No. 7. UNESCO, Paris.

Zektser, I.S. and Everett, L.G. 2004. *Groundwater Resources of the World and their Use,* IHP-VI, Series on Groundwater No. 6, UNESCO. (Section 6.4 – Human Activities impact on groundwater resources and their use).

We must treat each and every swamp, river basin, river and tributary, forest and field with the greatest care, for all these things are the elements of a very complex system that serves to preserve water reservoirs – and that represents the river of life.

Mikhail Gorbachev

CHAPTER 5

Coastal and Freshwater Ecosystems

By
UNEP
*(United Nations
Environment
Programme)*

Betsiboka River, Madagascar

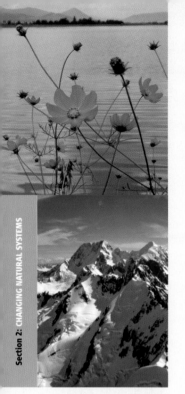

Section 2: CHANGING NATURAL SYSTEMS

Key messages:

Coastal and freshwater ecosystems are deteriorating in many areas and at a faster rate than any other ecosystem. Such changes are caused by intertwined factors, making it difficult to identify the problems early on. While progress in integrating these various factors in managing water and ecosystems has been made in some places, the majority of the world and its inhabitants increasingly suffers from a lack of priority given to environmental protection.

■ Humans depend upon healthy aquatic ecosystems for drinking water, food security and a wide range of environmental goods and services. Aquatic biodiversity is also extremely rich, with high levels of endemic species, and is very sensitive to environmental degradation and overexploitation.

■ Aquatic ecosystems and species are deteriorating rapidly in many areas. This is having an immediate impact on the livelihoods of some of the world's most vulnerable human communities by reducing protein sources for food, availability of clean water, and potential for income generation.

■ People in regions with highly variable climatic conditions are particularly vulnerable to droughts and floods and the resulting deteriorating condition of freshwater ecosystems. Coastal lowland areas, where population densities are usually very high and coastal habitats are fragile, are most likely to be affected by sea level rise in future.

■ The conservation of biodiversity (species, habitats and ecosystem functions) must become an integral part of all water resource management programmes. This will assist poverty reduction strategies by ensuring the sustainability of aquatic ecosystems for future generations.

■ Ecosystem approaches constitute a fundamental element of Integrated Water Resource Management (IWRM) and are essential for safeguarding and balancing the needs and requirements of water resources among different stakeholder groups and ecosystems. Ecosystem approaches are the subject of global and regional targets and policy initiatives, but they have yet to be implemented in practice. This requires awareness raising, tools and methodologies to monitor and negotiate the trade-offs involved in such broad-scale approaches.

■ Our understanding of the properties and functions of many aquatic ecosystems is seriously hampered by inadequate data. Enhanced monitoring efforts are required to provide a better assessment of the status, conditions and trends of global water ecosystems, habitats and species.

Top to bottom:
Ticti reservoir, Mexico

Franz Joseph Glacier,
New Zealand

Heavy rains in the
province of Misiones,
Argentina, carry off
significant quantities of
ferruginous earth into the
River Uruguay

Part 1. Ecosystems and their Capacity to Provide Goods and Services

Human population growth and the expansion of economic activities are collectively placing huge demands on coastal and freshwater ecosystems

The majority of us live in temperate and subtropical regions centred around the coast or inland water systems. Coastal waters, rivers, lakes, wetlands, aquifers and other inland water systems such as swamps and fens have in consequence been subjected to disproportionate human-induced pressures. These include construction along coastlines for harbours and urban expansion, alteration of river systems for navigation and water storage, drainage of wetlands to increase farmland, overexploitation of fisheries, and multiple sources of pollution. Human population growth and the expansion of economic activities are collectively placing huge demands on coastal and freshwater ecosystems. Water withdrawals, for instance, have increased sixfold since the 1900s, which is twice the rate of population growth (WMO, 1997). In addition, the quality of many water bodies is declining due to increased pollution from agriculture, industry, tourism, urban runoff and domestic sewage.

Desertification is also spreading as a consequence of the misuse of water resources, not only in Africa and Central Asia, but increasingly in other regions, such as in California and southern Europe. The dramatic shrinking of the Aral Sea in Central Asia and its consequences for biodiversity and human well-being have been well documented (UNEP, 2004b; Kreutzberg-Mukhina, 2004). There are many other water crises that have received less attention, such as the serious soil erosion and groundwater depletion occurring in parts of Spain and the eutrophication of many coastal waters as a result of intensive farming. In other regions, the problem may soon be one of too much water, threatening many low-lying coastal and floodplain areas. Predictions of the impacts of melting ice caps and increased discharge from Arctic rivers due to global warming remain uncertain, although it is clear that they will change the fragile Arctic Ocean ecosystem, with potentially devastating consequences further afield, especially along often highly populated coastlines (ACIA, 2004).

While many of the world's coastal and freshwater ecosystems are continuing to deteriorate at alarming rates, the reversal of these trends and the improvement of water quality in other areas indicate that this decline is neither inevitable nor always irreversible. The management of water and land resources requires a comprehensive understanding and careful consideration of ecosystem functions and interactions. The application of such knowledge in an integrated approach to land use and water management is often referred to as an 'ecosystem approach', and such a holistic response to the challenges facing the world's water resources is at the heart of international agreements and programmes like the Convention on Biological Diversity (CBD), the Global Programme of Action (GPA) for the Protection of the

Marine Environment from Land-based Activities and the World Summit on Sustainable Development (WSSD).

The ecosystem approach, a key element of integrated water resources management (IWRM) (GWP, 2003), is a strategy for the integrated management of land, water and living resources which promotes conservation and sustainable use in an equitable way (CBD, 2000). There is no single way to implement the ecosystem approach, as it depends upon local, provincial, national, regional and global conditions. **Box 5.1** discusses one of the many systems in which an ecosystem approach should be implemented to solve a current ecosystem crisis.

IWRM is a systematic participatory planning and implementation process for the sustainable management of water, land and coastal resources, which promotes coordinated development and is based on credible science. It involves the participation of stakeholders who determine equitable resource allocation and the sharing of economic benefits and monitoring within set objectives in order to ensure the sustainability of vital ecosystems. It is also a process that promotes the coordinated development and management of water, land and related resources in order to maximize the resultant economic and social welfare in an equitable manner without compromising the sustainability of vital ecosystems (GWP, 2000).

Integrated water resources management considers the following:

■ **The hydrological cycle in its entirety:** downstream and upstream interests are taken into account (basin-wide, also across national borders), as well as surface and groundwater sources and, most importantly, rainfall.

There is a growing recognition of the need for a sustained global effort to meet the immense challenges of managing the Earth's water resources

■ **The full range of sectoral interests:** integrated development and management implies close coordination between institutions that are often sectorally focused, the involvement of stakeholders in decision-making, and taking into account those stakeholders without a voice (such as the environment).

■ **Future needs:** as legitimate claims to the water resource, such as future generations – sustainability (Institute of Water and Sanitation Development of Zimbabwe, 1997).

■ **The management of water as a resource:** as well as the governance framework for provision of water services to stakeholders.

There is a growing recognition of the need for a sustained global effort to meet the immense challenges of managing the Earth's water resources. At the 2002 WSSD, participating countries committed themselves to halving the proportion of people who lack access to safe drinking water and sanitation by 2015 (Millennium Development Goal Target 10) and significantly reducing the rate of biodiversity loss in aquatic ecosystems by 2010. Reconciling these two goals constitutes a major challenge. The implementation of IWRM schemes on regional and local scales; the increasing use of ecosystem approaches focusing on river basins and their interaction with coastal zones; the decommissioning of dams in North America and Europe; and the many different river and wetland restoration projects taking place throughout the world all suggest that these commitments are starting to be taken seriously, although change is slow and not happening in every part of the world.

Indicators of ecosystems processes and functions are essential to the proper assessment of watershed resources by evaluating the pressures, state, driving forces and responses to change. Unfortunately, the necessary knowledge and data needed to develop and interpret indicators are often lacking. Data remain seriously incomplete and unharmonized at a global level, making detection and resolution of problems extremely difficult (see **Chapter 13**). Even more worrying is the fact that hydrographic and water quality monitoring networks have deteriorated in several parts of the world, further hampering the accurate assessment of global water resources. The Global Environment Monitoring System for freshwater (GEMS/Water), led by the United Nations

Environmental Programme (UNEP), maintains a water quality database with information from over 1,500 monitoring stations covering 112 watercourses from more than eighty countries. However, most stations contain only patchy and sporadic information, making the detection of long-term trends in water quality difficult. Information on groundwater resources is even less complete because of the difficulty and the costs of obtaining accurate measurements. Furthermore, there are few systematic epidemiological studies that allow us to understand the impacts of low quality or contaminated water supplies on human health and well-being (see **Chapter 6**). Existing data and information on aquatic species and the extent and condition of their habitats are also limited and fragmented. Although there are some groups (e.g. water-birds and amphibians) for which trend data do exist, they remain the exception. Given the need for information to manage resources in an integrated fashion, and the reality of the dearth of data and monitoring, we have to depend on indicators that measure drivers of change, which are currently quite clear and, for the most part, easier to assess and monitor. This is especially true in those countries where resources for extensive fieldwork and capacity are limited. For example, using data on the extent of agriculture in a watershed, or the size and location of dams, we can draw some conclusions about the relative degree of alteration or stress affecting a system. These geospatial indicators are often called proxies or surrogates, because they are indicators of current threat and give only indirect information about actual ecological integrity.

It is clear that if we continue to ignore ecosystem processes and functions, human activities will lead to the continuing degradation of coastal and freshwater ecosystems, as well as the loss of biodiversity and a consequential decline in human well-being.

This chapter addresses the current condition of coastal and freshwater ecosystems and their capacity to deliver an array of life-supporting ecosystem goods and services. It looks critically at several different management approaches and policies, and concludes by identifying some of the challenges that human society must face in attempting to achieve the Millennium Development Goals (MDGs) related to water resources and other international targets, such as those for biodiversity and climate change.

BOX 5.1: LAKE VICTORIA: AN ECOSYSTEM IN DECLINE

Lake Victoria is the second largest lake in the world. It supports a very productive freshwater fishery with annual fish yields exceeding 500,000 tonnes, with a value of US $400 million. In addition, the lake provides freshwater for irrigation, hydroelectric power, recreation and transport (see **Chapter 14**).

In some parts of the basin, its population density is well over 100 people per square kilometre (km[2]) (Cohen et al., 1996) and has been increasing at the rate of 3 to 4 percent per year. The lake faces considerable pressure from a variety of natural or anthropogenic causes. It has undergone enormous environmental changes within the last forty years, caused by human activities such as overfishing, siltation from deforested watersheds, erosion from poor agricultural practices, introduction of alien species and industrial pollution. Seasonal variability attributed to climate change has also been noted. These combinations of natural and anthropogenic causes have led to rapidly evolving changes in the lake that seriously threaten its ecosystem function and

dependent livelihoods. Three immediate causes of eutrophication so far identified include enhanced effluent discharge, from untreated municipal sewage, runoff and storm water, chemical pollutions from industries (such as small-scale mining in some parts of Tanzania, Panpaper Limited in Kenya, etc.) and agrochemicals from farms, including the expanding flower farms currently threatening internationally important wetland sites, such as Uganda's Lutembe Bay, which hosts large numbers of wintering waterbirds of global significance, including almost the entire population of the White-winged Tern (*Chlidonias leucopterus*).

Today, nearly half of the lake experiences prolonged anoxia (lack of oxygen) spells for several months of the year, whereas four decades ago anoxia was sporadic and localized. As a result, algal biomass concentration is almost five times greater in the surface waters today than reported in the 1960s, which indicates higher rates of photosynthesis. Also, water transparency values have decreased to one-third and the

silica concentration to one-tenth of what they were about forty years ago.

The lake was until recently home to over 600 endemic haplochromine cichlid fish (not all yet formally catalogued), as well as a number of other non-cichlid species. The extinction of species of haplochromine cichlid fish in the lake, primarily as a result of the introduction of the Nile perch (*Lates niloticus*), remains the single most dramatic event of vertebrate extinction attributable to specific human activities. More than 100 fish species have been driven into extinction since 1960. The lake's fisheries are currently dominated by three commercial species: the Nile perch, a non-native species introduced in the 1950s which now makes up 80 percent of the lake's fish population; the other 20 percent is formed by the indigenous tilapia (*Oreochromis niloticus* from the *Cichlidae* family) and dagaa (*Rastrineobola argentea*).

Sources: World Lakes Network (www.worldlakes.org/); Cohen et al., 1996; Kiremire, 1997; Verschuren et al., 2002; Dodman and Diagana, 2003; Hecky, 1993; Mugidde, 1993; Lehman, 1996; Johnson et al., 1996.

Algonquin National Park, Canada

Water hyacinth accidentally introduced in Lake Victoria from Latin America is having a huge impact on the lake's natural ecosystem

Part 2. The Environmental and Social Importance of Ecosystems

The first *United Nations World Water Development Report* (WWDR1) noted that a healthy and unpolluted natural environment is essential to human well-being and sustainable development, stressing that aquatic ecosystems and their dependent species are an integral part of our lives and provide a resource base that helps us to meet a multitude of human and ecosystem needs (UN-WWAP, 2003). These goods and services include water for human consumption, food production, irrigation, energy generation, regulating services (e.g. flood mitigation, water filtration, aquifer recharge and nutrient cycling), and transport and recreational services. Their value is irreplaceable, and they are an important part of the water, energy, health, agriculture and biodiversity (WEHAB) sectors, which are essential for poverty alleviation and socio-economic development. The ecosystem approach further focuses on coastal and freshwater ecosystems, which are stepping stones for migratory birds and fish species and provide global environmental services that underpin the natural functions of the Earth. But, as WWDR1 shows, these ecosystems are under severe pressures that threaten their ability to meet the multiple and growing demands placed upon them.

2a. Goods and services

All ecosystems, aquatic and terrestrial, play a role in regulating the way water flows through the landscape, highlighting the need to better understand the relations between them and manage them in an integrated way. Forests absorb precipitation and regulate streamflow, while wetlands act as sponges, absorbing excess water in times of heavy rain and high tides and releasing water slowly during dry periods. Aquatic ecosystems play a number of vital roles in human society: regulating climatic extremes, providing food resources and, in the case of freshwater, sustaining agricultural production. Many other human life-supporting goods and services (see **Table 5.2**) are derived from aquatic ecosystems, including:

- hydrological regulation of floods, availability and supply of water during dry periods
- sediment retention, water purification, and waste disposal
- recharge of groundwater supplies
- drinking water and sanitation for large populations
- irrigation water for crops and drinking water for livestock
- coastline protection
- climate change mitigation through greenhouse gas absorption and impact buffering
- recreation and tourism
- cultural and spiritual values
- a range of goods such as fibres, timber, animal fodder and other food products
- transport routes – sometimes the only accessible routes
- hydroelectric and mechanical power.

The ability of any particular aquatic ecosystem to supply the range of services listed above depends upon a variety of factors, such as the type of ecosystem, the presence of key species, management interventions, the location of human communities and the surrounding climate and topography. Few sites have the capacity to provide all of the above services. Whereas recreation opportunities may depend only on the presence of clean water, for example, the provision of fish for food usually depends upon the presence of a fully functioning food chain to sustain the fish populations. Generally speaking, the more biologically diverse an ecosystem is, the greater the range of services that can be derived from it. There is some evidence from aquatic systems that a rich regional species pool is probably needed to maintain ecosystem stability in the face of a changing environment (UNEP Millennium Ecosystem Assessment, 2006).

Aquatic ecosystems refer not only to coastal waters, rivers and lakes, but also to a complex and interconnected system of permanent and temporary habitats, with a high degree of seasonal variation. Temporary habitats play a key role in the overall value of water ecosystems. For example, coastal estuaries and river floodplains are among the most productive ecosystems on Earth (Junk et al., 1989). Some, such as the Amazon floodplain, stretch over thousands of kilometres, while others may be only a few metres wide. Seasonal variation is vital for the integrity of such ecosystems, as many fish depend on the seasonal inundation of river floodplains for breeding or feeding. The extent of flooding is in these cases positively correlated with fish catches (Welcomme, 1979). Many tropical freshwater wetlands have a low nutrient status, as in the black and clear floodplains in Amazonia (Furch, 2000). In these systems, high biodiversity is not an indicator of high productivity, but rather of quick and efficient nutrient recycling. These habitats are particularly vulnerable to overexploitation. Estuaries and river floodplains also have an important role in dissipating high tides and river flows and preventing flood damage and coastline erosion. In many countries, river floodplains also serve as nutrient-rich sites for agriculture.

The precise value of many of these services (particularly their monetary value) remain poorly understood. However, direct use values of water buffering (e.g. flood prevention) alone have been estimated at US $350 billion at 1994 prices, and recreational values at US $304 billion (Constanza et al., 1997). It was estimated that reef habitats provide human beings with living resources, such as fish, and services, such as tourism returns and coastal protection, worth about US $375 billion each year (Constanza et al., 1997). Economic losses from degradation can also be serious. One example is coastal erosion, which results from altered currents and sediment loads caused by changes in coastal and upstream land use. The beaches of Tangiers in Morocco, for instance, largely disappeared in the 1990s after new ports were built. The destination lost 53 percent of its international tourist night-stays and substantial tourism income, estimated at about US $20 million per year (Blue Plan, 2005). **Figure 5.1** roughly summarizes some of the estimates for marine ecosystems.

Studies of particular wetlands (coastal or inland) provide a fragmented picture of their physical and economic benefits in different regions of the world (see **Table 5.1**). For example, the value of using wetlands for waste treatment in Kampala, Uganda was estimated at US $2,000 to 4,000 per hectare (ha) (Emerton et al., 1999) (see **Chapter 14**). Zambia's wetlands were estimated to be around US $16.7 million per year, of which US $4.2 million was generated by the Barotse floodplain alone (Turpie et al., 1999).

2b. Fisheries

Fish are among the greatest and most obvious benefits that human societies derive from aquatic ecosystems. In 2001, the reported global marine capture amounted to 85 million tonnes of fish, crustaceans and molluscs (Fishstat, 2002). Unlike high-seas fisheries, however, coastal and inland fisheries are often dominated by small-

Figure 5.1: Estimated mean value of marine biomes

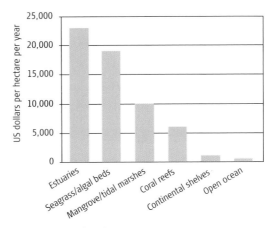

Source: Constanza et al., 1998.

scale and subsistence operations from the poorest sectors of society, for whom a catch provides a vital source of livelihood and affordable protein. Inland capture fishery production reported to the UN Food and Agriculture Organization (FAO) by 150 countries in 2001 indicated a total global production of 8.7 million tonnes, with the greatest continental catches reported in Asia (5.8 million tonnes) and Africa (2.1 million tonnes), mostly in developing or transition economies where fish production has rapidly increased over the past ten to fifteen years (FAO, 2002). Seven countries reported inland fisheries as their only source of fish, with twenty more considering inland fisheries extremely important, accounting for between 81 and 99 percent of their total fish production (Kura et al., 2004).

FAO (2002) recently estimated that marine and inland fisheries and aquaculture provide 16 percent of the global animal protein intake (see **Chapter 7**). This number exceeds 25 percent in the poorest countries, reaching up to 90 percent in some isolated rural areas. In the Upper Amazon Basin, for example, fish are reported to provide the majority of animal protein consumed by local households, with more than 200 kilograms (kg) of fish consumption per year per person (Batista et al., 1998). Fish are particularly important in communities relying primarily on a few staple foods such as rice, wheat, maize and cassava, which are deficient in essential nutrients and can be supplemented by fish (Thilsted et al., 1997). In areas where other economic opportunities may be declining, small-scale fisheries and related processing and trading offer an economic alternative for an increasing number of unskilled

Aquatic ecosystems refer not only to coastal waters, rivers and lakes, but also to a complex and interconnected system of permanent and temporary habitats, with a high degree of seasonal variation

Table 5.1: Estimated value of selected wetlands in Africa and Asia

Location	Value in million US $/ha/year	Services	Source
Bangladesh: Hail Haor	649	Crops, fisheries, plants, flood control, recreation, transportation, water quality and supplies, existence values	Colavito, 2002
Cambodia: Koh Kong Province mangroves	2 32	Carbon sequestration Storm protection	Bann, 1997
Cambodia: Ream National Park	59	Crops, fishing, plant use, hunting	Emerton et al., 2002
Cameroon: Waza Logone floodplain	3,000	Plant resources, grazing, crops, water supplies, fisheries	IUCN, 2001
Fiji: mangroves	158 5,820	Forestry, fisheries, crops Water purification	Lal, 1990
India: Bhoj urban wetland	1,206	Water quality and supplies, resource use, amenity and recreational values, crop cultivation	Verma, 2001
Indonesia: mangroves	86	Forest products and fisheries	Burbridge and Maragos, 1985
Japan: Kushiro National Park	1,400	Recreational and amenity values	Kuriyama, 1998
Kenya: Lake Nakuru National Park	400–800	Recreational value of wildlife viewing	Navrud and Mungatana, 1994
Republic of Korea: coastal wetlands	22,000	Fishery production and habitat, waste treatment, aesthetic functions	Lee, 1998
Malawi: Lower Shire wetlands	123	Plant resources, hunting, crops, grazing	Turpie et al., 1999
Malaysia: mangroves	35	Forest products	Hamilton et al., 1989
Mozambique: Zambezi Delta coastal wetlands	9	Plant resources, hunting, crops, grazing	Turpie et al., 1999
Namibia: Chobe-Caprivi wetlands	22	Plant resources, hunting, crops, grazing	Turpie et al., 1999
Nigeria: Hadejia-Nguru floodplain	2	Doum palm utilization, firewood, potash, agriculture	Eaton and Sarch, 1997
Nigeria: Hadejia-Nguru floodplain	20	Groundwater recharge for domestic consumption	Acharya, 1998
Philippines: Pagbilao mangroves	211	Forestry and fisheries	Janssen and Padilla, 1996
Sri Lanka: Muthurajawela urban marsh	2,600	Water supplies, wastewater treatment, flood attenuation, support to downstream fisheries	Emerton and Kekulandala, 2002
Thailand mangroves	165	Coastline protection	Christensen, 1982
Thailand: Surat Thani mangroves	77	Coastline protection	Sathirathai, 1998
Uganda: Nakivubo urban wetland	2,155	Wastewater treatment	Emerton et al., 1999
Uganda: Pallisa District wetlands	485	Crops, grazing, fisheries, plant use, sand and clay, maintenance of soil fertility, water supplies and quality	Karanja et al., 2001
Zambia: Barotse floodplain	16	Plant resources, hunting, crops, grazing	Turpie et al., 1999

Below: Las Huertas, Mexico

Bottom: Anawilundawa, Sri Lanka

labourers. In the Lake Chad Basin, for example, fish provide a source of income that is reinvested in farming (Béné et al., 2003).

Much of the increase in fisheries production is the result of enhancement efforts such as fish stocking and the introduction of non-native fish species in lakes and rivers (Kura et al., 2004), although the latter can in turn create environmental problems as discussed below. In 2001, aquaculture produced 37.9 million tonnes of fishery products, or nearly 40 percent of the world's total fish consumption, valued at US $55.7 billion (FAO, 2002). Aquaculture is the fastest-growing food production sector in the world, with freshwater finfish alone accounting for over 50 percent of global production. Asia, especially China, dominates inland fishery production. China produced close to 15 million tonnes of fish (about one-quarter of the world's total catch) in 2001, mostly carp for domestic consumption. Other leading inland aquaculture-producing countries include Bangladesh, Cambodia, Egypt, India, Indonesia, Myanmar, Tanzania, Thailand and Uganda (Kura et al., 2004).

Inland fishing is almost entirely dominated by small-scale and subsistence operations. In China alone, more than 80 percent of the 12 million reported fishermen are engaged in inland capture fishing and aquaculture (Miao and Yuan, 2001). In the Lower Mekong River Basin, which covers part of Cambodia, Laos, Thailand and Viet Nam, a recent study estimated that 40 million rural farmers are also engaged in fishing, at least seasonally (Kura et al., 2004). This is also true in Africa. In the major river basins and lakes in West and Central Africa, FAO (2003) estimated that fisheries employ 227,000 fishermen, producing 569,100 tonnes of fish products per year, with a value of US $295.17 million and a potential value of nearly US $750 million (Neiland et al., 2004).

All of these benefits depend on the continuation of healthy, functioning aquatic ecosystems. Unfortunately, many hydrological systems are currently being modified and damaged, resulting in a decline in biodiversity and a consequent loss of many of the services mentioned. It should be noted, furthermore, that information on inland fishery production is notoriously poor, particularly for subsistence fisheries, since catches are often grossly underestimated by national governments (Kura et al., 2004). FAO estimates under-reporting by a factor of three or four (FAO, 1999 and 2001). Despite their key role in providing nutrition to the poorest and most vulnerable members of society, coastal and inland fisheries frequently suffer from poor management, competition from industrial fishing and degradation from land-based activities, such as deforestation, pollution and upstream development (Kura et al., 2004).

Even though inland and coastal harvests continue to increase, maintained mainly by aquaculture expansion, most coastal and freshwater systems are stressed by overfishing, habitat loss and degradation, the introduction and presence of invasive species, pollution, and the disruption of river flows by dams and other diversions (FAO, 1999 and Revenga et al., 2000). This degradation threatens not only the biodiversity of riverine and lacustrine ecosystems, but also the food security and livelihood of millions of people – particularly those of poor rural and coastal communities in the developing world. The following section provides a brief overview of the status of freshwater and coastal ecosystems around the world.

...most coastal and freshwater systems are stressed by overfishing, habitat loss and degradation, the introduction and presence of invasive species, pollution and the disruption of river flows by dams and other diversions...

School of freshwater fish in the State of Mexico, Mexico

Part 3. Status of and Trends in Ecosystems and Biodiversity

Freshwater and coastal ecosystems comprise a range of highly productive habitats, such as lagoons, estuaries, lakes, rivers, floodplains, small streams, ponds, springs, aquifers and wetlands. The term 'wetland' describes a particular group of aquatic habitats representing a variety of shallow, vegetated systems, such as bogs, marshes, swamps, floodplains, coastal lagoons, estuaries, coral reefs and seagrass beds, where the shallowest sites are often transitional areas and can be seasonally or intermittently flooded (Groombridge and Jenkins, 1998).[1]

3a. Status of coastal and freshwater ecosystems

Proximity to water bodies has been an incentive for the location of human settlements for millennia, and the human alteration of coastlines, rivers, lakes and wetlands has gone hand in hand with social and economic development. Coastal and freshwater ecosystems have suffered multiple pressures, often undergoing degradation in small, incremental steps that are difficult to recognize. General analyses and reviews over the past two decades have identified a range of pressures that cause adverse change in these ecosystems (Allison, 2004; Revenga and Kura, 2003; Revenga et al., 2000; Groombridge and Jenkins, 1998; McAllister et al., 1997; Abramovitz, 1996; Bryant et al., 1998; Burke et al., 2001). These show physical alteration, habitat degradation and destruction, water withdrawal, overexploitation, pollution and the introduction of non-native species to be the leading causes of aquatic species decline and ecosystem degradation (see also **Table 5.2**). Rarely is a given species or habitat imperilled as a result of a single threat, and it is often impossible to decipher the intertwined effects of the many disturbances occurring within a given river basin (Malmqvist and Rundle, 2002). This 'creeping' nature of degradation not only makes it difficult to identify serious problems early on, but also allows people to get used to degradation as it is occurring, so that over time degraded ecosystems become accepted as the norm (Glantz, 1999).

A variety of attempts have been made to assess the global extent and distribution of aquatic habitats. However, estimates vary considerably, depending on the type of source material used. And while inventories of coastal zones, river basins and lakes do exist, there are no good data sets or indicators at the global level which track changes in conditions over time. Unfortunately, there are no unequivocally accepted global measures or indicators that demonstrate the changes in the overall extent of

wetlands. Finlayson and Davidson (1999) concluded that the information available is too patchy and inconsistent to provide a precise picture of global change. An often quoted estimate is that about 50 percent of the wetlands that existed in 1900 had been lost by the late 1990s as a result of the conversion of land to agriculture (Myers, 1997). However, this figure remains largely speculative.

Accurate information does exist, however, for some continents and regions. Junk (2002) recognized twenty-four major aquatic ecosystems in Africa, eight of which are strongly subjected to large-scale irrigation, with devastating environmental effects and losses of ecosystem services and biodiversity. If human population density is brought into the equation, south and Southeast Asian wetlands can be considered among the most severely degraded (see **Box 5.2**).

A global survey is beyond the scope of a chapter of this length. In the following section we first summarize some key global trends in wetland species and related ecosystem goods and services and then examine the range of pressures currently affecting aquatic ecosystems. Lastly, we discuss some specific examples of changes to the status of particular wetland habitats. Again, freshwater systems are covered in more detail than coastal systems, even though all these systems are interlinked and provide many specific goods and services.

3b. Global trends in key species

Species richness in relation to the extent of habitat is extremely high for many coastal and freshwater groups. It has been estimated, for example, that 12 percent of all animal species live in freshwater ecosystems (Abramovitz, 1996), while virtually all terrestrial species depend upon such ecosystems for their survival. In Europe, for example, 25 percent of birds and 11 percent of mammals use freshwater wetlands as their

---

1. The Ramsar Convention on Wetlands defines wetlands as 'areas of marsh, fen, peatland or water, whether natural or artificial, permanent or temporary, with water that is static or flowing, fresh, brackish or salt, including areas of marine water the depth of which at low tide does not exceed six metres'.

BOX 5.2: THREATENED SOUTH AND SOUTHEAST ASIAN WETLANDS

Given high population densities, increased rates of deforestation (particularly in Indonesia) and the large degree of ecosystem fragmentation in India, which has more than 4,000 dams, Southeast Asia's wetlands are probably the most degraded in the world. This is reflected in the rapid decrease or local extinction of large grazing wetland species, such as the Indian rhinoceros (*Rhinoceros unicornis*) and swamp deer (*Cervus duvauceli,*

C. eldi, C. schomburgki), and the high number of globally threatened fish, amphibians, water turtles and bird species in the region. In addition, more than half of the region's coral reefs, the most species-rich on Earth, are at high risk, primarily from coastal development and fishing-related pressures. Southeast Asia's mangroves, also the most biodiverse in the world, are under increasing pressure from timber industries, aquaculture and conversion

to agriculture despite their widely documented importance for coastal protection, water purification, carbon dioxide absorption, and as breeding and nursing grounds for many valuable subsistence and commercial fish species (see also **Chapter 14**).

Sources: IUCN, 2003a; IUCN et al., 2004; Bryant et al., 1998; Burke et al., 2001.

---

main breeding and feeding areas (EEA, 1995). Of the approximately 25,000 ray-finned (*Actinopterygii*) fish species described to date, 41 percent are considered primarily freshwater species. Individual freshwater systems can be extremely important in supporting high numbers of endemic species. According to the Ramsar Convention on Wetlands, Lake Tanganyika in Central Africa, for example, supports 632 endemic animal species. It is also important to note that the resilience of ecosystems increases with biodiversity, thus providing a relevant linkage between management and biodiversity conservation.

As for coastal waters, Conservation International (CI) has identified twenty-five biodiversity 'hotspots' around the world, twenty-three of which are at least partially located within coastal zones, mainly in Asia, the Caribbean, Africa and South America (UNEP, 2005). Coral reefs alone, representing only 0.2 percent of the total area of oceans (Bryant et al., 1998), harbour more than 25 percent of all known marine fish, with some reefs reaching densities of around 1,000 species per square metre, especially in parts of the Pacific and Indian Oceans (Tibbets, 2004). Semi-enclosed seas can also have a wealth of endemic flora and fauna. The Mediterranean, for example, contains 7 percent of the world's known marine species, although it covers only 0.8 percent of the ocean surface: 694 species of marine vertebrates have been recorded (580 fish, 21 mammals, 48 sharks, 36 rays and 5 turtles) and 1,289 marine plant taxons (Blue Plan, 2005).

Serious concern for the global status of aquatic biodiversity was raised in the early 1990s (e.g. Moyle and Leidy, 1992), focusing mainly on data relating to the conservation status of fish. Most of the relatively few global reviews have

appeared only during the past decade (Abramowitz, 1996; McAllister et al., 1997; Groombridge and Jenkins, 1998; Revenga et al., 1998; Revenga et al., 2000). These still rely heavily on information relating to fish, but draw from available case studies of other groups (e.g. molluscs in US waters), and also deal in increasing detail with threat factors and their sources. For instance, FAO's 2004 assessment of marine fish stocks for which information is available, concludes that about half of the stocks (52 percent) were fully exploited, 16 percent were overexploited, and 7 percent were depleted. Only about one-quarter were either underexploited (3 percent), moderately exploited (21 percent) or recovering from previous exploitation (1 percent).

Pressures on aquatic ecosystems have caused a severe decline in the condition of species, with more freshwater species threatened with extinction than in either terrestrial or marine environments (WRI et al., 2000; Revenga et al., 2000; Loh et al., 2004). Available indices tend to support the hypothesis that freshwater species are more threatened by human activities than species in other realms. The Living Planet Index (LPI) developed by UNEP's World Conservation Monitoring Centre (WCMC) and the World Wide Fund for Nature (WWF) is based on trends in populations of vertebrate species. The LPI found that on average freshwater species populations fell by about 50 percent between 1970 and 2000, representing a sharper decline than measured in either terrestrial or marine biomes. Furthermore, freshwater species declined most sharply in the Neotropical and Australasian realms (see **Figure 5.2**). However, this does not mean that marine species are in good condition. The Marine Species Population Index recorded a decline of about 35 percent over the same period.

Pressures on aquatic ecosystems have caused a severe decline in the condition of species. Freshwater species are more threatened with extinction than in terrestrial or marine environments

Figure 5.2: Living Planet Index, 1970–2000

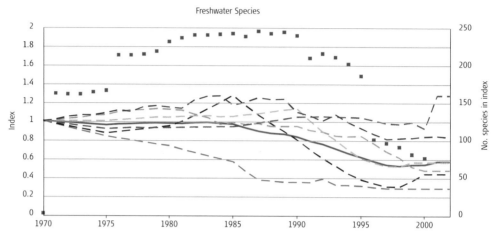

Freshwater Species

- – Australasia
- – Afrotropical
- – Indo-Malayan
- – Nearctic
- – Neotropical
- – Palearctic
— Freshwater Index
■ N

Source: Loh et al., 2004.

Other similar measures that reflect the level of threat to freshwater species include conservation status assessments, such as those compiled by the Species Survival Commission of the World Conservation Union (IUCN) and BirdLife International through the Red List of Threatened Animals and its derived red list indices. According to the 2003 IUCN Red List, 3,011 freshwater species are listed as threatened or extinct. Of these, 1,039 are fish and 1,856 are amphibian. Among other freshwater groups, four of the five river dolphins and two of the three manatees are threatened, as are several smaller aquatic mammals. About 40 freshwater turtles, more than 400 inland water crustaceans and hundreds of bivalve and gastropod molluscs are also listed as being threatened with extinction. However, the accuracy of available information tends to decline with lower taxa. Figures for crustaceans and molluscs, for example, may not reflect a true picture of the current global situation.

All of the world's amphibian species were recently assessed for the first time (IUCN et al., 2004), providing important new insights into the condition of this large faunistic group. The survey shows that amphibians are experiencing declines unprecedented in modern times, with nearly one-third (32 percent) of the world's 5,743 amphibian species threatened and 168 species already thought to have become extinct. Most amphibians are dependent on freshwater habitats during their larval stages (with the exception of arboreal species), and almost all species are highly sensitive to changes in habitat and water quality. The decrease in amphibian abundance and their threat status around the world are of major concern from a global biodiversity perspective. At least 43 percent of all amphibian species are declining in population, indicating that the number of threatened species can be expected to rise in the near future. The largest numbers of threatened species occur in Latin America. Although habitat loss and fragmentation clearly pose the greatest threat to amphibians, a newly recognized fungal disease is seriously affecting an increasing number of species, which might have developed in response to the increasing global

Figure 5.3: Trends in waterbird populations in the African-Eurasian (AEWA) regions

Note: The Agreement on the Conservation of African-Eurasian Migratory Waterbirds (AEWA) covers 117 countries from Europe, parts of Asia and Canada, the Middle East and Africa. In fact, the geographical area covered by the AEWA stretches from the northern reaches of Canada and the Russian Federation to the southernmost tip of Africa. The available evidence suggests that aquatic habitats and species are suffering a disproportionate decline in comparison to other habitats.

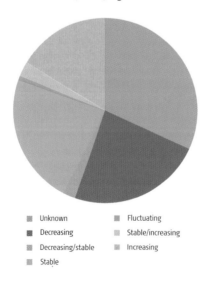

- ■ Unknown
- ■ Decreasing
- ■ Decreasing/stable
- ■ Stable
- ■ Fluctuating
- ■ Stable/increasing
- ■ Increasing

Source: Wetlands International, 2002.

eutrophication of aquatic ecosystems. Perhaps most disturbing is the fact that many species are declining for unknown reasons – complicating efforts to design and implement effective conservation strategies.

Birds have generally been recorded in more detail and over longer periods than any other group of species. Although there are limitations in using them as general indicators, the relative richness of data means they are often the best available proxy to suggest overall trends in biodiversity. The latest trend estimates from BirdLife International (Butchart et al., 2004) confirm that waterbird species are facing disproportionately serious problems. Some 22 percent of the world's seabirds alone are threatened species (WWF/IUCN, 2001). Additionally, global population estimates for waterbirds compiled by Wetlands International (2002) show a declining trend in the African-Eurasian flyway region (see **Figure 5.3**).

Part 4. Pressures and Impacts

Most aquatic ecosystems are vulnerable to a range of human activities. The likely impact of these activities varies from place to place and according to the type of habitat involved. Table 5.2 summarizes some of the key pressures with respect to different coastal and freshwater ecosystem types, as well as some of the goods and services that these ecosystems supply. Some specific pressures are discussed in greater detail below.

4a. Habitat alteration
Many aquatic ecosystems have undergone major alterations as a result of deliberate habitat change, either directly or through changes to nearby habitats. Various forms of land use changes have a major influence on water resources and ecosystems around the world (UNEP, 2004b). Several such changes are briefly described below.

Increased suspended loads
Increased concentrations of suspended solids in coastal waters, rivers and lakes resulting from human activity can cause significant changes in habitats. Examples include intensive agriculture, deforestation, road construction, urbanization, tourism, mining, dredging for harbours and shipping lanes, and gravel pit operations. Increased particulate matter in water leads to higher levels of turbidity, and thus reduces photosynthesis. In inland waters, it can fill downstream reservoirs faster than anticipated (UNEP, 2002b). As the suspended (occasionally polluted or even toxic) material settles out of the water column, the habitat for benthic organisms can change in ways that reduce biodiversity (Cobb et al., 1996). Some freshwater mammals are threatened with extinction because of increased silt loads in rivers, including the Spot-necked Otter (*Lutra maculicollis*) in South Africa, the Malagasy Web-footed Tenrec (*Limnogale mergulus*) in Madagascar, and the Giant Otter Shrew (*Potamogale velox*) in Cameroon (Revenga and Kura, 2003). Near shore coastal habitats are particularly impacted by suspended solids. Deltas, mangrove forests,

beaches and other coastal habitats are affected by altered currents and sediment delivery – to the benefit of some locations and the detriment of others (UNEP, 2002b). Coral reefs, mangrove forests and sea-grass beds may be smothered and deprived of light due to increases in sediment loads, thereby degrading important breeding and nursing grounds for many commercially valuable and subsistence fish species (Spalding et al., 2002). Fish populations are affected, both through reduced sources of food and by direct physical effects – such as clogging and abrasion of gills, behavioural changes (e.g. movement and migration), reduced resistance to disease, blanketing of spawning grounds and other habitat changes – and physical constraints that prevent functional egg and fry development (Singleton, 1985). Similarly, invertebrate communities are impacted if photosynthetic communities (e.g. periphyton) are affected. The direct invertebrate effects of suspended solids include smothering, clogging of interstices in gravel and cobble beds affecting microhabitats, abrasion of respiratory surfaces, and interference of food intake for filter-feeding species (Singleton, 1985).

Drainage and conversion of wetlands
Uncontrolled or poorly regulated wetland drainage has created severe threats to aquatic ecosystems and species in some parts of the world, with impacts sometimes affecting entire river basins or coastal habitats. While some drainage is often an essential step in agriculture and coastal development, done in the name of ensuring local

Increased concentrations of suspended solids in coastal waters, rivers and lakes resulting from human activity can cause significant changes in habitats

Table 5.2: Major threats to coastal and freshwater ecosystems and services

Ecosystem	Goods and services	Threats
Rivers	Many environmental, economic (e.g. fish, water supplies, transport, disposal, biological cleaning, climate regulation, etc.), religious and spiritual values	Reclamation, drainage, regulation of flow including dam construction, hydroelectric power, pollution, deforestation, soil erosion and degradation, climate change and alien invasive species
Estuaries	High biodiversity, fish, waterfowl, sedimentation, buffer zones, biological cleaning, recreation	Reclamation, drainage, irrigation, hydroelectric power, regulation of water flow, dams and dykes, pollution, agricultural intensification, deforestation, soil erosion/degradation, overexploitation of fish and other food species, climate change, waterborne disease control, and alien invasive species
Coral reefs	High species diversity, coastal protection, biological cleaning, tourism	Climate change, suspended solids from coastal construction, upstream agriculture and logging, tourism; nutrients from untreated sewage and agricultural runoff; pollution from industrial discharges, urban, agricultural and landfills runoff, mining
Mangroves	High species diversity, coastal protection, water purification, CO_2 absorption, breeding and nursing grounds for commercial fish species, source of firewood and timber, coastal protection, tourism	Cutting for firewood and building materials, timber industry, road construction, reclamation for aquaculture, agricultural, urban and industrial areas, tourism developments, and sea-level rise
Sea-grass beds	High species diversity, nursing grounds for commercial fish species, coastal protection, water purification, CO_2 absorption, sediment stabilization	Dredging for harbours, ports and shipping lanes, fishing by benthic trawling, aquaculture, coastal pollution, and clearance for beaches and other tourism developments and facilities
Inland deltas	Water supplies, sediment and nutrient retention, recreation	Drainage, irrigation, regulation of water flow, pollution, agricultural intensification, deforestation, soil erosion/degradation, overexploitation of fish and other food species, climate change
Floodplains	High productivity, high fish and fibre productivity, flood buffers, fire protection, carbon storage, recreation, groundwater recharge	Reclamation, drainage, irrigation, hydroelectric power, regulation of water flow, dams and dykes, pollution, agricultural intensification, deforestation, soil erosion/degradation, overexploitation of fish and other food species, climate change, waterborne disease control, and alien invasive species
Lakes	Water supplies, fibre, fish, waterfowl, recreation, groundwater recharge, religious and spiritual values	Pollution, agricultural intensification, eutrophication, deforestation, soil erosion/degradation, overexploitation of fish and other food species, climate change, waterborne disease control, and alien invasive species
Freshwater marshes	Flood buffers, carbon storage, reed, willow, food and fibre, purification	Drainage, regulation of water flow, dams and dykes, pollution, agricultural intensification, soil erosion/degradation, overexploitation of fish and other food species, and waterborne disease control
Raised bogs	Carbon storage, fossil fuels, purification	Reclamation, drainage, regulation of water flow, pollution, agricultural intensification, eutrophication and climate change
Fen mires	Carbon storage, pastoralism, willow, reed, groundwater recharge	Reclamation, drainage, regulation of water flow, pollution, agricultural intensification, and climate change
Alpine meadows	Species diversity, husbandry, pastoralism, recreation, groundwater recharge	Drainage, agriculture and climate change
Tundra wetlands	Carbon storage, climate regulation, water flow, subsistence hunting and herding, groundwater recharge	Pollution, climate change, overexploitation of fish and other food species
Forest swamps/ shrubs	Timber and fibre, biological cleaning, sanitation, flood buffers, groundwater recharge, purification	Deforestation, soil erosion, degradation and pollution
Groundwater aquifers	Water reservoirs, water storage, storage of nutrients	Irrigation, pollution, agricultural intensification, eutrophication, deforestation, soil erosion/degradation, overexploitation of food species and waterborne disease control
Freshwater springs and oases	Water and food supplies, stop-over sites for migratory species, recreation, religious and spiritual values	Irrigation, agricultural intensification, pollution, overexploitation of fish and other food species, and alien invasive species
Wet grasslands	Carbon storage, food supply, flood buffers (mostly on floodplains), groundwater recharge	Regulation of water flow, drainage, agricultural intensification, eutrophication, overexploitation of food species, and climate change
Ponds, gravel pits, drainage channels	Water supplies, recreation	Pollution, eutrophication and overexploitation of fish and other food species

Source: UNEP and UNEP-WCMC, 2004.

livelihoods, many such efforts usually bring short-term economic gains while neglecting the long-term impact on local communities. Draining wetlands can have serious effects on their natural regulatory functions, causing not only species and habitat loss but also significant detrimental impacts on human populations through increased and unpredictable droughts and flooding, and erosion and saline intrusions along coastlines.

The Pripyat River, for example, between Ukraine and Belarus previously had about 25 percent of its basin covered by peatlands, the subsequent clearance of which resulted in a long-term decline in river water quality (Bragg and Lindsay, 2003). Temperate wetlands, including peatlands, have been heavily modified by conversion to agriculture and other land uses in Western Europe, where many countries have lost more than 90 percent of their wetlands. Much of the wet grassland in Europe has also diminished due to drainage and land conversion. In England and Wales, for instance, less than 20 percent of traditional wet grasslands remained by the late 1990s. Similarly, wet grassland decreased in northern Germany on average by more than 50 percent between 1945 and the early 1990s, with devastating effects on biodiversity, as well as on water holding and carbon storage capacity. In Eastern Europe, socio-economic changes after 1990 led to the abandonment of agriculture on many wet grasslands in northern Russia, Poland and the Baltic States, thereby allowing them to develop into bushy wetlands with little or no drainage. If this trend continues, degraded fen mires and other habitats sensitive to intensive land use could regenerate and once again provide reservoirs of clean water, carbon storage and other services. When wetlands are drained, the natural flow of sediments also changes, with various impacts in habitats. It is sometimes possible, however, to reverse such changes.

Deforestation

Forests are often highly diverse systems, and water flowing through forested catchments is generally of high quality. However, these areas are also very sensitive to changes in land use, and any conversion of forest, including the loss of biomass and biodiversity (Krebs, 1978; Tischler, 1979), can disrupt the dynamics of water flow and recharge functions. Apart from the potential changes in water quality, quantity and flow continuity, deforestation also often results in increased sediment loads, with various impacts on downstream and coastline habitats.

Deforestation in the 1990s was estimated at a net loss of 14.6 million ha per year (taking into account

reforestation), or 4.2 percent of the world's natural forests (FAO, 2001). According to a recent report by the World Bank (Dudley and Stolton, 2003), much of the world's drinking water comes from catchments that are, or would naturally be, forested. This report also found that a third of the world's largest 100 cities rely on forests in protected areas for a substantial proportion of their drinking water, demonstrating that metropolitan authorities are increasingly recognizing the importance of the link between forests and water supplies (see **Chapter 12**). It is clear that forests often provide the basis for the integrated management of water resources, although precise effects vary from place to place; a topic that has been controversial among hydrologists. Knowledge of the type and age of trees, soil conditions and user needs can help determine what kind of forest management policies will be most beneficial in a given situation.

Agricultural land-use changes

Agriculture is the largest user of freshwater. Irrigation is responsible for almost 70 percent of all water withdrawals, involving around 250 million ha of land (Millennium Project, 2004), particularly in arid lands and in the larger rice-growing areas of the world. As a result, some rivers (e.g. the Colorado in the US, and the Nile in Africa) have no discharge to the sea during certain periods of the year (Postel, 1995). This, in and of itself, causes a range of downstream and coastal problems and in some cases accelerates the salinization of soils in irrigated regions and aquifers near coastlines. No global figures for salinization exist, but in the early 1990s a World Bank study estimated that up to 2 million ha of land were being withdrawn from agriculture each year due to water logging and salinization (Umali, 1993). While the majority of the world's crops are still grown in rainfed farmlands, 17 percent of the world's cultivated land currently produces 40 percent of its food (Wood et al., 2000), with an increasing trend towards irrigation.

Intensified agricultural practices that rely on the application of soluble fertilizers and pesticides can result in increased nutrient runoff – one of the major causes of deterioration in water quality. In extreme cases severe eutrophication and harmful algal blooms can result in both inland and coastal waters, leading to hypoxia, a condition where rapid algal growth depletes oxygen as it decomposes. Besides severely impacting human uses of water, eutrophication can cause major changes in aquatic food webs and ecosystem productivity.

Toxic algae warning sign on boating lake in Portishead, UK

Intensified agricultural practices that rely on the application of soluble fertilizers and pesticides can result in increased nutrient runoff – one of the major causes of deterioration in water quality

In addition, silt being washed into water from ploughed land and changes in the way stream and river banks are managed can damage fish spawning grounds and coastal habitats. For instance, the drainage of Dartmoor and Bodmin Moor in the UK has damaged salmon spawning in the Tamar, Fowey and Camel rivers, which together are worth a total of about US $27.3 million in rod fisheries.

Nutrients from agricultural runoff, aquaculture operations, and human and industrial wastes – including atmospheric depositions – can cause severe eutrophication and changes in the trophic conditions of coastal waters, rivers, lakes, reservoirs and wetlands. Nitrogen and phosphorus compounds are usually the major nutrients responsible for increases in unnaturally rapid growths of algae and other plants, which are symptomatic of eutrophic water bodies. Besides severely impacting human uses of water, eutrophication can cause major changes in aquatic food webs and ecosystem productivity. The die-off of excessive plant matter can lead to the deoxygenation of the water, killing many aquatic species and affecting chemical cycles that fuel biological productivity. Bacteria and other micro-

organisms require oxygen to decompose pollutants that enter aquatic systems.

Biological oxygen demand (BOD) is a measure of the quantity of oxygen necessary for biological oxidation of waterborne substances and therefore an indicator of organic pollution. Some aquatic species are particularly susceptible to declines in oxygen concentrations and thus to pollution from sewage or fertilizers. For example, salmon (*Salmonidae*) species require dissolved oxygen concentrations greater than 5 milligrams per litre (mg/L) and cyprinids and members of the Carp (*Cyprinus carpio*) family, more than 2 mg/L (Gleick et al., 2001). When nutrient levels increase, the delicate balance between corals and algae is also destroyed. The algae may overgrow and smother the corals, thereby affecting the marine organisms that depend on them. This may in turn affect humans who depend on these marine resources for their livelihoods. **Map 5.1** shows the distribution and changes in BOD for the regions of the world and major river basins. Oxygen-depleted coastal waters are also widespread along the eastern and southern coasts of North America, the

Map 5.1: Biological oxygen demand (BOD) for major watersheds by region, 1979–90 and 1991–2003

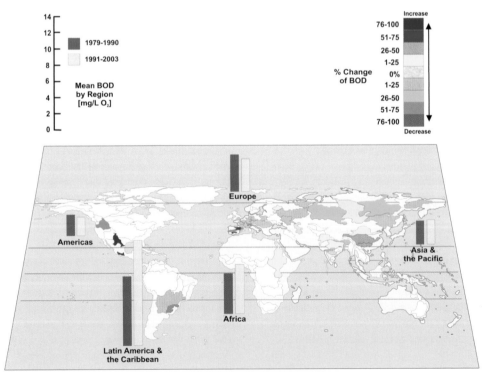

Note: Coloured areas on the map indicate percentage change and histograms the mean concentration changes by region. BOD is a measure of the amount of dissolved oxygen consumed as a result of decay of organic matter in the water column or at the sediment–water interface. Pollution in the form of municipal and industrial effluents is often high in organic matter and, thus, BOD is an indicator of ecosystem stress from municipal and industrial pollution. Watersheds in white indicate insufficient data in one of the time periods to calculate % change.

Source: Based on global water monitoring data maintained by the UNEP GEMS/Water programme, www.gemswater.org

southern coasts of Japan and China, and large parts of the many, often semi-enclosed, seas surrounding Europe (various sources compiled in UNEP, 2004b).

High concentrations of nitrate in water make it unusable for drinking purposes. A comparison of dissolved nitrogen concentrations in eighty-two major watersheds from the late 1970s (see **Map 5.2**) indicates that twenty-five watersheds had increased nitrate concentrations, thirteen had lower concentrations, probably due to improved nutrient control programmes for wastes, and the remainder showed no significant change or had insufficient data for accurate assessments to be made. The results suggest that, while conditions appear to be deteriorating in more areas than improving, significant improvements can be achieved if there is sufficient political will to improve wastewater treatment and modify agricultural policy.

In a global assessment of the human impacts on phosphorus leaching and its relation to eutrophication, it was found that even though households and industry tend

to be the most significant sources, the mining of phosphorus, and its subsequent use in fertilizers, animal feeds and other products is altering the global phosphorus cycle, causing it to accumulate in soils in some locations (Bennett et al., 2001). This can increase phosphorus runoff and subsequent loads to aquatic ecosystems, inland and coastal alike. The authors estimated that phosphorus storage in soils and aquatic systems was 75 percent greater than pre-industrial levels. In agricultural areas, the rate of phosphorus accumulation appears to be decreasing in developed nations and increasing in developing nations. As phosphorus is the key to biological production in most aquatic systems, in the future eutrophication problems are likely to increase in developing countries. Stored soil phosphorus can be transported to aquatic systems during storms and other events, which means there will be an inevitable lag before management actions taken to control eutrophication have a significant effect. Furthermore, phosphorus can build up in lake and coastal sediments. As this phosphorus can be remobilized under certain circumstances, it may threaten severe eutrophication in the future.

A waterfall in Sri Lanka

Map 5.2: Inorganic nitrogen concentrations for major watersheds by region, 1979–90 and 1991–2003

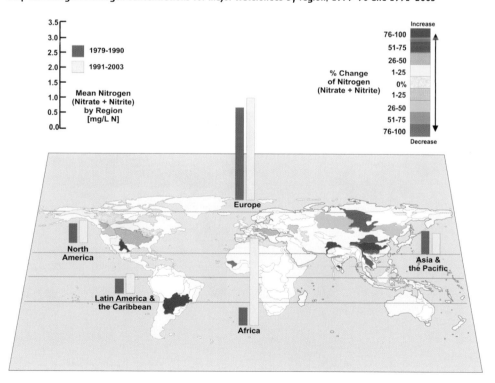

Note: Inorganic nitrogen, measured as nitrate + nitrite, is an indicator of trophic stress on ecosystems as a result of human activities. Inorganic nitrogen can enter aquatic ecosystems through agricultural activities, in the form of runoff from fertilizer applications, as well as from industrial and municipal processes. Nitrogen, in conjunction with phosphorus, controls the growth of plants and algae in aquatic systems and elevated levels of these nutrients can lead to overly productive or eutrophic conditions that can impair ecosystem health. Watersheds in white indicate insufficient data in one of the time periods to calculate % change.

Source: Figure generated based global water monitoring data maintained by the UNEP GEMS/Water programme, at www.gemswater.org

*Once species
become extinct,
they can never
be recovered,
with severe
potential effects
on entire food
webs and
ecological
processes*

4b. Fragmentation and flow regulation (dams and reservoirs)

While it is difficult to obtain a single and absolute measure of the condition of freshwater ecosystems, some indicators are available that can help to illustrate their overall status. The fragmentation and flow regulation indicator is a case in point. This indicator provides a measure of the degree to which freshwater systems have been altered by the construction of dams and reservoirs. According to the World Commission on Dams (WCD, 2000), most of the world's large dams were built during the second half of the twentieth century and, as of 2000, between 160 and 320 new large dams are still being built every year. Today, there are more than 45,000 dams of over 15 metres (m) high, with about 15 percent of the total annual river runoff being sequestered behind dams (Gornitz, 2000). Almost half of the existing large dams (22,000) are in China, followed by the United States with 6,390 (WCD, 2000).

Dams play a major role in fragmenting and modifying aquatic habitats, transforming lotic (flowing) ecosystems into lentic (still) and semi-lentic ecosystems, altering the flow of matter and energy and establishing barriers to the movement of migratory species. Waterfalls, rapids, riparian vegetation and wetlands can all disappear when rivers are regulated or impounded (Dynesius and Nilsson, 1994). These habitats are essential feeding and breeding grounds for many aquatic and terrestrial species and also contribute significantly to maintaining other vital ecosystem services, including water purification. The

fragmentation indicator presented here suggests that many unique riverine habitats have been fragmented or even eliminated. Given that habitat loss is the leading cause of species extinction in freshwater ecosystems, this indicator gives a measure of the risk that many freshwater species face. Once species go extinct, they can never be recovered, with severe potential effects on entire food webs and ecological processes.

This fragmentation and flow regulation indicator was developed by Umeå University in Sweden, in collaboration with the World Resources Institute (Nilsson et al., 2005). It assesses 292 of the world's largest river systems, which comprise approximately 60 percent of the world's river runoff, and occupy more than half (54 percent) of the world's land area. A large river system (LRS) is defined as a river system that has a river channel section with a virgin mean annual discharge (VMAD, the river discharge before any significant direct human manipulations) of at least 350 cubic metres per second (m^3/s) anywhere in its catchment (Dynesius and Nilsson, 1994). Results of the analysis (see **Map 5.3**) show that there are 105 strongly affected, 68 moderately affected, and 119 unaffected LRSs. Unaffected river systems are those without dams in their catchments, although dams in tributaries may not disqualify a river from being classified as 'unaffected' if flow regulation is less than 2 percent of the VMAD. A river system is never considered unaffected if there are dams in the main channel, and is never classified as strongly affected if

Map. 5.3: Fragmentation and flow regulation by Large River System (LRS)

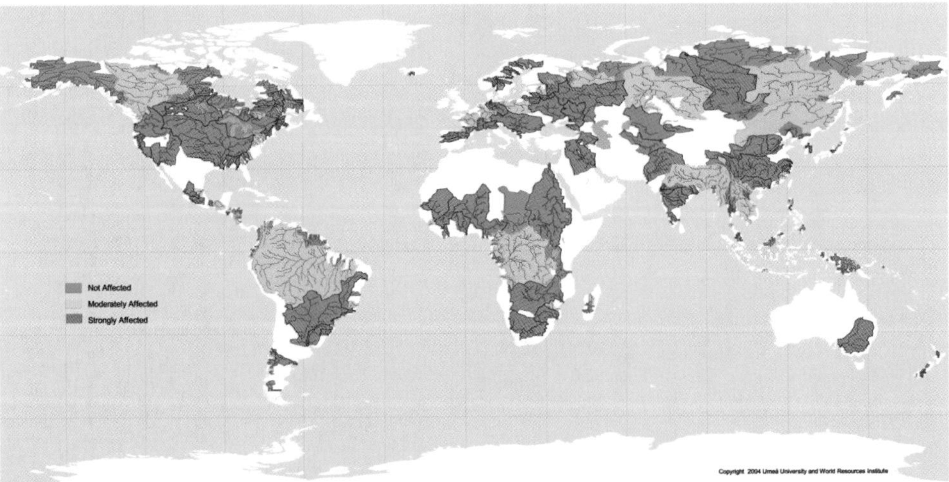

Note: This map presents the results of the river fragmentation and flow regulation indicator. Of the 292 of the world's LRSs, 173 are either strongly or moderately affected by dams; while 119 are considered unaffected. In terms of areas, strongly affected systems constitute the majority (52 percent or about 4,367 km^2) of total LRS catchment area. Grey colour represents potential LRSs in Indonesia and Malaysia that were not assessed due to lack of data.

Not Affected
Moderately Affected
Strongly Affected

Source: Nilsson et al., 2005.

there are no dams in the main channel. All river systems with no more than one-quarter of their main channel length left without dams are considered strongly affected.

The world's two river systems with the largest discharges, the Amazonas-Orinoco and the Congo rivers, are moderately affected, while the third largest, the Yangtze River in China, is strongly affected by fragmentation and altered flows. The largest river remaining unaffected by fragmentation and altered flows is the Yukon River system in Alaska. The other unaffected river systems are mainly smaller catchments in areas with low population densities, such as catchments surrounding the Hudson Bay in Canada and others in southern Chile and Argentina as well as in northern Siberia. Although fewer in number, river systems classified as moderately affected represent, on average, both the largest basins and those with higher discharges. On the other hand, strongly affected systems constitute the majority (52 percent) of total LRS catchment areas, despite contributing less VMAD per system.

At the continental level, Europe has the smallest number (four) and smallest proportion (10 percent) of free-flowing or unaffected large river systems. The highest number (forty) of unaffected LRSs is found in North and Central America, whereas Australasia contains the highest proportion (74 percent) of unaffected systems. In South America, the unaffected systems are on average smaller than affected systems both in discharge and catchment area. The situation is similar in Africa. For example, the moderately affected Congo River (Central Africa) system contributes 51 percent of total African LRS runoff.

This indicator does not address the within-basin distribution of impacts, which can be significant in large basins. For example, the Mackenzie (Northwest Territories, Canada) and the Amazonas-Orinoco systems, which are moderately affected, include extensive, virtually pristine areas as well as strongly affected areas. This within-basin variation is likely to have significant ecological implications. Furthermore, the data used are conservative and represent minimum values, implying that the LRS at a global scale may be more affected than depicted. An example can be seen with the Brahmaputra, a river thought to have more dams in Tibet than official sources report. If true, this would increase the Ganges-Brahmaputra system (Tibet, China, Bangladesh and India) classification to a higher fragmentation level. If irrigation pressure, planned dams and dams under construction are taken into account, the current fragmentation classifications will also change.

Despite gains in information, there are still data gaps that limit our understanding of the relationships between impacts on LRSs and ecosystem conditions. Most river systems in Indonesia, for example, are omitted, along with several in Malaysia, because reliable data are not available. This is particularly unfortunate because the region harbours some of the most unique and rich species assemblages on the planet, representing great conservation potential.

When the fragmentation and flow regulation indicator is correlated with terrestrial biome distribution, as classified by Olson et al. (2001), the analysis shows that unaffected LRS catchments are most represented in large biomes: tundra; boreal forests; tropical and subtropical moist broadleaf forests; and tropical and subtropical grasslands, savannahs and shrublands (see **Figure 5.4**). In fact, tropical and subtropical moist broadleaf forests and boreal forests contain low proportions of strongly affected river systems in terms of area. Smaller biomes retain little or no unaffected large river systems. Strongly affected systems are dominant in three biomes – temperate broadleaf and mixed forests; temperate grasslands, savannahs and shrublands; and flooded grasslands and savannahs – each of which retain less than 1 percent of their total surface area as unaffected LRSs. An important result is that strongly affected catchments alone constitute 80 percent of LRS area in deserts and xeric shrublands, and 99 percent for Mediterranean forests, woodlands and scrubs – highlighting the pressure on these ecosystems from altered river catchments and water abstraction. Furthermore, the eight most biogeographically diverse LRSs[2], spanning seven or more biomes each, are all moderately or strongly impacted.

Dams are often promoted as a means of meeting water and energy needs and supporting economic growth. We can therefore anticipate that the demand for large dams will continue to increase, particularly in regions with high water demand driven by growing populations and agricultural needs. The latest accounts of current dam development support this hypothesis. There are currently 270 dams over 60 m high planned or under construction around the world. Of the LRSs assessed, 46 presently have large dams planned or under construction, with anywhere between 1 and 49 new dams per basin (WWF and WRI, 2004). In addition, interbasin exchange of dam benefits may play a strong role in future decisions about dam construction. For example, over thirteen dams are planned or proposed for the currently unaffected Salween River (Tibet, China and Myanmar), the most imminent of

Approximately nine rivers are at risk of entering a higher impact class: from unaffected to affected, or from moderately to strongly affected

2. The eight most biogeographically diverse LRSs are the Amazonas-Orinoco Basins in South America; the Zambezi in Africa; the Amur, Ob and Yenisey in northern Asia (Russia, Mongolia); and the Irrawaddi, Ganges-Brahmaputra and Indus in Asia.

Figure 5.4: Fragmentation and flow regulation by biome type

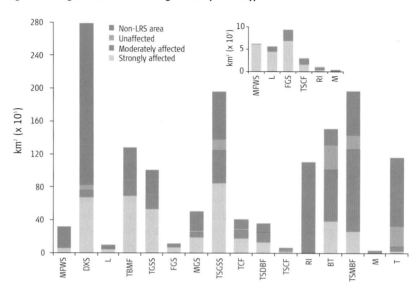

Note: The figure represents the distribution of surface area within each of the world's sixteen terrestrial biomes as belonging to unaffected, moderately affected or strongly affected LRSs. Biomes are listed in descending order from left to right by proportion of strongly affected area within LRS-covered area; the inset presents increased resolution of impact class distribution for six biomes with little LRS-covered area. MFWS = Mediterranean forests, woodlands and scrub; DXS = desert xeric shrubs; L = lakes; TBMF = temperate broadleaf mixed forests; TGSS = temperate grasslands, savannahs and shrublands; FGS = flooded grasslands and savannahs; MGS = montane grasslands and shrublands; TSGSS = tropical and subtropical grasslands, savannahs and shrublands; TCF = temperate conifer forests; TSDBF = tropical and subtropical dry broadleaf forests; TSCF = tropical and subtropical coniferous forests; RI = rock and ice; BT = boreal forests/taiga; TSMBF = tropical and subtropical moist broadleaf forests; M = mangroves; T = tundra. Grey colour represents non-LRS area, including potential LRSs in Indonesia and Malaysia that were not assessed because of lack of data.

Source: Nilsson et al., 2005.

which (the Tasang Dam on the main stem) is based on international and interbasin benefits and would alone make the Salween moderately affected.

Almost half of the new dams are located on just four rivers: forty-nine on the Yangtze (China), twenty-six on the Rio de la Plata (Argentina and Uruguay), twenty-six on the Tigris-Euphrates (Iraq, Syrian Arab Republic and Turkey), and twenty-five on the Ganges-Brahmaputra (WWF and WRI, 2004). In addition to the Salween, new dams are planned for several other unaffected LRSs, including the Cá and the Agusan rivers in Southeast Asia, and the Jequitinhonha in South America (see **Chapter 14**). Approximately nine rivers are at risk of entering a higher impact class: from unaffected to affected, or from moderately to strongly affected. Some of the impacts of these new dams may be limited by adopting recommendations from the WCD. Understanding and maintaining ecosystem functions in river systems where new dams are planned can be partly achieved by considering and balancing all the social, environmental and economic forces surrounding each dam proposal.

Fragmentation poses serious problems that can, in some cases, more than counteract any of the advantages of dams (see **Box 5.3**). The foregoing analysis identifies three biomes dominated by strongly affected LRSs (temperate broadleaf and mixed forests; temperate grasslands, savannahs and shrublands; and flooded grasslands and savannahs) that deserve immediate action to mitigate the impacts of existing alterations to flow regimes.

A reasonable goal might be to protect remaining unaffected basins from dam construction, as most of the unaffected basins are relatively small and free-flowing because their location and form have not made damming feasible, thereby making them easier to protect. For example, rivers with long, gently-sloping runs instead of large falls do not meet most hydroelectric requirements (see **Chapter 9**). Rivers on large plains, rather than distinct valleys, are also poor sites for constructing impoundments.

The possibility of decommissioning more dams might be considered, especially those that are old and no longer

BOX 5.3: DAMS AND THEIR ALTERNATIVES

The World Commission on Dams report proposed a new decision-making framework to improve the planning and management of dams and their alternatives. One of its strategic priorities, 'sustaining rivers and livelihoods', addresses the need for a basin-wide understanding of ecosystem functions and the livelihoods that depend on them, as well as adopting options and decision-making to avoid impacts, followed by minimization and mitigation of harm to the health and integrity of the river systems. Environmental flows are water flow allocations that are reserved for the river to sustain their ecological functions and species. While the core values and strategic priorities put forward in the report are widely accepted by all major stakeholders, the full set of recommendations, including policy principles and guidelines, has been the subject of dispute among some stakeholder groups and governments, which has limited the extent to which the recommendations have been applied on a global scale. The UNEP Dams and Development Project aims to promote dialogue on improving decision-making, planning and management of dams and their alternatives on the basis of WCD core values and strategic priorities.

Source: WCD, 2000.

fulfil their original purpose, those that have created serious environmental impacts, and those where endangered species and ecosystems face a high risk of extinction. These issues are now beginning to be addressed in some locations. In the US, for example, more dams are being removed each year than are being built. By 2000, 465 dams had been decommissioned in the US, the majority of which were followed by successful ecological and fisheries restoration (Postel and Richter, 2003). In some cases, improved dam operations, such as the implementation of fish ladders, the occasional flooding of downstream areas and the maintaining of minimum river flows, may be more feasible than removal and can also help to restore key habitats.

4c. Pollution

In addition to the pollution that generally accompanies agricultural intensification, aquatic ecosystems are affected by a wide range of pollutants that are leached from the soil, released directly into waterways or deposited from the atmosphere. Domestic and industrial effluents can seriously impact aquatic ecosystems, particularly in developing countries where wastewater treatment is minimal or non-existent and untreated effluents are often discharged directly into waterways. An estimated 80 percent of the pollutants entering coastal waters, mostly from land-based sources, are transported via rivers, and there are clear links between upstream river basins and associated coastal zones (UNEP, 2004b). Furthermore, at least eight of the ten regions defined by UNEP's Regional Seas Programmes[3] with sufficient data report that over 50 percent of their wastewater is still discharged into coastal and freshwaters untreated; and for five of these, it is over 80 percent (UNEP/GPA, 2004). Untreated sewage from municipal sources and animal wastes from agricultural

activities also add high concentrations of carbon-rich organic material to these pollutant loads.

Even in the developed world, industrial effluents can have significant negative impacts on aquatic ecosystems (see **Chapter 8**). In the US alone, it is estimated that industry generates about 36.3 billion kg of hazardous organo-pollutants each year, with only about 10 percent disposed of in an environmentally responsible manner (Reddy and Mathew, 2001). Concentrations of organochlorine pesticides, such as DDT and BHC[4], have been declining over the past decade in some countries' surface waters, as regulations to curtail their use have been put in place. Such compounds are the focus of major global studies (e.g. Li and Macdonald, 2005; Ueno et al., 2003), because they are harmful to aquatic biota, persistent in ecosystems, and their derivatives can bio-accumulate in food chains, having potentially significant impacts on animals at the top of these chains. Studies undertaken in the northern rivers of Russia clearly show the degree of decline in both river water quality and Burbot fish (*Lota lota*) (see **Figure 5.5**) (Zhulidov et al., 2002). Similarly, BHC concentrations in China have exhibited a significant decline over time. However, because of their persistence, the impacts of DDT and other organochlorines continue to be seen for many years after their use has been discontinued.

In recent years, there has been a growing concern about the impacts that personal care products[5] and pharmaceuticals are having on water quality and the productivity of aquatic systems and ecological functioning – through the disruption of endocrine systems in fish, for example (UN-WWAP, 2003). Between the 1940s and 1984, it is estimated that over 1 million tonnes of antibiotics were released into the biosphere (Mazel and

3. See www.unep.org/regionalseas/About/default.asp for more information about this programme.

4. Dichloro diphenyl trichloroethane and hexachlorocyclohexane, respectively.

5. This refers to a wide range of products used to, e.g., soften your water, boost the cleaning power of laundry detergents and other household products, skin protecting lotions, deodorants, compounds used to prevent things like shampoos and conditioners from spoiling after purchase, and increasing the protective power of sunscreens.

6. Invasive alien species (IAS) are defined as 'an alien species (species, subspecies or lower taxon, introduced outside its natural past or present distribution; and includes any part, gametes, seeds, eggs or propagules of such species that might survive and subsequently reproduce), whose introduction and/or spread threaten biological diversity' (CBD Decision VI/23).

Davis, 2003). As few studies have been undertaken to quantify the effects of personal care products and pharmaceuticals on components of aquatic ecosystems (e.g. effects on algal assemblages in freshwater, Wilson et al., 2003), very little is known generally about their distribution, fate and effects on aquatic systems and potable water supplies (Jones et al., 2005; Sharpe, 2003).

4d. Invasive species

Invasive alien species (IAS) are thought to be the second most important cause of biodiversity loss in freshwater systems after habitat loss and degradation. However in some lake ecosystems, they are now considered by some to be the primary cause of biodiversity loss (Ciruna et al., 2004).[6]

There are many ways in which invasive species can become established in an ecosystem, as a result of political, demographic, cultural, socio-economic or ecological factors. Introductions can either be intentional – through introduction of 'exotic' plants and organisms into gardens or waterways, or government-sanctioned releases of organisms for propagation or harvest – or

unintentional, as a result of the escape of aquaculture operations or the accidental transport of organisms attached to boats, structures, garbage or in ballast water.

There appears to be some correlation between levels of human activity, trade, ecological integrity and the resistance of ecosystems to invasion from introduced species (Ciruna et al., 2004). Where ecosystem functions have been degraded, there is generally a greater susceptibility to invasions. Once an IAS has established in a new region, it can cause severe damage to local species and habitats. **Table 5.3** indicates the scale of introductions in different regions.

In Mexico (see **Chapter 14**), for example, of the 500 or so known fish species, 167 are considered to be at risk and, of this total, 76 are thought to be threatened by invasive alien species (Ciruna et al., 2004). Changes in biodiversity through predation and competition for resources can lead to decreased local biodiversity. The Nile Perch (*Lates niloticus*) was originally introduced into Lake Victoria in 1954 to counteract the impacts of overfishing of native

Figure 5.5: Declines in the concentrations of organic contaminants in Russian and Chinese rivers

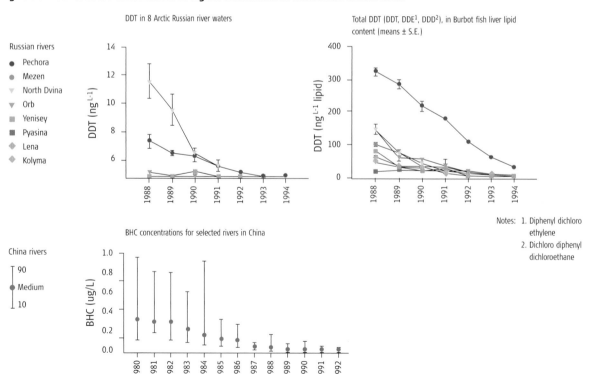

Sources: Russian data: Zhulidov et al., 2002; Chinese data: GEMS/Water www.gemswater.org

fish stocks (as seen in **Box 5.1**). However, aggressive competition and predation by this species on the native fish has since resulted in the apparent disappearance of up to 132 endemic fish (Stiassny, 2005). The introduction of the Atlantic comb jellyfish that caused the collapse of fisheries in the enclosed Black Sea is another well documented example of the detrimental effects of IAS (e.g. in UNEP, 2002b). And in the semi-enclosed Mediterranean, the accidentally introduced algae (*Caulerpa taxifolia*) now affects six western Mediterranean and Adriatic countries, covering 13,000 ha of the sea floor over 180 km of shoreline, where it has colonized precious sea-grass beds. IUCN classifies it as one of the 100 most dangerous invading species (Blue Plan, 2005).

Dominant invasive species can cause a rapid decrease in the productivity of the ecosystem. The water hyacinth (*Eichhornia crassipes*) is one of the most aggressive and fastest growing aquatic weeds in the world. Originating in South America, it is now present in more than fifty countries, primarily as a result of its introduction as an attractive ornamental plant. Within a matter of days, water hyacinth infestations can block waterways, preventing the passage of boats and interrupting economic activities, as well as dramatically reducing the availability of light and oxygen in the water – usually killing off endemic species in the process (Lowe et al., 2004).

4e. Climate change

The current and likely future impacts of climate change on coastal and freshwater ecosystems, as outlined by the Intergovernmental Panel on Climate Change (IPCC), are not yet fully understood. Sea-level rise, higher temperatures, greater carbon dioxide concentrations in seawater, increased droughts and floods, and increasingly frequent extreme weather events are all anticipated, with major implications for aquatic ecosystems. Warmer water, combined with anticipated changes in ocean currents, could have a devastating impact on water ecosystems and species diversity. One potential result is a reduction in the upwelling of nutrients, which would in turn reduce productivity in key fishing areas. Decreased growth may also be seen in coral reefs, with high concentrations of carbon dioxide in the water, impairing the deposition of limestone required for coral skeletons. A significant sea-level rise will cause some low-lying coastal areas to become completely submerged, while others will increasingly face high but short-lived water levels. These anticipated changes will have major impacts on coastal habitats and populations. Coastal zones harbour

Table 5.3: Introductions of invasive species by region

Region	Percentage of total recorded invasive species introductions
Europe	25.1
Asia	16.4
Africa	14.7
Oceania	14.7
South and Central America	14.1
Middle East	8.4
North America	6.3

Source: Ciruna et al., 2004.

approximately 38 percent of the global population and nine of the ten most densely populated cities in the world. The most vulnerable coastal nations, as recently assessed by UNEP through a vulnerability index, are Bangladesh, China, India, the Netherlands, Pakistan, the Philippines, the United States and the small island developing states, especially Barbados, Fiji, Haiti, the Maldives and the Seychelles (UNEP, 2005; see also **Chapter 10**).

More detailed scenarios for such areas as North America (Schindler, 1997) and southern Africa (Hulme, 1996) predict major changes, particularly for dynamic shallow-surface water systems, which will in turn affect their biodiversity and the livelihood of the populations that depend on them.

On a global level, polar and arid systems appear to be the most vulnerable to climate change (see also **Chapter 4**). Polar systems store the vast majority of freshwater, and most scenarios suggest they are likely to develop a considerably increased discharge of water, driven by higher temperatures in both the polar regions and particularly in the Arctic (ACIA, 2004). Arid regions are also expected to experience drastic changes.

While global warming may increase productivity in some regions and habitats, the overall predictions are that the impacts of climate change on aquatic ecosystems will be detrimental. Coastal wetlands such as mangroves and coral reefs (Southeast Asia), coastal lagoons (Africa and Europe) and river deltas (the Nile, Niger and Congo in Africa; the Ganges and Mekong in Asia) will be seriously affected by rising water levels, as well as other coastal lowland areas with an elevation of less than 0.5 m (UNEP, 2002c).

The crash of the European eel population (*Anguilla rostrata*) is an example of the detrimental effects of

The current and likely future impacts of climate change on coastal and freshwater are not yet fully understood

BOX 5.4: BIODIVERSITY IN LAKE CHAD

Lake Chad is 250 metres above sea level and its drainage basin is shared by Cameroon, Chad, Niger and Nigeria. As one of the largest wetlands in Africa, it hosts a biodiversity of global significance. These wetlands were once home to many large mammals, including elephants, hippopotami, gazelle, hyenas, cheetahs and wild dogs and also provided a habitat for millions of migratory birds. The lake supports fish populations that feed local communities and provide an important export trade, with 95 percent of the catch going to Nigeria in a trade worth an estimated US $25 million a year. During the severe drought periods of the 1970s, 1980s and early 1990s, Lake Chad shrunk significantly, from approximately 23,000 km^2 in 1963 to less than 2,000 km^2 in the mid-1980s. The key reasons for this phenomenon have been

identified as overgrazing, deforestation contributing to a drier climate and large, unsustainable irrigation projects in Cameroon, Chad, Niger and Nigeria which have diverted water from the lake and the Chari and Logone rivers. Since the 1990s, the lake levels have started to rise as rainfall has increased. However, the Intergovernmental Panel on Climate Change (IPCC) predicts reduced rainfall and increased desertification in the Sahel near Lake Chad, and drought frequencies are likely to increase again. The size of the affected region and the duration of the phenomenon are unprecedented. The changes have contributed to a widespread lack of water, major crop failures, livestock deaths, a collapse of local fisheries, rising soil salinity, and increasing poverty throughout the region. *Alestes naremoze*, a fish species that once

made up approximately 80 percent of the catch, is now rare due to the disappearance of its natural spawning beds. Sarch and Birkett report an annual fish catch in the Lake Chad Basin between 1986 and 1989 of 56,000 tonnes, compared to an annual catch of 243,000 tonnes between 1970 and 1977. The implications do not stop in the Sahel.

The decline in migratory bird species, such as the Central European wet grassland waders, including Ruff (*Philomachus pugnax*) and Black-tailed Godwit (*Limosa limosa*), is also related to the changes in conditions in Lake Chad and other wetland areas in the Sahel zone (see **Map 5.4**).

Sources: UNEP, 2004a, 2004c; Nami, 2002; Coe and Foley, 2001; FEWS, 2003; IPCC, 2001; Sarch and Birkett, 2000; Zöckler, 2002.

Map. 5.4: Levels of Lake Chad 1963–2001

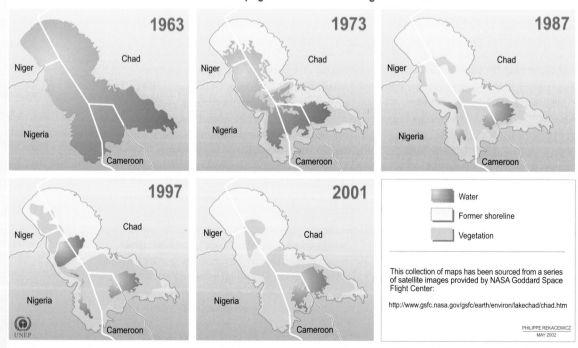

Source: UNEP, 2002c, 2004c.

BOX 5.5: DRAMATIC DECLINE OF THE ARAL SEA

In Central Asia, the Aral Sea has also declined dramatically in recent decades, with devastating consequences for both biodiversity and human well-being. **Map 5.5** shows the impact of highly intensive agriculture on the Aral Sea Basin, including the construction of ninety-four water reservoirs and 24,000 km of channels on the Amu Darya and Syr Darya rivers to support the irrigation of 7 million ha of agricultural land. As a direct result of these developments, the volume of water in the Aral Basin has been reduced by 75 percent since 1960. This loss of water, combined with the impact of excessive chemicals from agricultural runoff, has caused a collapse in the Aral Sea fishing industry, a loss

of biodiversity and wildlife habitat in the area's rich wetlands and deltas, and an increase in human pulmonary diseases and infant mortality resulting from the high toxicity of the salt concentrations in the exposed seabed. Whereas in 1959 the Aral Sea's fisheries produced almost 50,000 tonnes of fish, by 1994 the annual catch was only 5,000 tonnes. Biodiversity has also declined, with many local species extinctions in the region. The sensitive Turgay riverine forests, once a hotbed of biodiversity, have been reduced to marginal fragments in four nature reserves in Uzbekistan.

Sources: UNESCO, 2000; Postel, 1999; Kreutzberg-Mukhina, 2004.

Stranded boats on the exposed, former seabed of the Aral Sea

Map. 5.5: Major irrigation areas in the Aral Sea Basin

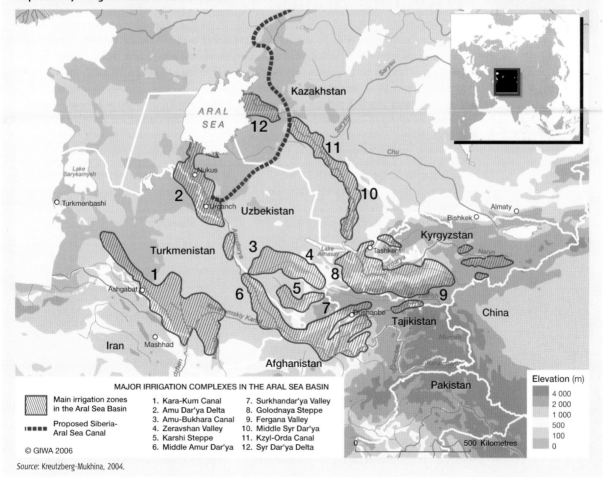

MAJOR IRRIGATION COMPLEXES IN THE ARAL SEA BASIN

Main irrigation zones in the Aral Sea Basin

Proposed Siberia-Aral Sea Canal

© GIWA 2006

1. Kara-Kum Canal	7. Surkhandar'ya Valley
2. Amu Dar'ya Delta	8. Golodnaya Steppe
3. Amu-Bukhara Canal	9. Fergana Valley
4. Zeravshan Valley	10. Middle Syr Dar'ya
5. Karshi Steppe	11. Kzyl-Orda Canal
6. Middle Amur Dar'ya	12. Syr Dar'ya Delta

Elevation (m)
4 000
2 000
1 000
500
100
0

0   500 Kilometres

Source: Kreutzberg-Mukhina, 2004.

climate change at the species level. The European eel fishery, which once sustained 25,000 fishermen, has systematically declined over the past thirty years, due in part to climate change and the weakening of the Gulf Stream. By the mid-1980s, the number of new glass eels (eel juveniles) entering European rivers had declined by 90 percent. Recent figures indicate that this level has now dropped to 1 percent of former levels (Dekker, 2003). While the major causes for this drastic decline are believed to be overfishing of eel juveniles for aquaculture operations (particularly in Japan), freshwater habitat loss and degradation, pollution, disease, and changes in climate and ocean currents are also contributing to reducing the number of juveniles (Dekker, 2003).

Scientists now believe that the glass eels may be unable to reach Europe because the Gulf Stream has slowed so much that they cannot survive for a sufficiently long period to make the 5,000-mile journey (Brown, 2004).

Dryland areas are naturally vulnerable to water stress. Changes have been well documented for sites such as Lake Chad and the Aral Sea (UNEP, 2002b), which are examined in **Boxes 5.4** and **5.5**, because they illustrate the extreme effects of current ecological changes. For many dryland waters, however, there is little information, including for much of Central Asia, the Middle East, and most parts of Africa.

Part 5. Policy and Management Responses: Implementing the Ecosystem Approach

This chapter has thus far discussed some of the most serious problems facing the world's coastal and freshwater ecosystems, from both a social and an environmental perspective. While recent improvements in some areas suggest that the situation is far from hopeless, failure to address these problems will have immediate social and economic costs and long-term – in some cases irreversible – impacts on biodiversity. It is perhaps not too alarmist to talk about a crisis currently facing water resources management. The following section considers some of the current and potential responses to this crisis.

According to the Global Water Partnership (GWP), current problems of water management often stem from the lack of integrating ecosystem functions and processes into natural resource management efforts. The ecosystem approach is not sufficiently implemented. The GWP also stresses that current management efforts suffer from a widespread lack of governance structures and legal frameworks for integrating the policies that can have a positive impact on the management of water resources. Good water governance exists where the responsible government bodies establish effective policies and legal frameworks to allocate and manage water in ways that are responsive to national and relevant international, social and economic needs, and to the long-term sustainability of the resource base. Such policies need to recognize the finite and sensitive nature of water resources, incorporate notions of the sustainable use of aquatic systems, and negotiate and develop partnerships with relevant stakeholder groups. In this way, government policies can be supported by the population, rather than

being opposed and resisted. This suggests that current water management frequently fails to meet these ideals (Rast and Holland, 2003), thus jeopardizing many of the goods and services provided to humanity by healthy aquatic ecosystems.

This is perhaps not surprising; negotiations over freshwater are among the oldest and most intractable problems relating to the use of the Earth's natural resources (see **Chapter 11**). Choices and trade-offs must often be made between a range of potential benefits that could be derived from an aquatic ecosystem. If a particular aquatic system is managed to maximize fisheries production, for example, then benefits that could be derived from the diversion of water for irrigation are likely to be reduced. Different needs have to be balanced within the natural limitations and functions of the ecosystem and between local communities (see **Chapter 12**), and local needs have to be balanced with those of more distant users, who may be far downstream or in recipient coastal areas (see

Chapter 11). Achieving the sustainable use of our readily available water resources in order to equitably share and value them is the reason for developing approaches associated with Integrated Water Resources Management.

5a. IWRM and its implementation challenges

It is becoming increasingly accepted that the most effective approach to the sustainable use of aquatic ecosystems is embodied in the concept of IWRM. A primary difference between the traditional sectoral approach to water management and IWRM is that the latter makes the link between water resources and human activities throughout the hydrological cycle and allows ecological and socio-economic issues to be considered within an ecosystem approach.

IWRM specifically considers the relationships between freshwater and coastal zones, along with other interactions between freshwater, land use and development. It seeks to reduce the negative impacts of development in a river basin through, for example, the use of alternative land-use practices that mitigate damage while maintaining economic and social benefits (Falkenmark et al., 1999; GWP, 2000). At the same time, integrated coastal zone management (ICZM) is widely accepted as the most appropriate policy framework for the coastal–marine interface, while the integrated coastal area and river basin management (ICARM) merges the two. The close link between freshwater and coastal ecosystems is further recognized by the Convention on the Protection and Use of Transboundary Watercourses and International Lakes, the Global Programme of Action (GPA) for the Protection of the Marine Environment from Land-based Activities, and the European Union (EU) Water Framework Directive (UNEP, 2004b).

Some governments and international development and conservation organizations use the integrated river basin management (IRBM) approach, a similar concept to IWRM, which considers the river or lake basin/aquifer as the ecologically defined management unit. The application of IRBM can therefore take place at a variety of scales, depending upon the size of the river basin, ranging from small catchments of a few square kilometres to major national basins (e.g. the Loire and Vistula in Europe), as well as transboundary basins where allocation and pollution issues cross international borders (e.g. Lake Chad and the Danube, Oder and Rhine river basins). Special institutional and governance structures, such as river basin organizations or authorities, have sometimes been established to set operational and legal frameworks

within which freshwater resources can be managed to meet different stakeholder interests (see **Chapter 14**).

Although IWRM is simple to envision, experience over the past decade suggests that it is difficult to implement effectively because of the need to integrate a complex, and often competing, combination of elements. These elements can include the following (GWP, 2000):

- land and water issues
- freshwater bodies and downstream coastal zones
- water consumed in the direct production of biomass as opposed to that flowing in rivers and aquifers (green versus blue water)
- surface-water and groundwater resources
- water quantity and quality
- differing water interests upstream and downstream.

IWRM becomes even more complex when transboundary water systems are involved, because this situation frequently requires that one or more countries subordinate some of their national interests in favour of their neighbours' needs (see **Chapter 11**). The World Lake Vision Committee has identified the lack of proper accountability on the part of citizens and governments as one of the most significant root causes of unsustainable water use, along with a general lack of accountable environmental stewardship, inadequate stakeholder participation, and inappropriate or ineffective governmental institutions and regulatory mechanisms (World Lake Vision Committee, 2003).

A series of regional workshops, convened by UNEP in various developing countries to address and redress the fact that IWRM schemes have frequently experienced serious problems, concluded that some of the major barriers to implementing IWRM (Rast, 1999) include the following:

- lack of proper coordination of management activities
- lack of appropriate management tools
- inability to integrate water resources policies
- institutional fragmentation
- insufficiently trained or qualified manpower
- shortfalls in funding
- inadequate public awareness
- limited involvement by communities, non-governmental organizations (NGOs) and the private sector.

Attempts to resolve some of these issues have included the establishment of freshwater and regional seas

...negotiations over freshwater are among the oldest and most intractable problems relating to the use of the Earth's natural resources

Looking beyond freshwaters to consider links with coastal waters has often proved difficult, partly because river managers are often water engineers, concerned with issues of water quantity and quality, food production and flood management

agreements at local, basin or regional levels (e.g. in the Mekong, the Black Sea and the Danube, the Mediterranean, and Lake Chad). While such initiatives have met with some success, they often still lack the policy tools necessary to promote long-term integrated water resources management.

UNEP has proposed four overall principles for the development of such approaches (UNEP, 2004b):

- **An adaptable management structure:** effective institutional management structures must incorporate a certain degree of flexibility, allowing for public input, changing basin priorities, and the incorporation of new information and monitoring technologies. The adaptability of management structures must also extend to non-signatory riparian countries (i.e. those within the same hydrological system) by incorporating provisions that address their needs, rights and potential accession.

- **Clear and flexible criteria for water allocations and quality:** water allocations, which are often at the heart of most water disputes, are a function of water quantity and quality, as well as political fiat. Effective institutions must therefore identify clear allocation schedules and water quality standards, which simultaneously provide for extreme hydrological events, new understandings of basin dynamics, and changing societal values and aquatic ecosystem needs. Riparian states may also consider prioritizing uses throughout the basin. Establishing catchment-wide water precedents may not only help to avert inter-riparian conflicts over water uses, but also to protect the environmental health of the basin as a whole.

- **Equitable distribution of benefits:** this concept, subtly yet powerfully different from equitable water use or allocation, is at the root of some of the world's most successful water management institutions, a noteworthy example being the US-Canada International Joint Commission (IJC, 1998). The idea concerns the distribution of the benefits derived from water use – whether from hydropower, agriculture, economic development, aesthetics, or the preservation of healthy aquatic ecosystems – rather than equal distribution of the water itself. Distributing water-use benefits allows for positive-sum agreements, whereas dividing the water itself among competing users may only allow for winners and losers (see **Chapter 12**).

- **Detailed conflict resolution mechanisms:** many basins may continue to experience disputes even after a treaty is negotiated and signed. Therefore, incorporating clear mechanisms for resolving conflicts is a prerequisite for effective long-term basin management for sustainable water use (UNEP, 2002a; see **Chapter 11**).

Other examples of integrated approaches are recommended for applying water management efforts in areas of water scarcity, as well as those with water abundance. The recommendations of a World Bank report (Abdel-Dayam et al., 2004) on agricultural development include the following:

- Evolving institutions for the governance, management and financing of agricultural drainage, as well as (re)designing physical interventions and technical infrastructure from the perspective of multi-functionality and plurality of values.

- Drafting policies that create environments conducive to change and empower actors to make the necessary changes.

Fortunately, IWRM is becoming increasingly accepted and ingrained in the planning and decision-making processes of water managers and policy-makers. It has become clear that there are great similarities between management issues in coastal and river areas, and the concept of IWRM has been extended from its initial freshwater focus to establish appropriate links to coastal waters. One major impetus towards the adoption of IWRM is the Plan for Implementation from the 2002 WSSD, through which participating governments agreed to develop IWRM and water-efficiency plans by 2005 (see **Chapter 2**). To this end, the GWP is also promulgating principles and approaches to assist governments to meet this deadline (see **Chapter 1**).

Looking beyond freshwaters to consider links with coastal waters has often proved difficult, partly because river managers are often water engineers, concerned with issues of water quantity and quality, food production and flood management. In contrast, much of the concern regarding coastal zones is focused on the impacts of land-based activities on downstream coastal areas. UNEP has promoted this management link since 1999 through its ICARM programme. The FreshCo Partnership was launched at the World Summit on Sustainable

BOX 5.6: **THE ECOSYSTEM APPROACH IN ACTION**

Quito Catchment Conservation Fund, Ecuador

About 80 percent of Quito's drinking water comes from two protected areas, the Cayambe Coca Ecological Reserve and the Antisana Ecological Reserve. A nominal water use fee on citizens of Quito together with 1 percent of revenues of hydroelectric companies and contributions expected from other sources in the future is used to finance conservation of the reserves.

The Komadugu-Yobe Integrated Management Project, Nigeria

To combat increasing tensions among local stakeholders about scarce water resources in the North Nigerian River Basin, the Nigerian National Council on Water Resources established the Hadejia-Jama'are-Komadugu-Yobe Coordination Committee in 1999, with support from the Komadugu-Yobe Integrated Management Project. The project will establish a framework for broad-based and informed decision-making, based upon agreed principles for equitable use and sustainable management of the Komadugu-Yobe Basin.

European Union Water Framework Directive

This Directive, adopted in 2000, stipulates that EU governments should adopt IRBM to achieve 'good' or 'high' ecological status in all water bodies (coastal and inland) by 2015 (see **Chapter 14**). Status is assessed by indicators measuring departure from natural or pristine conditions for any given category of water body. Data are evaluated against five categories of ecological status, ranging from 'high', representing the absence of anthropogenic disturbance in all variables, or only very minor alteration, to 'bad', reflecting an extensive departure from natural conditions.

Sources: Echavarria, 1997; IUCN, 2003b; EU, 2000.

Development (FreshCo Partnership, 2002)[7] to provide further impetus to this process.

IWRM is increasingly being recognized by the global community, and has become the subject of international commitments and targets that are starting to develop necessary legal and policy frameworks. There is now an urgent need to move beyond these preliminary steps to widespread implementation, and doing this effectively will require developing a series of tools and methodologies or adapting those already used in different biomes and situations. Of equal importance, the partnerships – between governments, communities, NGOs, industry interests and research groups – must move beyond general commitments to specific actions and active, flexible and durable working arrangements. In pursuing these partnerships, it will be essential that the case for conservation be rooted in the most reliable and accessible information on aquatic ecosystems, particularly with respect to their values, uses and flow requirements, and how these properties may vary between basins and ecosystems. Examples for implementing this ecosystem approach are provided in **Chapter 1**.

As demand for water grows with the need for hard decisions to meet this demand, so too will conservation need to be rooted in the soundest possible science.

None of these goals will be easy to achieve, but the fact that political commitment is growing along with the recognition of the urgency of these needs provides some cause for cautious optimism.

5b. Protecting and restoring habitats

One management response to pressures on wetlands is to protect a certain proportion through protected areas, such as national parks or wilderness areas. While most such areas are primarily designated to protect their biodiversity values, they can also serve other beneficial functions, such as the protection of fish breeding stocks or of coastlines, flood mitigation and the maintenance of water purity. Protected areas can be an important component within watersheds and coastal ecosystems managed under IWRM approaches. According to the latest report on the state of the world's protected areas (Chape et al., 2004), some 12.7 percent of wetlands are contained within protected areas recognized by the IUCN. Because these figures are based on remote sensing data, smaller wetland protected areas and wetlands classified as other kinds of habitat, such as forests or grassland, may have been missed, and the real degree of protection may be higher (see **Figures 5.6** and **5.7**). If tropical moist forest is included, of which a large proportion is regularly flooded, the value potentially increases to almost 20 percent.

7. See www.ucc-water.org/ freshco/ for more information.

Figure 5.6: Surface area and degree of protection of major terrestrial habitats

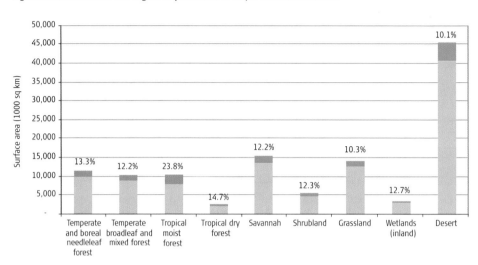

Source: Chape et al., 2004.

Figure 5.7: Distribution and degree of protection of wetland habitats by region

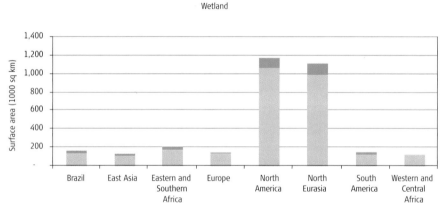

Source: Chape et al., 2004.

In addition to fully protected area status, a number of other important instruments exist to help safeguard freshwater and coastal ecosystems, while integrating their protective functions with other forms of sustainable development.

Ramsar sites

The Ramsar Convention on Wetlands, signed in Ramsar, Iran, in 1971, is one of the oldest intergovernmental treaties explicitly dedicated to the conservation of wetlands. It provides a framework for national action and international cooperation for the preservation and wise use of wetlands and their resources (see **Chapter 12**). There are presently 144 Contracting Parties to the

Convention, with 1,420 wetland sites, totalling 123.9 million ha, designated for inclusion in the Ramsar List of Wetlands of International Importance (See **Figure 5.8**). While many Ramsar sites are also officially protected areas, the Ramsar designation has been used as a 'softer' instrument to generate commitments to sustainable use and guarantee some degree of protection without necessarily ruling out all forms of sustainable development. According to a recent study by the World Bank, the designation of aquatic ecosystems as Ramsar sites is likely to have improved the conservation prospects of these sites for a variety of reasons, including increased awareness of their ecological importance, increased

Figure 5.8: Total area designated as Ramsar sites (1974–2004)

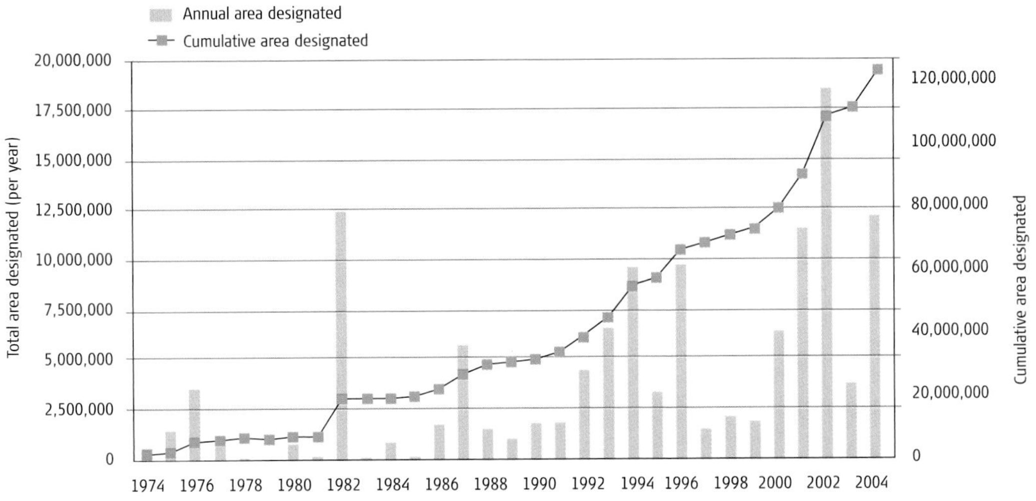

Source: Wetlands International, 2005: www.wetlands.org/RDB/global/AreaTrend.html

conservation funding (both international and domestic), increased participation by local stakeholders in conservation, and reduction of threats (Castro et al., 2002).

African-Eurasian Migratory Waterbird Agreement (AEWA)

Migratory bird species are particularly vulnerable to water degradation and habitat fragmentation. They utilize several different types of habitats, depending on the stage they have reached in their lifecycle and the possibility of unhindered movement between habitats. The Convention on Migratory Species (CMS) is another treaty with particular relevance to aquatic ecosystems, through its role in sustaining migration routes for waterbird species. The Convention helps to inform national and international agreements on the protection of migratory birds, fish and coastal species, such as whales and marine turtles, often also including official or voluntary protection of habitats. The AEWA is the largest agreement developed so far under the CMS, which came into force on 1 November 1999. The agreement covers 235 species of birds that are ecologically dependent upon wetlands for at least part of their annual cycle. The AEWA aims to enhance waterbird protection by establishing a site network and promoting the sustainable use of waterfowl and their habitats (Boere, 2003).

5c. Ecosystem restoration

Efforts to address the serious degradation problems facing many freshwater and coastal ecosystems are already demonstrating that ecosystem restoration is indeed possible. To date, most restoration activities have been initiated and carried out by NGOs, although an increasing number of governments and local communities are now undertaking such projects. Restoration has become a central activity in modern environmental management, as well as a growing stage for long-term sustainable management systems. However, it should be noted that restoration is not a panacea for poor management. Such projects are generally very costly, and some habitats are extremely difficult, if not impossible, to restore. As an example, raised bogs and fen mires are, at best, very difficult to restore, and in some cases, restoration is not possible because irreversible soil changes have occurred. The goal of preventing damage must still be the most important management objective, although restoration remains an important option when damage has already occurred. **Box 5.7** briefly describes some current restoration projects, both large and small, in the very different conditions that characterize Africa, Asia and Europe.

Many more examples of successful freshwater and marine restoration schemes could be cited, from large-scale government initiatives such as the restoration of the

BOX 5.7: RESTORED ECOSYSTEMS AND LIVELIHOODS

Mauritania

The Diawling Delta had been virtually destroyed by a combination of years of low rainfall and the construction of a dam in 1985, leading to an ecological crisis, loss of wetland-dependent livelihoods and the mass migration of its inhabitants. Restoration started in 1991, when the government declared 16,000 ha as a national park. Although the population was initially hostile to the declaration, acceptance of this designation grew as protected area managers and the World Conservation Union (IUCN) staff worked with communities to restore the region's biodiversity and local livelihood options. There is now an agreed management plan, increased management capacity and procedures to resolve resource conflicts. The restoration area covers 50,000 ha, larger than the national park area itself. Artificial flooding began in 1994, with the aim of reconstructing former flooding patterns and saltwater inflow, which has restored a diverse delta ecosystem, with fish catches rising from less than 1,000 kg in 1992 to over 113,000 kg in 1998. Seeds of the restored acacia trees are used in the tanning industry, and the indigenous women once again produce famous traditional mats from *Sporobulus robustus*, a brackish floodplain grass that again grows abundantly. Bird counts have risen from a meagre 2,000 in 1992 to over 35,000 waterbirds in 1998. The total value added to the region's economy as a result of this effort is approximately US $1 million per year.

Mozambique

The Kafue Flats, named after the Kafue River, a tributary of the Zambezi, consist of extensive savannah wetland of approximately 5,600 km², where the natural flooding regime was dramatically altered by two dams built in the 1970s. The dams had reduced the flooded area, changed the timing of the flooding, affected wetland productivity and reduced water resources, grazing areas and wildlife and fish populations, as well as the potential for tourism. The World Wide Fund for Nature (WWF) has been working with the Ministry of Energy and Water Development, and the Zambia Electricity Supply Cooperation, in a project that aims to restore the natural flooding regime of the Kafue and, in turn, restore wetland productivity and biodiversity. This can only be achieved by changing the operation rules of both dams. At the end of Phase I in 2002, an integrated river basin management (IRBM) strategy for the Flats was accepted by the Zambian Government. The first results of improvement are anticipated in the next few years.

The Netherlands

In 1982 the Dutch Government funded the creation of small ponds to replace the loss or simplification of natural wetlands and traditional canals. About 600 ponds were created in the first year, immediately colonized by the most common amphibian species, with some also being colonized by rarer species. Several thousand more ponds have been created throughout the country in subsequent years, and an International Pond Project has been launched under the auspices of the EU LIFE fund.

Germany

In order to reverse serious losses in natural habitats, the German Government implemented a programme to conserve and restore sites of national concern. By 2002, a total of fifty-three projects had been initiated, with forty-two of these relating to freshwater and coastal wetland habitats, conserving and restoring more than 180,000 ha. With NGOs or local communities as partners, a large proportion of the approximately US $400 million budget has been used over the past ten years for land purchases, which are often necessary before restoration can begin. In all of these projects, restoration work has started and mostly been completed, covering river stretches, wetland areas and fen mires.

United Kingdom

Restoration does not have to involve governments or international grants, as demonstrated by the increasing number of private reserves and restoration areas in many countries. One British farmer has converted 65 ha of former sugar-beet and wheat crops near the River Cam in East Anglia into a rich, diverse wetland, with meres, reedbeds and wet grassland areas to enrich local biodiversity. Since restoration work began in 1995, the project has resulted in the successful recovery and re-establishment of seventy-nine species of breeding waterbirds, including a rapid increase in the numbers of Lapwings (*Vanellus vanellus*), the establishment of Avocets (*Recurvirostra avocetta*), a colony of Common Terns (*Sterna hirundo*), and the successful breeding of endangered Bitterns (*Botaurus stellaris*) and Marsh Harriers (*Circus aeruginosus*).

Japan

Kushiro River, a Class A river originating in Kussharo Lake in Akan National Park in eastern Hokkaido, winds gently through the expansive Kushiro Swamp before reaching the Pacific Ocean. Its total length is 154 km with a basin dimension of 2,510 km², and 180,000 inhabitants in the area, approximately 75,000 of whom live in a flood area. Since salmon and trout run the river and artificial salmon hatching is operated, it is an important river for conservation. Kushiro Swamp, located downstream on the Kushiro River, is the biggest swamp in Japan: 18,000 ha, 5,012 ha of which are registered as a natural monument and 7,863 ha of which are designated under the Ramsar Convention. Located near an urban area, the swamp serves as an important flood barrier as well as a scenic tourist attraction. In the past fifty years, however, this area has decreased by 20 percent, as the swamp slowly gives way to alder forests. The Ministry of Land, Infrastructure and Transport has now joined forces with the Hokkaido Prefectural Government to explore various innovative options for restoring the river and swamp to their former glory. Ongoing research focuses on stream restoration, swamp vegetation management and control of sediment flows.

Sources: Hamerlynck and Duvail, 2003; WWF Mozambique, 2003; Stumpel, 1998; Scherfose et al., 2001; Hokkaido Regional Developement Bureau, 2003; Cadbury, 2003.

Everglades in the US and Australia's Great Barrier Reef, to local-level efforts in the coastal zones of the Mediterranean and Caribbean seas and the Pacific, Indian and Atlantic oceans. In all of these cases, the key aspects that have contributed to their success are as follows:

- the involvement of all stakeholders (government, community groups, environmental organizations, private sector, scientists and others) at all decision levels
- cross-sectoral planning and management (environment, development, agriculture, forestry, urban planning, tourism, public works, etc.)
- an appropriate landscape scale
- sufficient funding (e.g. using tourism returns)
- long-term planning.

Activities can vary widely and complement each other:

- closing damaging industries
- establishing 'no-take' sanctuaries
- banning illegal fishing
- promoting alternative livelihoods through micro-enterprise development
- launching public awareness campaigns
- supporting environmental clean-up
- developing disaster response strategies for oil spills and the like
- regulating tourism
- creating and maintaining mooring schemes.

Positive results do not come overnight, but experiences obtained over the past decade are very promising (UNEP, 2002a, 2004b; WRI, 2000; Bryant et al., 1998; Blue Plan, 2005).

Part 6. Facing Challenges and Managing Trade-offs

Accepted indicators, such as trends in the status of biodiversity, and pollution measurements, such as BOD and nitrate concentrations in water, indicate the continuing deterioration of our coastal and freshwater ecosystems. In addition, the global indicator of river fragmentation and flow regulation presented in this chapter shows that most of the large river systems are strongly or moderately affected by dams and altered flows. For the past decade, despite efforts to minimize or reverse these trends, aquatic ecosystems have continued to deteriorate – with freshwater systems declining at an even faster rate than marine or terrestrial ecosystems. Some specific habitats, such as freshwaters in arid areas and semi-enclosed seas, have been particularly affected. Furthermore, due to a widespread lack of comprehensive and coordinated monitoring programmes, our understanding of the status of many of these ecosystems remains poor or incomplete. The best available information currently focuses on coral reefs, waterbirds, amphibians and marine commercial fish, although even this information remains largely incomplete.

As demonstrated in this chapter, ecosystem changes do not only concern those interested in biodiversity, but also have direct and immediate impacts on human societies in terms of lost services such as drinking water, food production, employment opportunities, and recreational and aesthetic values. The poorest members of society generally suffer the most when coastal and freshwater are degraded, undermining national and international efforts at poverty alleviation (see **Chapter 1**).

It is critical that we recognize the direct links between the loss of biodiversity and ecosystem degradation and the loss of ecosystem resilience. Biodiversity and the

conservation of coastal and freshwater ecosystems are not separate issues from sustaining clean water and food security, but rather an integral part of the same agenda. As such, they must become an integral part of all future plans for water management and restoration.

These problems are now generally recognized by most governments, intergovernmental agencies, non-governmental organizations, major industries and – most crucially – by the communities directly involved. More importantly, there is general agreement regarding the appropriate way to move forward, based on the ecosystem approach and the harmonization of

It is critical that we recognize the direct links between the loss of biodiversity and ecosystem degradation and the loss of ecosystem resilience

conservation and development through what has become known as Integrated Water Resources Management.

However, despite all the fine words about this approach, the reality is that it has at most been implemented only locally, and even then often reluctantly. Many stakeholders remain ill-informed and unconvinced about the value of the ecosystem approach, and many communities continue to shun long-term benefits and values in favour of destructive short-term gains. Better public information and incentives for stakeholders to act in an 'environmentally friendly' manner are becoming vital if we are to improve the current situation.

The next challenge for national governments and the international community is to start implementing these approaches on a broader scale. Successful implementation will depend on winning the acceptance of the ecosystem approach by the majority of stakeholders and developing and providing implementing tools and applying methodologies to facilitate IWRM, many of which are related to managing the inevitable trade-offs involved. Rather than arguing about whether dams are good or bad, for instance, we need more robust criteria for deciding when they are, or are not, likely to produce net benefits, and how to best build dams to achieve societal needs while sustaining ecosystem functions. Similar tools are needed for weighing the trade-offs of different approaches to agricultural production and

Global warming has caused the Vatnajokull glacier in Iceland to retreat, revealing this spectacular lagoon

tourism. Better monitoring and evaluation systems are also needed to ensure that the impacts of management actions can be tracked over time and adjustments made as necessary. Those involved in managing water resources and ecosystems also need participatory and conflict resolution skills, in addition to greater technical expertise. The relevant organizations, strategies and frameworks for methodologies and partnerships are already being developed in many cases, although the urgency of the problems means that many of these efforts must be further accelerated. A clear, unequivocal lead from governments and the international community is needed to ensure that the good work developing in this critically important area is maintained and enhanced in the future.

Although data on biodiversity and water quality exist for some species groups, habitats and regions, there are still large gaps in the information available on many species, and very little information is available on the extent and quality of aquatic ecosystems. If the global community is serious about monitoring indicators that accurately describe the status of these ecosystems, habitats, species and their protection, in order to evaluate progress towards the WSSD's and Convention of Biological Diversity's 2010 target of reducing the rate of biodiversity loss, considerable improvements in the data quality, formats and geographical coverage are urgently required. Indeed, the ecosystem indicators presented today are only ever as good as the data that support them.

References and Websites

Abdel-Dayam, S., Hoevenaars, J., Mollinga, P. P., Scheumann, W., Slootweg, R. and van Steenbergen, F. 2004. *Reclaiming Drainage: Toward an Integrated Approach.* Washington, DC, World Bank.

Abramovitz, J. N. 1996. *Imperilled Waters, Impoverished Future: The Decline of Freshwater Ecosystems.* Worldwatch Paper No. 128, Washington DC, Worldwatch Institute.

Acharya, G. 1998. Valuing the hidden hydrological services of wetland ecosystems. Stockholm paper presented at 4th Workshop of the Global Economics Network, Wetlands: Landscape and Institutional Perspectives.

ACIA (Arctic Climate Impact Assessment). 2004. *An Assessment of Consequences of Climate Variability and Change and the Effects of Increased UV in the Arctic Region.* Anchorage, ACIA, Cambridge University Press.

Allison, E. 2004. The fisheries sector, livelihoods and poverty reduction in Eastern and Southern Africa. F. Ellis and A. Freeman (eds), *Rural Livelihoods and Poverty Reduction Policies.* London, Routledge.

Bann, C. 1997. An economic analysis of alternative mangrove management strategies in Koh Kong Province, Cambodia. EEPSEA (Economy and Environment Programme for Southeast Asia) Research Report Series, 1, Singapore, pp. 1–72.

Batista, V. S., Inhamuns, A. J., Freitas, C. E. C. and D. Freire-Brasil. 1998. Characterisation of the fishery in river communities in the low-Solimões/high-Amazon region. *Fisheries Management and Ecology*, Vol. 1, No. 5, pp. 419–35.

Belfiore, S. 2003. The growth of integrated coastal management and the role of indicators in integrated coastal management. *Ocean & Coastal Management*, Vol. 46, pp. 225–34.

Béné, C., Neiland, A.E., Jolley, T., Ovie, S., Sule, O., Ladu, B., Mindjimba, K., Belal, E., Tiotsop, F., Baba, M., Dara, L., Zakara, A., and Quensiere, J. 2003. Inland fisheries, poverty, and rural livelihoods in the Lake Chad basin. *Journal of Asian and African Studies*, Vol. 38, pp. 17–51.

Bennett, E. M., Carpenter, S. R. and Caraco, N. E. 2001. Human impact on erodable phosphorus and eutrophication: A global perspective. *BioScience*, Vol. 51, No. 3, pp. 227–34.

Blue Plan. 2005 (in prep). *Report on Environment and Development in the Mediterranean* (working title). Sophia-Antipolis, France, Blue Plan Regional Activity Centre of the Mediterranean Action Plan (MAP).

Boere, G. 2003. Global activities on the conservation, management and sustainable use of migratowaterbirds: an integrated flyway/ecosystem approach. *WSG Bulletin*, Vol. 100, pp. 96–101.

Bootsma, H. A. and Hecky, R. E. 1993. Conservation of the Great Lakes: A limnological perspective. *Conserv. Biol.*, Vol. 7, pp. 644–55.

Bragg, O. and Lindsay, R. 2003. *Strategy and Action Plan for Mire and Peatland Conservation in Central Europe.* Wetlands International, Publication 18.

Brown, P. 2004. Freshwater fish stocks revived, but climate change blamed for eel's decline. *The Guardian*, 25 August 2004.

Bryant, D. Burke, L., McManus, J. and Spalding, M. 1998. *Reefs at Risk: A Map-Based Indicator of Threats to the World's Coral Reefs.* Washington DC, WRI.

Burbridge, P. R. and Maragos, J. E. 1985. *Analysis of Environmental Assessment and Coastal Resources Management Needs (Indonesia).* Washington DC, International Institute for Environment and Development.

Burke, L., Kura, Y., Kassem, K., Revenga, C., Spalding, M., and McAllister, D. 2001. *Pilot Analysis of Global Ecosystems: Coastal Ecosystems.* Washington, DC, World Resources Institute.

Butchart, S. H. M., Stattersfield, A. J., Bennun, L. A., Shutes, S. M., Akçakaya, H. R., Baillie, J. E. M., Stuart, S. N., Hilton-Taylor, C. and Mace, G. M. 2004. Measuring global trends in the status of biodiversity: Red list indices for birds. *PLoS Biology*, Vol. 2, No. 12.

Cadbury, C. J. 2003. Arable to wetland: Restoring habitat for birds in the Cambridgeshire fens. *Cambridgeshire Bird Report*, Vol. 76, pp. 133–52.

Castro, G., Chomitz, K. and Thomas, T. S. 2002. *The Ramsar Convention: Measuring its Effectiveness for Conserving Wetlands of International Importance.* World Bank and World Wild Life Fund. www.ramsar.org/cop8_doc_37_e.htm

CBD (Convention on Biological Diversity). 2002. COP Decisions VI/23Alien species that threatens ecosystems, habitats or species. www.biodiv.org/decisions/?m=cop-06

——. 2000. COP Decision V/6: Ecosystem Approach. www.biodiv.org/decisions/?m=cop-05

Chape, S. J., Harrison, M., Spalding, M. and Lysenko, I. 2004. Measuring the extent and effectiveness of protected areas as an indicator for meeting global biodiversity targets. *Philosophical Transactions: Biological Sciences*, London, Royal Society.

Christensen, B. 1982. *Management and Utilization of Mangroves in Asia and the Pacific.* FAO Environment Paper No. 3.

Ciruna, K. A., Meyerson, L. A. and Gutierrez, A. 2004. The ecological and socio-economic impacts of invasive alien species in inland water ecosystems. Report to the Convention on Biological Diversity on behalf of the Global Invasive Species Programme, Washington DC.

Cobb, D. G., Galloway, T. D. and Flannagan, J. F. 1996. Effects of discharge and substrate stability on density and species composition of stream insects. *Canadian Journal of Fisheries and Aquatic Sciences*, Vol. 49, pp. 1788–95.

Coe, M. T. and Foley, J. A. 2001. Human and natural impacts on the water resources of the Lake Chad Basin. *Journal of Geophysical Research*, Vol. 106.

Cohen, A. S., Kaufman, L. and Ogutu-Ohwayo, R. 1996. Anthropogenic threats, impacts and conservation strategies in the African Great lakes: A review.*The Limnology, Climatology and Paleoclimatology of the East African Lake.* T. C. Johnson, and E. Odada (eds). Toronto, Gordon and Breach, pp. 575–624.

Colavito, L. 2002. Wetland economic valuation using a bioeconomic model: The case of Hail haor, Bangladesh, Paper presented at the Workshop on Conservation and Sustainable Use of Wetlands: Learning from the World. Kathmandu, Nepal, IUCN-The World Conservation Union.

Constanza, R., d'Arge, R. and de Groot, R. 1997. The value of the world's ecosystems services and natural capital. *Nature*, Vol. 387, pp. 253–60.

Constanza, R., d'Arge, R., de Groot, R., Farber, S., Grasso, M., Hannon, B., Limburg, K., Naeem, S., O'Neill, R. V., Paruelo, J., Raskin, R. G., Sutton, P. and Van der Belt, M. 1998. The value of the world's ecosystems services and natural capital. *Ecological Economics*, Vol. 1, No. 25, pp. 3–15.

David, L. J., Golubev, G. N. and Nakayama, M. 1988. The environmental management of large international basins: The EMINWA Programme of UNEP. *Water Resources Development*, Vol. 4, pp. 103–07.

Dekker, W. 2003. Eel stocks dangerously close to collapse. Copenhagen, Denmark, International Council for the Exploration of the Sea. www.ices.dk/marineworld/eel.asp

Dodman, T. and Diagana, C. H. 2003. *African Waterbird Census.* Wetlands International, *Global Series*, 16.

Dudley, N. and Stolton, S. 2003. *Running Pure: The Importance of Forest Protected Areas to Drinking Water.* World Bank/WWF Alliance for Forest Conservation and Sustainable Use.

Dynesius, M. and Nilsson, C. 1994. Fragmentation and flow regulation of river systems in the northern third of the world. *Science*, Vol. 266, pp. 753–62.

Eaton, D. and Sarch, T. M. 1997. The economic importance of wild resources in the Hadejia-Nguru Wetlands. Collaborative Research in the Economics of Environment and Development (CREED) Working Paper #13. London, International Institute for Environment and Development (IIED).

Echavarria, M. 1997. Agua!: Juntos Podremos Cuidarla! Estudio de Caso Para un Fondo Para la Conservacion de las Cuencas Hidrograficas Para Quito, Ecuador. Quito, Ecuador, The Nature Conservancy/USAID.

EEA (European Environment Agency). 1995. Europe's water: An indicator based assessment. European Environment Agency Topic Report 1/2003.

Emerton, L., Iyango, L., Luwum, P. and Malinga, A. 1999. *The Economic Value of Nakivubo Urban Wetland.* Kampala, Uganda, Uganda National

Wetlands Programme and Nairobi, IUCN-The World Conservation Union, Eastern Africa Regional Office.

Emerton, L. and Kekulandala, B. 2002. *Assessment of the Economic Value of Muthurajawela Wetland.* IUCN-The World Conservation Union, Colombo, Sri Lanka Country Office and Regional Environmental Economics Programme.

Emerton, L., Seilava, R. and Pearith, H. 2002. *Bokor, Kirirom, Kep and Ream National Parks, Cambodia: Case Studies of Economic and Development Linkages, Field Study Report, Review of Protected Areas and their Role in the Socio-Economic Development of the Four Countries of the Lower Mekong Region.* Brisbane, International Centre for Environmental Management and Karachi, IUCN-The World Conservation Union, Regional Environmental Economics Programme.

EU (European Union). 2000. Directive 2000/60/EC of the European Parliament and of the Council of 23 October 2000, establishing a framework for community action in the field of water policy, *Official Journal,* L 327, 22/12/2000 P. 0001–0073.

Falkenmark, M., Andersson, L., Castensson, R. and Sundblad, K. 1999. *Water: A Reflection of Land Use: Options for Counteracting Land and Water Mismanagement.* Stockholm, Swedish Natural Science Research Council.

FAO (Food and Agriculture Organisation of the United Nations). 2004. *Review of the State of World Marine Fishery Resources.* (FAO Fisheries Technical Paper 457). Rome, FAO.

——. 2002. *The State of World Fisheries and Aquaculture – Fisheries Resources: Trends in Production, Utilization and Trade.* Rome, FAO.

——. 2001. *The State of World Fisheries and Aquaculture 2000.* Rome, FAO, Fisheries Department.

——. 1999. *Review of the State of World Fishery Resources: Inland Fisheries.* FAO Fisheries Circular No. 942, Rome, FAO Fisheries Department.

Finlayson, C. M. and Davidson, N. C. 1999. Global review of wetland resources and priorities for wetland inventory: Summary report. C. M. Finlayson and A. G. Spiers (eds), *Global Review of Wetland Resources and Priorities for Wetland Inventory,* CD-ROM, Canberra, Australia, Supervising Scientist Report 144.

Fishstat. 2002. *Fishstat Plus, Universal Software for Fisheries Statistical Time Series.* FAO Fisheries, Software version 2.3.

FreshCo Partnership. 2002. *A partnership on linking Integrated Water Resources Management (IWRM) and Integrated Coastal Zone Management (ICZM).* www.ucc-water.org/freshco/

Furch, K. 2000. Evaluation of groundwater input as major source of solutes in an Amazon floodplain lake during the low water period. *Verhandlungen der Internationale Vereinigung für Theoretische und Angewandte Limnologie* [Proceedings of the

International Union for Theoretical and Applied Limnology], Vol. 27, pp. 412–15.

Gillespie, G. and Hines, H. 1999. Status of temperate riverine frogs in south-eastern Australia. A. Campbell (ed.), *Declines and Disappearances of Australian Frogs.* Canberra, Environment Australia, pp. 109–30.

Glantz, M. H. (ed.). 1999. *Creeping Environmental Problems and Sustainable Development in the Aral Sea Basin.* Cambridge, UK, Cambridge University Press.

Gleick, P. H., Singh, A. and Shi, H. 2001. *Threats to the World's Freshwater Resources.* Oakland, California, Pacific Institute for Studies in Development, Environment and Security.

Gornitz, V. 2000. Impoundment, groundwater mining and other hydrologic transformations: Impacts on global sea level rise. B. C. Douglas, M. S. Kearney and S. P. Leatherman (eds), *Sea Level Rise: History and Consequences.* San Diego, Academic Press, pp. 97–119.

Groombridge, B. and Jenkins, M. 1998. Freshwater biodiversity: A preliminary global assessment. WCMC Biodiversity Series No. 8.

GWP (Global Water Partnership). 2003. *Water Management and Eco Systems: Living with Change.* Global Water Partnership, Technical Advisory Committee Background Paper No. 9, Stockholm. www.gwpforum.org/gwp/library/TEC%209.pdf

——. 2000. *Integrated Water Resources Management.* Global Water Partnership, Technical Advisory Committee Background Paper No. 4, Stockholm. www.gwpforum.org/gwp/library/TACNO4.pdf

Hamerlynck, O. and Duvail, S. 2003. *The Rehabilitation of the Delta of the Senegal River in Mauritania.* Nouakchott, Mauritania, IUCN. www.iucn.org/themes/wetlands/pdf/diawling/Diawling_GB.pdf

Hamilton, L. S., Dixon, J. A. and Miller, G. O. 1989. Mangrove forests: An undervalued resource of the land and of the sea. E. Mann Borgese, N. Ginsburg and J. R. Morgan (eds), *Ocean Yearbook 8,* Chicago, USA, The University of Chicago Press, pp. 254–88.

Hecky, R. E., 1993. The eutropication of Lake Victoria. *Verhandlungen der Internationale Vereinigung für Theoretische und Angewandte Limnologie* [Proceedings of the International Union for Theoretical and Applied Limnology], Vol. 25, pp. 39–48.

Hokkaido Regional Development Bureau. 2003. Restoration of Kushiro River. *River Restoration,* seminar textbook. Foundation for Riverfront Improvement and Restoration. www.rfc.or.jp/rivernetwork/pdf/en/07kushiro_en.pdf

Hulme, M. 1996. *Climate Change and Southern Africa: An Exploration of Some Potential Impacts and Implications for the SADC Region.* Gland, Switzerland. Climatic Research Unit, University of East Anglia, Norwich, United Kingdom and WWF International, Gland, Switzerland.

IJC (International Joint Commission). 1998. The International Joint Commission and the Boundary Waters Treaty of 1909. Washington. DC, International Joint Commission.

Institute of Water and Sanitation Development, Zimbabwe. 1997. *Assessment of Integrated Water Resources Management activities in the Southern Africa Region.* A preliminary inventory. www.thewaterpage.com/IWRM_Zimbabwe.htm

IPCC (International Panel of Climate Change). 2001. *Climate Change 2001 – Impacts, Adaptation, and Vulnerability.* Cambridge, UK, Cambridge University Press.

IUCN (World Conservation Union) Water and Nature Initiative. Conservation International and Nature Serve. 2004. Global Amphibian Assessment. www.globalamphibians.org

——. 2003a. *Red List of Globally Threatened Species.* Cambridge, United Kingdom.

——. 2003b. Integrated management of the Komadugu-Yobe River Basin (Nigeria Water and Nature Initiative). Online at: www.waterandnature.org/d1.html

——. 2001. *Economic Value of Reinundation of the Waza Logone Floodplain, Cameroon.* Projet de conservation et de développement de la région de Waza-Logone, Maroua.

Janssen, R. and Padilla, J. E. 1996. Valuation and evaluation of management alternatives for the Pagbilao mangrove forest. Collaborative Research in the Economics of Environment and Development Working Paper Series No. 9.

Johnson, T. C., Schol, C. A, Talbot M. R., Kelts, K., Ricketts, R. D., Ngobi, G., Beuning, K. R. M., Ssemanda, I., and McGill, J. W. 1996. Late Pleistocene desiccation of lake Victoria and rapid evolution of cichlid fishes. *Science,* Vol. 273, pp. 1091–93.

Jones, O. A, Lester, J. N. and Voulvoulis, N. 2005. Pharmaceuticals: a threat to drinking water? *Trends in Biotechnology,* Vol. 23, pp. 163–67.

Junk, W., Bayley, P. B. and Sparks, R. E. 1989. The flood pulse concept in river-floodplain systems. *Canadian Journal of Fisheries and Aquatic Sciences,* No. 106, pp. 110–27.

Junk, W. 2002. Long-term environmental trends and the future of tropical wetlands. *Environmental Conservation,* Vol. 29. pp. 414–35.

Karanja, F., Emerton, L., Mafumbo, J. and Kakuru, W. 2001. *Assessment of the Economic Value of Pallisa District Wetlands.* Kampala, Uganda. Biodiversity Economics Programme for Eastern Africa, IUCN-The World Conservation Union and Uganda National Wetlands Programme.

Kiremire, B. T. 1997. The status of chemicals in Uganda and a survey of disposal methods. Presentation given at the American Chemical Society US Africa Workshop on Environmental Chemistry and Water Quality, Feb 1997, Mbarara, Uganda.

Krebs, C. J. 1978. *Ecology: The Experimental Analysis of Distribution and Abundance*, 2nd edn. New York, Harper & Row Publishers.

Kreutzberg-Mukhina, E. 2004. Effect of drought on waterfowl in the Aral Sea region: Monitoring of anseriformes at the Sudochie Wetland. 202.

Kura, Y., Revenga, C., Hoshino, E. and Greg, M. 2004. *Fishing for Answers: Making Sense of the Global Fish Crisis*. Washington DC, World Resources Institute.

Kuriyama, K. 1998. Measuring the value of the ecosystem in the Kushiro Wetland: An empirical study of choice experiments. Forest Economics and Policy Working paper, No. 9802.

Lal, P. N. 1990. *Ecological Economic Analysis of Mangrove Conservation: A Case Study from Fiji*. UNESCO Mangrove Occasional Paper, No. 6.

Lee, H. D. 1998. Use and value of coastal wetlands in Korea. *Intercoast Network Newsletter*, No. 32, pp. 7-8.

Lehman, J. T. 1996. Pelagic food webs of the east African great lakes. T. C. Johnson and E. Odada (eds), *The Limnology, Climatology and Paleoclimatology of the East African lakes*. Toronto, Gordon and Breach, pp. 281-301.

Li, Y. F. and Macdonald, R. W. 2005. Sources and pathways of selected organochlorine pesticides to the Arctic and the effect of pathway divergence on HCH trends in biota: a review. *Science of the Total Environment*, Vol. 342, pp. 87-106.

Loh, J., Randers, J. Jenkins, M., Kapos, V., Bernal, J., Smith, K., Lacambra, C. and Phipps, E. (eds). 2004. *Living Planet Report 2004*. Gland, Switzerland, World Wide Fund International.

Lowe, S., Browne, M. and Boudjelas, S. 2004. 100 of the world's worst invasive alien species: A selection from the global invasive species database. IUCN, Auckland, New Zealand. www.issg.org/booklet.pdf

Malmqvist, B. and Rundle, S. 2002. Threats to the running water ecosystems of the world. *Environmental Conservation*, Vol. 29, pp. 134-53.

Mazel, D. and Davis, C. B. 2003. Antibiotic resistance in microbes. *Cellular and Molecular Life Sciences*, Vol. 56, pp. 742-54.

McAllister, D. E., Hamilton, A. L. and Harvey, B. H. 1997. Global freshwater biodiversity: Striving for the integrity of freshwater ecosystems. *Sea Wind*, Vol. 11, No. 3. Special issue (July-September 1997).

Miao, W. and Yuan, X. 2001. Development and present status of inland fisheries and aquaculture in China. Unpublished paper prepared for the World Resources Institute, Qitang, Wuxi, China, Chinese Academy of Fishery Sciences.

Millennium Project, 2004. Interim Report, Task Force Water and Sanitation. www.unmillenniumproject.org/documents/tf7interim.pdf

Moyle, P. B. and Leidy, R. A. 1992. Loss of biodiversity in aquatic ecosystems: Evidence from fish fauna. P. L. Fielder et al. (eds), *Conservation Biology: The Theory and Practice of Nature Conservation and Preservation and Management*. New York and London, Chapman and Hall.

Mugidde, R., 1993. The increase in phytoplankton productivity and biomass in lake Victoria (Uganda). *Verhandlungen der Internationale Vereinigung für Theoretische und Angewandte Limnologie* [Proceedings of the International Union for Theoretical and Applied Limnology], Vol. 25, pp. 846-49.

Myers, N. 1997. The rich diversity of biodiversity issues. M. L. Reaka-Kudla, D. E. Wilson, and E. O. Wilson (eds), *Biodiversity II: Understanding and Protecting Our Biological Resources*. Washington DC, Joseph Henry Press, pp. 125-38.

Nami, B. 2002. *Environmental Degradation of the Lake Chad Basin: Implications for Food Security*.

Navrud, S. and Mungatana, E. D. 1994. Environmental valuation in developing countries: The recreational value of wildlife viewing. *Ecological Economics*, Vol. 11, pp. 135-51.

Neiland, A. E., Béné, C., Bennett, E., Turpie, J., Chong, C. K., Thorpe, A., Ahmed, M., Valmonte-Santos, R. A. and Balasubramanian, H. 2004. *River Fisheries Valuation: A Global Synthesis and Critical Review*. Penang, Malaysia, WorldFish Center and Comprehensive Assessment of Water in Agriculture.

Nilsson, C., Reidy, C. A., Dynesius, M. and Revenga, C. 2005. Fragmentation and Flow Regulation of the world's large river systems. *Science*, Vol. 308, No. 5720, 405-08.

Ogutu-Ohwayo, R. 1990. The decline of the native species of Lake Victoria and Kyoga (East Africa) and the impact of introduced species, especially the Nile perch, *Lates niloticus*, and the Nile tilapia, *Oreochromis niloticus*. *Env. Biol. Fish*, Vol. 27, pp. 81-96.

Olson D. M., Dinerstein, E., Wikramanayake, E. D., Burgess, N. D., Powell, G. V. N., Underwood, E. C., D'amico, J. A., Itoua, I., Strand, H. E., Morrison, J. C., Loucks, C. J., Allnutt, T. F., Ricketts, T. H., Kura, Y., Lamoreux, J. F., Wettengel, W. W., Hedao, P. and Kassem, K. 2001. Terrestrial ecoregions of the world: A new map of life on Earth. *BioScience*, Vol. 51, No. 11, pp. 933-38.

Postel, S. 1999. *Pillar of Sand: Can the Irrigation Miracle Last?* World Watch Institute, Washington DC.
——. 1995. Where have all the rivers one? *World Watch*, Vol. 8, No. 3, pp. 9-19.

Postel, S. and Richter, B. 2003. *Rivers for Life: Managing Water for People and Nature*. Washington DC, Island Press.

Rast, W. 1999. Overview of the status of implementation of the freshwater objectives of Agenda 21 on a regional basis. *Sustainable Development International*, Vol. 1, pp. 53-57.

Rast, W. and Holland, M. M. 2003. Sustainable freshwater resources: Achieving secure water supplies. M. M. Holland, E. R. Blood and L. R. Shaffer (eds), *Achieving Sustainable Freshwater Resources, A Web of Connections*. Washington DC, Island Press, pp. 283-315.

Reddy, C. A. and Mathew, Z. 2001. Bioremediation potential of white rot fungi. G. M. Gadd (ed.), *Fungi in bioremediation*. Cambridge, UK, Cambridge University Press.

Revenga, C., Murray, S., Abramovitz, J. and Hammond, A. 1998. *Watersheds of the World: Ecological Value and Vulnerability*. Washington DC, World Resources Institute and Worldwatch Institute.

Revenga, C., Brunner, J., Hinninger, N., Kassem, K. and Payne, R. 2000. *Pilot Analysis of Global Ecosystems: Freshwater Systems*. Washington DC, World Resources Institute.

Revenga, C. and Kura, Y. 2003. *Status and Trends of Biodiversity of Inland Water Ecosystems*. Technical Series No. 11. Montreal, Secretariat of the Convention on Biological Diversity.

Sarch, M. T. and Birkett, C. M. 2000. Fishing and farming at Lake Chad: Responses of lake level fluctuations. *The Geographical Journal*, Vol. 166, No. 2, pp. 156-72.

Sathirathai, S. 1998. Economic valuation of mangroves and the roles of local communities in the conservation of natural resources: Case study of Surat Thani, south of Thailand. EEPSEA Research Report.

Scheffer, M. 1998. *Ecology of Shallow Lakes*. Vol. 1, London, Chapman and Hall, pp. 1-357.

Scheffer, M., Carpenter, S. R., Foley, J. A., Folke, C. and Walker, B. 2001. Catastrophic shifts in ecosystems. *Nature*, Vol. 413, pp. 591-96.

Scherfose, V., Boye, P., Forst, R., Hagius, A., Klär, C., Niclas, G. and Steer, U. 2001. Naturschutzgrossgebiete des Bundes. [Large-scale conservation projects of national interest]. *Natur und Landschaft*, Vol. 76, pp. 389-97. (in German).

Schindler, D. W. 1997. Widespread effects of climatic warming on freshwater ecosystems in North America. *Hydrological Processes*, Vol. 11, pp. 1043-67.

Sharpe, M. 2003. High on pollution: drugs as environmental contaminants. *Journal of Environmental Monitoring*, Vol. 5, pp. 42N-46N.

Singleton, H. J. 1985. Water quality criteria for particulate matter. Technical appendix. Victoria, British Columbia, Canada, British Columbia Ministry of the Environment Lands and Parks.

Spalding, M. D., Taylor, M. L., Ravilius, C. and Green, E. P. 2002. *The Global Distribution and Status of Seagreass Ecosystems, 2002*. London, UK, UNEP/WCMC.

Stiassny, M. 2005. Personal communication with Dr. Melanie Stiassny, American Museum of Natural History, New York.

Stumpel, A. H. P. 1998. The creation and restoration of ponds as a habitat for threatened amphibians. Delbaere (ed.), *Facts and Figures on Europe's Biodiversity*. ECNC, 1998-1999.

Sverdrup-Jensen. 2002. Fisheries in the Lower Mekong Basin: Status and perspectives. MRC Technical Paper

No. 6, Phnom Penh, Cambodia, Mekong River Commission.

Thilsted, S. H., Roos, N. and Hassan, N. 1997. The role of small indigenous species in food nutrition security in Bangladesh. Paper presented at the International Consultation on Fisheries Policy Research in Developing Countries, Hirtshals, Denmark, 2–5 June 1997.

Tibbetts, J. 2004. The state of the oceans, Part 2: Delving deeper into the sea's bounty. *Environmental Health Perspectives*, Vol. 112, No. 8, June 2004.

Tischler, W. 1979. *Einführung in die Ökologie*, 2nd edn. Stuttgart and New York, Spektrum Akademischer Verlag (in German).

Turpie, J., Smith, B., Emerton, L. and Barnes, J. 1999. *Economic Valuation of the Zambezi Basin Wetlands*, Harare, Zimbabwe. IUCN-The World Conservation Union, Regional Office for Southern Africa.

Ueno, D., Takahashi, S., Tanaka, H., Subramanian, A. N., Fillmann, G., Nakata, H., Lam, P. K. S., Zheng, J., Muctar, M., Prudente, M., Chung, K. H. and Tanabe, S. 2003. Global pollution monitoring of PCBs and organochlorine pesticides using skipjack tuna as a bioindicator. *Archives of Environmental Contamination and Toxicology*, Vol. 45, pp. 378–89.

Umali, D. L. 1993. Irrigation-induced salinity: A growing problem for development and the environment. World Bank Technical Paper 215, Washington D.C. www-wds.worldbank.org/servlet/ WDSContentServer/WDSP/IB/1993/08/01/ 000009265_3970311124344/Rendered/PDF/ multi_page.pdf

Umeå University and WRI (World Resource Institute). 2004. *Fragmentation and Flow Regulation Indicator*. Umeå, Sweden, Umeå University and Washington DC, World Resources Institute.

UN-WWAP (United Nations World Water Assessment Programme). 2003. *Water for People, Water for Life. World Water Development Report*. Paris, UNESCO and London, Berghahn Books.

——. 2000. UN Millennium Development Goals, United Nations, New York. www.un.org/millenniumgoals/

——. 1992. *Report of the United Nations Conference on Environment and Development*. Resolutions Adopted by the Conference. United Nations, A/CONF.151/26/Rev.1 Vol. 1, pp. 275–314.

UNEP (United Nations Environment Programme). 2006. Marine and Coastal Ecosystems and Human Wellbeing: A Synthesis Report Based on the Findings of the Millennium Ecosystem Assessment. UNEP.

——. 2005. Assessing Coastal Vulnerability: Developing a Global Index for Measuring Risk. Nairobi, Kenya. DEWA, UNEP.

——. 2004a. *Lake Chad: Sustainable Use of Land and Water in the Sahel*. Environmental Change Analysis Series. DEWA, UNEP. Nairobi, Kenya.

——. 2004b. *GEO Yearbook 2003*. London, Earthscan Publications Ltd.

——. 2004c. Fortnam, M. P. and Oguntula, J. A. (eds), *Lake Chad Basin, GIWA regional assessment 43*. Kalmar, Sweden, University of Kalmar.

——. 2002a. *Atlas of International Freshwater Agreements*, Stevenage, England, Earthprint.

——. 2002b. *Global Environment Outlook 3: Past, present and future perspectives*. London, Earthscan Publications Ltd.

——. 2002c. *Vital Water Graphics: An Overview of the State of the World's Fresh and Marine Waters*. Nairobi, Kenya, UNEP.

UNEP/GPA (United Nations Environmental Programme/Global Programme for Action for the Protection of the Marine Environment from Land-based Activities). 2004. *Water Supply and Sanitation Coverage in UNEP Regional Seas. Need for Regional Wastewater Emission Targets (WET)*. Section III: An inventory of regional specific data and the feasibility of developing regional wastewater emission targets (WET). The Hague, the Netherlands, UNEP/GPA.

UNESCO (United Nations Educational, Scientific, and Cultural Organisation). 2000. *Water Related Vision for the Aral Sea Basin for the Year 2025*. Paris, France, UNESCO.

Verma, M. 2001. Economic valuation of Bhoj wetland for sustainable use. Report prepared for India: Environmental Management Capacity Building Technical Assistance Project, Bhopal, Indian Institute of Forest Management.

Verschurem, D., Johnson, T. C., Kling, H. J., Edgington, D. N., Leavitt, P. R., Brown, E. T., Talbot, M. R. and Hecky, R. E. 2002. History and timing of human impact on Lake Victoria, East Africa. *Proceedings of the Royal Society of London B*. Vol. 269, pp. 289–94.

Watson, B., Walker, N., Hodges, L. and Worden, A. 1996. Effectiveness of peripheral level of detail degradation when used with head-mounted displays. Technical Report 96–04, Graphics, Visualization & Usability (GVU) Center, Georgia Institute of Technology.

Welcomme, R. L. 1979. *Fisheries Ecology of Floodplain Rivers*. London, Longman.

——. 2005. Annual History of Ramsar Site Designations. www.wetlands.org/RDB/global/Designations.html and www.wetlands.org/RDB/global/AreaTrend.html

Wilson, B. A., Smith, V. H., Denoyelles, F. Jr., and Larive, C. K. 2003. Effects of three pharmaceutical and personal care products on natural freshwater algal assemblages. *Environmental Science & Technology*, Vol. 37, pp. 1713–9.

Wood, S., Sebastian, K., and Scherr, S. 2000. *Pilot Analysis of Global Ecosystems: Agroecosystems Technical Report*. Washington DC: World Resources Institute and International Food Policy Research Institute.

World Commission on Dams (WCD). 2000 *Dams and development : A new framework for decision-making*, Earthscan Publ., London, UK.

World Lake Vision Committee. 2003. *World Lake Vision: A Call to Action*. World Lake Vision Committee (International Lake Environment Committee, International Environment Technology Centre, United Nations Environment Programme and Shiga, Japan Prefectural Government).

WMO (World Meteorological Organization). 1997. *Comprehensive Assessment of the Freshwater Resources of the World*. UN, UNDP, UNEP, FAO, UNESCO, WMO, UNIDO, World Bank, SEI. WMO, Geneva, Switzerland.

——. 1992. *Dublin Statement and Report of the Conference*. International Conference on Water and the Environment: Development Issues for the 21st Century. Geneva, Switzerland, WMO.

WRI (World Resources Institute), UNDP (United Nations Development Programme), UNEP (United Nations Environment Programme) and World Bank. 2000. *World Resources 2000–2001: People and ecosystems: The fraying web of life*. Washington DC, WRI.

WWF (World Wide Fund for Nature). 2003. *Managing Rivers Wisely: Kafue Flats Case study, Mozambique*. www.panda.org/downloads/freshwater/ mrwkafueflatscasestudy.pdf

WWF and WRI (World Resources Institute). 2004. *Rivers at Risk: Dams and the Future of Freshwater Ecosystems*. www.panda.org/downloads/ freshwater/riversatriskfullreport.pdf

WWF/IUCN (Global Conservation Organization/World Conservation Union). 2001. *The Status of Natural Resources on the High Seas*. Gland, Switzerland

Zhulidov, A. V., Robarts, R. D., Headley, J. V., Liber, K., Zhulidov, D. A., Zhulidova, O. V. and Pavlov, D. F. 2002. Levels of DDT and hexachlorocyclohexane in burbot (*Lota lota*) from Russian Arctic rivers. *The Science of the Total Environment*, Vol. 292, pp. 231–46.

Zöckler, C. 2002. A comparison between tundra and wet grassland breeding waders with special reference to the ruff (*Philomachus pugnax*). *Schriftenreihe Landschaftspflege und Naturschutz*, Vol. 74.

FEWS (Famine Early Warning Systems Network). 2003. www.fews.net
Ramsar 2005. www.ramsar.org
Regional Ecosystem Office. 2005. Definitions N-Z. www.reo.gov/general/definitions_n-z.htm#R
UNEP-GEMS Water (Global Environment Monitoring System). 2004. www.gemswater.org
Wetlands International. 2005. Annual History of Ramsar Site Designations.
www.wetlands.org/RDB/global/Designations.html and www.wetlands.org/RDB/global/AreaTrend.html
World Lakes Organisation: www.worldlakes.org/

SECTION 3

Challenges for Well-being and Development

The provision of adequate drinking water is just one aspect of the role played by water in meeting basic needs and contributing to development. Having enough water to cover domestic hygiene needs promotes better health and well-being. Sanitation facilities help to ensure the safe disposal of human waste and reduce disease and death. Adequate water supplies improve the prospects of new livelihood activities, including agriculture, that are otherwise denied and which are often a key step out of poverty. Industry at all scales needs reliable water resources to prosper and grow. Water also plays a key role in energy generation and transportation.

We must examine the current conditions of and the different demands being placed on water for food, human health, industry and energy, as increased competition will demand integrated responses in order to ensure that there is enough water of adequate quality to meet each of these needs in a sustainable manner.

Global Map 5: *Domestic and Industrial Water Use*
Global Map 6: *Sediment Trapping by Large Dams and Reservoirs*

Chapter 6 – **Protecting and Promoting Human Health**
(WHO & UNICEF)

This chapter reviews the main components of the water cycle and provides an overview of the geographical distribution of the world's total water resources, their variability, the impacts of climate change and the challenges associated with assessing the resource.

Chapter 7 – **Water for Food, Agriculture and Rural Livelihoods**
(FAO & IFAD)

The demand for food is not negotiable. As the largest consumer of freshwater, the agriculture sector faces a critical challenge: producing more food of better quality while using less water per unit of output, and reducing its negative impacts on the complex aquatic ecosystems on which our survival depends. Better water management leads to more stable production and increased productivity, which in turn enhance the livelihoods and reduce the vulnerability of rural populations. This chapter examines the challenges of feeding a growing population and balancing its water needs with other uses, while contributing to sustainable development in rural areas.

Chapter 8 – **Water and Industry**
(UNIDO)

Despite industry's need for clean water, industrial pollution is damaging and destroying freshwater ecosystems in many areas, compromising water security for both individual consumers and industries. This chapter focuses on industry's impact on the water environment in routine water withdrawal and wastewater discharge, analysing a broad range of regulatory instruments and voluntary initiatives that could improve water productivity, industrial profitability and environmental protection.

Chapter 9 – **Water and Energy**
(UNIDO)

To be sustainable, economic development needs an adequate and steady supply of energy. Today's changing contexts require the consideration of a range of strategies to incorporate hydropower generation and other renewable forms of energy production to improve energy security while minimizing climate-changing emissions. This chapter stresses the need for the cooperative management of the energy and water sectors to ensure sustainable and sufficient supply of both energy and water.

Domestic and Industrial Water Use

Freshwater is critical to the ever-growing urban populations around the world, as well as the industrial base upon which these modern societies are based. With rapid urban growth – often poorly managed – the delivery of adequate, clean, and reliable supplies of freshwater becomes an important development challenge. Calculating contemporary domestic and industrial water use is today based on educated guesswork, for many countries lack comprehensive and standardized survey systems to determine water use. Decaying, poorly managed, and leaky delivery systems add to the difficulty. In the maps below, reported water withdrawals by country (WRI, 1998) were used to estimate domestic and industrial water use. The reporting year of national water use statistics differed from country to country. To make up for this inconsistency, regional water use trends reported in Shiklomanov (1996) were used to extrapolate national water use to a common year, in this case, the year 2000.

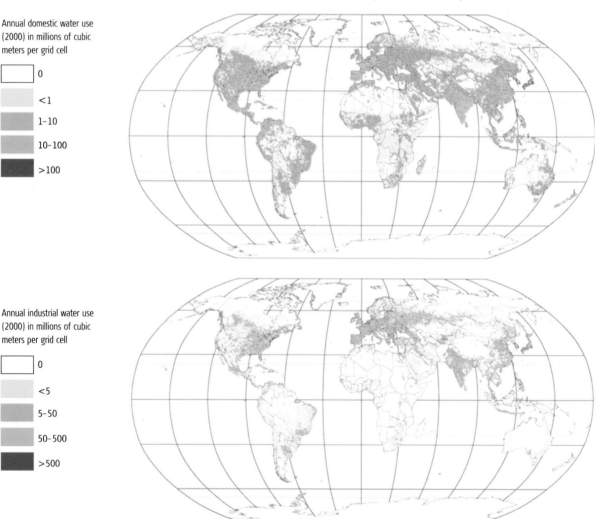

Annual domestic water use (2000) in millions of cubic meters per grid cell

- 0
- <1
- 1–10
- 10–100
- >100

Annual industrial water use (2000) in millions of cubic meters per grid cell

- 0
- <5
- 5–50
- 50–500
- >500

As shown in the maps above, a broad spectrum of water use arises, with high levels associated with dense settlement and advanced levels of economic development. Maps of water use such as these can be linked with those depicting water supply to define patterns of water scarcity and stress.

Source: Water Systems Analysis Group, University of New Hampshire. Datasets available for download at http://wwdrii.sr.unh.edu/

Sediment Trapping by Large Dams and Reservoirs

Dams and reservoirs create what are essentially large holding tanks that intercept and efficiently trap incoming particulate matter. Despite widespread increases in soil erosion from poor land management upstream, the construction of dams and reservoirs has made a significant impact on the transport of sediment destined for the world's coastal zones (Walling and Fang, 2003). At least 30 percent of continental sediment now fails to reach the oceans (Vörösmarty et al., 2003; Syvitski et al., 2005). Today, several large river basins, such as the Colorado and the Nile, show nearly complete trapping due to reservoir construction and flow diversions. Given that most of the major reservoirs of the world have been constructed only over the last 50 years, by any measure of global change, the impact of these structures has been substantial, rapid and unprecedented. The map, based on information from large, registered reservoirs only, illustrates the variable efficiency of sediment trapping worldwide. The additional impact of many smaller but unregistered impoundments numbering around 800,000 is unknown (McCully, 1996).

Sediment trapping efficiency

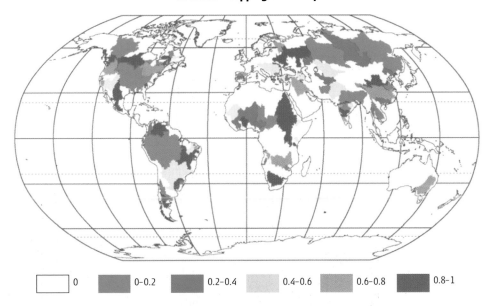

| | 0 | | 0–0.2 | | 0.2–0.4 | | 0.4–0.6 | | 0.6–0.8 | | 0.8–1 |

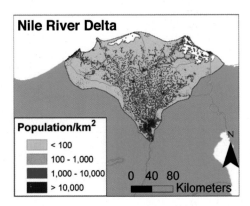

Nile River Delta

Population/km²
- < 100
- 100 - 1,000
- 1,000 - 10,000
- > 10,000

0 40 80
Kilometers

The impact of interrupted sediment flows is of more than academic interest. Siltation of reservoirs means a loss of water storage capacity and a shorter lifespan or costly maintenance for expensive infrastructure designed to support hydropower generation, irrigation, or domestic and industrial uses. The delivery of adequate supplies of freshwater and nutrient-rich sediment is critical to sustaining coastal ecosystems and preventing coastal erosion, such as in the Nile Delta (right). A recent sample of 40 deltas around the world shows that more than 75% are threatened predominantly by the upstream loss of sediment and only secondarily by global sea level rise. In this sample alone some 10 million people in coastal landscapes risk being flooded (Ericson et al., 2006).

Source: Water Systems Analysis Group, University of New Hampshire. Datasets available for download at http://wwdrii.sr.unh.edu/

By means of water we give life to everything
The Koran, Book of The Prophets 21:30

CHAPTER 6

Protecting and Promoting Human Health

By

WHO
(World Health Organization)

UNICEF
(United Nations Children's Fund)

6

Key messages:

Water-related diseases, including diarrhoea, are a leading cause of death in children of developing countries. However, they can be prevented and controlled by improving access to safe drinking water and sanitation, as well as domestic and personal hygiene. Yet progress remains very slow, especially in the provision of adequate sanitation in sub-Saharan Africa and South-East Asia. An integrated approach to human health and water resources management is urgently required. This should be characterized by flexible planning and implementation, analysis of the cost-effectiveness of local options, a significant reallocation of resources to drinking water, sanitation and hygiene, and attention to the most vulnerable groups in urban and rural settings. This is essential to save the lives of millions and ensure considerable long-term economic benefits.

■ Infectious diseases, especially diarrhoea followed by malaria, continue to dominate the global burden of water-related disease. Substantial progress has been made in reducing the mortality associated with diarrhoeal disease but morbidity remains essentially unchanged, while the burden of malaria is increasing.

■ Globally, the drinking water target set by MDG 7 is on schedule, but the sanitation target will not be met by 2015 without extra inputs and efforts. In sub-Saharan Africa, the trends observed since 1990 indicate that neither of the targets will be met by 2015.

■ The water-related disease burden and the relative efficiency of water interventions are key criteria in water/health decision-making. Disability-Adjusted Life Years (DALYs) and cost and effectiveness of interventions should be put upfront as key decision-making criteria.

■ Understanding of drinking water quality as it relates to health has evolved from rigid standards applied at the treatment facility to a process of risk assessment and management from catchment to consumer.

■ The importance of both accelerated access to safe water and adequate sanitation, and better Integrated Water Resources Management (IWRM) practices to achieving most MDG targets, need a higher profile. This can be achieved by refining and publicizing the correlations between water indicators and the indicators for childhood illness/mortality and nutritional status.

Above: Exterior and interior of a toilet block in Dar es Salaam, Tanzania

Below: A teacher assists a young girl to wash her hands with soap and clean water after using a sanitary latrine in a school in North Khway Ye village in Myanmar. Some 3,800 children die every day from diseases associated with lack of access to safe drinking water, inadequate sanitation and poor hygiene

Part 1. Human Health in Water Development

A range of water conditions and parameters essentially determine the health status of communities. Human health, therefore, cuts across all sectors responsible for water resources development, management and use.

In water for domestic uses, the focus is on the lack of access to sufficient supplies of safe drinking water, adequate sanitation, and the promotion of hygiene practices, all in relation to diarrhoeal and other water-related diseases. While infectious diseases are of principal concern, other health risks may be important under specific circumstances.

In water for food and energy, the focus is on the hydrological changes caused by dam construction (see **Chapter 5**) and irrigation development (see **Chapter 7**) and the ensuing transmission risks of vector-borne diseases, such as malaria, schistosomiasis, filariasis and Japanese encephalitis. The impact of irrigated crop production on the nutritional status of communities varies. On the whole, it is positive, but there may be vulnerable groups whose nutritional status declines with the introduction of irrigation, which shifts the economic balance from subsistence to cash crops. Over time, irrigation development may result in significant improvements in the economic status of communities, allowing better access to health services and, indirectly, an improved community health status. Increased energy generation through hydropower development benefits large segments of the population, the urban populations often disproportionately so; communities vulnerable to adverse health impacts live in the vicinity of dams and reservoirs (see also **Chapters 5 and 9**).

In water for ecosystems, the evidence base for associations between natural ecologies, biodiversity conservation and human health still requires substantial development. It may be safely assumed that many of the environmental services provided by wetlands, for example, are important to sustaining the health of communities that depend on these ecosystems for their livelihood. Yet, in specific settings, such as wetlands, there may also be health risks related to water-associated vector-borne diseases, sanitation-related diseases and impaired access to health services. However, health can be a key motivator in mobilizing communities to participate in nature conservation and environmental management.

Improvements in access to safe drinking water, adequate sanitation and hygiene have an impact on wider development issues, a fact that has been summarized by the Water Supply and Sanitation Collaborative Council (WSSCC, 2004) for the Millennium Development Goals (MDGs) (see **Table 6.1**).

Human health cuts across all water issues. Community health status is, therefore, the ultimate indicator of the success or failure of integrated water resources development and management. During and after water resources development, both negative and positive health effects can occur. Documented cases of adverse health impacts in the wake of water resources development abound; a recent example comes from Rajasthan, India (see **Box 6.1**).

...there may be vulnerable groups whose nutritional status declines with the introduction of irrigation...

BOX 6.1: THE EMERGENCE OF MALARIA IN INDIA'S THAR DESERT

The massive Indira Gandhi Nahar Pariyojana project is changing the face of the Thar Desert in Rajasthan, and will eventually irrigate 1.9 million hectares (ha) of arable land. Halfway through the project development, the number of locally transmitted malaria cases has risen from a few thousand to 300,000 a year. Key environmental changes include a rise in groundwater levels, more surface water bodies, changed water-retention properties of the soils, and an elevated relative humidity. The associated mosquito species succession from *Anopheles stephensi* to *A. culicifacies* has resulted in intensified transmission, which has shifted from seasonal to perennial. Between 1980 and 1995, the proportion

of Rajasthani malaria cases registered in the desert districts grew from 14.1 percent to 53.3 percent, and the share of *Plasmodium falciparum* cases (the most virulent malaria parasite species) rose from 11.6 percent to 62.5 percent. On the other hand, the extension of the canal system has made large quantities of water available for domestic use. Regrettably, the trend is that the ingenious traditional water supply systems of desert villages, consisting of small earthen underground reservoirs, are abandoned as soon as irrigation water becomes available. The increased quantities of water available for domestic use are nonetheless likely to provide important health benefits, despite the increased number of cases of malaria.

The two major environmental risk factors (seepage water collections from the canals, and pools of uncontrolled surplus run-off water) can be reduced by forestation and land reclamation, and by meticulous application of wet and dry irrigation management techniques meticulously. Such measures must be backed up by raising awareness among farmers and irrigation managers, and establishing effective institutional arrangements between health and irrigation authorities. Their application will greatly reduce (but not eliminate) the need for standard malaria control measures, such as case detection and treatment, and the use of insecticide-treated mosquito nets.

Source: Tyagi, 2004.

Section 3: CHALLENGES FOR WELL-BEING & DEVELOPMENT

Table 6.1: The relationship between the Millennium Development Goals (MDGs) and water, sanitation and hygiene

MDGs and their targets	The water, sanitation and hygiene perspective
Goal 1. Eradicate extreme poverty and hunger Target 1: Halve, between 1990 and 2015, the proportion of people whose income is less than US $1 a day. Target 2: Halve, between 1990 and 2015, the proportion of people who suffer from hunger.	■ The security of household livelihoods rests on the health of its members; adults who are ill themselves or must care for sick children are less productive. ■ Illnesses caused by unsafe drinking water and inadequate sanitation generate high health costs relative to income for the poor. ■ Healthy people are better able to absorb nutrients in food than those suffering from water-related diseases, particularly helminth[1] infections, which rob their hosts of calories. ■ Access to safe drinking water and adequate sanitation helps reduce household expenditures on health care. ■ The time lost because of long-distance water collection and poor health contributes to poverty and reduced food security.
Goal 2. Achieve universal primary education Target 3: Ensure that, by 2015, children everywhere, boys and girls alike, will be able to complete a full course of primary schooling.	■ Promotion of a healthy school environment is an essential element of ensuring universal access to education. School enrolment, attendance, retention and performance are improved; teacher placement is improved. ■ Improved health and reduced water-carrying burdens improve school attendance, especially among girls. ■ Separate school sanitation facilities for girls and boys increases girls' attendance, especially after they enter adolescence.
Goal 3. Promote gender equality and empower women Target 4: Eliminate gender disparity in primary and secondary education, preferably by 2005, and to all levels of education no later than 2015.	■ Sanitation improvement allows women and girls to enjoy private, dignified sanitation, instead of embarrassment, humiliation and fear from open defecation. ■ Access to safe drinking water and sanitation reduces the burden on women and girls from looking after sick children or siblings and from water carrying, giving them more time for productive endeavours, (adult) education and leisure. ■ Water sources and sanitation facilities closer to home reduce the risk of assault for women and girls when collecting water or searching for privacy.
Goal 4. Reduce child mortality Target 5: Reduce by two-thirds, between 1990 and 2015, the under-5 mortality rate.	■ Improved access to sanitation, safe drinking water sources and improved hygiene behaviour lead to a sharp decline in infant and child morbidity and mortality from diarrhoeal diseases. ■ Better nutrition and a reduced number of episodes of illness lead to the physical and mental growth of children.
Goal 5. Improve maternal health Target 6: Reduce by three-quarters, between 1990 and 2015, the maternal mortality ratio.	■ Good health and hygiene increase chances of a healthy pregnancy. ■ Safe drinking water and basic sanitation are needed in health-care facilities to ensure basic hygiene practices following delivery. ■ Accessible sources of water reduce labour burdens and health problems resulting from carrying water, thereby reducing maternal mortality risks.

Table 6.1: *continued*

MDGs and their targets	The water, sanitation and hygiene perspective
Goal 6. Combat HIV/AIDS, malaria and other diseases Target 7: Have halted by 2015 and begun to reverse the spread of HIV/AIDS. Target 8: Have halted by 2015 and begun to reverse the incidence of malaria and other major diseases.	■ Reliability of drinking water supplies and improved water management in human settlements contribute to reducing malaria and dengue fever transmission risks. ■ A reduction in stagnant water around tap points translates into less breeding places for mosquitoes. ■ Less pressure by other infections on the immune system of HIV/AIDS sufferers allows for better health. ■ Better, more hygienic and dignified possibilities to take care of ill people lift their burden. ■ Safe drinking water and basic sanitation help prevent water-related diseases, including diarrhoeal diseases, schistosomiasis, filariasis, trachoma and intestinal helminth infections.[2]
Goal 7. Ensure environmental sustainability Target 9: Integrate the principles of sustainable development into country policies and programmes and reverse the loss of environmental resources. Target 10: Halve by 2015 the proportion of people without sustainable access to safe drinking water and basic sanitation. Target 11: Achieve significant improvement in lives of at least 100 million slum dwellers by 2020.	■ Adequate treatment and disposal of wastewater result in a sharp decrease in environmental contamination by faeces, which contributes to better ecosystem conservation and less pressure on scarce freshwater resources. ■ Careful use of water resources prevents contamination of groundwater and helps minimize the cost of water treatment. ■ Better health is linked to a reduction in poverty, which in turn helps to put less strain on natural resources.
Goal 8. Develop a global partnership for development Target 12: Develop further an open, rule-based, predictable, non-discriminatory trading and financial system. Targets 13 and 14: Address special needs of less developed countries, landlocked and small island developing countries. Target 15: Deal comprehensively with the debt problems of developing countries through national and international measures in order to make debt sustainable in the long term. Target 16: In cooperation with developing countries, develop and implement strategies for decent and productive work for youth. Target 17: In cooperation with pharmaceutical companies, provide access to affordable essential drugs in developing countries. Target 18: In cooperation with the private sector, make available the benefits of new technologies, especially information and communications.	■ Development agendas and partnerships should recognize the fundamental role that safe drinking water and basic sanitation play in economic and social development. ■ Countries that illustrate improved access to and quality of safe drinking water and sanitation are more attractive, boosting tourism and national image. ■ These countries have more options for employment creation, as water supply and sanitation provision is labour intensive. ■ Safe drinking water and better sanitation provide a better chance for completing schooling, which leads to higher youth employment. ■ Including health impact assessment in water resources development planning prevents the transfer of hidden costs to the health sector.

1. Helminths are parasitic intestinal worms that include tapeworms, hookworms, whipworms and roundworms.
2. Human schistosomiasis is a chronic, usually tropical, disease caused by infection with parasitic blood flukes that have certain aquatic snail species as their intermediate host. Depending on the *Schistosoma* species, the infection will lead to disorders of the liver or urinary system. Filariasis is a parasitic disease caused by thread-like worms, which are transmitted by mosquitoes and invade the lymphatic vessels causing chronic swelling of the lower extremities. Trachoma is a contagious infection of the eye caused by a bacteria-like organism and can cause damage to the cornea leading to visual impairment and blindness.

Source: Adapted from WSSCC, 2004.

The DALY is a powerful tool for assisting policy-makers in sector-wide decision-making

Part 2. Update of the Burden of Water-related Diseases

The basic measures of disease frequency are incidence (new cases that occur in a population over time) and prevalence (existing cases in a population at a certain point in time). In principle, disease incidence data could be obtained from routine government health information systems. However, such data remain scant, inaccurate and often unreliable and fail to attribute diseases to specific social and environmental determinants. Data collected directly at the user/household level are generally more reliable. These data are mostly obtained through cross-sectional surveys that provide prevalence, not incidence, figures. For example, the proportion of people surveyed with helminth eggs in their stools provides an estimate of the prevalence of helminth infection. If there is prospective surveillance of large populations, direct incidence figures can be available. For diarrhoea, there are now sufficient results from longitudinal studies to make it possible to arrive at reliable global estimates of incidence (Kosek et al., 2003). For some diseases, such as scabies, no studies have been done and reliable global data, whether incidence or prevalence, are currently unavailable.

Mortality rates express the incidence of death in a particular population during a defined period of time. Mortality data are more widely available than comparable information on disease incidence rates (morbidity) and functional disabilities. Many of the water-related diseases affect children in particular, and the database for estimating child mortality is unquestionably much better developed than that for adult mortality (Murray and Lopez, 1994a). This provides a strong rationale for reporting mortality data for the under-5 years age group separately.

The first *UN World Water Development Report* described the methodology for assessing water-related health impacts at the global level (UN-WWAP, 2003; Prüss, et al., 2002). It referred to the Disability-Adjusted Life Year (DALY), a summary measure of population health, as an important indicator for assessing the disease burden associated with, for example, environmental exposures, and evaluating options for public health interventions. The DALY was developed under the *Global Burden of Disease* study (Murray and Lopez, 1996) and was a major step towards rational information-based health policy. One DALY represents the loss of one healthy life year. For each disease, DALYs are calculated at the global and regional levels as the discounted sum of years lost due to premature mortality and the years lost due to disability for incident cases of the ill-health condition.

The DALY is a powerful tool for assisting policy-makers in sector-wide decision-making for prioritizing health activities on the basis of cost-effectiveness analysis. It is therefore the

unit of choice for monitoring the burden of disease over time and across populations in relation to improvements in water supply and sanitation. Although not yet utilized for this purpose, it would also be a good measure in health impact assessment (HIA) of hydraulic infrastructure development projects, such as dams and irrigation schemes.

DALY estimates depend on the availability of data of sufficient quality as well as on assigning a certain class of severity of disability to each disease (Murray and Lopez, 1994b, 1996). This is based mainly on expert opinion and partly on empirical population valuations from surveys, such as those documented in recent *World Health Reports* (WHO 2003a, 2004a). In the case of certain water-related infectious diseases, such as intestinal helminth infections and schistosomiasis, DALYs are estimated on the basis of the number of new individuals infected with an associated low disability weight[1] (i.e. less severe). Once better data on clinical signs and symptoms associated with such diseases become available from community-based epidemiological studies, more appropriate disability weights can be assigned, and the DALY estimates will become more empirically based and less modelled. Furthermore, global estimates have to be validated by locally measured epidemiological data. For example, a study by Würthwein et al. (2001) in an area of Burkina Faso found much higher percentages of total burden of disease (mortality and morbidity) caused by malaria, diarrhoea, intestinal helminth infections and malnutrition than the *Global Burden of Disease* study (Murray and Lopez, 1996). Such differences are bound to have important implications

1. In the DALY methodology a disease is only included in terms of the disability it causes, whether temporary or permanent. Every health status gets a disability weight, varying between 0 (perfect health) and 1 (death).

for local health planning and decision-making. It has been stated that the normative value choices in the DALY on disability weighting, age weighting and discounting[2] tend to underestimate the disease burden attributed to young populations and communicable diseases. This goes against efforts to target diseases that are prevalent in poor populations (Arnesen and Kapiriri, 2004).

Clearly, water-related diseases continue to impose a large burden on health, especially in Africa and Asia (see **Chapter 14**). Globally, diarrhoeal diseases and malaria accounted, respectively, for 4 percent and 3 percent of DALYs lost and 1.8 and 1.3 million deaths in the year 2002. This burden is almost entirely concentrated in the group under 5 years of age. While the burden of

2. Based on the reality that individuals prefer benefits now rather than in the future, future life years are discounted.

Table 6.2: Global burden of disease: Deaths by age, gender, region and cause for the year 2002

Cause	Total number of deaths (thousands)	0–4 years (%)	Gender Male (%)	Gender Female (%)	Region[1] AFR (%)	SEAR (%)	WPR (%)	EMR (%)	AMR (%)	EUR (%)
All causes	57,029	18	52	48	19	26	21	7	10	17
Diarrhoeal disease	1,798	90	52	48	39	34	9	14	3	1
Malaria	1,272	90	48	52	89	5	1	5	0	0
Schistosomiasis	15	0	65	35	8	2	23	61	6	0
Lymphatic filariasis[2]	0	n/a[3]	n/a	n/a	n/a	n/a	n/a	n/a	n/a	n/a
Onchocerciasis	0	n/a	n/a	n/a	n/a	n/a	n/a	n/a	n/a	n/a
Dengue	19	22	45	55	1	63	20	5	11	0
Japanese encephalitis	14	36	49	51	0	61	21	17	0	0
Trachoma	0	n/a	n/a	n/a	n/a	n/a	n/a	n/a	n/a	n/a
Intestinal nematode infections	12	23	50	50	32	36	10	8	13	0
Protein-energy malnutrition[4]	260	57	50	50	40	26	5	10	16	2
Drowning[5]	382	15	69	31	17	26	35	7	6	10

Table 6.3: Global burden of disease: DALYs[6] by age, gender, region and cause for the year 2002

Cause	Total DALY (thousands)	0–4 years (%)	Gender Male (%)	Gender Female (%)	Region[1] AFR (%)	SEAR (%)	WPR (%)	EMR (%)	AMR (%)	EUR (%)
All causes	1,490,126	29	52	48	24	29	18	9	10	10
Diarrhoeal disease	61,966	91	52	48	38	33	11	14	4	1
Malaria	46,486	91	48	52	88	6	1	5	0	0
Schistosomiasis	1,702	1	60	40	78	0	3	13	4	0
Lymphatic filariasis[2]	5,777	4	76	24	35	56	7	2	0	0
Onchocerciasis	484	4	58	42	97	0	0	2	0	0
Dengue	616	23	45	55	1	62	21	5	11	0
Japanese encephalitis	709	37	48	52	0	43	45	12	0	0
Trachoma	2,329	0	26	74	52	7	17	16	7	0
Intestinal nematode infections	2,951	18	50	50	39	27	21	8	6	0
Protein-energy malnutrition[4]	16,910	88	51	49	34	36	11	12	6	1
Drowning[5]	10,840	19	69	31	18	25	35	7	6	8

1. WHO defines the regions of the world as follows:
   AFR – Africa south of the Sahara
   SEAR – South-East Asia (includes India)
   WPR – Western Pacific (includes China)
   EMR – Eastern Mediterranean (includes Sudan, Afghanistan, Pakistan)
   AMR – the Americas
   EUR – Europe (includes Central Asian republics)
2. Lymphatic filariasis, onchocerciasis and trachoma are diseases that are not fatal but that can lead to considerable disability (see Table 6.3).
3. For mortality rates that are zero, there can be no percentage.
4. Malnutrition is both a medical and a social disorder. It can occur as a primary disorder (with consequences for the susceptibility to infectious diseases) or as

a secondary disorder, prompted by infectious diseases, many of which are water-related.
5. Drowning is a major, non-communicable water-associated health problem.
6. The Disability-Adjusted Life Year is a summary measure of population health. One DALY represents a lost year of healthy life and is used to estimate the gap between the current health of a population and an ideal situation where everyone in that population would live into old age in full health.

Source: WHO, 2004a.

diarrhoea is distributed over both Africa and South Asia, malaria is largely a burden on children under the age of 5 in Africa. Africa accounts for more than half of the world's burden of onchocerciasis[3] (97 percent), malaria (88 percent), schistosomiasis (78 percent), and trachoma (52 percent). The World Health Organization (WHO) Region of South-East Asia accounts for more than half of the world's burden of dengue (62 percent) and lymphatic filariasis (56 percent). **Tables 6.2** and **6.3** provide estimates of the global burden of disease of the major water-related diseases for the year 2002, expressed in number of deaths and DALYs. These are based on data from the 2004 edition of the *World Health Report* (WHO, 2004a).

Diarrhoea and many other water-related diseases could eventually be controlled in a sustainable way by universal access to safe water and adequate sanitation, improved hygiene and optimal water management practices. In the short term, the control of many water-related diseases depends to a large extent (and puts a heavy burden) on the health care delivery system, which is responsible for oral rehydration therapy to prevent deaths from diarrhoea, insecticide-treated mosquito nets and chemotherapy to prevent and treat malaria and individual or mass drug treatment for the various helminth infections. There have been ongoing disease elimination programmes against some important water-related

diseases, notably Guinea worm infection, onchocerciasis, lymphatic filariasis and trachoma. These programmes, and those aimed at controlling intestinal helminth infections and schistosomiasis, are based on the mass treatment of at-risk populations. Low-cost, safe and effective drugs are available, but there are problems with respect to the insufficient capacity of health care delivery systems. This also applies to malaria control, where prompt treatment of patients and promotion of insecticide-treated nets (ITNs) is the backbone of the current strategy.

The following sections focus on the most important water-related diseases for which global data are available. There are many more water-related infectious as well as non-infectious diseases for which no data are available, and these cannot therefore be used to monitor progress in water-related development activities.

2a. Diseases related to lack of access to safe drinking water, poor sanitation and insufficient hygiene

Diarrhoeal diseases

It is estimated that, on average, each child under 5 years of age in a developing country suffers from three episodes of diarrhoea per year, with little change over the years (Kosek et al., 2003). While the number of cases has remained very high, substantial progress has been made in reducing the mortality associated with diarrhoeal disease. According to the *Global Burden of Disease* study by Murray and Lopez (1996), 2.9 million people died of diarrhoea in 1990, compared to 1.8 million in 2002, a decline of 37 percent. DALYs lost to diarrhoea went down with the same percentage from 99 million to 62 million. The reduction in mortality is probably due to improved case management, particularly oral rehydration therapy (ORT) (Victora et al., 2000). Despite this reduction, diarrhoeal diseases remain the leading cause of death from water-related diseases in children, accounting for 21 percent of all deaths of children under 5 in developing countries (Parashar et al., 2003). The increasing use of ORT from the early 1980s onwards is likely to have had its greatest impact on mortality due to dehydration from acute watery diarrhoea, such as that caused by rotavirus infection.[4] Persistent diarrhoea (episodes lasting fourteen days or longer, often associated with under-nutrition) and dysentery may now cause an increasing proportion of the remaining diarrhoeal deaths. No recent numbers

3. Onchocerciasis is a tropical parasitic disease caused by infection with filarial worms of the genus *Onchocerca* that after long and intense exposure can cause skin lesions and blindness. The worms are transmitted by *Simulium* blackflies that breed in the rapids and highly oxygenated parts of rivers, hence the popular name 'river blindness'.

4. Rotavirus gets its name from the Latin word for wheel, because of the wheel-like appearance of the virus under the electron microscope. Discovered in 1973, it is a leading cause of gastroenteritis and acute diarrhoea in young children.

Hundreds of professional launderers do their wash every day in the creek located at the entrance of the tropical forest of Le Banco (designated a national park in 1953) in Abidjan, Côte d'Ivoire

are available but based on a large review of studies between 1966 and 1997, the annual number of dysentery episodes caused by *Shigella*[5] throughout the world was estimated to be 164.7 million, of which 163.2 million were in developing countries (causing 1.1 million deaths) and 1.5 million in industrialized countries. A total of 69 percent of all episodes and 61 percent of all deaths attributable to shigellosis involved children under 5 years of age (Kotloff et al., 1999). Since the beginning of the 1990s large outbreaks of dysentery caused by *Shigella* have been reported with high case fatalities, first from Central Africa and later from other parts of the continent. While other pathogens, such as viruses, are more common causes of diarrhoea, *Shigella* is responsible for most deaths. This has important implications for control measures, as simple hygiene measures, especially handwashing after defecation, are very effective in its prevention and control (see **Box 6.2**).

Amoebiasis[6] is the second most important cause of dysentery and results in an estimated 100,000 deaths each year (WHO/PAHO/UNESCO, 1997). Two forms of *Entamoeba histolytica*[7] may be found in the stools of those carrying the infection: cysts and trophozoites. People with cysts can infect others, but they can be perfectly healthy themselves. Only the trophozoite, which is the motile form, is a sign of active infection. The cysts can be very persistent in the environment. In recent years, there has been an increased recognition of the protozoan parasite *Cryptosporidium parvum* as the cause of water-borne disease outbreaks, especially in the industrialized countries. The cysts are resistant to chlorine that is used for drinking water disinfection. *Cryptosporidium* and other protozoal infections are an important cause of chronic diarrhoea in patients infected with HIV. In the developing world, where highly effective antiretroviral treatment remains unaffordable, protozoa-related diarrhoea continues, by and large, to be a major cause of morbidity and mortality in HIV-infected individuals (Lean and Pollok, 2003).

In the early 1990s, cholera was concentrated in the Americas with 400,000 cases and 4,000 deaths in 1991. From the late 1990s onwards the problem shifted to Africa, where between 100,000 and 200,000 cases are officially reported each year: in 2002 a total of 123,986 cases with 3,763 deaths were reported. The actual number of cases is considered to be much higher. Poor surveillance systems and frequent under-reporting, often motivated by fear of trade sanctions and lost tourism, are the root causes.

5. Bacteria of the genus *Shigella* often cause dysentery.

6. Amoebiasis is an infection from a protozoan parasite (*Entamoeba histolytica*), which can lead to the destruction of the intestinal mucosa and, on penetrating the intestinal wall, may affect other organs, particularly the liver.

7. *Entamoeba* is a genus of dysentery-causing amoebae.

BOX 6.2: **CONTROL OF DIARRHOEAL DISEASES**

Outbreaks of diarrhoea, such as those caused by cholera, draw a lot of attention and often result in the mobilization of resources and policy changes. However, it is the day-to-day diarrhoea of small children that causes the great majority of deaths each year.

Diarrhoea is caused by a wide variety of microorganisms, including viruses, bacteria and protozoa. Rotavirus is the most common cause of watery diarrhoea in children in developed as well as developing countries. The primary pathway of rotavirus transmission is faecal-oral and infection can occur through ingestion of faecally contaminated water or food and contact with contaminated surfaces. An important cause of diarrhoea, especially in developing countries, is *Shigella*; infection with this bacterium often leads to bloody diarrhoea (dysentery). Typical for *Shigella* is the very small infective dose; therefore, it can spread easily from person to person.

The mainstay of diarrhoeal disease case management is oral rehydration therapy (ORT) to prevent dehydration. The discovery of ORT was an important public health advance of the twentieth century and has saved many lives. ORT is, however, most effective against acute watery diarrhoea and has less effect in preventing death due to dysentery.

Treatment of shigellosis usually involves antibiotics, in addition to ORT. Unfortunately, most *Shigella* bacteria have developed resistance to common antibiotics.

Simple hygiene measures are very effective in control and prevention of shigellosis, especially handwashing after defecation. Measures to improve the quality of drinking water, for example by boiling or adding chlorine to the water, are important for the prevention of rotavirus transmission, but are unlikely to have an impact on the transmission of *Shigella*.

Preventing the contamination of human fingers, legs of flies, water and food by the sanitary disposal of faeces would have an impact on both *Shigella* and rotavirus transmission. This provides a strong rationale for placing sanitation top of the agenda to combat diarrhoea.

Sources: Kotloff et al., 1999; Victora et al., 2000; Parashar et al., 2003.

Globally, iron-deficiency anaemia is the most common micronutrient disorder, known to be associated with high maternal mortality and morbidity

This woman's hands bear the marks of arsenic poisoning through drinking water

A number of water-related pathogens have emerged as new problems in developing as well as industrialized countries. These include Hepatitis E, *Escherichia coli* O157, and *Legionella pneumophila*, which can colonize water systems in buildings.

Typhoid fever is not a diarrhoeal disease, but it is associated with poor water supply, sanitation and hygiene. The global burden for the year 2000 was estimated at 21.6 million cases (with 216,510 deaths), half of which were from the WHO South-East Asia region (Crump et al., 2004).

Intestinal helminth infections

The roundworm (*Ascaris*), the whipworm (*Trichuris*), and hookworms (*Ancylostoma* and *Necator*) are mainly transmitted through soil that is contaminated with human faeces and are, therefore, directly related to the level of sanitary facilities. These soil-transmitted helminths flourish where poverty, inadequate sanitation and minimal health care prevail. In 1947, it was estimated that 1.5 billion people were infected with these worms. Fifty years later, this figure had increased to 3.5 billion. Taking account of the population increase, the proportion of the world population infected with these parasites remains virtually unchanged despite all the advances in medicine and technology (Chan, 1997). The clinical importance of a worm infection very much depends on the worm-burden. Above a certain number of worms, there are detrimental effects on physical fitness, growth development and school performance. In addition, hookworm infections cause blood loss from the intestine and are recognized as a major contributor to iron deficiency anaemia in adolescent girls and women of childbearing age. Globally, iron-deficiency anaemia is the most common micronutrient disorder, known to be associated with high maternal mortality and morbidity.

The reduction in DALYs lost from intestinal worm infections between 2000 and 2002, as reported in the

subsequent *World Health Reports*, can be attributed to an adjustment in the calculations and does not necessarily reflect a real reduction in the number of cases. Had symptoms and effects of disease been taken into account consistently, the estimated burden of disease would be much higher. It was estimated at 39 million DALYs for the year 1990 (see **Table 6.4**).

A recent update of the infection prevalences (see **Table 6.5**) shows that this has declined markedly in the Americas and Asia, but prevalence rates in Africa remain stagnant. This study (de Silva et al., 2003) also demonstrates the strong and reciprocal links between poverty and helminth infections, in particular hookworm infection.

Periodic drug treatment of school-age children living in areas of high endemicity is the control measure to obtain immediate benefits (WHO, 2002a). Long-term sustainable control will only be obtained by safe disposal of human faeces. The provision of culturally acceptable sanitary facilities for disposal of excreta and their proper use are necessary components to be included in any programme aimed at controlling intestinal parasites. In poor urban areas, sewerage and rainwater drainage can have a significant effect on the intensity of intestinal helminth infections by reducing transmission in the public domain (Moraes et al., 2004).

Skin and eye infections

Many infectious skin and eye diseases are related to poor hygiene and inadequate water supplies. Once enough water is available and used for personal and domestic hygiene, the prevalence of these diseases diminishes and they are therefore often classified as water-washed diseases. Trachoma is the leading cause of preventable blindness in the world, with an estimated 146 million cases, 6 million of which have caused actual blindness. The disease is related to poverty, illiteracy and unhygienic, crowded living conditions, particularly in dry dusty areas. Eye-seeking flies are important in the

Table 6.4: Estimated global burden of disease associated with soil-transmitted intestinal helminth infections, 1990

Helminth	Number of infections (millions)	Morbidity (cases, millions)	Mortality (deaths per year, thousands)	DALYs lost (millions)
Ascaris	1,450	350	60	10.5
Trichuris	1,050	220	10	6.4
Hookworm	1,300	150	65	22.1

Source: WHO, 2002a.

Table 6.5: Global estimates of prevalence and the number of cases of soil-transmitted helminth infections by region and age group, 2003

Helminth	Population (millions)		Infection prevalence (%)	Estimated number of infections (millions) Age groups (years)				
	At risk	Total		0–4	5–9	10–14	>15	Total
Ascaris								
LAC	514	530	16	8	10	10	56	84
SSA	571	683	25	28	28	25	92	173
MENA	158	313	7	3	3	3	14	23
SAS	338	363	27	13	15	13	56	97
India	808	1,027	14	15	18	17	89	140
EAP	560	564	36	20	25	25	134	204
China	1,262	1,295	39	35	44	51	371	501
Total	4,211	4,775	26	122	143	144	812	1,221
Trichuris								
LAC	523	530	19	10	12	12	66	100
SSA	516	683	24	26	27	23	86	162
MENA	52	313	2	1	1	1	4	7
SAS	188	363	20	10	11	10	43	74
India	398	1,027	7	8	9	9	47	73
EAP	533	564	28	16	19	19	105	159
China	1,002	1,295	17	15	19	22	163	220
Total	3,212	4,775	17	86	98	96	514	795
Hookworm								
LAC	346	530	10	1	3	5	41	50
SSA	646	683	29	9	18	29	142	198
MENA	73	313	3	0	1	1	8	10
SAS	188	363	16	2	5	8	44	59
India	534	1,027	7	2	5	8	56	71
EAP	512	564	26	4	9	16	120	149
China	897	1,295	16	3	9	18	173	203
Total	3,195	4,775	15	21	50	85	584	740

Abbreviated regions are as follows:
LAC – Latin America and the Caribbean
SSA – sub-Saharan Africa
MENA – Middle East and North Africa

SAS – South Asia
EAP – East Asia and the Pacific Islands

Source: de Silva et al., 2003.

Trachoma is the leading cause of preventable blindness in the world, with an estimated 146 million cases, 6 million of which have caused actual blindness

transmission and are associated with poor environmental sanitation. The provision of pit latrines in villages in Gambia resulted in a significant reduction of fly-eye contact and trachoma prevalence (Emerson et al., 2004). The main burden is in sub-Saharan Africa, with focal areas in the eastern Mediterranean and South and Central Asia.

There is sufficient scientific evidence to support the notion that with improved hygiene and access to water and sanitation, trachoma will disappear from these areas as it has from Europe and North America (Mecaskey et al., 2003).

2b. Vector-borne diseases associated with water

Water is the breeding site for many disease vectors that play a key role in the spread of disease-causing organisms. Malaria, Japanese encephalitis, filariasis and schistosomiasis are major vector-borne diseases associated with water resources development.

In 2003, WHO commissioned from the Swiss Tropical Institute[8] a number of systematic literature reviews focusing on the association between water resources development and four vector-borne diseases (malaria, lymphatic filariasis, Japanese encephalitis and schistosomiasis). The research led to global estimates of people at risk of these diseases in irrigation schemes and

8. www.sti.ch

In many parts of Africa the population faces intense year-round malaria transmission, resulting in a high disease burden, especially among children...

near dam sites, and provided evidence of the impact of water resources development on these diseases in different WHO sub-regions (Erlanger et al, 2005; Keiser et al., 2005a,b; Steinman et al., in press). The low level of association between water resources development and malaria and schistosomiasis in sub-Saharan Africa, where the estimated burden of these two diseases is highest, reflects the limited level of development of this continent's water resources potential rather than a lack of association (see **Chapter 14**). The at-risk population for Japanese encephalitis in rice irrigation schemes is highest in South Asia. While only 5.9 percent of the global population at risk of schistosomiasis lives in the western Pacific region (mainly China and the Philippines), relatively substantial parts of the population at risk living in irrigated areas or near dams (14.4 percent and 23.8 percent, respectively) are found in that region (see **Table 6.6**).

Malaria

Malaria remains one of the most important public health problems at a global level, causing illness in more than 300 million people each year. Its share of the global burden of disease has increased over the past few years and now stands at 46.5 million DALYs, 3.1 percent of the world's total. This is an increase of 23 percent, compared with the year 1990. Mortality increased by 27 percent from 926,000 in 1990 to 1,272,000 in 2002. The majority of the burden of malaria is concentrated in sub-Saharan Africa. In many parts of Africa the population

faces intense year-round malaria transmission, resulting in a high disease burden, especially among children below 5 years of age and pregnant women. In all malaria-endemic countries in Africa, on average 30 percent of all out-patient clinic visits are for malaria (WHO/UNICEF, 2003). In these same countries, between 20 percent and 50 percent of all hospital admissions are malaria-related. International efforts to reduce the malaria burden are coordinated by the WHO-led Roll Back Malaria (RBM) initiative, which was launched in 1998. The main strategy is to promote prompt diagnosis and treatment, and the use of insecticide-treated nets (ITNs).

Malaria control is hampered by a number of constraints. Vector mosquitoes are becoming increasingly resistant to insecticides and malaria parasites to inexpensive drugs. Climate and environmental change, population movements and behavioural change have helped malaria gain new grounds in many parts of the developing world. The difficulties in achieving a high coverage of ITNs among the vulnerable groups are a major issue, especially in Africa. In addition, operational constraints limit effective re-impregnation of ITNs. Most importantly, the countries facing severe malaria problems have an underdeveloped health care sector that is limited in its potential to implement the established strategies, particularly those related to ensuring early diagnosis and treatment, disease monitoring and community involvement in control activities.

Table 6.6: Global estimates of people at risk of four vector-borne diseases

Estimated numbers of	Malaria (million)	Lymphatic filariasis (million)	Japanese encephalitis (million)	Schistosomiasis (million)
People at risk globally	>2,000	>2,000	1,900	779
People at risk near irrigation schemes, globally	851.3	213	180–220	63
People at risk near dams, globally	18.3	n.a.	n.a.	42
People at risk in urban settings (no access to improved sanitation)		395	n.a.	
People at risk near dams and irrigation schemes, sub-Saharan Africa	9.4	n.a.	n.a.	39
People at risk near dams and irrigation schemes, excluding sub-Saharan Africa	860.3	n.a.	n.a.	66
People at risk near dams and irrigation schemes, Western Pacific	n.a.*	n.a.	n.a.	40
People at risk near irrigation schemes, South East Asia and Western Pacific	n.a.*	n.a.	132 (in irrigated areas) } SE Asia 167 (in rice irrigated areas) } 921 (in irrigated areas) } W. Pacific 36 (in rice irrigated areas) }	

*Not segregated to this level.

Sources: Erlanger et al., 2005; Keiser et al., 2005a,b; Steinman et al., in press; www.who.int/water_sanitation_health/resources/envmanagement/en/index.html

Water management for malaria control

Water resources development projects, especially irrigation systems, can provide the ecological conditions suited to the propagation of malaria vectors. The relationship between malaria and water resources development is, however, highly situation-specific, depending on the ecology, biology and efficiency of local vectors, people's behaviour and climate. The opportunities for malaria vector breeding are often associated with faulty irrigation design, maintenance or water management practices. The case of irrigation-related malaria in the Thar Desert is described in **Box 6.1**.

In Africa, but also in parts of Asia, several empirical studies have shown the counter-intuitive result of no intensification of malaria transmission in association with irrigation development and increased mosquito vector densities; socio-economic, behavioural and vector ecological factors may all play a role in this phenomenon, dubbed the 'paddy paradox' (Ijumba and Lindsay, 2001; Klinkenberg et al., 2004). Studies in West Africa on rice irrigation and farmers' health showed that irrigation altered the transmission pattern but did not increase the burden of malaria (Sissoko et al., 2004). It was also documented that irrigated rice cultivation attracted young families, improved women's income and positively affected treatment-seeking behaviour by shortening the delay between disease and initiation of treatment.[9]

Globally, it is estimated that only 18.9 million people (most of whom are in India) live close enough to large dams to be at risk of malaria transmitted by mosquitoes associated with man-made reservoirs (Keiser et al., 2005a). The population living close to irrigation sites in malaria endemic areas is much larger and has been estimated at 851.5 million (see **Chapters 7 and 8**). However, in Africa, where the main burden of malaria rests, only 9.4 million people live near large dams and irrigation schemes. Hardly any information is available on the impact of small dams, of which there are many hundreds of thousands in malaria endemic areas in Africa and elsewhere. Cumulatively, these could well be more important for malaria transmission than large dams and irrigation schemes. The potential for the further expansion of small dams is considerable, particularly in sub-Saharan Africa. There is therefore a pressing need for strategic health impact assessment as part of the planning of small dams that should encompass a broad approach towards health, including issues of equity and well-being (Keiser et al., 2005a).

The role of the aquatic environment as an essential condition for malaria transmission was recognized long ago. Environmental management methods were used for malaria control, especially in Asia, Central America and the Caribbean, Europe and the US (Konradsen et al., 2004; Keiser and Utzinger, 2005). A lack of scientific evidence of effectiveness, uncertainty about the present-day feasibility of implementation and remaining vertical vector-control structures prevent environmental management methods from playing a more important role in present-day malaria control. The joint World Health Organization (WHO), Food and Agriculture Organization (FAO), United Nations Environment Programme (UNEP) Panel of Experts on Environmental Management for Vector Control (PEEM) has played a central role in research and capacity-building in this field since the early 1980s. Recently, international research initiatives have focused on possibilities for reducing malaria as part of an ecosystem approach to human health, by looking at the relationship between all components of an ecosystem in order to define and assess priority problems that affect the health and livelihood of people and environmental sustainability.[10]

The Consultative Group on International Agricultural Research's (CGIAR) Systemwide Initiative on Malaria and Agriculture (SIMA) looks at the interaction of people with land, water and crops as they farm existing agricultural areas or develop new areas for farming. This is expected to lead to the identification of specific environmental management measures for the reduction of the disease transmission potential. In the absence of an effective vaccine, treatment of patients and promotion of insecticide-treated nets will remain the main evidence-based strategies for malaria control. But even in the African context, vector control (largely by indoor house-spraying with residual insecticides) and proper management of the environment is increasingly recognized as an indispensable part of malaria control (see the recent work done in Sri Lanka discussed in **Box 6.3**). In low transmission areas such as in many parts of Asia and in the latitudinal and altitudinal fringes of malaria distribution in Africa, environmental management is re-emerging as an important component of an integrated approach to malaria control. In such areas, it is also important that health impact assessments be part of the planning process of hydraulic infrastructure projects, in order to identify, qualify and possibly quantify adverse health effects at the earliest possible stage and suggest preventive solutions (Lindsay et al., 2004). In rural areas of Africa where mosquito breeding places are diffuse and

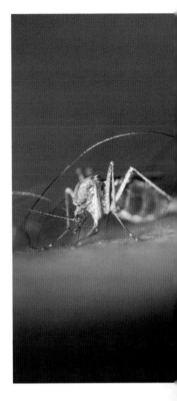

Mosquitoes are becoming increasingly resistant to insecticides and malaria parasites to inexpensive drugs

9. See www.warda.cgiar.org/research/health for more information

10. For more information, go to www.idrc.ca/ecohealth

In rural areas of Africa where mosquito breeding places are diffuse and varied, there may be little scope for environmental control measures

varied, there may be little scope for environmental control measures. The situation is different in African cities. In urban and peri-urban areas, breeding sites can be detected more easily than in rural areas, and environmental management is proposed as a main feature for an integrated control approach (Keiser et al., 2004). This can have an important impact on the overall malaria burden. According to different plausible scenarios, an estimated 25 to 100 million malaria cases occur in African cities.

Filarial infections

Mosquito-borne lymphatic filariasis is rarely life-threatening but causes widespread and chronic suffering, disability and social stigma. Globally, an estimated 119 million people are infected, with 40 million suffering from severe chronic disease. More than 40 percent of those infected live in India and 30 percent in Africa. In India alone, the disease causes losses of US $1 billion annually (Erlanger et al., 2005). The current Global Programme to Eliminate Lymphatic Filariasis (GPELF), led by WHO, is based on mass drug administration of the entire population at risk. It has been argued that vector control is an essential supplement that will add to the sustainability of these efforts. The programme provides significant opportunities to deliver public health benefits in the broader scope of intestinal helminth infections, malaria and dengue (Molyneux, 2003).

Urban vectors of lymphatic filariasis (*Culex* species) breed in organically polluted waters, such as blocked drains and sewers. An estimated 394 million urban dwellers, mainly in South Asia, are at risk of lymphatic filariasis if lack of access to improved sanitation is considered a key determinant (Erlanger et al., 2005). Urban improvement, including proper sanitation, a robust drainage infrastructure and environmental management to minimize mosquito-breeding places, has been shown to contribute significantly to the reduction of transmission risks. In the rural areas of Africa, where anopheline mosquitoes are the vectors, an estimated 213 million people are at risk because of their proximity to irrigation schemes (Erlanger et al., 2005). While densities of the mosquito vector are often much higher in irrigated areas as compared to irrigation-free sites, relatively few studies have been done aimed at linking water resources development and filarial disease. More research is needed to better define the potential of water management approaches for vector control in this connection. In rural areas of Africa, the vector of lymphatic filariasis also transmits malaria. Therefore, vector control activities such as implemented under the Roll Back Malaria initiative can be expected to reduce the transmission of malaria as well as lymphatic filariasis (Manga, 2002). In India, the vectors of lymphatic filariasis and malaria are different, but vector control, including breeding-site reduction and environmental

---

BOX 6.3: MALARIA CONTROL THROUGH STREAM WATER MANAGEMENT

From a global perspective, the use of environmental and engineering-based control interventions that make the water environment less conducive for vector-breeding plays a limited role in current malaria control efforts. However, research from around the world has shown the potential for using environmentally based control interventions as a component of an integrated control programme. Some of the interventions being field-tested today are based on the approaches used during the first half of the twentieth century, while others have come about through the use of modern technologies in an attempt to counter the new challenges resulting from large-scale changes in the freshwater environment. The experiences clearly point to the site-specific nature of the interventions,

reflecting the different patterns of transmission, disease-vector ecology and the local capacities available for implementation. Extensive field research and close collaboration between the water management and health sectors provide opportunities for a significant contribution to malaria control.

Recent work in Sri Lanka assessed options for the control of malaria vectors through different water management practices in irrigation conveyance canals and streams. The approach was based on the use of existing irrigation structures regulating the water levels in the waterways and was aimed at eliminating the principal breeding sites of the most important malaria vector in the country. Overall, the result

demonstrated a high potential for effective vector control by feasible changes in irrigation and stream water management, resulting from periodic fortnightly releases from upstream reservoirs, to eliminate mosquito breeding sites and render the habitat less conducive to *Anopheles culicifacies* breeding for some time after the water release (see **Chapter 14**). The approach followed did not result in a loss of water, since the water was captured in reservoirs downstream. The designated water management strategy was far cheaper than the use of chemical larvicides.

Sources: Konradsen et al., 1998, 1999; Matsuno et al., 1999; Keiser and Utzinger., 2005.

management, can have an impact on different vector species and both diseases (Prasittisuk, 2002).

Guinea worm infection is unique in that it is the only communicable disease that is transmitted exclusively through drinking water containing infected intermediate hosts. These are small crustaceans infected with the parasite *Drancunculus medinensis*, which causes the disease in humans. Thus, it is the only disease that can be prevented entirely by protecting supplies of drinking water. Guinea worm infection is about to be eliminated by improvements in water supply. In Africa, the number of cases has declined from 3.5 million in 1986 to 35,000 in 2003 (WHO/UNICEF, 2004). The majority of remaining cases are in Sudan, where many areas are inaccessible to eradication efforts due to ongoing civil conflicts.

Schistosomiasis

Schistosomiasis (bilharzia) is contracted by humans through contact with water infested with the free-swimming larval stages of parasitic worms (*cercariae*) that penetrate the skin and develop in the human body to maturity. Parasite eggs leave the human body with excreta. They hatch in freshwater and infect aquatic snail intermediate hosts. Within the snails they develop into *cercariae*, which are, in turn, released into the water to infect new human hosts. Transmission can take place in almost any type of habitat from large lakes or rivers to small seasonal ponds or streams. Man-made water bodies, including irrigation schemes, are particularly important, as the human population density is usually high around these and water contact patterns are intense. The disease occurs in seventy-four countries in Africa, South America and Asia, with an estimated 200 million people infected, 85 percent of whom live in sub-Saharan Africa.

Schistosomiasis is a chronic, debilitating parasitic disease, which may cause damage to the bladder, liver and intestines, lowers the resistance of the infected person to other diseases and often results in retarded growth and reduced physical and cognitive functions in children. The current estimate of the global burden of disease due to schistosomiasis as presented in the *World Health Report* is based on the number of people infected with an associated low disability weight because many infections do not result in clinical disease. With better data on morbidity and mortality now becoming available, DALYs due to schistosomiasis should be recalculated (Crompton et al., 2003). Recent estimates from sub-Saharan Africa indicate that 280,000 deaths per year can

be attributed to schistosomiasis, much higher than the 15,000 listed by the Global Burden of Disease Initiative (van der Werf et al., 2003).

The key element in the current control strategy is the regular treatment of at-risk populations, especially school children, with the drug praziquantel. This has to be combined with improvements in sanitation, which will prevent eggs from entering the environment. Contact with infested water has successfully been reduced by improving water supplies and providing laundry and shower facilities and footbridges. It has been stated that linking schistosomiasis control to improvements in water supply and sanitation has the potential to ensure long-term control and, in many instances, elimination of the disease (Utzinger et al., 2003). Results from national control programmes in endemic countries such as Brazil, China and Egypt are encouraging (see **Chapter 14**). However, there is currently little or no schistosomiasis control in sub-Saharan Africa (Engels and Chitsulo, 2003). Of the estimated population at risk (779 million globally), some 105 million live close to dams and irrigation schemes (Steinmann, in press). Proportionally, a high percentage of these live in the western Pacific region (China and Philippines).

The introduction or spread of schistosomiasis has been documented in relation to the construction of large dams and irrigation systems. In these settings, it is important to combine mass chemotherapy and improvements in water supply and sanitation with snail control. Reductions in snail populations can be achieved by various engineering means, including proper drainage, canal lining, removal of aquatic vegetation from canals, regular flushing of canals, increasing the flow velocity, drying of irrigation systems and changing water levels in reservoirs. Such often capital-demanding interventions should be focused on locations where water contact is intense. Of even greater importance for increased schistosomiasis transmission in the future could be the thousands of small dams that are being built on the African continent for agriculture, livestock, and drinking water supply.

Arboviral infections

Japanese encephalitis: Restricted to the Asian region, Japanese encephalitis (JE) is closely associated with irrigated rice ecosystems, where the *Culex* mosquito vectors prefer to breed. Transmission risks are greatly enhanced where pig rearing is practised as a source of food and income generation: pigs are amplifying hosts of

Globally, an estimated 119 million people are infected with lymphatic filariasis, with 40 million suffering from severe chronic disease

---

BOX 6.4: A NEW STRATEGY AGAINST *AEDES AEGYPTI* IN VIET NAM

The Australian Centre for International and Tropical Health and Nutrition and the General Department of Preventive Medicine and HIV-AIDS control of the Vietnamese Ministry of Health have progressively deployed a new strategy for the control of the container breeding mosquito, *A. aegypti*. It incorporates four elements:

■ a combined vertical and horizontal approach that depends on community understanding

■ prioritized control according to the larval productivity of major habitat types

■ use of predacious copepods of the genus *Mesocyclops* as a biological control agent

■ community activities of health volunteers, schools and the public.

Between 1998 and 2004, the strategy achieved the elimination of vectors from thirty-two out of thirty-seven communities, covering 309,730 people. As a result, no dengue cases have been detected in any of the communities since 2002, and the deployment of the strategy indicates so far its applicability and sustainability wherever large water storage containers are major sources of the vector.

Source: Kay and Nam, 2005.

Dengue ranks as the most important mosquito-borne viral disease in the world. In the last fifty years, its incidence has increased thirty-fold. An estimated 2.5 billion people are at risk in over 100 endemic countries

the JE virus. It is the leading cause of viral encephalitis in Asia with 30,000 to 50,000 clinical cases reported annually and an estimated global burden of 709,000 DALYs lost in 2002; under-reporting is, however, considerable. Vaccination initiatives are the mainstay of JE outbreak control, but water management methods have been used to control the mosquito vector, especially the alternate wet and dry method of cultivating rice (van der Hoek et al., 2001a; Keiser et al., 2005b).

Dengue: Dengue ranks as the most important mosquito-borne viral disease in the world. In the last fifty years, its incidence has increased thirty-fold. An estimated 2.5 billion people are at risk in over 100 endemic countries. Up to 50 million infections occur annually, with 500,000 cases of dengue haemorrhagic fever and 22,000 deaths, mainly among children. Prior to 1970, only nine countries had experienced cases of dengue haemorrhagic fever (DHF); since then the number has increased more than fourfold and continues to rise. In 2001, the Americas alone reported over 652,212 cases of dengue (of which 15,500 were DHF), nearly double the number of cases reported for the same region in 1995. Dengue is present in urban and suburban areas in the Americas, South and South-East Asia, the East coast of Africa, the Eastern Mediterranean and the Western Pacific regions. In South and South-East Asia, it has also spread to rural areas. In areas of high endemicity, dengue hemorrhagic fever considerably increases the disease burden caused by this virus. Several factors have combined to produce epidemiological conditions in developing countries in the tropics and subtropics that favour viral transmission by the *Aedes* mosquito vectors: rapid population growth, rural–urban migration, inadequate basic urban infrastructure (e.g. unreliable water supply leading

householders to store water in containers close to homes), promotion of inappropriate drinking water storage containers and increase in volume of solid waste, such as discarded plastic containers and other abandoned items which, following rains, provide larval habitats in urban areas.[11] **Box 6.4** shows a strategy developed in Viet Nam to control dengue transmission.

2c. Non-communicable water-associated health issues

Among the ill-health conditions that are water-associated but not caused by infectious agents, two stand out globally based on the burden of disease: drowning and the effects of long-term exposure to inorganic chemicals that occur naturally in the groundwater sources of drinking water (see **Box 6.5**). Anthropogenic chemical pollution of surface waters, mainly by industry and agricultural runoff, is a health hazard, but the impacts on health (for example, malignant tumours) generally occur only after extended periods of exposure and are difficult to attribute accurately to specific environmental or lifestyle factors.

Drowning is a significant problem worldwide, particularly in children under the age of fifteen, with 97 percent of all drowning incidents occurring in low- and middle-income countries (Peden and McGee, 2003). In 2002, an estimated 382,000 people drowned worldwide, which translates into the loss of more than 10 million DALYs. The risk of drowning is not just linked to recreational water use, but is also important in the context of natural disasters (for example, in low-lying areas when protective hydraulic works have not been adequately maintained). It is an occupational health hazard for sailors, fishermen and other professions. Obviously, the death toll of the Indian Ocean tsunami of 26 December 2004 will contribute to a sharp increase in the

11. Based on www.who.int/csr/ disease/dengue/impact/en/

mortality caused by drowning as an exceptional peak resulting from a phenomenon for which early warning is key (see **Chapters 1, 10 and 14**).

The strategy to improve rural drinking water supply by installing low-cost hand pumps that draw groundwater uncontaminated by disease-causing microbes, has been applied on a large scale in several countries. In Bangladesh alone, more than 4 million tubewells have been installed over the past twenty years to provide drinking water to 95 percent of the population. It is believed to have contributed significantly to the reduction of the burden of diarrhoea over the same period. Arsenic occurs naturally in groundwater, and excess exposure to arsenic in drinking water may result in a significant risk of skin lesions and cancer (WHO, 2004b). The high concentrations of arsenic in many Bangladeshi tubewells occur in an erratic pattern, and the scale of the problem has only become known in recent years. The full effects of arsenic poisoning will only become apparent at a later stage: deep wells have been in use since the late 1970s, and cancer has a long latency period (Yoshida et al., 2004). However, a return to surface water would inevitably result in an increase in diarrhoeal disease (Lokuge et al., 2004). In India, an estimated 66 million people rely on groundwater with fluoride concentrations exceeding WHO recommended norms for their drinking water needs. While arsenic is toxic and carcinogenic, an excess of fluoride leads to the mottling of teeth and, in severe cases, crippling skeletal deformities, as well as other health problems. In addition to the Indian subcontinent and China, clinical forms of dental and skeletal fluorosis are particularly common along the East African Rift Valley (WHO/IWA, 2006; see also **Chapter 14**).

The situation in Bangladesh and other areas, including parts of India, China and East Africa, calls for a pragmatic combination of practical, affordable and sustainable water supply programmes aimed at minimizing the combined risk to health posed by diarrhoeal disease, fluoride, arsenic and other chemical contaminants that may be present in the environment. Installing filters or other devices in millions of tubewells to remove arsenic and fluoride is an almost impossible task. Yet, it is imperative that water from each and every tubewell be tested for arsenic and fluoride in affected areas, before it is made available for consumption. Even in affected villages, one or more pumps could provide water with permissible arsenic and fluoride levels. In other cases, there might be no choice but to use surface water sources that are also used for agriculture and other uses with suitable treatment. This calls for clear, integrated policies on the joint use of surface and groundwater resources, as illustrated in **Box 6.6**.

There are many other chemicals that can cause health problems. However, at the global level these are not as important as fluoride or arsenic. Nitrate pollution of groundwater is a major environmental issue in developing as well as industrialized countries (see **Chapter 5**). Nevertheless, a recent review of the global burden of disease related to nitrate in drinking water concluded that nitrate is just one of the factors that play a role in the sometimes complex causal web underlying methaemoglobinemia ('blue baby syndrome').[12] Given the apparently low incidence of this condition and the complex nature of the role of nitrates and individual behaviour, it is currently inappropriate to attempt to link disease incidence with drinking water nitrate levels (Fewtrell, 2004).

The *Guidelines for Drinking-water Quality* (WHO, 2004b) establishes values for the concentrations of substances

12. This condition can occur when large amounts of nitrates in water are ingested by an infant and converted to nitrite by the digestive system. The nitrite changes the oxygen-carrying blood protein. As a result, body tissues may be deprived of oxygen, causing the infant to develop a blue colouration.

BOX 6.5: RECREATIONAL WATER USE, POLLUTION AND HEALTH

Drowning is not the only risk associated with recreational water use. Swimming may be exposed to health hazards at many places where raw or partially treated sewage is pumped into seas every day. Chemical contamination of seas and rivers arises principally from direct waste discharge (e.g. industrial effluent) or chemical spills and is typically local or regional in nature. Little is known about the adverse effects of exposure to chemical contaminants, but exposure from recreational water is likely to be a very small fraction of total exposure. Recreational waters in the tropics and subtropics pose special hazards, not just from some of the local aquatic or amphibious predators such as crocodiles, but also from the causative agents of a number of tropical diseases, especially schistosomiasis. WHO produces international norms on recreational water use and health in the form of guidelines. Volume 1 of the WHO Guidelines for Safe Recreational Water Environments addresses the health aspects of coastal and fresh waters.

Source: www.who.int/water_sanitation_health/ bathing/en/

BOX 6.6: AVAILABILITY AND QUALITY OF DRINKING WATER IN THE RUHUNA BASINS, SRI LANKA

The Ruhuna Basins in southern Sri Lanka were described as a pilot case study in the first *World Water Development Report* (UN-WWAP, 2003; see also **Chapter 14**). Recent studies in the area have confirmed the observation that seepage from irrigation canals and reservoirs is indispensable for maintaining water levels in shallow wells that people use for drinking. Canal seepage accounted for more than half of groundwater recharge, and canal closure resulted in groundwater levels decreasing by 1 to 3 metres within a few days, leading to the drying-up of many shallow wells and problems of access to domestic water supplies for farmers. To make agricultural water use more efficient, several canals were lined with concrete and this has reduced seepage and further restricted the availability of water for domestic use.

To improve the drinking water supply for the people that settled in newly developed irrigated

areas in the Ruhuna Basins, many tubewells were constructed in order to exploit deeper groundwater resources. However, a large proportion of these wells are not used by the local population, because the water is unpalatable, due to salt or other chemicals. Water quality testing showed that surface water was polluted by faecal matter, therefore presenting a potential risk for faecal-orally transmitted diseases, especially diarrhoea, if used for drinking. Shallow wells had lower levels of pollution, and tubewells had the lowest level of faecal indicator bacteria, often meeting the zero pathogen criterion referred to in the WHO guidelines for Drinking-water Quality. While the tubewell water from deeper aquifers was of good bacteriological quality, the water generally had high contents of iron, salt and fluoride. Prevalence of dental fluorosis among 14-year-old students in the area was 43 percent.

In basins such as these, providers of drinking water are faced with a dilemma. Taking availability and biological and chemical water quality into account, shallow wells seem to be the best water source for domestic purposes, especially those protected by a wall from surface inflow. However, irrigation rehabilitation programmes that include the lining of canals are threatening this source of drinking water. Residents may then be forced to look for alternative water sources and may have to revert to untreated surface water from larger canals and reservoirs if shallow wells fail. This stresses the need for a governance structure with intersectoral and integrated planning, development and management of water resources to ensure that the needs of at least the most important stakeholder group in the system – farmers – are met.

Sources: Boelee and van der Hoek, 2002; van der Hoek et al., 2003; Rajasooriyar, 2003.

above which toxic effects may occur; most chemicals included in the guidelines' listing are of health concern only after extended exposure of years rather than months. In addition to fluoride and arsenic, the guidelines give values for a number of other naturally occuring inorganic substances, including barium, boron, chromium, manganese, molybdenum, selenium and uranium. In relation to industrial waste, the guidelines list three inorganic chemicals (cadmium, cyanide and mercury) and

give values for some twenty organic substances of importance. Most pollutants originating from agricultural activities are pesticides and a large part of them (or their residues) have never been detected in drinking water, while another substantive part occurs at concentrations well below those at which toxic effects may occur. Finally, the guidelines single out the cyanobacterial toxins produced by many species of *Cyanobacteria* occurring naturally in lakes, reservoirs, ponds and slow-flowing rivers.

Part 3. Progress towards the MDG Targets on Water, Sanitation and Health

In 1990, 77 percent of the world's population used improved drinking water sources. Considerable progress was made between 1990 and 2002, with about 1.1 billion people gaining access to improved sources. Global coverage in 2002 reached 83 percent, keeping the world on track to achieve the MDG target; however, there are great regional disparities. Table 6.1 reviews the water and sanitation targets in relation to the MDGs.

New toilet installed thanks to a water and sanitation project in Bara Bari village, Bangladesh

3a. Status of MDG 7: Drinking water and sanitation targets

The region that made the greatest progress towards sustained access to safe drinking water is South Asia, where coverage increased from 71 percent to 84 percent between 1990 and 2002. This jump was fuelled primarily by increased access to improved water sources in India, home to over 1 billion people. Coverage in sub-Saharan Africa increased from 49 percent to 58 percent between 1990 and 2002. Yet this falls short of the progress needed to achieve the MDG target of 75 percent coverage by 2015 (see **Map 6.1**). Nevertheless, there are a number of success stories in water supply, sanitation and hygiene in sub-Saharan Africa, some of which are reported in the Blue Gold Series of the World Bank Water and Sanitation Programme (see also **Chapter 14**).[13]

Global sanitation coverage rose from 49 percent in 1990 to 58 percent in 2002. Still, some 2.6 billion people – half of the developing world and 2 billion of whom live in rural areas – live without improved sanitation. Sanitation coverage in developing countries (49 percent) is only half that of the developed world (98 percent). Major progress was made in South Asia between 1990 and 2002. Yet, more than 60 percent of the region's population still did not have access to sanitation in 2002. In sub-Saharan Africa, sanitation coverage in 2002 was a mere 36 percent, up 4 percent from 1990. Over half of those without improved sanitation – nearly 1.5 billion people – live in China and India.

To halve the proportion of people without improved sanitation, global coverage needs to grow to 75 percent by 2015, from a starting point of 49 percent in 1990. However, if the 1990–2002 trend continues, the world will fall short of the sanitation target by more than half a billion people. In other words, close to 2.4 billion people will be without improved sanitation in 2015, almost as many as there are today. The proportion of the world's population with improved sanitation has increased by just 9 percent since 1990, a rate far slower than that required to meet the MDGs. The widening gap between progress and target (see **Figure 6.1**) signals that the world will meet its sanitation goal only if there is a dramatic acceleration in the provision of services.

3b. Status of the other MDG targets with respect to water-related health issues

Some of the indicators for monitoring progress towards meeting the MDGs are especially relevant in connection with water-related diseases. WHO and the United Nations Children's Fund (UNICEF) are responsible for providing the UN Statistics Division with relevant international statistics and analyses of quantitative and time-bound indicators directly linked to water and sanitation. Data sets and information on water supply and sanitation coverage are derived from their Joint Monitoring Programme (JMP). In addition, progress towards the MDGs is monitored by a number of indicators that are health-related but that cut across different sectors. While there is progress in many parts of the world with respect to targets on child mortality, nutrition and water-related infectious diseases, the situation remains extremely worrisome in sub-Saharan Africa.

Figure 6.1: Projected population without access to improved sanitation

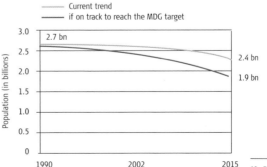

Source: WHO/UNICEF, 2004.

13. For more information, go to www.wsp.org/08_BlueGold.asp

Map 6.1: Coverage with improved drinking water sources, 2002

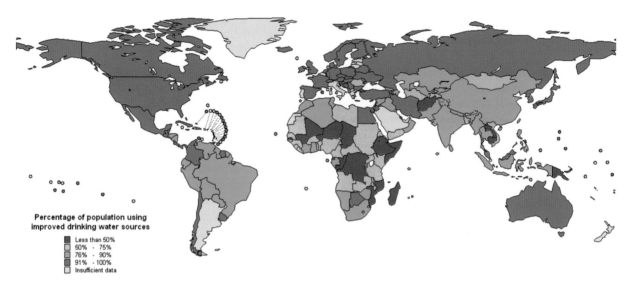

Percentage of population using
improved drinking water sources

- Less than 50%
- 50% - 75%
- 76% - 90%
- 91% - 100%
- Insufficient data

Source: WHO/UNICEF, 2004. The boundaries shown on this map do not imply the expression of any opinion whatsoever on the part of the World Health Organization concerning the legal status of any country, territory, city or area or of its authorities, or concerning the delimitation of its frontiers or boundaries. Dotted lines on maps represent approximate border lines for which there may not yet be full agreement.

Section 3: CHALLENGES FOR WELL-BEING & DEVELOPMENT

A significant part of child mortality rates can be attributed to water-associated diseases

■ *MDG Target 2: Halve, between 1990 and 2015, the proportion of people who suffer from hunger*

One of the two indicators for monitoring progress towards achieving this target is the prevalence of underweight children under 5 years of age.

It is unlikely that the MDG target of reducing 1990-level prevalence of underweight children by 50 percent in the year 2015 can be met, mainly due to the deteriorating situation in Africa (de Onis et al., 2004). Worldwide, the percentage of underweight children has been projected to decline from 26.5 percent in 1990 to 17.6 percent in 2015, a decrease of 34 percent. However, in Africa, the rate was expected to increase from 24 percent to 26.8 percent. In developing countries, stunting has fallen progressively from 47 percent in 1980 to 33 percent in 2000, but with very little, if any, progress in large parts of Africa (de Onis et al., 2000). Estimated trends indicate that overall stunting rates in developing countries will continue to decrease to 16.3 percent in 2020 (de Onis and Blössner, 2003). The great majority of stunted children live in South Asia and sub-Saharan Africa, where only minor improvements are expected.

■ *MDG Target 5: Reduce by two-thirds, between 1990 and 2015, the under-5 mortality rate*

Progress in reducing child mortality is low. No country in sub-Saharan Africa is making enough progress to reach this target. The developing world only achieved a 2.5 percent average annual decrease during the 1990s, well short of the target of 4.2 percent (UNDP, 2003). A significant part of this mortality rate can be attributed to water-associated diseases.

■ *MDG Target 8: Have halted by 2015 and begun to reverse the incidence of malaria and other major diseases*

Throughout sub-Saharan Africa, the decrease in under-5 mortality from all combined causes, apparent during the 1970s and 1980s, levelled off in the 1990s, perhaps partially as a result of increased malaria mortality (WHO/UNICEF, 2003).

Map 6.2: Coverage with improved sanitation, 2002

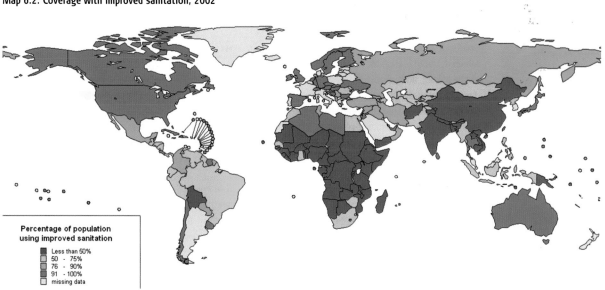

Percentage of population
using improved sanitation

- Less than 50%
- 50 - 75%
- 76 - 90%
- 91 - 100%
- missing data

Source: WHO/UNICEF, 2004. The boundaries shown on this map do not imply the expression of any opinion whatsoever on the part of the World Health Organization concerning the legal status of any country, territory, city or area or of its authorities, or concerning the delimitation of its frontiers or boundaries. Dotted lines on maps represent approximate border lines for which there may not yet be full agreement.

Part 4. Indicators

Good indicators must fulfil a number of criteria. They must have a scientific evidence base, be policy-relevant, make it possible to monitor progress towards internationally agreed targets (such as the targets of the MDGs), and reliable data necessary to compute the indicator values should be available in the public domain. Several indicators relevant to water and health are well-defined, well-established and backed by databases with global coverage that are updated at least on an annual basis. Examples include access to safe drinking water and adequate sanitation, under the WHO/UNICEF Joint Monitoring Programme (JMP); global burden of specific diseases, expressed in deaths and DALYs (WHO's *World Health Reports*); child mortality (UNICEF); and nutritional status (WHO Global Database on Child Growth and Malnutrition).

4a. Burden of water-related diseases

Databases on the number of deaths and DALYs by cause (disease), age, gender and region are maintained by WHO.[14] The major water-related diseases and hazards whose mortality rate and burden can be monitored in this way include diarrhoea, malaria, schistosomiasis, lymphatic filariasis, onchocerciasis, dengue, Japanese encephalitis, trachoma, intestinal helminth infections (separate for *Ascaris*, *Trichuris* and hookworm), and drowning. Some water-related diseases of interest are separately accounted for or not included, notably cholera, typhoid fever and Guinea worm disease. For these diseases, and for selected other diseases like diarrhoea and malaria as well, it is still useful to report the direct measures of disease frequency (incidence or prevalence) when data are available. Changing epidemiological patterns, with important implications for planning appropriate cost-effective interventions, make it preferable to segregate reported diarrhoea figures for watery diarrhoea, persistent diarrhoea and dysentery.

In the context of ongoing mass treatment campaigns, increasing numbers of baseline prevalence data will be generated for intestinal helminth infections. Such mass treatment campaigns will result in immediate prevalence

reductions. Over longer periods of time, the prevalence of intestinal helminth infections can be an important indicator for monitoring the impact of improvements in sanitation, so long as it is possible to control for other confounding factors, especially treatment. Spatial differences in prevalences following extended periods of mass treatment will indicate particular environmental risk factors linked to sanitation conditions and specific behaviours.

4b. Access to improved drinking water and sanitation: Standards and definitions

The question of what exactly constitutes access to safe drinking water and basic sanitation has been a topic of debate in recent years. Since the publication of the 2000 coverage estimates for access to improved facilities, produced by WHO, UNICEF and their Joint Monitoring Programme (WHO/UNICEF, 2000), in which definitions based on an expert consensus are presented, several publications have attempted to provide alternative definitions (see UN Millennium Project, 2004b).

JMP, responsible for monitoring progress towards the MDG targets, has used proxy indicators to estimate the number of people with and without access to safe

Table 6.7: Classification of improved and unimproved drinking water sources

Improved sources of drinking water	Unimproved sources of drinking water
Piped water (into dwelling, yard or plot)	Unprotected dug well
Public tap/standpipe	Unprotected spring
Tubwell/borehole	Vendor-provided water
Protected dug well	Tanker truck water
Protected spring	Surface water (river, stream, dam, lake, pond, canal, irrigation channel)
Rainwater collection	
Bottled water*	

*Bottled water is considered an 'improved' source of drinking water only where there is a secondary source that is 'improved'.

Source: WHO/UNICEF, 2005; www.wssinfo.org

14. See www.who.int/evidence/bod for more information.

Section 3: CHALLENGES FOR WELL-BEING & DEVELOPMENT

Table 6.8: Classification of improved and unimproved sanitation facilities

Improved sanitation facilities	Unimproved sanitation facilities
Flush/pour flush to: piped sewer system	Public or shared latrine
septic tank	
pit (latrine)	
Ventilated Improved pit latrine	Pit latrine without slab or open pit
Pit latrine with slab	Hanging toilet/hanging latrine
Composting toilet	Bucket latrine
	No facilities (so people use any area, for example, a field)

Source: WHO/UNICEF, 2005; www.wssinfo.org

drinking water and basic sanitation. These are the official indicators for monitoring the MDG targets; these proxy indicators for access are defined as the type of facility that people use to obtain their drinking water and meet their sanitation needs. JMP categorized these facilities as 'improved' or 'unimproved' (**Tables 6.7** and **6.8**). People relying on an improved source as their main source of drinking water are counted as having access to safe drinking water, while those using an improved sanitation facility are counted as having access to sanitation.

Specifically, the JMP definition for monitoring the proportion of the population with access to an improved drinking water source is as follows:

> An improved drinking water source is more likely to provide safe drinking water than a not-improved drinking water source, by nature of its construction, which protects the water source from external contamination particularly with faecal matter.

The JMP definition for monitoring the proportion of the population with access to basic sanitation is expressed in terms of the proportion of a population that uses an improved sanitation facility, defined as a facility that hygienically separates excreta from human contact.

Rather than providing an all-inclusive definition of what constitutes access to safe drinking water and basic sanitation, the categorization into improved and unimproved facilities was a necessary step to make the available data comparable between countries and within countries over time. This allows JMP to monitor progress, using the best available nationally representative population-based data obtained from household surveys (see **Box 6.7**). **Chapter 3** gives more details on the accuracy of local surveys compared to national censuses in urban areas.

JMP estimates do not always reflect whether or not an improved source provides drinking water of an acceptable quality; nor do they take into account accessibility of the drinking water source (in terms of the distance or time) or the affordability of drinking water. Issues of intermittence, reliability or seasonality are not reflected either. For access to basic sanitation, JMP monitors the number of people using different types of sanitation facilities, not taking into account whether or not they provide privacy and dignity or whether or not they are used by all household members at all times. Nor does the outcome of the monitoring process reflect the actual level of hygiene or cleanliness of the facility. This type of information is generally not collected at national level.

Nonetheless, using the categorization 'improved/ unimproved', JMP has a yardstick for measuring progress and change over time. It allows a reasonably accurate estimate of the number of people without access to any type of improved facility: the have-nots on which the MDGs focus.

However, access to safe drinking water and sanitation needs to be better defined. Howard and Bartram (2003) propose four access categories (see **Table 6.9**), based on the relationship between accessibility expressed in time or distance and the likely quantities of water collected or used. The four categories are: no access, basic access, intermediate access and optimal access. Global access, as monitored by JMP, corresponds to the level of basic access.

The definitions applied by WHO and UNICEF constitute a pragmatic approach to a complex global monitoring need and ensure consistency, replicability and a focus on those without access.

Over the years, a number of comprehensive definitions of access have been formulated. Such definitions and the accompanying standards may serve in the planning or design of new drinking water and sanitation services. The related indicators are specific, objective and measurable on an individual, setting-specific basis. However, when

Over longer periods of time, the prevalence of intestinal helminth infections can be an important indicator for monitoring the impact of improvements in sanitation...

Women and children collecting water for domestic use from a public water fountain, India

BOX 6.7: USER-BASED VERSUS PROVIDER-BASED DATA

Since 2000, JMP coverage estimates have been based on user data derived from nationally representative household surveys and national censuses. This marks an important shift away from the approach of using data originating from governments in the 1990s, which became possible after the introduction of the five-yearly Multiple Cluster Indicator Surveys (MICS) by UNICEF in sixty-four countries in 1995. Together with results of the Demographic and Health Surveys (DHS), or data from national censuses and other sources, including WHO's Water, Sanitation and Health Programme, this provides a large enough knowledge base to calculate coverage estimates supported by evidence-based datasets.

Why are data derived from household surveys better than those provided by governments or water utilities? The latter suffer from variations in the interpretation of what constitutes access. This complicates comparability between countries and even within a country over time. Often only those facilities that are constructed under government programmes or by water utility companies are counted. Facilities constructed by households, NGOs or the private sector may be partially or totally excluded. Water providers are inclined to report progressively on the number of facilities constructed and do not take into account facilities that are not used or that have fallen into disrepair. Household surveys, on the contrary, record, at a given point in time, the facilities people actually use – broken facilities are not counted.

Informal settlements and slums, even those that are home to hundreds of thousands of people, frequently do not appear in official government statistics because of questions of tenure or land ownership. In general, access to such areas tends to be poor and when not counted, a significant over-reporting of coverage will result. Household surveys usually do survey peri-urban areas when they fall into one of the selected sampling clusters, thus providing a better picture of the actual situation (see also **Chapter 3**). Household surveys including national censuses together provide the most reliable, nationally representative, comparable data, and they are available for almost every country in the world.

Source: www.wssinfo.org

used as a benchmark to assess globally whether or not existing services meet the required standards, the feasibility of measuring such indicators declines sharply and becomes a bottleneck for the frequent monitoring of progress and trends.

An example is provided by the lack of agreement on what exactly constitutes domestic and personal hygiene. The debate is centred around activities like bathing and clothes washing. Such activities usually require amounts of water equal to or larger than the amount used for all other basic personal and domestic water needs combined. In rural areas, bathing and clothes washing often takes place at the source or water point or in rivers or streams. In urban slum areas or during emergency situations, this might not be possible or desirable due to limited water availability, privacy concerns or public health concerns of contaminating the water source. Howard and Bartram (2003) argue that an improved source should provide adequate quantities for bathing and clothes washing as well, but recognize that the quantity per person required corresponds to the level of intermediate access and not to the level of basic access. It should be recalled that basic access is the current global standard for access. Drinking water for domestic and personal hygiene therefore does not necessarily include the use for extensive bathing and clothes washing.

4c. Water quality

The three principal international guidelines on water quality of relevance to human health are as follows:[15]

■ *Guidelines for Drinking-Water Quality*

■ *Guidelines for the Safe Use of Wastewater, Excreta and Greywater*[16]

■ *Guidelines for Safe Recreational Water Environments*.

These guidelines are addressed primarily to water and health regulators, policy-makers and their advisors, to assist in the development of national standards. For a long time, in the absence of good epidemiological studies, guidelines and standards for water-related hazards were based on the technical feasibility of providing treatment and took a 'no or very low' risk approach. However, setting targets that are too high can be counter-productive: they may be ignored if they are not attainable. National standards should therefore reflect national conditions, priorities and capacities to improve water supplies. All the recently developed guidelines are based on risk-assessment methods. This considers the risk for disease, not just the presence or absence of pathogens or chemicals in the water.

15. All of these guidelines are available online at www.who.int/water_sanitation_health/norms/

16. In four volumes: (1) Policy and regulatory aspects; (2) Wastewater use in agriculture; (3) Wastewater and excreta use in aquaculture; (4) Excreta and greywater use in agriculture. All of these are available at www.who.int/water_sanitation_health/norms/

Table 6.9: Requirements for water service levels and health implications

Service level	Access measure (distance or time)	Needs met	Level of health concern
No access – quantity collected often below 5 litres (L) per capita per day	More than 1,000 metres (m) or 30 minutes total collection time	Consumption cannot be assured Hygiene not possible (unless practised at the source)	Very high
Basic access – average quantity unlikely to exceed 20 L per capita per day	Between 100 and 1,000 m or 5 to 30 minutes total collection time	Consumption should be assured Handwashing and basic food hygiene possible; laundry and bathing difficult to assure unless carried out at source	High
Intermediate access – average quantity about 50 L per capita per day	Water delivered through one tap on plot or within 100 m or 5 minutes total collection time	Consumption assured All basic personal and food hygiene assured; laundry and bathing should also be assured	Low
Optimal access – average quantity 100 L per capita per day	Water supplied through multiple taps continuously	Consumption: all needs met Hygiene: all needs should be met	Very low

Source: Howard and Bartram, 2003.

...there is increasing recognition that a few key chemicals, notably fluoride and arsenic, cause large-scale health effects

Drinking water quality

An important recent event was the publication of the third edition of the *Guidelines for Drinking Water Quality* (WHO, 2004b). These guidelines are widely accepted in industrialized and developing countries. Recent developments in microbial risk assessment and its linkages to risk management are taken into account. Increased attention is paid to effective preventative management through a 'framework for drinking water safety', including 'water safety plans' (see **Box 6.8**). The guidelines pay attention to the adequacy of supply, which is not only determined by water quality but also by water quantity, accessibility, affordability and continuity. The importance of water quality at the point of use (within the house) is emphasized, while previously, quality guidelines tended to refer only to the source of the drinking water.

There is agreement that the best available indicator of faecal pollution of individual drinking water sources is *Escherichia coli* (or thermo-tolerant coliform bacteria). The presence of *E. coli* provides conclusive evidence of recent faecal pollution, but its absence does not automatically prove that the water is safe. There is certainly a need for additional indicators, especially for protozoa such as *Cryptosporidium parvum*. To date, no water quality standards regarding *Cryptosporidium* oocysts have been established, and the minimum concentration of oocysts in drinking water leading to

clinical illness in healthy individuals has not been conclusively defined.

Drinking water quality guidelines have always included permissible levels of chemical substances. Chemicals in drinking water can be naturally occurring or originate from pollution by agricultural activities (fertilizer, pesticides), human settlements and industrial activities. While the revised WHO guidelines state that microbial hazards continue to be a priority concern in both developed and developing countries, there is increasing recognition that a few key chemicals, notably fluoride and arsenic, cause large-scale health effects. For a risk analysis, information from the catchment on naturally occurring chemicals is essential. If chemicals such as fluoride or arsenic are present in unusually high concentrations in rocks, soil or groundwater, there is an elevated risk for public health. In many countries, the development of appropriate risk management strategies is hampered by a lack of information on the presence and concentrations of chemicals in drinking water and the lack of information on disease cases. In the case of chemical hazards with high measurable disease burden, the target would be to reduce the occurrence of disease cases. If the disease burden is low, it cannot be directly measured by public health surveillance systems, and quantitative risk assessment methods can be applied (see also **Chapter 10**).

Section 3: CHALLENGES FOR WELL-BEING & DEVELOPMENT

BOX 6.8: WATER SAFETY PLANS (WSPs)

To ensure that drinking water is safe, a comprehensive strategy that looks at risks and risk management at all stages in water supply (WHO, 2004b), from catchment to consumer, is needed. In the WHO *Guidelines for Drinking-Water Quality*, such approaches are called water safety plans (WSPs). WSPs have built-in quality control at each step of the process, from source to tap, and represent a paradigm shift in drinking water management, which previously tended to focus on the detection of contamination that had already taken place. WSP approaches exist for large (piped) supplies and smaller community or household supplies in developed and developing countries. The objectives of WSPs are the minimization of contamination of source waters, the reduction or removal of contamination through treatment processes and the prevention of contamination during storage, distribution and handling of drinking water. This is achieved by an assessment of the entire drinking-water supply chain, effective operational monitoring, and management plans.

Source: WHO, 2005; Davison et al., 2005.

Wastewater use in agriculture and aquaculture

With the increasing scarcity of freshwater resources available to agriculture, the use of urban wastewater in agriculture will increase, especially in arid and semi-arid regions. Wastewater is often the only reliable source of water for farmers in peri-urban areas, and it is widely used in urban and peri-urban areas, in both treated and untreated forms. A nationwide survey in Pakistan showed that an estimated 25 percent of all vegetables grown in the country are irrigated with untreated urban wastewater and that these vegetables, cultivated close to the urban markets, were considerably cheaper than the vegetables imported from different regions of Pakistan (Ensink et al., 2004). Likewise, 60 percent of the vegetables consumed in Dakar, Senegal are grown with a mixture of groundwater and untreated wastewater within the city limits (Faruqui et al., 2004). In this context, the use of wastewater for peri-urban agriculture provides an opportunity and a resource for livelihood generation.

The major challenge is to optimize the benefits of wastewater as a resource (both the water and the nutrients it contains) and to minimize the negative impacts on human health. There is sufficient epidemiological evidence that infection with intestinal helminths poses the major human health risk associated with the agricultural use of untreated urban wastewater. In those countries where sewage and excreta are used to feed fish, there are important risks for infection with flukes. Foodborne trematode (fluke) infections are a serious and growing public health problem, with an estimated 40 million people affected worldwide. Transmission to humans occurs mostly via consumption of raw freshwater fish and aquatic plants. A recent analysis indicates that residents in endemic areas living close to freshwater bodies more than double their risk of infection, and it is speculated that the exponential growth of aquaculture is the major contributing factor to this emerging disease trend (Keiser and Utzinger, 2005).

Mitigating health risks while maximizing benefits requires holistic approaches that involve all stakeholders in a process to enhance knowledge sharing, promote realistic measures for hygiene and sanitation improvement, generate income, produce food for better livelihoods and sustain the strengthening of water and sanitation services at household and community levels.

For the protection of public health in this context, WHO has developed updated *Guidelines for the Safe Use of Wastewater, Excreta and Greywater* (WHO, 2006a–d). They define an acceptable and realistic level of public health protection, which can be achieved through a combination of setting microbial water quality targets and implementing health protection measures, such as crop restriction, application techniques and irrigation timing. This approach is flexible and is applicable to both industrialized and less-developed countries. Countries can choose to meet the health target level by wastewater treatment alone, or through a combination of partial wastewater treatment and additional health protection measures.

In adopting wastewater use guidelines for national standards, policy-makers should consider what is feasible and appropriate in the context of their national situation. They should use a risk-benefit approach that carefully weighs the benefits to household food security, nutrition and local economic development against possible negative health impacts. The revised guidelines call for a progressive implementation of measures and incremental improvements in the public health situation.

4d. Child mortality

Children under the age of 5 are the most affected by poor water supply and sanitation. Diarrhoea is one of the directly preventable causes of under-5 mortality. Child mortality is the result of a complex web of determinants at many levels. The fundamental determinant is poverty, and an underlying determinant is under-nutrition. The under-5 mortality rate has become a key indicator of health and social development. It can be seen as a cross-cutting indicator for several of the challenge areas and for achieving the MDG targets.

There is sufficient evidence that improvements in water supply, sanitation and hygiene result in fewer cases of diarrhoea and lower overall child mortality. To obtain the maximum possible reduction in child mortality, these improvements would have to be combined, however, with other preventive interventions (breastfeeding, vitamin A supplementation) and treatment interventions (oral rehydration therapy and zinc) (Jones et al., 2003). This combination of interventions could save more than 1.8 million children under the age of 5 each year, which is 88 percent of the annual under-5 mortality due to diarrhoea.[17]

The infant mortality rate is a less suitable indicator than the under-five mortality rate in the context of water-related diseases, since only a small proportion of deaths in the neonatal period (first twenty-eight days of life) can be attributed to water-related diseases (Black et al., 2003). In the first six months of life, children are, to some extent, protected against diarrhoea (if they are being breastfed) and malaria. It is only towards the end of the first year of life that infectious diseases due to poor water, sanitation and hygiene take their huge toll on children's health.

4e. Nutritional status

Nutritional status is probably the single most informative indicator of the overall health of a population (see also **Chapter 7**). For evaluating the impact of water supply and sanitation interventions, nutritional status is as important and appropriate a measure as the incidence of diarrhoeal disease. Anthropometric measurements are well defined, and are easily and inexpensively performed. Data on childhood under-nutrition are available from the WHO Global Database on Child Growth and Malnutrition,[18] which is based on nationally representative anthropometric surveys. It is a good example of international collaboration in standardizing indicators and data collection systems (de Onis and Blössner, 2003).

One of the indicators for monitoring progress towards the MDG targets is the prevalence of underweight children under 5 years of age. Underweight (low weight-for-age) reflects the effects of acute as well as chronic under-nutrition. Weight-for-age is a composite indicator of height-for-age and weight-for-height, which makes its interpretation difficult. Stunting (low height-for-age) reflects chronic under-nutrition and is an indicator of the cumulative effects of standard of living, women's educational level, access to food, access to water supply and sanitation, and burden of infectious diseases. Stunting is a good indicator to monitor the long-term impact of improvements in water supply, sanitation and hygiene, provided it is possible to correct for confounding variables.

17. Child mortality data are available online from UNICEF at www.childinfo.org/cmr/revis/db2.htm

18. Available online at www.who.int/nutgrowthdb/

Part 5. Comparative Risk Assessment

Most water-related diseases have multiple risk factors. This raises a number of questions: What part of the burden of disease is attributable to inadequate water supply and sanitation? What would be the health gains of improvements in water supply and sanitation? Similar questions may be posed for water management in agriculture: What burden of disease can be attributed to poor water management, and what are the health benefits of improved water management?

To answer these questions, epidemiological measurements are needed that quantify the public health relevance of important risk factors. The population-attributable risk provides a measure of the amount of disease in the whole population, which is attributable to a certain level of exposure (risk to health), assuming that the association between exposure and the disease is one of cause-and-effect. The known attributable risks for a disease often add up to more than 100 percent, because some risk factors act through other more proximal factors, such as under-nutrition. The potential impact fraction expresses the proportion of disease that could be eliminated by reducing exposure. Risk

...washing hands with soap can reduce the risk of diarrhoeal diseases by 42 to 47 percent, and the promotion of handwashing might save a million lives per year

assessment methods using these measures were the subject of the *World Health Report 2002* (WHO, 2002b), which stated that approximately 3.1 percent of annual deaths (1.7 million) and 3.7 percent of DALYs (54.2 million) worldwide are attributable to unsafe water, sanitation and hygiene. The decrease in the burden of disease due to poor water, sanitation and hygiene, from 6.8 percent in 1990 to 3.7 percent in 2000, is partly due to a decline in mortality associated with global diarrhoeal disease.

The Comparative Risk Assessment (CRA) module of the *Global Burden of Disease* study aims to assess risk factors in a unified framework. It provides a vision of potential gains in population health by reducing exposure to a risk factor or a group of risk factors. This has provided sufficient evidence that in the poorest regions of the world, unsafe water, sanitation and hygiene are major contributors to loss of healthy life, expressed in DALYs (Ezzati et al., 2002). Globally, 88 percent of DALYs lost due to diarrhoea can be attributed to unsafe water, poor sanitation and lack of hygiene, while 92 to 94 percent of DALYs lost due to diarrhoea can be attributable to the joint effects of unsafe water, sanitation and hygiene; underweight; vitamin A deficiency; and zinc deficiency (Ezzati et al., 2003).

Further analysis by the CRA Collaborating Group in multiple age and exposure categories, or along a continuum of exposures, showed that globally, a considerable proportion of the disease burden attributable to major risk factors occurred among those with only moderately raised risk levels, not the extremes (Rodgers et al., 2004). This is consistent with the fundamental axiom in disease prevention across risk factors: 'A large number of people exposed to a small risk may generate many more cases than a small number exposed to high risk' (Rose, 1992). It follows that population-based strategies that seek to shift the whole

distribution of risk factors have the potential to substantially reduce total disease burden, possibly over long time periods if the interventions alter the underlying risk behaviours or their socio-economic causes (Rodgers et al., 2004).

The relative importance of the availability of drinking water, quality of drinking water, sanitation and hygiene behaviour for the occurrence of diarrhoeal diseases continues to be a subject of discussion. Many of the viral, bacterial and protozoan pathogens that cause diarrhoea can be transmitted through the ingestion of contaminated water. Accordingly, water supply utilities and programmes aim to remove these pathogens before the drinking water is provided to consumers. The importance of drinking water quality for the transmission of diarrhoeal diseases was challenged when several reviews in the 1980s and 1990s showed that increasing the quantity of water available for personal and domestic hygiene and ensuring the safe disposal of excreta led to greater reductions in diarrhoeal morbidity than improving drinking water quality (Esrey et al., 1991). Safe excreta disposal and handwashing after defecation would stop the transmission by preventing faecal pathogens from getting into the environment. If these primary barriers were in place, then secondary barriers such as removing faecal pathogens from drinking water would be less important. As the debate continues and setting-specific case studies tip the balance in one direction or the other, one thing is clear improvements in access to safe water will only provide real health benefits if sanitation facilities are improved at the same time. In this respect, it is alarming that global sanitation coverage has only increased from 49 percent in 1990 to 58 percent in 2002, lagging behind the successful increase in global coverage for access to safe drinking water, which is now 83 percent and on track to meet the 2015 MDG target (WHO/UNICEF, 2004).

BOX 6.9: BENEFITS OF IMPROVED SANITATION

The best way to prevent faecal-orally transmitted diseases such as diarrhoea is the sanitary disposal of human faeces in pit latrines or other improved sanitation facilities. Improved sanitation is also the only long-term sustainable option for controlling intestinal worms and schistosomiasis.

Improved sanitation has important additional benefits, especially to women. In many cultures, the only time when women or girls can defecate, if they have no latrine, is after dark. The walk to the defecation field, often in the dark, is when women run the greatest risk of sexual harassment and assault. The lack of adequate, separate sanitary facilities in schools is one of

the main factors preventing girls from attending school, particularly when menstruating. In Bangladesh, a gender-sensitive school sanitation programme increased girls' enrolment by 11 percent.

Source: www.lboro.ac.uk/well/resources/ fact-sheets/fact-sheets-htm/wps.htm

Based on current evidence, washing hands with soap can reduce the risk of diarrhoeal diseases by 42 to 47 percent, and the promotion of handwashing might save a million lives per year (Curtis and Cairncross, 2003). Handwashing promotion should become an intervention of choice. Hygiene depends on the quantity of water that people have available, and one has to realize that in many areas, handwashing after defecation or before preparing food seems like a luxury when the water has to be fetched from a water point far away.

Recently, there has been increased attention to the fact that drinking water, even if it is of good quality, can become contaminated between the point of collection and the home, and in the domestic environment, when children dip their faecally polluted hands in a household water container, for example. The water quality of drinking water sources might not be associated with the occurrence of diarrhoea (Jensen et al., 2004), because it does not reflect the water quality at the point of use. There is increasing evidence that simple, low-cost interventions at the household and community level are capable of improving the microbial quality of domestically stored water and of reducing the associated risks of diarrhoeal disease and death (Clasen and Cairncross, 2004; Sobsey, 2002). This has led to the creation of the WHO-coordinated International

Network to Promote Household Water Treatment and Safe Storage, providing a framework for global collaboration of UN and bilateral agencies, governments, NGOs, research institutions and the private sector committed to improve household water management as a component in water, sanitation and hygiene programmes.

Epidemiological studies have established a causal relationship between malnutrition and diarrhoea. Malnourished children experience higher risks of mortality associated with diarrhoea (Rice et al., 2000). This is especially true for persistent diarrhoea and dysentery that now account for the majority of deaths related to diarrhoea in the developing world. About 61 percent of deaths due to diarrhoea in young children are attributable to underweight (low weight-for-age) (Caulfield et al., 2004). About 15 percent of the global disease burden can be attributed to the joint effects of childhood and maternal underweight or micronutrient deficiencies. In terms of DALYs, in 1990, under-nutrition was the single leading global cause of health loss, estimated at 140 million DALYs (9.5 percent of total) attributable to underweight (Ezzati et al., 2002). Although the prevalence of underweight has decreased in most regions of the world, it has increased in sub-Saharan Africa.

Washing hands with soap can reduce the risk of diarrhoeal diseases by 42 to 47 percent

Part 6. Governance

Challenges with respect to water availability, water quality and sanitation are intertwined with challenges on food security, urbanization and environmental degradation. They stand in the way of poverty reduction and sustainable development. Providing for effective and sustainable water supply and sanitation services requires adequate governance structures and includes a commitment to good governance. In certain countries, broad policy and institutional constraints are greater obstacles than resource and technological constraints. Often, policy and institutional reforms are needed, and these would have to: (1) balance the competition for water between different uses and users; (2) implement a genuinely integrated approach to water resources management; and (3) establish effective governance institutions and institutional arrangements conducive to such an integrated management approach.

6a. Economic evaluation of interventions

Global disease control priorities should be based on the global burden of disease and the availability of cost-effective interventions. In the long term, many environmental health interventions have proved to be cost-effective compared to medical interventions. Water connections in rural areas have been estimated to cost US $35 per DALY saved, hygiene behaviour change US $20

per DALY saved, and malaria control US $35–75 per DALY saved (Listorti and Doumani, 2001).

Hutton (2002) was commissioned by WHO to test a number of intervention scenarios and concluded that cost-benefit ratio is high when all benefits are included, standing at an economic benefit of around US $3 to 6 per US $1 invested for most regions and for most interventions. Time saving

...estimated benefits of US $3 to 34 per US $1 invested if the water and sanitation MDG targets were achieved...

was found to be an important component in the overall benefits from water and sanitation improvements. When valued in monetary terms, using the minimum wage as a proxy for annual time savings, such savings outweigh the annual costs of the interventions.

The scenario scoring highest in actually reducing the burden of water-related disease to nearly zero is that where universal access to piped water and sewerage connections is provided, with an estimated cost of US $850 to 7,800 per DALY averted (Evans et al., 2004; Rijsberman, 2004). This is above income levels in developing countries. In the scenario that provides for low-cost technologies (standpipes and latrines, as opposed to piped water and sewerage connections to individual homes), the cost would improve to US $280 to 2,600 per DALY averted, if disinfection at the point of use is added.

A further analysis considering disease burden averted, costs to the health services and to individual households reduced, and opportunity costs (attending ill family members, fetching water) avoided arrived at estimated benefits of US $3 to 34 per US $1 invested if the water and sanitation MDG targets were achieved, with, on the whole, benefits from sanitation investments being greater than those from water interventions. In aggregate, the total annual economic benefits of meeting the MDG targets on water supply and sanitation accrue to US $84 billion (Hutton and Haller, 2004). While global estimates for the additional annual investment to meet the MDG water and sanitation targets all arrive at about US $11 billion, meeting the targets translates into 322 million

working days per year gained at a value of US $750 million (SIWI/WHO, 2005).

Based on a study in Burkina Faso, the cost of implementing a large-scale hygiene promotion programme was estimated at US $26.9 per case of diarrhoea averted (Borghi et al., 2002). Cost-effectiveness of a latrine revision programme in Kabul, Afghanistan ranged from US $1,800 to 4,100 per death due to diarrhoea averted, depending on age and payer perspective (Meddings et al., 2004). Fattal et al. (2004) estimated the cost of treating raw sewage used for direct irrigation to meet the WHO 1989 norms for safe irrigation of vegetables eaten raw with untreated wastewater at about US $125 per case of disease prevented.

6b. Water scarcity: Bridging the gaps between the different sectors

While the looming freshwater crisis is getting a lot of attention from water resources policy- and decision-makers, the provision of domestic water to rural populations is often not perceived as a problem in this context. Policy documents on integrated water resources management (IWRM), whether from governments or donor organizations, give first priority to water supply for agricultural production purposes in water allocation decisions; the domestic uses are only a small fraction of the total amount of freshwater utilized in a country. For example, the agriculture sector of the South Asian region receives about 96 percent of the total diversions. Even in sub-Saharan Africa, with a much less developed irrigation infrastructure than Asia, 84 percent of total water diversions is used in agriculture.

BOX 6.10: DOMESTIC USE OF IRRIGATION WATER

Millions of people around the world rely on surface irrigation water for most of their domestic needs. This is especially true of irrigation communities living in areas with low rainfall, under-developed drinking water supply systems, and in regions with low groundwater tables or unusable groundwater, due to high concentrations of salt or hazardous chemicals. In such circumstances, the way in which irrigation water is managed has a tremendous bearing on the health of the populations living in these areas. Unfortunately, irrigation water management is based entirely on crop

requirements and not on domestic water needs. Therefore, when decisions for water allocation are made, domestic uses are rarely taken into account. Also, with the looming freshwater crises, there is increasing pressure on the irrigation sector to make water use in agriculture more efficient. In this process, the non-agricultural uses of irrigation water need to be considered.

Studies in Punjab, Pakistan have documented the links between availability of irrigation water for domestic use and its impact on diarrhoea and the

nutritional status of children. It was concluded that irrigation water management has a clear impact on human health and that bridging the gap between the irrigation and domestic water supply sectors could provide great health benefits by taking into account the domestic water availability when managing irrigation water. In the same study, it was found that using irrigation seepage water as a safe source for domestic supplies was a possible option.

Source: Van der Hoek et al., 2001b; 2002a, b.

The difference between credible high and low estimates of the water globally required for agriculture in 2025 is in the order of 600 cubic kilometres (km^3) – more than is estimated to be required for all domestic uses. This has created a widely prevalent notion that a small diversion from the irrigation sector could fulfil the demands of a growing population for domestic water supply. In reality, this reallocation of water between sectors can be very difficult, and truly integrated water management is constrained by the traditional sectoral thinking and priorities set by professionals in the various disciplines and the existing power structure. The main concern of public health officials and researchers is the increasing deterioration of water quality due to industrial and urban waste, agricultural runoff and insufficient investments in the domestic water supply infrastructure.

This global concern for water quality is, to a large extent, a reflection of the very high quality standards traditionally imposed on drinking water by institutions and professionals in industrialized countries. On the other hand, the managers of water for agricultural production see their responsibilities largely confined to the provision of water in time and space in accordance with the cropping cycle requirements. Few irrigation managers would see it as part of their mandate to supply water for domestic use. To water planners, domestic uses in rural areas concern only a small fraction of the total amount of freshwater utilized and are therefore easily overlooked. This may lead to the situation that high investments have been made to mobilize freshwater into an area, without considering other uses than irrigation.

6c. Multiple uses of water

In many areas, the most readily available surface water is from irrigation canals and reservoirs. It has not been sufficiently recognized that apart from irrigating crops, irrigation water is used for many other purposes, including drinking, cooking, livestock rearing, aquaculture and wildlife. Washing clothes and bathing are probably the most frequently observed domestic uses of irrigation systems throughout the world. When there is a poor supply of domestic water from underground sources, but abundant supply for agricultural production, irrigation water from canals and reservoirs can be the only source of water for domestic use. In a few cases, such uses have been considered in the design of irrigation systems, but as a general rule, designers and engineers have tended to focus exclusively on water use in crop production. On the other hand, providers of domestic water rarely

consider the usage of irrigation water as an option, because the conventional strategy has been to utilize groundwater, not surface water for domestic purposes.

As a result, the non-agricultural household uses of irrigation water have neither been systematically documented, nor have the possibilities they offer been seriously explored. A large gap therefore remains between what happens in irrigation schemes (what people do) and what is taken into account in water resources planning and policies. With increasing focus on improved water use efficiency within irrigation systems, there is a risk that recognized uses of water (crop irrigation) will be prioritized to the detriment of other valuable but non-recognized uses, such as domestic needs (see **Box 6.6** for an example in Sri Lanka). There is a critical need, therefore, to understand the health dimensions of the multiple uses of irrigation water, the determinants of its use, the realistic alternatives, and the consequences of these uses in order to promote informed water policy formulation (see **Box 6.10**).

6d. Falling groundwater levels

The over exploitation of groundwater for agricultural and industrial purposes renders the availability of shallow groundwater for drinking and domestic purposes increasingly problematic. In some of the major breadbaskets of Asia, such as the Punjab in India and the North China Plain, water tables are falling 2 to 3 metres a year. The wealthier farmers can continue to drill deeper tubewells with larger, more expensive pumps, but poor farmers are unable to do so. The problem of falling groundwater levels is now seen by all stakeholders as a threat to food security. What has received less attention is that it also causes the shallow drinking water wells of poor communities to run dry. Deepening these wells is very costly and beyond the resources of the poor. In coastal areas such as the State of Gujarat, India, over-pumping causes salt water to invade freshwater aquifers, making them unsuitable for drinking. Over-pumping has also been linked to the contamination of drinking water with arsenic. Clearly, pumping groundwater has become a key policy issue that can only be dealt with in the context of IWRM.

6e. Poverty Reduction Strategy Papers

One of the main instruments for national governments in their attempts to reduce poverty are the Poverty Reduction Strategy Papers (PRSPs), which give clarity and direction to all the development work in a country. These

The main concern of public health officials and researchers is the increasing deterioration of water quality due to industrial and urban waste...

BOX 6.11: SUCCESSFUL WATER SUPPLY IN PHNOM PENH, CAMBODIA

Cambodia is one of the poorest countries in South-East Asia. It is still recovering from decades of conflict, and all sectors, including the health sector, require reconstruction. The life of most people in Cambodia is still defined by poverty and a very high burden of disease with a government health care system that is ill-equipped to deal with a range of health problems. Access to improved drinking water sources (estimated at 34 percent in 2002) is extremely low, even by developing country standards (WHO/UNICEF, 2004). In the capital, Phnom Penh, the water supply and drainage systems have deteriorated over the years due to war, poor management and lack of maintenance. This problem has been worsened by the rapid growth of the urban population. However, successful rehabilitation projects have taken place with foreign aid and technical assistance. Since 1993, the Phnom Penh Water Supply Authority (PPWSA) has increased its distribution network from serving 40 percent of the Phnom Penh population to over 80 percent. Non-revenue water – the result of leaks, mismeasurement, illegal connections and illegal sales – has been reduced to 22 percent (from 72 percent) and collections are at almost 99 percent with full cost recovery achieved. By mid-2004, it is predicted that the water supply capacity in the city will have increased to 235,000 cubic metres per day. This is now considered a success story for which the PPWSA was awarded the Water Prize of the Asian Development Bank.

Charging for water and the relative roles of public versus private management are controversial issues (see **Chapters 2 and 12**). Phnom Penh provides a rare example of an efficient water-delivery system in a large city run by a public body.

Source: www.adb.org/Documents/News/2004/nr2004012.asp.

---

are country-owned development strategies demanded by the World Bank and the International Monetary Fund of countries that want to be eligible for loans. Reducing an excessively-high disease burden will have a positive economic impact, and strategies on communicable disease control and child health can be seen as evidence of a pro-poor approach. A review of twenty-one PRSPs found that all of them included strategies on communicable disease control, child health and water and sanitation improvement (WHO, 2004c). However, the emphasis was overwhelmingly on government delivery of health services to reach health goals without examination of the role of non-government providers and other sectors. Furthermore, quantifiable targets were mostly not mentioned, making it difficult to link PRSP indicators with the MDGs. One of the overarching criticisms of the PRSPs from NGO sources has been that participation – the widely proclaimed centrepiece of national ownership of the PRSPs – is poorly implemented (UN Millennium Project, 2004a).

Part 7. Water for Life: Making it Happen

With respect to human health, this second edition of the World Water Development Report consolidates our new and updated insights into the diverse nature and broad scope of conditions where the development, management and use of water resources are associated with community health status. The concept of burden of disease, expressed in Disability-Adjusted Life Years lost, has strengthened its position as a universal indicator of that status with valid applications in economic evaluation as well as development planning. New tools have also become available to better estimate the costs and benefits of different options, particularly for improving access to drinking water and sanitation.

Water taps provided by relief organizations at the Virginia Newport high school in Monrovia, Liberia, where some of the 25,000 internally-displaced people had taken refuge

The basic driving forces of the water–health nexus have not changed in nature and include population expansion, rapid urbanization, globalization and increasing scarcity of good-quality freshwater resources. At the global policy level, the MDGs are exerting an increasingly marked pressure on both thinking about and acting on water-health issues; newly emerging economic realities (particularly the rapid developments in China and India) further modulate these pressures.

Positive and negative trends can be distinguished. The positive trends include:

- Global progress towards achieving the MDG target on drinking water.

- A significant reduction in mortality due to childhood diarrhoea.

- The availability of good indicators for monitoring progress towards achieving health-related MDG targets at the global and regional level.

- A significant evolution in approaches to managing the quality of drinking water, recreational waters and wastewater, from a technical no-risk concept to a comprehensive system of risk assessment and management.

- Greater recognition of health impact assessment as the critical starting point for a functional incorporation of human health considerations, especially into integrated water resources development and management.

Together, these trends will influence and improve the governance of water and health issues in the coming years. Authorities now can apply adaptive management and ensure optimal solutions in local settings. Decision-making will have a stronger evidence base, even though

the indicators used need further development and refinement. An example of this is provided by the new vision on the safe use of wastewater, excreta and greywater in agriculture and aquaculture that assesses and manages health risks and that balances health costs and benefits rather than applying rigid water quality standards. In many parts of the world rigid standard setting has proved to be neither feasible nor enforceable, whereas through water safety plans, through safe household water management and storage or through safe use of wastewater, governments can achieve solid and sustainable progress.

On the downside, the following constraints and bottlenecks can be observed:

- Lack of progress towards achieving the MDG sanitation target left 2.6 billion people without access to improved sanitation at the end of 2002.

- The significant increase in the absolute number of people without access to an improved drinking water source and improved sanitation, in both urban and rural areas, since 1990 as exclusively experienced in sub-Saharan Africa.

- The problematic health situation (with no signs of improvement) in sub-Saharan Africa, as reflected in practically all indicators, and in particular by the increasing malaria burden.

- Lack of progress in the implementation of the IWRM concept specifically, and in the realization of intersectoral action for health in general.

- An inadequate evidence base needed to advocate for increased investment in urban sewage treatment, resulting from a lack of indicators and mechanisms for monitoring the sewerage discharge and the added burden of disease for people downstream.

Despite the general acceptance of the Integrated Water Resources Management concept, the different water use sectors still by and large fail to coordinate their planning and to collaborate at the implementation phase, with a range of predictable, and therefore in many cases preventable, adverse consequences for human health. Like the more generic concept of intersectoral collaboration, IWRM is embraced by all, but funded by few. The innovative proposals of the World Commission on Dams for improved planning procedures and best practice in implementation, in the broader governance context (of generic value to all water resources development and transferable to water and health issues), have received insufficient follow-up and seem all but forgotten.

Inadequate funding also continues to bog down efforts to achieve the sanitation target. While there are several factors to which this can be attributed, the continued lagging behind of support for sanitation projects (as compared to drinking water projects) deserves special mention.

Growing challenges in the health sector range from drug resistance in important parasitic and bacterial pathogens to newly emerging diseases (with SARS and the H1N5 avian influenza virus as recent examples) underscore the need for water resources development, management and use to take human health into consideration in a far more comprehensive and integrated manner.

The following **recommendations** aim to strengthen the positive trends and help counter the constraints:

■ Re-focus a much more broadly supported programme of development aid and technical assistance on meeting the MDG drinking water and sanitation targets especially in rural areas that still lag far behind urban areas, but also in peri-urban and slum areas that are likely to absorb most of the urban population increase in the coming decade.

■ Increase investments in sanitation coverage and improvement worldwide, ensuring a progressively expanding portion for sewerage and proper maintenance.

■ Increase investments for meeting the MDG drinking water and sanitation targets in sub-Saharan Africa.

■ Refine the correlations between water indicators and the indicators for childhood illness/mortality and nutritional status, the importance for accelerated access to safe water and adequate sanitation, and better Integrated Water Resources Management (IWRM) practices.

■ Promote intervention studies that provide scientific information and help strengthen the evidence base on the effectiveness of environmental management methods for control of water-associated vector-borne diseases, and develop a toolkit for environmental managers in this area.

■ Make the multiple uses and multiple users of water the starting point of planning, developing and managing water resources at the river basin level, and promote the principle of subsidiarity in the governance of water resources.

■ Introduce the use of available tools for estimating costs and benefits of different drinking water and sanitation options initially at the national and subsequently at lower levels of governance.

Signs of the times: a camel drinking bottled water in Petra, Jordan

References and Websites

Arnesen, T. and Kapiriri, L. 2004. Can the value choices in DALYs influence global priority-setting? *Health Policy*, Vol. 70, pp. 137–49.

Black, R. E., Morris, S. S. and Bryce, J. 2003. Where and why are 10 million children dying each year? *Lancet*, Vol. 361, pp. 2226–34.

Blumenthal, U. J., Mara, D. D., Peasey, A., Ruiz-Palacios, G. and Stott, R. 2000. Guidelines for the microbiological quality of treated wastewater used in agriculture: recommendations for revising WHO guidelines. *Bulletin of the World Health Organization*, Vol. 78, pp. 1104–16.

Boelee, E. and van der Hoek, W. 2002. Impact of irrigation on drinking water availability in Sri Lanka / Impact de l'irrigation sur la disponibilité de l'eau potable au Sri Lanka. ICID-CIID 18th Congress on Irrigation and Drainage, 21–28 July, Montreal, Canada. Q. 51, R. 5.04. International Commission on Irrigation and Drainage.

Borghi, J., Guinness, L., Ouedraogo, J. and Curtis, V. 2002. Is hygiene promotion cost-effective? A case study in Burkina Faso. *Tropical Medicine and International Health*, Vol. 7, pp. 960–9.

Caulfield, L. E., de Onis, M., Blössner, M. and Black, R. E. 2004. Undernutrition as an underlying cause of child deaths associated with diarrhea, pneumonia, malaria, and measles. *American Journal of Clinical Nutrition*, Vol. 80, pp. 193–8.

Chan, M. S. 1997. The global burden of intestinal nematode infections – Fifty years on. *Parasitology Today*, Vol. 13, pp. 438–43.

Clasen, T. F. and Cairncross, S. 2004. Household water management: refining the dominant paradigm. *Tropical Medicine and International Health*, Vol. 9, pp. 187–91.

Crompton, D. W. T., Engels, D., Montresor, A, Neira M. P. and Savioli, L. 2003. Action starts now to control disease due to schistosomiasis and soil-transmitted helminthiasis. *Acta Tropica*, Vol. 86, pp. 121–4.

Crump, J. A., Luby, S. P. and Mintz, E. D. 2004. The global burden of typhoid fever. *Bulletin of the World Health Organization*, Vol. 82, pp. 346–53.

Curtis, V. and Cairncross, S. 2003. Effect of washing hands with soap on diarrhoea risk in the community: a systematic review. *Lancet Infectious Diseases*, Vol. 3, pp. 275–81.

Davison, A., Howard, G., Stevens, M., Callan, P., Fewtrell, L., Deere, D. and Bartram, J. 2005. *Water Safety Plans: Managing Drinking-water Quality from Catchment to Consumer*. Geneva, WHO.

De Onis, M. and Blössner, M. 2003. The World Health Organization Global Database on Child Growth and Malnutrition: methodology and applications. *International Journal of Epidemiology*, Vol. 32, pp. 518–26.

De Onis, M., Blössner, M., Borghi, E., Frongillo, E. A. and Morris, R. 2004. Estimates of global prevalence of childhood underweight in 1990 and 2015. *Journal of the American Medical Association*, Vol. 291, pp. 2600–6.

De Onis, M., Frongillo, E. A. and Blössner, M. 2000. Is malnutrition declining? An analysis of changes in levels of child malnutrition since 1980. *Bulletin of the World Health Organization*, Vol. 78, pp. 1222–33.

De Silva, N. R., Brooker, S., Hotez, P. J., Montresor, A, Engels, D. and Savioli, L. 2003. Soil-transmitted helminth infections: updating the global picture. *Trends in Parasitology*, Vol 19, pp 547–51.

Emerson, P. M., Lindsay, S. W., Alexander, N., Bah, M., Dibba, S. M., Faal, H. B., Lowe, K. O., McAdam, K. P., Ratcliffe, A. A., Walraven, G. E. and Bailey, R. L. 2004. Role of flies and provision of latrines in trachoma control: cluster-randomised controlled trial. *Lancet*, Vol. 363, pp. 1093–8.

Engels, D. and Chitsulo, L. 2003. Schistosomiasis. D. W. T. Crompton, A. Montresor, M. C. Nesheim and L. Savioli (eds) 2004. *Controlling Disease due to Helminth Infections*. Geneva, WHO.

Ensink, J. H. J., Mahmood, T., van der Hoek, W. and Raschid-Sally, L. 2004. A nationwide assessment of wastewater use in Pakistan: An obscure activity or a vitally important one? *Water Policy*, Vol. 6, pp. 197–206.

Erlanger, T. E., Keiser, J., Caldas de Castro, M., Bos, R., Singer, B. H., Tanner, M. and Utzinger, J. 2005. Effect of water resource development and management on lymphatic filariasis, and estimates of populations at risk. *American Journal of Tropical Medicine and Hygiene*, Vol. 73(3): 523–33.

Esrey, S. A., Potash, J. B., Roberts, L. and Shiff, C. 1991. Effects of improved water supply and sanitation on ascariasis, diarrhoea, dracunculiasis, hookworm infection, schistosomiasis, and trachoma. *Bulletin of the World Health Organization*, Vol. 69, pp. 609–21.

Evans, B., Hutton, G. and Haller, L. 2004. Closing the sanitation gap – the case for better public funding of sanitation and hygiene. Background paper for the Roundtable on Sustainable Development, 9–10 March, 2004, Paris, OECD.

Ezzati, M., Lopez, A. D., Rodgers, A., Vander Hoorn, S., Murray, C. J. L. and Comparative Risk Assessment Collaborating Group. 2002. Selected major risk factors and global and regional burden of disease. *Lancet*, Vol. 360, pp. 1347–60.

Ezzati, M., Vander Hoorn, S., Rodgers, A., Lopez, A. D., Mathers, C. D., Murray, C. J. L. and Comparative Risk Assessment Collaborating Group. 2003. Estimates of global and regional potential health gains from reducing multiple major risk factors. *Lancet*, Vol. 362, pp. 271–80.

Faruqui, N., Niang, S. and Redwood, M. 2004. Untreated wastewater reuse in market gardens: a case study of Dakar, Senegal. C. A. Scott, N. I. Faruqui and L. Raschid-Sally (eds), *Wastewater Use in Irrigated Agriculture: Confronting the Livelihood and Environmental Realities*, pp. 113–25. Wallingford, CAB International.

Fattal, B., Lampert, Y. and Shuval, H. 2004. A fresh look at microbial guidelines for wastewater irrigation in agriculture: a risk-assessment and cost-effectiveness approach. C. A. Scott, N. I. Faruqui, L. Raschid-Sally, (eds). *Wastewater Use in Irrigated Agriculture: Confronting the Livelihood and Environmental Realities*. Wallingford, CAB International Publishing.

Fewtrell, L. 2004. Drinking-water nitrate, methemoglobinemia, and global burden of disease: a discussion. *Environmental Health Perspectives*, Vol. 112, pp. 1371–4.

Howard, G. and Bartram, J. 2003. *Domestic Water Quantity, Service Level and Health*. WHO/SDE/WSH/03.02. Geneva, WHO.

Hutton, G. 2002. *Evaluation of the Global Non-Health Costs and Benefits of Water and Sanitation Interventions*. Basel, Swiss Tropical Institute.

Hutton, G. and Haller, L. 2004. Evaluation of costs and benefits of water and sanitation improvements at the global level. Document WHO/SDE/WSH/04.04. Geneva, WHO.

Ijumba, J. N. and Lindsay S. W. 2001. Impact of irrigation on malaria in Africa: Paddies paradox. *Medical and Veterinary Entomology*, Vol. 15, pp. 1–11.

Jensen, P.K., Jayasinghe, G., van der Hoek, W., Cairncross, S. and Dalsgaard, A. 2004. Is there an association between bacteriological drinking water quality and childhood diarrhoea in developing countries? *Tropical Medicine and International Health*, Vol. 9, pp. 1210–15.

Jones, G., Steketee, R. W., Black, R. E., Bhutta, Z. A., Morris, S. S., and Bellagio Child Survival Study Group. 2003. How many child deaths can we prevent this year? *Lancet*, Vol. 362, pp. 65–71.

Kay, B.H. and Nam, Vu Sinh, 2005. New strategy against *Aedes aegypti* in Viet Nam. *Lancet*, Vol. 365, pp. 613–17.

Keiser, J., Caldas de Castro, M., Maltese, M. F., Bos, R., Tanner, M., Singer, B. H. and Utzinger, J. 2005a. The effect of irrigation and large dams on the burden of malaria on global and regional scale. *American Journal of Tropical Medicine and Hygiene*, Vol. 72, pp. 392–406.

Keiser, J., Maltese, M. F., Erlanger, T. E., Bos, R., Tanner, M., Singer, B. H. and Utzinger, J. 2005b. Effect of irrigated rice agriculture on Japanese encephalitis and opportunities for integrated vector management. *Acta Tropica*, Vol. 95, pp. 40–57.

Keiser, J., Singer, B. H. and Utzinger, J. 2005c. Reducing the burden of malaria in different settings with environmental management: a systematic review. *Lancet Infectious Diseases*, Vol. 5, pp. 695–707.

Keiser, J. and Utzinger, J. 2005. Food-borne trematodiasis: An emerging public health problem. *Journal of Emerging Infectious Diseases*, Vol. 11: 1507–14.

Keiser, J., Utzinger, J., Caldas de Castro, M., Smith, T. A., Tanner, M. and Singer, B. H. 2004. Urbanization in sub-Saharan Africa and implications for malaria control. *American Journal of Tropical Medicine and Hygiene*, Vol. 71 (Suppl. 2), pp. 118–27.

Klinkenberg, E., van der Hoek, W. and Amerasinghe, F. P. 2004. A malaria risk analysis in an irrigated area in Sri Lanka. *Acta Tropica*, Vol. 89, pp. 215–25.

Konradsen, F., Matsuno, Y., Amerasinghe, F. P., Amerasinghe, P. H. and van der Hoek, W. 1998. *Anopheles culicifacies* breeding in Sri Lanka and options for control through water management. *Acta Tropica*, Vol. 71, pp. 131–8.

Konradsen, F., Steele, P., Perera, D., van der Hoek, W., Amerasinghe, P. H. and Amerasinghe, F. P. 1999. Cost of malaria control in Sri Lanka. *Bulletin of the World Health Organization*, Vol. 77, pp. 301–9.

Konradsen, F., van der Hoek, W., Amerasinghe, F. P., Mutero, C. and Boelee, E. 2004. Engineering and malaria control: Learning from the past 100 years. *Acta Tropica*, Vol. 89, pp. 99–108.

Kosek, M., Bern, C. and Guerrant, R. L. 2003. The global burden of diarrhoeal disease, as estimated from studies published between 1992 and 2000. *Bulletin of the World Health Organization*, Vol. 81, pp. 197–204.

Kotloff, K. L., Winickoff, J. P., Ivanoff, B., Clemens, J. D., Swerdlow, D. L., Sansonetti, P. J., Adak, G. K. and Levine, M. M. 1999. Global burden of *Shigella* infections: Implications for vaccine development and implementation of control strategies. *Bulletin of the World Health Organization*, Vol. 77, pp. 651–66.

Lean, S. and Pollok, R. C. G. 2003. Management of protozoa-related diarrhoea remains a major cause of morbidity and mortality in HIV-infected individuals. *Review of Anti-infective Therapy*, Vol. 1, pp. 455–69.

Lindsay, S., Kirby, M., Baris, E. and Bos, R. 2004. *Environmental Management for Malaria Control in the East Asia and Pacific (EAP) Region.* HNP Discussion Paper, Washington, DC, World Bank.

Listorti, J. A. and Doumani, F. M. 2001. *Environmental Health: Bridging the Gaps.* World Bank Discussion Paper No. 422. Washington, DC, World Bank. www.worldbank.org/afr/environmentalhealth/

Lokuge, K. M., Smith, W., Caldwell, B., Dear, K. and Milton, A. H. 2004. The effect of arsenic mitigation interventions on disease burden in Bangladesh. *Environmental Health Perspectives*, Vol. 112, pp. 1172–7.

Manga, L. 2002. Vector-control synergies, between 'roll back malaria' and the Global Programme to Eliminate Lymphatic Filariasis, in the African Region. *Annals of Tropical Medicine and Parasitology*, Vol. 96, Supplement 2, pp. 129–32.

Matsuno, Y., Konradsen, F., Tasumi, M., van der Hoek, W., Amerasinghe, F. P. and Amerasinghe, P. H. 1999. Control of malaria mosquito breeding through irrigation water management. *International Journal of Water Resources Development*, Vol. 15, pp. 93–105.

Mecaskey, J. W., Knirsch, C. A., Kumaresan, J. A. and Cook, J. A. 2003. The possibility of eliminating blinding trachoma. *Lancet Infectious Diseases*, Vol. 3, pp. 728–34.

Meddings, D. R., Ronald, L. A., Marion, S., Pinera, J. F. and Oppliger, A. 2004. Cost effectiveness of a latrine revision programme in Kabul, Afghanistan. *Bulletin of the World Health Organization*, Vol. 82, pp. 281–9.

Molyneux, D. 2003. Lymphatic filariasis (elephantiasis) elimination: a public health success and development opportunity. *Filaria Journal*, Vol. 2, 13 (www.filariajournal.com/content/2/1/13)

Moraes, L. R. S., Cancio, J. A. and Cairncross, S. 2004. Impact of drainage and sewerage on intestinal nematode infections in poor urban areas in Salvador, Brazil. *Transactions of the Royal Society of Tropical Medicine and Hygiene*, Vol. 98, pp. 197–204.

Murray, C. J. L. and Lopez, A. D. (eds). 1996. *The Global Burden of Disease.* Boston, Harvard University Press.

——. 1994a. Global and regional cause-of-death patterns in 1990. *Bulletin of the World Health Organization*, Vol. 72, pp. 47–480.

——. 1994b. Quantifying disability: data, methods and results. *Bulletin of the World Health Organization*, Vol. 72, pp. 481–94.

Parashar, U. D., Bresee, J. S. and Glass, R. I. 2003. The global burden of diarrhoeal disease in children. *Bulletin of the World Health Organization*, Vol. 81, p. 236.

Peden, M. M. and McGee, K. 2003. The epidemiology of drowning worldwide. *Injury Control and Safety Promotion*, Vol. 10, pp. 195–9.

Prasittisuk, C. 2002. Vector-control synergies, between 'roll back malaria' and the Global Programme to Eliminate Lymphatic Filariasis, in South-east Asia. *Annals of Tropical Medicine and Parasitology*, Vol. 96, Supplement 2, pp. 133–7.

Prüss, A., Kay, D., Fewtrell, L. and Bartram, J. 2002. Estimating the burden of disease from water, sanitation, and hygiene at a global level. *Environmental Health Perspectives*, Vol. 110, pp. 537–42.

Rajasooriyar, L. 2003. A study of the hydrochemistry of the Uda Walawa Basin, Sri Lanka, and the factors that influence groundwater quality. Ph.D. thesis, University of East Anglia, UK.

Rice, A. L., Sacco, L., Hyder, A. and Black, R. E. 2000. Malnutrition as an underlying cause of childhood deaths associated with infectious diseases in developing countries. *Bulletin of the World Health Organization*, Vol. 78, pp. 1207–21.

Rijsberman, F. 2004. *The Water Challenge.* Copenhagen Consensus Challenge Paper. Copenhagen, Environmental Assessment Institute.

Rodgers, A., Ezzati, M., Vander Hoorn, S., Lopez, A. D., Ruey-Bin Lin, Murray, C. J. L. and Comparative Risk Assessment Collaborating Group. 2004. Distribution of major health risks: findings from the Global Burden of Disease Study. *PLOS Medicine*, Vol. 1, Issue 1, e27.

Rose, G. 1992. *The Strategy of Preventive Medicine.* Oxford, Oxford University Press.

Saadé, C., Bateman, M. and Bendahmane, D. B. 2001. *The Story of a Successful Public-Private Partnership in Central America: Handwashing for Diarrheal Disease Prevention.* Arlington, VA, Basic Support for Child Survival Project, EHP, UNICEF, USAID, World Bank.

Scott, C. A., Faruqui, N. I. and Raschid-Sally, L. (eds). 2004. *Wastewater Use in Irrigated Agriculture: Confronting the Livelihood and Environmental Realities.* Wallingford: Cabi Publishing.

Shordt, K., van Wijk, C., Brikké, F. and Hesselbarth, S. 2004. *Monitoring Millennium Development Goals for Water and Sanitation. A Review of Experiences and Challenges.* Delft, IRC.

Sissoko, M. S., Dicko, A., Briët, O. J. T., Sissoko, M., Sagara, I., Keita, H. D., Sogoba, M., Rogier, C., Touré, Y. T. and Doumbo, O. K. 2004. Malaria incidence in relation to rice cultivation in the irrigated Sahel of Mali. *Acta Tropica*, Vol. 89, pp. 161–70.

SIWI and WHO (Stockholm International Water Institute and World Health Organization). 2005. *Making water a part of economic development.* Report commissioned for the 13th session of the Commission on Sustainable Development, Stockholm and Geneva, SIWI and WHO.

Sobsey, M.D. 2002. *Managing Water in the Home: Accelerated Health Gains from Improved Water Supply.* Geneva, WHO, report ref: WHO/SDE/WSH/02.07.

Steinmann, P., Keiser, J., Bos, R., Tanner, M. and Utzinger, J. In press. Schistosomiasis and water resource development: Systematic review, meta analysis and estimates of people at risk. *Lancet Infectious Diseases.*

Tyagi, B.K. 2004. A review of the emergence of *Plasmodium falciparum*-dominated malaria in the irrigated areas of the Thar Desert, India. *Acta Tropica*, Vol. 89, pp. 227–39.

UN Millennium Project. 2004a. *Interim Report of Task Force 4 on Child Health and Maternal Health.*

UN Millennium Project. 2004b. *Monitoring Target 10 and Beyond: Keeping Track of Water Resources for the Millennium Development Goals.* Issues Paper prepared for CSD 12. Millennium Task Force on Water and Sanitation.

UNDP (United Nations Development Programme). 2003. *Human Development Report, 2003. The Millennium Development Goals: a Compact Among Nations to End Human Poverty.* New York, Oxford University Press, 2003.

UN-WWAP (United Nations World Water Development Programme). 2003. *The United Nations World Water Development Report: Water for People, Water for Life.* Paris and London, UNESCO and Berghahn Books.

Utzinger, J., Bergquist, R., Shu-Hua, X., Singer, B. H. and Tanner, M. 2003. Sustainable schistosomiasis control: The way forward. *Lancet*, Vol. 362, pp. 1932–4.

Van der Hoek, W., Sakthivadivel, R., Renshaw, M., Silver, J. B., Birley, M. H. and Konradsen, F. 2001a. *Alternate Wet / Dry Irrigation in Rice Cultivation: A Practical Way to Save Water and Control Malaria and Japanese Encephalitis?* Research Report 47. Colombo, IWMI.

Van der Hoek, W., Konradsen, F., Ensink, J. H. J., Mudasser, M. and Jensen, P. K. 2001b. Irrigation water as a source of drinking water: is safe use possible? *Tropical Medicine and International Health*, Vol. 6, pp. 46–54.

Van der Hoek, W., Feenstra, S. G. and Konradsen, F. 2002a. Availability of irrigation water for domestic use in Pakistan: its impact on prevalence of diarrhoea and nutritional status of children. *Journal of Health, Population and Nutrition*, Vol. 20, pp. 77–84.

Van der Hoek, W., Boelee, E. and Konradsen, F. 2002b. Irrigation, domestic water supply and human health. *Encyclopedia of Life Support Systems: Knowledge for Sustainable Development* (EOLSS), Oxford, EOLSS Publishers.

Van der Hoek, W., Ekanayake, L., Rajasooriyar, L. and Karunaratne, R. 2003. Source of drinking water and other risk factors for dental fluorosis in Sri Lanka. *International Journal of Environmental Health Research*, Vol. 13, pp. 285–93.

Van der Werf, M. J., de Vlas, S. J., Brooker, S., Looman, C. W. N., Nagelkerke, N. J. D., Habbema, J. D. F. and Engels, D. 2003. Quantification of clinical morbidity associated with schistosome infection in sub-Saharan Africa. *Acta Tropica*, Vol. 86, pp. 125–39.

Victora, C. G., Bryce, J., Fontaine, O. and Monasch, R. 2000. Reducing deaths from diarrhoea through oral rehydration therapy. *Bulletin of the World Health Organization*, Vol. 78, pp. 1246–55.

WHO (World Health Organization). 2006a. *Guidelines for the Safe Use of Wastewater, Excreta and Greywater. Volume 1: Policy and Regulatory Aspects.* Geneva, WHO. World Health Organization,

——. 2006b. *Guidelines for the Safe Use of Wastewater, Excreta and Greywater. Volume 2: Wastewater Use in Agriculture.* Geneva, WHO.

——. 2006c. *Guidelines for the Safe Use of Wastewater, Excreta and Greywater. Volume 3: Wastewater and Excreta Use in Aquaculture.* Geneva, WHO.

——. 2006d. *Guidelines for the Safe Use of Wastewater, Excreta and Greywater. Volume 4: Excreta and Greywater Use in Agriculture.* Geneva, WHO.

——. 2005. Water safety plans: managing drinking-water quality from catchment to consumer, prepared by Annette Davison, Guy Howard, Melita Stevens, Phil Callan, Lorna Fewtrell, Dan Deere and Jamie Bartram. *WHO publication WHO/SDE/WSH/05.06*, Geneva, WHO. www.who.int/water_sanitation_health/dwq/wsp0506/en/index.html

——. 2004a. *World Health Report 2004. Changing History.* Geneva, WHO. www.who.int/whr/2004

——. 2004b. *Guidelines for Drinking-water Quality. Third Edition. Volume 1. Recommendations.* Geneva, WHO. www.who.int/water_sanitation_health/dwq/en

——. 2004c. *PRSPs: Their Significance for Health: Second Synthesis Report.* WHO/HDP/PRSP/04.1, Geneva, WHO.

——. 2003a. *World Health Report 2003. Shaping the Future.* Geneva, WHO. www.who.int/whr/2003

——. 2003b. *Guidelines for Safe Recreational Water Environments. Volume 1: Coastal and Fresh Waters.* Geneva, WHO.

——. 2002a. *Prevention and Control of Schistosomiasis and Soil-Transmitted Helminthiasis.* WHO Technical Report Series 912, Geneva.

——. 2002b. *World Health Report 2002. Reducing Risk, Promoting Healthy Life.* www.who.int/whr/2002

WHO/IWA (World Health Organization/International Water Association). 2006. *Fluoride in drinking water.* Geneva, World Health Organization.

WHO/PAHO/UNESCO (World Health Organization/Pan-American Health Organization/United Nations Educational, Scientific and Cultural Organization). 1997. A consultation with experts on amoebiasis. Mexico City, Mexico 28–29 January

1997. *Epidemiological Bulletin – Pan American Health Organization*, Vol.18, pp. 13–4.

WHO/UNICEF (World Health Organization/United Nations Children's Fund). 2005. Water for Life: Making it Happen. Geneva.

——. 2004. *Meeting the MDG Drinking Water and Sanitation Target. A Mid-term Assessment of Progress.* New York, Geneva.

——. 2003. *Africa Malaria Report.* WHO/CDS/MAL/2003.1093.

——. 2000. *Global Water Supply and Sanitation Assessment 2000 Report.* New York, Geneva.

WSSCC (Water Supply and Sanitation Collaborative Council). 2004. *Resource Pack on the Water and Sanitation Millennium Development Goals.* Geneva, WSSCC.

Würthwein, R., Gbangou, A., Sauerborn, R. and Schmidt, C. M. 2001. Measuring the local burden of disease. A study of years of life lost in sub-Saharan Africa. *International Journal of Epidemiology*, Vol. 30, pp. 501–8.

Yoshida, T., Yamauchi, H. and Fan Sun, G. 2004. Chronic health effects in people exposed to arsenic via the drinking water: dose – response relationships in review. *Toxicology and Applied Pharmacology*, Vol. 198, pp. 243–52.

World Health Organization: www.who.int

    WHO Water, Sanitation and Health: www.who.int/water_sanitation_health

    WHO Evidence and Information for Health Policy: www.who.int/evidence

WHO Regional Offices

    Africa: www.afro.who.int/wsh/index.html

    Americas: www.paho.org/Project.asp?SEL=TP&LNG=ENG&ID=86

    Eastern Mediterranean/Centre for Environmental Health Activities: www.emro.who.int/ceha/community.asp

    Europe: www.euro.who.int/healthtopics/HT2ndLvlPage?HTCode=drinking_water   www.euro.who.int/ecehrome

    South East Asia: www.searo.who.int

    Western Pacific: www.wpro.who.int/health_topics/water_sanitation_and_hygiene

United Nations Children's Fund: www.unicef.org

    UNICEF Water, Environment and Sanitation: www.unicef.org/wes

Monitoring programmes

    WHO/UNICEF Joint Monitoring Programme for Water Supply and Sanitation: www.wssinfo.org

    UNICEF Monitoring the Situation of Women and Children: www.childinfo.org

Water Supply and Sanitation Collaborative Council: www.wsscc.org

World Bank: www.worldbank.org

    World Bank Water Supply and Sanitation Programme: www.worldbank.org/watsan

    World Bank Water and Sanitation Blue Gold Series: www.wsp.org/08_BlueGold.asp

Asian Development Bank: www.adb.org

Water for All Programme: www.adb.org/Water

Health, Nutrition and Population Programme: www.adb.org/Health

International Development Research Centre (IDRC Canada) Ecohealth Programme: www.idrc.ca/ecohealth

International Water and Sanitation Centre, the Netherlands: www.irc.nl

International Water Association: www.iwahq.org.uk

International Commission for Irrigation and Drainage: www.icid.org

WHO Collaborating Centres in Water, Sanitation and Health:
   Water Quality and Health Bureau, Health Canada: www.hc-sc.gc.ca/waterquality
   Office national de l'Eau potable (ONEP), Morocco: www.onep.org.ma
   Institute of Environmental Engineering and Research: www.uet.edu.pk/Departments/Environmental/environmental_main.htm
   DBL Institute for Health Research and Development, Denmark: www.dblnet.dk
   DHL Water and Environment Denmark: www.dhi.dk
   Institute for Water, Soil and Air Hygiene, Federal Environment Agency, Germany: www.umweltbundesamt.de
   Institute for Hygiene and Public Health, Bonn, Germany: www.meb.uni-bonn.de/hygiene
   Institute for Water Pollution Control (VITUKI), Hungary: www.vituki.hu
   University of Surrey, School of Engineering: www.surrey.ac.uk/eng
   National Centre for Environmental Toxicology: www.wrcplc.co.uk/asp/business_areas.asp#ncet
   British Geological Survey, Groundwater Systems and Water Quality Programme: www.bgs.ac.uk
   International Water Management Institute, Sri Lanka: www.iwmi.cgiar.org
   Faculty of Tropical Medicine, Mahidol University, Bangkok, Thailand: www.tm.mahidol.ac.th
   Asian Institute of Technology, Urban Environmental Management Programme: www.ait.ac.th
   Queensland Institute for Medical Research, Mosquito Control Laboratory, Australia:
   www.qimr.edu.au/research/labs/briank/index.html
   National Institute of Public Health, Department of Water Supply Engineering, Japan: www.niph.go.jp
   Centre regional pour l'Eau potable et l'Assainissement à faible Coût, Burkina Faso: www.reseaucrepa.org

Future Harvest centres associated with the CGIAR doing research on water management/health
   International Food Policy Research Centre: www.ifpri.org/events/seminars/2005/20050623AgHealth.htm
   Africa Rice Centre (formerly: West African Rice Development Association): www.warda.cgiar.org/research/health
   International Water Management Institute (IWMI): www.iwmi.org

Starvation is the characteristic of some people not having enough food to eat. It is not the characteristic of there being not enough food to eat.

Amartya Sen

CHAPTER 7

Water
for Food,
Agriculture
and Rural
Livelihoods

By

FAO
*(Food and
Agriculture
Organization of the
United Nations)*

IFAD
*(International Fund
for Agricultural
Development)*

Rice terraces, China

Key messages:

Above: Farmer ploughing a rice field, Indonesia

Below: Wollo women diverting a stream to irrigate land, Ethiopia

Bottom: Rendille livestock enclosures, Kenya.

Bottom right: Fruit and vegetable market, Jordan

In the context of demographic growth, increased competition for water and improved attention to environmental issues too often left out by agricultural policies, water for food remains a core issue – that can no longer be tackled through a narrow sectoral approach. New forms of water management in agriculture, including irrigation, are to be explored and implemented, in order to focus on livelihoods rather than just on productivity.

■ To satisfy the growing demand for food between 2000 and 2030, production of food crops in developing countries is projected to increase by 67 percent. At the same time, a continuing rise in productivity should make it possible to restrain the increase in water use for agriculture to about 14 percent.

■ As competition for water increases among different sectors, irrigated agriculture needs to be carefully examined to discern where society can benefit most effectively from its application. Access to natural resources needs to be negotiated with other users in a transparent fashion in order to achieve optimal uses under conditions of growing scarcity.

■ Farmers are at the centre of any process of change and need to be encouraged and guided, through appropriate incentives and governance practices, to conserve natural ecosystems and their biodiversity and minimize their negative impact, a goal that will only be achieved if the appropriate policies are in place.

■ Irrigation institutions must respond to the needs of farmers, ensuring more reliable delivery of water, increasing transparency in its management and balancing efficiency and equity in access to water. This will not only require changes in attitudes, but also well targeted investments in infrastructure modernization, institutional restructuring and upgrading of the technical capacities of farmers and water managers.

■ The agriculture sector faces a complex challenge: producing more food of better quality while using less water per unit of output; providing rural people with resources and opportunities to live a healthy and productive life; applying clean technologies that ensure environmental sustainability; and contributing in a productive way to the local and national economy.

■ Action is needed now to adapt agricultural and rural development policies, accelerate changes in irrigation governance and, through adequate water laws and institutions, support the integration of the social, economic and environmental needs of rural populations.

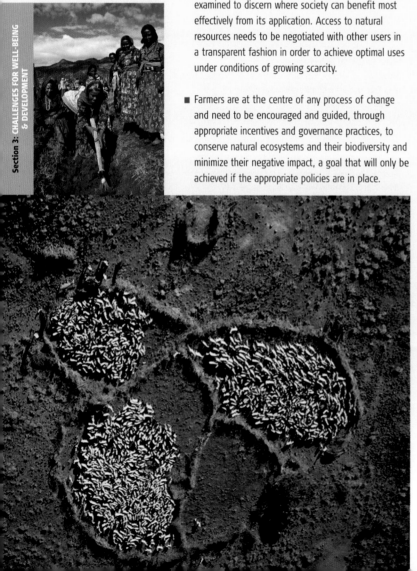

Part 1. Water's Role in Agriculture

During the second half of the twentieth century, the global food system responded to a twofold increase in the world's population by more than doubling food production, and this in an environment of decreasing prices for agricultural products. During the same period, the group of developing countries increased per capita food consumption by 30 percent and nutritional situations improved accordingly. In addition, agriculture continued producing non-food crops, including cotton, rubber, beverage crops and industrial oils. However, while feeding the world and producing a diverse range of commodities, agriculture also confirmed its position as the biggest user of water on the globe. Irrigation now claims close to 70 percent of all freshwater appropriated for human use.

In the absence of competition for raw water, and with little initial concern for environmental implications, agriculture has been able to capture large quantities of freshwater and ensure its claim to freshwater use. In the near future, the need to produce and process more food for the world's growing population will translate into an increased demand for irrigation. However, agriculture is now increasingly obliged to accommodate its claims on water within a complex framework in which social, economic and environmental objectives have to be negotiated with other sectors. The basis for well-informed negotiation hinges upon the degree of effective governance that can be found within the respective social, economic and environmental sectors.

Food production, whether on a large commercial farm, a homestead garden plot or a fish pond, is a local activity. However, the decisions underlying the way in which food is produced are increasingly beyond the reach and influence of local communities and of local agricultural organizations. Prices and specifications for agricultural export products are decided at far away market places. National governments in many developing countries have discriminated against the rural sector in order to favour urban constituencies, and rich countries have subsidized their agricultural exports, with dire consequences for undercapitalized rural producers operating in an environment of poor or non-existing physical, financial, educational and health infrastructure. At present, about 13 percent of the world's population does not have access to enough food to live a healthy and productive life, yet the ability, technology and resources needed to produce enough food for every man, woman and child in the world do currently exist. Lack of health, financial or natural resources such as land and water, and lack of skills to link productive activities with remote markets and ensure employment, are all intimately related to poverty.

Table 7.1 presents the various levels of governance linked to agricultural water management. Water governance issues emerge around water allocation and distribution, but governance aspects other than those concerning water are equally important. Secure tenure of sufficient land is fundamental to water governance, as is the availability of supporting infrastructure related to distribution and marketing. Market access is critical to income generation. Clearly, the management of rural water, including irrigation facilities, calls for some form of local governance. Recent trends towards increased responsibility of water users present new challenges for such arrangements. Again, beyond water management, the whole system of food governance arises when the implementation of national food policies (through subsidies, taxes, tariffs and even food aid in some instances) distorts markets and marginalizes the rural poor.

With current trends towards international trade liberalization, the complexity of these governance problems and their interconnection is growing. Water governance and related food system issues have to be examined from local, national and global perspectives. This chapter reviews the main links between water, food production, markets and rural livelihoods, as well as their implications in terms of governance at all levels.

1a. The water variable in agriculture
Feeding the world's population: from need to surplus
For adequate nutrition, a person's daily diet should be complete (in energy terms) and balanced (in nutritional terms). The indicator used as a proxy to assess the nutritional situation of a population is the average dietary energy intake in the form of kilocalories per person per day (kcal/person/d). The dietary energy supply (DES) value of 2,800 kcal/person/d is taken as a threshold to

...agriculture is now increasingly obliged to accommodate its claims on water within a complex framework in which social, economic and environmental objectives have to be negotiated with other sectors

Section 3: CHALLENGES FOR WELL-BEING & DEVELOPMENT

Table 7.1: Governance issues at different levels in agricultural water management

Level	Water	Land	Infrastructure	Market-Services
Farmer	Access to water: water rights; water markets	Access to land: land tenure; size of farm holdings	Access to affordable technology, including irrigation	Access to production inputs and markets
Farmer groups	Water rights; equity; water distribution; accountability	–	Management authority (irrigation schemes)	Farmer cooperatives, unions, meteorological forecasting
Irrigation service	Reliability, equity and flexibility of irrigation service delivery	Crop patterns and licensing	System management and maintenance; cost recovery; transparency; accountability	Farm roads maintenance and other scheme infrastructures
Local government	Water licensing (nepotism); conflict resolution	Land-use planning	Decentralization; development of new infrastructure (including markets)	Market infrastructure and transport; access to finance; market information
Basin authority	Sectoral water allocation; water quality management; water conservation (financial incentives)	Soil conservation; watershed protection	Main hydraulic infrastructure planning; development and management (corruption)	–
National government	Water policy and legislation; institutional arrangements	Land-use policy and legislation; cadastre; land-use planning	Policies and legislation on: decentralization; infrastructure development planning; cost recovery; financing mechanisms for infrastructure; access to finance for local stakeholders	Policies and legislation on: food security; agriculture (subsidies); rural development; trade (tariffs, subsidies); food self-sufficiency; rural finance
Regional level	Transboundary water; security of supply	–	Transboundary water shared infrastructure	Regional trade agreements
Global level	International security and solidarity	–	–	Agricultural subsidies and tariffs

In developed countries, food consumption has kept increasing strongly and their population is now increasingly stricken by obesity

national food security: below this level, countries are likely to suffer severe chronic undernutrition problems in their population. However, even in conditions of sufficient national food supply (above 2,800 kcal/person/d), a part of the population may be suffering from undernutrition because individuals and households lack the capacity to access the food they need, either through self-production or purchase. **Box 7.1** briefly discusses this matter.

Based on DES/person/d as a food security indicator, **Figure 7.1** shows for the world and for each region the levels of food consumption in 1965 and 1998, with projections for 2030. The nutritional situation has improved everywhere and, towards the turn of the century, the world average has reached the 2,800 kcal/person/d 'graduation' level. This can be argued to mean that universal food security – food for all – is within grasp. The sub-Saharan Africa and South Asia

regions, where food consumption is the lowest, have improved their situation but their level is still low. The developing countries of East and Southeast Asia, as well as Latin America and the Caribbean, are close below or above the threshold level. In developed countries, food consumption has kept increasing strongly and their population is now increasingly stricken by obesity and growing food wastage. In the details, it is interesting to note that most of the progress in the food situation has occurred in some of the most populous developing countries, in particular Brazil, China and Indonesia. Significant progress has also been made in India, Nigeria and Pakistan. There remain, however, thirty nations with a very low level of food intake (below 2,200 kcal/person/d), plagued by both the inability to produce enough food and to earn foreign exchange to import the food that their population needs. Of these thirty countries, twenty are located in sub-Saharan Africa.

BOX 7.1 MEASURING HUNGER AND UNDERNUTRITION

The method used by the Food and Agriculture Organization (FAO) of the United Nations (UN) to monitor the state of the world's food insecurity is based on information from national food balance sheets, adjusted through household income and expenditure surveys. These surveys allow the construction of a distribution function of dietary energy consumption on a per-person basis. The proportion of undernourished people in the total population is defined as that part of the distribution lying below a minimum level of energy requirement, which is derived taking into account sex, age and type of activity.

Other methods for measuring hunger and undernutrition involve income and expenditure survey data, adequacy of dietary intake, child nutritional status based on anthropometric surveys, and qualitative methods for measuring people's perception of food insecurity and hunger. None of these methods alone suffices to capture all the aspects and dimensions of food insecurity.

Sources: FAO, 2003b; Mason, 2002.

Figure 7.1: Per capita food consumption by region, 1965–2030

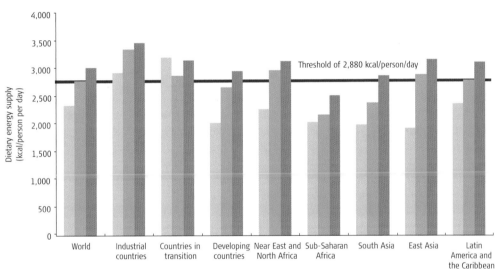

Note: The nutritional status of most regions of the world has improved. Sub-Saharan Africa and South Asia have also improved their nutritional level, but they lag behind and are host to the majority of the undernourished people in the world.

Source: FAO, 2003a.

Agriculture: Crops and livestock products

Drinking water intake typically varies between 2 and 3 litres per person per day (L/person/d). In addition, there are domestic water requirements for personal and household hygiene and related requirements, which are quantified at 30 to 300 L/person/d, according to standard of living and quality of water supply. Producing food requires much more water: from 2,000 to 5,000 L/person/d, depending on diet and climate differences and the efficiency of local food production systems. Most of the water used to produce food or other crops comes from rain that is stored in the soil (so-called green water), where it is captured by crop roots. Irrigation is practised in places and times where rainwater is insufficient for adequately supplying water to crops. It provides a guaranteed supply of water and protects against droughts and dry spells. Globally, rainfall provides about 90 percent of the water used by crops. Although it covers only 10 percent of the water used in agriculture, irrigation claims 70 percent of all the freshwater (so-called blue water) used for human consumption and so comes under heavy scrutiny when discussing freshwater governance. The concept of green water and blue water is further discussed below (see also **Chapter 4**).

The production of food commodities presents different demands on producers and the environment, including the quantity and quality of the water required.

Harvesting fish, State of Mexico

Young girl sorting freshly harvested tomatoes for packing, Honduras

Agriculture, including livestock production and aquaculture, produces most of the food we eat. Direct food harvest from inland and marine environments, including capture fisheries[1], plays a vital role in the livelihood of many rural and indigenous people. However, its total quantity is only a fraction of global food supply (**Figure 7.2**). In addition to food and beverage commodities, agriculture also produces non-food commodities including fibre (such as cotton), and industrial oil.

Food consumption habits are strongly linked to local environment and production capacities, but through the years, there have been major changes in food consumption patterns, which depend largely on factors such as income. **Figure 7.3** shows changes in diet composition from the 1960s through the recent past and projected to 2030. Rice, wheat and other cereals are a major component of the human diet. They represent, in terms of dietary energy supply, more than half of all food consumed. However, the relative weight of cereals in the human diet tends to decline as income increases, signalling an increase in the consumption of vegetable oils and meat. While the demand for cereals intended for direct human consumption is decreasing, about one-third of all cereals is now used for animal feed, and agriculture keeps responding to this strong aggregate demand. The oil crops sector is one of the most dynamic in the world: since the 1970s, oil crops have provided 20 percent of the increase in caloric intake in developing countries.

Livestock production accounts for some 40 percent of the gross value of agricultural production, and its role is continually growing as consumers adopt a diet richer in meat and dairy products. Current trends (**Figure 7.3**) show strong growth in the meat component of diets. Livestock production is the world's largest user of agricultural land, through grazing and consumption of fodder and feed grains. Although overgrazing and ensuing land erosion has caused concern, it is worth noting that grazing natural pasture is a very efficient way of using rainwater to produce food that is rich in protein. A fairly recent development within the livestock sector is the shift of feed production from natural pasture and hay to cereals. Contrary to grazing, when feed grains are produced using irrigation, the shift in livestock production methods has an impact on water demand. Looking at another facet of this complex question, in traditional livestock management, animal manure is an important element in the process of returning nutrients to the soil, and

lack of manure is a cause of soil degradation. Conversely, intensive methods of livestock production result in the production of large amounts of manure that all too often are discharged into surface waters or left on the land to seep into the groundwater, threatening water supplies needed for potable water and other high-value uses.

Food commodities: Fish

Capture fisheries and aquaculture are critical to food security among poor communities in inland and coastal areas. Fish, including shellfish and crustaceans, provides about 16 percent of animal proteins consumed worldwide and is a valuable source of minerals and essential fatty acids in human nutrition. Fish consumption, however, varies greatly among regions, and average values for fish consumption have little meaning. As an environmental product, capture fisheries, both marine and inland, are the largest source of 'wild' food. Because of the open access conditions that prevail in fisheries, overexploitation has led to the depletion of many stocks. Conflicts are frequent between native and community-based small-scale subsistence fishers, recreational fishers and large-scale or industrialized fishers who emphasize commercial and income-producing purposes. More significantly, fisherfolk also suffer from encroachment by, and conflicts with, other resource users and sectors, including urban and industrial development, tourism, agriculture and energy sectors (see **Chapter 11**).

While fish production from marine capture fisheries has not increased since the 1990s, production from inland capture fisheries has shown modest but steady increases. Inland fishery catches are believed to be greatly under-reported and could be twice as high as records show (FAO, 2003d). Freshwater fisheries, in particular the unreported fraction of strategic relevance for poor people, tend to be undervalued or completely neglected in water development projects. This is especially the case in the estimation of the adverse effects of aquatic pollution and habitat degradation that result from the unsustainable use of water resources in fishing communities. There are nevertheless significant opportunities for considering and implementing measures (see **Box 7.2**) that integrate fisheries and agriculture activities in order to enhance fish production and food security (FAO, 1998).

Aquaculture in freshwater, brackish water and seawater has increased its production from less than 5 million metric tonnes (t) in the 1960s to close to 50 million t in

---

1. Fisheries are usually divided into capture fisheries and aquaculture, where capture fisheries refers to the direct harvesting of fish and other aquatic organisms from their natural environment, while aquaculture can be defined as the farming of aquatic organisms with some sort of intervention in the rearing process to enhance their production.

Figure 7.2: Main sources of global food supply, 2002

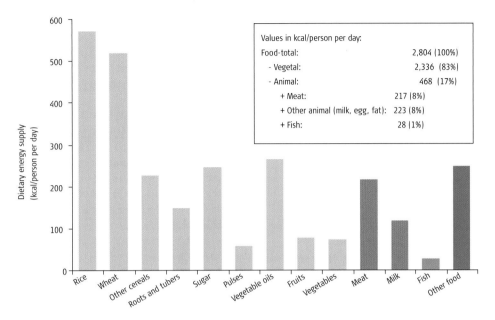

Note: Cereals, in particular rice and wheat, dominate food supply and provide the largest share of energy to the world's population. Although the livestock and fisheries sectors remain marginal in global terms, they play an important role in the supply of proteins. These global figures hide a large geographical variability in people's dietary energy supply.

Source: FAOSTAT, accessed in 2005.

Figure 7.3: Dietary changes in developing countries, 1965–2030

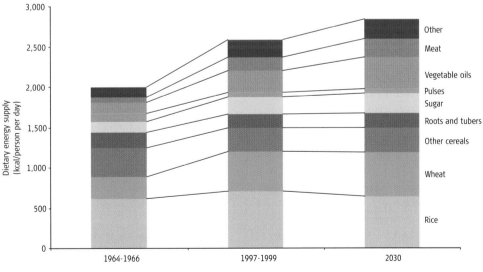

Note: While rice, wheat and other cereals remain the main component of the human diet, their relative weight tends to decline as income increases, compensated by an increase in the consumption of vegetable oils and meat. The oil crops sector is one of the most dynamic in the world: since the 1970s, oil crops have provided 20 percent of the increase in caloric intake in developing countries.

Source: FAO, 2003a.

Of particular relevance is the rapid growth of aquaculture production in China – an average of 11.5 percent per year in the last thirty years, compared to 7 percent per year for the rest of the world

2000. This trend is bound to continue, driven both by market demand for high-priced fish (including shellfish) and by the ability of aquaculture to produce low-priced but nutritious fish for local consumption. Of particular relevance is the rapid growth of aquaculture production in China – an average of 11.5 percent per year in the last thirty years, compared to 7 percent per year for the rest of the world. When properly managed and integrated into other agricultural activities, aquaculture constitutes a valuable way to increase the productivity of water. But as it intensifies, new environmental challenges arise, and aquaculture can be the cause of water pollution and consequent damage to natural systems.

Competition in the use of water for agriculture and for fisheries is a key issue. Agricultural water use is strongly consumptive and results in reduced river outflow that adversely affects fish habitat. Intensive agriculture tends to release agrochemicals applied in excess to groundwater and rivers thus adding to the deleterious, often lethal, effects of water pollution on fisheries.

At the core of the requirements of fisheries are: maintaining aquatic habitats, sustaining river flow to keep ecosystems healthy and abating pollution upstream; for agriculture, it matters crucially to have water securely available for use during the irrigation season. In an integrated resources management (IRM) context, local and basin-wide interest groups representing agriculture and fisheries can negotiate and trade off their benefits and duties to construct viable compromises. In practice, however, settling complex contentious matters in an integrated context meets difficulties because the prospective negotiating partners are likely to be unequal in knowledge, financial and political clout, and in negotiating skills, and strong governance support is therefore needed to foster the process and guarantee fair and equitable outcomes.

Sources of water in agriculture: Rainfed and irrigated agriculture

The bulk of the world's agricultural production is rainfed, not irrigated. Claims that agricultural production is threatened by global water shortages usually fail to note that most of the world's food production does not rely on freshwater withdrawals at all and does not necessarily accelerate the naturally occurring rates of evapotranspiration. The largest share of the water uptake by plants is transpired back into the atmosphere through plants' leaves. In addition to its energy dissipating role, the

transpiration process is necessary for lifting nutrients from -photosynthesis takes place. If soil moisture levels fall below the wilting point, plant growth slows and eventually stops, and the potential crop yield is not fulfilled. Irrigation aims at ensuring that enough moisture is available at all times during the plant's life cycle to satisfy its water demand, thus supporting maximum crop yields.

The concept of 'blue' and 'green' water has been used for quite some time to distinguish between two fundamentally different elements of the water cycle (see **Chapter 4**). When atmospheric precipitation reaches the ground, it divides into several sections, which pursue the terrestrial part of the hydrological cycle along different paths. Out of a total annual amount of 110,000 cubic kilometres (km^3) of precipitation on the land surface, about 40,000 km^3 is converted into surface runoff and aquifer recharge (blue water) and an estimated 70,000 km^3 is stored in the soil and later returns to the atmosphere through evaporation and plant transpiration (green water, see **Figure 7.4**). Blue water is the freshwater that sustains aquatic ecosystems in rivers and lakes; it can also be applied to drinking or domestic purposes, to industry or hydropower or to irrigated agriculture. Rainfed agriculture uses only green water. Irrigation uses blue water in addition to green water to maintain adequate soil moisture levels, allowing the crop plants to absorb the water and fulfil their crop yield potential. The green water/blue water concept has proven to be useful in supporting a more comprehensive vision of the issues related to water management, particularly in reference to agriculture (Ringersma et al., 2003). It is estimated that crop production takes up 13 percent (9,000 km^3 per year) of the green water delivered to the soil by precipitation, the remaining 87 percent being used by the non-domesticated vegetal world, including forests and rangeland. While irrigation currently withdraws about 2,300 km^3 of freshwater per year from rivers and aquifers, only about 900 km^3 is effectively consumed by crops (this issue is addressed in more detail later in the section on water use efficiency).

Out of the world's total land area of 13 billion hectares (ha), 12 percent is cultivated, and an estimated 27 percent is used for pasture. The 1.5 billion ha of cropland include 277 million ha of irrigated land, representing 18 percent of cropland. In population terms, cropland amounts to a global average of 0.25 ha per person. **Figure 7.5** shows the evolution of cropland compared to population between 1960 and 2000, illustrating the huge productivity increase

Sprinkler irrigation on an experimental field of asparagus, Brazil

Figure 7.4: Blue and green water in the hydrological cycle

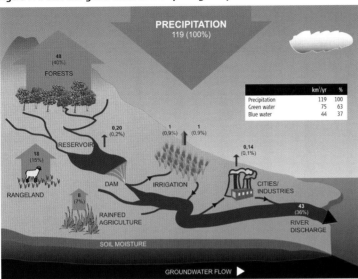

	km³/yr	%
Precipitation	119	100
Green water	75	63
Blue water	44	37

Note: About two-thirds of the water falling on land is evaporated from the ground or transpired by the vegetation (forests, rangeland, cropland) and rainfed agriculture uses about 8 percent of the green water. Irrigation uses both green water and blue water. Although the share of water used in irrigation is marginal in the global hydrological cycle, irrigated agriculture is the single most important user of blue water, leading in an increasing number of places to severe shortage and competition for water.

Source: adapted from Shiklomanov, 2000; FAO, 2002b; Ringersma et al., 2003; Rockström, 1999.

of agriculture during that period. The intensification of agricultural production made it possible to limit the expansion of agricultural land to a few percentage points as the population was more than doubling.

The options for increasing the amount of land dedicated to agriculture at the expense of natural forests and savannahs are limited, because land has to be both suitable and available for conversion to agriculture. The food needs of a growing population are therefore largely served through increased land productivity, meaning harvesting a larger quantity of crops from the existing agricultural land surface. Irrigation lifts the constraint on crop productivity that stems from insufficient and unreliable rainfall. During the second half

of the twentieth century, irrigation development became a core part of the strategy to feed a doubling world population, with a consequent increase in the quantity of water utilized for this purpose. **Map 7.1** shows the location and extent of irrigation in the world in 2000 and the relative importance of irrigation in national agriculture, a distribution that strongly correlates with climatic conditions.

Trends and projections

According to FAO projections, food demand in 2030 is expected to be 55 percent higher than in 1998, taking into account increases in both population and per capita food intake. To respond to this demand, global food production should increase at an annual rate of

BOX 7.2: INTEGRATED RESOURCES MANAGEMENT IN SUPPORT OF FOOD PRODUCTION IN RURAL AREAS

Integrated approaches to resource management (IRM) are challenging but provide significant avenues for enhancing fish production as well as the productivity of inland water bodies. In particular, opportunities exist for the enhanced integration of inland fisheries and aquaculture into agricultural development planning, especially irrigation, adding value to shared

resources. Basic features of good IRM include the formation of extensive partnerships and the close involvement of local interests. It implies improved cross-sectoral cooperation among the agriculture, forestry and fisheries sub-sectors.

Increased efforts are needed to provide better targeted technical assistance and policy

guidance on IRM for sustainable production of fish and other food at the local level. Such IRM efforts are also often required at regional and international levels, and further assistance is necessary in regional decision-making for the transboundary management of shared river and lake basins (see **Chapter 11**).

Figure 7.5: Evolution of cropland, 1961–2000

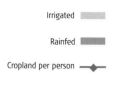

Irrigated

Rainfed

Cropland per person

Note: In forty years, cropland has increased only slightly while population was more than doubling, leading to a sharp reduction in the amount of land needed to produce food for one person. These rapid increases in productivity were obtained through intensification of agricultural production, in which irrigation has played an important role.

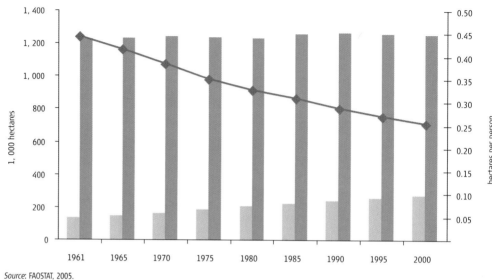

Source: FAOSTAT, 2005.

A farmer and his cattle return home across rice terraces, Indonesia

1.4 percent. This growth would occur mostly in developing countries, where about 80 percent of the projected growth in crop production will come from intensification in the form of yield increases (67 percent) and increased cropping intensity (12 percent). The remaining 20 percent will come from cropland expansion in some countries of sub-Saharan Africa, Latin America and East Asia that still have land potential (FAO, 2003a).

In 2030, irrigated agriculture in ninety-three developing countries would account for over 70 percent of the projected increase in cereal production. In these countries, the area equipped for irrigation is expected to expand by 20 percent (40 million ha) between 1998 and 2030. This projected increase in irrigated land is less than half of the increase of the preceding period (100 million ha). Thanks to increased cropping intensity, the area of harvested crops in irrigation is expected to increase by 34 percent by 2030. In the same period, the amount of freshwater that will be appropriated for irrigation is expected to grow by about 14 percent to 2,420 km³ in 2030. Compared to the projected 34 percent increase in harvested irrigated area and the 55 percent increase in food production, the 14 percent increase in water withdrawal for irrigation is modest. Irrigation in the ninety-three developing countries, aggregated as a group, still claims a relatively small part of their total water resources. At the local level, however, where there

are already water shortages, such as in the Near East and North Africa, growing competition between agriculture, cities and industries will exacerbate water scarcity, and it is likely that the share of freshwater available for agriculture will decrease (Faurès et al., 2003). In countries and regions facing serious water scarcity problems, the gap between demand and production will grow, forcing them to rely increasingly on importing food to satisfy domestic needs. Already today, several countries like Egypt or Jordan have a structural food deficit and cannot produce the food they need to satisfy domestic demand.

1b. Drivers of change in agricultural production

Changing patterns of demography, food production, food demand and diets

Global demographic projections point to declining rates in population growth. Deceleration of demographic growth and gradual saturation in per capita food consumption will contribute to a slowing growth of food demand. Nevertheless, the expected absolute annual increments in population growth continue to be large, of the order of 76 million people per year at present and 53 million people per year towards 2030. Almost all of this population growth will take place in developing countries, with large regional differences. These countries have to find an adequate mix of policies stimulating local food

Map 7.1: Distribution of areas under irrigation in the world, 2000

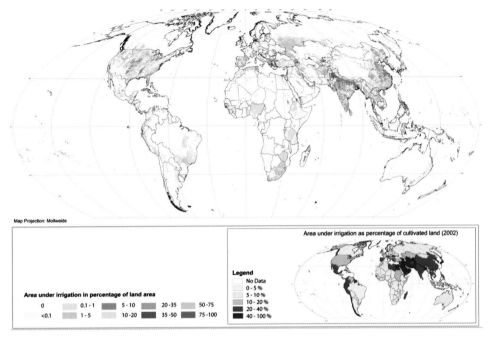

Map Projection: Mollweide

Area under irrigation in percentage of land area

| 0 | 0.1 - 1 | 5 - 10 | 20 - 35 | 50 - 75 |
| <0.1 | 1 - 5 | 10 - 20 | 35 - 50 | 75 - 100 |

Area under irrigation as percentage of cultivated land (2002)

Legend
No Data
0 - 5 %
5 - 10 %
10 - 20 %
20 - 40 %
40 - 100 %

Note: Irrigation is concentrated in arid and semi-arid areas, where it represents a significant share of cropland, and in the humid tropics of Southeast Asia, where it made it possible to move from one to two or even three harvests of rice per year.

Source: Siebert et al., 2005; FAO/AQUASTAT, 2005.

production, income generation for the poorer, mostly rural, segments of the population and generation of foreign exchange to import the complement of food needed to satisfy domestic food requirements.

Agricultural commodity supply and demand are also changing (Schmidhuber, 2003). In addition to the required quantity, many factors in changing food demand and production patterns, sometimes cancelling out one another, influence agricultural production and the way inputs are managed. The distribution of bulk grain has become more reliable and 'just-in-time', allowing world reserves to be progressively reduced over the past decades from about four months to less than three months of global demand (FAO, 2005). Food commodities are produced, conditioned, refrigerated and transported over increasing distances at the cost of energy and environmental degradation.

Meat demand has been shifting towards poultry, and the world is now consuming more poultry meat than bovine meat. Given that poultry has a much better conversion rate of cereals into meat (two to one) than cattle

(between five and seven to one), this shift releases some of the pressure projected on the cereal sector and water demand for the irrigated cereal production.

As diets diversify and become healthier and better balanced, the demand for fresh vegetables and fruits increases. These goods are produced under intensive farming methods, including the use of greenhouses and irrigation for timely year-round production following exacting specifications. The controlled agro-ecological environment under which vegetables and fruits are produced also allows for accurate water control with minimum wastage. However, this form of agriculture is only possible under full control of water, which should be available on demand and in good quality. Many irrigation systems are not equipped with the necessary storage, conveyance and control systems and do not have the capacity to deliver water under these stringent conditions.

In countries and regions facing serious water scarcity problems, the gap between demand and production of food will grow, forcing them to rely increasingly on importing food to satisfy domestic needs

Fishermen, Turkey

In arid areas, return flow from agriculture itself and multiple reuses of water lead to progressive degradation in water quality

Urbanization

Growing urban centres have a strong impact on nearby rural economies. Newly urbanized people tend to change their food consumption habits with a preference for foods that are easier and less time-consuming to prepare (for example preferring rice to millet). Urban markets are larger and more diverse than rural markets and often lead rural-urban migrants to take advantage of opportunities in informal activities such as producing, processing and trading food products. In addition to creating urban livelihoods, the rural-urban links established through urbanization open markets for rural products and thus also improve rural livelihoods. However, cities are rapidly increasing their claim on water, which is often satisfied at the expense of nearby rural areas. Furthermore, urban centres represent a source of water pollution that has profound impacts on downstream agriculture and aquatic ecosystems.

Growing urbanization also has significant impacts on food markets. In recent decades, a handful of transnational corporations have gained increasing control over trade, processing and sales of food, and the thirty largest supermarket chains now account for one-third of food sales worldwide (FAO, 2004a). This progression has been particularly fast during the last decade in South America and East Asia. Smallholders face difficulties in meeting the requirements imposed by supermarkets that increasingly prefer to contract limited numbers of suppliers. They often need to invest substantially in irrigation, greenhouses and storage to meet the standards for supply quality and reliability.

Impact of climate change

The average temperature of the earth's surface has risen by 0.6 °C since the late 1800s. It is expected to increase by another 1.4 to 5.8 °C by the year 2100, and the sea level may rise from 9 to 88 cm during the same period (IPCC, 2001). Climate change is expected to have significant impact on agriculture and food production patterns through three major factors: global warming, change in rainfall patterns and the increase in carbon dioxide (CO_2) concentration in the atmosphere. A temperature increase of more than 2.5 °C could affect global food supply and contribute to higher food prices. The impact on crop yields would vary considerably from one region to another. Heat stress, shifting monsoons and drier conditions can reduce yields by as much as one-third in the tropics and subtropics. Dry continental areas, such as central Asia and the African Sahel, would be expected to experience drier and hotter climates, whereas longer growing seasons and increased

rains might boost productivity in temperate regions. Higher temperatures would also influence production patterns, shifting production ranges of specific crops towards the poles. A similar expansion of the range of pests increases the risk of crop losses.

In a scenario of moderate climate change (a temperature increase of less than 2 °C), gradual adaptation of cultivars and agricultural practices could occur with no expected major impact on food production in tropical areas. However, regional impact would vary widely, affecting the production capacity of some countries. Those most vulnerable to these changes are the poor and landless in rural areas dependent on isolated rainfed agricultural systems in semi-arid and arid regions. The changes in the hydrological cycle and rainfall patterns – more precipitation, more frequent intense rainfall events and more evaporation – would affect soil moisture and increase erosion (see also **Chapter 4**). In drought-prone areas, the number and duration of dry spells would be expected to increase, affecting crop production. It is generally admitted that higher levels of CO_2 in the atmosphere could stimulate photosynthesis and contribute to an increase in crop productivity. This is particularly true for C3 crops that include wheat, rice, soybeans, barley, cassava and potato, for which a CO_2 concentration increase of 50 percent leads to a 15 percent increase in potential production. At the same time, as most weeds are also C3 plants, they will also become more aggressive. C4 crops, which include several tropical crops like maize, sugar cane, sorghum and millet, as well as many pasture and forage grasses, are less responsive to higher levels of CO_2.

Increased water scarcity and competition for water

Historically, irrigation represents between 70 and 80 percent of all water uses, with some countries using 90 percent or more for irrigation. This percentage is changing as more and more countries face water shortages. It is estimated that over 1 billion people now live in countries and regions where there is insufficient water to meet food and other material needs. By 2030, over 60 percent of the population will live in urban areas (UN, 2004), claiming an increasing share of water abstraction.

Much of this water will have to come from agriculture; of all freshwater use sectors, agriculture in most cases shows the lowest return on water in economic terms. As the stress on water resources increases, competition grows between irrigation agencies fighting to retain their power and cities needing to satisfy the needs of their rapidly growing populations (Johnson III et al., 2002). Water stress and the pressing need to renegotiate intersectoral allocations are usually factors that force changes in the way water is managed in agriculture. Declining water quality adds to the stress on supply. In developing countries, water diverted to cities is often released after use without adequate treatment. In arid areas, return flow from agriculture itself and multiple reuses of water lead to a rapid degradation in quality. In particular, combined problems of water pollution and water scarcity can have disastrous effects on fish populations and habitats.

Above: Nomad women drawing water from a well during a sandstorm, Mauritania

Part 2. How Agriculture Can Respond to the Changing Nature of Demand for Water

A farmer watering crops using watering-cans, Thailand

In the global debate about increasing water scarcity, agriculture is often associated with the image of inefficient, wasteful water use. This image is conveyed by poor performance in terms of 'water use efficiency', a term that was defined as the ratio between the irrigation water absorbed by the plants and the amount of water actually withdrawn from its source for the purpose of irrigation. FAO has estimated that overall water use efficiency in irrigation ranges around 38 percent in developing countries and has projected only a minor increase in overall water use efficiency in the forthcoming decades (FAO, 2003a). The word 'efficiency', when its value is significantly below 100 percent, implies that water is being wasted. However, from a water balance perspective, water not taken up and transpired by the crop plants, even if unnecessarily withdrawn from its natural course, is not necessarily wasted. Unused water may be used further downstream in the irrigation scheme, it can flow back to the river or it can contribute to the recharge of aquifers. Renewable freshwater is only effectively 'lost' when it evaporates from the soil, is fatally polluted or when it joins a saltwater body.

2a. Raising water productivity in agriculture

This fact does not, by itself, deny justification of programmes aimed at increasing water use efficiency in irrigation; in most cases it is better for water to be left in its natural course rather than being extracted. The adoption of water-saving technologies and improved water management is justified in terms of better equity within irrigation schemes, higher reliability of water service, reduced waterlogging and reduced energy cost in cases when pumping is required, when it can be demonstrated that excess water is actually lost to the sea or to salty depressions, and when water withdrawal jeopardizes the sustainability of the ecosystems. In all cases, the overall implications in terms of water balance must be clearly understood (Seckler et al., 2003).

From water use efficiency to water productivity

Rather than water use efficiency, the concept of water productivity is now widely accepted as a measure of performance in agricultural water use. By definition, productivity represents the output of any production process expressed per unit of a given input, in this case water. In agriculture, several types of output can be considered. In a strict commodity production vision, the output is usually expressed in volumes or value of a given agricultural production. **Figure 7.6** shows water requirements and economic return for a selection of crops in Cyprus. However, productivity calculations are increasingly being extended to assess the water value of other outputs, including the social and environmental services provided by irrigation (Molden et al., 2003).

Figure 7.6: Water productivity of different crops, Cyprus

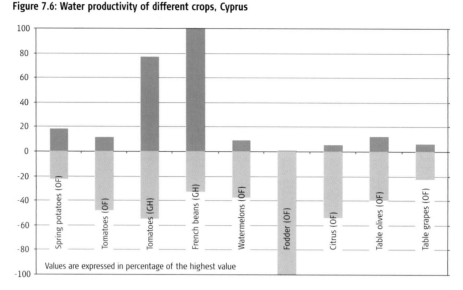

■ Economic return per m³ of water (Cy£)
■ Water requirements (m³/ha)

OF=On field
GH=Greenhouse

Note: The productivity of water in agriculture can be expressed in economic terms. This figure shows how water productivity changes from one crop to another in a given environment.

Values are expressed in percentage of the highest value

Source: FAO and WDD, 2002.

Section 3: CHALLENGES FOR WELL-BEING & DEVELOPMENT

In the last forty years, water productivity – expressed in kilograms of crop produced per cubit metre of water – has nearly doubled; that is, only half as much water is needed today than was needed to produce the same quantity of food in the 1960s (Renault, 2003). This remarkable improvement in crop water productivity is the result of increased yields for all major crops, thanks to breeding programmes that developed crop varieties with increasing harvest index, better physiological adaptation, and deeper rooting systems leading to better drought resistance. Water productivity of modern rice varieties, for example, is three times greater than the water productivity of traditional varieties, and progress in extending these achievements to other crops has been considerable (Bennett, 2003).

Boosting rainfed agriculture

Rainfed farming represents 82 percent of cropland and the bulk of the world's agricultural production. It is practised mostly in temperate climates and humid and sub-humid tropics. In rainfed farming, the outcome of farming operations depends on rainfall patterns and associated climatic phenomena. Both droughts and excess precipitation can lead to a partial or total loss of agricultural production. In the future, rainfed agriculture is projected to continue providing a large part of food production. There is still a large 'yield gap' to be closed, meaning that the production yields achieved under experimental station conditions are at present significantly larger than the average yields obtained in

practice. **Figure 7.7** shows that in India, for instance, long-term on-station experiments have indicated a steady gap between improved and traditional technologies, with a yield ratio of one to four in 2001 (Wani et al., 2003). Such a gap demonstrates that in most cases, production is more constrained by market opportunities and food prices than by production capacity.

Various cropping systems (combining appropriate soil fertility management and water conservation practices) contribute to boosting rainfed agriculture by reducing soil evaporation and increasing nutrient availability for plants during the growing season. Crop residues, or even specific crop cover management and mulching, also help to conserve moisture by limiting evaporation and runoff (Rockström et al., 2003). A long list of effective practices could be written, as farmers and researchers have developed and will continue to invent systems that help to mitigate the effects of drought at all scales and contribute to increasing water efficiency in rainfed conditions.

The potential of biotechnology

Biotechnology can be defined as any technological application that uses biological systems and living organisms to make or modify products or processes for a specific use. This definition covers many of the tools and techniques that are commonplace in agriculture and food production. While there is little controversy about many aspects of biotechnology, including traditional breeding, genetically modified organisms (GMOs) have become

Traditional irrigation system, Oman

Figure 7.7: Grain yield under improved and traditional technologies, 1977–2001, Andhra Pradesh, India

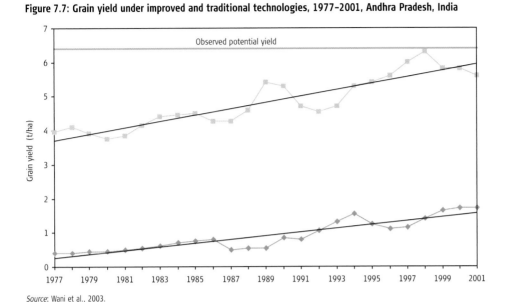

Source: Wani et al., 2003.

Note: Long-term monitoring experiments show that yields increase steadily, but that a significant gap remains between yields obtained by farmers and potential yields, even in rainfed areas.

the target of a very intensive and emotionally charged debate. The key difference between traditional breeding and modern genetic modification is that in the latter, genetically modified plants or animals are created using genes isolated from other living organisms, or from across the border between species and between animal and vegetal organisms. Current research focuses on the nutritional value, new products (including biodegradable plastics) and agronomic characteristics like salt and drought tolerance (FAO, 2002a; 2003c; 2004b).

The impact that biotechnology may have on water resources management is hard to assess. It is certain that yield increases of main crops will continue to enhance crop water productivity. Development of drought- and salinity-resistant crops could be relevant in the context of water scarcity; however, experts disagree on the possibility of achieving important progress with such crops in the near future: some success is being reported in tackling drought and salinity tolerance, but to date, there have been no big breakthroughs in developing such crops (FAO, 2003c). The relevance of GMOs for the food security of the rural poor is also a subject of controversy. Many of the currently available technologies generated by conventional research methods have yet to reach the poorest farmers' fields, and there is no guarantee that new biotechnologies will fare better. Identifying small farmers' constraints to technology access and use continues to be an issue that the development community must address (FAO, 2003a; 2004a).

Virtual water and food trade

The concept of virtual water is attracting attention in relation to the analysis of trade flows and increasing water scarcity (Allan, 2003; see also **Chapters 8, 11** and **12**). Producing goods and services generally requires water. The water used in the production process of an agricultural or industrial product is called the virtual water contained in the product. For producing 1 kg of wheat we need for instance 1 to 2 m^3 of water. Producing livestock products requires even more water: producing 1 kg of cheese requires about 5 m^3 of water, and it takes about 16 m^3 of water to produce 1 kg of beef (Hoekstra, 2003). The concept makes it clear that, in a reasonably safe, interdependent and prosperous world, a country with limited water resources could depend on the import of agricultural products showing high levels of embedded water (e.g. meat) and apply its own water resources to produce other commodities of lower value in terms of water content (see **Table 7.2**). Conversely, a country with abundant water resources could benefit from its comparative natural advantage by exporting products that are high in embedded water.

Food trade analysis shows that most trade takes place between countries that show substantial endowments in water resources, indicating clearly that factors other than water drive international food trade. Yet an increasing number of arid countries that face water scarcity (Egypt, Tunisia, etc.) are progressively embracing policies aimed at increasing their dependency on the import of staple

Farmers working on the construction of contour bunds, Myanmar

Table 7.2 Virtual water content of selected products

Product	Litres of water per kilo of crop
Wheat	1, 150
Rice	2, 656
Maize	450
Potatoes	160
Soybeans	2, 300
Beef	15, 977
Pork	5, 906
Poultry	2, 828
Eggs	4, 657
Milk	865
Cheese	5, 288

Note: Virtual water is the total amount of water used in the production and processing of a given product.

Source: Adapted from Hoekstra, 2003.

During the twentieth century, the world population increased threefold, while water used in agriculture through irrigation increased sixfold and some major rivers approached an advanced level of water depletion

crops and thus releasing water for more financially productive uses. Such policies usually imply long-term trade agreements between importing and exporting countries and therefore tend to facilitate increased stability in international relations.

2b. Improving irrigation

During the twentieth century, the world population increased threefold, while water used in agriculture through irrigation increased sixfold and some major rivers approached an advanced level of water depletion. The 'green revolution' was based on a technology package comprising components of improved high-yielding varieties of cereals, irrigation, improved soil moisture utilization and the application of plant nutrients, pest control, and associated management skills. The use of these technology packages on good land in suitable socio-economic environments resulted in increased crop yields and increased incomes for millions of farmers, particularly in Asia. Statistics indicate that yields of rice, wheat and maize approximately doubled between the 1960s and the 1990s. The green revolution has been a major achievement, and its effects are continuing, but the need for systematic use of irrigation, mineral fertilizers and agrochemical-based pest and weed control created environmental and health problems.

Achievements and failures of irrigation

The aim of large-scale irrigation projects was to drive regional and national development through the participation of significant segments of populations in direct and indirect project benefits. This socio-economic objective justified the implementation, at substantial public cost, of hydraulic infrastructure, including dams and canals, financed mostly by national governments, with support from international lending institutions. However, irrigation project performance problems started to emerge as early as the mid-1960s: not all the irrigation areas created were actually irrigated; crop yields were below projections; maintenance was substandard, and rehabilitation too frequently required; in some areas, soils started to become salinized; the return on investments was lower than expected; and the benefits to poor people were fewer than foreseen when calling for public funding (Mollinga and Bolding, 2004).

Understanding of the causes of poor performances in irrigation has improved. Design faults, including missing or inadequate drainage infrastructure, were often observed and sometimes traced to the application of inappropriate design standards (i.e. calling on materials, equipment and skills that were not locally available). Today's economic context calls for changes in agricultural policies and practices that these schemes cannot easily accommodate. In the Indus River Basin, for example, large irrigation schemes were initially designed and built to spread and share water thinly and equitably in order to reach as many farmers as possible, with deliveries covering only part of water needs when the entire area is considered. These systems cannot accommodate current demands for crop diversification and intensification. Other dysfunctions can be traced to disregard for relevant socio-economic conditions, lack of consultations with stakeholders and target groups, and generally poor governance both at the level of the countries implementing the irrigating works and of the financing institutions and donors. In many cases, women were excluded from the benefits of irrigation, because, according to social traditions, they could not have access to land rights and/or would not be allocated water rights (see **Box 7.3**). Among typical governance problems in the irrigation sector are the capture of benefits, including the control of water by the most influential farmers at the expense of poorer smallholders, and control of irrigation systems by rent-seekers, usually well-connected with local decision-makers.

Institutional reforms in irrigation management

By the 1990s, the major agencies funding development were making their loans conditional on the adoption of reform packages that required a balanced fiscal budget, a reduced role for the state and a larger role for the private sector. These packages emphasized economic water pricing, financial autonomy for irrigation agencies and the devolution of management responsibilities to lower levels.

BOX 7.3: **THE ROLE OF WOMEN IN IRRIGATED FARMING IN SUB-SAHARAN AFRICA**

In most countries, access to irrigation water is mediated by race, social status and gender. In sub-Saharan Africa, a complex set of rights and obligations reflecting social and religious norms prevails within rural communities and dictates the division of labour between men and women farmers. Irrigation projects have often been implemented without considering existing social and cultural practices like the gendered division of labour and responsibilities.

In Burkina Faso, a case study showed that overall productivity increased when women and men were allocated small separate plots rather than larger household plots. Women proved to be good irrigation managers and preferred to work on their own plots. As they became economically less dependent upon their husbands, they were able to help support their relatives and increase their own opportunities for individual accumulation of wealth in the form of

livestock. The effects of having an individual plot also significantly improved the bargaining position of women within households.

Sources: FAO, 2002a; Rathgeber, 2003.

Among these reforms, irrigation management transfer (IMT) appears to be the most systematic and far-reaching effort so far. The philosophy behind IMT lies in the perception that increased ownership, representation and active participation of farmers in the operation and maintenance of irrigation systems would be more effective than publicly run systems and would create an incentive for farmers to be more responsible towards their common obligations. IMT is based on the principle of subsidiarity, which holds that no responsibility should be located at a higher level than necessary.

Several approaches were developed to reform irrigation institutions, often in combination with each other. Decentralization, devolution, privatization and the development of public-private partnerships for irrigation management are all possible elements of institutional reform packages being implemented at various levels in over fifty countries, including Australia, India, Mexico, the Philippines and Turkey. All of them imply substantial changes in the way water governance is being practised. **Figure 7.8** schematically presents the implications of these institutional packages in terms of ownership and management of water for a range of typical situations.

Cattle drinking from the riverbank, Ethiopia

Figure 7.8: Examples of institutional reforms and implications in terms of ownership and management

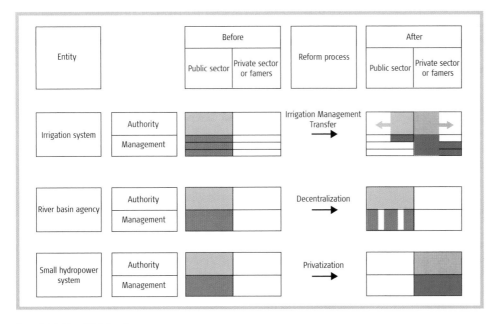

Note: Institutional reforms in irrigation management can take many forms. In most cases, they imply changes in the ownership, responsibility and authority over different parts of the irrigation system, land, water and infrastructures. The figure shows a sample of possible changes, and the new roles of governments, farmers and the private sector (including farmers' associations).

Source: Adapted from FAO, forthcoming.

BOX 7.4: MALI, OFFICE DU NIGER: THE SUCCESSFUL REFORM OF AN IRRIGATION MONOPOLY

In the 1960s, the Office du Niger, a state enterprise, managed 45,000 hectares (ha) of rice fields in the inland delta of the Niger River in Mali, producing over half of Mali's rice. The agency had a monopoly on rice marketing, and farmers were obliged to sell their paddy crops to the Office at a low price. The agency staff not only operated the irrigation network but also supplied inputs, provided extension, stored the paddy, operated the rice mills, ran a seed farm, produced farm equipment, managed guest houses and restaurants, staffed health clinics and ran literacy courses. The vertical integration of the agency was useful initially, but gradually

the central control it exercised became a source of inefficiency and lack of accountability. As rice yields declined to 2 metric tons per hectare (t/ha) and cropping intensity sank to 60 percent, national rice imports grew. Eventually, farmers refused to grow rice, and the project survived on fund infusions by the government. In turn, the government solicited support from donors for financial and technical assistance.

In 1993, a presidential delegate was appointed to draw up and implement a reform plan. Rice mills were sold and the seed farm, guest houses, equipment workshop and training centre were all

leased out, while the health clinics were transferred to the respective ministry and literacy and extension services were delivered for a fee. By 1996 the Office had been turned into a public enterprise with financial and administrative autonomy. Farmers were involved through joint committees established in each zone of the irrigation scheme. Paddy yields climbed to 6.5 tons/ha in places and cropping intensity climbed to 115 percent, contributing to improving the financial sustainability of the scheme (see also **Chapter 14**).

Source: Johnson III et al., 2002.

...reforms have often tended to disregard the need for funding the provision of public goods such as education, health and water services, which are all fundamental for development...

Successful institutional reforms require strong political backing and a willingness to shift responsibilities for delivering efficient and equitable irrigation services from government agencies to new, more representative institutions (see **Box 7.4**). Progress is slow, and reforms have often tended to disregard the need for funding the provision of public goods such as education, health and water services, which are all fundamental for development, poverty reduction and nutritional improvements in rural areas. Modalities for large-scale irrigation governance continue to be a subject of debate, and covering social, economic and environmental aspects proves difficult to attain (Mollinga and Bolding, 2004).

Modernization of large irrigation systems
Modernization of an irrigation system can be defined as the act of upgrading or improving the capacity of the system to respond appropriately to current water service demands, keeping in perspective future needs. The process involves institutional, organizational and technological changes. It implies changes at all operational levels of irrigation schemes, from water supply and conveyance to the farm level. Improvements in canal operation is generally a critical step in the process (Facon, 2005).

Modernization implies that a large part of the authority on irrigation management be transferred from government institutions to farmers and that farmers be in a position to decide on the level of service they want and are willing to pay for. The term, 'modernization', refers, therefore, not only to the rehabilitation, upgrading or transformation of physical

infrastructure in irrigation systems, but also to innovation or transformation in how irrigation systems are operated and managed. The concept is highly relevant to most of the large irrigation schemes in Asia, where rapid economic development poses new challenges to local agricultures.

Management reform of irrigation systems requires adequate and sustainable financing for ongoing operation and maintenance and, when needed, for the rehabilitation and upgrading of the infrastructure, including water gates and automated controls, which are a precondition for accurate and flexible water distribution. It also often includes demand management approaches in order to encourage efficient water allocation.

Water pricing in irrigation serves several purposes. Water charges refer to the collection of fees from water users, with the prime purpose of covering the cost of operation and maintenance, and, sometimes, to recover part of the investment costs. Water pricing, instead, has been advocated as an effective tool for reducing wastage and achieving more productive water use. Experience shows that such economic incentives alone have little chance for success and need to come as part of a comprehensive package of policies aimed at enhancing the productivity of irrigation systems: water distribution rules, effective local institutions and technological choices are essential complements to any attempt at conserving water in response to higher prices.

The roles of private and public sectors in financing irrigation

Figure 7.9 illustrates the decline in irrigation investment since the 1970s. This decline has been attributed to a conjunction of reasons, including poor technical, economic and social performance of large-scale irrigation systems and the increasing cost of irrigation development, as the best sites are already taken, and the price of food commodities is at an historic low. Public sector investment in irrigation now needs to be more strategic than in the past and explicitly geared towards growth and poverty-reduction outcomes. In this respect, the evaluation of the impact of irrigation cannot be reduced to increases in yields, outputs or the economic rate of return alone, but has to measure the impact of each marginal dollar of investment on poverty reduction (Lipton et al., 2003). Indeed, the most recent figures tend to show a renewed interest of major developing banks in agriculture and rural development as a response to the need to reach the Millennium Development Goals (MDGs) in view of the fact that the majority of the poor and hungry people live in rural areas. The recent report of the Commission for Africa (2005), the Report of the UN Millennium Project Task Force on Hunger (2004) and several recent commitments taken in the framework of the African Union and the New Partnership for Africa's Development (NEPAD) all call for more investments in water control in rural areas as an effective contribution to poverty eradication and reduction of malnutrition, particularly in sub-Saharan Africa.

Government agencies tend to systematically underestimate the role of the private sector in irrigation investment. Even in public irrigation schemes, farmers usually provide up to 50 percent of investment in irrigation development. Private sector investment aims to produce maximum revenue for the investor. As such, it is typically dynamic, responds well to market opportunities, is economically effective, and usually provides a substantial contribution to rural economies. It does not, however, take into account considerations of social equity and so does not directly target poverty reduction. In the future, public investment and policies must better recognize and support private initiatives in agricultural water management by providing appropriate incentives and an environment that favours investments by the private sector, including small farmers. In many places, land tenure reform and stable water rights are paramount to the involvement of the private sector in irrigation. The role of the public sector is also to ensure that private sector development benefits the largest possible number of rural people and is performed in a way that guarantees long-term environmental sustainability.

Irrigation technology: Moving towards more precision agriculture

Where agriculture moves away from subsistence farming

Computer-controlled irrigation in a salad factory, Germany

In many places, land tenure reform and stable water rights are paramount to the involvement of the private sector in irrigation

Figure 7.9: World Bank lending in irrigation, 1960–2005

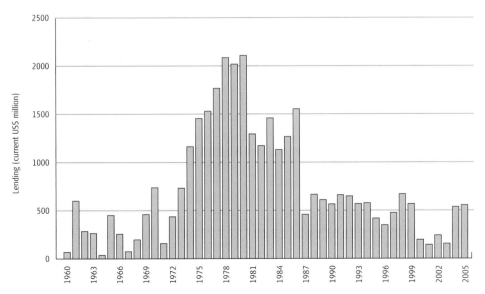

Source: based on World Bank data.

Note: World Bank lending in irrigation and drainage in constant dollars peaked between the mid-1970s to the mid-1980s, followed by a sharp decline, as a result of falling prices of main agricultural commodities, increased cost of new irrigation investments and progressive reduction in water availability. The most recent data, however, suggest a renewed interest in rural development, including water management in agriculture.

and farmers progressively shift self-sufficient activities into more business-oriented activities, irrigation is increasingly used for precision agriculture. Precision agriculture is an agricultural concept relying on the existence of in-field variability. It seeks to tailor agricultural practices to suit local conditions. Precision agriculture is well adapted to markets that demand delivery on a precise schedule of products subject to stringent specifications. Precision agriculture calls for optimal control of water deliveries and is an ideal condition for the application of pressurized irrigation technologies (sprinkler and localized irrigation). Localized irrigation finds its most rewarding applications in

horticulture and fruit tree production. Combined with automatic fertilizer application, or 'fertigation', it usually guarantees high returns on investments, reliability in the control of soil moisture, and reduced labour costs. When it is well managed, it can ensure an irrigation efficiency of close to 100 percent, thus contributing to minimizing water loss. Precision agriculture has a bright future in serving an increasing number of consumers in urban areas, but its application to low-cost staple food and commodities, representing the largest share of irrigation, is unlikely to materialize.

Part 3. Towards More Sustainable Agriculture

Agriculture has significant impacts on environment and people's health and too often, pursuing a narrow development goal of increased productivity has lead to the breakdown of the resilience of natural ecosystems. The negative impacts of water management in agriculture are related to land and water uses, in particular through encroachment on natural ecosystems, water extraction, erosion, or loss of soil biodiversity. The drainage and return of flows from irrigation often result in unwanted impacts, including loss of water quality. Inappropriate agricultural practices like excessive application of pesticides and fertilizers have direct impacts on water quality and affect people's health; waterlogging and salinization is also the result of inadequate planning and management of irrigation and drainage in agriculture. Finding alternative ways to alleviate these impacts is therefore essential to maintain the integrity and productivity of the ecosystems on which it depends and to create the conditions for agriculture to contribute, in a sustainable way, to food security, poverty alleviation and economic growth.

3a. Water storage and the evolution of groundwater-based economies

Irrigation backed by water storage has been conceived to provide a high degree of water security to reduce agricultural risk and encourage farmers to invest. Water stored in reservoirs is a secure asset on which farmers can rely. Surface water storage in reservoirs, however, contends with several problems, including the cost and liability of the impounding structure; the loss of water through evaporation; reservoir and canal sedimentation; river fragmentation and changes in river regimes; and the destruction of local livelihoods and resources. In the past, these costs (externalities) tended to be systematically underestimated while the potential benefits of dams were overestimated. The World Commission on Dams (WCD, 2000) represented significant progress in stating that all costs (social, economic and environmental) must be assessed against potential benefits derived from the construction of reservoirs.

Groundwater, where it is available, is a highly reliable source of water and provides an attractive alternative to surface storage. In the last few decades, groundwater has played a central role in enabling the transformation of rural communities from low productivity subsistence agriculture to more intensive forms of production. In contrast to surface water systems based on reservoirs and canals, users pump the aquifer as and when needed, and because groundwater extraction occurs through a pressurized system instead of open channels, precision application, which can greatly increase yields, is also possible. Except in places where energy is subsidized, groundwater productivity tends to be good, as pumping costs provide an incentive to saving water. For farming populations, groundwater access enables increases in production while reducing agricultural risk, enabling them to invest in more intensive forms of production and improve their livelihoods. Consequently, groundwater has played a particularly large role in yield and production increase. Evidence also suggests that groundwater access often plays a role in reducing rural poverty (Moench, 2001). Groundwater-based rural development has been central to major productivity gains in agriculture and the improvement of rural livelihood. But groundwater-based rural economies also show signs of vulnerability as aquifers are depleted (see also **Chapter 4**).

Managing groundwater

Private irrigation often uses groundwater, and its individualistic nature makes groundwater extraction difficult to control, leading to risks of anarchic exploitation and unequal water access. As water levels drop, competition among users increases, progressively excluding poorer farmers who cannot afford the cost of competitive well deepening, whereas those farmers who develop groundwater early, and those who already possess diverse forms of social or other capital, often consolidate their economic advantage. Managing a groundwater body consists mainly in ensuring that the extraction of water by a large number of individual users is performed in a controlled manner. This is technically feasible, but in reality meets major legal, institutional and social obstacles. Conventional bureaucratic controls based on well-licensing procedures, water pumping quotas and policing entail large transaction costs and have generally proven ineffective. It is a cause for concern that many aquifers continue to be drawn from until declining discharges, growing salinity and increasing pumping costs announce groundwater depletion (Burke and Moench, 2000).

Few examples exist of successful groundwater management. In Guanajuato State, Mexico, an experiment involving users at the local level has led to the creation of Aquifer Management Councils that have resulted in reduced transaction costs and progressive changes in users' behaviour. The sustainability and replicability of such a model remains to be ascertained (see also **Chapter 14**).

3b. Environment and water quality

Water abstractions for agriculture and other purposes modify the water balance and reduce the quantity of water that flows its natural course. The impact on the aquatic environment ranges from negligible to deleterious and deadly in extreme cases. The return of contaminated water into natural water bodies, when exceeding the natural recovery capacity of these systems, further reduces the amount of freshwater of adequate quality available for various uses and for sustaining the aquatic environment. Agriculture is a major cause of river depletion in the regions of the world sustained by irrigation and is the main source of nitrate pollution of groundwater and surface water, as well as a principal source of ammonia pollution. It is also a major contributor to the phosphate pollution of lakes and waterways and to the release of methane and nitrous

oxide into the atmosphere. The improper use of pesticides has detrimental impacts on the environment, resources and human health. (see **Chapters 5** and **6**).

While there is no question that agriculture must reduce the impact of its negative externalities, there is also wide recognition that agriculture must not necessarily be considered in opposition to natural ecosystems: positive externalities generated by agriculture go beyond the strict economic systems of crop production. In the same way that humans have for millennia drawn their food from the environment, agricultural systems that have proved sustainable in the long term can go far in the preservation of ecosystems and their biodiversity, while enhancing rural livelihoods. The twentieth century has been a time of productivity, based on the application of agronomic practices that favour a limited number of strategic crops. In this aspect, much indigenous knowledge has been neglected and partly lost in the drive to always apply agricultural systems that are high in energy inputs and low in species and cultivar diversity. As conditions now exist to release the pressure on natural resources, the trend, in numerous developed countries, is to transform agriculture into a landscape management practice, offering new avenues for better integration of agriculture with its environment (see also **Chapter 13**).

In addition to producing food and other goods for farm families and markets, sustainable agriculture also contributes to a range of public goods, such as clean water, wildlife, management of living aquatic resources, carbon sequestration in soils, flood protection and landscape quality. Certain non-food functions of sustainable agriculture cannot be produced by other sectors, including on-farm biodiversity, groundwater recharge, or social cohesion. Thus, what many see as an almost unbearable challenge for the agriculture sector – internalizing externalities – might in fact also be seen as a major opportunity to promote sustainable development in rural areas (FAO and MAFF, 2003).

Salinity, a hazard of irrigation in arid zones

Irrigation development has caused numerous cases of soil and water salinization, which is mostly restricted to arid and semi-arid areas, where about 40 percent of the world's irrigated land is located and where the rate of evapotranspiration is high. By withdrawing water from rivers for application on land, irrigation tends to accelerate the rate of accumulation of salts on land through evaporation and increase its concentration in rivers. Salinization is also likely to become a problem on poorly

Fields near Quito, Ecuador. The plateaus of Quito benefit from the humid, gentle climate of the Sierra, which favours the cultivation of cereals and potatoes

Heavy monsoon rains submerge villages, roads and fields in India

Salinization seriously affects 20 to 30 percent of the area under irrigation in arid and semi-arid zones

drained soil, when the groundwater level is close to the surface. In such cases, water tends to rise from the water table to the surface by capillary action and then evaporate from the soil surface, leaving salts accumulating at the surface. In more humid regions, rainfall usually provides enough leaching to prevent harmful salt accumulation.

No exact assessments are available on the extent and severity of salinization, but Smedema and Shiati (2002) suggest that it seriously affects 20 to 30 million ha worldwide, that is, about 25 percent of the area under irrigation in arid and semi-arid zones and about 10 percent of all areas under irrigation. Most of this is a legacy of large-scale water works developed from the 1950s. The spread of salinization in these 'old' irrigation areas is now drastically reduced. Current global estimates of the rate of extension of salinization are in the order of 0.25 to 0.50 million ha per year.

To a large degree, irrigation-induced land and river salinization is inherent to the practice of irrigation in arid and semi-arid areas. The adverse impacts of salinization can, to some extent, be prevented and mitigated, but large-scale development in arid regions will always represent a salinity hazard that goes beyond the single irrigation scheme and amplifies as one travels downstream. Of particular concern are the major rivers that have their sources in the Himalayas and flow into desert areas in Pakistan and Central Asia. Preventive action includes planning irrigation development better, avoiding highly saline areas and establishing drainage infrastructure. Among the measures that can be applied are the application of river basin-level salt balance models to predict and monitor the incidence of salinization and the interception and disposal of highly saline runoffs. Salt control programmes have contributed to arresting river salinization. However, significant lowering of salt concentrations in rivers would generally require radical measures such as a substantial reduction in land under irrigation.

Recycling: Achieving an adequate urban–rural balance of wastewater use

Cities produce large quantities of solid and liquid waste that are disposed of and more or less treated back into the environment. If this process is not adequately considered in all its steps and consequences – and in developing countries it generally is not – the impact can be devastating to the environment and to people living close to disposal sites, causing the disruption of ecosystems and putting people at risk of poisoning (see **Chapter 3**). Liquid waste (the contents of sewers) is discharged into rivers and coastal zones, where it may overtax the recovery capacity of natural water bodies, leading to the establishment of new, less desirable ecological systems (i.e. anaerobic systems).

Agriculture around cities is generally dynamic and well connected to markets, making profitable use of water. Wastewater provides users with a stable source of water with a high nutrient content. However, the use of untreated wastewater in agriculture poses risks to human health. Governments have usually responded to such risks by implementing strict regulations limiting or preventing the use of wastewater and advocating treatment before use. Full treatment, however, can be expensive, and achieving an adequate urban–rural balance and distribution of charges and benefits continues to elude governance. While the uncontrolled use of wastewater cannot be encouraged, unconditional restriction is not a practical option, as wastewater is too valuable a resource for farmers who have no other alternatives.

A more pragmatic approach to wastewater use in agriculture is now emerging. It includes enhanced monitoring, health protection and education, and alternative agricultural practices. This approach is being adopted in the revision of the WHO guidelines for wastewater use in agriculture (see **Box 7.5**).

BOX 7.5: REVISED WORLD HEALTH ORGANIZATION (WHO) GUIDELINES FOR THE SAFE USE OF WASTEWATER IN AGRICULTURE

WHO first published its 'Guidelines for the safe use of wastewater and excreta in agriculture and aquaculture' in 1989. A revision of these guidelines is currently under preparation. The revised WHO guidelines will incorporate a risk-benefit approach, in which the assessment of tolerable risks takes place before the setting of

health targets. This framework allows more flexibility to countries in adapting what would be available and achievable in the context of local social, economic and environmental factors. In addition, the interaction between wastewater use and poverty in the political context and international development targets is mentioned

with expanded sections on risk analysis and management, revised microbial guideline values and further elaboration of chemical contaminants, including pharmaceuticals and endocrine disrupting substances, health-impact assessment, and wastewater use planning strategies at sub-national levels.

Tonle Sap, in central Cambodia, is the largest freshwater lake in Southeast Asia. The annually flooded area includes a ring of freshwater mangrove forest, shrubs, grassland and rice fields. The Tonle Sap is also a rich fishing ground, with an estimated catch of 250,000 metric tonnes (t) per year on average.

Traditionally, the people living around the lake in areas subject to flooding have cultivated rice varieties that could cope with the high water level by elongating their stems up to five metres, with a maximum growth of 10 cm per day. Where the flooding is not as deep, normal wet rice varieties are transplanted into the fields once the flood has reached them. In some areas, rice is planted in the fields as floodwaters recede.

Licensed fishing lots occupy the most productive areas in terms of fish catch. There is a tendency to underestimate the importance of rice field fisheries in the Tonle Sap, because they tend to yield only small amounts of fish at a time, but this provision of fish is available for many people on a regular basis (it is estimated that fish consumption around the lake averages about 60 to 70 kg per person per year).

The Tonle Sap is also an important source of biodiversity. A survey carried out in 2001 identified seventy different species of fish and other organisms captured in rice field ecosystems for consumption as food and for other purposes. They include several species of fish, snake, turtle, crab, shrimp and amphibians,

all of them tradable on local markets. In the vegetal world, besides rice, thirteen plant species were recorded, of which six were marketed.

In conclusion, the Tonle Sap ecosystem is of major importance to the local population, not only for the supply of rice, but also for animal protein and vegetables. Development that focuses only on increasing yields of rice through intensification and the use of agrochemicals may provide more rice to eat, but may also eliminate many aquatic animals and vegetables harvested from and around the rice fields.

Source: Adapted from Balzer et al., 2002.

Wetlands: *Fragile ecosystems, sources of livelihood*

Wetlands are fragile ecosystems and an important source of biodiversity, with complex hydrological and livelihood support functions, including regulation, silt retention, grazing land, hunting, fishing and wood production. In the past, the attractive characteristics of wetlands for agricultural production (particularly their fertility and soil moisture) have led planners to undervalue their environmental and socio-economic functions and promote their conversion into agricultural production. Conversion of wetlands into farmland, largely a matter of the past in developed countries, is still actively underway in regions with high demographic growth that suffer from food insecurity, as in sub-Saharan Africa. Not all wetlands can be preserved, and research is needed to identify critical wetlands of particular importance for biodiversity, so that a critical core of wetlands can be preserved. The Ramsar Convention, which initially focused on wetland conservation to ensure the survival of migrating bird species, now works with its partners to promote a wise use of wetlands in general, emphasizing the needs of local populations and the complex livelihood support functions of the wetlands. Resolution VIII.34 of the 8th Conference of the Contracting Parties (2002) focuses on the necessary interactions between agriculture, wetlands and water resources management (see **Box 7.6**).

3c. Water to combat hunger and poverty in rural areas

The projections of total food demand suggest that per capita food consumption will continue to grow significantly, and the world average will approach 3,000 kcal in 2015, compared to 2,800 kcal around 2000. The world will be producing enough food for everyone, but its distribution will continue to be unequal. In absolute figures, the number of undernourished people in the world has been stagnating since the early 1990s, and was estimated at 850 million in 2000–02, of which 815 million were in developing countries (see **Map 7.2**, **Figures 7.10** and **7.11**). Projections show a decline to 610 million in 2015, which is progress, but still distant from the 1996 World Food Summit target of 400 million in 2015 (FAO, 1997). Although the Millennium Development Goal of reducing by half the proportion of people living in extreme poverty and hunger by the year 2015 is well within grasp, at present, 15,000 children under the age of five die every day as a consequence of chronic hunger and malnutrition (see **Chapters 1** and **6**).

Chronic hunger is a reflection of extreme poverty, as those affected by hunger do not have the resources needed to produce or buy food. Hunger is not only a result of poverty but also contributes to poverty by lowering labour productivity, reducing resistance to disease and depressing educational achievements.

Irrigation in the Eastern Cape, South Africa

Figure 7.10: Proportion of undernourished people in selected developing countries, 2000-02

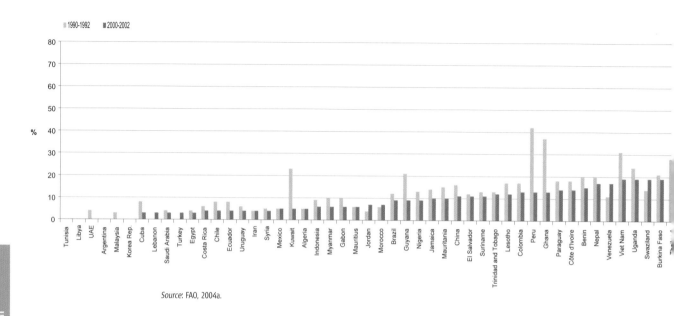

■ 1990-1992   ■ 2000-2002

Source: FAO, 2004a.

Map 7.2: Proportion of undernourished people in total population, 2000-02

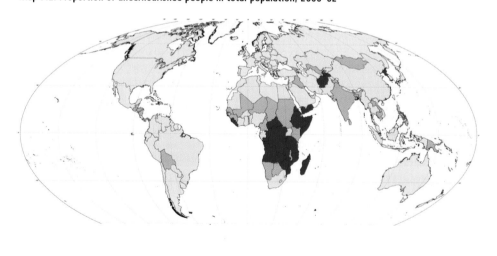

Note: While the world is progressively graduating out of poverty, sub-Saharan Africa remains plagued with high levels of undernutrition. Although less severe in relative terms, the situation in South Asia also deserves attention, as it holds the largest absolute number of undernourished people in the world.

Map Projection: Mollweide

No Data   < 5   5 - 20   20 - 35   35 - 100   **Percentage of total population undernourished**

Source: FAO, 2004a.

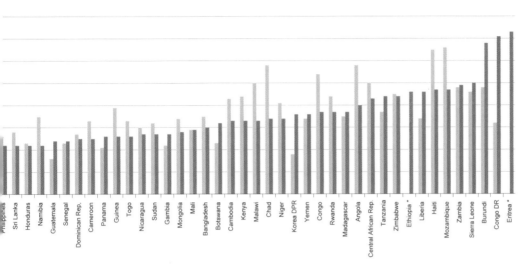

Note: This graph shows changes in the percentage of undernourished people over a ten-year period in ninety developing countries. It shows that relatively rapid changes can be obtained in food security when the right political decisions are taken. It also shows the negative impacts that civil unrest and wars can have on people's food security.

* Ethiopia and Eritrea: no separate data available for 1990–92; Afghanistan, Iraq, Papua New Guinea and Somalia: no data available for 2000–02.

Figure 7.11: Estimated and projected number of undernourished people by region, 1991–2030

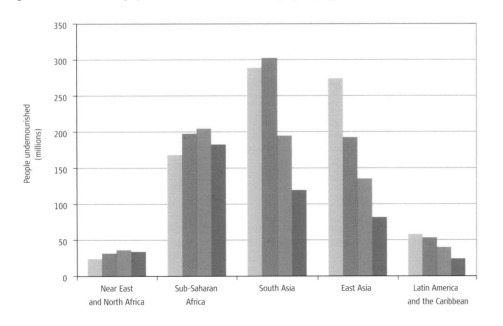

1990-92
1997-99
2015
2030

Note: Projections towards 2030 indicate a reduction in the total number of undernourished people in all developing regions, but increases are still expected between now and 2015 in the Near East and sub-Saharan Africa. Asia is expected to see the highest drop in the number of undernourished people as a result of steady economic growth.

Source: FAO, 2003a.

Access is a pivotal concept in the development of livelihoods, and is allied with the notion of entitlement. Poverty results from the failure to express such entitlements...

Decades of international concern about an ethically unacceptable global level of extreme poverty and hunger, and national and international policies and governance formulated in this spirit, have been insufficient to transform the livelihoods of the poor for the better. Forty years ago, there was hope that the green revolution, with its new high-yielding varieties of rice, wheat and maize, would bring world hunger to an end by increasing food supply. The green revolution boosted food production and, in relative terms, the global situation has improved. Nevertheless, universal food security has not been achieved, and the absolute number of chronically hungry people remains high. So what options are now available to eradicate hunger and poverty, and what role does water play in this endeavour?

Water in livelihood as a pathway out of poverty

The livelihoods approach to understanding and tackling poverty, its causes and consequences, is gaining momentum across the developing world and among the development partners. For water and food, it means a fundamental shift beyond considering water as a resource for increasing food production to focusing on people and the role that water plays in their livelihood strategies. This puts people at the centre of development and means that issues such as drought and secured access to water become problem-led

rather than discipline-led, leading to a focus on institutional and political barriers to water access and on physical infrastructure needed for its management.

At the heart of the livelihoods approach are the 'capital' assets of households, a particularly relevant approach in rural areas. These include not only natural assets, such as land and water, but also social, human, physical and financial assets, presenting a comprehensive view of the basis of livelihood, as opposed to the more classical approach that tends to address single issues separately (see **Table 7.3**). Within a sustainable livelihoods approach, water is treated as an economic good and as an asset that can be invested to generate benefits and income. To consider only the role of water in agricultural production is not sufficient; in a livelihoods approach, it is necessary to understand the impacts of improved water supplies on the socio-economic livelihood circumstances of households. The challenge for the future is to introduce this systemic approach in places where the majority of organizations and the professionals working in them are still driven by a sectoral approach (see **Chapter 12**).

Irrigation is a direct source of livelihood for hundreds of millions of the rural poor in developing countries because of the food, income options and indirect benefits it

Table 7.3: Shifting towards a livelihood-based approach in rural areas

Capital	Issue	Production-based approach	Livelihood-based approach
Physical capital	Infrastructure for rainfed and irrigation systems	Rainfed and irrigation farming systems improved to increase agricultural production.	Improves decision-making ability through better rainfed and irrigation farming systems. Removes risk and uncertainty including maintenance and management of natural capital stocks.
Social capital	Community approach needed to raising or managing other forms of capital, of crucial importance in irrigation management, water user associations (WUA), networks	Communities mobilized to establish water user associations (WUA) to improve agricultural water management.	Identifies poorest households and strengthens participation in, and influence on, community management systems; creates safety-nets within communities to ensure the poor have access to water; improves rights to land and water and establishes right to access by poor households within communities.
Natural capital	Land and water availability	Develops new and enhances existing water resources using physical and social assets.	Enhanced through training in catchment protection and maintaining natural environment.
Financial capital	Cash, credit, savings, animals	Develops individual or community-based tariffs and charges mechanisms for water use.	Secured through access to small-scale credit.
Human capital	Labour, knowledge (through education, experience)	Trains people in agricultural water management and promotes gender equity.	Knowledge of demand, responsive approaches, community self-assessment of needs, participatory monitoring, gender mainstreaming.

Source: Adapted from Nicol, 2000.

- Employment and income for landowners and the landless who benefit from new employment opportunities.

- Increased production options throughout the year, used for both home consumption and sale.

- Health improvements through access to safe domestic water supply and sanitation.

- Attraction of immigration and improved provision of services, such as education.

- Widening of social networks through participation in water committees.

- Boost to local economy and family welfare.

Source: Adapted from Vincent, 2001; Meinzen-Dick and Bakker, 1999; Zwarteveen 1996.

generates (Vincent, 2001). The anti-poverty effects of irrigation can be assessed on two levels: (1) production, related to the national or regional economy, and (2) livelihoods, related to the household and its well-being. The former has been the method traditionally used to assess irrigation impacts; conversely, a livelihoods approach to irrigation places adequate and secure livelihood aims before increased production. Negative impacts of irrigation systems and provision on livelihoods, such as water-borne infections, waterlogging and salinity, increases in land prices and in women's unpaid workload, displacement and disparity of benefits between inequity in irrigation water distribution, are outweighed in most cases by benefits (see **Box 7.7**).

The balanced achievement of these livelihoods benefits, without a disproportionate gap between those who lose and those who gain in irrigation processes, can only be reached if access to water or to the benefits it generates to third parties is secured by the poor and landless. Access is a pivotal concept in the development of livelihoods, and is allied with the notion of entitlement. Poverty results from the failure to express such entitlements, not from the lack of individual rights to the commodities at stake.

Despite difficulties in participatory irrigation management processes, a beneficial by-product has been the strengthening of social capital, increasingly accounted for in livelihood improvements (see **Box 7.8**). Increasing the positive impacts of water in supporting and enhancing the livelihood of the poor has three implications:

- recognizing the role and importance of water in non-agricultural uses in single purpose management systems and identifying complementarities among uses (Meinzen-Dick, 1997)

- supporting people's capacity to manage their water in a fair and sustainable manner (Vincent, 2001)

- engaging a policy move from production and health to sustainable livelihoods in water assessments (Nicol, 2000), that is, moving from supply-led to demand-responsive policies that take requirements and claims of user groups into account and make more efficient and equitable use of existing supplies (Winpenny, 1997).

In Sudan, the livelihood of more than 67,000 poor farming families is to be improved on a large irrigation scheme in the Gash watershed in the arid eastern part of the country. Set up in the 1920s in order to settle nomadic people, the project fell into decline in the 1970s. The management was fragmented and nepotistic, and farmers complained about its ineffectiveness in meeting their needs for social and economic development. Soon, production shifted to low-return subsistence crops, farmers stopped paying irrigation fees and the system fell into disrepair.

Traditional approaches to rehabilitating such projects usually focus on infrastructure repairs, with little room for adaptation. In the Gash Sustainable Livelihood Regeneration Project (GSLRP) (2004–12), the improvement of farmers' livelihood was selected as the first objective. Capacity development and institutional reforms have been designed to ensure that all stakeholders are involved in the decision-making process. This is seen as critical to the success of the project. New organizations are being set up to ensure that citizens gain more secure rights to

land and water by building on existing local community organizations. Efforts are underway to counter the strong tradition of supply-driven irrigation management, where farmers are tenants, and empower farmers to take on more management responsibility. This puts household livelihood, rather than infrastructure development, firmly at the core of future investments.

Source: IFAD, 2003.

In many parts of the world, poor farmers rank soil erosion and lack of soil fertility among the main constraints to improving crop yields – both of which are linked to water management. Technical solutions have long been available, yet the problems and the solutions do not appear to be connected and so the rate of adoption of good practices remains poor. The links between research, extension and poor farmers need to be strengthened in new and innovative ways that encourage two-way communication with farmers (IFAD, 2001).

Water mobilization targeted towards benefiting the poor can effectively contribute to reducing extreme poverty and hunger. Safe water supply improves personal health, the primary physical capital, thus facilitating the undertaking of gainful activities. Water availability sustains the natural ecosystems on which the livelihoods of the rural poor largely depend. Irrigation can reduce the risk of droughts and increase cropping intensities by 'extending' the wet season in the humid and tropical zones. Introducing irrigation technology can reduce household risks by raising incomes. Groundwater access often plays a particularly important role in reducing rural poverty. However, while irrigation is one of the success stories of the twentieth century, providing significant increases in food production, its poverty-reduction impact is not a foregone conclusion. Inequality in access to land and water resources, such as in southern Africa and Latin America, tends to exacerbate social inequities. If not properly managed, resources tend to end up in the hands of large influential farmers, thereby leaving almost none for small poor farmers to control (Lipton et al., 2003). Access to financial capital is also important; most often poor farmers have neither the money to invest in irrigation nor the collateral, such as land tenure rights, to obtain credit.

In the future, a purely sectoral approach to water management will no longer be possible...

Part 4. Governance Matters at All Levels in Agriculture

Agriculture requires that large quantities of water be taken up by crops from the soil in the root zone. The production of meat requires substantially more water, and fish production needs large quantities of clean water in ponds, rivers and estuaries. Globally, irrigated agriculture claims close to 70 percent of all freshwater withdrawn from its natural course, but this represents only about 10 percent of water used by agriculture – rainfall that replenishes soil moisture provides the larger part. However, irrigation has a strategic role in agriculture. Depending on various circumstances, irrigation helps to produce two to three times as much per hectare than non-irrigated agriculture. It is of crucial importance in boosting agricultural productivity and limiting horizontal expansion of cropland.

However, agriculture is now coming under much more scrutiny as competition for water between sectors increases. Degraded land and water systems, competition from other economic sectors and the need to conserve the integrity of aquatic ecosystems are progressively limiting water availability to agriculture and imposing cleaner production methods. In the future, a purely sectoral approach to water management will no longer be possible, and substantial adaptations of agricultural policies will be necessary to align production with overall river basins and aquifer management objectives.

As competition increases, irrigated agriculture will need to be systematically examined to discern where society can most effectively benefit from its application. Access to natural resources needs to be negotiated with other users in a transparent fashion in order to achieve optimal allocation and uses under conditions of growing demand for water.

The modernization of irrigated agriculture, through technological upgrading and institutional reform, will be essential in ensuring much-needed gains in water productivity. Irrigation institutions will have to respond to the needs of farmers, ensuring flexible and reliable delivery of water, increasing transparency in its management and balancing efficiency and equity in access to water. This will not only require changes in attitudes, but also well-targeted investments in infrastructure modernization, institutional restructuring and upgrading of the technical capacities of farmers and water managers.

Agriculture is under pressure to reduce its negative impact on the environment and other sectors, particularly when associated with the use of fertilizers and pesticides, as well as wasteful water use. However, there is currently a much wider recognition that better agricultural water management can also have a profoundly positive impact, reaching far beyond the strict economic system of crop production. Farmers are at the centre of any ecological process of change. They need to be encouraged and enabled, through appropriate incentives and governance practices, to conserve natural ecosystems and their biodiversity and minimize the negative impacts of agricultural production, a goal that will only be achieved if the appropriate policies are in place.

Farmers around the world are deeply affected by economic factors out of their control. Historically, governments in developing countries have tended to neglect agricultural development in favour of industrialization and national and urban activities. However, it is now more generally acknowledged that agriculture is the main engine of growth in many developing economies. Thirty countries, most of them in Africa, are highly dependent on agriculture, and progress in improving their food security situation depends on, more than any other factor, the development of local food production. In most cases, there is a need for substantial increase in investment in rural areas, where water management plays a central role in raising the productivity of agriculture and related rural activities.

At the same time, targeted policies are needed to address the causes of chronic hunger and poverty. To be effective, such policies need to focus on people and develop the assets they control. Titles to land and secure and equitable access to water and basic rural services (education, finance, etc.) are also needed if rural populations are to emerge from marginalization and integrate their farming activity in their region's economy.

The agriculture sector faces complex challenges: producing more food of better quality, while using less water per unit of output; providing rural people with resources and opportunities to live healthy and productive lives; applying clean technologies that ensure environmental sustainability; and contributing in a productive way to the local and national economy. Continuing 'business as usual' is unlikely to deliver the Millennium Development Goals on the path towards freeing humanity of extreme poverty and hunger and ensuring environmental sustainability. Action is needed now to adapt agricultural and rural development policies, accelerate changes in irrigation governance and, through adequate water laws and institutions, support the integration of the social, economic and environmental needs of rural populations.

Ultimately, the reduction of rural hunger and poverty depends on the decisions and actions of the farming community in developing countries – 500 million farm households. Their potential contribution will not be fully realized in the absence of a socio-economic environment that encourages, supports and protects their aspirations, ideas and initiatives.

A class of children from Shanghai, China, drew their vision of their environment for the 'Scroll around the world' project

Workers harvest carp raised in a stock pond, India

References and Websites

Allan, J. A. 2003. Virtual water – the water, food and trade nexus: useful concept or misleading metaphor? *Water International*, Vol. 28, pp. 4–11.

Balzer, T., Balzer, P. and Pon, S. 2002. Kampong Thom Province, Kingdom of Cambodia. M. Halwart, D. Bartley, and H. Guttman (eds), *Traditional Use and Availability of Aquatic Biodiversity in Rice-based Ecosystems*. CD-ROM, Rome, FAO.

Barker, R. and Molle, F. 2004. Evolution of Irrigation in South and Southeast Asia. *Comprehensive Assessment Research Report 5*. Colombo, Sri Lanka, International Water Management Institute.

Bennett, J. 2003. Opportunities for increasing water productivity of CGIAR crops through plant breeding and molecular biology. J. W. Kijne, R. Barker and D. Molden (eds), *Water Productivity in Agriculture: Limits and Opportunities for Improvement*. Wallingford, UK, CABI Publishing and IWMI.

Burke, J. and Moench, M. H. 2000. *Groundwater and Society: Resources, Tensions and Opportunities*. New York, UN DESA and ISET.

Commission for Africa. 2005. *Our Common Interest*. Report of the Commission for Africa. www.commissionforafrica.org

Dixon, J., Gulliver, A. and Gibbon, D. 2001. *Farming Systems and Poverty: Improving Farmers' Livelihoods in a Changing World*. Rome/Washington DC, FAO/World Bank.

Facon, T. 2005. Asian irrigation in transition – service orientation, institutional aspects and design/operation/infrastructure issues. G. Shivakoti, D. Vermillion, W. F. Lam, E. Ostrom, U. Pradhan and R. Yoder (eds), *Asian Irrigation in Transition: Responding to Challenges*. London, Sage Publications Ltd.

FAO (Food and Agriculture Organization of the United Nations). Forthcoming. *Irrigation Management Transfer: Worldwide Efforts and Results*. Rome, FAO.

——. 2005. *FAO Food Outlook*. Quarterly Report No. 1, April 2005, Global information and early warning system on food and agriculture (GIEWS). Rome, FAO.

——. 2004a. *The State of Food Insecurity in the World 2004*. Rome, FAO.

——. 2004b. *The State of Food and Agriculture 2003–2004: Agricultural Biotechnology: Meeting the Needs of the Poor?* Rome, FAO.

——. 2003a. *World Agriculture Towards 2015/2030: An FAO perspective*. Rome/London, FAO/Earthscan Publishers.

——. 2003b. International scientific symposium on measurement and assessment of food deprivation and undernutrition. Summary of Proceedings. 26–28 June 2002. Rome, FAO.

——. 2003c. *Report of the FAO Expert Consultation on Environmental Effects of Genetically Modified Crops*. 16–18 June 2003, Rome, FAO.——. 2003d. *Review of the State of World Fishery Resources: Inland Fisheries*. FAO Fisheries Circular, No. 942, Rev. 1, Rome, FAO.

——. 2002a. *The State of Food and Agriculture 2002*. Rome, FAO.

——. 2002b. *Crops and Drops: Making the Best Use of Water for Agriculture*. FAO, Rome.

——. 1999. Global issues and directions in inland fisheries. *Review of the State of World Fishery Resources: Inland Fisheries*. FAO Fisheries Circular. No. 942, Rev. 1. Rome, FAO.

——. 1998. Integrating fisheries and agriculture to enhance fish production and food security. *The State of Food and Agriculture, 1998*, No. 31, pp. 85–99. Rome, FAO Agriculture Series.

——. 1997. Report of the World Food Summit, 13–17 November 1996, Part 1. Rome, FAO.

FAO and MAFF (Ministry of Agriculture, Forestry and Fisheries of Japan). 2003. Issue paper for the Ministerial meeting on Water for food and agriculture. Rome, FAO and MAFF.

FAO and WDD (World Development Department of Cyprus). 2002. Reassessment of the water resources and demand of the island of Cyprus. Synthesis report. Rome/Nicosia, FAO/WDD.

FAO-IPTRID (International Programme for Technology and research in Irrigation and Drainage). 2003. The irrigation challenge: Increasing irrigation contribution to food security through higher water productivity from canal irrigation systems. Issue Paper 4. Rome, FAO and IPTRID.

Faurès, J. M., Hoogeveen, J. and Bruinsma, J. 2003. *The FAO Irrigated Area Forecast for 2030*. Rome, FAO www.fao.org/ag/agl/aglw/aquastat/reports/index.htm

Hoekstra, A. Y. (ed.) 2003. Virtual water trade, proceedings of the international expert meeting on virtual water trade, *Value of Water Research Report* No. 12. Delft, the Netherlands, IHE.

IFAD (International Fund for Agricultural Development). 2003. Republic of the Sudan, Gash sustainable livelihoods regeneration project, project document. Rome, IFAD.

——. 2001. *Rural Poverty Report 2001: The Challenge of Ending Rural Poverty*. Oxford, IFAD.

IPCC (International Panel on Climate Change). 2001. *Third Assessment Report – Climate Change 2001: Synthesis Report*. Geneva, IPCC.

IWMI (International Water Management Institute). 2003. *Confronting the Reality of Wastewater Use in Agriculture*. Water Policy Briefing No. 9. Colombo, Sri Lanka, IWMI.

Johnson III, S., Svendsen, M. and Gonzalez, F. 2002. Options for institutional reform in the irrigation sector. International Seminar on Participatory Irrigation Management, Beijing.

Kijne, J. W., Barker, R. and Molden, D. (eds). 2003. *Water Productivity in Agriculture: Limits and Opportunities for Improvement*. Wallingford, UK, CABI Publishing.

Lipton, M., Litchfield, J. and Faurès, J. M. 2003. The effects of irrigation on poverty: a framework for analysis. *Water Policy*, Vol. 5, No. 5/6, pp. 413–27.

Mason, J. B. 2002. Measuring hunger and malnutrition. Measurement and assessment of food deprivation and undernutrition. Proceedings of an international scientific symposium convened by FAO, 26–28 June, Rome.

Meinzen-Dick, R. 1997. Valuing the multiple uses of irrigation water. M. Kay, T. Frank and L. Smith (eds), *Water: Economics, Management and Demand*. London, E. & F. N. Spon.

Meinzen-Dick, R. and Bakker, M. 1999. Irrigation systems as multiple-use commons: Water use in Kirindi Oya, Sri Lanka. *Agriculture and Human Values*, No. 16, pp. 281–93.

Moench, M. 2001. Groundwater: Potential and Constraints. *2020 Vision Focus (Overcoming Water Scarcity and Quality Constraints)*, No. 9. Washington, DC, IFPRI.

Molden, D., Murray-Rust, H., Sakthivadivel R. and Makin, I. 2003. A water productivity framework for understanding and action. J. W. Kijne, R. Barker, and D. Molden (eds), *Water Productivity in Agriculture: Limits and Opportunities for Improvement*. Wallingford, UK, CABI Publishing and IWMI.

Mollinga, P. P. and Bolding, A. 2004. *The Politics of Irrigation Reform: Contested Policy Formulation and Implementation and Implementation in Asia, Africa and Latin America*. Aldershot, UK, Ashgate Publishing.

Nicol, A. 2000. Adopting a sustainable livelihoods approach to water projects: Implications for policy and practice. Working Paper 133, London, Overseas Development Institute.

Rathgeber, E. 2003. *Dry taps... Gender and Poverty in Water Resources Management*. Rome, FAO.

Renault, D. 2003. Value of virtual water in food: principles and virtues. A.Y. Hoekstra (ed.), *Virtual Water Trade, Proceedings of the International Expert Meeting on Virtual Water Trade*. Delft, the Netherlands, UNESCO-IHE.

Ringersma, J., Batjes, N. and Dent, D. 2003. *Green Water: Definitions and Data for Assessment*. Wageningen, the Netherlands, ISRIC.

Rockström, J. 1999. On-farm green water estimates as a tool for increased food production in water scarce regions. *Phys. Chem. Earth B*, Vol. 24, No. 4, pp. 375–83.

Rockström, J., Barron, J. and Fox, P. 2003. Water productivity in rain-fed agriculture: Challenges and opportunities for smallholder farmers in drought-prone tropical agro-ecosystems. J. W. Kijne, R. Barker and D. Molden (eds), *Water Productivity in Agriculture: Limits and Opportunities for Improvement*. Wallingford, UK, CABI Publishing.

Schmidhuber, J., 2003. The outlook for long-term changes in food consumption patterns: Concerns and policy options. Paper prepared for the FAO Scientific Workshop on Globalization of the Food System: Impacts on Food Security and Nutrition, 8–10 October 2003, Rome, FAO.

Section 3: CHALLENGES FOR WELL-BEING & DEVELOPMENT

Seckler, D., Molden, D. and Sakthivadivel, R. 2003. The concept of efficiency in water-resources management and policy. J. W. Kijne, R. Barker and D. Molden (eds), *Water Productivity in Agriculture: Limits and Opportunities for Improvement*. Wallingford, UK, CABI Publishing.

Shiklomanov, I. 2000. Appraisal and assessment of world water resources. *Water, International*, Vol. 25, No. 1, pp. 11–32, March 2000. IWRA.

Siebert, S., Döll, P., Feick, S. and Hoogeveen, J. 2005. *Global Map of Irrigated Areas*. Version 3.0, interactive map. Frankfurt/Rome, Johann Wolfgang Goethe University and FAO.

Smedema, L. K. and Shiati, K. 2002. Irrigation and salinity: A perspective review of the salinity hazards of irrigation development in the arid zone. *Irrigation and Drainage Systems*, Vol. 16, No. 2, pp. 161–74.

UN (United Nations). 2004. *World Population Monitoring 2003: Population, Education and Development*. Department of Economic and Social Affairs, Population Division. New York, United Nations.

UN (United Nations) Millennium Project Task Force on Hunger. 2004. Halving hunger by 2015: A framework for action. Interim report. Millennium project. New York, United Nations.

Vincent, L. 2001. Water and rural livelihoods. R. Meinzen-Dick and M. W. Rosegrant (eds), *2020 Vision Focus 9 (Overcoming Water Scarcity and Quality Constraints)*. Brief 5. Washington, DC, International Food Policy Research Institute.

Wani, S. P., Pathak, P., Sreedevi, T. K., Singh, H. P. and Singh, P. 2003. Efficient management of rainwater for increased crop productivity and groundwater recharge in Asia. J. W. Kijne, R. Barker and D. Molden (eds), *Water*

Productivity in Agriculture: Limits and Opportunities for Improvement. Wallingford, UK, CABI Publishing.

Winpenny, J. T. 1997. Demand management for efficient and equitable use. M. Kay, T. Frank and L. Smith (eds), *Water: Economics, Management and Demand*. London, E. & F.N. Spon.

WCD (World Commission on Dams). 2000. *Dams and Development, A New Framework for Decision-making: The Report of the World Commission on Dams*. London and Sterling, VA, Earthscan Publications Ltd.

Zwarteveen, M. Z. 1996. *A Plot of One's Own: Gender Relations and Irrigated Land Allocation Policies in Burkina Faso*. Washington, DC, The Consultative Group on International Agricultural Research (CGIAR).

CGIAR (Consultative Group on International Agriculture Research) – Challenge Program on Water and Food:
www.waterforfood.org/
One of the greatest challenges of our time is to provide food and environmental security. The CGIAR Challenge Program on Water and Food approaches this challenge from a research perspective.

Comprehensive Assessment of Water Management in Agriculture (CA): www.iwmi.cgiar.org/Assessment/Index.asp
A multi-partner assessment process hosted by International Water Management Institute (IWMI). The CA Synthesis Report will be released in August 2006. It will examine trends, conditions, challenges and responses in water management for agriculture in order to identify the most appropriate investments for enhancing food and environmental security over the next fifty years.

FAO-AQUASTAT: www.fao.org/ag/aquastat/
Global information system of water and agriculture. Provides users with comprehensive information on the state of agricultural water management across the world, with emphasis on developing countries and countries in transition (statistics, country profiles, maps and GIS).

FAO-FAOSTAT: faostat.external.fao.org/
Online multilingual database containing over 3 million time-series records covering international statistics in the areas of food production, prices, trade, land use, irrigation, forests, fisheries, etc.

FAO – Global Perspective Studies: www.fao.org/es/ESD/gstudies.htm
Includes the report, *World Agriculture: Towards 2015/2030*, which is FAO's latest assessment of the long-term outlook for the world's food supplies, nutrition and agriculture.

FAO – The State of World Fisheries and Aquaculture (SOFIA): www.fao.org/sof/sofia/index_en.htm
Published every two years with the purpose of providing policy-makers, civil society and those who derive their livelihood from the sector with a comprehensive, objective and global view of capture fisheries and aquaculture, including associated policy issues.

FAO – The State of Food Insecurity (SOFI): www.fao.org/sof/sofi/index_en.htm
Reports annually on global and national efforts to reduce by half the number of undernourished people in the world by the year 2015.

ICID (International Commission on Irrigation and Drainage): www.icid.org/
ICID is a non-profit organization dedicated to enhancing the worldwide supply of food and fibre by improving the productivity of irrigated and drained lands through the appropriate management of water and environment and the application of irrigation, drainage and flood management techniques.

IFAD (International Fund for Agricultural Development) – Rural Poverty: www.ifad.org/poverty/
In its *Rural Poverty Report 2001, The Challenge of Ending Rural Poverty*, the International Fund for Agricultural Development argues that, to be successful, poverty-reduction policies must focus on rural areas.

We are no longer able to think of ourselves as a species tossed about by larger forces – now we __are__ those larger forces.

Bill McKibben, *The End of Nature*

CHAPTER 8

Water and Industry

By

UNIDO
(United Nations Industrial Development Organization)

*Aerial view of the disposal of mine wastes into a water
body, Ishpeming, Michigan, US*

Top to bottom:
Industrial site in Grangemouth, Scotland

Dockside construction site in the US

Water treated on site at a rubber factory, Malaysia

Key messages:

For the majority of the world's population, a thriving economy and improvement in the quality of life are closely linked to better access to consumer goods. Growing local industries create much-needed jobs, so people have more disposable income to spend on manufactured products. This often comes at the cost of increasing volumes of dumped solid waste, deteriorating water quality, and increased air pollution, when industry discharges untreated wastes onto land and into water and air. However, the linkage between industry and pollution is not inevitable. The purpose of this chapter is to show that manufacturing activities can be both clean and profitable. Indeed, industry can lead the way in pricing water at its true value and conserving high-quality water resources. Governance has an important role to play in creating the conditions that promote healthy and sustainable industrial growth.

■ Industry is a significant engine of growth providing 48 percent of gross domestic product (GDP) in East Asia/Pacific, 26 percent of GDP in lower-income countries and 29 percent of GDP in higher-income countries, although this last figure is declining.

■ Much industrial activity in middle- and lower-income countries is accompanied by unnecessarily high levels of water consumption and water pollution.

■ Worldwide, the total rate of water withdrawals by industry is slowing, whereas the rate of water consumed is steadily increasing.

■ It is possible to decouple industrial development from environmental degradation, to radically reduce natural resource and energy consumption and, at the same time, to have clean and profitable industries.

■ A very wide range of regulatory instruments, voluntary initiatives, training and advice is available to help industrial managers improve water-use productivity and to reduce polluting emissions to very low levels. At the same time, these tools can aid production efficiency, reduce raw material consumption, facilitate recovery of valuable materials and permit a big expansion of re-use/recycling.

The total water withdrawal from surface water and groundwater by industry is usually much greater than the amount of water that is actually consumed

Part 1. Industry in an Economic Context

Industry is the engine of growth and socio-economic development in many developing countries. In the fast-growing East Asia and Pacific region, industry now provides 48 percent of the total gross domestic product (GDP), and this proportion is still increasing. In heavily indebted poor countries, the proportion of GDP provided by industry grew quickly from 22 percent to 26 percent between 1998 and 2002. In rich countries, by contrast, the proportion of GDP coming from the production of manufactured goods is slowly declining, currently providing some 29 percent of GDP, with services making up the bulk of the economy. Overall however, industrial production continues to grow worldwide, as economies grow (World Bank, 2003).

1a. Water use by industry

Water is used by industry in a myriad of ways: for cleaning, heating and cooling; for generating steam; for transporting dissolved substances or particulates; as a raw material; as a solvent; and as a constituent part of the product itself (e.g. in the beverage industry). The water that evaporates in the process must also be considered in accurate assessments as well as the water that remains in the product, by-products, and the solid wastes generated along the way. The balance is discharged after use as wastewater or effluent. The total water *withdrawal* from surface water and groundwater by industry is usually much greater than the amount of water that is actually consumed, as illustrated by the graphs in **Figures 8.1** and **8.2**. Industrial water use tends to be measured in terms of water withdrawal, not water consumption.

Following major growth between 1960 and 1980, water withdrawal for use by industry worldwide has pretty much stabilized. Industrial water withdrawal in Europe has actually been dropping since 1980, although industrial output continues to expand. In Asia, the growth in industrial water withdrawal was rapid up to 1990, and has since been growing much more slowly, despite the region's high growth in manufacturing output. As shown by these figures, the intensity of water use in industry is increasing in these regions, as is the value added by industry per unit of water used (see **Table 8.4** at end of chapter).

Once more information becomes available on environmental water flow requirements in many rivers and rainfed agriculture, a fuller picture may be presented of the allocation of water among all its various uses. It will also be necessary to analyse actual water use in terms of consumption by the various sectors (see **Figure 8.3**). The return flows from the different sectors to surface water and groundwater must be accurately depicted, with the inclusion of water reuse cycles (and water reclamation, see discussion below). Only then can a realistic water balance be prepared for a given river basin or country.

1b. Negative industrial impacts on the water environment

Frequently of greater concern than the actual volume of water used by industry is the negative impact of industry on the water environment. Water quality is deteriorating in many rivers worldwide, and the marine environment is also being affected by industrial pollution. How does this take place? Much of the water used by industry is usually disposed of 'to drain'. This can mean one of the following things:

■ direct disposal into a stream, canal or river, or to sea

■ disposal to sewer (which may be discharged, untreated, further downstream, or may be routed to the nearest municipal sewage treatment plant)

■ treatment by an on-site wastewater treatment plant, before being discharged to a watercourse or sewer treatment in a series of open ponds.

There are many instances of water reclamation (treating or processing wastewater to make it reusable), where industrial effluent is not returned immediately to the natural water cycle after use. It can be recycled or reused directly on-site, either before or after treatment. The water may also be treated and then reused by other industries nearby or agricultural or municipal users, as well as for cropland irrigation or local parks and gardens. All these possibilities for water reclamation and reuse are dependent on the quality of the discharge and are discussed in more detail in Part 3. Reclaimed water that has been treated can also help to conserve the water environment by being injected to replenish underground aquifers or prevent salt-water intrusion or by being discharged into a drought-stricken wetland.

Of major concern are the situations in which the industrial discharge is returned directly into the water cycle without adequate treatment. If the water is contaminated with

Figure 8.1: Trends in industrial water use by region, 1950-2000

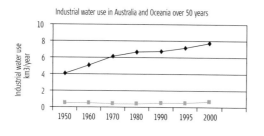

Industrial water use in Australia and Oceania over 50 years

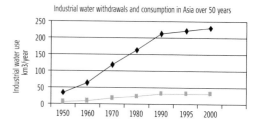

Industrial water withdrawals and consumption in Asia over 50 years

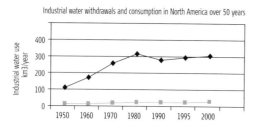

Industrial water withdrawals and consumption in North America over 50 years

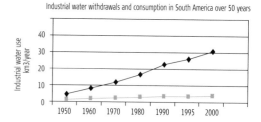

Industrial water withdrawals and consumption in South America over 50 years

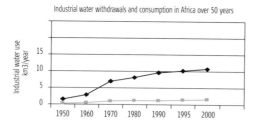

Industrial water withdrawals and consumption in Africa over 50 years

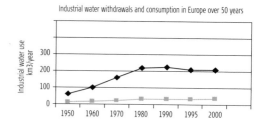

Industrial water withdrawals and consumption in Europe over 50 years

Note: Vertical scale varies among the graphs. Industrial water withdrawals in Africa and South America are still rising, albeit off a very low base. In Asia, North America and Europe, industrial water use accounts for the bulk of the global figure for industrial water withdrawals. Note that industrial water consumption is everywhere much lower than the volume of water withdrawn.

Source: Shiklomanov, 2000.

Legend:
◆ Withdrawals
■ Consumption

Figure 8.2: Total world industrial water use, 1950-2000

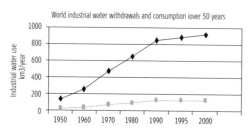

World industrial water withdrawals and consumption iover 50 years

Legend:
◆ Withdrawals
■ Consumption

Source: Shiklomanov, 2000.

Figure 8.3: Water use by industry vs. domestic use and agriculture

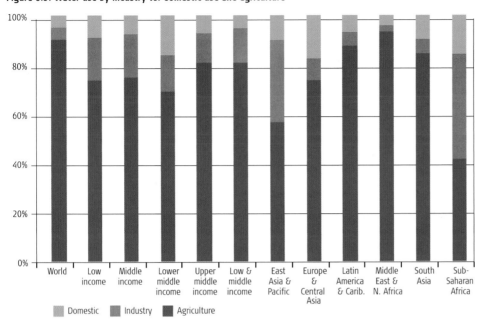

Domestic ■ Industry ■ Agriculture

Above: An Ijaw child shows off the oil that has damaged the communal forest around her village in the Delta region of Nigeria. The fish catch has dropped dramatically as a result of oil spillage from a nearby extraction pipe

Below: Wastewater from bleaching cotton in a mill, Ndola, Zambia

Note: There is increasing competition for water among the various water use sectors in many river basins. If we compare water use (i.e. water withdrawal) by industry to water use by other sectors, namely agriculture and domestic use, it is clear that globally, industry uses only a fraction of the amount of water used by agriculture. However, in East Asia and the Pacific, industrial water use has grown to a significant proportion of total use, in line with its significance to the economies of those countries. In sub-Saharan Africa, although overall water use is low, the water used by industry is a larger proportion of the total, because more agriculture is rainfed, rather than irrigated. These data exclude rainfed agriculture from the calculations of water use, and do not include environmental flow requirements as a water use category. In many catchment areas and river basins, environmental needs have not yet been calculated (see **Chapter 5**).

Source: World Bank, 2002.

heavy metals, chemicals or particulates, or loaded with organic matter, this obviously affects the quality of the receiving water body or aquifer. The sediments downstream from the industrial discharge can also be contaminated. Water that has a high organic content (called the biochemical oxygen demand, or BOD) often appears cloudy or foamy, and is characterized by the rapid growth of algae, bacteria and slime (see **Figures 8.4** and **8.5** and **Chapter 5**). The growth of these organisms depletes the level of oxygen in the water. It is more difficult for fish, insects, amphibians and many species of aquatic plants to live and breed in such oxygen-depleted water. If the water discharged is still hot, this 'thermal pollution' may also affect the aquatic ecosystems downstream, which have to adjust to a temperature that is higher than normal (see also **Chapter 9**).

A much larger volume of water may actually be affected than the volume of the industrial discharge itself. Industries and water quality regulators in some places still

rely on the so-called 'dilution effect' to disperse contaminants within the water environment to the point where they fall below harmful levels. In areas where industries are growing fast and more industrial plants are coming on-stream with many newly created discharge points, this approach can quickly result in polluted rivers and reservoirs. The toxicity levels and lack of oxygen in the water can damage or completely destroy the aquatic ecosystems downstream as well as lakes and dams, ultimately affecting riverine estuaries and marine coastal environments. In international river basins, routine pollution and polluting incidents such as industrial accidents and spillages may have transboundary effects. Significant pollution sources in river basins, such as large industrial plants, may be termed 'hot spots' and prioritized for clean-up within a river basin management plan (see **Box 8.1**).

It is important to consider not only the level or concentration of individual substances, but also their

combined effect. It is very expensive to monitor water quality for the presence of numerous chemicals, each of which must be tested for separately. By monitoring the populations of certain organisms, called indicator organisms (such as frogs, molluscs or certain insect species), it is possible to create a picture of how the water body is being affected over time. These eco-toxicological methods provide a more cost-effective way of assessing the impact of industrial discharges and are discussed in more detail in **Chapter 5** on ecosystems.

Direct human health impacts can result if the industrial discharge is located upstream of:

■ a recreational bathing and swimming area or commercial, recreational or subsistence fishing grounds

■ a point where farmers withdraw water in order to irrigate their crops

■ a point where a municipality withdraws water for domestic use

■ a point where people without a formal water supply withdraw water for drinking.

Many municipalities now find that the quality of the drinking water which they supply is compromised by industrial pollution. This raises water treatment costs for the water supply utility. Where the problem is variable freshwater quality, caused by irregular effluent discharges, the water treatment plant may not be able to cope adequately with the contaminants. In such cases the health of local people may be affected in the longer term, depending on the concentration and type of substances involved.

Two additional ways in which industries may more indirectly affect the water environment are through the following:

BOX 8.1: IDENTIFICATION, ASSESSMENT AND PRIORITIZATION OF POLLUTION 'HOT SPOTS'

The methodology for evaluating hot spots was developed within the framework of the Global Environment Facility (GEF) regional project preparing a Strategic Action Plan for the transboundary Dnieper River Basin, including areas of the three countries involved, namely Russia, Ukraine and Belarus. The objectives of the Strategic Action Plan are to facilitate the reduction of pollution in the river basin, and ultimately to contribute to the protection of the Black Sea.

As in many river basins in populated areas, there are thousands of pollution sources in the Dnieper River Basin. The Hot Spot methodology identifies, assesses and prioritizes the most significant sources of pollution, based on their impacts and characteristics. These include point sources, such as industrial and municipal effluents, and non-point sources, such as agricultural and urban runoff. Each contributes to human health risk and environmental degradation, including significant impacts to environmentally sensitive areas where biodiversity is threatened.

A multi-stage screening system, developed by the United Nations Industrial Development Organization (UNIDO), is used to identify priority hot spots. They are evaluated according to pollution control issues, water quality issues and biodiversity issues, as well as economic and employment criteria. Point sources of pollution are scored on a number of criteria under each of these general headings, which are then weighted according to their significance, before a total score is assigned. Non-point sources and areas that are difficult to characterize quantitatively (e.g. abandoned military facilities, or large tailings ponds) may still rank as hot spots but are described qualitatively, based upon the professional judgement of national experts. Finally, for a small group of priority hot spots, mitigation measures are proposed together with an estimation of implementation costs and a cost-benefit analysis.

An example of water quality issues scored includes the following criteria:

■ location of nearest municipal drinking water withdrawal downstream
■ influence of river quality on the nearest municipal drinking water withdrawal point
■ population being supplied by river water within 25 kilometres downstream of the hot spot
■ recreational bathing areas located near the hot spot
■ other aquatic recreational activities near the hot spot
■ any illnesses attributed to the recreational areas
■ hot spot directly identified as the source of illnesses
■ proximity of recreational fishing areas and sustainability
■ proximity of commercial fishing areas and sustainability
■ agricultural water utilization in proximity to the hot spot
■ sediment quality
■ proximity to national boundaries.

Source: UNIDO, 2003.

■ The leaching of chemicals from solid wastes: The solid
wastes generated by industrial activity may contain a
quantity of contaminated water or other liquids, which
gradually seep out once the waste is disposed of. In the
rain (or in groundwater, if the waste is buried in a
landfill), further chemicals may be leached or mobilized
from the solid waste over time. This leachate eventually
reaches a stream or an aquifer. Industrial dumpsites and
municipal landfill sites, if not adequately constructed,
are frequently found to generate such 'leachate plumes'
that can be significant pollution hot spots.

■ The atmospheric deposition of chemicals distributed
through air and rain pollution: Some industries emit
significant quantities of sulphur and nitrogen
compounds (SOx and NOx) into the atmosphere. These
may dissolve into raindrops, and fall as acid rain. Many
streams, rivers and lakes in Europe are more acid than
they would naturally be, due to this process. Other
compounds such as dioxins and furans may also be
released into the atmosphere from furnaces, and
thereby enter the water cycle.

More detailed information on industrial pollution in Europe
is available since the introduction of the European
Pollutant Emissions Register. All factories in the European
Union over a certain size are required to report their
emissions. **Figure 8.5** shows figures on total organic carbon
releases to water (a more accurate measure than the BOD
data available). **Table 8.1** shows the amount of benzene,
toluene, ethylbenzenes and xylenes being released annually
to the water environment, both directly and indirectly.
These toxic hydrocarbons are emitted by a range of
industries, from oil refineries to pharmaceutical plants.

1c. Natech disasters

Natech disasters are a new disaster category, identified
by the UN International Strategy on Disaster Reduction
(UN-ISDR) as a technological disaster triggered by a
natural hazard (see **Box 8.2**). In Europe, for instance,
there are many vulnerable installations close to rivers or
in earthquake-prone regions, which are vulnerable to
flooding or strong tremors (see **Chapter 10**).

For example, the magnitude of an earthquake in Turkey
that measured 7.0 on the Richter Scale in August 1999
triggered unprecedented multiple and simultaneous
hazardous materials releases, wreaking havoc on relief
operations assisting earthquake victims. In one incident,
the leakage of 6.5 million kilograms (kg) of toxic
acrylonitrile (ACN) contaminated air, soil and water,

**Figure 8.4: Industry shares of biological oxygen demand (BOD), by industrial sector
and in selected countries**

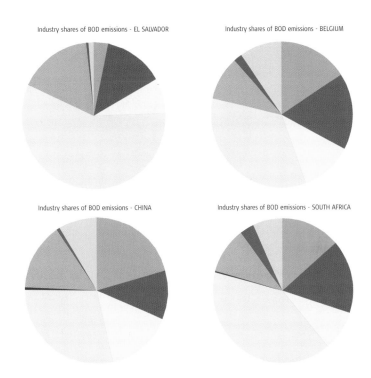

Primary metals
Paper and pulp
Chemicals
Food and beverages
Stone, ceramics and glass
Textiles
Wood
Other

Note: This figure shows the industry shares of organic pollution emissions
(using BOD as an indicator), by industrial sector in selected countries. The
data on BOD is the only pollution data available, and it is not very accurate
as it is calculated indirectly from employment data in the various industrial
sectors.

In less developed economies, such as in El Salvador, the food and beverage
industry generates the majority of the organically loaded effluent. In
developed countries such as Belgium, where the economy is more
diversified, effluent from the food and beverage industry is still significant,
but there is a wider spread of other contributing sectors. In China, the
primary metals sector, chemicals and textiles contribute the lion's share. In
all countries, the pulp and paper sector can be a significant polluter of the
aquatic environment, if untreated effluent is released.

Source: World Bank, 2002.

threatening residential areas. Automatic foam sprayers
were available at the industrial facility at the time of the
earthquake, which would normally have contained the
ACN release, but these were rendered useless due to a
lack of water and power. These technological disasters
posed additional health and psychological problems to an
already devastated population. Hence current industrial
risk management regulations should be carefully revised to
ensure that this kind of 'natech' risk is being addressed.

Figure 8.5: Release of total organic carbon (TOC) directly or indirectly to water in thirteen EU Member States, 2003

■ Industrial plants for pulp from timber or other fibrous materials and paper or board production (>20t/d) – 240,771,200.00kg

■ Installations for the disposal of nonhazard waste (>50t/d) and landfills (>10t/d) – 123,851,400.00kg

Basic organic chemicals – 41,957,238.80k

Others – 30,943.80kg

■ Slaughterhouses (>50t/d), plants for the production of milk (>200t/d), other anim raw materials (>75t/d) or vegetable raw materials (>300t/d) – 18,628,850.00kg

Source: EC, 2004.

Table 8.1: Release of benzene, toluene, ethylbenzene and xylenes directly or indirectly to water in eight EU Member States, 2003

Activity releasing benzene, toluene, ethylbenzene and xylenes	Directly to water (kg/year)	Indirectly to water (kg/year)
Combustion installations (> 50 MW)	967	2,830
Mineral oil and gas refineries	67,486	880
Coke ovens	390	–
Coal gasification and liquefaction plants	1,020	–
Metal industry and metal ore roasting or sintering installations	16,080	8,080
Basic organic chemicals	40,328	127,158
Basic inorganic chemicals or fertilisers	57,996	–
Biocides and explosives	6,170	365
Pharmaceutical products	1,282	7,550
Installations for the disposal or recovery of hazardous waste (>10 tons/day)	2,300	2,136
Plants for the pre-treatment of fibres or textiles (>10 tons/day)	–	707
Installations for surface treatment or products using organic solvents (>200 tons/year)	–	3,773
Total	**194,019**	**153,479**

Source: EC, 2004.

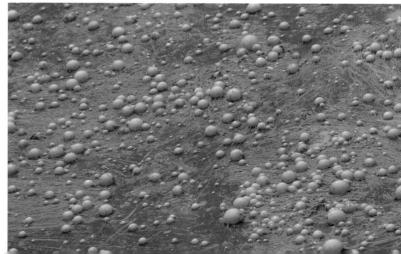

Section 3: CHALLENGES FOR WELL-BEING & DEVELOPMENT

BOX 8.2: INDUSTRIAL DISASTERS AROUND THE WORLD

The Tisza tailings dam disaster

On 30 January 2000, a breach in a tailings dam released some 100,000 cubic metres (m^3) of cyanide-rich tailings waste into the river system near Baia Mare in northwest Romania (see also **Chapter 14**). This spill released an estimated 50 to 100 tonnes (t) of cyanide, as well as heavy metals, into the Somes, Tisza and finally the Danube Rivers before reaching the Black Sea. Ice on the rivers and low water levels in Hungary delayed the dilution of the cyanide, increasing the risk to municipal water supplies. High concentrations of copper, zinc and lead, leached by the cyanide, compounded the problem. Impacts included:

■ contamination and interruption of the drinking water in twenty-four locations, affecting 2.5 million people
■ massive fish kill and destruction of aquatic species in the river systems
■ severe negative impact on socio-economic conditions of the local population
■ long-term reduction of revenue from tourism and canoeing
■ drop in real estate prices.

The Red Rhine Incident

In 1986 a fire destroyed a chemical store in Basel, Switzerland, near the borders of France and Germany. Chemicals reached the water in the Rhine River through the plant's sewage system when huge amounts of water (10,000-15,000 m^3) were used to fight the fire. The store contained large quantities of thirty-two different chemicals, including insecticides and raw ingredients, and the water implications were identified through the presence of red dye in one of the substances, which turned the river red. The main wave of chemicals destroyed eels, fish and insects, as well as habitats for small animals on the riverbanks. The total eel population was destroyed for 500 kilometres (km) downstream, from Basel in Switzerland down to Loreley in Germany. It took three months after the incident for the contaminant concentrations to drop to normal values. As a result of new regulations and precautions put in place following this incident, the permanent chemical load in the Rhine has been reduced, and information systems on potential incidents improved.

Arsenic contamination from mines in Thailand

Past mining activities caused heavy arsenic contamination of groundwater and topsoil over a 40 km^3 area in Nakhon Si Thammarat province, Thailand (see **Chapter 14**). The contamination was revealed in a study commissioned by the Japan International Cooperation Agency (JICA) in 2000. One conclusion of the study was that the contamination would last for the next thirty to fifty years. Testing of 1,000 samples showed arsenic contamination in some groundwater wells to be 50 to 100 times higher than the World Health Organization's guideline value for drinking water (0.01 milligrams per litre). Most people in the affected district stopped drinking well water in 1993, after the local health effects were found, and are now paying a very high cost for tap water.

Martin County, Kentucky, United States

On 11 October 2000, a coal tailings dam failed after the collapse of an underground mine beneath the slurry impoundment, and 950,000 m^3 of coal waste slurry was released into local streams. About 120 km of rivers and streams turned an iridescent black, causing a fish kill along the Tug Fork of the Big Sandy River and some of its tributaries. Towns along the Tug were forced to turn off their drinking water intakes.

Nandan County, Guangxi Province, China

Also in October 2000, after a tailings dam failure, at least 15 people were reported killed and 100 missing. More than 100 houses were destroyed by the tailing wave downstream.

Sebastião das Águas Claras, Nova Lima District, Minas Gerais, Brazil

On 22 June 2001, the failure of an iron mine waste dam caused a tailings wave to travel at least 6 km. Five mine workers died in the incident.

San Marcelino, Zambales, Philippines

In August 2002, at the Dizon Copper Silver Mines, after heavy rain, the overflow and spillway failure of two abandoned tailings dams caused some tailings to spill into Mapanuepe Lake and eventually into the St. Tomas River. By 11 September, several low-lying villages had been flooded with mine waste: 250 families were evacuated; no injuries were reported.

Source: WHO, 2004.

Part 2. Governance Issues and Sustainable Development in Industry Regulation

Environmental governance is central to ensuring that growing industries undertake an acceptable and affordable level of pollution control and environmental management. This section discusses a number of international conventions and multilateral environmental agreements (MEAs) that exist in order to regulate industries, and in particular those dealing with hazardous and toxic chemicals (see Table 8.2). It is also important to note that voluntary systems go a long way towards demonstrating that industries can be both clean and profitable. Various voluntary approaches have been developed over time and are discussed later in this chapter.

The 'polluter pays principle', or the 3Ps, was first widely discussed at the 1992 United Nations Conference on Environment and Development held in Rio de Janeiro, Brazil (Rio Principle 16). The principle was endorsed by the attending country representatives, and was also adopted by the Conference of the Parties of the Basel Convention among others (see section 2a below). Under the Basel Convention, the 3Ps states that the potential polluter must act to prevent pollution, and that those who cause pollution must pay for remedying the consequences of that pollution. With the development of the concept of Cleaner Production (see section 3b below), the 3Ps now stand for Pollution Prevention Pays: in other words, good environmental management need not be simply an extra cost for a company to bear, but can actually improve production processes, save money and resources, and make the company more efficient, more profitable and more competitive in the global marketplace.

The World Summit on Sustainable Development (WSSD), held in Johannesburg, South Africa, in August 2002 proposed a Plan of Implementation, which makes a strong link between the related goals of industrial development, poverty eradication and sustainable natural resource management[1] (see **Box 8.3**). The Johannesburg targets proposed for industry to build on what was defined in Goal 7 of the Millennium Development Goals (MDGs) in 2000 by doing the following:

■ ensuring environmental sustainability

■ integrating the principles of sustainable development into country policies and programmes

■ reversing the loss of environmental resources.

For countries adopting poverty-reduction policies, industrial growth is desirable in order to diversify their economy, create jobs and add value to primary products and raw materials being produced. However, it is very important that the necessary legal and institutional arrangements be in place to enable this growth to take place sustainably. Since water pollution can have significant transboundary effects, good environmental governance at the national level includes committing to international agreements and conventions on transboundary cooperation on shared waters (see **Chapter 11**).

2a. Best environmental practices and international standards for industry

Voluntary measures and self-regulation are the means whereby industries can demonstrate their commitment to improving the environment and monitoring their own performance. The extent of self-regulation tends to vary with the size of the enterprise, and the industrial sector in question. Consumers and media pressure can often influence the level of eco-awareness of companies in a particular sector. Women working through consumer organizations and environmental pressure groups have been particularly successful in bringing about good environmental practices in the companies making household products such as detergents. Eco-labelling is a growing practice whereby consumers can choose to buy certain products labelled to indicate that they are produced in a cleaner and more environmentally responsible way.

The international competitiveness of a company and its products in the global market is therefore often enhanced by its commitment to best environmental practices (BEP).[2] A company can show its high quality of self-regulation by seeking certification through ISO 14001, which is the current international environmental standard administered

1. See www.un.org/esa/sustdev/ documents/WSSD_POI_PD/ English/POIToc.htm

2. Best Environmental Practices (BEP) are guidelines that exist for all sectors of human society striving to co-exist with the natural environment, such as housing, infrastructure, industry and tourism. BEPs for industry include carrying out environmental impact assessments for new projects, environmental audits for existing projects and using best available technology.

by the International Organization for Standardization. By the end of 2002, nearly 50,000 companies in 118 countries had received this certification (see **Table 8.2**).

Some of the approaches to BEP, discussed below, have been made mandatory in national environmental law in some countries, but not in others. BEP can begin at the planning and design stage of a new industrial installation, with an environmental impact assessment (EIA), and be continued by putting in place an environmental management system (EMS) for the plant. Periodic or occasional environmental audits can be carried out during the plant's lifetime in order to assess the effectiveness of the environmental management system, and the plant's compliance with environmental regulations. The use of Best Available Technology (BAT) usually goes hand-in-hand with BEP (see **Box 8.4**).

Environmental Impact Assessment (EIA)

The EIA process is now required by law for new projects and significant extensions of existing projects in many countries. It covers a broad range of activities ranging from industrial to infrastructure projects. The process introduces procedural elements, such as the provision of an environmental impact statement and consultation with the public and environmental authorities, within the framework of development consent procedures for the activities covered.

...good environmental management need not be simply an extra cost for a company to bear...

BOX 8.3: INTERNATIONAL AGREEMENTS AND MULTILATERAL ENVIRONMENTAL AGREEMENTS (MEAs)

The Basel Convention on the Control of Transboundary Movements of Hazardous Wastes and their Disposal

Entered into force in May 1992. As of 28 May 2004, 159 states and the European Union were Parties to the Convention, which is an effective mechanism for addressing waste generation, movement, management and disposal. It plays a significant role in the safe management of chemicals. Recently, the Basel Convention joined with other existing international organizations in the creation of the Africa Stockpile Project, aimed at eliminating harmful stockpiles of pesticides on that continent. The Basel Convention is also working to create useful partnerships in areas as diverse as e-waste[1], biological and medical waste, and a global partnership aimed at addressing the stockpile of used oils in Africa.

The Rotterdam Convention on the Prior Informed Consent (PIC) Procedure for Certain Hazardous Chemicals and Pesticides in International Trade

A voluntary procedure from 1980 onwards, and is now mandatory in the 73 countries that are party to the Convention. This Convention entered into force in February 2004. A total of 27 hazardous chemicals are currently subject to the PIC Procedure. To put this into perspective, about 70,000 chemicals are currently on the market, with 1,500 new ones being added each year. This poses a significant challenge to governments, which must monitor and manage the use of these chemicals. The treaty helps countries to reduce the risks associated with the manufacture, trade and use of hazardous chemicals.

The Stockholm Convention on Persistent Organic Pollutants (POPs)

A global treaty, which entered into force in May 2004 and is designed to protect human health and the environment from persistent organic pollutants (POPs). POPs are chemicals that remain intact in the environment for long periods, become widely distributed geographically, accumulate in the fatty tissue of living organisms, and are toxic to humans and wildlife. They have been shown to cause cancer and to damage the nervous, reproductive and immune systems, as well as causing birth defects. At present, twelve hazardous chemicals, including DDT, dioxins and furans, are listed as POPs. In implementing the Convention, governments take measures to eliminate or reduce the release of POPs into the environment.

BOX 8.3: CONTINUED

The European Union Water Framework Directive (WFD) on Integrated River Basin Management for Europe

Adopted in October 2000, it coordinates the objectives of European water policy in order to protect all waters, including surface water and groundwater, using a river basin management approach. The WFD coordinates with all previous EU directives relating to water, including the **Integrated Pollution Prevention and Control Directive** (IPPC) of 1996, which addresses industrial installations with a high pollution potential. Such installations may only be operated if the operator holds a permit containing requirements for the protection of air, water and soil, waste minimization, accident prevention and, if necessary, site clean-up. These requirements must be based on the principle of **Best Available Techniques** (BAT) (see following section). The European Pollutant and Emissions Register, which has been compiled under the IPPC Directive, contains information on the emissions to air and water of nearly 10,000 industrial installations across Europe.[2]

The UNECE Convention on the Protection and Use of Transboundary Waters and International Lakes (UNECE Water Convention)

Intended to strengthen national measures for the protection and ecologically sound management of transboundary surface water and groundwater. It obliges Parties to prevent, control and reduce water pollution from point and non-point sources. More than 150 major rivers and 30 large lakes in the UNECE region run along or straddle the border between two or more countries. **The Convention entered into force in October 1996** and has been ratified by 34 countries and by the European Community. It is open for accession by all UN Member States.

The 1992 UNECE Convention on the Transboundary Effects of Industrial Accidents

Entered into force in April 2000, and 32 countries and the European Community are currently Parties to the Convention. This Convention cooperates with the UNECE Water Convention on issues related to the prevention of chemical accidents and the limitation of their impact on transboundary waters. In 2003 this resulted in the signing of the joint Protocol on Civil Liability and Compensation for Damage caused by Industrial Accidents on Transboundary Waters. The following work is being undertaken under both Conventions:

- An inventory of existing safety guidelines and best practices for the prevention of accidental transboundary water pollution[3]
- Safety guidelines and best practices for tailing dams, pipelines, and navigation of ships on rivers
- Alarm and notification systems
- International response exercises[4]
- Transboundary contingency planning.

1. E-waste is electronic and electrical waste including domestic computers and appliances.
2. The upgraded register, the European Pollutant Release and Transfer Register, should go online in 2009 and will then replace the present one, the European Pollutant and Emissions Register, EPER. Just like EPER, it will provide clear information about the level of specific pollutants, the quality of our local environment, emissions from specific industrial facilities and activities, and by country. But while EPER reports on 50 substances emitted to air and water, the PRTR will report on more than 90 substances released to air, water and land. The present register covers 56 industrial activities; the new one will cover 65. It will also have information on what the industrial installations do with their waste and waste water. The reporting cycle will be annual instead of every three years. What is more, the PRTR will compile reporting of pollution from diffuse sources such as road traffic, aviation, shipping and agriculture.
3. For more information see www.unece.org/env/teia/water/inventory.htm
4. For more information see www.unece.org/env/teia/response.htm

Table 8.2: Trends in ISO 14001 certification regionally and globally, 1997–2002

National standards institutes from individual countries have created the ISO 14000, which provides voluntary environmental management systems standards. The table below shows the number of companies in each region that have received the International Organization for Standardization (ISO) 14001 certification by December of any given year. Companies adhering to the ISO 14001 implement environmental management systems, conduct environmental audits, and evaluate their environmental performance. Their products adhere to environmental labelling standards, and waste streams are managed through life cycle assessments. However, the ISO does not require companies to provide public reports on their environmental performance.

The number of companies with ISO 14001 certification globally increased more than tenfold between 1997 (two years after the introduction of the standard) and 2002. Europe and the Far East dominate the statistics, with 47 percent and 36 percent respectively of all companies certified worldwide. (ISO 14001 replaced ISO 14000 in 1995).

Regions	Number of companies with ISO 14001 certification					
	1997	1998	1999	2000	2001	2002
North America	117	434	975	1,676	2,700	4,053
Share in percent	2.64	5.50	6.91	7.32	7.35	8.20
No. of countries	3	3	3	3	3	3
Europe	2,626	4,254	7,365	11,021	18,243	23,316
Share in percent	59.24	53.94	52.21	48.13	49.62	47.14
No. of countries/economies	25	29	32	36	41	44
Central and South America	98	144	309	556	681	1,418
Share in percent	2.21	1.83	2.19	2.43	1.86	2.87
No. of countries/economies	5	12	14	18	22	21
Africa/West Asia	73	138	337	651	923	1,355
Share in percent	1.65	1.75	2.39	2.84	2.51	2.74
No. of countries/economies	10	15	21	25	29	31
Australia/New Zealand	163	385	770	1,112	1,422	1,563
Share in percent	3.68	4.88	5.46	4.86	3.87	3.16
No. of countries	2	2	2	2	2	2
Far East	1,356	2,532	4,350	7,881	12,796	17,757
Share in percent	30.59	32.10	30.84	34.42	34.81	35.90
No. of countries/economies	10	11	12	14	16	17
World total	4,433	7,887	14,106	22,897	36,765	49,462
Number of countries/economies	55	72	84	98	112	118

This rapid worldwide increase in the number of companies certified is accompanied by a dramatic rise in environmental awareness of managers and workers, because there is a substantial element of training and capacity-building involved in the certification process. This capacity-building and development of the environmental knowledge base necessarily goes hand-in-hand with the introduction of new management systems and performance yardsticks in these companies (see **Chapter 13**).

Source: www.iso.org/iso/en/prods-services/otherpubs/iso14000/index.html

Above: Polluted water due to discharge of refuse from a sugar-mill in Nosy Bé Dzamandzar, Madagascar

The EU commissioned eighteen case studies of the EIA process to be reviewed across Europe in order to investigate its effectiveness. This review showed that the EIA process improved the implementation of the project itself and ensured the protection of the environment. In the majority of the case studies, all participants confirmed that the EIA process had assisted in decision-making in one or more of the following ways:

- the quality of the project design was improved in 83 percent of the case studies

- key environmental issues were identified in 94 percent of the case studies

- higher standards of mitigation were achieved than would otherwise have been expected in 83 percent of the case studies

- environmentally sensitive areas were avoided through relocation or redesign of the project in 56 percent the of the case studies

- a better framework for preparing conditions and legal agreements to govern future operation of the project was provided in 72 percent of the case studies

- environmental concerns were incorporated from an earlier stage in the design process in 61 percent of the case studies

- better decision-making was achieved in at least 61 percent of the case studies, due to the following:

  - a more systematic and structured framework for analysis
  - more objective and credible information
  - increased rigour in evaluating environmental information.

BOX 8.4: BEST ENVIRONMENTAL PRACTICES (BEPs)

Environmental Management Systems (EMS)

Aimed at achieving the organization's environmental policy, as defined by the top management. The system needs to describe various aspects, including setting responsibilities, defining environmental objectives, the means of achieving these, and the operational procedures, training needs, monitoring and communication systems that are to be used. The structure of an EMS is essentially an 'empty shell', within which the company defines its own unique ways of operating and establishing accountability. The most widely used EMS is that provided by the International Organization for Standardization within the ISO 14001 framework (see **Table 8.2**).

Environmental audits

These should be carried out periodically in order to assess the effectiveness of the management system in place and its conformity with the organization's environmental policy and programme. Quantitative technical audits can form part of this process, such as water and energy audits. An environmental audit must also assess the organization's compliance with relevant environmental regulatory requirements.

The European Union's Eco-Management and Audit Scheme (EMAS)

A voluntary scheme designed to promote continuous improvements of the environmental performance and compliance with all relevant regulatory requirements regarding the environment. To achieve this aim, industrial sites in Europe are required to use an environmental management system to monitor efficiency and to report on their achievements regarding environmental performance. They are also encouraged to seek ISO 14001 accreditation. EMAS statements are principally focused on improvements of environmental performance by describing the current environmental conditions and the operational aspects required at the site level to deliver continuous improvements in environmental performance. In regulatory

instruments, permits set the emission limit values for the company's activities. The licence stipulates a maximum load on the environment and what reduction of this load must be achieved. The EMAS environmental management system is an instrument that provides industries with a set of appropriate means for achieving an effective reduction in the load on the environment.

Best Available Technology (BAT)

The BAT concept is a useful standard-setting tool for emissions reduction in many industrial sectors. In the European Union (EU), however, the use of BAT is mandatory as part of the Integrated Pollution Prevention and Control (IPPC) Directive of 1996. Since 2000, all new industrial installations in the EU have been subject to the IPPC Directive and the BAT provisions. Imposing new and considerably tougher BAT rules on all existing industrial installations in the EU could jeopardize many European jobs, and therefore the IPPC Directive granted these installations an eleven-year transition period (i.e. to 2007). In many cases, BAT means quite radical environmental improvements, and sometimes it can be very costly for companies to adapt their existing plants to BAT.

'Best' is defined as that which is most effective in achieving a high general level of protection of the environment as a whole. 'Available' techniques are defined as those which can be implemented in the relevant industrial sector, under economically and technically viable conditions, taking into consideration the costs and benefits and whether or not the techniques are used or produced inside the country, as long as they are reasonably accessible to the operator. This includes both the technology used and the way in which the industrial installation is designed, built, maintained, operated and decommissioned. The following considerations should be taken into account when determining BAT:

- the use of low-waste technology
- the use of less hazardous substances
- the furthering of recovery and recycling of substances generated and used in the process and of waste, where appropriate
- comparable processes, facilities or methods of operation which have been tried with success on an industrial scale
- technological advances and changes in scientific knowledge and understanding
- the nature, effects and volume of the emissions concerned
- the commissioning dates for new or existing installations
- the length of time needed to introduce the best available technique
- the consumption and nature of raw materials (including water) used in the process and their energy efficiency
- the need to prevent or reduce to a minimum the overall impact of the emissions on the environment and the risks to it
- the need to prevent accidents and to minimize the consequences for the environment
- the use of information published by the EU or by international organizations.

Water can be saved either by cutting down on water input... or by water recycling and reuse...

Part 3. The Vision: Towards High Water Productivity and Zero Effluent Discharge

Both water quantity and quality need to be considered in the challenge of improving industrial water use. Where water *quantity* is concerned, it is useful to consider water productivity, in terms of the industrial value added per unit of water used (see Table 8.5 at end of chapter). The higher the water productivity, the greater the intrinsic value being placed on the water. In water-scarce regions, where there is competition for water among various users, water is likely to be allocated to the more highly productive uses. Industry achieves higher water productivity than agriculture, but as it is difficult to compare the water productivity of domestic use or environmental flow requirements, because the adequate data and economic instruments are not available to make such comparisons. Within industry, as in other sectors, it is important to strive towards greater water productivity.

Where water *quality* is concerned, zero effluent discharge is the ultimate goal, in order to avoid any releases of contaminants to the water environment. Zero effluent discharge entails water recycling, which also contributes to raising water productivity. If zero effluent discharge is not economically and technically feasible, there are some valuable intermediate strategies, which can be pursued to reduce pollution and to ensure that waste substances are recovered and water reused.

3a. Strategies for saving water and increasing industrial water productivity

Water auditing
Conducting a water audit of an industrial plant or manufacturing facility clearly shows where the water supplied to the plant is being used, how much is used in each process, and where it ultimately ends up. Rainwater that falls on the site, as well as the natural evaporation that occurs, should also be included in the audit. Once a water audit has been done, it is possible to draw a flow chart and show the water balance across the plant, or over individual units of the process. This is the first step in finding innovative ways to save water on an industrial site.

Water can be saved either by cutting down on water input, where it is being unnecessarily wasted, or by identifying water recycling and reuse opportunities, discussed in more detail below. On-site rainwater harvesting may also be considered, since this is preferable to allowing rainwater (which may have become contaminated) to simply run off into the stormwater system. Case studies from the same industrial sector can provide some ideas and general lessons on saving water, but each site needs to be audited and analysed individually.

Matching water quality to use requirements
In many instances, the water used in industry is of an unnecessarily high quality for the use to which it is put. The analogy in domestic water use is, for instance, using water of drinking quality in order to flush toilets or water the garden. Similarly, in industrial processes there are many applications where lower water quality could be used. This offers recycling opportunities. Often 50 percent or more of an industrial plant's water intake may be used for the purpose of process cooling, a need that can often be met with lower quality water. On the other hand, some industries (such as the pharmaceutical industry) require water of exceptionally high quality. In such processes, additional water treatment is carried out on the water received from the local water utility, or withdrawn from groundwater or surface waters, in order to further improve the water quality before it is used.

There are cases in industry where water is used inappropriately, where a completely different approach could be taken to save water in water-scarce areas. An example of this would be switching to using pneumatic or mechanical systems for transportation, instead of using water to move the products, as is often done in the poultry and other food industries.

Water recycling and on-site reuse
Water recycling is the primary means of saving water in an industrial application: taking wastewater that would otherwise be discharged and using it in a lower quality application (often after treatment). Each cubic metre (m^3) of water that is recycled on-site represents one cubic metre that will not have to be withdrawn from a surface water source or from groundwater. Water can even be used many times over. In such cases, where, for instance, a

The most common uses of reclaimed water are industrial cooling and power generation, followed by boiler feed and quenching

given cubic metre of water is used ten times in the process (a 'recycle ratio' of ten to one), this represents 9 m^3 that are not withdrawn from a freshwater source. Increased water savings can be made by raising the recycle ratio. The industrial water productivity of the product is thereby also greatly increased, as far less freshwater is used to produce the same quantity of product.

The way in which water recycling is done on-site must be governed by the principle of matching water quality to use requirements, as mentioned above. This is dependent on the nature of the manufacturing process, as well as on the degree of wastewater treatment carried out on the site. Processes such as heating, cooling and quenching are the most common applications for lower quality water. It can also be used as washdown water, and for site irrigation.

A second consideration in recycling industrial water is the cost of treating the wastewater to the required level, including the cost of new or additional pipes and pumps, as compared to the cost of 'raw' water supplies (freshwater). Where the quality of freshwater is declining locally, or where freshwater supplies are becoming unreliable due to water scarcity in the region (droughts or falling groundwater levels), on-site industrial water recycling becomes an increasingly attractive option. On-site

water recycling can be regarded as a component of industrial risk management, since it contributes to reducing the risk related to the unreliability of freshwater supplies.

For example, the micro-chip manufacturer, Intel, established the Corporate Industrial Water Management Group to improve water use efficiency at its major manufacturing sites, which use large amounts of highly treated water for chip cleaning. The group includes representatives from fabrication sites, corporate technology development experts, and regulatory compliance staff. Intel's initial goal was to offset by 2003 at least 25 percent of its total incoming freshwater supply needs by using recycled water and installing more efficient systems. In 2002, the company exceeded this goal by achieving 35 percent water savings through recycling water and efficiency gains.

Using reclaimed water

A more indirect means of recycling water occurs when an industrial enterprise reuses the wastewater produced by another industrial plant close by (with a treatment step in between, if necessary). Again, the principle of matching water quality to use requirements must be followed. The availability of wastewater, when needed, and its variability in terms of quality also need to be considered. For instance, an industrial plant could use wastewater from a

---

BOX 8.5: DEFINITIONS OF WATER RECLAMATION, REUSE AND RECYCLING

'Water reclamation', 'reuse' and 'recycling' should not be used interchangeably. In the wastewater treatment industry, the following definitions are used:

Water recycling normally involves only one use or user, and the effluent from the user is captured and redirected back into that use scheme. Water recycling is predominantly practised by industry.

Wastewater reclamation is the treatment or processing of wastewater (industrial or municipal) in order to make it reusable.

Reclaimed water is treated effluent suitable for an intended water reuse application.

Water reuse is the use of treated wastewater for beneficial purposes such as agricultural irrigation and industrial cooling. Water reuse can be done directly or indirectly.

Direct water reuse requires the existence of pipes or other conveyance facilities for delivering reclaimed water.

Indirect water reuse is the discharge of an effluent to receiving waters (a river, lake or wetland) for assimilation and further withdrawals downstream. This is recognized to be important and can be planned for, but does not constitute direct water reuse.

Water recycling and reuse has far-reaching benefits in industries beyond the mere requirement of complying with the effluent discharge permits:

1. Reduction in freshwater withdrawal and consumption
2. Minimization of wastewater discharge by reclaiming wastewater, thereby reducing clean-up costs and discharge liabilities
3. Recovery of valuable by-products
4. Improvement of the profit margin by cost reduction
5. Enhancement of corporate image, public acceptance and environmental responsibility.

Source: Asano and Visvanathan, 2001.

Above: This golf course in Arizona, US, is watered with recycled water from the city of Page

nearby municipal sewage treatment plant. The result is usually called reclaimed water and is sold to industry by municipalities in many countries, including Australia, South Africa and the US. The most common uses of reclaimed water are industrial cooling and power generation, followed by boiler feed and quenching. In such arrangements, the use of reclaimed water by industry eases the pressure on scarce water resources in the region.

In the metropolitan region of Durban, South Africa, an innovative public-private partnership has been supplying reclaimed water to industries since 1999 (see **Chapter 14**). The Southern Sewage Works of the Durban Metro Water Services treats over 100,000 m³/day of domestic and industrial effluent (through primary treatment only), prior to discharging it to sea through a long sea outfall. Projections showed that the capacity of the sea outfall would soon be reached, due to the growing population and industrial water discharges in the area. A secondary treatment plant with a capacity of 48,000 m³/day was built, which was allowed to discharge water into a canal that flows over the beach into the sea. A nearby paper mill then contracted to take 9,000 m³/day of the treated water. A local survey was undertaken, which found that further (tertiary) treatment would be required to sell reclaimed water to other industries in the area, which needed higher quality water than the paper mill. Since it was not economically feasible for the municipal water utility to construct and operate such a high-tech plant, the tertiary treatment works (which currently treats and sells up to 30,000 m³/day of reclaimed water to local industries) was built through a public-private partnership.

Agricultural irrigation and urban irrigation (of parks, sports fields and golf courses) are also major applications for reclaimed water, which is important since irrigation is usually the largest water user in any region (see **Chapter 7**). Israel currently reuses some 84 percent of its treated sewage effluent in agricultural irrigation. The World Health Organization has laid down guidelines for the use of reclaimed water in irrigation, as there may be health implications when reclaimed water is sprayed in the open (WHO, 2005). Reclaimed water can also be used to recharge aquifers, for instance to avoid saline water intrusion into the aquifer, or simply to augment the groundwater supply. In the Adelaide region of Australia, half of the city's water demand is met through reclaiming water by aquifer storage and recovery.

In construction applications, reclaimed water can be used for dust control, soil settling and compaction, aggregate washing and concrete production. Domestic applications for reclaimed water include fire fighting, car washing, toilet flushing and garden watering. Supplying reclaimed water in urban areas requires two sets of piping: one for potable water (drinking water) and the other for reclaimed water – termed 'dual reticulation'. The installation of dual reticulation is usually done in new housing developments, as laying it retrospectively may be prohibitively expensive. The Tokyo Metropolitan Government in Japan has long encouraged the fitting of new office blocks and apartments in Tokyo with dual reticulation (see also **Chapter 14**). There are even a few cities in arid regions, such as in Windhoek, Namibia, where reclaimed water is treated to a very high standard and then reused directly to augment the potable water supply.

Table 8.3: Wastewater treatment requirements as a function of end-use for industrial water supply

Industrial water use	Nitrogen and Phosphorus removal	Chemical precipitation	Filtration
Cooling tower makeup	Normally	Yes	Yes
Once through cooling			
- Turbine exhaust condensing	Sometimes	Seldom	Sometimes
- Direct contact cooling	Seldom	No	Sometimes
- Equipment and bearing cooling	Yes	Yes	Yes
Process water	Yes	Yes	Yes
Boiler feed water	Requires more extensive treatment; use of reclaimed wastewater generally not recommended		
Washdown water	Sometimes	Seldom	Yes
Site irrigation	No	No	Normally

Source: Asano and Visvanathan, 2001.

Both freshwater and reclaimed water can contain constituents that can cause problems, but their concentrations in reclaimed water are generally higher.

- **Scaling:** This refers to the formation of hard deposits on surfaces, which reduce the efficiency of heat transfer processes. Due to repetitive recycling of feed water in the cooling water, water lost by evaporation leads to increases in the concentration of mineral impurities such as calcium, magnesium, sodium, chloride and silica, which eventually lead to scale formation. Scale forming constituents can be eliminated using appropriate chemical precipitation techniques.

- **Corrosion:** Ammonia, which may be present in significant concentration in reclaimed municipal wastewater, is one of the prime causes of corrosion in many industrial water reuse installations. Dissolved oxygen and certain metals (manganese, iron, and aluminium) may also promote corrosion because of their relatively high oxidation potential. The corrosion can be controlled by adding chemical corrosion inhibitors.

- **Biological growth:** Slime and algal growth are common problems in reclaimed water due to a high nutrient content, which promotes biological growth. This growth can be controlled or eliminated by addition of biocides during the internal treatment process.

- **Foaming:** Associated with the presence of biodegradable detergents, foaming problems can be avoided by using anti-foaming chemicals.

- **Pathogenic organisms:** When reclaimed water is used in industry, the assurance of adequate disinfection is a primary concern for protecting the health of workers and plant operators. The most stringent requirement, similar to unrestricted reclaimed water use in food crop irrigation, would be appropriate if there exists a potential for human exposure to spray.

Minimizing virtual water in manufactured products

The concept of the virtual water trade has been mentioned in **Chapters 7, 11** and **12**, in relation to trade in crops and food products. The same concept applies in relation to manufactured products, where the virtual water of a particular finished product represents the volume of water that was used to produce it. This can be calculated in two ways: either as m^3/t of product, or as m^3/dollar of added value. By looking at the imports and exports of each type of product, it is possible to calculate the virtual water flows into and out of the region. One can also calculate the industrial water productivity of various products and sectors, in terms of the industrial value added per unit of water used (see **Table 8.4** at end of chapter). In water-scarce regions, it makes sense to focus on the manufacture of products that use little water, and to therefore only export products with a high water productivity. This minimizes the amount of virtual water that is exported. Conversely, water-intensive products and products with low water productivity, such as aluminium and beer, should be imported into water-scarce regions, as this represents a way of indirectly importing water.

Better policy instruments and economic incentives

Industrial water management strategies that intend to minimize water consumption and wastewater generation, and thereby improve water productivity, can be either internal or external to the enterprise itself. Internal strategies are those measures that are required to be taken at a factory level in order for water consumption and wastewater generation to be controlled, such as water recycling. These measures can be taken more or less independently of external strategies.

External strategies, on the other hand, are measures that are required at the industry level in the context of local, regional or national industrial water management. Generally, the factory management does not control these strategies, although in certain cases some measures are required at the factory level in response. The nature and number of a particular type of industry present in a locality or a region can significantly influence these strategies. Some of these strategies are summarized as follows:

- national water recycling and reuse policies

- grouping of industries in a particular site (industrial parks) coupled with combined treatment methods and reuse policies

- rationing the water use within industry, so that each process uses a defined quantity of water

- applying economic instruments such as penalties, water charges, subventions, credits and grants.

...the virtual water of a particular finished product represents the volume of water that was used to produce it

Pollution from a chemical factory upstream devastated fish farms along the Tuo River, a tributary of the Yangtze River, China

National water conservation policy, also called water demand management, is the key factor in water recycling and reuse in industries. It forms an important component of national water efficiency plans. In some developing countries, industry is not charged for water nor for wastewater services; in other words, industrial water withdrawals and wastewater discharges are still free and unregulated. Both regulation and the imposition of stepped water tariffs according to the volume of water used are key instruments for governments to use in these situations (see **Chapter 12**).

Compliance with stringent effluent requirements can force industries to implement new water-saving technologies to reduce effluent discharges and prevent pollution incidents. Fines for non-compliance and the threat of closure for repetition of non-compliance can also significantly achieve higher recycling and reuse. Higher charges for raw water can be applied to industries using large volumes of water. An example can be seen in Singapore, which levies a 15 percent water conservation tax on operations using more than a specified amount. New factories using more than 500 m^3 of water per month must apply for approval from the City Council during the planning phase. A fertilizer plant in Goa, India cut water demand by 50 percent over a six-year period in response to higher water prices. Dairy, pharmaceutical, and food processing industries in São Paulo, Brazil (see also **Chapter 14**), reduced water use per unit output by 62 percent, 49 percent and 42 percent, respectively (Kuylenstierna and Najlis, 1998).

Given proper incentives, it is generally found that industry can cut its water demand by 40 to 90 percent, even with existing techniques and practices (Asano and Visvanathan, 2001). However, water conservation policies need to be fair, feasible and enforceable. Economic incentives should be given to industry to comply with standards and policy and to reduce raw water intake and wastewater discharge. Such incentives could include subsidies for industries implementing innovative environmental technologies and financial and advisory support for industries that fund new research.

3b. Strategies and methodologies for reducing pollution: Paving the way to zero effluent discharge

No industrial plant operator sets out in the morning to pollute the environment. The objectives of a plant operator are to minimize production costs and to maximize the volume of production. Releases of pollutants into rivers and streams are carried out because this is typically a low-cost and low-tech option for waste and wastewater disposal. Frequently, once plant operators and managers become aware of the possibilities of cleaner production, the water and energy savings as well as the cost savings that can be made, their response is enthusiastic. In addition, the company as a whole can become more competitive in the global market by advertising its environmental policy and its strategies to reduce pollution. Some examples of how individual companies have benefited from implementing cleaner production methods are described in **Boxes 8.6** and **8.7**.

Strategies to reduce pollution include applying the principles of cleaner production in general. This is illustrated by a specific example of a UNIDO methodology for promoting cleaner production, the TEST (transfer of environmentally sound technology) strategy, which has been implemented in several river basins. Some of the specific processing techniques that can be applied to reducing water pollution on industrial plants include stream separation, raw material and energy recovery, and the reuse of waste and wastewater. End-of-pipe treatment technologies are also discussed briefly below, as these technologies are applicable both to recycling water, and to treating it prior to discharge back into the water environment. In the long run, the most desirable approach is that of aiming for zero effluent discharge, so that no industrial wastewater at all is discharged back directly into the environment.

Cleaner production

The principles of cleaner production include optimizing the use of resources used as inputs to the production process, such as raw materials, water and energy, and reducing to a minimum the generation of wastes. They are also concerned with the method of disposal of wastes and their impact on water, air and soil. A Cleaner Production Assessment (CPA) is a specific methodology for identifying areas of inefficient use of resources and poor management of wastes by focusing on the environmental aspects and impacts of industrial processes. This methodology (developed by UNEP) consists of the following five phases:

- **Phase I:** Planning and organization: obtain management commitment; establish a project team; develop policy, objectives and targets; plan the CPA.

- **Phase II:** Pre-assessment (qualitative review): prepare company description and flow chart; undertake a walk-through inspection; establish a focus.

- **Phase III:** Assessment (quantitative review): collect quantitative data; assess material balance; identify Cleaner Production opportunities; record and prioritize options.

- **Phase IV:** Evaluation and feasibility study: prepare preliminary evaluation; conduct technical evaluation; conduct economic evaluation; conduct environmental evaluation; select viable options.

- **Phase V:** Implementation and continuation: prepare an implementation plan; implement selected options; monitor performance; sustain Cleaner Production activities.

There is a network of National Cleaner Production Centres in twenty-seven countries, established by UNIDO, which assist in spreading the CPA methodology among companies in various industrial sectors. Many successful case studies of cleaner production are available, demonstrating how CPA can help companies become cleaner and more profitable.

Environmentally Sound Technologies (the TEST strategy)

Whereas CPAs are focused at the enterprise level, TEST is applied at a river basin level in order to scale up the benefits of clean production. The TEST strategy is a methodology that has been developed by UNIDO, which involves individual enterprises, national and local governments, and river basin organizations (see also **Chapter 14**). To date, it has been applied in the Danube and the Dnieper Basins with considerable success, helping to protect the Black Sea marine ecosystem (see **Box 8.8**). The methodology consists of the following aspects:

- identifying the pollution hot spots in the basin (see **Box 8.1**)

- introducing the principles of cleaner production and reducing the industrial pollution associated with certain enterprises, selected as demonstration sites, which are located at these pollution hot spots

- scaling up the TEST by training and equipping water managers, plant operators and local trainers and consultants with the principles and tools of the Cleaner Production Assessment (CPA), the Environmental Management System (EMS) and Environmental Management Accounting (EMA)

- supporting the relevant river basin organization in building awareness of the benefits of Cleaner Production and increasing industrial water productivity.

Given proper incentives, it is generally found that industry can cut its water demand by 40 to 90 percent...

BOX 8.6: CLEANER CHEESE PRODUCTION IN EL SALVADOR

A typical dairy company in El Salvador was using 10 litres (L) of milk and about 80 L of water in order to produce one kilogram (kg) of cheese. Nearly 9 L of whey were produced as a by-product and discharged into the wastewater. Whey is a highly concentrated organic liquid, containing proteins and lactose. Large dairy companies use ultra-filtration plants to produce pure lactose, additives for ice cream and other food products from this by-product. However, this technology is not affordable for small and medium-sized companies.

The solution proposed by the National Cleaner Production Centre in El Salvador was to process the whey in order to produce a marketable whey-fruit drink. Such drinks are available in the European market and are popular with consumers. No additional investment was required by the company in order to process the whey. The estimated benefits were found to include:

- 11.5 percent reduction in the volume of wastewater
- 40,000 milligrams per litre (mg/L) reduction in biochemical oxygen demand (BOD) level in wastewater
- 60,000 mg/L reduction in chemical oxygen demand (COD) level in wastewater
- US $60,000 annual savings in wastewater treatment costs.

Other dairy companies in El Salvador are starting to produce this product, and similar programmes are being developed in Guatemala and Mexico.

BOX 8.7: CLEANER BEER PRODUCTION IN CUBA

Tinima Brewery is the second largest brewery in Cuba, producing 47,600 cubic metres (m^3) of beer per year. In 2002, the National Cleaner Production Network responded to a request from the brewery's management for a cleaner production assessment (CPA) of the brewery.

Beer is produced by mixing, milling and boiling three main components: barley malt, sugar and water. This process produces a sugary liquid called wort, which is cooled, fermented and filtered in order to obtain the final product. A new technology was proposed to Tinima Brewery, in which the concentrated sugar syrup and water short-circuit the hot section and are added directly to the fermentation tank. This

means that the main volume of liquid does not pass through the hot section. The investment required was low, as these changes needed only some new pipeline arrangements.

This new technology was implemented at Tinima Brewery in 2003. The beer produced was found to be just as acceptable to consumers, and the technology is now approved for all breweries in Cuba. The following savings were achieved by Tinima Brewery as a result of the cleaner production technology:

- 74 percent reduction in cooling water consumption
- 7 percent reduction in total water consumption

- 11 percent reduction in the volume of wastewater
- 4 percent savings on sugar consumption (used as an additive to the beer)
- 3 percent savings on caustic cleaning solution consumption
- 50 percent savings on thermal energy consumption in the heating and evaporating stages
- 30 percent savings on thermal energy consumption in the cooling stages
- 12 percent savings in total electricity consumption
- 21 percent reduction in greenhouse gas emissions.

Source: UNIDO, 2004.

BOX 8.8: THE TEST STRATEGY IN THE DANUBE BASIN

The implementation of the TEST programme in the Danube River Basin began in May 2001 and was completed in December 2003, successfully introducing the TEST approach in seventeen companies in five countries (Bulgaria, Croatia, Hungary, Romania and Slovakia). The companies were identified as hot spots of industrial pollution, from various industrial sectors, including chemicals, food, machinery production, textiles, pulp and paper. Tangible results were achieved in terms of increased productivity and improved environmental performance. These results are used to show other enterprises in the basin that it is possible to reduce environmental impacts to acceptable levels while becoming more competitive.

There were 224 Type A (no cost or low cost) measures identified, of which 128 were implemented by the 17 participating enterprises. These were mostly 'good housekeeping' measures involving the following:

- process improvement (with small technological changes using existing equipment)
- raising the skills of operational staff
- revising laboratory procedures

- improving scheduling management and maintenance
- improving raw materials storage
- adjusting water consumption by reducing wastage and leaks, and improving process control.

There were 260 Type B (involving a relatively small investment with a short payback period) measures identified, of which 109 were implemented by the end of the project. The total investment undertaken by the 17 enterprises to implement the Type B measures was US $1,686,704, while the estimated financial savings as a result of these measures are US $1,277,570 per year.

It is interesting to note that the Bulgarian, Romanian and Slovak companies had the highest number of Type A and B measures identified. This can be explained by the fact that these companies used relatively outdated technology, thus many more measures could be identified to optimize the existing process. The environmental benefits were significant in terms of reduced consumption of natural resources (including

freshwater and energy), reduced wastewater discharges and pollution loads into the Danube River and its tributaries, as well as reduction of waste generation and air emissions. The total reduction in wastewater discharge into the Danube River Basin, achieved by the end of 2003 as a result of implementing Type A and B measures, was 4,590,000 m^3 per year. Pollution loads in the wastewater were reduced in most of the companies, including chemical oxygen demand (COD), biochemical oxygen demand (BOD), oily products, total suspended solids, heavy metals and toxic chemicals.

Finally, 141 Type C (requiring a significant financial investment) measures were identified, of which 38 were approved by top management of the various enterprises for implementation. The total investment required for the approved Type C measures is US $47,325,000 and they are scheduled for completion by 2007. The total additional reduction in wastewater discharges will be 7,863,000 m^3 per year, with estimated annual savings by the companies of US $5,362,000.

Source: UNIDO, 2004.

Stream separation

The principle of stream separation is a useful tool when assessing wastewater flows for treating the final discharge and when identifying flows of process water that may be recycled and reused. Wastewater containing a variety of contaminants is much more difficult and expensive to treat effectively than wastewater containing only one contaminant. Also, mixing a concentrated stream of effluent with a more diluted stream may result in much larger volumes of wastewater entering an expensive treatment process. The diluted stream alone may be suitable for discharge directly into a sewer, to be dealt with by a municipal wastewater treatment plant or may be suitable for on-site recycling for direct reuse in another part of the process. Treating the concentrated stream alone may become easier, because it may contain fewer contaminants and is likely to be cheaper, because the volume is much smaller. Now that a wide range of treatment technologies is available, stream separation may provide better and more cost-effective solutions in comparison to producing a single mixed effluent. The larger the enterprise, the more cost-effective this approach becomes. However, even small and medium-sized enterprises may benefit from considering stream separation.

Raw material and energy recovery from waste

An important aspect of reducing pollution is to look carefully at the solid and liquid waste generated by a given production process in order to calculate the quantities of unconverted raw material remaining in the waste streams. This unconverted raw material can potentially be reused. The feasibility of the recovery process can be assessed by determining the following:

- the cost of the separation process, which can recover the raw materials from the remainder of the waste
- the quantities of recoverable material
- the cost of the raw material
- the cost of waste disposal.

If the cost of the separation process is too high, or if the quantities of recoverable material are too small, material recovery becomes unprofitable. Similarly, very cheap raw materials and a low cost of waste disposal mitigate against the feasibility of the process.

Energy recovery may be possible in the same way from wastewater carrying waste heat. Once discharged, the waste heat becomes thermal pollution in the receiving water body. However, the heat could potentially be recovered and reused in another part of the process, or indeed in another enterprise nearby, which requires lower grade heat. Indirect energy recovery can be done through a biological form of wastewater treatment, in which anaerobic bacteria break down organic matter in the wastewater and produce methane (biogas). The methane may then be used to fire a boiler or to generate electricity (see **Chapter 9**).

Reuse of waste

The recycling of glass, paper and various types of plastics are the best-known examples of waste reuse. However, there are many cases in industry where used solvents, oils, concentrated wastewater containing starch, or various solid wastes can be traded for their residual value and reused. One innovative industrial park in Cape Town, South Africa, has set up a voluntary Waste Register for the companies located within the park (see **Chapter 14**). Each company is required to log its waste production, the quantity and the type of waste, as well as the types of raw materials used as inputs and whether these may be reclaimed materials. The Waste Register may be searched on the industrial park's website, so that companies can identify sources that fit their requirements. This approach results in savings on raw materials and waste disposal costs for the companies involved.

Wastewater treatment technologies

These technologies are applicable both to recycling water and treating it prior to discharge back into the water environment. End-of-pipe treatment is often applied prior to discharging wastewater into the sewage system, as municipalities can charge industries for accepting their wastewater, with a rising tariff according to the concentration of the discharge. Some of the technologies mentioned below result in the recovery of energy or raw materials from the wastewater, such as anaerobic treatment, which produces biogas, and sulphate removal, which produces gypsum.

Today, there is a very wide range of treatment technologies available. Wastewater treatment technologies typically fall into two broad categories:

- physical/chemical treatment (e.g. settling, filtration, reverse osmosis, adsorption, flocculation, chlorination) and biological treatment (aerobic or anaerobic treatment which remove organic matter)

- other more specialized processes such as phosphate reduction and sulphate removal.

...there are many cases in industry where used solvents, oils, concentrated wastewater containing starch, or various solid wastes can be traded for their residual value and reused

There are also land-based treatment methods, such as treatment in flow through wetlands, and various types of contained lagoons (most of which are both aerobic and anaerobic). Matching the type of treatment to the contaminants in the wastewater (coupled with stream separation as mentioned above) results in the most cost-effective solution. There is no technological constraint on the quality of the water which can be achieved through treatment, although there are inevitably cost constraints.

Achieving zero effluent discharge

Zero Effluent Discharge is a key target for both reducing water withdrawal by industry, and reducing pollution to the water environment. All the effluent that would normally be discharged is recycled or sold to another user, using the principles of stream separation, raw material and energy recovery from waste, and reuse of waste, which have been discussed above.

Reducing the volume of effluent discharged back into the water environment by industry is essential to closing the gap between water withdrawal and actual water consumption, which was noted in the first section of this chapter. Once no more water is discharged from an

industrial installation, its overall water consumption will equal its water withdrawal from source. In practice, this means that water withdrawal by industry will gradually decline, as levels of water recycling increase, down to the point where withdrawal equals consumption. This process has already begun, as has been shown by the declining water withdrawal by industry in Europe over the past twenty-five years (see **Figure 8.1** above).

The city of St Petersburg, Florida, US, is the first municipality in the world to have achieved zero effluent discharge to its surrounding surface waters. Situated on a bay that is a major tourist attraction, the city has laid an extensive dual-reticulation system for reclaimed water. All the generated domestic and industrial wastewater is treated to a high standard. The reclaimed water is then reused for irrigation and industrial cooling applications by thousands of customers, accounting for nearly half of the city's water needs of 190,000 m^3. By substituting reclaimed water for potable water in many applications, the city has eliminated the need for expansion of its potable water supply system until the year 2030. Equally important for the area, one sees no pollution of the beaches and marine ecosystems by municipal wastewater, and no unsightly sea outfalls.

Part 4. The 'Cradle-to-Cradle' Concept

There is a growing worldwide understanding that we need to find ways to make our industrialized society more conservative in its use of materials and resources. At present the uses of materials and other inputs to industrial processes (such as energy and water) are primarily linear. Resources are mined, products are manufactured, wastes are generated and then disposed of at the end of the process. Sooner or later, the products themselves end up as waste to be thrown away. By contrast, the 'cradle to cradle' concept is a vision of cyclical flows of materials: the materials that make up products are reused over and over again. This approach seeks to eliminate the whole concept of waste, bringing industrial production closer to the natural ecological process.

...a vision of cyclical flows... the materials that make up products are reused over and over again

In 2002, this concept was put forward by William McDonough and Michael Braungart in their book, *Cradle to Cradle: Remaking the Way We Make Things.* They argue that there need to be two major flows of materials: one of biodegradable materials, termed 'biological nutrients', and one of non-biodegradable materials (such as metals and plastics) termed 'technological nutrients'. The earth's natural chemical and biological cycles will ensure the recycling of the biodegradable materials, while society needs to put in place the systems necessary for the complete recycling of the 'technological nutrients'. For the recycling of materials, it is important that the two flows be kept distinct, otherwise each cycle will progressively contaminate the other. Hence the sourcing of materials for use in industry and the design of products become crucial: each product must be easily consigned to one or the other of the nutrient flows at the end of its lifespan so that cross-contamination does not occur between the two cycles. The authors argue that manufacturers need to change from becoming suppliers of a product, to suppliers of a service. A customer would use a product for a certain period, after which the manufacturer takes it back and breaks it down, reusing all of its materials to create new products.

What the cradle-to-cradle concept means for water is a complex question, as water is an essential part of both the biological and the technological cycles. However, water recycling is clearly the key. The basic principle must be refraining from contaminating the water that is returned to the biological cycle with any 'technological nutrients', or non-biodegradable materials, metals and solvents. Ideally, water should recirculate either within the biological cycle, or within the technological cycle. Implementing zero effluent discharge to the water environment would ensure that once used by industry, water stay within the technological cycle, except for that fraction that evaporates and is thereby returned to the biological cycle, free of contaminants. Preventing the atmospheric deposition of contaminants and the leaching of solid wastes into the water environment would also be essential to implementing the 'cradle-to-cradle' vision.

In the long term, the cradle-to-cradle vision could become the basis of almost all product design in manufacturing. This would result in a much cleaner and healthier world. Applications of the concept need to be found that can be implemented by industries without excessive cost in developing countries. This implies building both the necessary design capacity in those countries, and the capability to tap into global technology and information support networks. The experience of the companies that have already adopted the 'cradle-to-cradle' strategy shows that effective design not only generates positive externalities but can also make good business sense.

This chapter has largely focused on industry's impact on the water environment through its routine water withdrawal and wastewater discharge. However, solid waste disposal and air pollution by industry, as well as major industrial disasters, such as the Red Rhine incident, can also affect freshwater quality and thereby pose a risk to the water environment. Aquatic ecosystems in many places worldwide are being damaged or destroyed by industrial pollution, which in the case of fisheries has direct economic consequences. Many municipalities now find that the quality of the drinking water they supply is compromised by industrial pollution. This raises water treatment costs for the water supply utilities, which are then passed on to consumers through increased water rates. Where the problem is variable freshwater quality, caused by irregular effluent discharges, the water treatment plant may not treat the contaminants adequately. Hence companies requiring clean freshwater, as well as municipalities, are finding their water security

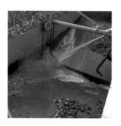

Washing sugarbeets as they are unloaded at a sugar factory, Antois-Picardy River Basin, France

...industries often lead the way towards a more sustainable society by implementing water recycling and putting environmental management systems in place...

increasingly affected by water shortages and deteriorating water quality.

Industries therefore have a dramatic effect on the state of the world's freshwater resources, both by the quantity of water that they consume and their potential to pollute the water environment by their waste discharge. Yet industries often lead the way towards a more sustainable society by implementing water recycling and putting environmental management systems in place in their factories and offices. As discussed, there has been an exponential increase over the past decade in the numbers of industrial companies worldwide seeking certification with ISO 14001, the international environmental standard. This demonstrates many companies' commitment to being

environmentally responsible as well as profitable, enhancing both their corporate image and their competitiveness.

As stated by the WSSD Plan of Implementation, poverty eradication and sustainable natural resource management are strongly linked. For countries adopting poverty-reduction policies, industrial growth is highly desirable in order to diversify the economy, create jobs, and add value to primary products and raw materials being produced. However, it is important that the necessary legal and institutional arrangements be in place to enable this growth to take place sustainably, keeping in mind that environmental commitment can be a highly efficient tool for enhancing profitability and competitiveness.

Table 8.4: Industrial water productivity by country, 2000/01

Country	Industrial value added (IVA): 2001* (billion constant 1995 US $) (1)	Industrial water use: 2000 (Km³/year) (2)	Population: 2000 (million) (3)	Industrial water productivity (IWP) (US $ IVA/m³) (4) = (1)÷(2)	IWP per capita (US $ IVA/m³/c) (5) = (4)÷(3)
Algeria	23.21	0.80	30.29	28.97	0.96
Angola	4.84	0.06	13.13	86.13	6.56
Argentina	69.13	2.76	37.03	25.07	0.68
Armenia	1.19	0.13	3.79	9.18	2.42
Australia	107.29	2.40	19.14	44.70	2.34
Austria	82.15	1.35	8.08	60.85	7.53
Azerbaijan	1.02	4.77	8.04	0.21	0.03
Bangladesh	13.10	0.52	137.44	25.25	0.18
Belarus	5.76	1.30	10.19	4.44	0.44
Benin	0.40	0.03	6.27	15.04	2.40
Bolivia	2.25	0.05	8.33	46.86	5.63
Botswana	3.41	0.03	1.54	127.97	83.05
Brazil	239.36	10.65	170.41	22.48	0.13
Bulgaria	3.48	8.21	7.95	0.42	0.05
Cambodia	0.79	0.02	13.10	35.14	2.68
Cameroon	2.67	0.08	14.88	33.69	2.26
Canada	205.98	31.57	30.76	6.52	0.21
Central African Republic	0.23	‹0.01	3.72	56.08	15.09
Chad	0.30	‹0.01	7.89	93.97	11.92
Chile	26.29	3.16	15.21	8.33	0.55
China	593.70	161.97	1,282.44	3.67	‹0.01
Colombia	25.24	0.40	42.11	62.36	1.48
Congo, Dem. Republic	0.86	0.06	50.95	14.66	0.29
Costa Rica	4.16	0.46	4.02	9.11	2.26
Côte d'Ivoire	2.40	0.11	16.01	21.79	1.36
Czech Republic	20.97	1.47	10.27	14.31	1.39
Denmark	44.90	0.32	5.32	138.59	26.05
Ecuador	7.18	0.90	12.65	7.96	0.63
Egypt	24.03	9.57	67.88	2.51	0.04

Table 8.4: *continued*

Country	Industrial value added (IVA): 2001* (billion constant 1995 US $) (1)	Industrial water use: 2000 (Km³/year) (2)	Population: 2000 (million) (3)	Industrial water productivity (IWP) (US $ IVA/m³) (4) = (1)÷(2)	IWP per capita (US $ IVA/m³/c) (5) = (4)÷(3)
El Salvador	3.39	0.20	6.28	16.89	2.69
Estonia	1.72	0.06	1.39	26.80	19.24
Ethiopia	0.77	0.15	62.91	5.30	0.08
Finland	53.22	2.07	5.17	25.66	4.96
France	430.02	29.76	59.24	14.45	0.24
Gabon	2.85	0.01	1.23	198.17	161.12
Germany	748.18	31.93	82.02	23.43	0.29
Ghana	2.04	0.08	19.31	26.52	1.37
Greece	28.18	0.25	10.61	114.44	10.79
Guatemala	3.53	0.27	11.39	13.20	1.16
Guinea	1.59	0.03	8.15	45.85	5.62
Honduras	1.32	0.10	6.42	13.63	2.12
Hungary	17.26	4.48	9.97	3.85	0.39
India	120.24	35.21	1,008.94	3.42	‹0.01
Indonesia	94.42	0.56	212.09	169.18	0.80
Iran, Islamic Republic	29.51	1.69	70.33	17.50	0.25
Italy	332.94	16.29	57.53	20.44	0.36
Jamaica	1.95	0.07	2.58	28.08	10.90
Japan	1,889.94	15.80	127.10	119.62	0.94
Jordan	1.80	0.04	4.91	40.27	8.20
Kazakhstan	8.39	5.78	16.17	1.45	0.09
Kenya	1.32	0.10	30.67	13.60	0.44
Korea, Republic	285.64	3.05	46.74	93.66	2.00
Kyrgyz Republic	0.36	0.31	4.92	1.17	0.24
Lao PDR	0.59	0.17	5.28	3.46	0.66
Latvia	1.88	0.10	2.42	19.60	8.10
Lebanon	2.52	0.01	3.50	333.78	95.47
Lesotho	0.43	0.02	2.04	19.31	9.49
Lithuania	2.48	0.04	3.70	60.34	16.33
Malawi	0.28	0.05	11.31	5.90	0.52
Malaysia	48.65	1.90	22.22	25.58	1.15
Mali	0.81	0.02	11.35	49.92	4.40
Mauritania	0.31	0.05	2.67	6.34	2.38
Mexico	99.69	4.29	98.87	23.25	0.24
Moldova	0.78	1.33	4.30	0.59	0.14
Mongolia	0.26	0.12	2.53	2.06	0.81
Morocco	13.36	0.20	29.88	66.51	2.23
Mozambique	1.50	0.01	18.29	102.65	5.61
Namibia	0.94	0.01	1.76	73.34	41.74
Netherlands	119.90	4.76	15.86	25.17	1.59
New Zealand	15.85	0.20	3.78	79.26	20.98
Nicaragua	0.48	0.03	5.07	14.34	2.83
Niger	0.39	0.01	10.83	31.69	2.93
Nigeria	14.31	0.81	113.86	17.65	0.16
Norway	49.05	1.46	4.47	33.56	7.51
Pakistan	15.71	3.47	141.26	4.53	0.03
Panama	1.49	0.04	2.86	34.47	12.07
Papua New Guinea	1.65	0.03	4.81	51.24	10.66

For countries adopting poverty-reduction policies, industrial growth is highly desirable in order to diversify the economy

Table 8.4: *continued*

Country	Industrial value added (IVA): 2001* (billion constant 1995 US $) (1)	Industrial water use: 2000 (Km³/year) (2)	Population: 2000 (million) (3)	Industrial water productivity (IWP) (US $ IVA/m³) (4) = (1)÷(2)	IWP per capita (US $ IVA/m³/c) (5) = (4)÷(3)
Paraguay	2.61	0.04	5.50	61.90	11.26
Peru	17.08	2.03	25.66	8.42	0.33
Philippines	28.07	2.69	75.65	10.42	0.14
Poland	50.65	12.75	38.61	3.97	0.10
Portugal	36.71	1.37	10.02	26.87	2.68
Romania	12.32	7.97	22.44	1.55	0.07
Russian Federation	139.79	48.66	145.49	2.87	0.02
Rwanda	0.37	0.01	7.61	35.11	4.61
Senegal	1.42	0.06	9.42	24.32	2.58
Sierra Leone	0.17	0.01	4.41	24.86	5.64
South Africa	51.35	1.61	43.31	31.99	0.74
Spain	208.17	6.60	39.91	31.54	0.79
Sri Lanka	4.11	0.31	18.92	13.33	0.70
Sweden	81.68	1.61	8.84	50.67	5.73
Syrian Arab Republic	3.18	0.36	16.19	8.74	0.54
Tajikistan	0.70	0.56	6.09	1.25	0.21
Tanzania	1.07	0.03	35.12	42.31	1.20
Thailand	69.52	2.14	62.81	32.46	0.52
Togo	0.33	0.01	4.53	25.22	5.57
Trinidad and Tobago	3.46	0.08	1.30	41.98	32.44
Tunisia	7.04	0.07	9.46	105.03	11.10
Turkey	50.00	4.11	66.67	12.18	0.18
Turkmenistan	3.46	0.19	4.74	18.34	3.87
Uganda	1.34	0.05	23.30	29.40	1.26
Ukraine	21.62	13.28	49.57	1.63	0.03
United Kingdom	340.03	7.19	59.63	47.28	0.79
United States	2,147.80	220.69	283.23	9.73	0.03
Uruguay	5.32	0.04	3.34	147.69	44.26
Uzbekistan	2.79	1.20	24.88	2.33	0.09
Venezuela, Bolivian Republic	31.69	0.59	24.17	53.82	2.23
Viet Nam	10.89	17.23	78.14	0.63	0.01
Yemen	1.86	0.04	18.35	43.74	2.38
Zambia	1.11	0.13	10.42	8.45	0.81
Zimbabwe	1.58	0.12	12.63	13.15	1.04

*For some countries only 2000 statistics are available.

Note: Values for IVA and population have been rounded to two decimal places. *Source:* World Bank, 2001; FAO, 2003.

References and Websites

Asano, T. and Visvanathan, C. 2001. Industries and water recycling and reuse. *Business and Industry – A Driving or Braking Force on the Road towards Water Security.* Founders Seminar, organized by Stockholm International Water Institute, Stockholm, Sweden, pp. 13–24.

EC (European Commission). 2004. *European Pollutant Emissions Register.* Luxemburg, Office for Official Publications of the European Communities.

ISO (International Organization for Standardization). 2004. The ISO Survey of ISO 9000 and ISO 14001 Certificates, Twelfth cycle: up to and including 31 December 2002, ISO, Geneva.

Kuylenstierna, J. and Najlis, P. 1998. The comprehensive assessment of the freshwater resources of the world - policy options for an integrated sustainable water future. *Water International,* Vol. 23, No.1, pp. 17–20.

Levine, A. D. and Asano, T. 2004. Recovering sustainable water from wastewater. *Environmental Science & Technology.* June, 2004, pp. 201-08.

McDonough, W. and Braungart, M. 2002. *Cradle to Cradle: Remaking the Way We Make Things.* New York, North Point Press.

Morrison, J. and Gleick, P. 2004. *Freshwater Resources: Managing the Risks Facing the Private Sector,* Pacific Institute, Oakland, California.

Shiklomanov, I.A. 1999. *World Water Resources and their Use.* Paris, UNESCO and the State Hydrological Institute, St Petersburg.

UN (United Nations). 2002. Johannesburg Plan of Implementation. New York, UN. www.un.org/esa/sustdev/documents/WSSD_POI_PD/English /POIToc.htm

UNECA (United Nations Economic Commission for Africa). 2002. The Way Forward. Addis Ababa, UNECA.

UNIDO (United Nations Industrial Development Organization). 2004. *Industry, Environment and the Diffusion of Environmentally Sound Technologies.* Annual Report, 2004. Vienna, UNIDO.

——. 2003. Identification, assessment and prioritisation of Pollution Hot Spots: UNIDO Methodology.

Unilever. 2003. *Unilever Environment Report.* Unilever, N.V. Netherlands, Unilever.

WHO (World Health Organization). 2006. Guidelines for the Safe Use of Wastewater, Excreta and Greywater. Geneva, WHO.

——. 2004. Guidelines for Drinking Water Quality. Geneva, WHO. www.who.int/water_sanitation_health/dwq/gdwq3/en/inde x.html

World Bank. 2003 *World Development Indicators.* New York, World Bank.

Basel Convention: www.basel.int/

Chronology of major tailings dams failures: www.wise-uranium.org/mdaf.html

The European Pollutant Emission Register: www.eper.cec.eu.int/eper/default.asp

The EU Water Framework Directive: europa.eu.int/comm/environment/water/water-framework/index_en.html

International Standards Organization: www.iso.org

Pacific Institute: www.pacinst.org

Rotterdam Convention: www.pic.int/

Stockholm Convention: www.pops.int/

UNECE Convention on the Protection and Use of Transboundary Watercourses and International Lakes (Water Convention): www.unece.org/env/water/

UNECE Convention on the Transboundary Effects of Industrial Accidents: www.unece.org/env/teia/welcome.htm

UNEP description of technological disasters: www.uneptie.org/pc/apell/disasters/lists/technological.html

UNEP Disaster database: www.uneptie.org/pc/apell/disasters/database/disastersdatabase.asp

UNEP Division of Technology, Industry and Economics: www.uneptie.org/

UNIDO: www.unido.org/

Unilever: www.unilever.com/environmentsociety/

World Resources Institute: www.wri.org/

World Water Resources and Their Use – a joint SHI/UNESCO product: webworld.unesco.org/water/ihp/db/shiklomanov/index.shtml

And what is a man without energy?
Nothing – nothing at all.

Mark Twain

CHAPTER 9

Water and Energy

By

UNIDO

(United Nations Industrial Development Organization)

Colorado River dam in Arizona, United States

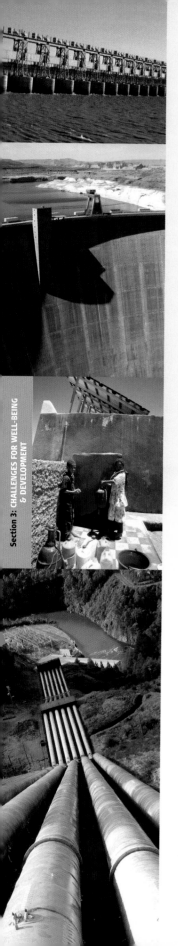

Key messages:

Water and energy are two highly interconnected sectors: energy is needed throughout the water system, from supplying water to its various users, including urban people, to collecting and treating wastewater. On the other hand, water is essential to producing energy, from hydropower to water cooling in power stations.

In the context of a growing world population, leading to increasing demands and competition for water and energy, it is time to integrate the management of these resources. This chapter takes stock of the various possibilities to be explored in order to enhance water and energy efficiency and ensure sustainable development.

While much progress has yet to be made for ensuring universal access to water supply and sanitation, even more progress is needed to provide electricity for all. In order to achieve these challenging and urgent targets, water supply and energy production systems both need improvements that do not jeopardize the environment.

■ There are very strong links between water and electrical power usage which at present are not fully taken into account in policy-making, management and operation of both water and electricity generation systems. The consequence is that many opportunities for both energy and water savings are being lost.

■ Access to electricity for many poor people in lower-income countries continues to lag a long way behind access to an improved water supply. Access to electricity plays a big role in poverty alleviation, improved health and socio-economic development. Accelerating access to electricity for the poor, although not one of the Millennium Development Goals (MDGs), was one of the targets set at the World Summit on Sustainable Development (WSSD) in Johannesburg in 2002.

■ Concern about the impact on the environment of traditional methods of electrical power generation is driving the introduction of a variety of non-polluting, renewable energy sources. However, economies of scale on large thermal and hydropower plants, existing transmission/distribution grids plus government subsidies for these traditional systems, put the renewable approaches at a cost disadvantage. A wide range of renewable electricity production options is now available, together with a growing range of incentives and economic instruments to promote their use and also to promote increased efficiency of energy usage.

■ Hydropower is available at different scales from very large systems to small systems. It is very flexible, permits rapid start up and can augment both thermal power plant base loads at peak times and compensate for fluctuating renewable supplies, as well as providing stand-alone generating capacity for smaller and remote communities. There is some controversy over whether large hydropower schemes are renewable power sources, but run-of-river systems are and there are now many options to increase sustainability.

■ The supply of water and wastewater services of all kinds to urban areas generally involves high electrical energy consumption. However, by taking a total system approach to energy management in these systems, including energy audits, it is possible to achieve big energy savings. Desalination of saline and brackish water for urban water supply is growing as technology improvements bring significant decreases in costs.

■ Experience has shown that the simultaneous analysis of water and energy use at the policy level can enable significant increase in productivity in the use of both resources. Water conservation can lead to large energy savings, as can taking full account of energy efficiency approaches in water policy decisions.

This page from top: Kut Al Amara dam, Iraq; Glen Canyon Dam, Arizona; Villagers draw water from a pump powered by solar panels, Tata, Morocco; Water pipeline transporting water up to a valley accumulation station

Right: The Blue Lagoon is an artificial lake fed by the surplus water drawn from the geothermal power station at Svartsengi, Iceland. Captured at 2,000 m below ground, the water reaches the surface at a temperature of 70°C, at which point it is used to heat neighbouring cities

Water and electricity use are inextricably linked. Large quantities of water are used for cooling in many electricity generating methods, such as coal and nuclear power stations. Hydropower, while not a consumptive use of water, often requires the construction of reservoirs and other large engineering works, which modify the aquatic environment. Conversely, large amounts of electrical power are used to pump water from its sources to the places it is used, especially in irrigated agriculture and municipal water systems.

Further links between the water and energy sectors are created by the frequent inefficiency and wastage in the way both resources are used. There are serious inefficiencies in many parts of the world in electricity generation, transmission, distribution and usage. Likewise, there are inefficiencies and leaks in water distribution systems. It follows that substantial efficiency gains in water use will reduce electric power requirements, which in turn will lead to more savings of water otherwise used in power generation.

A great deal of the infrastructure for both power and water in middle- and lower-income countries is poorly maintained. There is also a serious lack of the infrastructure needed to extend necessary power and water services to the many people presently unserved. Access to electricity for the poor lags far behind access to drinking water supply in many countries. For example, in sub-Saharan Africa, only 25 percent of the population have access to electricity, while 83 percent of the urban population and 46 percent of the rural population have access to a water supply (see **Table 9.5** at the end of this chapter).

Thus there is great pressure on governments in developing countries to build power stations and to deliver more electricity for domestic use and industrial development. Yet increasing power generation through burning coal, oil and gas presents its own set of sustainability issues linked to the generation of carbon dioxide (CO_2) and the greenhouse effect. The majority of electricity worldwide is generated by fossil fuel power stations from which emissions exacerbate the problems of climate variability and changes, raising the intensity of natural disasters, which mainly impact the poor. Moving away from a carbon-rich power-generating environment to more sustainable generation methods and reducing the inefficiencies mentioned earlier will help to alleviate this problem.

In the rapidly growing urban environments in developing countries, energy costs draw budgetary resources from other municipal functions, such as education, public transportation and health care. Without the provision of reliable sustainable energy supplies, it is unlikely that the Millennium Development Goals (MDGs) of reducing hunger, providing safe drinking water, providing sanitation and improving health will be achieved.

Access to electricity for the poor lags far behind access to drinking water supply in many countries...in sub-Saharan Africa, only 25 percent of the population have access to electricity...

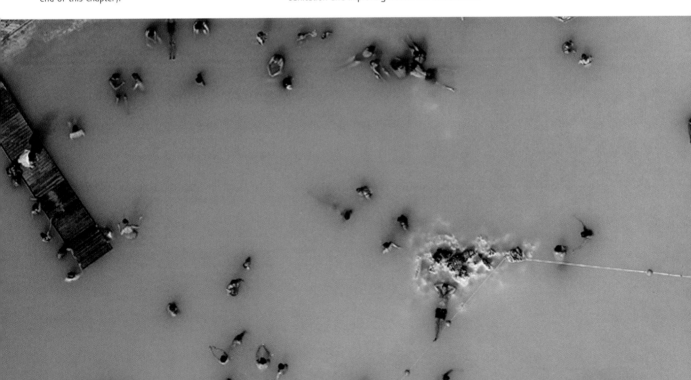

Part 1. Energy for Water Supply

In many countries, agricultural irrigation, groundwater pumping, interbasin transfers, and urban water supply and sanitation systems are major electricity users. Improving water use efficiency and introducing water conservation measures can therefore significantly reduce energy use. This section sets out to show how the two issues are interlinked, focusing upon urban water supply, and how the two systems should be co-managed, with future implications for both water and energy policies.

...the energy required by the end uses of water is far greater than in the other steps of the urban water cycle

1a. Energy use in water supply and sanitation services

Nearly all modern urban water and wastewater systems require energy in all phases of the treatment, delivery, collection, treatment and disposal cycle. Where historical systems once depended on surface water sources, gravity distribution systems and dilution for wastewater treatment, the water needs of growing urban areas need additional energy input to safeguard human health.

Extraction, conveyance and treatment

The first step of the urban water cycle is extraction, conveyance and treatment. The most widely used sources of potable water are surface sources and groundwater wells. The use of a particular source in a region depends on the availability and the cost of water extraction. Surface sources such as lakes, rivers and reservoirs typically require some treatment to achieve potable quality. The quality of the water body and the desired level and type of treatment are key variables in energy inputs required at this step. Groundwater sources have a more direct energy requirement, since energy is needed to pump the water up out of the ground, usually from bore. The amount of energy required by a pump and motor system to extract groundwater depends on the depth of the water table beneath the ground. It is important to note that water recycling and reuse, including a treatment step, is generally far less energy-intensive than developing any new physical source of water, other than local surface water.

Distribution

Distribution of potable water is often the most variable step in the urban water cycle. Ideally, the water source is at a higher elevation than the intended destination. In this case, gravity is used to distribute water and no energy input is required. In most cases, though, varied topography requires energy input through booster pumps to generate sufficient pressure in the system to distribute water to communities at higher elevations. Pumped storage,[1] which is further discussed later in this chapter, is often used at this stage to take advantage of off-peak

energy rates, converting pumping mechanical energy to potential energy by storing water at higher elevations. There are areas where conveying water can be highly energy intensive. Since water conservation saves all of the upstream energy inputs as well as the end-use energy inputs, water conservation in areas with energy intensive water supplies will save substantially more energy than water conservation in other areas.

Consumer end-use

Once water is delivered for consumer use, additional energy inputs come from heating and cooling water. Residential and commercial consumers heat water for bathing, radiant heating and dishwashing and cool water for air conditioning. Industrial consumers vary widely in their needs for heated and cooled water based on the industrial application and their process needs. However, the energy required by the end-uses of water is far greater than in the other steps of the urban water cycle. While there are efficiency improvements that can reduce the energy inputs required at each stage of the water use cycle, the greatest energy and water savings come from reducing water consumed by various end-uses (see **Box 9.1**). Water conservation at the end-use stage eliminates all of the upstream energy required to bring the water to the point of end-use, as well as all of the downstream energy that would otherwise be spent to collect, treat and dispose of this water.

Wastewater collection

Like distribution, wastewater collection is ideally done using gravity systems. When this is not possible, pumps are used to boost the wastewater to treatment facilities. In combined sanitary and storm-water sewers, precipitation affects the energy requirements of collection systems and heavy rains sometimes overwhelm the available infrastructure.

Wastewater treatment requires energy to remove contaminants and prepare the water for discharge or reuse. In aerobic wastewater treatment, the largest energy input is in the aeration system itself. Some types of wastewater

1. This involves pairs of reservoirs with a significant height difference. Water is pumped up when there is spare capacity in the network and then allowed to flow down again and generate power at times of peak demand.

treatment require very little energy (e.g. lagoons) but large amounts of land. In urban areas where land is scarce, more energy is required to treat large amounts of wastewater in a treatment plant requiring less land area. Opportunities exist to recover some of the energy embodied in the organic material present in wastewater, by recovering methane gas through anaerobic treatment and then using this fuel to power the treatment facility. Indeed, some wastewater treatment plants even provide electricity to the national grid.

1b. Approaches to energy and water efficiency
Because of the interconnectedness of water and energy, it is vital to manage them together rather than in isolation. The energy savings from water conservation and the water savings from energy efficiency are inextricably linked, and these linkages should be considered when determining the best course of action from an economic, social or environmental perspective. Energy efficiency in the water and wastewater industry saves money in operations and maintenance costs, reduces capital costs of new supply, improves solvency and operations capacity of water utilities, improves service coverage, reduces emissions and improves water quality, among a host of other related benefits.

In order to support larger efforts to reduce energy use in water and wastewater systems, larger-scale energy and water management should be entrusted to the local level for implementation. The term 'watergy' efficiency has

been coined to describe the combined water and energy efficiencies which are available to municipalities and water users (see **Box 9.2**).

Involvement of the energy utility provides the needed support for implementing energy efficiency measures and ensuring that efforts to reduce energy and water waste are sustainable as a business practice. Energy efficiency in any water utility never has a beginning or an end. To sustain its energy savings, a water utility must continue to monitor its energy use and set goals for improvement.

Identifying water/energy efficiency opportunities
Energy and water audits are used to identify areas of concern in water and wastewater systems. The boundaries of the system to be audited are usually chosen based on budgetary considerations and areas that are presumed to yield the largest energy savings for the investment.

Major areas that are frequently identified as water/energy savings opportunities in water supply systems include the following:

- Repairing leaks from valves, distribution pipes, etc. Many urban water distribution systems in developed cities were installed more than fifty years ago, and leaks caused by corrosion of pipe material or other problems

Small-scale hydro-energy generation in Lao Cai, Viet Nam

BOX 9.1: WATER CONSERVATION VERSUS ENERGY CONSERVATION

Energy intensity measures the amount of energy used per unit of water. Some water sources are more energy intensive than others; for instance, desalination requires more energy than wastewater recycling. Water conservation technology may either increase or decrease energy intensity. Yet when water planners make decisions, they should look not just at energy intensity, but also at the total energy used from source to tap. In the case of water conservation, some programmes may consume a lot of energy at one stage in the energy/water use cycle, but still decrease the amount of energy used overall. The following three examples illustrate the interplay between energy intensity and total energy use:

- Water conservation may increase energy intensity and increase total energy use: A particular irrigation technology could reduce water use by 5 percent but require so much energy to operate that it increases the energy intensity by 10 percent. This would increase total energy use by 4.5 percent.

- Water conservation may increase energy intensity and decrease total energy use: The average high-efficiency dishwasher increases the energy intensity of dishwashing by 30 percent, but reduces water use by 34 percent. As a result of using less water (and therefore less energy to convey water from the source to the dishwater) the net total energy needed to wash dishes declines by 14 percent.

- Water conservation may decrease energy intensity and decrease total energy use: The average high-efficiency clothes washer reduces water use by 29 percent, compared to average low-efficiency machines, and simultaneously lowers energy intensity by 27 percent. Energy intensity declines, because mechanical aspects of the machines are also improved. By reducing total water use as well as energy intensity, total energy use is reduced by 48 percent.

Source: NRDC, 2004.

Pumping systems are significant energy consumers in most water distribution systems, and the potential for substantial energy reduction exists in most water facilities throughout the world

Mobile solar water-heating equipment provided by an environmental centre in a slum area, Johannesburg, South Africa

can lead to the loss of significant amounts of potable water. Distribution system losses increase the energy intensity of water supply by requiring utilities to treat and convey water that will be lost. Losses vary significantly among urban water suppliers and range as high as 30 percent in developed cities. In developing cities, the proportion of water lost to leakage may be even higher, typically 40 to 60 percent. (Only about 2 percent of this lost water goes to unmetered uses, such as firefighting and construction.)

■ Correct sizing, design and maintenance of the pump and motor system. Pumping systems are significant energy consumers in most water distribution systems, and the potential for substantial energy reduction exists in most water facilities throughout the world. Small improvements in efficiency can be achieved by installing the most efficient pump equipment available. More significant, however, is the saving potential through optimizing the entire pumping system, including the pump, motor, drive, controls, piping, valves and any other ancillary equipment. Similarly, for large systems made up of multiple pumps in parallel or series, opportunities exist for optimizing control strategies. This type of optimization can be carried out through adopting the systems approach.

The systems approach

The cost-effective operation and maintenance of a pumping system requires attention not only to the needs of the individual pieces of equipment, but also to the system as a whole. A 'systems approach' considers both the supply and demand sides of the system and how they interact, essentially shifting the focus from individual components to total system performance. Often, operators are so focused on the immediate demands of the equipment that they overlook the broader question of how system parameters are affecting the equipment. For example, the frequent replacement of pump seals and bearings can keep a maintenance crew so busy that they overlook the system operating conditions that caused the problems in the first place. **Box 9.2** provides an example of the systems approach as applied in California.

1c. Desalination

Ninety-seven percent of the world's water is too salty for consumption or agriculture. Desalination is not a new

concept, as it has been practised since biblical times. However, the process typically consumes large quantities of energy in order to produce drinking water from seawater or polluted water, making energy cost the major determinant of the desalination cost. Hence desalination technology has tended to be used in water-scarce countries where energy is cheap and plentiful. (see **Table 9.1**) Some 65 percent of the world's desalination plants are located in the Arabian Gulf countries.

Desalination can be achieved either by removing salt from water, or by removing pure water from a saline or polluted source. For producing large quantities of freshwater from a saline source, it is necessary to remove the water from the salt. This process leaves behind a highly concentrated saline solution, or brine, which must be disposed of as a waste product, often in the sea.

Traditionally, thermal desalination or distillation has been the most commonly used technology for producing large quantities of freshwater from seawater. Different thermal desalination processes require different magnitudes and combinations of heat and electricity. The economic efficiency of desalination plants is improved by combining the purposes of power and water production. Most of the

Table 9.1: Volume of desalinated water produced, selected countries, 2002

Country	Desalinated water (million m³/year)
Kazakhstan	1,328.0
Saudi Arabia	714.0
United Arab Emirates	385.0
Kuwait	231.0
Qatar	98.6
Libyan Arab Jamahiriya	70.0
Algeria	64.0
Bahrain	44.1
Oman	34.0
Malta	31.4
Egypt	25.0
Yemen	10.0
Tunisia	8.3
Morocco	3.4
Iran, Islamic Rep.	2.9
Jordan	2.0
Mauritania	1.7
Turkey	0.5
Sudan	0.4
Somalia	0.1
Djibouti	0.1

Source: FAO's AQUASTAT, 2003.

BOX 9.2: ENERGY CONSERVATION IN THE MOULTON NIGUEL WATER DISTRICT, CALIFORNIA

The Moulton Niguel Water District, located in southern California, has a water system capacity of 181 million litres per day and a wastewater system capacity of 64 million litres per day. The water district began investigating energy efficiency measures when facing significant increases in energy costs. The water district staff used the systems approach when evaluating both their water and wastewater facilities. Changes implemented included the following:

■ installing an energy management system using programmable logic controllers that activate and de-activate pumps at seventy-seven district pumping stations to take advantage of off-peak electricity rates

■ installing variable frequency drives onto the wastewater pumps to reduce motor wear, improve control over lift station wastewater levels, and help prevent stagnant sewage in pipes

■ specifying high-efficiency (95 to 97 percent) electric motors for all new construction, while establishing a policy to replace existing motors as they fail or approach the end of their useful life.

The annual savings attributed to these efficiency improvements is over US $330,000, representing a reduction in electricity costs of approximately 25 percent.

Source: Alliance to Save Energy, (www.ase.org).

BOX 9.3: DESALINATION USING RENEWABLE ENERGY, GREECE

Research being carried out in Greece has linked a reverse osmosis unit for membrane desalination to a wind generator and a solar photovoltaic array, in order to create a unit that can be used in remote areas where there is no potable water and no electricity grid. The unit currently produces 130 litres per hour (L/h) of potable water from seawater containing approximately 37,000 parts per million (ppm) of total dissolved solids (TDS), while processing 1,000 L/h of seawater. The recovery ratio is approximately 15 percent, which is low compared to other systems. During a Greek summer, the unit can operate for an average of eight hours per day on solar energy, while in winter the operating time drops to an average of five hours per day; however, the operating time is boosted when wind energy is available. This is a promising technology for helping remote coastal areas to improve access to drinking water supplies.

Source: Martinot, 2004.

desalination plants operating in the Middle East and elsewhere are dual-purpose multistage flash distillation plants that produce both water and electricity, using oil as the energy source. However, oil price rises undermine the economic performance of these plants, even in the Arabian Gulf region. As a result, nuclear power is increasingly being considered as a viable energy source for thermal desalination plants, particularly in countries that have local uranium reserves. The advantages include fuel price stability and the long-term availability of the fuel, but these need to be balanced against the well-known drawbacks of high initial investment costs and the disposal of spent nuclear fuel.

Reverse osmosis (membrane desalination) is an electrically-driven process that uses special membranes through which water molecules may pass under pressure, leaving behind larger molecules, including salt. The capital cost of reverse osmosis units is dropping, and they are now the most common choice for new desalination plants. The water utility Thames Water in the UK is currently investing £300 million (US $539 million) in a reverse osmosis plant to treat water from the tidal estuary of the River Thames. The plant, due for completion in 2007, will serve 900,000 customers in London, producing up to 150,000 m^3/day of drinking quality water. This technology can be combined with renewable energy processes, as shown in **Box 9.3**.

Some countries, such as Spain, increasingly prefer the desalination option to environmentally damaging inter-basin transfers, in order to bring water to arid coastal areas. However, before advocating moving forward with large-scale desalination, it must be recognized that relatively little is yet known about its impact on marine and coastal environments. Few studies have been conducted on marine resource impacts from the large-scale desalination facilities in the Middle East. The range of potential adverse environmental impacts that may arise from new desalination facilities include the impacts from construction, waste discharge, injury and death of aquatic life from water intakes, and the secondary impacts of

Countries such as Spain are increasingly choosing the desalination option in preference to environmentally damaging inter-basin transfer...

Reverse osmosis desalination unit in the Virgin Islands, United States

By 2003 there had been about 29 million domestic solar water heaters installed worldwide, of which 21 million were in developing countries

increased energy consumption. These must all be explored and adequately addressed before the development of desalination facilities begins. In most cases, water conservation and water recycling offer cheaper and better alternatives. The heavy energy cost of desalination also suggests the need to consider desalination plants as an emergency water supply, to be used during water demand peaks or droughts, rather than as a base supply.

1d. Solar energy for water supply

Small and inexpensive solar units are now available for many water-related applications, including pumping, water purification and solar water heating.

Solar pumping

Solar power can be used to help achieve the MDG of providing safe and accessible drinking water in countries that have plenty of sunlight. The great potential of solar pumping is to bring freshwater to villages that have no electricity and pump groundwater. There are many different types of solar pumps now available for various applications. At present, sales of solar pumps are largely to developed countries, because the prices of the systems are still rather high, but they are dropping rapidly as demand grows.

Solar water purifiers

The simplest and cheapest solar water disinfection system has been named SoDis (Solar Disinfection), and is designed for use at the household level. It improves the microbiological quality of drinking water by using solar ultraviolet-A (UV-A) radiation and heat to inactivate the pathogens that cause diarrhoea. The system uses commonly available plastic soft drink bottles. Contaminated water is filled into the transparent plastic bottles and exposed to full sunlight for six hours. The water must be relatively clear, and the bottles must be clean and unscratched. The required heating can be achieved by placing the bottles on a corrugated iron sheet or on a rooftop.

A more sophisticated solar water disinfection system called Naiade has been developed for use in developing countries. It produces safe drinking water from polluted water in a sustainable manner, without the use of chemicals, by means of UV irradiation. The unit weighs 44 kg and can produce on average up to 2,000 litres per day of high-quality drinking water. Water from a well or surface source is poured into the unit, either by hand or pipe. The water passes through a sieve, which removes large impurities, then through two filters, which remove

microscopic particles (including nematodes), and finally under an ultraviolet lamp. The ultraviolet light kills bacteria, viruses and worm eggs. It can be activated by the use of an electric battery, by the connection to electricity mains or by using a 75-watt solar panel. Maintenance and management of the unit is simple: if the filters become blocked, they can be easily cleaned by hand, which needs to be done daily.

Heating water for domestic use

Solar thermal capacity for domestic hot water and space heating is growing rapidly. Worldwide, the sector grew by 16 percent in 2003, while in China it grew by 30 percent. Although some developing countries are located in warm or tropical climates where hot water is not of primary importance, in many areas, especially those that are mountainous, there is a considerable demand for hot water. Solar water heaters are especially useful in the tourism sector and the hotel industry, as well as laundries, hospitals and clinics. Where solar water heaters displace electrical ones, they play a significant role in reducing peak electricity demand and reduce the negative environmental impact of fossil fuel use. Appropriate policies and economic incentives need to be put in place to stimulate the spread of this technology.

By 2003 there were about 29 million domestic solar water heaters installed worldwide, 21 million of which were in developing countries. Several million are located in China and India, while Egypt and Turkey have hundreds of thousands of households served by solar water heaters. In Barbados there are over 35,000 solar water heaters installed (33 percent of all households). Each unit saves about 4,000 kWh per year. This represents a considerable foreign exchange savings on the import of diesel fuel for the island, in addition to avoiding carbon emissions. It has been calculated that these solar water heaters replace 30 to 35 MW of additional electric generating capacity that would otherwise have to be installed in Barbados.

The success of solar water heaters in Barbados was supported through various governance mechanisms. A 30 percent consumption tax was put on electric water heaters; furthermore, the cost of electricity is relatively high in Barbados, which is also an incentive. Homeowners can gain concessions on their mortgages by installing solar water heaters. In Australia, each solar water heater with an electricity equivalent of 1 MWh over its lifetime receives between ten and thirty-five green certificates. These certificates have an economic value (US $18 in 2002), since electricity suppliers are obliged to purchase a certain share of electricity from renewable energy sources, which they can prove by presenting a corresponding number of green certificates. In other countries, different means have been used to foster the use of solar water heaters, including direct grants. In Namibia, the government requires solar water heaters to be installed in the construction of all new government housing, while in India, the government has introduced accelerated depreciation for commercial and public applications of solar water heaters.

Part 2. Water for Energy Generation

Hydropower, and small hydropower (SHP) in particular, is recognized as a flexible and affordable renewable energy source. Its role in electricity generation, especially in rapidly developing countries, is crucial. The World Commission on Dams (WCD, 2002) focused attention internationally on the negative environmental and social impacts of large dams, which raised questions about the environmental sustainability of large hydropower projects. However, only about 25 percent of the world's large dams are involved in producing hydropower. The rest were built for other purposes, mainly for irrigation, but also for water storage, for recreation and for assisting in river transport. Conversely, many large hydropower projects are run-of-river projects, which do not necessitate the building of a dam, while the role of small, mini- and micro-hydropower schemes is becoming increasingly important in the energy security of many countries, led by the example of China. It is therefore important to disassociate a discussion of the role of hydropower from the debate over large dams, while not glossing over the environmental and social considerations involved in the choice of technology.

Electricity plays a key role in reducing poverty, promoting economic activities and improving quality of life, health, and education opportunities...

2a. Hydropower in context

Governments have a pressing need to provide, at an affordable price, the convenience and reliability offered by electricity. The role of energy, and electricity in particular, in meeting development targets was discussed in depth in the first edition of the *UN World Water Development Report*. Statistics show that for many developing countries, access to electricity lags behind access to an improved water supply (see **Table 9.5** at the end of the chapter). Although improving access to electricity is not one of the MDGs, it was a target at the Johannesburg Plan of Implementation adopted at the World Summit on Sustainable Development (WSSD) in 2002 (see **Box 9.4**). Electricity plays a key role in reducing poverty, promoting economic activities and improving quality of life, health, and education opportunities, especially for women and children.

Since 1970, as worldwide demand for electricity has steadily increased, governments have met this demand through increasing thermal (gas, oil, coal and nuclear), as well as hydropower generation capacity. Although the share of hydropower in total world energy supply was only 2.2 percent in 2002, hydropower accounted for 19 percent of all electric power generated (see **Figures 9.1** and **9.2**).

Over the same time period, there has been a perceptible increase in the use of other renewable energy sources (geothermal, solar photovoltaics, wind, and combined heat and power[2] [CHP]). **Table 9.2** shows the renewable power capacity in all countries and in developing countries in 2003. Environmental concerns, particularly over climate change and nuclear waste disposal, as well as safety and security of supply, have prompted governments to introduce policies aimed at accelerating the penetration of renewables and CHP (see **Box 9.5**). Total worldwide investment in renewable energy rose from US $6 billion in 1995 to approximately US $22 billion in 2003, and is increasing rapidly.

The economies of scale available to the thermal and hydropower options and the existence of transmission and distribution grids continue to give them a significant cost advantage when compared with renewables. Both the thermal and hydropower options, particularly when used together, offer the load-following capability and reliability demanded by electricity consumers. Subsidies of all types have historically been used worldwide to establish a top-down energy supply system favouring thermal and large hydropower generating plants of ever-increasing capacity. However, both thermal and large hydropower options bring

2. CHP is the simultaneous generation of electric power and steam used for heating.

BOX 9.4: WORLD SUMMIT ON SUSTAINABLE DEVELOPMENT: ENERGY TARGETS

■ 'Take joint actions and improve efforts to work together at all levels to improve access to reliable and affordable energy services for sustainable development sufficient to facilitate the achievement of the Millennium Development Goals, including the goal of halving the proportion of people in poverty by 2015, and as a means to generate other important services that mitigate poverty, bearing in mind that access to energy facilitates the eradication of poverty' (Target II.9).

■ 'Improve access to reliable, affordable, economically viable, socially acceptable and environmentally sound energy services and resources, taking into account national specificities and circumstances, through various means, such as enhanced rural electrification and decentralized energy systems, increased use of renewables, cleaner liquid and gaseous fuels and enhanced energy efficiency...' (Target II.9a).

■ 'Assist and facilitate on an accelerated basis ... the access of the poor to reliable, affordable, economically viable, socially acceptable and environmentally sound energy services, taking into account the instrumental role of developing national policies on energy for sustainable development, bearing in mind that in developing countries sharp increases in energy services are required to improve the standards of living of their populations and that energy services have positive impacts on poverty eradication and improve standards of living' (Target II.9g).

■ 'Diversify energy supply by developing advanced, cleaner, more efficient, affordable and cost-effective energy technologies, hydro included, and their transfer to developing countries on concessional terms as mutually agreed. With a sense of urgency, substantially increase the global share of renewable energy sources with the objective of increasing its

contribution to total energy supply... ensuring that energy policies are supportive to developing countries' efforts to eradicate poverty, and regularly evaluate available data to review progress to this end' (Target III.20e).

■ 'Assist developing countries in providing affordable energy to rural communities, particularly to reduce dependence on traditional fuel sources for cooking and heating, which affect the health of women and children' (Target VI.56d).

■ 'Deal effectively with energy problems in Africa, including through initiatives to... support Africa's efforts to implement NEPAD objectives on energy, which seek to secure access for at least 35 percent of the African population within twenty years, especially in rural areas' (Target VIII.62j).

Source: UN, 2002.

BOX 9.5: CLIMATE CHANGE AND ATMOSPHERIC POLLUTION: POWER GENERATION FROM FOSSIL FUELS

In the industrialized world, the future of fossil fuel-based electricity generation will be largely determined by requirements for reducing greenhouse gas emissions. Targets established under the Kyoto Protocol to the UN Framework Convention on Climate Change amount to an aggregate reduction shared among all Parties to the Protocol of at least 5 percent from 1990 levels by 2008–12. As its Kyoto commitment, the European Union (EU) agreed to an 8 percent reduction shared between its Member States. The EU has also established a plan whereby sources from which emissions are to be capped may trade their emissions allowances (the EU Emissions Trading Scheme). The majority of

capped emissions sources under this scheme are coal-based power plants.

The EU Large Combustion Plant Directive establishes limits for emissions to air for nitrogen oxides, sulphur dioxide and particulates (dust) from combustion plants with a thermal input of 50MW or more. Similar environmental legislation is in place in the United States, other Organisation for Economic Co-operation and Development (OECD) Member States and several developing nations. Limits on emissions of heavy metals and organic pollutants from fossil fuel combustion can also be expected. Today, flue gas filtration and other emissions control

technologies have enabled new fossil fuel-fired power plants to meet these requirements.

Unlike oil and gas, worldwide coal reserves are plentiful and sufficient for the next 200 years. Coal deposits are widely distributed geographically, and coal is traded internationally. Several developing nations can be expected to continue using coal for decades to come. China is adding 15-20 gigawatt-equivalents (GWe) of new coal-fired capacity each year. Decarbonization of fossil fuels, particularly coal, is being developed as an interim measure, together with carbon sequestration, bridging the gap towards a fully renewable energy system.

Table 9.2: Grid-based renewable power capacity in 2003

Generation type	Capacity in all countries (gigawatts)	Capacity in developing countries (gigawatts)
Small hydropower	56.0	33.0
Wind power	40.0	3.0
Biomass power*	35.0	18.0
Geothermal power	9.0	4.0
Solar photovoltaics (grid connected)	1.1	<0.1
Solar thermal power	0.4	0
TOTAL RENEWABLE POWER CAPACITY	**141.5**	**58.0**
For comparison: Large hydropower	674.0	303.0
Total electric power capacity	3,700.0	1,300.0

* Excluding municipal solid waste combustion and power from landfill gas.

Source: Adapted from Martinot, 2002.

Figure 9.1: Global generation of electricity by source, 1971–2001

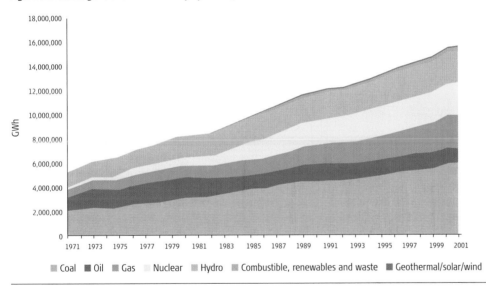

Source: International Energy Agency, 2004.

Figure 9.2: Total primary energy supply by source, 2002

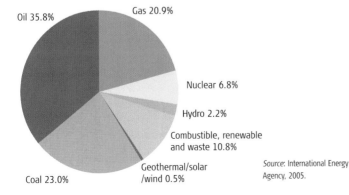

Source: International Energy Agency, 2005.

Solar power plant in Whitecliffs, Australia

...hydropower offers great benefits and holds a unique place in the range of energy options for electricity generation which are currently available

environmental problems and potentially unacceptable social consequences, such as the displacement of inhabitants for the construction of large dams, which now constrain their deployment. Hence more sustainable approaches, small-scale alternatives and distributed generation are gaining ground in many countries.

In 2001, hydropower generated 2,740 terawatt hours (TWh), or 19 percent of the world's electricity. This is the equivalent of 2.1 billion metric tons of CO_2 emissions, if that energy had been generated by oil, gas or coal power stations. The usage of hydropower varies greatly from country to country. Twenty-four countries generate more than 90 percent of their electricity through hydropower, whereas others generate none at all. Europe makes use of 75 percent of its hydropower potential, while Africa has developed only 7 percent. This is seen to be the possible future cornerstone of Africa's development, including significant export potential, with plans to establish a continent-wide electricity grid (see **Box 9.6**). At the end of this chapter, **Table 9.6** shows countries' capability for providing energy through hydropower.

There are several different types of hydropower, each suitable for different needs and circumstances:

- **Reservoirs:** This type of hydropower involves the construction of a dam (large or small) and the formation of a reservoir. Such construction is usually multi-purpose, both for water supply and for electricity production. This type of hydropower brings maximum flexibility of supply and maximum efficiency.

- **Pumped storage:** This involves pairs of reservoirs with a significant height difference. Water is pumped when there is spare capacity in the network and then allowed to flow down again and generate power at times of peak demand. It uses more power than it generates, but is essential as a flexible reserve and can make an electricity network more efficient.

- **Run-of-river:** This form of hydropower uses the stream's flow and has little or no reservoir capacity for storage or regulation. In social and environmental terms, it is seen as preferable to reservoir construction. Smaller hydropower schemes (including micro- and

---

BOX 9.6: THE DEVELOPMENT OF HYDROPOWER IN AFRICA

Africa is a heavy user of traditional (non-commercial) energy, namely biomass. Per capita electricity consumption is particularly low in Central, East and West Africa. In 2000, it was calculated that there were 7,730 megawatts (MW) of surplus installed generating capacity in the southern African grid. Hydropower presently provides 22 percent of electricity generation in Africa; nuclear power provides 2 percent, while thermal power stations provide 70 percent. However, reliance on hydropower is 80 percent or greater in Cameroon, the Democratic Republic of the Congo, Ghana, Mozambique, Rwanda, Uganda and Zambia. Because of the enormous potential of Africa's great rivers, particularly the Zambezi and the Congo, hydropower is seen as the motive force for future development in Africa.

Mozambique

Mphanda Nkuwa is a very large hydropower project by international standards. The planned capacity is 1,300 MW, making it slightly smaller than Cahora Bassa, which was commissioned in the mid-1970s and has an installed capacity of 2,075 MW. The site is located on the Zambezi River between Cahora Bassa and Tete. Mphanda Nkuwa is rated as one of the most attractive undeveloped hydropower projects in the world. The hydrological risk has been limited and well documented with long time-series of water flows. The geological risk is low and the dam site can be developed at a cost of US $640 per kilowatt of installed capacity. Since existing upstream dams, such as Cahora Bassa, Kariba and Kafue Gorge, regulate the Zambezi River, the project can be developed as a run-of-river hydropower plant requiring a small reservoir relative to its size, with very limited negative environmental impact.

Democratic Republic of Congo

The estimated cost of the 3,500 MW Inga III power station to be built on the Congo River is US $3,74 billion. The planned transmission lines of the so-called Western Corridor – with the termination points being up to 3,500 km from Inga – will require a further investment in the region of US $652 million. In addition, two 1,500 MW converter stations that will cost US $842 million are planned as part of the project. The aim is to have Inga III and the Western Corridor on line by 2015. This would open the way for further development of the Inga site, namely the Grand Inga project (a run-of-river project) with a generating capability of up to 39,000 MW. The economic feasibility of Grand Inga is seen to be dependent upon a continental market developing throughout Africa, with power ultimately being exported to North Africa and possibly even to Europe.

Source: UNECA, 2004.

pico-hydro) are usually run-of-river projects, but the technology is increasingly being applied in large schemes where the topography makes it feasible, as fast-flowing water is required.

Conventional hydropower (other than pumped storage) is vulnerable to droughts and seasonal fluctuations in rainfall. However, its value to the electricity system of a country is that it brings great flexibility. Electric power

cannot be stored, yet demand is constantly fluctuating in response to both predictable and unpredictable events.

Hydropower can be used at peak periods to supplement supply from less flexible thermal and nuclear power stations. Water can be held indefinitely in a reservoir and then released exactly when it is needed to produce power (pumped storage). It is particularly valuable in tandem with other renewable sources, such as solar or wind power, whose outputs wax and wane

---

BOX 9.7: SMALL HYDROPOWER IN CHINA

China is well known for its large hydropower schemes, such as the Xaolangdi power station on the Yellow River, which generates 1,800 megawatts (MW) of electricity, and the planned power station of the newly completed Three Gorges Dam on the Yangtze River, which will generate up to 16,000 MW. However, China has also attracted international attention due to its small hydropower (SHP) development, focusing on rural electrification.

Approximately one half of all commissioned SHP plants worldwide are located in China. Although the total feasible development potential of SHP is approximately 100,000 MW throughout China, the majority of the rapid recent development has been located in the southeast and southwest of the country. By the end of 2002, China had installed 28,489 MW of capacity through building 42,221 SHP plants. The unique features of China's SHP policy can be characterized by the following features:

- its decentralized approach
- its use of local grid structures (or mini-grids)
- the specific policies and strategies it adopts, particularly the overarching SHP policy of self-construction, self-management and self-utilization
- the popularization of a share-holding investment system
- its utilization of cost-effective SHP technology and equipment, the use of indigenous equipment manufacturers, and the prioritization of training.

SHP now accounts for about 30 percent of China's overall hydropower capacity. Three hundred million people in China now use SHP-derived electricity. The improvement in access to electricity in remote rural areas has been dramatic: 28 million people in China had no electricity in 2000, and that figure had dropped to 10.15 million by the end of 2002.

Nearly half of China's SHP generators are linked to local mini-grids (especially in mountainous areas), while only 10 percent are connected to the national grid. The remainder of SHP stations operate in isolation. Forty-four percent of China's SHP projects fall into the sub-category of micro-hydro, with capacity of less than 100 kW. Another 46 percent are mini-hydro, with a capacity up to 500 kW. The larger SHP projects – with a capacity between 500 kW and 25 MW – make up the remaining 10 percent of the projects, but account for 75 percent of the electricity output.

The construction of SHP-based local grids to serve specific rural supply areas is a unique electricity supply system developed by China. The rate of electrification in villages and rural households rose from 78.1 and 65.3 percent in 1985 to 97.7 and 97.5 percent in 2002, respectively. The quality of electricity supply was improved, and tariffs were reduced to be equal with that of urban centres. There are two types of SHP stations: those invested, owned and managed by local government and small Independent Power Producers (IPPs) developed with private investment. The recent

disposition of SHP stations in China, according to installed capacity, operation mode and ownership can be seen in **Table 9.3**.

The rapid development of SHP in China can be attributed to the following:

- **Preferential policies**: The Chinese Government introduced many preferential policies for SHP, such as tax reductions, soft loans/grants from government, encouragement of private firms to invest in SHP stations and policies protecting water supply areas and property ownership. The ratio of central government investment in SHP to that of the private/individual contribution is only 1:24.

- **Indigenous manufacturing capability**: In view of the fact that equipment costs form the largest percentage of the total cost of SHP development – unlike in large hydro, where civil works generally takes a higher proportion – the Chinese government decided to promote local manufacturing in order to reduce the overall cost of developing SHP stations.

- **Recognizing the advantages of SHP over large hydropower**: China has known for a long time that SHP has its own peculiar advantages that cannot be achieved through large hydropower generation.

Source: International Networking on Small Hydropower (www.inshp.org).

Pumped storage reservoirs are smaller than conventional reservoirs and less controversial, since they are less dependent on the topography

Table 9.3: Status of small hydropower stations in China in 2002

		SHP stations by Installed Capacity			
	Types	**Micro**	**Mini**	**Small**	**Total**
Station	Number	18,944	19,606	4,427	43,027
	percent	44.0	45.6	10.4	100
Installed capacity	MW	687	7,171	8,404	26,262
	percent	2.6	27.3	70.1	100
Annual output	GWh	1,860	20,245	65,036	87,141
	percent	2.1	23.2	74.7	100
		SHP stations by Operating Mode			
	Mode	**National Grid**	**Local Grid**	**Isolated**	**Total**
Station	Number	4,722	20,465	17,840	43,027
	percent	10.9	47.6	41.5	100
Installed capacity	MW	6,412	17,869	1,981	26,262
	percent	24.5	68.0	7.5	100
Annual output	GWh	20,097	60,792	6,252	87,141
	percent	23.1	69.8	7.2	100
		SHP stations by Ownership			
	Ownership	**State ownership**	**Others**	**Total**	
Station	Number	8,244.0	34,783	43,027	
	percent	19.2	80.8	100	
Installed capacity	MW	17,500	8,762	26,262	
	percent	66.6	33.4	100	
Annual output	GWh	62,954	24,187	87,141	
	percent	72.2	27.8	100	

Source: International Networking on Small Hydropower (www.inshp.org).

BOX 9.8: SMALL HYDROPOWER IN NEPAL

In Nepal, where almost 83 percent of the land is mountainous, grid extension is not usually cost-efficient, due to the high cost and the low load factor inherent in supplying power to remote and scattered settlements. Small-scale hydropower of less than 100 kW (micro-hydro), and less than 5 kW (pico-hydro) can be used in circumstances where hydrological conditions (availability and gradient of water flow) are adequate. Some 2,000 SHP generators represent a capacity of 13 MW as well as supplying mechanical power. Another forty small projects run by the Nepal Electricity Authority provide 19 MW of installed capacity. The Government, in its tenth five-year plan (2002-2007), has set itself a target of producing a further 10 MW of electricity from decentralized micro-hydro schemes, thus providing off-grid electricity to 12 percent of the population, mostly in rural mountainous areas, who are currently unserved.

Generally, micro- or pico-hydro electrification schemes in Nepal are privately or community-owned. Since it requires a significant amount of capital investment and organization to establish a micro-hydro scheme, it is more common for a community to get together and raise capital to build a scheme that serves their village than for a micro-hydro plant to be built privately. Community-owned schemes are also more likely to receive support from non-governmental organizations. However, a government subsidy is provided irrespective of who owns the project. A wide range of technology is used in micro- and pico-hydro schemes throughout the country. Very small hydropower schemes such as Peltric Sets, pioneered by a Nepalese manufacturer,

Kathmandu Metal Industries, are extremely popular because of their simplicity and low capital investment. The units are very small (often less than 2 kW) vertical shaft Pelton turbine and induction generation modular units, which require very little construction to install. Polythene pipes, generally used for water supply and irrigation purposes, guide water into the turbine from a canal, directly from the river or occasionally from a small reservoir. As of 2001, about 700 such projects were installed in various parts of Nepal. For larger electrification projects of up to 100 kW, Pelton and Crossflow are the most popular turbines. There are almost twenty manufacturers of micro- and pico-hydro turbines and other components in the country. All the required hardware is manufactured locally in Nepal.

Source: IT Power (www.itpower.co.uk).

with the weather. In addition, hydropower's fast response time enables it to meet sudden fluctuations almost instantaneously. Hence hydropower offers great benefits and holds a unique place in the range of energy options for electricity generation that are currently available.

2b. Focus on small hydropower (SHP)

There is no universal international consensus on the definition of what counts as small hydropower. A generally accepted definition is a hydropower plant up to 10 MW, but in the US and Brazil, for example, the limit is up to 30 MW. At the smaller end of the SHP scale, the definitions are subdivided: mini-hydro at less than 500 kW; micro-hydro at less than 100 kW, and pico-hydro at 10 kW or less.

SHP plants have considerable longevity, as has been demonstrated by the successful rehabilitation of numerous projects. The costs of a project are almost entirely in up-front capital, with fixed and predominantly small running and maintenance costs throughout a long lifetime. There can also be further benefits like greater control over flooding, irrigation, water storage and supply.

When SHP plants require a reservoir, it has been found that they use much more reservoir space per unit of power than larger hydropower plants. On average, plants of less than 100 MW capacity use 249 hectares per megawatt (ha/MW), while the biggest projects, producing between 3,000 and 18,000 MW, occupy only 32 ha/MW.

The introduction of electricity in remote communities in developing countries with difficult terrain has been possible only by way of decentralized small-scale hydropower schemes. Lighting of homes and surrounding areas is the major application of electricity generated from such projects, and provides both economic and social benefits. Examples of small hydropower in use in Asia can be seen in **Boxes 9.7** and **9.8** and **Table 9.3**. The use of SHP can contribute to poverty alleviation through sustainable socio-economic development, increasing employment opportunities for local people, improving rural living standards, and promoting environment-friendly development.

2c. Pumped storage

Pumped storage works like a giant rechargeable battery, a reserve source of power available at any time regardless of the weather. Pumped storage does not depend on rivers or rainfall as it uses the same water over and over again. When there is spare electrical capacity in an electricity system, at night, for example, it is used to pump water from a low reservoir to a high one. Then, at times of peak demand, the water is allowed to flow down again, generating extra power to supplement the grid. Pumped storage reservoirs are smaller than conventional reservoirs and less controversial, since they are less dependent on the topography. They are particularly effective in countries with limited water supplies, such as South Africa (see **Box 9.9**). Japan, the biggest user of pumped storage hydropower, has even been

The Atatürk Dam (Turkey) is the largest in a series of 22 dams and 19 hydroelectric stations built on the Euphrates and Tigris rivers

---

BOX 9.9: PALMIET PUMPED STORAGE SCHEME, SOUTH AFRICA

The Palmiet pumped storage scheme, just outside Cape Town, has two complementary functions: it provides a flexible electricity reserve for the South African national network and helps to supply freshwater to Cape Town. The project consists of two reservoirs with a height difference of 285 metres and a connecting conduit passing through a hydropower station with a reversible turbine.

Using surplus electricity from the national grid, the reversible turbine pumps water uphill through a two-kilometre chain of steel-lined tunnels to the higher reservoir at quiet periods every weekday, and for thirty-three hours during the weekend. Then, during

the working days when demand for power from South Africa's industries and people is at its highest, the water is allowed to flood back down through the turbines, generating electricity to pump back into the grid. The higher reservoir also has a separate outlet through which spare water can flow down the other side of the mountain range, into the large Steenbras reservoir, effectively transferring water from one catchment area to another. From there it is fed into the Cape Town city supply, contributing a total of 25 million cubic metres a year.

Built between 1983 and 1988, Palmiet has played a vital role in providing stability to South Africa's electricity supply. More than 90 percent

of the country's electricity comes from coal-fired power stations, which are relatively inflexible and cannot easily cope with demand fluctuations.

Palmiet, which has a capacity of 400 MW, and South Africa's other pumped storage plant, the 1,000 MW Drakensberg project, together account for only 1.5 percent of total electricity production. However, they help the system to absorb any shocks from breakdowns or surges in demand and allow the thermal power stations to run at constant, energy-efficient levels of output (see **Chapter 14**).

Source: International Hydropower Association (www.hydropower.org).

Section 3: CHALLENGES FOR WELL-BEING & DEVELOPMENT

BOX 9.10: HYDRO TASMANIA, AUSTRALIA

Hydro Tasmania, which is Australia's biggest hydropower producer with 2,300 MW of capacity, has a project to build wind farms generating 1,000 MW. The plan is dependent on the Basslink undersea cable connecting Tasmania with the rest of Australia and is scheduled to be completed around the beginning of 2006. Hydro Tasmania will then be able to use its joint wind-hydro production on the offshore territory to supply peak power to the industries of Victoria State, Australia's biggest electricity network, on the mainland. The state-owned company would be able to guarantee supply,

as its hydropower projects on Tasmania would provide backup if weather conditions made wind generation impossible.

The scheme was created by the Australian Government's Mandatory Renewable Energy Targets (MRET), designed as a first step towards reducing the country's greenhouse gas emissions and heavy dependence on coal-fired power stations. Under this initiative, Australia's regional authorities would have to obtain at least 2 percent of their energy from renewable sources by 2010. Those that fail would be fined

or would have to purchase tradable Renewable Energy Certificates from renewable power producers such as Hydro Tasmania.

The synergy between hydropower and other energy sources not only enables an increase in the penetration of renewables into the energy market, but by meeting peak demand, it also reduces the need for additional investment in base load generation.

Source: International Hydropower Association (www.hydropower.org).

---

experimenting with seawater pumped storage, but the technical difficulties have made this approach economically unappealing (see **Chapter 14**).

The world's pumped storage capacity amounted to 103 GW by 2003, which represents about 13 percent of total hydropower capacity. Japan and the US account for 24 and 20 percent of this amount, respectively. Italy, France and Germany also have substantial pumped storage capacities.

Pumped storage units can start up in very few minutes in an emergency to provide the necessary reserve capacity. This allows coal-fired stations to operate at constant levels of output, thus functioning more efficiently and reducing CO_2 emissions; however, pumped storage has an efficiency level of 70 to 75 percent and uses about a quarter more electric power than it creates. In a system with a substantial proportion of thermal plants, this is more than compensated by the increase in efficiency of the oil, gas and coal generators as well as the subsequent reduction in the amount of greenhouse gases they emit.

Pierre Bénite dam, France

2d. Sustainable hydropower solutions

There are three options for increasing electricity generation through hydropower, which are especially sustainable, and cost-effective: piggybacking an alternative energy source, adding hydropower capacity to existing infrastructure and extending the life and increasing the operating efficiency of existing hydro power projects.

Piggybacking alternative energy sources

Hydropower's flexibility and reliability of operation make it ideally suited to working in tandem with alternative energy sources, which means that it can play an essential role in the development of these young industries. Few of these can yet provide the steady, guaranteed supply of power an electricity network requires, but when piggybacked with hydropower, they can be effective suppliers of electricity to a system, providing financial incentives for developing these sectors.

When wind turbines or solar panels are injecting energy into a grid, hydropower units can reduce their own output and store extra water stocks in their reservoirs. These stocks can then be used to increase hydropower output and fill the gap when the wind drops or the sun is covered by clouds and input from these sources falls. This is well illustrated by the large combined wind and hydro project in Tasmania (see **Box 9.10**).

Adding hydropower capacity to existing infrastructure

It must be remembered that only about 25 percent of the world's dams are involved in producing hydropower. Water supply and energy policies were often poorly coordinated in the past, often with a reluctance to plan across sectors and cater for both uses. In Africa, the continent with the least developed hydropower potential, only about 7 percent of dams have hydropower as their main purpose. This leaves a window of opportunity for adding hydropower capacity to existing dams.

The hydropower industry now offers an array of different types of equipment suitable for this purpose. This is also relevant in parts of the world where hydropower potential is largely developed already, such as in Europe (75 percent) and North America (69 percent). An extra 20,000 MW of generating capacity could be added in the US by installing generating units in some 2,500 dams that at present have none. Many authorities are looking at new ways to add hydropower-generating capacity without building new dams (see **Box 9.11**). As the most suitable sites from a physical, political and financial point of view have been used, it becomes harder to win approval for any project on a new site, as hostility to dams is still strong among some environmental groups. In these conditions, adding generating capacity to existing dams is an attractive option.

Extending the life and improving the efficiency of hydropower schemes

The structural elements of a hydropower project, which tend to take up about 70 percent of the initial investment cost, have a projected life of about 100 years. On the equipment side, some refurbishment can be an attractive option after thirty years. Advances in hydro technology can justify the replacement of key components or even complete generating sets. Typically, generating equipment can be upgraded or replaced with more technologically advanced electro-mechanical equipment two or three times during the life of the project, making more effective use of the same flow of water.

A turbine commissioned in the 1970s, for example, might have a peak operating efficiency of 80 to 85 percent, whereas a modern turbine would raise this to 90 to 95 percent. The long life of hydro and extremely low running costs make even a modest improvement in output financially attractive. A number of techniques short of replacing the whole turbine can be used to increase output – using computerized testing and simulation, for example. Changing the shape of turbine runner blades has been effective, for example, at Arapuni in New Zealand, where productivity was increased considerably after an efficiency upgrade in 2002.

There are three principal ways of improving operating efficiency within existing hydropower projects, which allow for more electricity generation from the same scheme:

■ Improving water management and allowing plants to operate at their optimal level of efficiency, by adjusting flows to maximize the available 'head' (drop) at each site. The only costs may be the testing of equipment performance and staff training.

■ Installing equipment that is designed to have a higher efficiency over a wider range of water flows through the turbine. This is particularly significant for small projects for which the volume of water flow may vary sharply during rainy and dry seasons, and even during the same day, depending on rainfall.

■ Increasing the flow to the turbines and reducing losses, through minor changes to the hydraulic passages. This solution involves some civil engineering works. An example is the Manapouri scheme in New Zealand, which was completed in 1971. In 2002, a new 10-kilometre tailrace tunnel was commissioned to carry the water away from the turbines. The improved design of the tunnel enabled the output to be substantially increased.

2e. Environmental impacts of thermal power generation including water use

Where electricity transmission and distribution grids are established, thermal power is currently the major alternative to hydropower for base-load electrical power generation. In a thermal power plant, heat is generated from either the combustion of fossil fuel (coal, oil or gas), or through nuclear fission of radioactive material. The heat is used to raise steam, which generates electrical power by means of a steam turbine mechanically connected to an electricity generator. Water and steam circulate between the steam generator (boiler) and the steam turbine in a closed circuit.

Steam exiting the turbine must be condensed, and, since condensation employs cooling towers, where cooling water is lost due to evaporation, there are water resources issues associated with the deployment of thermal power generation. Where surface water is used for cooling and returned directly to the river or lake from where it is sourced (once-through cooling), the return water will be several degrees warmer, resulting in temperature changes capable of affecting aquatic ecosystems (see **Chapter 5**). In addition, a typical power plant using once-through cooling technology can kill tons of fish each year by trapping fish against intake screens or drawing fish into the facility.

Coal-fired power station in Bergheim, Germany

The long life of hydro and extremely low running costs make even a modest improvement in output financially attractive

---

BOX 9.11: HYDROPOWER GENERATION IN FREUDENAU, AUSTRIA

One example of creative thinking in hydropower generation is in Freudenau on the Danube in Austria. This run-of-river facility in the heart of Vienna was completed in 1998. The project is comprised of six giant Kaplan turbines, with installed capacity of 172 MW, enough to power about half the homes in the city. As well as generating hydropower, it provides flood protection, has helped to raise groundwater under Vienna and has restored water levels

in two blind arms of the river. It improves navigation in related canals and includes two large locks for river traffic on one of Europe's busiest waterways.

The owner, the state-controlled utility Verbund, later added modern Matrix turbines to the navigation locks. These sets of small turbines capture the energy from water that would normally be discharged through gates and

valves, producing an additional 5 MW of power. This is a small amount compared with the total output of the project, but it is significant in itself, especially as it could be added without the disruption of creating a fresh site for electricity generation.

Source: International Hydropower Association (www.hydropower.org).

Water availability is now driving the development of cooling technology for thermal power application

In a thermal plant, cooling is essential for efficient operation of the steam turbine and forecovery/recirculation of the highly purified water, which must be returned in closed circuit to the steam boiler. Power plant operating costs and lifetime (normally thirty to forty years) will be seriously affected if cooling water availability becomes constrained. Coastal plants can obtain cooling water from the sea. However, coal-fired plants are frequently located close to the coal deposit ('mine-mouth' plants) to reduce coal transportation costs, in which case they are heavily dependent on the availability of local cooling water.

In several major coal-producing countries, including China, parts of India and South Africa and the US, coal deposits are located in arid areas. In the US, about 40 percent of daily freshwater usage is for power generation. Most of this is returned to the source; about 2 percent is consumed/evaporated. Water availability is now driving the development of cooling technology for thermal power application.

It is important to distinguish between water diversion and consumptive use. The former indicates the amount of water removed from a water body (river or lake), the bulk of which is returned to the water basin albeit at an elevated temperature that gives rise to some environmental concerns. Consumptive use indicates the actual amount of resource loss, because this is the amount of water evaporating in the cooling process and not returned to its original source. Technologies are available to control water temperature discharged from the thermal power cooling systems, as well as to reduce consumptive use to virtually zero, but applying those technologies incurs additional costs.

In addition to impacts on the aquatic environment through the discharge of cooling water, fossil fuel (especially coal-based) power generation is also responsible for air pollution. Emissions of oxides of sulphur and nitrogen are responsible for acid rain, the deposition of which results in the degradation of ecosystems, as well as damage to agricultural production and to buildings. As a result of these measures, the use of low sulphur coal, the deployment of dust filters, flue gas desulphurization and nitrogen oxide control technologies are common practice at modern fossil fuel generating plants. Atmospheric emissions from coal combustion are now the major challenge confronting a continued deployment and development of coal-fuelled power plants.

The greenhouse gas emission reduction targets agreed under the Kyoto Protocol will limit releases of CO_2, the major greenhouse gas contributing to climate change. CO_2 emission mitigation will be considerably more expensive to implement than controls on acid deposition. This will also have a pronounced effect on the competitive position of coal in comparison with power sources that do not generate greenhouse gases, such as hydropower and nuclear power generation options (the latter of which also has environmental concerns linked to waste storage and possible accidents).

Table 9.4 lists the fifty countries with the world's highest carbon intensities of electricity production (WRI, 2004). Cutting-edge 'clean coal' technologies, including supercritical pulverized fuel, pressurized fluid bed and integrated gasification/combined cycle (IGCC) technologies, control CO_2 emissions by generating electricity at higher efficiencies. Older coal-based technologies have efficiencies in the range of 30 to

35 percent. New clean coal technologies, particularly IGCC, reduce CO_2 emissions per unit of power generated and have the potential to achieve efficiencies of 45 percent and above. Wide-scale application of CO_2 sequestration is likely to be accompanied by an accelerated use of IGCC technology. In several OECD countries, research is underway into the capture of CO_2 from power plant flue gas, transportation and CO_2 sequestration in depleted oil/gas reservoirs, deep saline aquifers, unmineable coal beds and in the deep ocean.

An interesting observation can be made by exploring the relationship between the carbon intensity of electricity production and the role of nuclear power in the electricity sector. In the group of twenty-five countries with the highest carbon intensity, only three countries have nuclear power in their electricity portfolios, each at a rather modest level. In the group of the next twenty-five countries, however, there are five countries with nuclear electricity, and in three of them, nuclear power provided around 30 percent of electricity in 2002. It is likely that countries with the economic means to invest in nuclear technology will increasingly turn to this solution as a means of reducing their dependence on fossil fuels, achieving energy security and reducing greenhouse gas emissions.

Table 9.4: Carbon intensity of electricity production in 2002

	Country	Grams of carbon per kilowatt hour		Country	Grams of carbon per kilowatt hour
1	Estonia	328.9	26	Czech Republic	206.8
2	Moldova	314.2	27	Singapore	206.7
3	Kazakhstan	309.0	28	Lebanon	200.3
4	Qatar	300.4	29	Romania	198.5
5	Poland	286.1	30	Bahrain	187.4
6	China	259.9	31	Trinidad and Tobago	185.3
7	Turkmenistan	245.8	32	Côte d'Ivoire	184.6
8	India	240.7	33	Algeria	183.4
9	Senegal	237.1	34	Kuwait	182.6
10	Malta	234.7	35	Morocco	180.3
11	Bosnia and Herzegovina	232.0	36	Jordan	179.0
12	Cyprus	231.5	37	Ireland	178.7
13	Belarus	229.9	38	Zimbabwe	175.8
14	South Africa	229.7	39	Libya	172.6
15	Serbia and Montenegro	227.6	40	Kenya	170.0
16	Oman	222.8	41	Indonesia	166.8
17	Togo	222.2	42	Hungary	166.3
18	United Arab Emirates	220.7	43	Nicaragua	166.1
19	Greece	220.1	44	Denmark	165.6
20	Israel	215.7	45	Latvia	162.0
21	Australia	215.6	46	Russian Federation	158.8
22	Cuba	214.9	47	Bulgaria	154.8
23	Azerbaijan	212.8	48	Bangladesh	152.2
24	Brunei	208.4	49	Iran	151.8
25	Uzbekistan	207.1	50	Iraq	148.8

Note: These data cover fossil fuel generation, hydropower, nuclear power, renewables and waste. The countries listed generate the highest amounts of greenhouse gases, per unit of electricity generated, and hence have the greatest potential for applying technological solutions in order to reduce their carbon releases. Untapped hydropower options are available in China and India, as has been already discussed, as for many other countries appearing in this table. However, for countries with a high reliance on gas and coal, technological improvements in thermal power generation will become necessary.

Source: WRI, 2004.

China, India and Turkey frequently argue that their electricity requirements for economic growth and social development outweigh the environmental concerns surrounding hydropower...

Part 3. Governance of Energy and Water Resources

In the past, some hydropower projects, particularly big reservoirs, have had a negative impact on their immediate surroundings. Damage to the local environment and inadequate provision for those affected in the area have contributed to the hostility shown by some environmental and human rights organizations towards the hydropower industry. The World Commission on Dams attempted to bring the various parties together, although its recommendations were not universally welcomed (WCD, 2000). More recent guidelines published by the International Hydropower Association (IHA) in 2004 have been broadly accepted throughout the large hydropower industry, in particular the core principles of equity, participatory decision-making and accountability.

3a. The continuing debate on large hydropower

IHA argues that equitable sharing of the benefits of any power project requires a careful balance between different stakeholders and interested parties. Hydropower uses renewable water supplies, not finite fossil fuels. In contrast to nuclear power, it leaves no toxic waste to threaten future generations, and in contrast to thermal power, it emits virtually no greenhouse gases. While the vast majority of a project's costs are borne at the start, the benefits continue for 100 years or more.

Furthermore, while any negative effects of a hydropower project are inevitably borne by the local community, the benefits – in reliable electricity supplies – are shared by everyone in the nation or region.

The key to managing changes lies in advance planning and consultation with all interested parties. The IHA Sustainability Guidelines state that hydro developers planning a project should try to minimize the following:

■ health dangers, particularly from water-borne diseases or malaria
■ loss of homes, farms and other livelihoods
■ disruption of community networks and loss of cultural identity
■ changes to biodiversity in the affected area.

They should try to maximize the following:

■ timely consultation at all levels
■ the flow of relevant information to all those affected
■ negotiated settlement of disputes
■ timely and adequate payment of any compensation.

Where people or communities have to be transferred to new sites, developers should do the following:

■ investigate possible alternative ways of doing the project
■ ensure adequate consultation with the people to be displaced throughout the project
■ guarantee equivalent or improved livelihoods at the new location
■ provide better living standards and public health at the new location.

Rapidly developing countries such as China, India and Turkey frequently argue that their electricity requirements for economic growth and social development outweigh the environmental concerns surrounding hydropower, and that support for large hydropower development is a pro-poor policy. This need was recognized in the Johannesburg Plan of Implementation (UN, 2002) where hydropower is included among the 'advanced, cleaner, more efficient, affordable and cost-effective energy technologies' required by developing countries. However, several non-governmental organizations are campaigning to have large hydropower excluded from global efforts to promote renewable energy. Among the arguments advanced for this position are the following:

■ including large hydro in renewables initiatives reduces the available funding for new renewable energy technologies
■ there is no technology transfer benefit from large hydro, which is a mature technology
■ large hydro projects often have major social and ecological impacts
■ large reservoirs can emit significant amounts of greenhouse gases from rotting organic matter
■ large hydro reservoirs are often rendered non-renewable by sedimentation.

This long-standing debate is still a major issue. Many large hydropower projects necessitate the construction of large dams. These are structures with a long life, which permanently alter the river downstream and affect a significant stretch of the river upstream. They are not, strictly speaking, renewable. However, as discussed in this chapter, there are also very large run-of-river hydropower projects, as well as small, mini and micro-hydropower projects, which are all renewable energy providers. It must also be remembered that the driving force for much new dam construction is irrigation, rather than hydropower generation.

The water/energy nexus can be better understood by distinguishing the issue of large dams from that of hydropower, except in the cases of certain hydropower projects that do require the construction of large new reservoirs. In these specific cases, greater transparency, accountability and oversight of the contractual process to ensure the exposure of corrupt practices are all necessary in order to promote social equity and good governance.

3b. Renewable energy and energy efficiency: Incentives and economic instruments

In the world's developed regions, electricity is delivered to the vast majority of consumers through vertically integrated utility industries based on central power generation. Over the past several decades the efforts of energy policy-makers, utility planners, regulators and generation technology developers have enabled this conventional power generation and supply system to keep pace with rising demand, but with social and environmental impacts that are increasingly considered unacceptable. The inertia within the power supply system – power plants and transmission/distribution systems have lifetimes of several decades – means that this trend will be difficult to change.

However, with a worldwide 30 percent annual growth rate, renewable-based generating capacity is currently increasing faster than the conventional power option. Accelerated interest in renewable energy can be traced to the 'oil crisis' of the 1970s, but a list of environmental concerns headed by global climate change is responsible for the recent surge in interest in clean energy.

In developing countries where affordable power is desperately needed, environmental concerns must be carefully weighed against urgent development needs. As seen earlier in this chapter, governments will be less responsive to objections to the construction of dams for large hydropower generation or to the deployment of new, greenhouse gas-emitting coal-fired power plants, when their priority is meeting rapidly growing electricity demand. Clearly, the transition to a fully sustainable, global energy supply system needs cooperative and innovative, if not radically new, policy-making.

International and national mechanisms implemented with the Kyoto Protocol

At the international level, the Clean Development Mechanism (CDM) and Joint Implementation (JI) measures established by the Kyoto Protocol seek to provide incentives for the use of low carbon-emitting and renewable energy technologies in developing countries, through the sale of carbon credits arising from clean energy investments. Given its proven track record of decades of successful experience, it is not surprising that hydropower projects are prominent in the current portfolio of CDM and JI projects. Multilateral initiatives establish emission reduction goals and a cooperative means of achieving them. However, they will need to be accompanied by national policies to stimulate a thriving market for renewable energy resources, such as wind, biomass, solar photovoltaics and hydropower as well as for combined heat and power (CHP) generation options. For example, feed-in tariffs oblige utilities to buy renewable electricity from any developer in their service area at tariffs set by government. These are generally a little lower than zthe electricity retail price, facilitating a good return on investment and assuring long-term support. Renewable Portfolio Standards (RPS) require the share of renewable power purchased by a utility to increase yearly to a given percentage. An RPS creates long-term stability and demand that establishes a flourishing renewables-based generation market. Within a given country, regional discrepancies arising from cost and availability of renewable power sources can be levelled out by means of tradable 'renewable energy certificates' (called 'green certificates' in Australia). Feed-in tariffs, RPSs and tradable certificates may need the further support of long-term and stable subsidies, such as investment tax credits and accelerated depreciation (see **Box 9.12**). In effect, renewable power markets need to be driven by a combination of demand- and supply-side measures capable of keeping the costs of electricity retailers and their retail customers at a minimum.

The case of rural electrification

Rural electrification is a special case. The provision of rural electrification through rural cooperatives was employed

Aerial close-up of Tucson Electric Power's cooling towers, Arizona, United States

BOX 9.12: RENEWABLE OBLIGATION CERTIFICATES: A POLICY INSTRUMENT PROMOTING RENEWABLE ENERGY

Small-scale hydropower development stands to benefit from policy instruments designed to promote renewable energy. The largest hydropower operator in the UK, Scottish and Southern Energy, is engaged in a €360 million (US $439 million) investment programme to upgrade its older projects over ten years. This is expected to increase UK hydropower output of some 5,000 GWh per year by 200 GWh. The programme was made possible by the British

Government's decision to allow the refurbishment of hydropower projects with a capacity of less than 20 MW to qualify for Renewable Obligation Certificates. The company had previously refurbished its larger hydro plants, for an increase in output of 6 percent at a cost of €60 million (US $73 million). Renewable Obligation Certificates were then introduced by the British Government to encourage the development of renewable energy. Each

electricity supplier has to produce a certain proportion of its power from sources qualifying for the Certificates, or face fines for every MW it produces. The first of the smaller plants to qualify was a 17 MW capacity plant at St Fillans, upgraded in 2002 for an 8 percent increase in output and a thirty-year extension to its life.

Source: UNIDO, 2004.

Although environmentally friendly, and free from expensive fuel costs, renewables are often intermittent and carry a burden of high capital investment

effectively in the industrialized world between the 1930s and the 1950s. This institutional model has been employed successfully in several developing countries. The high costs of grid extension, particularly to remote parts of many developing countries, mean than isolated rural communities are usually served by diesel-based mini-grids rather than centrally operated electricity distribution networks. Electricity produced by diesel sets can be two to three times the cost of grid power in urban areas, but still cost-effective relative to grid extension. Costs of maintenance and transporting diesel fuel are high. Greenhouse gas emissions per unit of generated power from a diesel engine are particularly high.

Where renewable energy resources, in the form of solar, wind, biomass, biogas and mini-hydro are available, their use can replace or supplement diesel. Although environmentally friendly, and free from expensive fuel costs, renewables are often intermittent and carry a burden of high capital investment. Policy reforms to make capital resources more readily available for small-scale rural energy investments are needed. Micro-financing is now almost a prerequisite for rural energy development projects. Micro-finance schemes are especially important for photovoltaic and other renewable energy-based technologies. There is also a need to stimulate local manufacturing of renewable energy equipment and gradually increase domestic content.

Rural electrification has to be seen in the broader context of rural development. Even though major barriers exist, the subsidized development and deployment of renewables-based mini-grids is proceeding in many rural parts of the developing world. With improved knowledge of rural development needs and a clearer understanding of the contribution of reliable, cost-effective and clean energy

to rural development, these projects should lead to replication, commercial support, the phasing-out of subsidies and the ultimate goal of reducing poverty.

Improving energy efficiency

In the same way that innovative policies are needed to overcome obstacles to the accelerated deployment of renewable energy technologies, new approaches are required to encourage energy users to take advantage of the enormous potential for improving end-use energy efficiency. Much of the world's future energy demand will have to be met by efficiency improvements.

The market for energy efficiency products and services is far from perfect, and information is neither widespread nor clear. The potential for industrial energy system optimization remains largely unrealized. Production, not energy efficiency, is the top priority of industrial manufacturers. Plant operating budgets and capital improvement budgets are accounted for separately, so that the consequences of purchasing less efficient equipment is not linked to increased operating costs, although these can be 80 percent or more of the lifecycle cost of the equipment. These disincentives can only be reversed through integrated policy-making, including changes in tax laws, and measures for incorporating life-cycle energy costs into bidding procedures for capital projects (see **Chapters 2** and **12**).

Whereas manufacturers of energy-consuming equipment have successfully improved the performance of individual components, such as pumps, compressors, fans and steam boilers, these components only provide a service to the user when operating as part of a system. There is scope to improve energy efficiency and reduce greenhouse gas

---

BOX 9.13: DISTRIBUTED GENERATION: POWER SUPPLY IN THE FUTURE

Electricity power generators – conventional and renewable alike – sell their product through distribution grids or networks. In the future, renewable energy will be provided from large numbers of individually small and frequently intermittent (such as combined heat and power [CHP] and wind) power generators. The network needs increasingly to be managed as an interlinking mesh, rather than a unidirectional funnel of energy, so that suppliers can continue to provide continuous and reliable electricity to their customers. To attain renewables targets in established grid systems, network operators will therefore need new tools and incentives, but this will by no means present insurmountable challenges. Indeed, the wider use of on-site generation can reinforce stressed networks. Today's grids support thermal and hydro central power stations delivering 'mass-produced'

electricity to meet an agglomerated load curve. To accommodate an ever-increasing penetration of individually small power producers, new grid operating procedures and protocols are required, many of which are already available.

Distributed generation (DG) offers a promising set of solutions and benefits. DG means producing power close to the customer using a network supplied by several small generators. Operating in parallel with the main grid, DG systems provide some or all of the power required by the user, while the grid either absorbs the surplus or provides the shortfall. On-site or local CHP generation not only reduces thermal energy losses from conventional plants, but also significantly decreases the network losses arising from moving electricity long distances from remote hydro and thermal

plants. From a cost perspective, investment in local generation also avoids the high expense of constructing long-distance transmission networks that can prove vulnerable to disruption.

The function of the network is therefore evolving from a supply role to that of a buffer. All network users should be required to pay their reasonable share for the construction and maintenance of those networks according to the use made of them and according to the services provided for network reinforcement. Achieving reasonable cost-recovery arrangements is complex but manageable, and is likely to provide a further incentive for investment in generation that reduces network use.

Source: WADE (www.localpower.org).

---

emissions across the entire industrial sector by improving the design and operation of the systems that deliver energy to the point of use. Pumping systems alone account for 20 percent of the world's electrical energy demand and range from 25 percent to 50 percent of total energy use in some industrial operations.

One way to increase the implementation and persistence of energy efficiency measures in the industrial sector would be for industry participants to incorporate their commitments to energy efficiency into the ISO 9000/14000 quality and environmental management system (see **Chapter 8**). ISO certification has become a significant trade facilitation vehicle for developing countries, with more than 155,000 plants participating in these countries as of December 2002. Tracking energy efficiency projects and milestones in their ISO quality and environmental management system will help each company to maintain a focus on its energy efficiency commitments, provide visibility for its achievements and provide a verification of their results for project funders. All of these measures will help stimulate a significantly higher level of activity in industrial energy efficiency programmes.

3c. Policy-making for co-management of water and energy resources

Policy-makers in the water and energy sectors need to

find better ways of integrating decisions between the two sectors in order to optimize benefits, address financing barriers and identify potential new partnerships. Inclusion of energy considerations can improve water resource management decisions and avoid potentially significant but unintended energy consequences. The key indicators of the success of such co-management would be the increased availability, acceptability and affordability of both water and energy services.

A recent report based upon three detailed case studies in California has shown clearly that including energy considerations in water management decisions can lead to major energy and money savings (NRDC, 2004). The case study analysis supports two primary recommendations for how policy-makers can begin to achieve these savings, which are generally applicable in many areas beyond California:

- Decision-makers should better integrate energy issues into water policy decision-making.
  Looking at energy use and water use simultaneously generates valuable insights that do not arise from separate policy analyses of water and energy issues.

The report makes the following recommendations:

- modify planning tools for water resources management to include energy use and costs

...including energy considerations in water management decisions can lead to major energy and money savings...

Governments that have ratified the Kyoto Protocol are bound to reduce greenhouse gas emissions and promote clean energy investments

■ improve coordination among resource management agencies to better identify and address the energy implications of water policy decisions

■ conduct an energy intensity analysis of water distribution systems and identifying regions and districts where large amounts of power are required to deliver water

■ develop partnerships designed to produce energy, economic and environmental benefits through voluntary water transfers away from the agricultural sector with a focus on dry-year transfers in locations where large water diversions upstream reduce downstream flows and downstream hydropower generation.

■ Both water and energy policy-makers need to give water conservation higher priority.

The amount of energy required for end use is the largest component of energy use in urban water supply. Hence, policy actions that affect the end uses of water may have much larger energy implications than policy actions that affect the mix of physical water sources. Conservation has much greater stronger energy-related economic and environmental benefits than has previously been recognized. In addition, the energy benefits of water conservation can generate air quality benefits as well as climate change benefits. The report recommends the following actions:

■ prioritize water conservation funding

■ enforce existing water conservation requirements

■ promote conservation through pricing strategies and water metering

■ offer water conservation incentives.

There are serious weaknesses in the pricing of both electricity and water in many parts of the world, which send the wrong signals to consumers about the need for conservation of both of these resources (see **Chapter 12**). In addition, regulatory regimes, where these exist, are frequently not sufficiently focused on the need for efficiency of use and conservation. Frequently the cultures within the electricity and water sectors of many countries are very different, and there is rarely the level of communication required to exploit the potential synergies of the two sectors. Availability of both energy and water is essential for human survival and national prosperity. In the globalized world of the twenty-first century, energy and water supply security will require governance regimes that are sensitive to environmental and social, as well as political and economic, considerations. In the many cases of countries where water/energy resource availability and the environmental and social consequences of using both are interrelated, a strong case can be made for policies and regulations that address water and energy simultaneously. Many inefficiencies in both sectors impact not only on poverty alleviation and socio-economic development but on other sectors of water and the environment at large. Governments need to recognize the very close connection between the two sectors, in order to maximize the benefits of synergies between them.

Table 9.5: Access to electricity and water in 2000

	Population access to improved water source		Electricity production kWh (billions)	Population access to electricity (%)
	Urban (%)	Rural (%)		
Afghanistan	19	11	–	2.0
Albania	99	95	4.9	–
Algeria	94	82	25.4	98.0
Angola	34	40	1.4	12.0
Argentina	97	73	89.0	94.6
Armenia	–	–	6.0	–
Australia	100	100	208.1	–
Austria	100	100	60.3	–
Azerbaijan	93	58	18.7	–
Bangladesh	99	97	15.8	20.4
Belarus	100	100	26.1	–
Belgium	–	–	82.7	–
Benin	74	55	0.1	22.0
Bolivia	95	64	4.0	60.4

Table 9.5: *continued*

	Population access to improved water source		Electricity production kWh (billions)	Population access to electricity (%)
	Urban (%)	Rural (%)		
Bosnia and Herzegovina	–	–	10.4	–
Botswana	100	90	–	22.0
Brazil	95	53	349.2	94.9
Bulgaria	100	100	40.6	–
Burkina Faso	66	37	–	13.0
Burundi	91	77	–	–
Cambodia	54	26	–	15.8
Cameroon	78	39	3.5	20.0
Canada	100	99	605.1	–
Central African Republic	89	57	–	–
Chad	31	26	–	–
Chile	99	58	41.3	99.0
China	94	66	1,355.6	98.6
Hong Kong, China	–	–	31.3	–
Colombia	99	70	44.0	81.0
Congo, Dem. Republic.	89	26	5.5	6.7
Congo	71	17	0.3	20.9
Costa Rica	99	92	6.9	95.7
Côte d'Ivoire	92	72	4.8	50.0
Croatia	–	–	10.7	–
Cuba	95	77	15.0	97.0
Czech Republic	–	–	72.9	–
Denmark	100	100	36.2	–
Dominican Republic	90	78	9.5	66.8
Ecuador	90	75	10.6	80.0
Egypt, Arab Rep.	99	96	75.7	93.8
El Salvador	91	64	3.9	70.8
Eritrea	63	42	–	17.0
Estonia	–	–	8.5	–
Ethiopia	81	12	1.7	4.7
Finland	100	100	70.0	–
France	–	–	535.8	–
Gabon	95	47	1.0	31.0
Gambia	80	53	–	–
Georgia	90	61	7.4	–
Germany	–	–	567.1	–
Ghana	91	62	7.2	45.0
Greece	–	–	53.4	–
Guatemala	98	88	6.0	66.7
Guinea	72	36	–	–
Guinea-Bissau	79	49	–	–
Haiti	49	45	0.5	34.0
Honduras	95	81	3.7	54.5
Hungary	100	98	35.0	–
India	95	79	542.3	43.0
Indonesia	90	69	92.6	53.4
Iran, Islamic Republic.	98	83	121.4	97.9
Iraq	96	48	33.7	95.0
Ireland	–	–	23.7	–
Israel	–	–	43.0	100.0
Italy	–	–	269.9	–
Jamaica	98	85	6.6	90.0

This solar voltaic panel's energy is used to pump water, Kabekel village, Gambia

Table 9.5: *continued*

	Population access to improved water source		Electricity production kWh (billions)	Population access to electricity (%)
	Urban (%)	Rural (%)		
Japan	–	–	1081.9	–
Jordan	100	84	7.4	95.0
Kazakhstan	98	82	51.6	–
Kenya	88	42	3.9	7.9
Korea, Dem. Republic.	100	100	31.6	20.0
Korea, Rep.	97	71	292.5	–
Kuwait	–	–	32.5	100.0
Kyrgyz Republic	98	66	14.9	–
Lao Peoples Dem. Republic.	61	29	–	–
Latvia	–	–	4.1	–
Lebanon	100	100	7.8	95.0
Lesotho	88	74	–	5.0
Libyan Arab Jamahiřiya	72	68	20.7	99.8
Lithuania	–	–	11.1	–
Madagascar	85	31	–	8.0
Malawi	95	44	–	5.0
Malaysia	–	94	69.2	96.9
Mali	74	61	–	–
Mauritania	34	40	–	–
Mauritius	100	100	–	100.0
Mexico	95	69	204.4	–
Moldova	97	88	3.3	–
Mongolia	77	30	–	90.0
Morocco	98	56	14.1	71.1
Mozambique	81	41	7.0	7.2
Myanmar	89	66	5.1	5.0
Namibia	100	67	1.4	34.0
Nepal	94	87	1.7	15.4
Netherlands	100	100	89.6	–
New Zealand	100	–	39.0	–
Nicaragua	91	59	2.3	48.0
Niger	70	56	–	–
Nigeria	78	49	15.8	40.0
Norway	100	100	142.4	–
Oman	41	30	9.1	94.0
Pakistan	95	87	68.1	52.9
Panama	99	79	4.7	76.1
Papua New Guinea	88	32	–	–
Paraguay	93	59	53.5	74.7
Peru	87	62	19.9	73.0
Philippines	91	79	45.3	87.4
Poland	–	–	143.2	–
Portugal	–	–	43.4	–
Romania	91	16	51.9	–
Russian Federation	100	96	876.5	–
Rwanda	60	40	–	–
Saudi Arabia	100	64	128.4	97.7
Senegal	92	65	1.5	30.1
Sierra Leone	75	46	–	–
Singapore	100	–	31.3	100.0
Slovak Republic	100	100	30.4	–
Slovenia	100	100	13.6	–

Table 9.5: *continued*

	Population access to improved water source		Electricity production kWh (billions)	Population access to electricity (%)
	Urban (%)	Rural (%)		
South Africa	99	73	207.8	66.1
Spain	–	–	221.7	–
Sri Lanka	98	70	6.8	62.0
Sudan	86	69	2.4	30.0
Sweden	100	100	145.9	–
Switzerland	100	100	66.0	–
Syrian Arab Republic	94	64	22.6	85.9
Tajikistan	93	47	14.2	–
Tanzania	90	57	2.3	10.5
Thailand	95	81	96.0	82.1
Togo	85	38	0.0	9.0
Trinidad and Tobago	–	–	5.5	99.0
Tunisia	92	58	10.6	94.6
Turkey	81	86	124.9	–
Turkmenistan	–	–	9.8	–
Uganda	80	47	–	3.7
Ukraine	100	94	171.4	–
United Arab Emirates	–	–	38.6	96.0
United Kingdom	100	100	372.2	–
United States of America	100	100	4,003.5	–
Uruguay	98	93	7.6	98.0
Uzbekistan	94	79	46.8	–
Venezuela, Bolivarian Republic	85	70	85.2	94.0
Viet Nam	95	72	26.6	75.8
Yemen	74	68	3.0	50.0
Yugoslavia, Fed. Republic	99	97	31.9	–
Zambia	88	48	7.8	12.0
Zimbabwe	100	73	7.0	39.7
World	**94**	**71**	**15,346.5**	**–**
Low-income countries	**90**	**70**	**11,44.7**	**37.4**
Middle-income countries	**95**	**70**	**4,777.2**	**94.0**
Lower middle-income countries	95	70	3,429.3	93.8
Upper middle-income countries	94	69	1,347.9	94.7
Low and middle-income countries	93	70	5,921.9	65.0
East Asia and Pacific	93	67	1,722.1	87.3
Europe and Central Asia	96	83	1,827.5	–
Latin America and Carib.	94	65	973.2	86.6
Middle East and N. Africa	96	78	481.9	90.4
South Asia	94	80	634.8	40.8
Sub-Saharan Africa	83	46	282.4	24.6
High income	–	–	9,424.6	–
Europe (European Monetary Union)	–	–	2,018.0	–

Source: World Bank, 2003.

Table 9.6: Hydropower: Capability at the end of 2002

	Gross theoretical capability TWh/yr	Technically exploitable capability TWh/yr	Economically exploitable capability TWh/yr
Algeria	12	5	–
Angola	> 150	108	65
Benin	2	1	–
Burkina Faso	1	n.a.	n.a.
Burundi	6	2	1
Cameroon	294	115	103
Central African Republic	7	3	–
Chad	n.a.	n.a.	–
Congo	> 125	> 50	–
Congo Dem. Republic	1,397	774	419
Côte d'Ivoire	46	12	2
Egypt	> 125	> 50	50
Ethiopia	650	> 260	260
Gabon	200	80	33
Ghana	17	11	7
Guinea	26	19	15
Guinea-Bissau	1	n.a.	n.a.
Kenya	> 30	9	–
Lesotho	5	2	–
Liberia	28	11	–
Madagascar	321	180	49
Malawi	15	6	–
Mali	> 12	> 5	–
Mauritius	n.a.	n.a.	–
Morocco	12	5	4
Mozambique	50	38	32
Namibia	25	10	2
Niger	> 3	> 1	1
Nigeria	43	32	30
Rwanda	1	n.a.	–
Senegal	11	4	2
Sierra Leone	17	7	–
Somalia	2	1	–
South Africa	73	14	5
Sudan	48	19	2
Swaziland	4	1	n.a.
Tanzania	39	20	3
Togo	4	2	–
Tunisia	1	n.a.	n.a.
Uganda	> 18	> 13	–
Zambia	52	29	11
Zimbabwe	19	18	–
Total Africa	**> 3,892**	**> 1,917**	**–**
Belize	1	n.a.	n.a.
Canada	1,284	948	522
Costa Rica	223	43	20
Cuba	3	1	–
Dominica	n.a.	n.a.	n.a.
Dominican Republic	50	9	6
El Salvador	7	5	2
Greenland	800	14	–
Grenada	n.a.	n.a.	n.a.

Table 9.6: *continued*

	Gross theoretical capability TWh/yr	Technically exploitable capability TWh/yr	Economically exploitable capability TWh/yr
Guatemala	54	22	–
Haiti	4	1	N
Honduras	16	7	–
Jamaica	1	n.a.	–
Mexico	135	49	32
Nicaragua	33	10	7
Panama	26	> 12	12
United States of America	4,485	1,752	501
Total North America	**7,122**	**> 2,873**	**–**
Argentina	172	130	–
Bolivia	178	126	50
Brazil	3,040	1,488	811
Chile	227	162	–
Colombia	1,000	200	140
Ecuador	167	134	106
Guyana	64	> 26	26
Paraguay	111	85	68
Peru	1,577	> 260	260
Surinam	32	13	–
Uruguay	32	10	–
Venezuela	320	246	130
Total South America	**6,920**	**> 2,880**	**–**
Armenia	22	8	6
Azerbaijan	44	16	7
Bangladesh	5	2	–
Bhutan	263	70	56
Cambodia	88	11	5
China	5,920	1,920	1,270
Cyprus	59	24	–
Georgia	139	68	32
India	2,638	660	–
Indonesia	2,147	402	40
Japan	718	136	114
Kazakhstan	163	62	27
Korea Republic	52	26	19
Kyrgyz Republic	163	99	55
Lao People's Dem. Republic	233	63	–
Malaysia	230	123	–
Mongolia	56	22	–
Myanmar	877	130	–
Nepal	727	394	221
Pakistan	307	263	–
Philippines	47	20	18
Russian Federation	2,295	1,670	852
Sri Lanka	9	7	5
Taiwan, China	103	20	8
Tajikistan	527	> 264	264
Thailand	18	16	15
Turkey	433	216	126
Turkmenistan	24	5	2
Uzbekistan	88	27	15
Viet Nam	300	100	90
Total Asia	**18,695**	**> 6,844**	**–**

Table 9.6: continued

	Gross theoretical capability TWh/yr	Technically exploitable capability TWh/yr	Economically exploitable capability TWh/yr
Albania	40	15	6
Austria	75	> 56	56
Belarus	7	3	1
Belgium	1	n.a.	n.a.
Bosnia and Herzogovina	60	24	19
Bulgaria	27	15	12
Croatia	10	9	8
Czech Republic	12	4	–
Denmark	n.a.	n.a.	n.a.
Estonia	2	n.a.	–
Faroe Islands	1	n.a.	n.a.
Finland	48	25	20
France	270	100	70
Germany	120	25	20
Greece	80	15	12
Hungary	7	5	–
Iceland	184	64	40
Ireland	1	1	1
Italy	340	105	65
Latvia	7	6	5
Lithuania	6	2	1
Luxembourg	n.a.	n.a.	n.a.
Macedonia, former Yugoslav Republic	9	6	–
Moldova	2	1	1
Netherlands	1	n.a.	n.a.
Norway	600	200	187
Poland	23	14	7
Portugal	32	25	20
Romania	70	40	30
Serbia and Montenegro	37	27	24
Slovakia	10	7	7
Slovenia	13	9	6
Spain	138	70	41
Sweden	176	130	90
Switzerland	144	41	35
Ukraine	45	24	17
United Kingdom	40	3	1
Total Europe	**2,638**	**> 1,071**	–
Iran, Islamic Republic	176	> 50	50
Iraq	225	90	67
Israel	125	50	–
Jordan	n.a.	n.a.	n.a.
Lebanon	2	1	n.a.
Syrian Arab Republic	5	4	4
Total Middle East	**533**	**> 195**	–
Australia	265	> 30	30
Fiji	3	1	–
French Polynesia	n.a.	n.a.	n.a.
New Caledonia	2	1	n.a.

Table 9.6: continued

	Gross theoretical capability TWh/yr	Technically exploitable capability TWh/yr	Economically exploitable capability TWh/yr
New Zealand	46	37	24
Papua New Guinea	175	49	15
Samoa	n.a.	n.a.	–
Solomon Islands	2	> 1	–
Total Oceania	**493**	**> 119**	**–**
TOTAL WORLD	**> 40,293**	**> 15,899**	**–**

n.a. = not applicable due to flat topography

– = information not available

Notes:

1. A quantification of hydropower capability is not available for Comoros, Equatorial Guinea, Mauritania, Réunion, São Tomé and Principe, Guadeloupe, Puerto Rico, St Vincent and the Grenadines, French Guiana, Afghanistan, Korea (Democratic People's Republic) and Palau.

2. As the data available on economically exploitable capability do not cover all countries, regional and global totals are not shown for this category.

Sources: *The International Journal on Hydropower and Dams*; International Hydropower Association Member Committees, 2003; *Hydropower Dams World Atlas 2003*.

References and Websites

Martinot, E. 2002. Indicators of investment and capacity for renewable energy. *Renewable Energy in the World*. Vol. Sept/Oct.

NRDC (National Resources Defense Council). 2004. *Energy down the Drain*. New York, NRDC.

UN (United Nations). 2002. Johannesburg Plan of Implementation. www.un.org/esa/sustdev/documents/WSSD_POI_PD/English/WSSD_PlanI mpl.pdf

UNECA (United Nations Economic Comission for Africa). 2004. *African Water Development Report*. Addis Ababa, UNECA.

US DOE (United States Department of Energy). 2004. *Improving Pumping System Performance: A Sourcebook for Industry*. Washington DC, US DOE, *2nd edition*.

World Bank. 2003. *World Development Indicators*. New York, World Bank.

WRI (World Resources Institute). Climate Analysis Indicators Tool, Washington, DC.

Alliance to Save Energy www.ase.org: www.watergy.org

FAO's AQUASTAT www.fao.org/ag/agl/aglw/aquastat/main/index.stm

International Energy Agency, Energy Statistics: www.iea.org/Textbase/stats/index.asp

International Energy Agency Coal Centre: www.iea-coal.org.uk

International Networking on Small Hydropower: www.inshp.org

International Hydropower Association: www.hydropower.org/

IT Power: www.itpower.co.uk

Naiade solar water purifiers: www.nedapnaiade.com/

Solar water Disinfection (SoDis): www.sodis.ch

UNIDO: www.unido.org/

WCD (World Commission on Dams): www.dams.org

World Resources Institute: Climate Analysis Indicators Tool, Data on Carbon Intensity to Electricity Production from 2002, available online at cait.wri.org/

World Resources Institute: www.wri.org/

SECTION 4
Management Responses and Stewardship

Balancing the increasing competition among the diverse and different water using sectors and the demands of upstream and downstream users – whether within or between countries – is a challenge in watersheds worldwide. Decisions on water allocations have to be made at different scales, based not only on the various demands for water, but taking into account its many values as well.

Though the urgency of many water problems means that effective actions are needed now, water management approaches must also be forward-looking in their ability to deal with changing contexts, such as climate variability and its impact on water-related hazards, namely floods and droughts. The capacity to adapt and to make wise decisions depends upon preparedness, which depends in turn on a sound knowledge base; the complexity of water issues requires a more effective policy framework that builds, maintains, extends and shares our knowledge and uses of water resources, and respects the values we place on them.

Global Map 7: *The Climate Moisture Index Coefficient of Variation*
Global Map 8: *The Water Reuse Index*

Chapter 10 – **Managing Risks: Securing the Gains of Development** (WMO & UN-ISDR)

The climate is changing, thus increasing the occurrence and intensity of water-related natural disasters and creating greater burdens on human and environmental development. Employing an integrated approach, this chapter explores some of the ways of better reducing human vulnerabilities and examines the recent developments in risk reduction strategies.

Chapter 11 – **Sharing Water** (UNESCO)

Increasing competition for water resources can have potentially divisive effects. Mechanisms for cooperation and shared governance among users must be further developed in order to ensure that the resource become a catalyst for cooperation and a medium for deterring political tensions, while encouraging equitable and sustainable development.

Chapter 12 – **Valuing and Charging for Water** (UNDESA)

Water has a range of values that must be recognized in selecting governance strategies. Valuation techniques inform decision-making for water allocation, which promote not only sustainable social, environmental and economic development but also transparency and accountability in governance. This chapter reviews techniques of economic valuation and the use of these tools in water policy development and charging for water services.

Chapter 13 – **Enhancing Knowledge and Capacity** (UNESCO)

The collection, dissemination and exchange of water-related data, information and know-how are imbalanced and, in many cases, deteriorating. It is now more urgent than ever to improve the state of knowledge concerning water-related issues through an effective global network of research, training and data collection and the implementation of more adaptive, informed and participatory approaches at all levels.

The Climate Moisture Index Coefficient of Variation (CMI-CV)

Water scarcity, in part, is determined by the availability of renewable fresh water supply. One useful measure of available water is the Climatic Moisture Index (CMI) (Willmott and Feddema, 1992), a measure of the balance between annual precipitation and evaporation and a function of climate. The CMI ranges from +1 to −1 with wet climates showing positive values and dry climates negative values. The variability of CMI over multiple years, critical to determining the reliability of water supplies, is measured by the Coefficient of Variation (CV), defined as the ratio of annual deviation to the long-term annual mean. A CMI-CV value < 0.25 is considered low variability, while 0.25 to 0.75 is moderate and > 0.75 high. Increased climate variability indicates larger year-to-year fluctuations, and hence, less predictability in the climate. As shown in the map below, variability is low in the most humid regions (i.e., the tropics) as well as the most arid regions (i.e., major deserts) of the world. Increased CMI variability often occurs along the interfaces between different climate zones, for instance, between the dry Sahelian region of North Africa and the humid tropical zone of southern west Africa, or in the Great Plains region of the United States. These areas are well known for periodic severe droughts and water scarcity.

Climate Moisture Index CV

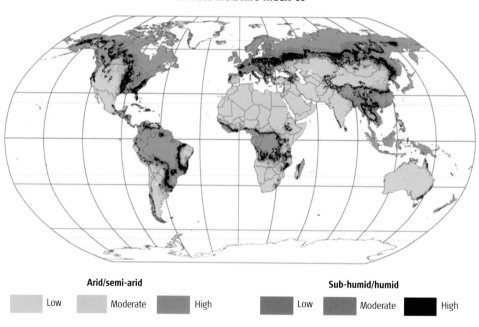

Arid/semi-arid: Low | Moderate | High
Sub-humid/humid: Low | Moderate | High

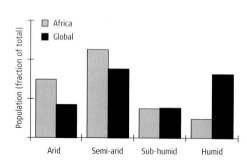

Water scarcity is fundamentally a problem of the distributions of climate and human society, which vary greatly around the world. Compared to the global proportion of 52% of total population living in arid or semi-arid regions, approximately 75% of all Africans live in such conditions (mean CMI-CV < 0; see inset). In addition, 20% of all Africans live in areas that experience high interannual climatic variability as expressed by a CMI-CV > 0.75 (Vörösmarty et al., 2005a). This explains why Africa suffers disproportionately from water scarcity and water stress compared to other continents.

Source: Water Systems Analysis Group, University of New Hampshire. Datasets available for download at http://wwdrii.sr.unh.edu/

The Water Reuse Index

Water use by humans is a recursive phenomenon, by which water is withdrawn, used and reused as it passes from upstream source areas downstream past agricultural, industrial and domestic users to the sea. The Water Reuse Index (WRI) provides a measure of pressure on river basin water resources (Vörösmarty et al., 2000, 2005a). Beginning at zero in the headwaters, the WRI can vary widely over the course of a river according to the pressures of different uses along its trajectory. If demand is high relative to the available flow (e.g. when encountering a city or major irrigation works), the WRI will move upwards (see graph below). If relatively low-use tributaries feed into the mainstream, the Index will decline. In many of the world's river

systems water reuse can exceed, sometimes greatly, natural river flow. With high values for this Index, we see increasing competition between water users – both nature and society – as well as pollution and potential public health problems. The WRI can shift markedly with climate variability. For example, for the Orange River in Africa (inset graph below), the relative water-use ratio remains well below 1.0 (i.e. 100 percent use of river flow) at mean annual flow conditions. However, water use becomes 'over-subscribed' by a factor of more than 10 under 30-year low flow (drought) conditions (Vörösmarty et al., 2005a). If water is to be delivered to all users, then it must be reused, flowing through canals, pipes and pumps more than ten times to satisfy all.

Water Reuse Index

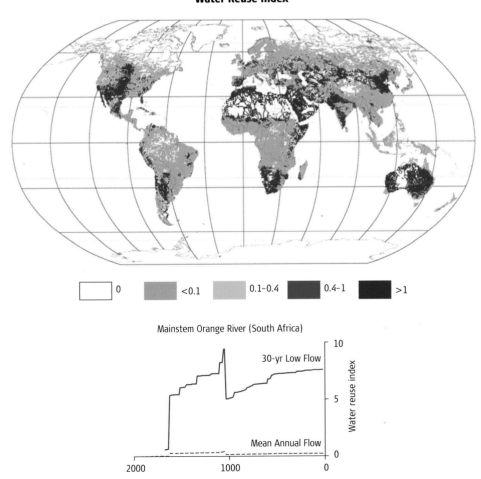

| | 0 | | <0.1 | | 0.1–0.4 | | 0.4–1 | | >1 |

Mainstem Orange River (South Africa)

30-yr Low Flow

Mean Annual Flow

Water reuse index

10

5

0

2000   1000   0

Source: Water Systems Analysis Group, University of New Hampshire. Datasets available for download at http://wwdrii.sr.unh.edu/

Better decision-making, improved planning, effective risk management, innovation in development and environmental protection activities – these are the human activities that can reduce the vulnerability of communities. To this end, risk assessment and disaster reduction should be integral parts of all sustainable development projects and policies.

Kofi Annan, *United Nations Secretary-General*

CHAPTER 10

Managing Risks: Securing the Gains of Development

By

WMO
*(World
Meteorological
Organization)*

UN-ISDR
*(Inter-Agency
Secretariat, United
Nations International
Strategy for Disaster
Reduction)*

*Left: Coastal destruction in the wake of the
26 December 2004 tsunami in Indonesia*

Key messages:

Over the past decade, there has been progress in risk management, thanks to scientific advancements and the recognition of the political, social and cultural dimensions of risk. Yet technical and organizational constraints remain high and slow down the design and implementation of efficient risk reduction policies.

■ Water-related disaster risk reduction calls for stronger integration of risk-related public policies, and improved cooperation among decision-makers, risk managers and water managers.

■ Indicators are needed to detect and monitor changes in the natural and social environment to provide a quantitative basis for the design of disaster risk reduction policies and to monitor the effectiveness of these policies.

■ Societies need to improve decision-making in situations of uncertainty, in order to better adapt to ongoing and future global changes, such as increased climate variability.

■ Based on an integrated policy and national risk management frameworks, institutional coordination and management mechanisms need to be strengthened. This could be achieved through the promotion of national disaster prevention forums including stakeholders.

■ Implementation plans need to be developed based on an integrated risk management policy. These plans need to be well-resourced and underpinned by a clear legislative framework.

■ Risk management strategies should consider the preparedness of societies to live and deal with risks, taking into account risk perception issues and emerging threats such as global warming and climate change.

Above: Flooding of the Danube in Budapest, Hungary
Floods in Viet Nam

Below: Hue, Viet Nam
Girl standing on dry, crackled ground following a drought in Turkey
Tornado damage in Osceola County, Florida, United States. This type of violent tornado, rare in Florida, is generally linked to the climatic phenomenon of El Niño

Section 4: MANAGEMENT RESPONSES & STEWARDSHIP

Part 1. Setting the Scene of Water-related Disasters

The number, scale and increasing impact of water-related disasters in recent years have resulted in massive losses of lives and livelihoods. Vulnerable societies throughout the world, developing countries in particular, are enduring the long-term negative social, economic and environmental consequences of these disasters.[1] These impacts were acknowledged in both the Johannesburg Plan of Implementation (JPoI)[2] and the Millennium Development Goals (MDGs) (UNDP, 2004).

Children collecting water during drought, India

Developing countries are disproportionately affected by disasters; their losses are about five times higher per unit of gross domestic product (GDP) than those of rich countries. These losses often offset years of hard-won social and economic development progress. Managing risks has therefore become a priority for alleviating poverty, ensuring socio-economic progress and securing the gains of development.

Extreme hydrometeorological events often interact with other water-related hazards. Other threats include pollution and chemical spills, aquifer depletion, land subsidence, salinization of arable land, marine intrusions, sea and storm surges, coastal flooding and water-borne diseases.[3] The scope of risk reduction policies and activities must therefore be broadened to include these multiple threats. Explicit links with other challenge areas of the *World Water Development Report* can also be made. These include health and sanitation (**Chapter 6**), ecosystems and biodiversity (**Chapter 5**), food security

(**Chapter 7**) and water quality (**Chapter 4**), as well as development and sharing of water resources (**Chapter 11**).

1a. Water-related hazards in the global disaster landscape

Statistics from the Center for Research on the Epidemiology of Disasters (CRED) in Belgium revealed that during the ten-year period from 1996 to 2005, about 80 percent of all natural disasters were of meteorological or hydrological origin. In the last decades, between 1960 and 2004 (see **Figure 10.1**), there has been a significant rise in water-related extreme events, such as floods and windstorms. Drought, as well as water-related and landslide events, have also increased across the same period equivalently to all other natural disasters (such as earthquakes, volcanoes, etc.). During the ten-year period 1995–2004, wave-surge disasters including storm-surges and the 2004 tsunami in the Indian Ocean threatened increasing numbers of people worldwide.

Figure 10.1: Global trend of water-related disasters by type of hazard, 1960–2004

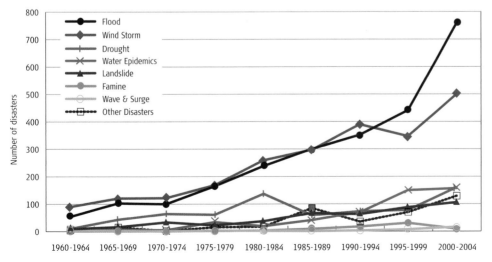

Source: Data from the Center for Epidemiology of Disasters (OFDA-CRED) in Louvain (Belgium). Analysis by the Public Works Research Institute (PWRI) in Tsukuba (Japan), 2005.[4]

1. For 1992–2001, losses from water-related disasters were estimated at US $446 billion, accounting for about 65% of economic loss due to all natural disasters (ISDR, 2004a).

2. World Conference on Sustainable Development, September 2002, Johannesburg, Republic of South Africa.

3. For risk-related terminology see the first edition of the *World Water Development Report* (UN-WWAP, 2003) and the International Strategy for Disaster Reduction (ISDR) global review, *Living with Risk* (ISDR, 2004a).

4. We acknowledge support from Dr T. Merabtene, and Dr Y. Junichi, at PWRI.

...the causes for disasters need to be analysed so as to guide investment in reconstruction, in particular for infrastructure and land use

The global distribution of water-related disasters shows important regional differences (see **Figure 10.2**), with a large share of events occurring in Asia. **Figure 10.3** offers an overview of the impact of water-related disasters in terms of numbers of deaths and people affected.

To some extent, the increase of water-related disasters shown in **Figure 10.3** can be explained by an increase in reporting activities. Likewise, the number of people affected by disasters and material losses can be attributed to population growth and increasing value of assets. In some cases, however, risk and disaster statistics are still difficult to produce. This is the case for instance when it comes to giving a clear definition of people 'affected' by a disaster, where health, sanitary, social and economic dimensions must be taken into account. Differences are also introduced when comparing developing and developed countries. Statistical difficulties also exist with the qualification of drought-related disasters.

1b. Disaster risk reduction at the international level

Milestone events during the United Nations (UN) International Decade for Natural Disaster Reduction (IDNDR, 1990–2000) and in the Yokohama Strategy and Plan of Action (UN/GA, 1994) have provided policy guidance and tools for the mitigation of natural disasters.

5. For details visit
   www.mrcmekong.org

6. See the final report of the
   World Conference on Disaster
   Reduction (18–22 January
   2005, Hyogo, Japan) at
   www.unisdr.org/wcdr

Based on a review of global disaster risk reduction initiatives, the Secretariat of the International Strategy for Disaster Reduction (ISDR) identified the principal limitations and challenges to the implementation of this strategy and action plan (ISDR, 2004a). These limitations have also become key areas for developing a relevant framework for action in disaster risk reduction for the International Decade 'Water for Life' 2005–2015 and include the following (WCDR, 2005):

- governance: organizational, legal and policy frameworks
- risk identification, assessment, monitoring and early warning
- knowledge management and education
- reduction of underlying risk factors
- preparedness for effective response and recovery.

These points are consistent with the priorities identified in two other major policy documents agreed upon by the international community: the Johannesburg Plan of Implementation (JPoI) and the Millennium Development Goals (MDGs). **Table 10.1** illustrates how the latter are related to risk reduction.

Complementary to the commitment of the international community, many countries have also engaged, bilaterally, regionally and internationally, in cooperative arrangements for water-related disaster risk reduction. This is, for instance, the case in the Mekong River Basin, where in 2001 riparian countries established a Flood Mitigation and Management Plan under the aegis of the Mekong River Commission.[5] In southern Africa, countries of the Southern African Development Community developed a web-based information system to monitor regional conditions when cyclones, floods and droughts occur in the region (see **Chapter 14**).

Strong linkages have been identified between poverty, high social vulnerability to and low capacity to cope with water-related hazards and disasters.[6] The next section discusses how risk management is a key issue of sustainable development.

1c. Linking disaster risk reduction and development planning

Water-related disasters disrupt economic development as well as the social fabric of vulnerable societies. This jeopardizes the accumulated gains in social and economic development and investments in better living conditions and quality of life. Disaster risk reduction policies and measures

Section 4: MANAGEMENT RESPONSES & STEWARDSHIP

Figure 10.2: Distribution of water-related disasters by region, 1990–2004

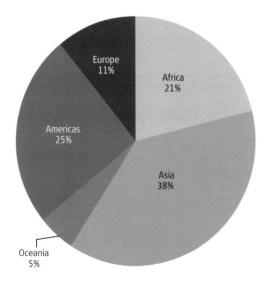

Europe 11%
Africa 21%
Americas 25%
Asia 38%
Oceania 5%

Source: Data from OFDA-CRED in Louvain (Belgium) and analysis by PWRI in Tsukuba (Japan), 2005.

Table 10.1: The Millennium Development Goals (MDGs) and disaster risk reduction

Millennium Development Goals (MDGs)	Related risk reduction aspects
MDG 1: Eradicating extreme poverty and hunger	Human vulnerability to natural hazards and poverty are largely codependent. Exposure to hazards plays a critical role in poverty-ridden areas. Hunger reduces individual capacity to cope with stress caused by disasters.
MDG 2: Achieving universal primary education	Educational attainment is a fundamental determinant of human vulnerability and social marginalization. Basic literacy and numeracy enable individuals to become more engaged in their society. Broadening participation in decision-making is key to disaster risk reduction.
MDG 3: Promoting gender equality and empowering women	Facilitating the participation of women and girls in the development process is a key priority. Women across the world play critical roles in shaping development. In some contexts, women may be more exposed to natural hazards. At the same time, women are often more likely than men to participate in communal actions to reduce risk and enhance development.
MDG 4: Reducing child mortality	Children under five years of age are particularly vulnerable to the impacts of environmental hazards, ranging from the everyday risks of inadequate sanitation and drinking water to death and injury as a result of catastrophic events and their aftermath. Post-traumatic psychological disorders are also a major issue.
MDG 5: Improving maternal health	As environmental hazard stress or shock erodes the savings and capacities of households and families, marginal people within these social groups are most at risk. In many cases, it is women, girls or the elderly who are the least entitled to household or family assets. Reducing drains on household assets through risk reduction will contribute to enhancing maternal health.
MDG 6: Combating HIV/AIDS, malaria and other diseases	Interactions between epidemiological status and human vulnerability to subsequent stresses and shocks are well documented. For example, rural populations affected by HIV/AIDS are less able to cope with the stress of drought. Likewise, individuals living with chronic or terminal diseases are more vulnerable to emergency situations.
MDG 7: Ensuring environmental sustainability	Major disasters, or the accumulation of risk from regular and persistent but smaller events, can wipe out any hope of sustainability for urban or rural environments. Again, the equation works both ways. Increasing destruction due to landslides, floods and other disasters related to environmental and land-use patterns are a clear signal that massive challenges remain in achieving this goal.
MDG 8: Developing a global partnership for development	Efforts to enhance sustainable development and reduce human vulnerability to natural hazards are hampered by national debt burdens, terms of international trade, the high price of necessary drugs, lack of access to new technology and new hazards associated with global climate change, among other hurdles. Building a global partnership for development would contribute to disaster risk reduction.

Source: Adapted from UNDP, 2004.

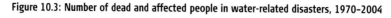

Figure 10.3: Number of dead and affected people in water-related disasters, 1970–2004

Note: The disasters reported in this figure include floods, windstorms, landslides, avalanches, droughts, famines, water-related epidemics and technological water-related disasters (such as traffic accidents due to water). This figure indicates a possible improvement in crisis management, disaster relief and humanitarian activities, while simultaneously illustrating that the number of people living in hazard-prone areas is increasing. The alarming increase of affected people since the start of the twenty-first century is notable: from 2000 to 2004 (four-year data), 1,942 water-related disasters claiming the lives of 427,045 people and more than 1.5 billion affected people were reported in the CRED disaster database.

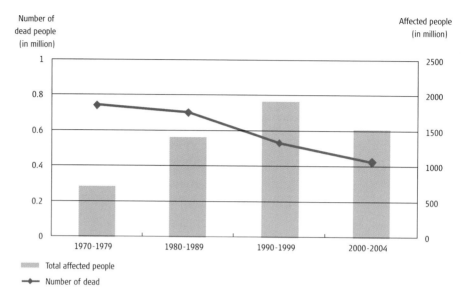

Source: Data from OFDA-CRED in Louvain (Belgium) and analysis done in 2005 by Public Works Research Institute (PWRI) in Tsukuba (Japan).

need to be designed in such a way that they are consistent with integrated long-term development objectives and implementation plans. Therefore, managing water-related risks is a matter of governance. In particular, post-disaster relief and reconstruction activities need to be improved with the long-term objective of 'building back better'. This means that the causes for disasters need to be analysed so as to guide investment in reconstruction, in particular for infrastructure and land use. Limiting the extent of damage and reducing vulnerability are two interrelated objectives of the risk management cycle (**Figure 10.4**).

There is now international acknowledgement that efforts to reduce disaster risks must be systematically integrated into policies, plans and programmes for sustainable development and poverty reduction (Abramovitz, 2001 in ISDR, 2004a). At the local level, for instance, disaster risk reduction efforts should assist communities not only to recover from disasters but also to move above the poverty line. Land-use planning is another example of integrated policies that can help reduce disaster risk, which should take account of the positive social and economic aspects of flooding, including sediment provision for soil fertility, environmental flow maintenance, and ecosystem maintenance.[7] The value of wetlands for flood protection has also been increasingly recognized as a complement to structural measures.[8]

As part of national and regional plans for sustainable development, risk assessment is needed to anticipate possible impacts of global changes on water resources. It is now widely recognized that climate variability and change have the potential to threaten sustainable development (IPCC, 2001).

A convergence of interests is emerging, and efforts have been made since the 1990s to develop cooperative actions with the objective of integrating climate-related coping strategies into disaster risk reduction and poverty reduction efforts. For instance, the Co-operative Programme on Water and Climate aims to improve the capacity to cope with the impacts of the increasing variability of the world's climate.[9]

However, despite these national and international investments and efforts, limitations still remain in current disaster risk reduction activities.

1d. Limitations in risk reduction: lessons from current practice

A recent study conducted by the World Meteorological Organization (WMO) identified challenge areas in risk management (adapted from WMO, 2004):

- Challenge areas related to scientific observations and improved methodologies:

Section 4: MANAGEMENT RESPONSES & STEWARDSHIP

7. For details, see the Integrated Flood Management concept (IFM) of the World Meteorological Organization-Global Water Partnership (WMO-GWP) Associated Programme for Flood Management (APFM) at www.apfm.info/

8. See, for instance, the activities of the World Conservation Union (IUCN) at iucn.org/themes/wetlands/

9. For more information, visit www.waterandclimate.org/home.asp

- improving the quantity and accuracy of data in order to map hazards and assess impacts
- making Geographic Information Systems (GIS) more user-friendly
- quantifying uncertainties related to forecasting hydrometeorological extremes
- building up and disseminating knowledge of the effects of climate variability and change
- further developing robust vulnerability assessment methods
- incorporating integrated environmental strategies in risk management.

▨ Challenge areas related to social and political issues:
- building risk management frameworks that reflect an integrated approach to risk management
- promoting the inclusion of risk management aspects in transboundary agreements
- enhancing public participation in risk management programmes and activities.

Sustainable development, poverty reduction, appropriate governance and disaster risk reduction are interconnected objectives, as reflected by the evolution of risk reduction approaches detailed in the next section.

Figure 10.4: The risk management cycle

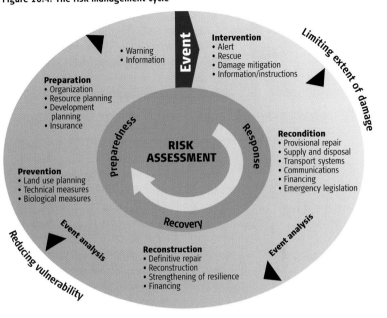

Source: Adapted from Swiss Civil Protection. This mitigation-crisis-rehabilitation cycle shows the challenges of post-disaster reconstruction.

Flooding in Tana River Valley, Kenya, due to extended and unseasonal rain

Part 2. Risk Management Frameworks

The first *UN World Water Development Report* (WWDR1) showed that over the past decade, there has been a shift from crisis management, which was mainly responsive in nature, to proactive risk management and strategies orientated towards disaster prevention. The basic characteristics of these different approaches appear in Table 10.2.

Fighting against rising flood waters in Germany

2a. Risk management over time: From response to integration

In recent years, the understanding of water-related disaster risks has improved, thanks to advances in modelling and forecasting of physical processes, such as climate variability and change, and the progressive inclusion of social and environmental dimensions in assessment (Viljoen et al., 2001). This has been useful in identifying the social and economic factors of disasters, such as the value of assets exposed to hazards, livelihood functions, social vulnerability, coping capacity, cultural dimensions and the role played by the insurance sector (for theory and examples of practice, see Dercon, 2004).

The recognition of the social dimensions of risk and disasters has fostered research and action in developing participatory processes for risk management (WHO, 1989; McDaniels et al., 1999; Parker, 2004; ISDR, 2004a). The objective is that all relevant institutions and stakeholders that are exposed to risk have the opportunity to share their experiences and concerns in the decision-making process (see **Box 10.1** for details).

Such participatory approaches have been implemented in many countries. In the Netherlands, for instance, participatory planning for flood management has been successfully tested (Frijters and Leentvaar, 2001). In France, Germany and Poland, the European Union funded a project involving floodplain communities in the design of a flood information system based on information technologies.[10] In Cambodia, the non-governmental organization Action Against Hunger and the Cambodian Red Cross have developed, since 1998, a community-based project of early warning for Mekong floods (Affeltranger and Lictevout, 2005).

As a support tool to these community-based processes, rapid developments in modern communication technologies can help record and disseminate experience, convey professional knowledge and contribute to decision-making processes. Information and knowledge, both of which are institutional and community-based,[11]

are integral to the design and successful implementation of risk reduction policies (see also **Chapter 13**).

2b. Managing risk-related knowledge and information wisely: Preventing data loss

Water- and risk-related data are needed to support multi-hazard approaches, design risk-related indicators, operate efficient warning systems, develop awareness-raising programmes and enable institutions to adapt to environmental and social changes. Availability of and access to data are therefore essential for hazard analysis and vulnerability assessment (ISDR, 2004a). However, risk-related knowledge and information are often unavailable or missing. Difficulties include a loss of institutional memory and limited access to data and information.

Water- and risk-related data accessibility problems include the following (ISDR, 2004a):

- Data is restricted for presumed security purposes.
- Inadequate cross-sector communication exists about the existence of data.
- Dissemination of information is not considered a priority by the organization.
- Information is maintained in non-standardized formats.
- Existing information is costly to convert into more readily accessible formats.
- Data compilers have not consulted users about their data requirements.
- Information for women's advocacy organizations and other community-based groups is not readily available, and gender-specific data is not consistently gathered or disseminated.

Risk knowledge and know-how can also be lost over time for various reasons, including lack of funding for database maintenance, lack of information-sharing among administrations, loss of institutional memory when civil servants retire or leave office for a job in the private sector. Lost knowledge and know-how include overviews of hydrometeorological processes in river basins, the

---

10. For more information, visit www.ist-osiris.org

11. Institutional knowledge includes the expertise of civil servants; official statistics and databases; and hazard and risk mapping resources. Community-based knowledge includes past flood experience; empirical knowledge; and coping and adaptive capacities.

Table 10.2: Response-based versus prevention-oriented strategies to disaster risk reduction

Response-based strategies (relief)		Integrated strategies (prevention, mitigation and relief)
1. Primary focus on hazards and disaster events	**Emphasis**	1. Primary focus on vulnerability and risk issues
2. Single, event-based scenarios		2. Dynamic, multiple risk issues and development scenarios
3. Basic responsibility to respond to an event		3. Fundamental need to assess, monitor and update exposure to changing conditions
4. Often fixed, location-specific conditions		4. Extended changing and shared regional or local variations
5. Responsibility concentrated in a single authority or agency	**Operations**	5. Involves multiple authorities, interests, actors
6. Command and control, directed operations		6. Situation-specific functions, free association
7. Established hierarchical relationships		7. Shifting, fluid and tangential relationships
8. Often focused on hardware and equipment		8. Dependent on related practices, abilities and knowledge base
9. Dependent on specialized expertise		9. Specialized expertise, squared with public views and priorities
10. Urgent, immediate and short-term periods in outlook, planning, attention, returns	**Time horizons**	10. Comparative, moderate and long-term periods in outlook, planning, values, returns
11. Rapidly changing, dynamic information usage, often conflicting or sensitive		11. Accumulated, historical, layered, updated or comparative use of information
12. Primary, authorized or singular information sources, need for definitive facts		12. Open or public information, multiple, diverse or changing sources, differing perspectives and points of view
13. Directed, 'need to know' basis of information dissemination, availability	**Information use and management**	13. Multiple use, shared exchange, inter-sectoral use of information
14. Operational, or public information based on use of communications		14. Matrix, nodal communication
15. In-out or vertical flows of information		15. Dispersed, lateral flows of information
16. Relates to matters of public security, safety	**Social, political rationale**	16. Matters of public interest, investment and safety

Source: ISDR, 2001.

Left: 2nd wave of the 26 December 2004 tsunami, Sri Lanka

Right: Desertification in Chott El-Djerid, Tunisia

BOX 10.1: VALUE OF STAKEHOLDER PARTICIPATION IN DISASTER RISK REDUCTION

Before the disaster:
- improved hazard assessment by relying on local traditional or scientific knowledge
- improved vulnerability analysis by identifying risk perceptions and hidden weaknesses
- identification of most vulnerable groups and prioritization of mitigation investment
- assessment of self-help capacity: awareness, knowledge and resources
- assessment of the information needs of flood-prone communities

- improved social understanding and ownership of official mitigation strategies.

During the disaster:
- communal good-neighbourhood protective behaviour and helping capacity for relief
- improved understanding of warnings and disaster management activities
- improved trust in official authorities and relief forces.

After the disaster:
- improved commitment in post-crisis feedback analysis and debriefing activities
- coordinated reconstruction using the concept of 'building back better'.

Source: Affeltranger, 2002.

The loss of risk-related information and knowledge is a threat to the sustainability of institutions and organizations responsible for disaster risk reduction

location of stored data, station maintenance, the operation of models and checking forecasts for consistency. The loss of risk-related information and knowledge is a hindrance to institutions and organizations responsible for disaster risk reduction.

Options for preventing this loss of institutional memory include improvement of legislation on the management of information produced by national administrations, clear allocation of duties regarding information management and custodianship, internal procedures and total-quality approaches.

Data and information for hazard analysis
As a primary input for hazard analysis and identification of trends in hazards, high quality comprehensive data and information are crucial in support of research in natural and man-made processes that govern the frequency and magnitude of hazards. Such data are the base for comprehensive risk assessments that are essential for planning and monitoring purposes (WMO, 1999). The analysis of past data, and of extreme events in particular, is helpful for quantifying disaster trends and impacts in terms of exposure and human and economic loss.

Data collection and management remains a key challenge to successful hazard analysis. This is particularly the case when dealing with extreme hydrometeorological events about which there are limited available data, especially in developing countries, where historical data series are often missing or incomplete. Reasons for this can be technical, financial or political. Climate variability and change are other sources of uncertainty in terms of the reliability of past series of water-related data.

These various factors usually result in forecasting uncertainty, poor modelling results and inadequate warnings, as well as biased hazard mapping, which in turn can result in major losses when applied to land-use and urban planning purposes. Lack of data can then lead to higher risks.

Data and information for vulnerability assessment
Efficient vulnerability assessment requires availability of and access to data on attributes of groups and individuals, including socio-economic class, ethnicity, gender, age and religion. These features can be used to help differentiate and rank the level of vulnerability to various social groups and subgroups.

While hazard mapping has been improved by the wider use of GIS techniques, the inclusion of social, economic and environmental variables into GIS models still remains a major challenge (ISDR, 2004a). Indeed, assigning quantifiable values to the social and economic dimensions of vulnerability is not always possible. The various and usually interrelated scales of vulnerability patterns also make spatial representation, mapping and visualization difficult. In addition, the level of data quality and detail required for GIS analysis is often incompatible with available information produced or provided by national administrations.

When necessary, low-technology approaches, such as paper maps, billboards and drawings, can offer a local-level, cheaper alternative to GIS-based techniques. However, the use of GIS for the analysis of vulnerability level and coping capacity is a rapidly developing field.[12]

12. For an example of application, see the OSIRIS Project at www.ist-osiris.org

---

BOX 10.2: EXAMPLE OF MULTI-HAZARD ASSESSMENT IN COSTA RICA

Turrialba is a city of 33,000 people, which is located in the central part of Costa Rica and regularly affected by flooding, landslides and earthquakes. In order to assist the local emergency commission and the municipality, the United Nations Educational, Scientific and Cultural Organization (UNESCO) sponsored a project in capacity-building for natural disaster reduction. A multi-hazard risk assessment of the city was conducted based on a Geographic Information System (GIS) application for risk assessment and management. The city's cadastral database was used in combination with various hazard maps for different return periods to generate vulnerability maps for the city. Cost maps were combined with vulnerability maps and individual hazard maps to obtain graphs of probability and resulting loss values. The resulting database is an example of a tool that local authorities can use to assess the effects of different mitigation measures and conduct cost-benefit analyses.

Source: ITC, 2005 and ISDR, 2004a.

2c. Advances in water-related risk management: Examples of good practices

The following selections of good practices have been chosen in terms of their relevance to various aspects of disaster risk reduction. The first example offers a holistic approach to the design of risk reduction policies. The second example addresses the design of multi-hazard approaches, while the third example considers the involvement of stakeholders in the design of warning systems.

Integrated flood management: A holistic approach to policy design

Water-related disaster mitigation should be seen as a key component of integrated water resources management (IWRM) and can be found in the following principles (APFM, 2003):

- managing the water cycle as a whole (basin-wide, including underground resources)
- integrating land and water management (including water allocation and land-use patterns)
- adopting a sound mix of flood management strategies (structural and non-structural)
- ensuring a participatory approach (involving all relevant stakeholders in decision-making)
- adopting integrated hazard management approaches
- breaking the poverty cycle through improved risk management.

Managing extreme water-related events must be linked to water resources management. Training programmes, tools and awareness-raising material used for water resources management should include a risk-related component, such that an integrated assessment of land-use changes and floodplain restoration can be developed in parallel to structural measures for flood protection (Brouwer et al., 2001). Integrated flood management requires a holistic analysis of social, economic and ecological services by floods, as shown by projects developed in England on watershed management (DEFRA, 2003).

Multi-hazard approaches: Integrating existing mechanisms and tools

A multi-hazard approach to early warning, forecasting, preparedness and response, notably through the use of existing observational and telecommunication systems, is the ideal method for saving lives and protecting infrastructure (Grabs, 2005). Such multi-hazard approaches to warning also help to design warning messages that fit the various hazards imposed on a human settlement or a community. Such messages are more likely to provide decision-makers and other stakeholders with helpful information to cope with natural hazards.

The multi-hazard approach, advocated by WMO, also promotes improved cooperation and coordination of national agencies responsible for development planning, disaster mitigation and water resources management, including National Hydrological and Meteorological Services (NHMSs). National Platforms for Disaster Risk Reduction (NPs) are a major implementation instrument, with the following objectives:

- increased collaboration and coordination among national stakeholders
- increased knowledge and visibility of national situations at the regional and international levels
- increased levels of knowledge and skills in the global risk reduction community
- national leadership and commitment to the sustainability of the National Platforms for Disaster Risk Reduction.
- increased national counterparts to help strategy implementation

Water-related disaster mitigation should be seen as a key component of integrated water resources management...

BOX 10.3: MANUALS FOR COMMUNITY-BASED FLOOD MANAGEMENT: PROJECT IN BANGLADESH, INDIA AND NEPAL

The Community Approaches to Flood Management project, developed by WMO and its partners, has developed country-wide manuals on community flood management on the basis of information provided by selected flood-prone communities in Bangladesh, India and Nepal. Through field research, including Participatory Rapid Appraisal (PRA), it was first ascertained

which activities had been undertaken, individually and collectively, at various stages – before, during and after floods have occurred – with a view to reducing lost lives and destroyed livelihoods and the suffering caused by floods. Once drafted, the manuals were reviewed by the selected communities during workshops and subsequently adopted. This approach, when

implemented in selected flood-prone areas, has proven effective at improving the flood management capacity of the communities concerned and reducing their flood vulnerability.

Source: Unpublished project report (as of April 2005), WMO Commission for Hydrology and Water Resources Management.

BOX 10.4: METHODOLOGY BEHIND THE DISASTER RISK INDEX (DRI)

A mortality-based index was developed in order to enable comparisons of countries hit by different hazards types, such as droughts versus floods. The other reason for such a choice was that data on mortality is the most complete and the most reliable (the Emergency Disasters Database from CRED was used for this purpose). Other parameters, such as economic losses, number of injured or losses of livelihood, all suffer from either a lack of data or a lack of comparative potential, if not both. The formula used for risk estimation was based on the UN definition of 1979, which states that the risk

results from three components: the *hazard occurrence probability*, the *elements at risk* (in this case the population) and their *vulnerability*. By multiplying the frequency of hazards by the population affected, the *physical exposure* was obtained. This figure represents the average number of people affected yearly by a specific hazard. The first task was to find all the requested geophysical data and then model the different hazards in order to obtain the frequency for earthquakes, drought, floods and cyclones for each location on the globe. The model for population distribution, developed by

the Center for International Earth Science Information Network, was multiplied by frequency to compute the physical exposure. This already normalized the differences between populations highly affected by a selected hazard and those less frequently affected.

Note: The United Nations Environmental Programme's (UNEP) Global Resource Information Database started a process to update the DRI methodology in April 2005.

Source: UNDP, 2004.[16]

■ strengthened credibility across different institutions and interest groups
■ increased commitment to help the most vulnerable.[13]

Box 10.2 provides an example of multi-hazard assessment in Costa Rica.

User-based design of warning systems for floods and drought

Designing efficient flood warning systems poses technical, organizational and social challenges: technical constraints include a lack of data, modelling inadequacy and differing flood types, and organizational constraints include weak dissemination of information and institutional deficiencies in the coordination of joint measures for risk management and disaster prevention. Social and cultural limitations include a poor understanding of warning, limited ownership, conflicting information sources and resistance to follow guidance and instructions.

The efficiency of warning systems for water-related disasters was found to be greatly improved by the early involvement of stakeholders in the design of the warning system (McDaniels et al., 1999; Vari, 2004). The objective is to design a warning message that will be most useful to people confronting an impending hazard. Participatory design of warning strategies has been successfully implemented in many developed and developing countries. These approaches aim at involving warning receivers in the various development phases of a warning system, including forms and contents of the message, dissemination channel and options for feedback. (Affeltranger, 2002; Parker, 2004; Affeltranger and Lictevout, 2005). **Box 10.3** provides an example of a community-based approach to flood management.

The development of warning systems for droughts is another challenge for risk managers and water managers. Early warnings of drought help farmers select appropriate crops and irrigation schedules and

13. For more information, see ISDR's Guiding Principles – National Platforms for Disaster Risk Reduction www.unisdr.org/eng/country-inform/docs/Guiding%20Principles%20for%20NP.pdf

methods, thus contributing to food security. Timely warning also provides water managers with a chance to allocate available water resources based on rational priority criteria.

Several initiatives have been developed to improve drought-related information management and warning activities. For instance, at the request of twenty-four countries in eastern and southern Africa, WMO established two Drought Monitoring Centres (DMCs), in Nairobi, Kenya (see **Chapter 14**), and Harare, Zimbabwe, in 1989 with financial support from the United Nations Development Programme (UNDP). The main objective of the centres is to contribute to early warning systems and the mitigation of adverse impacts of extreme climatic events on agricultural production.[14]

14. For more information, visit www.drought.unl.edu/monitor/ EWS/ch11_Ambenje.pdf

Part 3. Indicators for Risk Management

Indicators are needed to inform the design of disaster risk reduction policies and monitor the implementation and assessment of these policies. Indicators help identify patterns in disaster losses, as well as underlying physical, social or economic trends that influence hazard, vulnerability and risk patterns. Such risk factors include environmental degradation, population growth and the increasing value of assets in flood-prone areas and risk perception. Quantifiable indicators in particular are needed when decisions involve trade-offs between development options with varying degrees of risk.

The development of indicators for water-related risk management is a relatively new field. In water-related risk management, risk-based indicators remain scarce and suffer from limitations in terms of conceptual design, paucity of data and largely insufficient robustness. There is a clear need to further develop indicators for risk management and encourage governments and relevant national and international organizations to provide the necessary data on which these indicators are built. Such data should be of high quality and supplied on a regular basis to enable the development of long-term indicators, especially for monitoring purposes.

Below are three examples of indicators selected to demonstrate their actual or potential applications on global, regional and national scales (see also **Chapter 1**). These indicators are in different testing and application stages. Some are already undergoing a revision of their science base and robustness, underlining the necessity for further research and development on their concepts and applicability.

3a. Disaster Risk Index (DRI)

This index[15] has been developed to help compare disaster-risk country situations, based on a quantitative approach to disaster impacts. Natural hazards assessed by this index include floods, cyclones, earthquakes and droughts. This index allows global ranking on the basis of relative vulnerability of nations (UNDP-BCPR, 2004) (**Box 10.4**).

Indicators used for the DRI aim at grasping the socio-economic dimensions of risks. They include the Human Development Index (HDI), the number of physicians per 1,000 inhabitants, the rate of urban growth, etc. The results showed surprisingly high correlations.[17] This analysis provides a useful and neutral tool for the evaluation of countries facing natural hazard risks. UNDP hopes that this tool will help countries with both high vulnerability and high exposure to adopt more risk reduction measures. **Maps 10.1** and **10.2** and **Figures 10.5** and **10.6** show the DRI graphic results as applied to floods and droughts.

3b. Climate Vulnerability Index (CVI)

Developed for a range of scales (from community to national and regional levels), this indicator links climate variability and change, water availability, and socio-economic factors (Sullivan and Meigh, 2005).[18] The assessment of risk in relation to water resources is strongly dependent on people's vulnerability to water-related hazards. In addition, the uncertainty generated by climate variability and change plays an important role.

The CVI identifies a range of social, economic, environmental and physical factors relevant to vulnerability (see **Table 10.3**) and incorporates them into an integrated index.

15. This index was developed by the United Nations Development Programme's Bureau for Crisis Prevention and Recovery (UNDP/BCPR), based on research done by United Nations Environmental Programme's Global Resource Information Database (UNEP/GRID, Geneva).

16. We acknowledge the contribution of Dr Pascal Peduzzi (UNEP/GRID, Geneva) in the drafting of this section.

17. A web-based interactive tool for comparing countries is provided at: gridca.grid.unep.ch/undp/. The location of frequency and physical exposure can be visualised at: grid.unep.ch/preview

18. For this section, we acknowledge the support of Dr Caroline Sullivan, from the Centre for Ecology and Hydrology (CEH) in the UK.

Map 10.1: Physical exposure and relative vulnerability to floods, 1980-2000

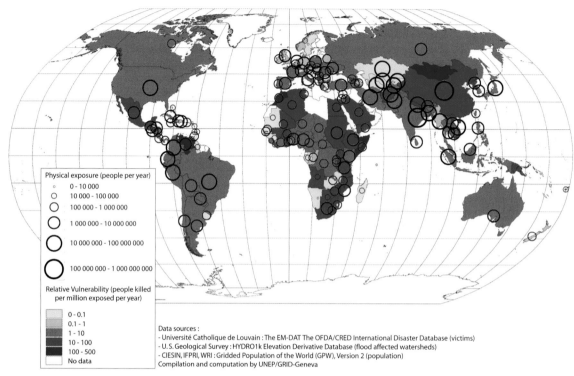

Physical exposure (people per year)
- 0 - 10 000
- 10 000 - 100 000
- 100 000 - 1 000 000
- 1 000 000 - 10 000 000
- 10 000 000 - 100 000 000
- 100 000 000 - 1 000 000 000

Relative Vulnerability (people killed per million exposed per year)
- 0 - 0.1
- 0.1 - 1
- 1 - 10
- 10 - 100
- 100 - 500
- No data

Data sources :
- Université Catholique de Louvain : The EM-DAT The OFDA/CRED International Disaster Database (victims)
- U.S. Geological Survey : HYDRO1k Elevation Derivative Database (flood affected watersheds)
- CIESIN, IFPRI, WRI : Gridded Population of the World (GPW), Version 2 (population)
Compilation and computation by UNEP/GRID-Geneva

Source: UNDP, 2004.

Figure 10.5: Relative vulnerability for floods

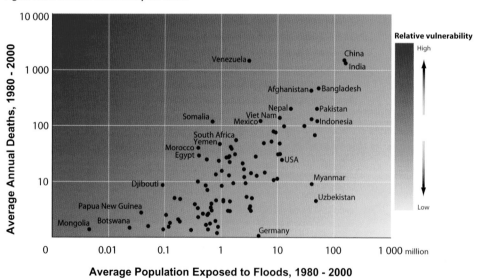

Data source: The EM-DAT OFDA/CRED International Disaster Database and UNEP/GRID-Geneva.

Source: UNDP, 2004.

Map 10.2: Physical exposure and relative vulnerability to droughts, 1980–2000

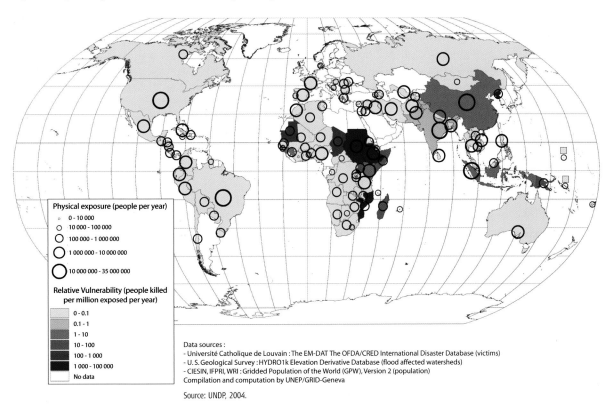

Physical exposure (people per year)

- 0 - 10 000
- 10 000 - 100 000
- 100 000 - 1 000 000
- 1 000 000 - 10 000 000
- 10 000 000 - 35 000 000

Relative Vulnerability (people killed
per million exposed per year)

- 0 - 0.1
- 0.1 - 1
- 1 - 10
- 10 - 100
- 100 - 1 000
- 1 000 - 100 000
- No data

Data sources :
- Université Catholique de Louvain : The EM-DAT The OFDA/CRED International Disaster Database (victims)
- U.S. Geological Survey : HYDRO1k Elevation Derivative Database (flood affected watersheds)
- CIESIN, IFPRI, WRI : Gridded Population of the World (GPW), Version 2 (population)
Compilation and computation by UNEP/GRID-Geneva

Source: UNDP, 2004.

Figure 10.6: Relative vulnerability for droughts

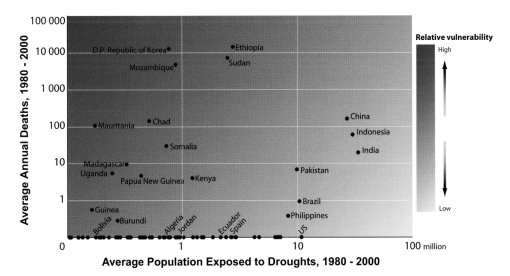

Data source: The EM-DAT OFDA/CRED International Disaster Database and UNEP/GRID-Geneva.

Source: UNDP, 2004.

Indicators are needed to assess the effectiveness of current policies for risk reduction and to explore other policy options

Based on a series of subcomponents, the six major components are combined using a composite index approach (similar to the HDI). The resulting scores range on a scale from 0 to 100, where 100 represents the highest level of vulnerability. The CVI can be applied at a scale more appropriate for resource management and disaster mitigation. **Map 10.3** shows an example of CVI application in Latin America.

The CVI provides a powerful technique for systematically expressing human vulnerability in relation to water resources, both for current conditions and for future scenarios. It can therefore help risk managers and water managers develop a warning system for water scarcity and possible drought events.

3c. Indicator on flood policy effectiveness

Indicators are needed to assess the effectiveness of current policies for risk reduction and to explore other policy options.[19] The indicator developed by the Public Works Research Institute (PWRI) in Japan has provided a clearer picture of goals and expected outcomes of a risk index for assessing policy effectiveness.[20]

Based on the DPSIR Framework (Driving forces, Pressure, State, Impact, Response), **Table 10.4** shows examples of indicators used to assess the effectiveness of flood countermeasures.

When applied to river basins in Japan, the selected indicators reflect policy effectiveness towards priority objective functions (see **Figure 10.7** for details and **Chapter 14**).

Table 10.3: Major components of the Climate Vulnerability Index (CVI)

Geospatial	Includes a range of factors specific to the location being examined that are likely to give rise to vulnerability (such as population density and dependence on imported food), slope and temperature, etc.
Resources	The physical availability of both surface water and groundwater, taking into account temporal variability and quality as well as the total amount of water.
Access	The extent of access to water for human use, including access to water for industry and irrigation.
Use	Water use efficiency for domestic, agricultural and industrial purposes.
Capacity	The effectiveness of people's ability to manage water.
Environment	A measure of how water use impacts on environmental integrity and on ecosystem goods and services provided by aquatic habitats.

Source: Sullivan and Meigh, 2005.

Table 10.4: Effectiveness of flood countermeasures: Examples of indicators

Target (T)	Framework	Perspective	Indicators
	Driving forces (**D**)	Indicators on water use and pressures on water systems that would trigger disasters as a result of socio-economic conditions (poverty, urbanization, etc.) and human activities	▪ Change in precipitation ▪ Change in river peak discharge
Targets (MDGs): 2015 reduce by 50% – the proportion of the population threatened by water-related hazards – the total losses in economic values	Pressure (**P**)	Change in state as a result of pressure	▪ Increase in land cover by urbanization ▪ Increase of population in flood-prone area
	Impact (**I**)	Impact of the driving forces and pressures on social and economic state	▪ Vulnerability of property ▪ Casualties and affected people ▪ Inundated areas ▪ Total economic loss
	Response (**R**)	Response (measures) to address changes in DPSIR	▪ Transition in allocated budget (i.e. investment) for flood risk mitigation

Source: Public Works Research Institute (PWRI) in Tsukuba (Japan) 2005.

19. This section is based on the contribution of Dr Tarek Merabtene and Dr Y. Junichi from PWRI Institute (Tsukuba, Japan).

20. For details on the PWRI Index see www.unesco.pwri.go.jp

Map 10.3: The Climate Variability Index (CVI) as applied to Peru at national and subnational scales

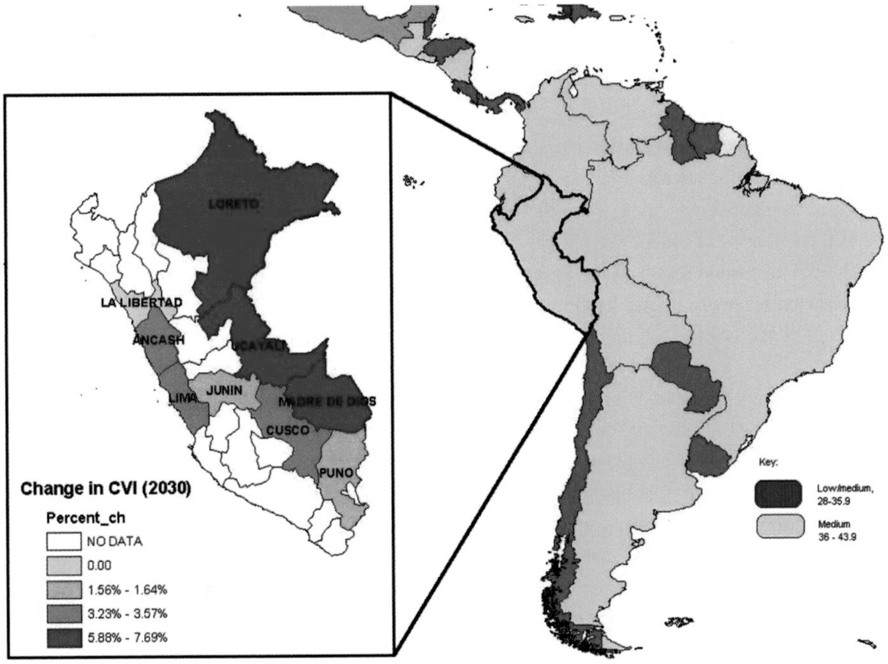

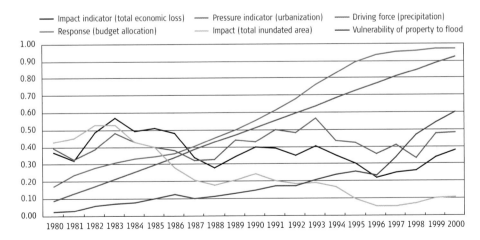

Source: Sullivan and Meigh, 2005.

The CVI provides a technique for expressing human vulnerability in relation to water resources, both for current conditions and for future scenarios

Figure 10.7: PWRI Risk Index: Case study basin in Japan, 1980–2000

—— Impact indicator (total economic loss) —— Pressure indicator (urbanization) —— Driving force (precipitation)
—— Response (budget allocation) —— Impact (total inundated area) —— Vulnerability of property to flood

Japan case study five years moving average.

Source: PWRI in Tsukuba (Japan), 2005.

Note: This graphic displays the various indicators composing the PWRI Risk Index. Each indicator refers to a particular dimension of public policies for flood mitigation. The figure shows that as a result of the chosen flood policy, flooded areas were decreasing significantly over the period of observation, while on the other hand, economic losses were still rising and overall vulnerability of people to floods has not been decreased. This type of information is valuable as a basis for causal analysis of this situation and the possible review of the flood management policy.

Part 4. From Frameworks to Policies

The design of a risk management framework is a prerequisite step to the successful development of risk reduction policies.

4a. Risk management frameworks

Risk management frameworks are meant to address the multiple goals of disaster risk reduction in a way that is consistent with the planning of social and economic development. Such frameworks also guide the design of a sound legislation basis, a necessary step for ensuring good governance of risk reduction activities.

Based on an extensive, global review of disaster risk reduction initiatives, the ISDR Secretariat designed a framework for disaster risk reduction, which provides a design concept for the development of risk management policies. The framework shows that treating interrelated issues such as 'knowledge development', 'political commitment' and 'application of risk reduction measures', involves a wide range of public policy issues (see **Figure 10.8** for details and **Chapter 14**).

The elements presented in the ISDR framework also advocate the development of a solid institutional background for disaster risk reduction policies.

21. Much of this section has been adapted from Plate, 2002.

4b. Risk management: A matter of legislation and policy

The legal basis for risk reduction policies is critical for transparent decision-making and allocating public funding for disaster mitigation. Examples include legislation, land-use planning regulations, building codes, inter-administration cooperation and operation rules for reservoirs. In some cases, the adoption of a new law on water-related disaster risk reduction has been fostered by the occurrence of a disaster or by a noticeable change in the natural environment. Under these circumstances, a new law concerning countermeasures against flood damages in urban areas was enacted in Japan in June 2003 (see **Figure 10.9** for details).

Disaster risk reduction policies also need to be consistent with existing policies in other sectors that have risk-related components (see **Table 10.5** for details).

4c. Example of practice: Flood risk management

Flood risk management[21] includes the planning of natural, technical or social systems, in order to reduce flood risk. Risk management therefore involves the value system of a given society, because it aims at balancing the desired state of the environment and the demands placed on it, while managing where trade-offs are best made.

Risk management actually takes place on three different levels: the operational level (see **Figure 10.10**), which is associated with the operation of existing systems; the project planning level, which is used when a new project or a revision of an existing project is planned (see **Figure 10.11**); and the project design level, which is embedded into the second level and describes the process of reaching an optimal solution for the project.

In the operation of an existing flood protection system, risk management involves a series of actions including the process of risk analysis, which provides the basis for long-term management decisions for the flood protection system. Continuous improvement of the system requires a reassessment of the existing risks and an evaluation of the hazards, making use of state-of-the-art data, information and assessment tools.

Table 10.5: Public policies with water-related risk components

Public policy	Risk-related aspect or impact
Development planning	▪ Social and economic activities
	▪ Poverty reduction
Land-use planning	▪ Urban sprawl in flood-prone areas
	▪ Exposure of the most vulnerable groups
Water resources management	▪ Upstream/downstream flow of water
	▪ Environmental flow management
	▪ Drought warning and management
Agriculture and forestry	▪ Erosion and sedimentation
	▪ Concentration time of river basin
Civil defence and the military	▪ Relief response capacity
	▪ Warning and crisis communication
Public health	▪ Emergency relief response capacity
	▪ Water-borne disease management
Education	▪ Awareness-raising campaigns
	▪ Learning self-protective behaviours
	▪ Academic research and staff training
Diplomacy	▪ Cooperation for water sharing
	▪ Exchange of data for forecast/warning
	▪ International basin management

Note: This table is indicative and should be adapted to the characteristics of regional, national and local situations.

Figure 10.8: Framework for disaster risk reduction

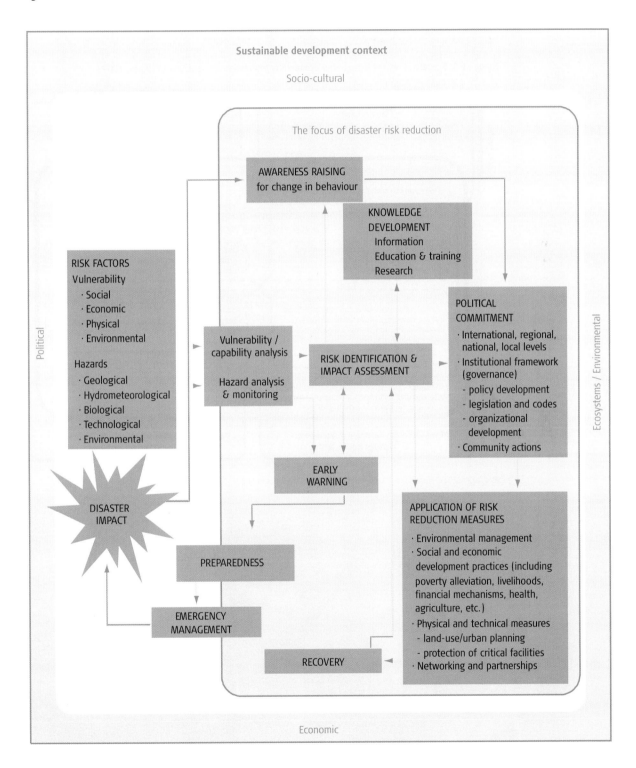

Source: ISDR, 2004a.

...in some cases the adoption of a new law has been fostered by the occurrence of disaster

The project planning aspect of risk management is summarized in **Figure 10.11**. This figure basically consists of two parts: risk assessment, which yields the basis for decisions on which solution to use, and the implementation phase, which involves a great deal of activities ranging from the fundamental decision to move forward to the studied complexity design and construction.

Figure 10.9: Framework of the Designated Urban River Inundation Prevention Act (Japan, 2003)

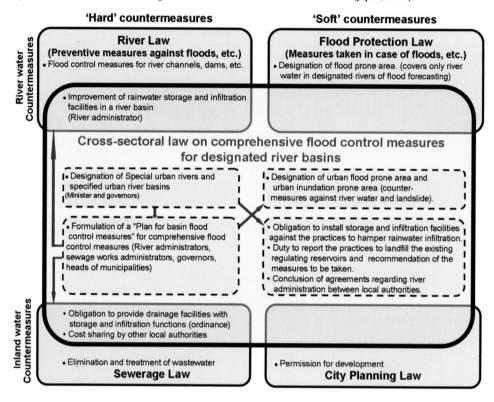

Note: The central box in this graphic figures the logical connections existing between the new law and the existing laws (River, Flood Protection, Sewerage and City Planning).

Source: Ministry of Land, Infrastructure and Transport (MLIT), Japan. Graphic provided by PWRI Institute (Tsukuba, Japan).

Figure 10.10: Risk management at the operational level

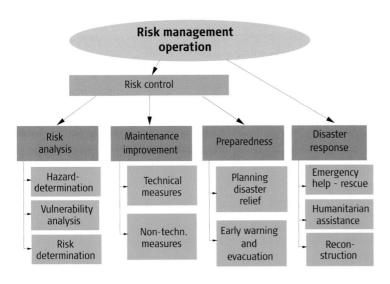

Source: Plate, 2002.

Figure 10.11: Risk management at the project planning level

Note: Although not appearing in this graphic, a 'public participation' dimension is needed at all levels. Involving stakeholders in project planning is essential to the social ownership of the process.

Source: Plate, 2002.

Above: The Indonesian coast, between Banda Aceh and Meulaboh, after the earthquake and tsunami of 26 December 2004

Below: Inundated houses during a flood in New Bethlehem, Pennsylvania, United States

Left: Floating market, Mekong Delta, Viet Nam

Floating market, Mekong Delta, Viet Nam

Part 5. Strategies for the Future

The future of disaster risk reduction depends heavily on the capacity of societies to cope with changes in the nature of water-related hazards, and in the nature of social vulnerability. This section therefore advocates flexibility in the design and implementation of disaster risk reduction policies and activities.

5a. Climate variability and change: Consequences for risk reduction

Climate variability and change are natural trends aggravated by the emission of natural and man-made greenhouse gases. The International Panel on Climate Change (IPCC) has noted that 'regional changes in climate have already affected hydrological systems and terrestrial and marine ecosystems', and that 'the rising socio-economic costs related to weather damage and to these regional variations suggest increasing vulnerability to climate change'. This in turn is projected to 'increase threats to human health, particularly among lower-income populations and within tropical and subtropical countries' (IPCC, 2001).[22]

Inhabitants of small islands and low-lying coastal areas are particularly at risk of severe social and economic effects from rising sea-levels, storm surges and tsunamis.[23] There are also severe threats to the freshwater resources on many of these islands, due to climate variability and change.

It is now widely recognized that climate change poses a major threat to sustainable development. The extreme vulnerability of certain societies to present and future climate risks necessitates integrating climate change issues in the planning of social and economic development. As disaster reduction has been recognized as a developmental issue, a convergence of interests to better manage risks related to climate and disasters for sustainable development is emerging. **Box 10.5** provides an example of climate change impacts and governmental response in Uganda.

Disaster risk reduction and uncertainty

Dealing with uncertainty in water-related risk management is not a new topic. Both natural and social scientists, as well as decision-makers, risk managers and water managers, have been dealing with this issue for decades. However, the challenge is to devise disaster risk reduction policies and strategies that can be adapted to uncertain changes in the environment, which are influencing both natural processes (e.g. global warming) and social systems (e.g. demographic pressure).

For instance, and in addition to limitations in forecasting accuracy, climate variability and change are additional sources of uncertainty for decision-makers and risk managers, potentially reducing the effectiveness of risk reduction measures. Limitations also include inaccurate hazard mapping, biased land-use planning and inefficient warning systems.

In addition to increased investments in natural and social science research, a way to reduce uncertainty is to improve the information exchange between the climate and risk management communities. This is, for instance, the aim of the Disaster Reduction and Climate Change (DR+CC) Infolink, an initiative that stimulates linkages and information exchanges between the disaster reduction and climate change communities.[24]

5b. Advocating adaptive risk reduction strategies

As explained above, the issue of climate variability and change needs to be treated as a cross-cutting issue related to governance issues including the following topics:

- climate variability change: changing rainfall and hazard patterns (frequency, magnitude, etc.)
- land degradation: deforestation, erosion, sedimentation in rivers, landslides
- migration and demographic pressure, uncontrolled urbanization
- poverty: loss of livelihoods, financial capacity for rehabilitation, illness, weakness, health
- loss of knowledge: migration to hazard-prone areas, lack of risk-related experience
- governance: failing States, corruption, political fragmentation, etc.

In a context of potentially increased uncertainty, successful disaster risk reduction strategies need to be adaptive, stressing, for instance, resilience to changes in the recurrence periods and duration of floods and droughts, in terms of exposure to water-related risks and to changes in patterns of social vulnerability.

22. For additional information, IPCC, 2001; findings of the World Water Agenda; MunichRe Topics Geo Annual Review 2003; 10 Year Review process of Barbados Plan of Action for SIDS, etc.

23. For more information on tsunamis, visit www.tsunamiwave.info On the 2004 Indian Ocean Tsunami Warning and Mitigation System, visit ioc3.unesco.org/indotsunami/ and see **Chapter 1**.

24. For details, see www.unisdr.org/eng/risk-reduction/climate-change/rd-cch-infolink1-03-eng.htm#n1

BOX 10.5: CLIMATE CHANGE AND DISASTER PREPAREDNESS IN UGANDA

Climate in Uganda, particularly rainfall, has been erratic since the early 1990s (see **Chapter 14**). The incidence, duration and amount of rainfall have all exhibited abnormal departures from long-term means. While rainfall in some years was far short of long-term means, thereby causing droughts, in other years it was excessive and produced catastrophic floods. The heaviest rains in recent years were recorded in 1994 and were associated with the El Niño phenomenon. This led

to sharp rises in lake levels, widespread flooding, washing away of roads and bridges, extensive soil erosion and landslides. In Lake Kyoga, rising water levels caused the detachment of previously firmly anchored floating papyrus swamps, which in turn caused a near total blockage in the lake. The blockage caused a further rise in lake levels and led to partial inundation of marginal homesteads and farmlands, the spread of water-borne diseases and the disruption of economic activities

around the lake shores. In order to respond to these risks, the Government designed a National Strategy for Disaster Preparedness and Management. This strategy aims at creating an integrated and multi-sectoral strategy to address these threats.

Source: Uganda National Water Development Report, 2004 from the World Water Assessment Programme, March 2005, Personal Communication.

BOX 10.6: PROJECTED IMPACTS OF CLIMATE CHANGE IN THE RHINE RIVER BASIN

■ Water supply: Demand for irrigation water will increase, which may lead to critical supply conditions in summer months. Drinking water supply may be constrained during summer months due to extreme low flows and reduced aquifer recharge.

■ Floods: Winter peak floods in alpine rivers will increase, but major changes in the flood condition of small catchments in the middle hill section of the Rhine are not anticipated using the present distribution of precipitation patterns. In the main stem of the Rhine River, an increased risk of winter floods is anticipated. Based on a design period of

1,250 years, the design flood may increase by 5 to 8 percent by the year 2050 in the lower stretch of the Rhine River.

■ Low flows: More frequent low flows have negative impacts on inland navigation, energy supply and the ecology of wetlands along the Rhine River. Use of processed water for industrial purposes and cooling water for thermal power plants will be constrained due to low flows and limits on warming up the river water. Low flows have a direct impact on the costs of shipping on the river.

■ Natural disasters: Due to a shift of the 0°C isothermal line in the Alps, an increased frequency of mudflows and slope failures is expected, which can cause dangerous flash floods.

■ Winter tourism in the Alps: By the year 2020 the decrease of winter sport potential in the Swiss Alps will be dramatic. In addition, it is expected that cumulated losses in income generation from winter tourism will be in the range from 1.8 to 2.3 billion Swiss francs (US $1.4 to 1.8 billion) by 2030–2050.

Source: Grabs, 1997.

All aspects of water-related risk management need to be considered in an adaptive perspective:

■ Adaptive risk reduction can be achieved through a society's capacity to devise new legislation and revise institutional integration accordingly. For instance, new public and private partners can be introduced into the National Platforms for disaster reduction.

■ A better response to changing conditions also requires a more flexible decision-making process. This can be the case for the chains of command and response from the forecasting services down to civil defence agencies and local instructions for the public. These objectives clearly require improved access to and circulation of information for decision-makers and other key players.

■ A capacity to anticipate changes in risk and disaster patterns requires a further development of risk-related indicators to monitor environmental and social changes.

A scenario-based study of hydrological impacts of climate change is also an important option for introducing flexibility in risk reduction policies and actions. See **Box 10.6** for an example on the Rhine River Basin.

5c. Vulnerability assessment: An insight into human security
Kofi Annan (2005) has recently stated that 'Human Security can no longer be understood in purely military terms. Rather, it must encompass economic development, social justice, environmental protection,

...climate variability and change are additional sources of uncertainty for decision-makers and risk managers...

BOX 10.7: COMMUNITY RISK ASSESSMENT BASED ON VULNERABILITY AND RESILIENCE

■ **Contextual aspects:** analysis of current and predicted demographics, recent hazard events, economic conditions, political structures and issues, geophysical location, environmental conditions, access/distribution of information and traditional knowledge, community involvement, organizations and management capacity, linkages with other regional/national bodies, critical infrastructures and systems

■ **Highly vulnerable social groups:** infants, children, elderly, economically disadvantaged, intellectually, psychologically and physically disabled, single-parent families, new

immigrants and visitors, socially/physically isolated, seriously ill, poorly sheltered

■ **Identifying basic social needs/values:** sustaining life, physical and mental well-being, safety and security, home/shelter, food and water, sanitary facilities, social links, information, sustaining livelihoods, maintaining social values/ethics

■ **Increasing capacities/reducing vulnerability:** positive economic and social trends, access to productive livelihoods, sound family and social structures, good governance, established regional/national networks, participatory

community structures and management, suitable physical and service infrastructures, local plans and arrangements, financial and material resources reservation, shared community values/goals, environmental resilience

■ **Practical assessment methods:** constructive frameworks and data sources including local experts, focus groups, census data, surveys and questionnaires, outreach programmes, historical records, maps, environmental profiles.

Source: ISDR, 2004a.

Figure 10.12: Pressure and Release (PAR) model in vulnerability analysis

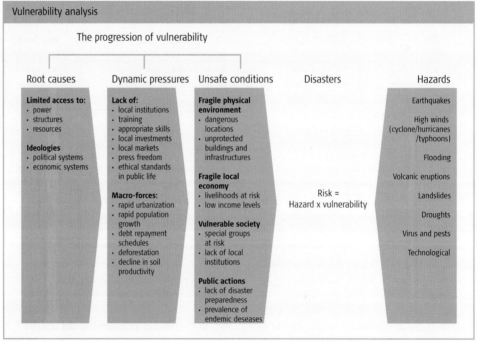

Note: 'Hazards' are the natural or man-made physical processing threatening social systems. In some cases (e.g. land degradation and landslides), hazards, characteristics are influenced by social practices. Besides, vulnerability level differs depending on social groups. For instance, some of these have been pushed to settle in marginal, hazard-prone areas, therefore increasing their exposure to hazards

Source: Blaikie et al., 1994.

democratization, disarmament, and respect for human rights and the rule of law'. Safeguarding human security requires a new approach for a better understanding of many interrelated social, political, economic, technological and environmental variables. These dimensions of human security are also key factors that influence the severity of impact generated by environmental deterioration and hydrometeorological extreme events.

Vulnerability is recognized as a central concept of human security and risk management. One definition of vulnerability is 'the conditions determined by physical, social, economic, and environmental factors or processes, which increase the susceptibility of a community to the impact of hazards' (ISDR, 2004a). In a wider perspective, however, the definition and effective assessment of vulnerability need to include more dynamic elements, such as social processes of exposure and responses to natural disasters.

Conceptual frameworks and models that provide a basis for vulnerability analysis in relation to specific hazards

have been developed. These models link dynamic processes at different scales and access to resources with vulnerability conditions. The Pressure and Release (PAR) model (see **Figure 10.12**) provides a good basis for the analysis and further identification of specific vulnerable conditions (Blaikie et al., 1994).

The basis for the PAR model is that a disaster is at the intersection of two opposing forces (Wisner et al., 1994): those processes generating vulnerability on one side, and the natural hazard event (or sometimes a slowly unfolding natural process) on the other.

In a risk management framework, vulnerability is also treated as a matter of scale, because individual vulnerability can be very different from vulnerability at the scale of communities (see **Box 10.7**), municipalities, regions or nations. Likewise, vulnerability is goal-specific as it involves activities such as knowledge management, awareness raising, risk perception, warning systems and communication mechanisms. Both features advocate for a strong community-based approach when designing, implementing and assessing disaster risk reduction strategies.

New Orleans residents walking through flood water in search of gasoline after the passing of hurricane Katrina in August 2005

Part 6. The Way Forward

The six key messages identified at the beginning of the chapter are specifically aimed at decision-makers, risk managers and water managers. The importance of establishing an integrated risk management policy has been stressed throughout the chapter, with the need to seek a sturdy framework from which implementation plans can stem. The World Conference on Disaster Reduction organized by UN/ISDR in Kobe, Japan, in January 2005 was of particular importance in providing a venue for reviewing disaster risk reduction strategy and its primary output, the Hyogo Framework for Action, proposes building a new strategy for the next ten years.

6a. The Hyogo Framework for Action 2005–15

The World Conference on Disaster Reduction (WCDR), held in January 2005 in Kobe, Japan, provided essential recommendations to decision-makers and risk managers. Although it dealt with all kinds of natural hazards, its framework provides very relevant guidance for water-related disaster risk reduction.

National delegates to the WCDR and international organizations both agreed on the following key challenge areas for developing a relevant framework for action for the International Decade 'Water for Life' 2005–15 (ISDR, 2005):

- governance: organizational, legal and policy frameworks; public participation
- risk identification, assessment, monitoring and early warning
- knowledge management and education
- reducing underlying risk factors
- preparedness for effective response and recovery.

WCDR participants also recognized the need to enhance international and regional cooperation, as well as assistance in the field of disaster risk reduction, by the following means for advanced international and regional cooperation in disaster risk reduction (ISDR, 2005):

The Indonesian coast, between Banda Aceh and Meulaboh, after the earthquake and tsunami of 26 December 2004

25. Revised Version Oct 2005.

BOX 10.8: HIGHLIGHTS OF THE HYOGO FRAMEWORK FOR ACTION 2005-2015

■ Ensure that the reduction of disaster risk from hydrometeorological events is a national and a local priority. An IWRM approach is needed, together with a strong institutional basis for implementation: national institutional and legislative frameworks, resources, community participation.

■ Identify, assess and monitor hydro-meteorological disaster risk and enhance early warning: national and local risk assessments, early warning systems, capacity building, regional and emerging risks. Improve regional and international cooperation for hazard assessment and data sharing.

■ Use knowledge, innovation and education to build a culture of safety and resilience at all levels: information management and exchange, education and training, research, public awareness. Foster applied research in technical and social aspects of hydro-meteorological hazards, risks and disasters.

■ Reduce the underlying risk factors: environmental and natural resources management; social and economic development practices, water resources management and development, land-use planning and other technical measures.

■ Strengthen disaster preparedness for effective response at all levels.

Source: ISDR, 2005.

■ transfer of knowledge, technology and expertise to enhance capacity building for disaster risk reduction
■ sharing of research findings, lessons learned and best practices
■ compilation of information on disaster risk and impact at all scales in a way that can inform sustainable development and disaster risk reduction
■ appropriate support to enhance governance for disaster risk reduction, for awareness-raising initiatives and for capacity-development measures at all levels in order to improve the disaster resilience of developing countries
■ consideration of the impact of disasters on the debt sustainability of heavily indebted countries
■ financial assistance to reduce existing risks and avoid the generation of new risks.

The Hyogo Framework for Action 2005–2015 sets a useful road map for the design of improved risk management frameworks and implementation plans. Finally, and in view of the practical implementation of the recommendations made above, WCDR identified key actions to improve disaster risk reduction. See **Box 10.8** for details, from a water-related disaster perspective.

6b. Conclusions

The future of living with water-related risks lies in the capability of societies to anticipate and adapt to changes occurring in their natural and social environment.

Improved management of risk-related knowledge and information is therefore a first and necessary step in that direction. There is a need to support further investment in data collection and analysis and modelling capacities,

as well as in indicator development. Indicators are essential to identifying and monitoring underlying trends in disasters, hazards, vulnerability and risk.

It is necessary that risk-related knowledge be made available to decision-makers, risk managers and water managers. Access to information is vital for the design of comprehensive risk management frameworks. Integrated policies for risk reduction need a sound governance framework, which includes a good legislation basis and efficient cooperation among the various administrations and institutions involved.

Disaster risk reduction is a key component of IWRM and sustainable development. Consequently, disaster risk reduction objectives need to be integrated into social and economic development planning. Moreover, risk reduction policies need to be consistent with other risk-oriented policies of different organizational entities such as different ministries or line departments and agencies. Risk assessment is therefore an important step on the route to sustainable development. At the local level, the involvement of stakeholders in the design, implementation and assessment of policies must be ensured.

Global processes, such as climate variability and change, increase the level of uncertainty for both water-related physical processes and the social processes of exposure to hazards, vulnerability and adaptation to change. Again, theoretical and applied research in the natural and social sciences needs to receive additional financial support with the purpose of improving our understanding of the physical and social processes leading to increased vulnerability. Climate variability and change is a strong incentive for advocating more adaptive policies for disaster risk reduction.

The main points raised in this chapter also show explicit links to challenge areas – and related chapters – of the *World Water Development Report*:

First, the need to incorporate disaster risk planning into national policies for socio-economic development confirms the governance dimension of risk management (see **Chapter 2**). This dimension is itself related to vulnerability assessment for human settlements – in particular when it comes to marginal communities and smaller social groups.

Second, managing the aftermath of water-related disasters requires continued investment in epidemiology and public health, as well as in water and sanitation development (see **Chapter 6**). Providing these resources to water users should however integrate features of the water cycle. These include: ecosystem functions, pollution and consequences of climate variability and change (see **Chapters 4 and 5**). In particular, managing risks cannot

be separated from food security issues, such as livelihood functions of riverine environments (see **Chapter 7**).

Third, controversies related to water resources development such as hydropower, show that the management of water-related risks is related to the broader debate on energy security, policies and technical choices (see **Chapters 8** and **9**). This dimension is particularly acute on transboundary river basins, where risk management very much depends on the institutional choices made for sharing water resources and conflict avoidance (see **Chapter 11**).

These various, interrelated dimensions of risk management all point to the common issue of knowledge management. Despite an increasing volume of environmental data produced worldwide, technologies for analysing water-related information remain insufficient – especially in developing countries, where information exchange remains very low. One of the key challenges related to risk management is the adequate sharing of water-related data and information, both within and among countries.

Access to information is vital for the design of comprehensive risk management frameworks

References and Websites

Affeltranger, B. 2004. Flood forecasting on the Mekong: The politics of hydrological data. Paper presented at the Canadian Consortium for Asia-Pacific Security Studies, 9–12 December 2004, Quebec, Canada.

——. 2002. User-based design of efficient flood warnings. Proceedings of the International Workshop on Flood Forecasting and Early Warning Systems (FFEWS), Mekong River Commission Secretariat, Phnom Penh, Cambodia, February–March 2002.

Affeltranger, B. and Lictevout, E. 2005 (in press). Community-based development of flood warning systems in Cambodia. F. Lasserre and A. Brun (eds) *Local Level Management of Water Resources: Principles and Challenges*. Quebec, Canada, Presses Université du Québec.

Annan, K. 2005. Towards a culture of peace: Letters to future generations. www.unesco.org/opi2/lettres/TextAnglais/AnnanE.html

APFM (Associated Programme on Flood Management). 2003. Integrated Flood Management (IFM) Concept Paper. WMO-GWP APFM, Geneva, WMO. www.apfm.info/

Blaikie, P., Wisner, B., Cannon, T. and Davis, I. 1994. *At Risk: Natural Hazards, People's Vulnerability and Disasters*, 2nd edn. London, Routledge.

Brouwer, van Ek, Boeters and Bouma. 2001. Living with floods: An integrated assessment of land use changes and floodplain restoration as alternative flood protection measures in the Netherlands. CSERGE Working Paper ECM 01-06. www.uea.ac.uk/env/cserge/pub/wp/ecm/ecm_2001_06.pdf

Burton, I., Kates, R. W. and White, G. F. 1993. *The Environment as Hazard*, 2nd edn. New York/London, Guilford Press.

DEFRA (Department for Environment, Food and Rural Affairs, UK). 2003. Wetlands, land use change and flood management, Joint statement, www.defra.gov.uk/environ/fcd/policy/wetlands/Wetlands3.pdf.

Dercon, S. (ed.) 2004. *Insurance Against Poverty*. Oxford, Oxford University Press.

Eikenberg, Chr. 1998. *Journalistenhandbuch zum Katastrophenmanagement*. 5th edn. Bonn, German IDNDR-Committee.

Frijters and Leentvaar. 2001. Participatory planning for flood management in the Netherlands. www.unescap.org/esd/water/disaster/2001/netherlands.doc

Grabs, W. (ed.) 1997. Impact of climate change on hydrological regimes and water resources management in the Rhine basin. CHR Report No. I-16, Lelystad.

IPCC (Intergovernmental Panel on Climate Change). 2001. *Scientific Assessment of Climate Change, Summary for Policymakers, Climate Change 2001: Synthesis Report of the IPCC Third Assessment Report*. XVIII Session of the IPCC, Wembley, United Kingdom, 24–29 September 2001.

ISDR (International Strategy for Disaster Reduction). 2005. Hyogo Framework for Action 2005–2015, Building the Resilience of Nations and Communities to Disasters, Advance Copy www.unisdr.org/wcdr/

——. 2004a. *Living With Risk: A Global Review of Disaster Reduction Initiatives*. 2nd edn. Geneva, UN-ISDR.

——. 2004b. Second World Conference on Disaster Reduction, January 2005, Kobe, Japan, First Announcement Flyer, 20 February 2004, Geneva, UN-ISDR.

——. 2003. Drought, Living With Risk: An Integrated Approach to Reducing Social Vulnerability to Drought. Report of the Ad Hoc Discussion Group, April 2003, Geneva, ISDR.

——. 2001. *Living With Risk: A Global Review of Disaster Reduction Initiatives*. 1st edn. Geneva, UN-ISDR.

McDaniels, T. L., Gregory, R. S. and Fields, D. 1999. Democratizing risk management: Successful public involvement in local water management decisions. *Risk Analysis*, Vol. 19, No. 3.

Merabtene, T. and Yoshitani, J., 2005. Technical Report on Global Trends of Water-related Disasters. PWRI Technical Memorandum, ISSN 0386-5878, No 3985, pp. 124. Also available online at unesco.pwri.go.jp

Parker, D. J. 2004. Designing flood forecasting, warning and response systems from a societal perspective. *Meteorologische Zeitschrift*, Vol. 13, No. 1, pp. 5–11.

Plate, E. J. 2002. Flood risk and flood management. *Journal of Hydrology*, Vol. 267, pp. 2–11.

Sullivan, C. A. and Meigh, J. R. 2005. Targeting attention on local vulnerabilities using an integrated index approach: The example of the Climate Vulnerability Index. *Water Sciences and Technology*, (Special Issue on Climate Change) Vol. 51, No. 5, pp. 69–78.

Sullivan, C. A., Meigh, J. R. and Acreman, M. C. 2002. Scoping study on the identification of hot spots: Areas of high vulnerability to climatic variability and change identified using a Climate Vulnerability Index. Report to the Dialogue on Water and Climate, Centre for Ecology and Hydrology, Wallingford, UK.

UN/GA (United Nations General Assembly). 1994. Yokohama Strategy and Plan of Action, International Decade for Natural Disaster Reduction.

www.unisdr.org/eng/about_isdr/bd-yokohama-strat-eng.htm

UNDP (United Nations Development Programme). 2004. *Reducing Disaster Risk: A Challenge for Development*. New York, UNDP/BCPR Bureau for Crisis Prevention and Recovery. www.undp.org/bcpr/disred/rdr.htm

UNDP-BCPR (United Nations Development Programme – Bureau for Crisis Prevention and Recovery). 2004. *Reducing Disaster Risk: A Challenge for Development*. New York, UNDP-BCPR. www.undp.org/bcpr/disred/rdr.htm

UN-WWAP (United Nations – World Water Assessment Programme). 2003. *The United Nations World Water Development Report: Water for People, Water for Life*. Paris/Oxford, UNESCO, Berghahn Books.

Vari, A. 2004. Hungarian experiences with public participation in water management. *Water International*, Vol. 29, No. 3, pp. 329–37.

Viljoen, M. F., du Plessis, L. A. and Booysen, H. J. 2001. Extending flood damage assessment methodology to include sociological and environmental dimensions. *Water SA*, Vol. 27, No. 4, Oct. 2001, pp. 517–21.

WCDR (World Conference on Disaster Reduction). 2005. Hyogo Framework for Action 2005–2015: Building the Resilience of Nations and Communities to Disasters, Extract from the final report of the WCDR (A/CONF.206/6).

WHO (World Health Organization). 1989. *Le personnel local de santé et la communauté face aux catastrophes naturelles* [Local health personnel and the community faced with natural catastrophes]. Technical Guide with the International Federation of Red Cross/Red Crescent, Geneva, World Health Organization.

WMO (World Meteorological Organization). 2005. Climate, water and disasters: A call for a millennium development target, Keynote Address, CSD-13 Side Event, 14 April 2005, New York.

——. 2004. Practices, approaches and methods in risk management related to flooding and climate variability. Web-based study, Contribution to the Project on Risk Management, WMO Commission for Hydrology (CHy), August 2004, Geneva, WMO.

——. 1999. Comprehensive Risk Assessment for Natural Hazards. WMO/TD No. 955, Geneva.

UN Agencies

International Strategy for Disaster Reduction (UN-ISDR): www.unisdr.org/

UNESCO Intergovernmental Oceanagraphic Commission (IOC): ioc.unesco.org

United Nations Development Programme, Disaster Reduction Unit, Bureau for Crisis
Prevention and Recovery (UNDP-DRU-BCPR): www.undp.org/bcpr/disred/rdr.htm

United Nations Environment Programme, Division of Early Warning and Assessment, Global
Resource Information Database (UNEP-DEWA-GRID): www.grid.unep.ch/

United Nations University Institute for Environment and Human Security (UNU-EHS):
www.ehs.unu.edu/

World Meteorological Organisation (WMO): www.wmo.ch/index-en.html

International NGOs

Action Against Hunger (AAH): www.actionagainsthunger.org/

International Federation of Red Cross and Red Crescent Societies (IFRC): /www.ifrc.org/

Research Centres & Universities

Centre for Ecology and Hydrology (UK): www.ceh.ac.uk/

Centre for Research on the Epidemiology of Disasters (CRED, Belgium): www.cred.be/

Flood Hazard Research Center (FHRC, UK): www.fhrc.mdx.ac.uk/

Natural Hazards Center (USA): www.colorado.edu/hazards/

Public Works Research Institute (Japan): www.pwri.go.jp/eindex.htm

Other Organizations

Asian Disaster Preparedness Center (ADPC, Thailand): www.adpc.net/

National Oceanic and Atmospheric Administration (NOAA, USA): www.noaa.gov/

Water is not a commercial product like any other, but rather a heritage that must be protected, defended and treated as such.

European Commission Water Framework Directive

CHAPTER 11
Sharing Water

By

UNESCO
*(United Nations
Educational,
Scientific and Cultural
Organization)*

*Geothermal power plant with bathers enjoying
geothermally-heated water, Blue Lagoon, Iceland*

Key messages:

The emerging water culture is about sharing water: integrated water resources managements (IWRM) looks for a more effective and equitable management of the resource through increased cooperation. Bringing together institutions leading with surface water and aquifer resources, calling for new legislative agreements all over the world, increasing public participation and exploring alternative dispute resolutions are all part of the process.

■ Sharing water resources constitutes a major part of integrated water resources management (IWRM).

■ There is a need to further expand special indicators for measuring efficient, effective and equitable water sharing.

■ Increasing complexity and interdependence regionally, nationally and internationally requires new approaches to shared water systems.

■ There is a need for developing new knowledge and new capabilities in order to understand aquifers and the difficulties of underground boundaries that are difficult to define.

■ There is a need to concentrate on the implementation of mechanisms for conflict avoidance and conflict management.

Above: A man-made entrance to an underground aquifer in Quintana Roo, Mexico

Right: Itaipu dam and hydro-electricity power station on the river Parana, Brazil / Paraguay

Below: Tea plantation in Kerala, India

Section 4: MANAGEMENT RESPONSES & STEWARDSHIP

Part 1. Towards Integration and Cooperation

The comprehensiveness of water resource planning and sharing has been the subject of much controversy and debate. It has been widely recognized that in order to maximize the benefits from any water resource project, a more systematic analysis of the broader environment is needed. In addition to a broadening of traditional management approaches, there needs to be increased sensitivity to decision-making that involves multi-purpose actions and multi-user considerations.

A proposed framework for sharing water would mean taking the following issues into account:

- natural conditions (e.g. aridity and global changes)
- variety of uses (irrigation, hydropower, flood control, municipal uses, water quality, effluent control, etc.)
- various sources of supply (surface water, groundwater and mixed sources)
- upstream/downstream considerations
- socio-demographic conditions (population composition and growth, urbanization, industrialization, etc.).

The mismatch between political boundaries and natural river basins has become a focal point for the difficulties of joint planning, allocation of costs and benefits, advantages of scale and other integrated water management issues and is usually referred to as transboundary (the terms transnational, trans-state and international have also been used), which refers to any water system that transcends administrative or political boundaries, which often do not coincide with river basins' or watersheds' natural boundaries (see **Chapter 4**).

The time lag between the implementation and impact of management decisions – sometimes measured in decades – significantly reduces the power of contemporary water resource institutions. Efforts to implement more integrated shared water resources management are confronted with continuous changes in values, structural transformations in society and environment, as well as climatic anomalies and other exogenous shifts. These transformations have created a context of complexity, turbulence and vulnerability. The emerging water sharing paradigm attempts to bring together the above concerns with cross-cutting sustainability criteria, such as social equity, economic efficiency and environmental integrity.

Access to adequate water is becoming a highly contested issue, which is further complicated by traditional values

and customs, cultural and religious considerations, historical factors and geographical variations. As for sharing the resources of an aquifer system, in which upstream-downstream relationships do not apply, current thinking is moving away from 'equitable utilization', a remarkably vague notion, given the predominance of slow responding storage overflows, towards ensuring the sound functioning and integrity of the aquifer system.[1]

1a. Setting the context

Sharing water is essential to meeting the goals of equity, efficiency and environmental integrity and answering the more complex questions that stem from broader challenges, such as the issue of overall security. Water sharing mechanisms (i.e. new institutional arrangements) help us adapt to these challenges through structural changes (specific organizations, joint engineering structures, etc.) and more resilient political institutions.

In 2002, UNESCO and the Organization of American States (OAS) launched the International Shared Aquifer Resource Management (ISARM) project for the Americas, which organized three workshops, in 2003, 2004 and 2005, to present the data gathered on transboundary groundwater in North, Central and South America and highlight the need to follow up on this cooperative project. The UNESCO-IHP ISARM project initiated transboundary aquifer resources inventories, covering the Americas (sixty-five aquifers; see **Map 11.1** and **Table 11.1**) and Africa (thirty-eight aquifers) as well as a recent update including the Balkan countries (forty-seven aquifers) and plans to extend coverage to Asia and the Pacific.[2]
Table 11.1 provides detailed information on shared aquifers located in Central and South America. To date, the UNESCO-ISARM project has inventoried over 150 shared aquifer systems with boundaries that do not correspond to those of surface basins. Progress in the consolidation of these newly created inventories has resulted in unprecedented development in global transboundary aquifer resources assessment.

Efforts to implement more integrated shared water resources management are confronted with continuous changes in values

1. The integrity of an aquifer can be destroyed if, for example, saline intrusion invades to such an extent that the aquifer system ceases functioning and cannot be effectively rejuvenated.

2. A publication on the achievements of the project is under preparation. Maps for these regions can be found on the CD-ROM accompanying the book and at www.unesco.org/water/wwap

The depletion of national water resources, recurring droughts and expanding socio-economic demands have all fuelled confrontations and forced international exchanges and cooperation

Map 11.1: Transboundary aquifers of the Americas (in progress)

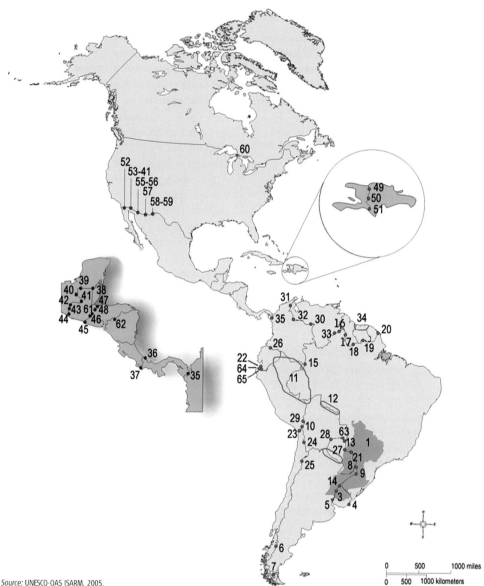

Source: UNESCO-OAS ISARM. 2005.

Underlying such broad considerations are apprehensions about the prospects for achieving the necessary cooperation for managing shared water systems, owing, for example, to persistent national sovereignty demands and further political fragmentation in many regions, despite cooperative efforts (see **Chapter 2**). Upstream states and regions lack incentives to enter into conflict resolution negotiations and other cooperative mechanisms driven by principles of comprehensive sustainable river development.

The geopolitical nature of water – a function of both geography and technology – produces different and complex cultural, historical and ecological adaptations, as well as varying power to use resources. The depletion of national water resources, recurring droughts and expanding socio-economic demands have all fuelled confrontations and forced international exchanges and cooperation. (This has generally been the case for surface waters, which are more visible, though attention is now also turning to transboundary aquifers.) There are more

Table 11.1: Transboundary aquifers of the Americas (in progress)

Map Ref.	Transboundary aquifers	Countries	Country number
	NORTH AMERICA		
52	Tijuana	Mexico-United States	2
53	Valle de Mexicali	Mexico-United States	2
54	Valle San Luis - Rio Colorado (Yuma)	Mexico-United States	2
55	Rio Santa Cruz	Mexico-United States	2
56	Nogales	Mexico-United States	2
57	Rio San Pedro	Mexico-United States	2
58	Conejos - Medanos	Mexico-United States	2
59	Bolson (Valle de Juarez)	Mexico-United States	2
60	Cambrian - Ordovician	Canada-United States	2
	CENTRAL AMERICA		
36	Sixaola	Costa Rica-Panama	2
37	Coto	Costa Rica-Panama	2
38	Hondo San Pedro	Guatemala-Mexico	2
39	San Pedro	Guatemala-Mexico	2
40	Usamancita	Guatemala-Mexico	2
41	Chixoy - Xaclbal	Guatemala-Mexico	2
42	Selegua - Cuilco	Guatemala-Mexico	2
43	Coatan - Suchiate	Guatemala-Mexico	2
44	Bajo Suchiate	Guatemala-Mexico	2
45	Cuenca La Paz (Ahuachapan-Las Chinamas)	El Salvador-Guatemala	2
46	Alto-Paz - Ostua/Metapan	El Salvador-Guatemala	2
47	Motagua Norte	Guatemala-Honduras	2
48	Motagua Sur	Guatemala-Honduras	2
61	Olopa	Guatemala-Honduras	2
62	Rio Negro	Honduras-Nicaragua	2
	CARIBBEAN		
49	Artibonito	Haiti-Republica Dominicana	2
50	Masacre	Haiti-Republica Dominicana	2
51	Pedernales	Haiti-Republica Dominicana	2
	SOUTH AMERICA		
1	Guarani	Argentina-Brazil-Paraguay-Uruguay	4
2	Yrenda-Toba -Tarijeno	Argentina-Bolivia-Paraguay	3
3	Salto Chico - Salto Chico	Argentina-Uruguay	2
4	Litoraneo-Chuy	Brazil-Uruguay	2
5	Litoral - Sistema Acuífero en Areniscas Cretácicas	Argentina-Uruguay	2
6	Probable	Argentina-Chile	2
7	El Condor	Argentina-Chile	2
8	Caiua	Argentina-Brazil-Paraguay	3
9	Serra Geral; Serra Geral-Arapey	Argentina-Brazil-Paraguay-Uruguay	4
10	Ignimbritas Cordillera Occidental	Bolivia-Peru	2
11	Solimoes	Bolivia-Brazil-Colombia-Ecuador-Peru	5
12	Jaci Parana y Parecis	Bolivia-Brazil	2
13	Pantanal	Bolivia-Brazil-Paraguay	3
14	Permianos	Brazil-Uruguay	2
15	Ica	Brazil-Colombia	2
16	Sedimentos Paleo-Proterozoicos	Brazil-Guyana-Venezuela	3
17	Serra do Tucano	Brazil-Guyana	2
18	Boa Vista	Brazil-Guyana	2
19	Sem Denominacao	Brazil-Surinam	2
20	Costeiro	Brazil-Guyana (F)	2
21	Furnas e Altos Gracas	Brazil-Paraguay	2
22	Zarumilla - Machala	Ecuador-Peru	2
23	Concordia - Caplina	Chile-Peru	2
24	Ascotan - Silala - Ollague	Bolivia-Chile	2
25	Puna	Argentina-Chile	2
26	Tulcan	Colombia-Ecuador	2
27	Coronel Oviedo Basamento Cristalino	Brazil-Paraguay	2
28	Agua Dulce Palmar de las Islas	Bolivia-Paraguay	2
29	Titicaca	Bolivia-Peru	2
30	Arauca	Colombia-Venezuela	2
31	Guajira	Colombia-Venezuela	2
32	San Antonio Urena Santander	Colombia-Venezuela	2
33	Sedimentos Grupo Roraima	Brazil-Venezuela	2
34	Zanderji; Coesewijne; A-sand	Guyana-Surinam	2
35	Jurado	Colombia-Panamá	2
63	Rio Negro-Itapucumi	Bolivia-Paraguay	2
64	Tumbes - Puyango	Ecuador-Peru	2
65	Chira - Catamayo	Ecuador-Peru	2

Source: UNESCO-OAS ISARM. 2005.

On an international scale, sharing and cooperative efforts can provide benefits that far exceed those that attempt to maximize individual and national self-interests...

than 3,800 unilateral, bilateral or multilateral declarations or conventions on water: 286 are treaties, with 61 referring to over 200 international river basins.[3] Such agreements, which serve to emphasize the importance of cooperation in many shared water settings, are expected to expand in the future. A new impetus to adopt transboundary aquifer agreements could also arise from the anticipated publication of the United Nations (UN) International Law Commission's (ILC) draft Convention on transboundary aquifers (see **Box 11.3**).

Vulnerability is increasingly discussed in the literature of environmental change, where it is associated with the shift in environmental studies from impact analysis to crisis assessment and vulnerability evaluation. Such assessment steps reflect the large number of variables involved; cumulative, interactive, synergistic and unexpected consequences, as well as multiple sources of threats. Moreover, vulnerability has been tied (especially in transboundary water systems) to security in all its forms – from food security, economic security and political security, all the way down to individual security. This dynamic evolution also coincides with the evolution from simple linear models to more complex non-linear feedback approaches. When combined with volatility and greater resilience to risks, the powerful new theme of expanding the timeframe of analysis and assessment emerges.

1b. The emerging water use paradigm

Traditional reactive crisis approaches were replaced by risk assessment and other proactive strategies at the beginning of the twenty-first century. These new approaches call for anticipatory action and multi-stakeholder involvement. Rapid socio-economic changes, socio-political upheavals and the transitions witnessed during the turbulent decades of the 1980s and 1990s underscored the need for a greater emphasis on environmental challenges – from the search for sustainable development and the promotion of integrated planning and governance to the attempt to combine structural and non-structural solutions to persistent water resources problems and transnational interdependencies.

In this setting of increasing complexity, interdependence and vulnerability, there is an urgent need for intergovernmental integration of the following issues:

■ *hydrological interdependencies*: in terms of both uses (agricultural, urban, industrial and recreational) and water regimes (surface water and groundwater, quality and quantity)

■ *political interdependencies*: both in terms of horizontal coordination in space and vertical cooperation between levels of government units

■ *transboundary interdependencies*: representing both social and hydrological trans-state interdependencies

■ *exogenous interdependencies*: the most notable of which are the potentially dramatic impacts and consequences of climatic change.

It is important to recognize water as a catalyst for cooperation; regions with shared international waters are often subject to water-related controversies. When coupled with reasonable and equitable utilization of the resource, cooperative efforts allow for more effective approaches to allocating and sharing water. However, cooperation is not simply an abstract term for peaceful coexistence, but also an important mechanism for managing natural resources by addressing the underlying historical, political, economic and cultural causes of water-stressed environments. It emphasizes the need for combining technological capabilities with political will and is an important part of international agreements, proclamations at water conferences and in millennial declarations, future scenarios and goal-oriented planning efforts, as well as in environmental law, conventions and regulatory provisions. On an international scale, cooperative efforts can provide benefits that far exceed those that attempt to maximize individual and national self-interests (Sadoff and Grey, 2002).

Since complexity, interdependence and rapidly changing socio-economic conditions each increase the likelihood of water conflict, we are faced with considering more complex models for understanding cooperation and contestation. Looking beyond the environmental debates and differing perspectives (optimism/pessimism, society/individualism, market/common good, etc.), we can see the broad outlines of a newly emerging paradigm for sharing water. This new paradigm emphasizes integrated

3. For more information, see Wolf et al., 2003 and www.unesco.org/water/wwap/pccp/

management, the duty to cooperate, equitable utilization, sustainable use, minimization of harm and true cost, in addition to public participation (EC, 2005).

This new water culture paradigm recognizes the inter-generational, inter-spatial and inter-species differentiations in allocating water resources. We must work towards setting up protocols for environmental protection, regulatory reform and sustainable use, such as the National Environmental Policy Act in the United States (US) and the Water Framework Directive (WFD) in the European Union (EU) (see **Chapter 14**), which can lead to more detailed practices, such as treaties and bilateral agreements, which, in turn, prescribe implementable action and monitoring performance mechanisms. Also needed is the allocation of finances to support the transaction costs of transforming contemporary institutions.

Many international conferences and other fora, including the UN's *Comprehensive Assessment of the Freshwater Resources of the World* (Kjellén and McGranahan, 1997),[4] have warned that we must fundamentally alter the way we think about and manage water. They have also made it clear that we must embrace new policies that are not only comprehensive, participatory and anticipatory, but also environmentally sound. Sound shared water management should promote intergovernmental dialogue and address long-term goals and objectives.

This shifting emphasis in water sharing has led to greater attention to cooperation rather than conflict, the latter including conflict prevention, management and resolution. Equally important is the emphasis on intra-state approaches, which address competing and conflicting uses of water through the concept of subsidiarity, or relegating responsibility to the lowest appropriate level of governance and decision-making. Other complementary approaches reinforce the need for capacity-building, the creation of an enabling environment and the mobilization of finances, as well as citizen participation. However, despite these positive trends, there remain many challenges to sharing water.

1c. The challenges to sharing water

Water resources are unequally distributed, and water scarcity and abundance are further affected by political changes, mismanagement and climatic anomalies. These create massive upheavals, demographic transformations and uneven development efforts, all of which, in turn, contribute to significant socio-economic differentiations. Ecological degradation and political instability can produce conflict or be catalysts for cooperation. At the same time, competition for water is also manifested in the demands between different uses – urban versus rural, present uses versus future demands, competing regions, water quantity versus water quality and water concerns versus other social priorities. Past research has stressed the following types of conflicts:

■ *direct* (competing and conflicting demands)

■ *indirect* (migration, environmental refugees or seasonal high peak demands from tourism)

■ *structural sources* that emphasize a broader socio-economic context, such as limited institutional and social capacity, fragmented authority, transboundary interdependencies, insufficient public participation, etc.

The above list supports the observation that more emphasis has been placed on conflict, with less importance given to efforts towards the peaceful sharing of water and long-term cooperation.

Geography suggests that – by virtue of physical unity and regardless of political divisions – a river basin should be developed and managed as an indivisible whole. Moving water ties land together, and interference with its movement has repercussions elsewhere in the basin. While geographic ties prescribe the unitary development of river basins and aquifer systems, politics, culture and history distort this process. The nation state covets its sovereignty and attempts to maximize benefits for itself. However, this state-central behaviour can generate international friction and even lead to conflict. We are, therefore, faced with a situation in which states confronted with limited choices tend to adopt a non-cooperative stance. But an increasing amount of literature argues that conflict is not the inevitable result of scarcity (Carius et al., 2004). A number of variables, such as cultural traditions, the degree of social cleavages, the nature of institutions and ideologies about or perceptions of the environment, can lessen the possibility of conflict due to water scarcity.

...we must embrace new policies that are not only comprehensive, participatory and anticipatory, but also environmentally sound

4. See also Guerquin et al., 2003; Cosgrove and Rijsberman, 2000; 3rd World Water Forum, 2003.

BOX 11.1: SHARED AQUIFERS BETWEEN ARGENTINA, BOLIVIA AND PARAGUAY

The Yrenda-Toba-Tarijeño aquifer system occupies about 300,000 square kilometres (km²), located mostly in the Gran Chaco Americano region. Its recharge zone, located in Argentina and Bolivia, determines groundwater flow towards the east and crosses national boundaries, emerging in low-lying lands and draining into a series of streams that discharge into the Paraguayan-Argentine Chaco and eventually into the Parana River in Paraguay.

The livelihood of the 1 million indigenous people in the region is closely linked to the aquifer's surface area. Increasing pressure on scarce water resources, poor land quality and soil degradation is causing alarm. The natural water quality transition (fresh in Bolivia, to brackish and saline in Paraguay and Argentina) may be changing.

There are many pressures on the land in the region, which have arisen from the expansion of poorly planned mechanized agriculture, which has in turn led to land degradation, the decline of wetlands and the deterioration of water quality. Increased rain intensity from anticipated climate change could trigger erosion, and re-sedimentation in recharge zones could inhibit aquifer infiltration from stream beds. Due to poor awareness and divergent regulations, current aquifer management by institutions in the sharing countries is inadequate. Therefore, coordination for the long-term management and protection of the recharge zones, as well as the discharge zones, is lacking.

A case study by the UNESCO International Shared Aquifer Resource Management Programme (UNESCO-ISARM) is part of a Plata Basin project financed by Global Environment Facility (GEF). The case study's activities focus on raising awareness of the aquifer system, as well as ensuring the sustainability of its resources, the lifeline of the local population and the aquifer-dependent environment. The project will help further develop engaged and strengthened institutions that practise sound aquifer management and offer educational and technical support to the community.

Source: www.isarm.net.

High altitude landscape at the border between Argentina and Bolivia

Part 2. Water and Geopolitics

Given the interdependencies of water resource uses, spatial variations and surface water and groundwater, as well as upstream and downstream differentiations, the need to develop mechanisms for the sustainable sharing of water is obvious. Attention to environmental security exemplifies the growing regional and global environmental concerns that could also lead to new forms of conflict.

2a. Trends in geopolitical developments

History shows few outright transboundary water-related conflicts. Although strong competition does occasionally occur between users, such as in the Tigris-Euphrates Basin, in the Jordan Basin and the Paraná-La Plata Basin (see **Box 11.2** for an example in southern India), there is an increasing trend towards inter-state collaboration (as in the case of the Nile), as well as cooperation through increased public participation, non-governmental organizations (NGOs), the common search for alternative water sources and the collaborative spirit of international water conferences, arbitration mechanisms and mediating agents (see **Box 11.3**). Efforts like the Division of Early Warning and Assessment (DEWA), UNESCO's From Potential Conflict to Cooperation Potential (PCCP) and ISARM have been developing case studies on the management of transboundary water resources, illustrating the impressive range of examples of water as a catalyst for peace and cooperative capacity-building. Many programmes – financed through the International Waters focus area of the Global Environment Facility (GEF) in Eastern Europe – are working together to develop cooperative frameworks and encourage the development and implementation of policies that support the equitable use of water and the sound functioning of other water-related natural resources.

BOX 11.2: CAUVERY RIVER DISPUTE IN SOUTHERN INDIA

In India, the federal government plays a mediating role in river water disputes. The Inter-State Water Disputes Act of 1956 requires the government to encourage states to settle disputes through dialogue. If that does not work, a tribunal is to be constituted. After a hearing, the tribunal makes a binding judgement.

The Cauvery Basin in southern India has 75,000 square kilometres (km^2) of area spread over four riparian states: Karnataka, Kerala, Tamil Nadu and Pondichery. The basin is mainly drained by the 780 km-long, rain-fed, perennial Cauvery River, which flows from west to east into the Bay of Bengal. In addition to being a major source of irrigation and hydroelectric power, the Cauvery River is an important water supply source for Bangalore, a centre for information technology and the software industry.

When a dam project was developed by the upstream state of Mysore (now in Kerala), two agreements were made (in 1892 and in 1924) detailing how the river waters were to be shared. The agreement was open for review once it expired, but no agreement has been reached between the two main riparians, Kerala and Tamil Nadu, since the 1970s. A tribunal was constituted in 1990, and an interim judgement was passed in 1991. The tribunal is expected to make its final decision soon.

The dispute is based on the fact that the demand for irrigation far exceeds the irrigation potential of the river. In drought years, this leads to a flash-point. The monsoon pattern is peculiar: the southwest monsoon brings rains to the upstream areas in June and July. The downstream and delta regions depend mainly on the weaker northeast monsoon (September-October). In the Cauvery Delta in Tamil Nadu, three crops are grown annually, but the summer crop depends on the timely release of waters from upstream areas. However, upstream farmers argue that it is unfair to be forced to share their water in summer when demand for water is at its highest. The downstream farmers argue that historically they have grown three crops and hence, their livelihoods crucially depend on maintaining the sharing scenario as accepted in the 1924 agreement.

There have been attempts to promote citizen efforts towards conciliation through people-to-people dialogues, and most recently, to form a 'Cauvery family'. Such efforts should help in encouraging informed dialogue and building trust. Collective action theory suggests that it is possible for riparian states to voluntarily reach self-enforcing agreements, provided the costs and benefits are considered in a transparent manner and sustainable development priorities are given primacy.

Table 11.2: Timeline of geopolitical developments: Inter-state water-related conflicts and cooperation since 2002

2002	● In early 2002, Friends of the Earth Middle East launched the Good Water Neighbors project to raise awareness about regional water and environmental issues. A variety of cooperative programmes have been set up in Jordan, Palestine and Israel to promote exchange of information and ideas between different communities in the region. These programmes have also furthered the campaign to protect the Jordan River, which brings stakeholders from the entire region together to work on sustaining the flow of this important river.
2003	● During the Joint River Commission (JRC) talks between India and Bangladesh – held in September 2003 – India agreed to involve Bangladesh in future discussions on the controversial US$ 200 billion (EUR 172 billion) river-linking project which will bring water for irrigation from the Ganges, Brahmaputra and Meghna river basins to Haryana and Gujarat. In February 2004, Bangladesh called on the Indian government to assess the impact on both the environment and biodiversity before it begins implementing the project.
	● In October 2003, the United States and Mexico came to an agreement over water used for irrigation. Mexico – in water debt to the United States due to prior agreements – agreed to release water from its reservoirs to relieve Texan farmers of the local drought.
	● In November 2003, the Limpopo Basin Permanent Technical Committee created the Limpopo Watercourse Commission (LIMCOM). The objective of the commission is to facilitate capacity-building to better manage the shared water resources in the basin states of South Africa, Botswana, Mozambique and Zimbabwe.
	● Iran signed a contract with Kuwait on 13 December, 2003 to provide the country with drinking water for almost 30 years. Three hundred million cubic metres per year will be pumped to Kuwait through a 540 kilometres pipeline, to be built at an estimated cost of US$ 2 billion.
2004	● In 2004, Kazakhstan warned of a potential environmental disaster in reaction to China's plans to divert water from the Irtysh and Ili Rivers. Similar concerns were expressed in reaction to Russian scientists' suggestion to revive an old Soviet plan to divert Siberia's Ob and Irtysh Rivers in order to replenish the Amudarya and Syrdarya Rivers. Kazakhstan and China have signed a transboundary water management agreement, but a joint commission has failed to address their concerns.
	● The Dniester-Odra Project – launched in 2004 – is an Eco-TIRAS project in partnership with Polish and Ukrainian NGOs, which promotes cooperation between local NGOs as well as state and local governments in large European river basins. The focus of the project is to share knowledge of integrated transboundary water management between the Dniester and Odra Rivers.
	● In 2004, countries sharing the Amazon River Basin – Brazil, Bolivia, Colombia, Ecuador, Guyana, Surinam, Peru and Venezuela – renewed their commitment to contain environmental damage and protect the planet's biggest reserve of freshwater. Representatives of these countries agreed to create three working groups to address the impact of pollution on the Amazon at a meeting of the Amazonian Cooperation Treaty Organization (OTCA) in Rio de Janeiro, Brazil.
	● Despite sporadic challenges, the riparians of the Nile Basin continue work on the Nile Basin Initiative (NBI). The initiative – formally launched in 1999 – is an important project geared towards cooperative development and institutional capacity-building for the entire East Africa region as well as all riparian states: Burundi, Congo, Egypt, Eritrea, Ethiopia, Kenya, Rwanda, Sudan, Tanzania and Uganda. In late May 2004, the Nile Basin Transboundary Environmental Action Programme was implemented; the first of eight on-the-ground projects to be initiated by the Shared Vision Programme of NBI.
	● On 13 July 2004, seven of the eight riparian nations of the Zambezi River signed the Zambezi Watercourse Convention (ZAMCOM). The signatory states have initiated the ratification process and the convention will enter into force when two-thirds of the signatories have ratified it, to occur most probably by the end of 2005.
	● Improving Water Governance in the Volta Basin – a project initiated by the World Conservation Union (IUCN)/the Swedish International Development Cooperation Agency (SIDA) in September 2004 – aims to aid Burkina Faso and Ghana to define shared principles and construct a framework for international cooperation in the management of the Volta Basin.
2005	● In early 2005, the European Union launched the Water and Environmental Resources in Regional Development (WERRD) project which has the objective of examining policies in improving livelihoods in international river basins. The project is currently being carried out in the Okavango River with the participation of Botswana, England, Namibia, South Africa and Sweden.
	● In 2005, Bolivia and Chile came to an agreement in sharing the groundwater of the Silala Aquifer – a potent source of conflict for years. The status of the Silala River – a disputed body of water flowing from Bolivia to Chile – however, is yet to be decided. Talks between the two countries have stopped due to the interim government in Bolivia, formed after the resignation of President Carlos Mesa.
	● In 2005, the governments of Honduras and Nicaragua approached the Organization for American States (OAS) for financial and technical assistance to improve their border relations after the Rio Negro was diverted as a result of Hurricane Mitch.
	● In April 2005, Friends of the Earth Canada and Friends of the Earth US called on the Canadian government to take the United States to the International Court of Justice in The Hague to stop the completion of the Devils Lake Emergency Outlet in North Dakota. According to the groups, the completion of the outlet threatens to move polluted waters and invasive species from the United States into a Canadian river that flows into Hudson's Bay.
	● After several years of attempting to stop India from constructing the Baglihar Dam on the Chenab River, claiming that it is in violation of the Indus Waters Treaty of 1960, Pakistan invoked the dispute resolution mechanism outlined in the Treaty, marking the first time the mechanism was invoked by either party. The Treaty states that the original broker of the Treaty, the World Bank, should appoint a neutral arbitrator to resolve the issue.
	● In May 2005, Green Cross International launched the La Plata Dialogues 'Water for Life' through its Water for Peace program, in cooperation with Itaipu Bi-National and the Intergovernmental Coordination Committee of the Plata Basin Countries (CIC). These high-level talks included stakeholders from all sectors of society to identify projects for improving the management of the La Plata River Basin.
	● In June 2005, Guinea rejoined the *Organisation pour la mise en valeur du fleuve Sénégal* (OMVS, the Senegal River Basin Organization) and thereby brought together all four countries of the river basin – Guinea, Mali, Mauritania and Senegal – for the first time in more than 30 years.
	● A Memorandum of Understanding between Israel, Jordan, and the Palestinian Authority, signed in July 2005, agreed to a two-year study to investigate the social and environmental impact of transporting large quantities of water through 200 kilometres of piping from a small canal on the Red Sea to the Dead Sea. The World Bank facilitated the $15 million agreement.
	● In 2002, UNESCO and the Organization of American States (OAS) launched the International Shared Aquifer Resource Management (ISARM) project for the Americas. The project – International Transboundary Aquifers of the Americas – organized three workshops in 2003, 2004 and 2005 presenting the data gathered on transboundary groundwater in North, Central and South America and highlighting the need to follow-up on this cooperative project.
	● In October 2004 UNESCO and the Aristotle University of Tessaloniki, Greece, presented in a joint workshop the draft inventory of internationally shared aquifers in southeast Europe. These were the first results of the ISARM-MED project. The results are published in this chapter.
	● In 2005, UNESCO and FAO published a review of all existing treaties and other legal instruments related to the use of groundwater resources, in a publication entitled: *Groundwater in International Law: Compilation of Treaties and Other Legal Instruments*.
	● In 2005, the UN International Law Commission continues work on the draft legal instrument on transboundary groundwater resources begun in 2002.

2b. The case of aquifer systems

The equally important challenge of demarcating aquifer systems, as well as the need to manage them through appropriate inter-state compacts or multilateral agreements, has begun in earnest. Unlike inter-state surface water compacts (legal agreements specially made for allocating and managing shared water systems among various water users and uses), the focus of the aquifer compact is the design of a joint resource allocation and management system that would ensure the sound functioning of the systems through appropriate policies adopted by all the overlying countries. Although pilot studies to develop such mechanisms and approaches have been initiated, it is still too early to deliver firm guidelines (UNESCO-IHP/ISARM, 2001).

New inter-state aquifer agreements have to be carefully crafted to fit retroactively into existing inter-state river agreements in which the groundwater component may have been given insufficient prominence. Reviews and reassessments, coupled with new agreements, offer the possibility of managing groundwater resources in a holistic way, as opposed to disparate groundwater administrations based on artificial state boundaries. Administrative bodies could be created to address present and future disputes over claims of excessive groundwater withdrawal between neighbouring states or basins, provided that the institutional transaction costs can be covered.

Parallel conceptual and methodological advances have raised the questions of indicators, both in terms of flash points and red flags for conflict and obstacles to cooperation. In addition to the traditional sources of conflict outlined above, especially for transboundary conflicts, much literature also emphasizes rivers forming a shared boundary, human action triggering disruptions like dams, power asymmetries, unilateral basin development and extreme hydrological events like droughts and floods. On the other hand, the shifting interest in conflict management and resolution has also helped support efforts that aim to better understand obstacles to and mechanisms for cooperation. An interesting conceptual scheme by Sadoff and Grey (2002) exemplifies types of cooperation by distinguishing between increasing benefits *to the river* (the ecological river), increasing benefits *from the river* (the economic river), reducing *costs because of the river* (the political river) and increasing benefits *beyond the river* (the catalytic river). A similar conceptual scheme for aquifers could also be formulated.

All such considerations point to two conclusions: first, water conflicts are intertwined with other, larger socio-political issues, and second, the development of indicators and early warning signs for preventing and mitigating conflict must be coupled with parallel indicator systems of cooperation that complement the challenge of water sharing in both intra- and inter-state situations (see **Table 11.2**).

2c. Water regimes and hydrodiplomacy

The notion of water regimes also helps to better delineate water sharing challenges. Water regimes imply a specific set of rules, institutions and practices, as well as relationships of power, position and interest. Such regimes exemplify a specific hydroculture of established cultural and socio-political traditions, attitudes and practices. For example, one can think of river basins as water regimes where voluntary cooperation has emerged over time (as in the case of the Columbia basin between the US and Canada), or where external incentives, such as foreign investment and even threats by a major power in the basins, as in the Nile River Basin and the Mekong, have contributed to cooperation. It is interesting to note that such a culture is lacking for groundwater resources, except in regions where cultural norms and traditions have brought cooperative forces together (qanats or *kharez*,[5] for example) and where land management and groundwater sustainability go hand-in-hand.[6]

Supporting water sharing efforts is also a general principle of conduct in international law, treaties, binding acts and judgments of international courts that shape the rules and procedures of shared transboundary waters. The five major legal principles that shape hydrodiplomacy, including intra-state practices, are as follows:

■ the principle of international water and the concept of an international watercourse

■ the principle of reasonable and equitable utilization, which has generated perpetual debates and interpretations of the terms 'reasonableness' and 'equity'

■ the obligation not to cause significant harm and the exercise of due diligence in the utilization of an international watercourse

■ the principle of notification and negotiations on planned measures

■ the duty to cooperate, including regular data exchange

Reviews and reassessments, coupled with new agreements, offer the possibility of managing groundwater resources in a holistic way...

5. A qanat is a traditional Middle Eastern subsurface network of tunnels and wells used to transport water from an elevated mountainous area downwards to the ground surface. *Kharez* is the term used in Baluchistan, which is a region on the border areas of Pakistan, Iran and Afghanistan.

6. See the activities of the recently created UNESCO Centre on Qanats and other hydraulic structures in Yazd, Iran.

Harvest in Two Buttes, Colorado, United States

The draft articles prepared by UN-ILC on the use of transboundary aquifers are developing similar though alternate principles for groundwater resources that encourage aquifer system states to focus on the integrity of the functions of the aquifer systems.

The legally binding and far-reaching WFD of the European Commission (EC, 2000) established a detailed process for community action in water policy that accentuates many normative aspects (social preferences, goals and established practices) of valuing and sharing water. It also placed a renewed emphasis on public participation. Existing legal approaches to water resources law (from the 1966 non-binding Helsinki Rules on the Uses of Waters and International Rivers to the International Law Association Berlin Conference of 2004) have now been further elaborated.[7] At the same time, declarations, international law organizations' drafts, the creation of the World Water Council and international water conferences (such as the Fourth World Water Forum in Mexico City in 2006) all aim to expand the spatial envelope (from narrowly local to national and transnational, if not global, geographical units) and accentuate integrated approaches to sharing water.

Progress in the twenty-first century will require an institutional order of cooperation, comprehensive management principles and sharing of experiences gained through the practice of ecosystemic principles in water resource projects. Paths towards further effectiveness for some authors (Rogers and Kordab, 2004) entail boosting governmental *concern*; enhancing the *contractual* and bargaining environment and, finally, building national *capacity*.

There is widespread interest in a paradigm shift to a new model that questions the traditional methods of governing water resources, as well as an ongoing debate as to what this new paradigm entails. This involves the search for new judicial norms, flexible institutions, demand-driven water policies, new concepts of water types (blue and green water, see **Chapter 4**, or virtual water, see **Chapter 12**), as well as sustainability, transparency and public participation. Conflict prevention and similar concepts of interdependence in other efforts to share water resources in a sustainable manner are also pivotal.

Noble as these goals may be, there have been repeated warnings about the difficulties of generating international

agreements given the obstacles to cooperation, which include the increasing split between the North and the South, the persistence of national sovereignty and the lack of sufficient incentives to bring nations to the table for sustained negotiations.

2d. Water sharing and the public good

The use of the terms 'reasonable', 'equitable' and 'sustainable' in the last three decades illustrates the increasing emphasis on water sharing as a public good. It also stresses the importance of accommodating competing demands and expresses the underlying wish to manage water by hydrological boundaries rather than by administrative or political borders. The complexity of the physical river system, the interdependence of surface water with groundwater, is not a new phenomenon. But the spate of activities and declarations in international water meetings have served to redefine efforts of purposeful water sharing and laid the ground for new upstream-downstream institutional arrangements. The Global Water Partnership's (GWP) effort to clarify and utilize toolboxes (specific methodological guidelines for defining ethical variables and relevant measurable indicators) is now joined by other institutional mobilization efforts, methodological advances and mechanisms for measuring performance and output.

The desire to maintain communal control and support public participation in upstream-downstream relations further expands the recognition of water as a public good and also points out the danger of commodifying water by distinguishing between the value, price and cost of supplying water[8] (see **Chapter 12**).

In articulating general principles of integrated shared water resources, one must also raise questions about the criteria for evaluating institutional performance. Here, in addition to the standard literature, the early UN Water Conference in Mar del Plata (1977) promulgated a set of performance criteria by blending practical experiences from the social sciences. Criteria include a cluster of institutional characteristics: clarity of water ownership, legal authority capable of enforcing decisions, transparent national policies and mechanisms for coordination. They also include other institutional performance criteria, such as consideration of alternatives, incorporation of externalities, responsiveness to national and local priorities and expeditious movement from planning to implementation.

---

7. This is especially evident in articles 10–16 dealing with internationally shared water.

8. The now famous essay, 'Tragedy of the Commons' (Hardin, 1968), is the metaphor for the problem of sharing this public good, and the shift from exploitation for growth to the preservation of ecosystem health.

A good example of capacity-building comes from the WFD (2000), which refers to five cross-cutting principles for implementation:

- opportunities for integrated approaches between different sectors (the environment and agriculture, for example)
- scale of intervention through a distinction between large and smaller river basins
- timing in terms of early implementation
- participation and encouragement of building on traditions of public or stakeholder involvement
- capacity, or the historical existence of strong technical and scientific traditions or expertise.

The right to water is already recognized in several legal or political instruments (see **Table 11.3**). It guarantees access to water, without discrimination, in a permanent and sustainable manner – and at a socially and economically acceptable cost. It also addresses the issues of subsidiarity, solidarity, and cooperation. Finally, it takes into account the interests of disadvantaged populations and the importance of decision-making at local levels.

2e. Institutions, procedures and regulatory principles

Institutions are defined as established and organized procedures; water institutions represent established values, norms and practices that provide a policy, legal and administrative framework for sharing water.

A variety of actors – local institutions, NGOs, research institutions, private sector participants, donors, riparian government institutions and transnational river basins – face problems of vertical and horizontal integration, not only in water resources projects, but also within and between other resource management entities and organizations.

Cutting across this complex setting are the theoretical problems of legal doctrines, as well as the power of all encompassing international agreements. The UN Convention on the Law of the Non-navigational Uses of International Watercourses, as well as the non-binding Helsinki Rules, have adopted the principles of limited territorial sovereignty with equitable and reasonable

Table 11.3 The right to water timeline

Several acts, declarations, conventions and constitutions make explicit or implicit provisions for the right to water:

1949-1977	■ The international humanitarian law applicable to armed conflicts and human rights law: - *Convention (I) for the Amelioration of the Condition of the Wounded and Sick in Armed Forces in the Field.* Geneva, 12 August 1949. - *Convention (II) for the Amelioration of the Condition of Wounded, Sick and Shipwrecked Members of Armed Forces at Sea.* Geneva, 12 August 1949. - *Convention (III) relative to the Treatment of Prisoners of War.* Geneva, 12 August 1949. - *Convention (IV) relative to the Protection of Civilian Persons in Time of War.* Geneva, 12 August 1949. - *Protocol Additional to the Geneva Conventions of 12 August 1949, and relating to the Protection of Victims of International Armed Conflicts (Protocol I),* 8 June 1977. - *Protocol Additional to the Geneva Conventions of 12 August 1949, and relating to the Protection of Victims of Non-International Armed Conflicts (Protocol II),* 8 June 1977.
1979	■ The 1979 *Declaration on the Elimination of All Forms of Discrimination Against Women,* adopted by the UN General Assembly on 18 December 1979 (resolution 34/180), entered into force on 3 September 1981.
1989	■ The 1989 *Declaration on the Rights of the Child,* adopted by the General Assembly and opened for signature, ratification and accession by on 20 November 1989 (resolution 44/25), entered into force on 2 September 1990.
1997	■ The 1997 *Convention on the Law of the Non-navigational Uses of International Watercourses,* adopted by the UN General Assembly on 21 May 1997 (Resolution 51/229).
2000	■ Resolution A/RES/54175 of the UN General Assembly of 2000: *The Right to Development.*
2001-02	■ General Comment No.15 of the Committee on Economic Social and Cultural Rights of November 2002: *The Right to Water* (arts. 11 and 12 of the International Covenant on Economic, Social and Cultural Rights).
	■ The Council of Europe and the European Parliament declared themselves in favour of this right, successively in 2001 and 2002.
2003	■ The Johannesburg Declaration and the documents produced at the Third World Water Forum (Kyoto 2003) include the right to basic sanitation as part of the right to water.

BOX 11.3: TRANSBOUNDARY AQUIFERS ON THE AGENDA OF THE INTERNATIONAL LAW COMMISSION

In 2002, the International Law Commission (ILC) of the United Nations incorporated in its work programme the topic of shared natural resources, including groundwater, oil and gas. In his *First Report on Outlines* submitted in 2003, the Special Rapporteur presented the background of the topic at the ILC, and indicated his intention to start with groundwaters. He also presented an addendum technical in nature and containing general information on groundwater (importance, characteristics, uses, causes of degradation, non-renewable).

In his *Second Report on Shared Natural Resources: Transboundary Groundwaters*, of 2004, the Special Rapporteur expressed his decision to consider all groundwaters, and introduced the concept of aquifers. He presented an outline of a future framework instrument for transboundary aquifers and seven draft articles dealing with aquifers' scope and

definition, the obligation not to cause harm, the general obligation to cooperate, the regular exchange of data and information and the relationship between different kinds of uses. The addendum contained case studies and models of several types of transboundary aquifers.

In his third report of 2005, the Special Rapporteur proposes a complete set of draft articles for an instrument on the law of transboundary aquifers. In addition to the previous draft articles, the Special Rapporteur introduces the distinction between recharging and non-recharging aquifers; articles on bilateral and regional arrangements in relation to other conventions; the need for equitable and reasonable utilization; monitoring; protection; preservation and resource management (with provisions on ecosystems, recharge and discharge zones); and activities affecting other states. The report was discussed at the fifty-

seventh session of the ILC (May-July 2005), and was generally well received. A working group was set up to review all the draft articles and will be reconvened at the ILC's next session in 2006 to complete its work. The third report was discussed at the last session of the sixth Committee of the UN General Assembly in October and November 2005. Member State delegates presented their comments and expressed their appreciation of the work being achieved by the Special Rapporteur.

An ad hoc groundwater multidisciplinary experts group was established, and meetings were organized in Paris and Tokyo. Experts convened by UNESCO-IHP participated in briefings on aquifers and the science of hydrogeology for the members of the ILC in Geneva and the members of the sixth Committee in New York.

Source: UN, 2005; 2004; 2003.

utilization. They currently make up the accepted doctrine in international water law (at least for surface water). In the draft articles submitted by the Special Rapporteur on transboundary aquifers, the principle of equitable and reasonable utilization was also introduced and is currently being debated at the International Law Commission (see **Box 11.3**). The doctrines of absolute territorial sovereignty and absolute territorial integrity are increasingly becoming outdated.

River basin planning and management have been a long honoured tradition – from the development of the Tennessee Valley Authority in the US in the 1930s to the Senegal development plans of more recent years. The common thread underlying these efforts has been the development of water resources for a variety of uses. The existing literature raises the important question of whether such integrative regional water plans can fit within the geographic limits of a river basin or watershed, and if so, how water can be shared in an equitable manner. Can such joint management take place in the vast expanses of the Aral, Nile, Amazon, Mekong and the Parana-LaPlata basins? Or should it be restricted to regional, specific socio-political conflicts and well-defined

geographic, cultural, environmental and economic boundaries?

What implications does this have for the long-term process of sharing water? First, we need to recognize the difficulties associated with legalistic approaches that tend to emphasize conduct rather than formal governance, especially in cases where there is no agreed-upon river regime. Second, existing legal approaches can be supplemented by flexible mechanisms, such as second track diplomacy (environmental diplomacy or hydrodiplomacy); alternative dispute resolution (ADR)[9] through international bodies or in the spirit of Agenda 21, a technical professional or an independent panel of experts (water ombudsmen or a water cooperation facility); and public participation and mobilization.

The operational terms for the above are complementarity and implementation. However, there are three troublesome problems in multilateral agencies: the historical and cultural inertia of past differences and practices, the calculation of all costs involved in the development of shared waters and incorporating social and environmental concerns related to effective water sharing. Cooperation and conflict are also

9. Alternative dispute resolution (sometimes referred to as appropriate dispute resolution) is a general term, used to define a set of approaches and techniques aimed at resolving disputes in a non-confrontational way. It covers a broad spectrum of approaches, from party-to-party engagement in negotiations as the most direct way to reach a mutually accepted resolution, to arbitration and adjudication at the other end, where an external party imposes a solution. Somewhere along the axis of ADR approaches between these two extremes lies mediation, a process by which a third party aids the disputants to reach a mutually agreed solution (Shamir, 2003).

expressions of the need to accommodate geographic realities and hydrogeology into the social context of shared water. Donors who provide incentives for the creation of voluntary agreements are necessary for a true community of riparians, in addition to their willingness to come together in joint institutional mechanisms.

Part 3. Preventing, Managing and Resolving Shared Water Conflicts

There is a long history of water-related disputes, from conflicts over access to adequate water supplies to deliberate attacks on water systems. As growing populations, urbanization and economic development will all require more water for agricultural, municipal and industrial uses, there is a risk that such contestations will increase. At the same time, water availability may be coming up to what Falkenmark (1999) has described as the 'water barrier', a level of supply below which serious constraints to development arise. These limits may be further stretched by potential climatic anomalies, which in turn, could intensify regional conflicts between upstream and downstream users.

There are three distinct phases in water conflicts: conflict creation, management and resolution. In the first phase, the emphasis is on diagnosis, anticipation and prevention, problem architecture and joint fact-finding. The second phase represents a trust-building stage through mechanisms such as mediation, arbitration and neutral expert fact-finding. Finally, conflict resolution involves consensus-building and the depolarization of conflicting interests through public fact-finding processes or adjudication.[10]

The search for a typology for conflicts and appropriate responses has led to several conceptual schemes. One conclusion is that we should pay particular attention to international river basins and aquifer systems, where confrontations and conflicts can be far-reaching. Sources of potential water conflicts include the following:

- scarcity (permanent and temporary)
- differences in goals and objectives
- complex social and historical factors (including pre-existing antagonisms)
- misunderstandings or ignorance of circumstances and data
- asymmetric power between localities, regions or nations
- significant data gaps or questions of validity and reliability
- specific hydro-political issues at stake (dam construction or diversion of water)

- non-cooperative settings and value conflicts, especially in terms of water mythology, culture and water symbolism.

Alternative dispute resolution mechanisms have become important means for resolving conflicts. The search for alternatives to legal institutions for arbitration was prompted not only by the saturation of legal mandates, but also by increasing litigation and confrontation. Mediation – a compromised discussion between disputants aided by a neutral third party – has become a viable alternative to adversarial processes. The gamut of adjudication, arbitration, mediation, conciliation and even principled negotiation illustrates several more alternative processes of dispute resolution. Public participation and negotiation can be tools in maximizing agreement not only about the nature of a problem but also about the desirability of specific outcomes.

The outcome of conflicts for all water regimes will also depend on other variables: the number of actors involved, external factors, preventive rather than corrective emphasis in potential cooperation, information available to all parties concerned, bureaucratic rigidity, lack of relevant institutions, historical animosities, etc. It is difficult to determine which variables are most important and how they interact with larger processes, such as demographic and socio-political pressures or with resource scarcity and environmental degradation.

10. Training courses on conflict management and cooperation-building have been instrumental: see UNESCO's PC-CP project (webworld.unesco.org/ water/wwap/pccp/cd/ educ_tools.html) and the World Bank's 'Shared Water, Shared Future' course workbook and the International Water Academy's Building a Curriculum for Training in Water Conflict Resolution, Prevention, and Mediation (www.thewateracademy.org/ OppActivities/index_ main.html), among others.

Relevant indicators – as well as the combination of data and sound judgment – are needed in order to establish a baseline of the status of shared water resources...

Section 4: MANAGEMENT RESPONSES & STEWARDSHIP

11. The approach followed was to provide a set of indices, or potential indicators, which were then examined against the recent literature. These indicators were the basis for discussion at the 'Indicators of water conflict and cooperation' workshop convened in Paris in November 2004 by UNESCO-PCCP.

3a. The search for relevant indicators in water sharing

Recent literature has highlighted the need to move from descriptive studies towards the creation of measurable indicators that measure the performance of shared water systems, monitor the process of equitable sharing and provide the mechanisms for monitoring both the current state and changes in interdependent water systems, in addition to gaining more realistic insights through field studies (Carius et al., 2004; Gleick, 2005; Millennium Ecosystems Assessment, 2003).

Theoretical approaches must be counterbalanced by practical examples of water sharing at every level. Although there is great difficulty in measuring the performance, process and product of water sharing through quantitative and qualitative indicators, they can help us to develop the critical thresholds; articulate the interesting differences, as well as significant trends and developments; and put forward the data necessary for balanced decision-making.

Relevant indicators – as well as the combination of data and sound judgment – are needed in order to establish a baseline of the status of shared water resources in order to discern the nature and rate of changes, provide a solid base for understanding the process of equitable allocation and offer early warnings of emerging problems.

It is important to indicate the broad complementary goals of indicators: policy relevance as well as technical credibility and relevance. Specific criteria for indicator selection include analytical soundness (expressed in scientific and technical terms); measurability (readily available, adequately documented, valid and reliable data and information); and utility to users (simple to interpret, showing trends over time, responsive to change, comparable and providing threshold or reference value against which one can assess ultimate significance).

The critical question is how to combine metric data with more qualitative approaches that offer ordinal data, at best. The question between quantitative and qualitative measurement reflects the difficulties of the availability, validity and reliability of existing data and information, not to mention the challenges of statistical manipulation. The first *World Water Development Report* (2003) noted that beyond the general commentary of other studies' indicators, there is no empirical derivation of testing. This chapter has preliminary empirical results of indicators

measuring potential conflict or cooperation in international basins. By identifying sets of parameters that appear to be interrelated and using these sets for advanced statistical analyses, two findings stand out. First, most cases show cooperation rather than conflict; and second, rapid changes – in institutional capacity or in the physical system – have historically been at the root of most water conflicts. These changes were measured by three indicators: internationalized basins (i.e. in newly independent states); basins that include unilateral developments (and the absence of cooperative regimes); and cases where basin states show hostility over non-water issues (Wolf et al., 2003).

Important methodological and policy questions for cumulative, interactive, synergistic effects and consequences remain. In addition, many of the water sharing indicators intersect with other dimensions of water discussed in this volume.

The list of potential indicators can be summarized across the following dimensions:[11]

■ **Operational/administrative interdependencies for sharing water**
- number of international basins and transboundary aquifers
- dependency on inflow from other river basins
- serious impact on upstream water diversions and impoundments
- impact on groundwater ecosystems
- upstream and downstream integrative mechanisms
- systematic considerations of water users and uses interdependencies
- high water stress/scarcity/poverty conditions
- basin-wide operational water planning and management
- surface/groundwater conjunctive use
- number of treaties/cooperative events.

■ **Cooperation/conflict**
- existing conflict accommodation and resolution mechanisms
- significant number of water treaties or conventions
- economic, scientific or industrial agreements
- cooperative events involving transboundary rivers
- unilateral projects, highly centralized water megaprojects

– existence of laws and regulations for fair water allocation
– stakeholders' involvement and participation mechanisms
– publication of joint inventories of transboundary resources
– effectiveness of community-based management
– newly internationalized river basins.

■ **Vulnerability/fragility**
– high degree of rivalries, disputes and contestation within and between countries or areas
– ratio of water demand to supply
– environmental and social fragility, non-robust social system
– diminishing water quality and degraded groundwater dependent ecosystems
– poverty, lack of good sanitary conditions
– extreme hydrological events and periodic water disasters (flood and droughts)
– demand changes (sectoral) and distribution
– dependence on hydroelectricity.

■ **Sustainability/development**
– expressed and implementable water conservation measures
– competence for dealing with and managing water-related conflicts
– desire for and implementation of balanced environmental policies
– capacity to recover the true costs of water projects
– importance of virtual water in food trade
– unaccounted-for water
– integrated resources water management (IWRM).

The underlying concern, as far as interdependencies are concerned, has to do with types of conflict and cooperation, over water or over efforts to adapt to scarcity. Ohlsson (1999) makes distinctions between stages like supply management challenges, end-use efficiency and allocative efficiency, while using first-order and second-order disputes. Here concerns with upstream-downstream relations also appear, as well as the varieties of water shortage, such as scarcity, stress and poverty. Also included in this category are a large number of groundwater indicators, such as total resource, recharge rates, total abstraction, depletion rates and risks and composite measurements of conjunctive water use.

The cooperation/conflict dimension raises questions about institutional mechanisms and conflict resolution efforts. The conflict management techniques outlined earlier support a cooperative context in conflict prevention, management and resolution. Some even suggest friendship/hostility indicators and supportive water cultures for cooperative efforts. **Box 11.4** shows how traditional societies have put water at the core of their values. Others emphasize conflict-processing capabilities, including risks of victimization or critical thresholds in forecasting political crises, social and geographical spread of conflict, and increasing competition over water and land distribution.

The vulnerability/fragility dimension emphasizes volatility and turbulence in terms of the surrounding environment and society. The larger issues of poverty, sanitation, environmental degradation and a lack of social resilience, community mobilization, preparedness and disaster absorption institutions are all indications of the tenuous fabric of environmental and social structures. Questions of environmental security, risks and the inability to adapt to threats and disasters – referred to as the index of human insecurity in the combination of environment, economy, society and institutions – also belong in this category.

The sustainability/development dimension emphasizes not only cleavages in expectations and achievements, but the current preoccupation with balances between environment, economy and society. It also addresses questions of growth and carrying capacity as well as the underlying debates of survival and fulfilment. A vast array of socio-economic indicators on income distribution, environmental damages, effects on cultural heritage, freedom of action, resilience, adaptability, recuperability, environmental integrity and IWRM are likewise considered.

It is important to emphasize the dynamics of the coordination/cooperation/collaboration continuum, which stresses attempts to minimize a catastrophic water crisis preoccupation, and forces us to explicate pragmatic cases of environmental conflict management, stakeholders' engagement and trade-off considerations more clearly. Articulating and testing a few central indicators in various settings is also important. These indicators form the water interdependency indicator, exemplified by the amount of water inflow from other river basins; the cooperation indicator, measured by the number of significant joint projects, treaties or other formal agreements; the vulnerability indicator, resulting from the

The underlying concern has to do with types of conflict and cooperation – over water or over efforts to adapt to scarcity

BOX 11.4: TRADITIONAL APPROACHES TO WISE WATER SHARING

Modern statutory law often overrides and marginalizes traditional methods of managing natural resources, usually to the detriment of the rural poor. A technocratic approach to water as an exploitable resource has become the standard outlook in modern states. Traditional societies, by contrast, consider water and other natural resources as embedded in a holistic world view, deeply rooted in the traditional lives of their communities. Water is, therefore, not only of economic and social importance but also of cultural and spiritual importance. It is the focus of community building and has a broad range of non-economic values that go far

beyond the modern utilitarian perception. It is this broader understanding of water that is the basis for traditional mechanisms of cooperative water management and governance, which often differ considerably from modern state-centred approaches. Yet these mechanisms offer sophisticated methods for water-related conflict resolution.

Traditional laws and rights should be included in efforts to share and govern water wisely. A participatory approach that takes the inclusion and empowerment of local stakeholders seriously should combine traditional methods of

water governance and conflict resolution with modern mechanisms. Questions that we need to examine more thoroughly include the following: How can traditional and modern approaches be harmonized? How and to what extent can traditional mechanisms be transferred from the local to the national and international levels? Can traditional ways contribute to non-state-centred modes of natural resource governance? Wise water governance can only occur if local indigenous water management knowledge is tapped.

Winnowing rice in Madurai, Tamil Nadu, India

ratio of water demand and supply; the fragility indicator, measured in terms of environmental deterioration and social unrest (especially poverty and rivalries), within and between countries; and the development indicator, summarized by competence/commitment for managing water-related conflicts.

3b. Capacity-building and institutional mobilization

As world attention turns to questions of sustainable development, the restoration of degraded environments and the creation of cooperative arrangements for shared water resources, it is clear that institution-building, comprehensive management and alternative dispute resolution efforts will be central endeavours for many years to come (see **Box 11.5**). One of the main challenges is bringing together surface water and groundwater management institutions – the former dominated by irrigation interests and the latter dominated by mineral resource interests. Degraded water resources and their potential impacts on international security also provide opportunities for cooperative institutions and transnational cooperation.

The arid environments of the Middle East, the fragmented entities of the Balkans and other volatile regions will force us to consider, once again, the pivotal role of water as an agent of peace. To what extent will states, multinational corporations, NGOs and existing international bodies respond to sharing power and implementing action that promotes ecological

interdependence and addresses other environmental challenges? Traditional areas of contestation – historical areas of concern like the Middle East, the Nile River and the Himalayan Basins – remain priorities. The transformation of the former Soviet Union and its newly independent states has opened up new areas of concern, such as the emergence of new states in Eastern Europe and the Balkans and enclosed seas such as the Aral and Caspian seas. The Tumen River in Asia and the Lauca and Parana/La Plata rivers in Latin America are also water systems of growing concern.

Future sources of conflict are likely to be diverse and reflect a combination of internal and external considerations, as well as the larger conditions of environmental change, such as climate and land-use questions. Resource degradation and depletion, in addition to political confrontations and climate-based environmental changes, become the backdrop for potential sources of discord, which are difficult to manage with the tools that are currently available.

3c. Mechanisms for cooperation and crisis avoidance

Increasing sensitivity about the need to integrate competitive demands and stakeholders' interests, in addition to the evolving need for political accommodation and the proactive stance in avoiding conflict, have all contributed to a shift from confrontation to cooperation, from monologue to dialogue and from dissent to consensus.

BOX 11.5: MAJOR INITIATIVES IN BUILDING INSTITUTIONAL CAPACITY TIMELINE

Launched in 2002, the Consultative Group on International Agricultural Research (CGIAR) Challenge Programme includes a project on Transboundary Water Policies and Institutions. The five major themes make reference to nine benchmark basins, most of which are international.

In April 2003, the Woodrow Wilson International Center for Scholars launched a Working Group for a three-year project, Navigating Peace: Forging New Water Partnerships. Meetings have been held regularly and a report of the project is forthcoming.

The United Nations Economic Commission for Europe (UNECE) finalized the agreement on the Protocol on Civil Liability and Compensation for Damage Caused by the Transboundary Effects of Industrial Accidents on Transboundary Waters, which was formally adopted and signed by twenty-two countries at the Ministerial Conference Environment for Europe in Kiev,

Ukraine, in May 2003. The Protocol will give individuals affected by the transboundary impact of industrial accidents on international watercourses – fishermen and operators of downstream waterworks – a legal claim for adequate and prompt compensation.

In November 2003, the parties of the UNECE Convention opened up the possibility of acceding the Convention and its articles on Protection and Use of Transboundary Watercourses and International Lakes to countries outside the UNECE region. This will allow others to use the Convention's legal framework and benefit from the experience in transboundary water cooperation that has been gained, since it was signed in Helsinki in 1992.

In August 2004, the International Law Association (ILA) developed the Berlin Rules, an update of the 1966 Helsinki Rules, which aim to codify customary international water law.

In December 2004, UNESCO launched the Water Cooperation Facility, which promotes cooperation, peace and prosperity in developing and managing transboundary waters, in cooperation with the World Water Council, the Universities Partnership for Transboundary Waters and many others involved in the management of shared water resources.

In 2005, the International Network of Basin Organizations (INBO) launched the TwinBasin project, which enables the mobility of staff between twinned Basin Organizations and capitalizes on the knowledge thus acquired.

In May 2005, the Euphrates-Tigris Initiative for Cooperation (ETIC) was established in Kent, Ohio, US in order to facilitate cooperation within the Euphrates-Tigris System and in the riparian countries of Syria, Turkey and Iraq in the domains of technical, social and economically sustainable development.

Although there are many conflict resolution methods available, there is no standard legal structure for proceeding. A series of non-legally binding voluntary methods – such as arbitration, adjudication, negotiation, mediation, enquiry and conciliation – are used to settle disputes. Positive evidence from these efforts and other voluntary mechanisms of ADR allow us to be realistically optimistic about our ability to resolve future potential conflicts. In 2004, an international conference was organized by UNESCO in Zaragoza, Spain on the use of these techniques in finding solutions to difficult shared water resources.

Collaboration in addressing environmental disputes involves three phases: *problem setting* (problem definition or problem architecture); *direction setting* (predominantly negotiations over substantive problems); and *implementation* (systematic management of inter-organizational relations and monitoring of agreements). All proposed alternative processes (direct negotiations, good offices mediation, conciliation, etc.) could allow the parties to reach a more timely and appropriate resolution of disputes.

Furthermore, in many countries citizen involvement has been formally incorporated into decision-making processes and is an important means of social regulation, as well as a key component of integrated natural resources management. The current emphasis on public participation is not only an attempt to fight the problem of elitism in planning, but also part of the commitment to address public demands, involve all stakeholders and increase awareness about the nature of water resources development, as well as the potential for conflicts within and between countries. Instructive here is the CABRI-Volga project, an international coordination action aimed at facilitating cooperation in large river basins in the EU, Russia and Central Asia.

Public involvement can range from simply supplying public information releases to undertaking joint planning and decision-making. A proactive public can lead to conflict management and increased consensus, and in international cases, a reinforcement of the spirit of transnational commons. Public *awareness* involves one-way information by alerting the community to issues.

Table 11.4 Recent international conferences of interest

■ Second Israeli-Palestinian International Conference on 'Water for Life', Turkey, October 2004, sponsored by the Israel-Palestine Center for Research and Information.

■ 'Water Conflicts and Spiritual Transformation: A Dialogue', Vatican City, 13-15 October 2004, co-sponsored by Oregon State University, Pacific Institute, International Water Academy, and the Pontifical Academy of Sciences

■ Second International Conference, Zaragoza, Spain, October 2004, organized by UNESCO's PCCP project. The conference brought together water managers, decision makers, students, trainers and a broad range of other stakeholders involved in the management of shared water resources. A series of interactive role-plays enhanced participants' conflict management skills and improved their knowledge of selected basins.

■ 'Water as a Source of Conflict and Cooperation: Exploring the Potential', Tufts University, Boston, Massachusetts, United States, February 2005. This working group brought together some the brightest minds in the field of international waters, to discuss whether or not cooperation over water resources acts as a catalyst for cooperation in other areas (www.tufts.edu/water/WorkshopLogin.html for more information.)

■ The Third Biennial International Waters Conference, Brazil, June 2005, organized by the Global Environment Facility (GEF). The primary objective of the meeting was to 'foster knowledge sharing and collaboration between participating governments, GEF International Waters projects, GEF Implementing and Executing Agencies, donor partners, and the private and non-profit sectors.' This idea was proposed by, and supported by the Dushanbe International Fresh Water Forum held 29 August - 1 September 2003. Following the UN General Assembly initiative,

■ The Government of Tajikistan organized an International Conference on Regional Cooperation on Transboundary River Basins in Dushanbe, on 30 May – 1 June 2005. This event was organized as a follow-up of the United Nations General Assembly's Resolution A/RES/58/217, dated 23 December 2003, declaring the period from 2005 to 2015 the International Decade of Action 'Water for Life', upon the recommendations of the President of the Republic of Tajikistan, and the President of the International Fund for Saving the Aral Sea.

Source: Prepared by Aaron T. Wolf, Department of Geosciences, Oregon State University and Joshua T. Newton, The Fletcher School of Law and Diplomacy, Tufts University

A proactive public can lead to conflict management and increased consensus

Public *involvement* implies two-way communication and is a means of engaging community members in information exchange and dialogue. Finally, public *participation*, the most intense form of interaction between authorities, experts and citizens, implies shared leadership, truly joint planning and a democratic delegation of power.

In the last two decades, the simultaneous growth of participatory democracy and expertise in decision-making have been advocated, although it is difficult to maximize both value preferences. There is a distinction between the idealized conceptions of citizen participation and the harsh demands of public policy-making, especially in transboundary considerations. Again, participation is a double-edged sword: planners and decision-makers must be open to collaborating with citizens, while citizens must be active and competent in planning and negotiation. It remains true that the broader the base of citizen participation, the more potential influence on managing conflict there is.

3d. Contentious water sharing and environmental security

Water sharing also raises other challenges: the quest for environmental security and cooperative water agreements that deal with the more strategic issues of conflict prevention. What was previously a concern mainly with

overt military conflicts has now expanded to incorporate environmental disasters, such as chemical spills in the Danube and the destruction of the Aral Sea. These cases broaden our definition of security to incorporate contested water bodies that, according to the UN, are issues of human security and involve human life and dignity. A growing body of literature addresses water conflicts and hydrodiplomacy, not only as strategic socio-political issues, but also as environmental interventions that affect claims on water bodies and groundwater-dependent ecosystems.

In the end, reducing the risk of conflicts and promoting equitable water sharing require regional and international approaches. However, there are limited international mechanisms for environmental security issues concerned with access to water and the restoration of polluted habitats (e.g. the restoration of rivers in Eastern Europe polluted by military bases during the Cold War).

Concern about volatile and stressed regions, as well as water-scarce environments and discussions of complex shared waters, have raised questions about potential future contestation areas, critical thresholds and red flags.[12] Perhaps even more to the point, terrorist attacks have also increased the concern about security of local water supplies.

12. Brauch et al. (2003) offer a rich conceptual discussion of the emerging difference between traditional and environmental security.

Insightful works are linking water resources to vulnerability, a function of many factors that include economic and political conditions, water availability, population growth, climate variability and the extent to which a source of water supply is shared. 'Regions at risk' are suggested as a result of basic qualitative calculations. The scarcity of water is replacing that of oil as a flashpoint for conflict between high-risk countries (Brauch et al., 2003; Gleditsch, 1997). On the other hand, shortages can also stimulate cooperative solutions or international intervention for profit management. Extreme hydrological events – droughts or floods, institutional problems and expanding populations – are exacerbating problems in these regions.

Regions can therefore be characterized by three particular vulnerabilities:

- ecological vulnerability: arid regions and regions of limited resources

- economic vulnerability: concerned with past practices of traditional exploitation and state economics

- social vulnerability: over-utilizing resources, as well as complex social economic and ecological forces affecting an area's natural equilibrium.

Hence the terms 'fragility', 'volatility' and 'carrying capacity'[13] have become indicators of conflict or cooperation in shared water systems.

Concern about water privatization and civil unrest, which can also lead to conflict, has increased with the acquisition of national water companies by multinational corporations. The commodification of water has raised questions about poverty alleviation, water markets' effect on local economies and the search for a water democracy (see **Chapter 12**). Human rights issues, visionary declarations and the centrality of water equity all raise questions of fairness, distributive justice and the responsibilities of international communities vis-à-vis water sharing.[14]

Finally, the role of virtual water needs further attention as well. A broad indicator – a water footprint – links virtual water and world trade: via the sum of domestic water use and virtual water, we can then consider how

13. Carrying capacity is the measure of an environment's ability to sustain itself and its ecosystem.

14. *Ethics and Water*, a series of short volumes compiled by UNESCO, raises fundamental questions about our ethical and moral duties in sharing non-renewable groundwater resources and creating cooperative mechanisms. Available online at webworld.unesco.org/ihp_db/ publications/GenericList_ themes.asp. See also Cosgrove, 2000.

Map 11.2: National water footprints around the world, 2004

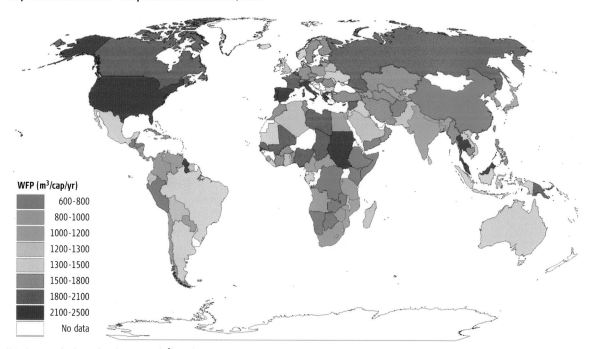

WFP (m³/cap/yr)

- 600-800
- 800-1000
- 1000-1200
- 1200-1300
- 1300-1500
- 1500-1800
- 1800-2100
- 2100-2500
- No data

Note: Average national water footprint per capita (m³/capita/yr). Green means that the nation's water footprint is equal to or smaller than global average. Countries with red have a water footprint beyond the global average.

Source: Chapagain and Hoekstra, 2004.

water used for the production of export commodities on the global market can contribute significantly to the changes in local and regional water systems (**Box 11.6**). It has been noted, for example, that since Japan consumes large quantities of American cereals and soybeans, it might be suggested that this in turn leads to the mining of aquifers (Ogallala, for example) and further water use of rivers in North America. **Map 11.2** shows national water footprints around the world. The concept

Below: Iguazu Falls, Brazil

of virtual water was first defined by Allan (2003) as the 'water embedded in commodities'. In terms of global trade, not only does it raise awareness about water interdependencies, but it can also serve as a means for improving water efficiency (see **Map 11.3** on water savings around the globe and **Map 11.4** on net virtual water imports). In addition, it can be an indicator of sharing water, as well as a sign of contributing to water security in water-poor regions.

BOX 11.6: VIRTUAL WATER AND THE WATER FOOTPRINT

International virtual water flows

The International trade of commodities implies flows of virtual water over large distances, where virtual water should be understood as the volume of water required to produce a commodity. Virtual water flows between nations can be estimated from statistics on international product trade and estimates of the virtual water content of products. The global volume of virtual water flows related to the international trade in commodities is 1.6 trillion m^3/yr. About 80 percent of these virtual water flows relate to the trade in agricultural products, while the remainder is related to industrial product trade. An estimated 16 percent of global water use is not for producing domestically consumed products, but rather products for export. With the increasing globalization of trade, global water inter-dependencies and overseas externalities are likely to increase. At the same time, the liberalization

of trade creates opportunities to increase global water use efficiency (see **Chapter 12**).

Globally, water is saved if agricultural products are traded from regions with high water productivity to those with low water productivity. At present, if importing countries produced all imported agricultural products domestically, they would require 1.6 trillion m^3 of water per year; however, the products are being produced with only 1.2 trillion m^3/yr in the exporting countries, saving global water resources by 352 billion m^3/yr.

The water footprint

The water footprint shows the extent and locations of water use in relation to consumption. The water footprint of a country is defined as the volume of water needed for the production of the goods and services consumed by the inhabitants of the country. The internal water footprint is the volume

of water used from domestic water resources, whereas the external water footprint is the water used in other countries. Water footprints of individuals or nations can be estimated by multiplying the volumes of goods consumed by their respective water requirement. The US appears to have an average water footprint of 2,480 cubic metres per capita per year (m^3/cap/yr), while China has an average footprint of 700 m^3/cap/yr. The global average water footprint is 1,240 m^3/cap/yr. The four major factors that determine the water footprint of a country are volume of consumption (related to the gross national income); consumption patterns (e.g. high versus low meat consumption); climate (growth conditions); and agricultural practice (water use efficiency).

Sources: Chapagain and Hoekstra, 2004; Chapagain, et al., 2005.

Map 11.3: Water savings around the world

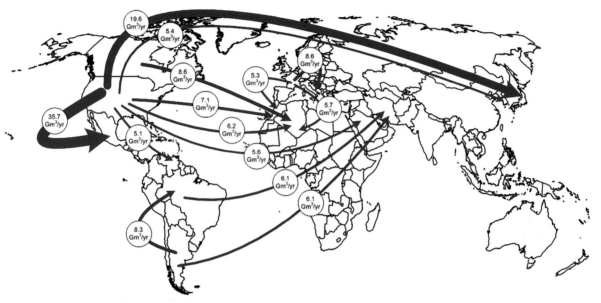

Global water saving = 352 x 10^9 m^3/yr

Note: Global water savings (>5.0 Gm^3/yr) associated with international trade of agricultural products. Period 1997-2001. The arrows represent the trade flows. The numbers show the global water savings, calculated as the trade volume (ton/yr) multiplied by the difference between water requirement (m^3/ton) in the importing country and water requirement (m^3/ton) in the exporting country. Global water savings occur if an exporting country requires less water per ton of product than an importing country.

Source: Chapagain, et al., 2005.

Map 11.4: Net virtual water imports around the world

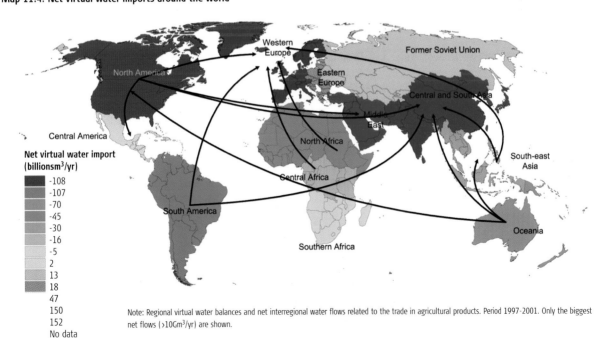

Note: Regional virtual water balances and net interregional water flows related to the trade in agricultural products. Period 1997-2001. Only the biggest net flows (>10Gm^3/yr) are shown.

Source: Chapagain and Hoekstra, 2004.

*...water
transcends
national
boundaries and
must be
managed
cooperatively
and equitably,
using the best
science
available*

Part 4. Conclusions and Recommendations: The Dynamics of Cooperation

Benefits to sharing can be identified at any level of cooperation; from initial description or analysis, through series of negotiations and, finally, to concrete, implementable steps. Benefits and costs can be calculated under alternative management, development scenarios and institutional analysis. Capacity-building can delineate the best form of cooperative agreements. At the same time, no supranational agency has the mandate to handle transboundary water disputes, even though third-party mediation boards have helped to bring contending parties together.[15]

A useful way of articulating the dynamics of cooperation is to summarize them under a continuum:

■ *Coordination* relates to sharing of information, communication, as well as some preliminary regional assessments.

■ *Cooperation* elevates the level of contact through joint projects, exchange of scholars and researchers, active planning, adaptation of national plans to capture regional costs and benefits, contingency alternatives or joint water flow forecasts.

■ *Collaboration* implies formalized agreements, continuous interaction, integrated river basin management, joint institutions, river basin commissions, permanent secretariat and staff and other forms of joint management.

One can see the basis for indicators and indices that could measure efficiency, effectiveness and equity in evaluating and monitoring performance, processes and results of comprehensive water sharing efforts. At the same time, we also need to recognize levels of water sharing ranging from *macro-level* or transboundary river basin management (overall water allocation), to *meso-level* or country river basin (emphasizing water use allocation) and *micro-level*, which refers to multi-objective, multi-purpose and multi-stakeholder water activities, such as irrigation districts and water users' associations.

Macro-scale cooperation aims to identify water issues that act as barriers to implementing national water efforts, such as Millennium Development Goals (MDGs) that fall within the capabilities of particular nations. At the national or meso-scale, the appropriate emphasis centres around the achievement of water use priorities

expressed in National Water Policies, such as access to urban water, sanitation and agricultural development. Local and focused activities on a smaller scale as well as specific projects and programmes are found at the micro-level.

For transboundary waters, there are two key interrelated water sharing issues:

■ how to induce affected parties to discuss cooperation on joint water management where no prior agreement exists (in terms of identifying common interests, initiatives, future collaboration, and implementable action)

■ for those countries that have agreements, treaties and established coordination mechanisms, deciding how to hold the parties accountable for implementation, which enforcement mechanisms and sanctions exist and how existing agreements can be strengthened and modified.

Again for transboundary cases, we can summarize coordination, cooperation and collaboration dynamics as part of a three-step process. The first step is to outline incentives and accepted cooperative mechanisms for water sharing at the international level. The second step is to reinforce the mechanisms by referring to different paradigms of equity and fairness that run through the variety of declarations, treaties, conventions and frameworks. The third step is to develop success indicators that reflect the results of implementation.

This chapter has continued to expand the discussion on the increasing complexity, volatility and vulnerability of water resources in a fast-changing socio-economic and environmental context. A general framework for sharing water resources tends to outline sources of tension, competing demands and the mismatch between political

---

15. Between 2003-06 a number of institutes and centres were established, all contributing to facilitating the management of shared water resources: the International Centre for Water Hazard and Risk Management (ICHARM) in Tsukuba, Japan; the International IHP-HELP Centre for Water Law, Policy and Science in Dundee, United Kingdom; the Regional Water Centre for Arid and Semi-arid Zones of Latin America and the Caribbean (CAZALAC) in La serena, Chile; the European Regional Centre for Echohydrology in Lodz, Poland (all at www.unesco.org/water/ihp/partners.shtml); UNESCO-IHE Institute for Water Education (www.unesco-ihe.org); Regional Centre on Urban Water Management (RCUWM) in Teheran, Iran (www.rcuwm.org.ir/).

boundaries, natural river basins and aquifer systems. At the same time, it is also important to note the positive role of cooperative efforts brought about by recent legislative and institutional developments, as well as recognition of the need for capacity-building, governance, and equitable and sustainable sharing of water. Also notable are the widespread international agreements towards peaceful settlement of differences and the concern about the consequences of present water use trends and developments.

Reflecting on recent developments since 2003, one can see noteworthy progress in water sharing. To start with, the case studies series, published by UNESCO's PCCP programme, new programmes and centres, and a series of volumes on water security have raised awareness about potential conflicts as well as the crucial role of cooperation. Similarly, there has been a steady stream of major meetings on IWRM, integrated river basin management, the application of WFD and other global water initiatives on comprehensive approaches, public participation and the sustainable management of water. Research centres have increased their efforts in extensive data collection, modelling, indicator development and practical applications of findings. UNESCO's Water Cooperation Facility (WCF) initiative, the World Water Council and other partners, for example, will turn attention to cooperation in managing transboundary waters. Visionary declarations in various international conferences, coupled with detailed guidelines for participatory assessment methods and multi-stakeholder manuals and additional water sharing agreements, illustrate the spirit and practice of the past several years. Varady and Iles-Shih (2005) point out that such initiatives have proliferated largely because of the belief that water transcends national boundaries and must be managed cooperatively and equitably, using the best science available.

An increasing number of examples of successful cooperation illustrate yet again that interdependence and changing socio-political conditions do not only produce conflicts, but also lead to collaboration. A body of formal guidelines and lessons learned from expanding sharing efforts – both nationally and internationally – are now available.

It is important to reiterate the distinction in incentives between coordination, cooperation, and collaboration. Indeed, the challenge for cooperation and the risk for

conflict raise two overarching issues: namely, how to induce the affected parties to discuss cooperation on joint water management when no prior agreements exist and how, when such treaties do exist, to hold the parties accountable for implementation, promote enforcement mechanisms and continuously monitor both performance and acceptable outcomes.

The great difficulty in measuring water systems performance, levels of significance, critical thresholds and comparability of data and measurement over time is repeated in all recent literature. There are emerging agreements about specific indicators, but they tend to be evasive and difficult to pin down when there is an attempt to describe socio-political and institutional dimensions. Improving techniques in ADR and public participation complement more reasonable and acceptable solutions to water challenges. Political will and commitment are important preconditions for successful water sharing. This also implies the coordination of water resources with other natural resources, especially land-use controls and comprehensive planning and management.

Political will and commitment are important preconditions for successful water sharing

Feluccas on the Nile river near Philae, Egypt

We can hope that shared water management becomes a realistic and thoughtful instrument for a balanced approach...

A review of recent literature and ongoing practices reveals the following lessons learned:

■ Cooperation, not conflict, is the norm in inter-state water relations in an increasing number of water courses.

■ Treaties, agreements and the principles of international water law help crystallize mechanisms for conflict management and dispute resolution.

■ Strengthening institutional mechanisms and legal frameworks for IWRM is needed to ensure that associated transaction costs can be covered.

■ More emphasis should be placed on building capacity in terms of IWRM and conflict prevention. Involving excluded or weaker groups, especially at the local level, may prevent them from developing grievances.

■ Adopting alternative dispute resolution mechanisms and confidence-increasing measures among affected parties is necessary.

■ Data, information and comparative indicators can provide an arena of focused disagreement and help concentrate the debate on concrete points of contention, as is the case for Regional Water Data Banks Project and the recent efforts of UNEP's Global Environment Monitoring System.

■ New paradigms of complexity and interdependence need not imply only optimal solutions and complicated models. Reasonable approximations are parts of necessary trade-offs, as are efforts to balance the relationship between ideal futures and real, or pragmatic, considerations of unfolding changes and practical solutions.

From security studies to management schemes and from administrative guidelines to conventions and bilateral agreements, a recurrent theme is the importance of the role of law and informal ties in reducing conflict and increasing cooperation. This new emphasis favours the development of contingency preparedness for continuous changes, building resilience into vulnerable systems, the ability to cope realistically with the challenges of upstream/downstream interdependencies, long-term planning and diversity and flexibility in thinking and practice. Such a combination of management would be based on regional cooperation principles, focusing on river basins and aquifer systems, with an emphasis on ecosystemic and social needs. It would focus on functionally interrelated natural resources problems, reduce potential points of friction and stress in advance and eliminate conflicting demands through risk management and vulnerability assessment. We can, then, hope that shared water management becomes a realistic and thoughtful instrument for a balanced approach and a useful tool for managing long-standing competitions, confrontations and potential outright water conflicts.

Section 4: MANAGEMENT RESPONSES & STEWARDSHIP

References and Websites

Allan, J. A. 2003. Virtual Water – the Water, Food, and Trade Nexus: Useful Concept or Misleading Metaphor? *Water International*, Vol. 28, No. 1.

Appelgren, B. (ed.). 2004. *ISARM Africa: Managing Shared Aquifer Resources in Africa*. Proceedings of the International Workshop, Tripoli, Libya, 2–4 June. IHP *Groundwater Series* No. 8. Paris, UNESCO-IHP.

Bayarsaihan, T. and McKinney, D. 2002. Past Experience and Future Challenges: Cooperation in Shared Water Resources in Central Asia. Paper presented at Asian Development Bank Workshop in Almaty, Kazakhstan, 26–28 September 2002. ADB, *Water for All Series 12* online at www.adb.org/Documents/Books/Water_for_all_Series/Past_And_Future/default.asp

Brauch, H. G., Liotta, P. H. Marquina, A. Rogers, P. F. and El-Sayed Selim, M. (eds). 2003. *Security and Environment in the Mediterranean: Conceptualizing Security and Environmental Conflicts*. Berlin, Springer.

Bruch, C.E. 2003. New Tools for Governing International Watercourses. *Global Environmental Change*, Vol. 14. pp. 15–23.

Burchi, S. and Kerstin, M., 2005. *Groundwater in International Law: Compilation of Treaties and Other Legal Instruments*. Rome, FAO/UNESCO.

Carius, A., Dabelco, G. D. and Wolf, A. T. 2004. Water, Conflict, and Cooperation. *ECSP Report, Issue 10*, pp. 60–66.

Chapagain, A. K. and Hoekstra, A. Y. 2004. *Water Footprints of Nations, Volume 1: Main Report*. Value of Water Research Report Series No. 16., Delft, UNESCO-IHE.

Chapagain, A. K., Hoekstra, A. Y. and Savenije. 2005. *"Saving water through global trade"*, Value of Water Research Report Series No. 17, UNESCO-IHE, Delft, The Netherlands

Clarke, R. and J. King. 2004. *The Atlas of Water*. London, Earthscan Publications Ltd.

Cosgrove William J. (ed.). 2003. UNESCO Technical Document in Hydrology, PCCP series, Vol. 29.

Cosgrove William J. and Rijsberman Frank R. 2000. *World Water Vision: Making Water Everybody's Business*. London, Earthscan.

Creighton, J. L. 2004. Designing Effective Public Participation Programs: A U.S. Perspective: A Water Forum Contribution. *Water International*, Vol. 29, No. 3, pp. 384–91.

EC (European Commission). 2005. European Decalartion for a New Water Culture, Madrid 18 February 2005. www.unizar.es/fnca/euwater/index2.php?x=3&idioma=en

——. 2000. Directive 2000/60/EC of the European Parliament and of the Council of 23 October 2000 establishing a framework for Community action in the field of water policy. *Official Journal of the European Communities*, Brussels, EC.

Falkenmark, M. 1999. Forward to the future: A conceptual framework for water dependence. *Ambio* Vol. 28, No. 4, pp. 356–61.

Gleditsch N. P. (ed.). 1997. *Conflict and the Environment*. Dordrecht, Kluwer Academic.

Gleick, P., Cain, N., Haasz, D., Henges-Jeck, C., Hunt, C., Kiparsky, M., Moench, M., Palaniappan, M., Srinivasan, V., Wolff, G. 2005. *The World's Water 2004–2005: The Biennial Report on Freshwater Resources*. Washington DC, Island Press.

Guerquin, F., Ahmed, T., Hua, M., Ozbilen, V. and Schuttelar. 2003. *World Water Actions : Making Water Flow for All*. World Water Council, Japan Water Resources Association, UNESCO.

Hardin, G. 1968. The tragedy of the commons, *Science*, Vol. 162, No. 3859, pp. 1243–48.

ILA (International Law Association). 2004. Water Resources Law. Paper presented at the Berlin Conference, 4–21 August 2004.

International Water Academy. 2002 Building a Curriculum for Training in Water Conflict Resolution, Prevention, and Mediation. www.thewateracademy.org/OppActivities/index_main.html Millennium Ecosystem Assessment. 2005. *Ecosystems and Human Well-Being*. Washington DC, Island Press and World Resources Institute.

Ohlsson, L. 1999. *Environment, Scarcity, and Conflict – A Study of Malthusian Concerns*. Department of Peace and Development Research, Gothenburg University.

Rogers, P. and Kordab, I. 2004. 'Conflict Resolution in Water Resources Management: Ronald Coase meets Vilfredo Pareto.' Presented at Symposium on Challenges Facing Water Resources Management in Arid and Semi-Arid Regions, American University of Beirut, 7–9 October, 2004.

Sadoff, C.W. and D. Grey. 2002. 'Beyond the River: the benefits of cooperation on international rivers' *Water Policy* 4 (2002) pp. 389–403.

Shamir Yona 2003 UNESCO Technical Document in Hydrology, PCCP series, 2003, Vol. 7.

World Water Council. 2003. *The 3rd World Water Forum Final Report*. Marseille, World Water Council.

UN (United Nations). 2005. *Third Report on Shared Natural Resources*. UN Doc A/CN.4/551 New York, UN.

——. 2004. *Second Report on Shared Natural Resources: Transboundary Groundwaters*. UN Doc. A/CN.4/539, New York, UN.

——. 2003. *First Report on Outlines*. UN Doc. A/CN.4/533, New York, UN.

——. 1977. Convention on the Law of the Non-navigational Uses of International Watercourses. UN Doc. A/51/869, New York, UN.

UNECE (United Nations Economic Commission for Europe). 2000. Guidelines on Monitoring and Assessment of Transboundary Groundwaters. Lelystad, UNECE Task Force on Monitoring and Assessment, under the Convention of the Protection and Use of Transboundary Watercourses and International Lakes (Helsinki 1992).

UNESCO/ISARM (United Nations Educational, Scientific and Cultural Organization/International Shared Aquifer Resource Management). 2001. *International Shared Aquifer Resource Management: A Framework Document*. IHP Groundwater Series No. 1. Paris, UNESCO.

Varady, Robert G. and Matthew Iles-Shih (forthcoming) 'Global Water Initiatives: What do the Experts Think?' in A.K.Biswas (ed) Impacts of Mega-Conferences on Global Water Development and Management. Springer Verlag.

Wolf, A, Yoffe, S. and Giordano, M. 2003. International waters: Identifying basins at risk. *UNESCO Technical Document in Hydrology, PCCP Series*, Vol. 20.

European Union water policy: www.europa.eu.int/comm/environment/water/

Global Water Partnership (GWP): www.gwpforum.org/servlet/PSP

Integrated River Basin Management (IRBM): www.panda.org/about_wwf/what_we_do/freshwater/our_solutions/rivers/irbm/index.cfm

International Network of Basin Organizations (INBO): www.riob.org/

International Rivers Network (IRN): www.irn.org

International Water Management Institute (IWMI): www.iwmi.cgiar.org

River Basin Initiative (RBI): www.riverbasin.org/ev_en.php

Stockholm International Water Institute (SIWI): www.siwi.org

United Nations Environmental Programme (UNEP): www.unep.org/themes/freshwater

United Nations Educational, Scientific and Cultural Organization (UNESCO) water portal: www.unesco.org/water

World Bank: www.worldbank.org/water.htm

World Water Council (WWC): www.worldwatercouncil.org

Organization of American States (OAS)/UNESCO: www.oas.org/usde/isarm/ISARM_index.htm

INWEB: www.inweb.gr/workshops/documents/groundwater_final_report.html

European Declaration for a New Water Culture: www.unizar.es/fnca/euwater/index2.php?x=3&idioma=en

Nowadays people know the price of everything and the value of nothing.

Oscar Wilde

CHAPTER 12

Valuing and Charging for Water

By

UNDESA
*(UN Department of
Economic and Social
Affairs)*

Key messages:

Because of the unique characteristics and socio-cultural importance of water, attempts to value water, or more specifically water services, in monetary terms is both difficult and, to some people, inappropriate. Nevertheless, economic valuation – the process of attaching a monetary metric to water services – is an increasingly important tool for policy-makers and planners faced with difficult decisions regarding the allocation and development of freshwater resources. With market prices unable to capture the full spectrum of costs and benefits associated with water services, economists have developed special techniques to estimate water's non-market values. Two important occasions when these tools are employed are assessments of alternative government strategies and tariff-setting. In this chapter, we examine valuation tools, explain how they are used, and explore underlying social, economic and environmental principles that condition their application. Finally we look at the emerging issues including private-sector participation, 'virtual water' trade, and payments for environmental services, which are playing an ever more prominent role in the debate on the allocation and development of scarce freshwater resources.

Top to bottom:

*Men and women bathing in the Ganges, India
above: Irrigated paddies, Viet Nam*

Women washing clothes at the Ralwala spring in Siaya district, Kenya. The spring serves 30 households

*Public water pump in Amboseli Reserve, Kenya
below: Public water pump in Amboseli Reserve, Kenyaenya*

■ Given its unique, life-sustaining properties and multiple roles, water embodies a bundle of social, cultural, environmental and economic values. All of these must be taken into consideration in the selection of water-related policies or programmes if the goals of integrated water resources management (IWRM) are to be realized.

■ Public policy analysis employing economic valuation provide a rational and systematic means of assessing and weighing the outcomes of different water policies options and initiatives and can assist stakeholders, planners and policy-makers to understand the trade-offs associated with different governance options.

■ Charging for water services – household, commercial, industrial and agricultural – requires, firstly, consensus on the underlying principles and objectives (e.g. full cost recovery, protecting the needs of the poor and the marginalized, etc.); secondly, a thorough, systematic analysis of all costs and perceived benefits; and thirdly, a tariff structure that endeavours to maximize governance objectives within prevailing socio-economic conditions.

■ Public-private partnerships, though not appropriate to all situations, can play a significant role in developing cost-efficient water service systems. Government authorities, however, must be open to a variety of initiatives, including local enterprise, public-private partnerships, community participation and water markets, and must take an active regulatory role in ensuring that societal goals are met with regard to social equity and environmental sustainability – as well as economic efficiency.

■ There is a great need not only for planners and policy-makers who understand the advantages and limitations of economic valuation techniques and their potential role in informing decisions regarding water resources management, but also for technicians who can clearly express these concepts, utilize these tools, and assist stakeholders in expressing their values and preferences. In this way, economic valuation can contribute more fully to information sharing and transparency, all of which are important for good governance.

Part 1. Understanding the Multi-faceted Value of Water

Water is vital for all life on this planet, but is also essential for food production, many manufacturing processes, hydropower generation, and the service sector. The value of water varies for different users depending on the ability to pay, the use to which the water will be put, access to alternative supplies, and the variety of social, cultural and environmental values associated with the resource. Acknowledging the totality and interdependence of water-related values important to stakeholders and water users is critical to realizing Integrated Water Resources Management (IWRM). Understanding the distinction between the *value* of water – determined by its socio-cultural significance and the broad spectrum of direct and indirect benefits it provides – the *price* of water, as charged to consumers, and the *cost* of water as derived from the expense of providing water to consumers, is a critical first step to understanding the role of economic valuation in water governance and management.

The Dublin principle[1] to treat water as an economic good follows a growing consensus on the need to maximize benefits across a range of water uses. Still, the importance of ensuring equitable access and meeting the needs of the poor and disadvantaged members of society is widely recognized. How to finance this task remains a key challenge. While higher-income countries move toward systems of water tariffs based on full cost recovery and metered service, low-income countries struggle to cover basic operating costs and, for the most part, still tolerate various systems of subsidies as many users are unwilling or unable to pay for water services. According to the World Bank, pro-poor policies relying on cross-subsidization have created an inefficient and unsustainable water services sector with serious impacts on the environment in many countries. Similarly, the 'polluter pays' principle, like the 'user pays' principle, although broadly accepted, suffers from poor enforcement due to a weak governance environment. While some countries favour decentralization and management transfer as a way to relieve cost burdens, others see private sector participation as a means of achieving better services and improved cost recovery. Pricing or tariff-setting is widely supported in the financial community both to raise the needed investment capital and to curb inefficient use. None of these options are without problems.

Difficulties associated with decentralization often stem from political weakness and lack of institutional capacity at the local level. Half-hearted support by national and international organizations for community-driven development of water services has also been a problem in some areas. Private sector involvement, often touted

as a key to solving financial problems in this sector, remains limited in many areas while the transfer of management models from one region to another has met with mixed results. Pricing, expected to serve a variety of objectives, including cost recovery, more prudent use of water, distributive justice and assured supplies for poor, has generally led to rising prices and a decline in water use in some countries. Many would argue that the poor would be better served by more focused tariff systems, which would be gradually introduced and underpinned by a minimal level of free service, or complimentary vouchers for water service rather than cross-subsidization.

Although economic valuation is recognized by many as an important tool in water management and substantial efforts have already been made in clarifying concepts associated with this technique, valuing water remains a controversial issue. Many stakeholders still feel that economic valuation is incapable of fully capturing the many social, cultural and environmental values of water. However, the variety of innovative initiatives attempted worldwide illustrates an increasing sensitivity to local needs and a growing understanding that the development and management of water resources must be a shared responsibility.

Although economic valuation is recognized by many as an important tool in water management... valuing water remains a controversial issue

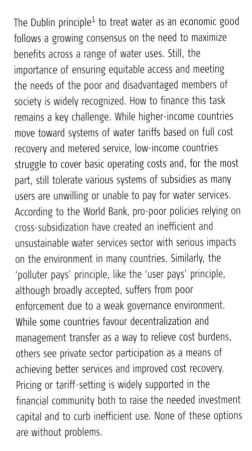

---

1. See Chapter 1 for definition.

Part 2. The Socio-cultural Context

We speak of a crisis in water management because in many places the available freshwater is insufficient to meet all demands. As discussed in previous chapters, demand for water is increasing because, despite falling fertility rates in many countries, the world's population continues to increase while freshwater water supplies remain constant. Meanwhile, economic growth in many countries, especially in India and China, has increased disposable income and instigated lifestyle changes that have often meant increased water consumption. Agriculture and industry, also growing in response to these changes, require water for production, processing and transport, while electric utilities look increasingly to hydropower to meet 'clean' energy demands. Urbanization, a seemingly unstoppable trend worldwide, intensifies the pressure.

As the competition for water resources accelerates, one becomes increasingly aware of water uses in different areas. The essential value of water is universally recognized: without water there is no life. For many

ecosystems, such as rivers, lakes and wetlands, freshwater is the defining element. The utility of water as a raw material, as a solvent, and as a source of kinetic energy has long been recognized. The role of water in

---

BOX 12.1 THE HIGH COST OF BOTTLED WATER

Over the last decade, sales of bottled water have increased dramatically to become what is estimated to be a US $100 billion industry (Gleick et al., 2004). From 1999 to 2004, global bottled water consumption grew from approximately 26 billion gallons to over 40 billion gallons (IBWA, 2005). In several cities of the developing world, demand for bottled water often stems from the fact that the municipal water supplies – if available at all – fail to meet basic criteria for drinking water quality. But companies manufacturing bottled water are also generating large revenues in developed countries. Bottled water sales in the United States in 2004 – higher than in any other country – totalled over US $9 billion for 6.8 billion gallons of water, that is, enough water to meet the annual physiological needs of a population the size of Cambodia (IBWA, 2005). Countries in the top ten list of bottled water consumers include Mexico, China, Brazil, Italy, Germany, France, Indonesia, Spain and India.

When asked why they are willing to pay so much for bottled water when they have access to tap water, consumers often list concerns about the safety of tap water as a major reason for preferring bottled water (NRDC, 1999). While

most companies market this product on the basis that it is safer than tap water, various studies indicate that bottled water regulations are in fact inadequate to ensure purity or safety. The World Health Organization (WHO, 2000) warns that bottled water can actually have a greater bacterial count than municipal water. In many countries, the manufacturers themselves are responsible for product sampling and safety testing. In the United States, for example, the standards by which bottled water is graded (regulated by the Food and Drug Administration) are actually lower than those for tap water (regulated by the Environmental Protection Agency) (Gleick et al., 2004).

The explosive increase in bottled water sales raises important questions related not only to health, but also to the social and environmental implications of the phenomenon. It remains to be seen, for example, how the growth of this industry will affect the extension and upkeep of municipal water services upon which the poor depend. In fact, those most likely to need alternative, clean water sources are also those least likely to be able to afford the high cost of bottled water. In China, where roughly 70 percent of rivers and lakes are polluted, the

largest demand for bottled water comes from city dwellers, for in rural areas people are too poor to pay for this alternative (Yardley, 2005).

Most water bottles are meant to be recyclable. However, only 20 percent of polyethylene terephthalate (PET), the substance used for water bottles, is actually recycled (Gleick et al., 2004). In Greece, it is estimated that 1 billion plastic drinking water bottles are thrown away each year (BBC, 2005). In addition, the PET manufacturing process releases harmful chemical emissions that compromise air quality.

Where safe tap water is temporarily unavailable, bottled water can provide an effective short-term solution for meeting a population's needs. But as noted above, the massive growth in sales of bottled water worldwide comes at a cost. A better appreciation of how people value water may help us understand how the bottled water phenomenon is impacting society's health, economic and environmental goals.

Sources: BBC, 2005; Gleick et al., 2004; IBWA, 2005; NRDC, 1999; WHO, 2000; Yardley, 2005.

BOX 12.2: VALUING WATER, VALUING WOMEN

In most, if not all, developing countries, collecting water for the family is women's work. While water for drinking and cooking must be carried home, dishes, clothes and often children may be carried to the water source for washing. Women and girls are often seen queuing with their water pots at all-too-scarce taps, then walking long distances home balancing them on their heads or hips. If the water is contaminated and a family member falls ill, it is often the woman who must care for them. Children in particular are vulnerable. In parts of the world where AIDS is rampant, individuals with weakened immune systems also easily fall prey to pathogens in the water supply.

The low status of women in many societies means that their contribution – in terms of the time and energy spent, for example, in fetching water – is considered to be of little value. In economic terms, the opportunity cost[2] of their labour is perceived as near zero. Where

women have been given access to education and to money-making work, such as handicrafts production, and are permitted to sell their products and to earn income for the family, their social as well as financial position improved dramatically.

Attitudes toward water-collecting can also shift. The time women spend collecting water, especially when simple and readily available technological alternatives exist, looks very different and far more costly to the family and

society as a whole, when women have income-earning opportunities. Thus, investments made to improve access to safe drinking water are both a reflection of the value placed on water for human well-being and the value accorded women. Providing regular and dependable access to safe drinking water is one way of improving the position of women as well as society as a whole.

Source: UNICEF/WHO, 2004.

human health is, of course, critical. Recently we have seen the growth in bottled water consumption, which although a necessity in some cases, is also a growing trend in places where safe and inexpensive water is readily available on tap (see **Box 12.1**).

As a physical, emotional and cultural life-giving element, water must be considered as more than just an economic resource. Sharing water is an ethical imperative as well as an expression of human identity and solidarity (see **Chapter 11**). Accordingly, the high value placed on water can be found in the cosmologies and religions and the tangible and intangible heritage of the world's various cultures. The unique place water holds in human life has ensured it an elevated social and cultural position, as witnessed by the key role water plays in the rituals of all major faiths. The proposition that water is a human right alongside the increasing competition between water users has resulted in water becoming a political issue in many regions (see **Chapter 2**). The amount of time spent in collecting water – a task mainly performed by women and children – is increasing in many areas. Water supply

must, therefore, also be viewed as a social issue and, more specifically, a gender issue (see **Box 12.2**).

Restored interest in ethnic and cultural heritage in many societies around the world has lead to a revival of numerous traditional rituals, festivals and social customs, many of which feature water as a key element. Thus, the tradition of social bathing endures, for example, in Turkey and Japan. Water sports too play an important role; currently nearly one-third of Olympic sports use water, snow or ice. Many archaeological sites – the Roman aqueducts, the Angkor ruins, the Ifugao and Inca terraces, among others – are monuments to ancient societies' ingenuity in water engineering. Listing these historic sites on the roster of World Heritage protected cultural properties is in effect formal recognition of the high value that the international community accords these locations (see **Map 12.1**).

Water splashing at a festival in the Dai ethnic Minority Village, China

2. Opportunity cost is defined as the maximum worth of a good or input among possible alternative uses (OMB, 1992).

Map 12.1: World Heritage Sites with important water-related value

1. Venice and its Lagoon, Italy
2. The Old Bridge of Mostar, Bosnia and Herzegovina
3. Rice Terraces of the Cordilleras, the Philippines
4. Angkor, Cambodia
5. Nubian Monuments, Egypt
6. Heritage Medal for City of Potosi, Bolivia
7. Xochimilco and Mexico City, Mexico

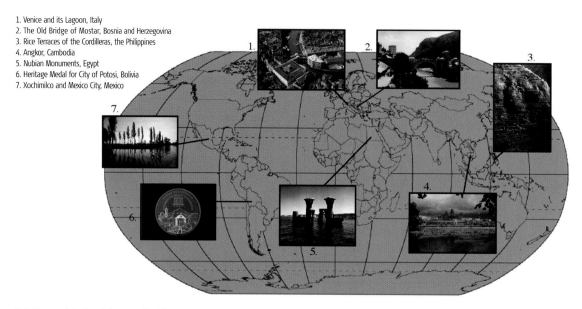

Note: These are just a few of the water-related sites on the World Heritage Cultural Properties list. The list contains more than 628 cultural sites that have been cited as having 'outstanding universal value to humanity'. With the 1972 Convention concerning the Protection of the World Cultural and Natural Heritage, UNESCO seeks to encourage the identification and protection of cultural and natural heritage around the world so that it may be enjoyed by all peoples. More information on this programme and these and many other sites can be found at whc.unesco.org.

Map 12.2: Ramsar sites with important water-related value

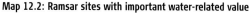

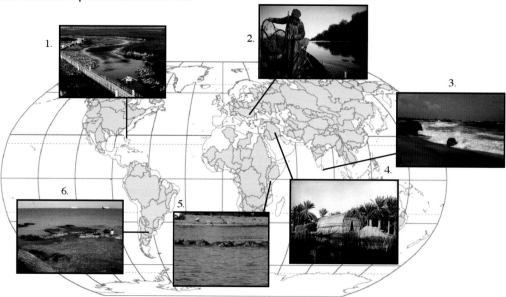

1. Everglades National Park, US
2. Danube Delta, Romania
3. Bundala, Sri Lanka
4. Marshlands, Iraq
5. Lake Naivasha, Kenya
6. Rio del Plata, Uruguay

Note: The Convention on Wetlands, signed in Ramsar, Iran, in 1971, is an intergovernmental treaty providing the framework for national action and international cooperation for the conservation and wise use of wetlands and their resources. There are presently 146 Contracting Parties to the Convention, with 1,459 wetland sites, totalling 125.4 million hectares designated for inclusion in the Ramsar List of Wetlands of International Importance.

Source: www.ramsar.org/key_sitelist.htm

A growing appreciation of ecological processes is developing in many countries as society's experience of pollution and other environmental disasters increases. The wholesale destruction of the natural environment, as in the case of the Aral Sea, is no longer socially or politically acceptable. Increasingly, society has come to value healthy ecosystems and accordingly has instituted legally-binding accords to better protect the environment. The Ramsar Convention, setting aside more than 1,400 wetland sites around the world for preservation and protection (see **Map 12.2**), is a testimony to the international recognition of the environmental, social and economic importance, and correspondingly the value, accorded to these special ecosystems.

The interaction between people and their environment is embedded in their culture. The ways in which water is conceived and valued, allocated and managed, used and abused, worshipped or desecrated, are influenced by the cultural and environmental context. Cultural diversity – the result of millennia of human development – includes a treasure trove of practical approaches to water husbandry. Indigenous knowledge holders can be invaluable partners for scientists in the quest for solutions to the challenges facing planners and practitioners pursuing IWRM (see **Chapter 13**). Given their fundamental role in human life and society, water-related traditions and practices are deep-rooted and thus changing them could have significant social impact. The strong cultural and social dimensions of water management and use must be understood in all their diversity if we are to find sustainable solutions to global water problems. Valuing water, including sustaining and fostering water-related cultural diversity, heritage and knowledge, is critical to enhancing our ability to adapt in a changing world. Economic valuation of water resources must be recognized as existing within this larger and more complex context of valuing water.

Angkor, Cambodia

Cultural diversity... includes a treasure trove of practical approaches to water husbandry

Part 3. The Role of Valuation

In formal models of policy analysis, valuation is the process of assessing significance against the projected results of proposed strategies. Values, in this sense, are weights assigned to outcomes of specific policy proposals, and can vary depending on which services are being valued, on the location of the services and other circumstances. Valuation assesses situations *with* versus situations *without* governance intervention, in other words, *incremental* gains and losses.

Essentially a tool for policy-makers and planners, public policy evaluation (including benefit-cost analysis), is a process of 'reasoned' decision-making and involves several logical, sequential steps. First, identify the relevant social goals. Second, characterize the perceived problems (defined as discrepancies between goals and the current situation). The third step is to specify alternative courses of action (governance strategies) to achieve the agreed goals. Fourth, predict and evaluate the predicted outcomes, both costs and benefits. The fifth and final step is to choose and implement the strategy that seems best able to achieve society's goals (MacRae and Whittington, 1997).

Economic valuation thus refers to the application of special techniques to determine the economic value (demand or willingness to pay)[3] of water services for purposes of informing policy decisions regarding the management and allocation of water resources. In the context of selecting governance strategies, valuation is seldom applied to water itself, but rather to the consequences of proposed policy initiatives. The beneficial and adverse effects of proposed initiatives (or the benefits and costs) are identified, and values (monetary assessments) assigned to these effects. In the context of public water supply policy, charging, for instance, is a governance strategy, and the choice of how much of the cost of water supply to recover, and from whom, is an important policy issue. Non-specialists sometimes incorrectly equate the observed price, or charge to the user, with economic value. Although tariff-setting must reflect both cost and value considerations, it

3. Willingness to pay is the maximum amount an individual would be willing to pay, or give up, in order to secure a change in the provision of a good or service (OMB, 1992).

must be remembered that the upper level of charges is limited by willingness to pay, rather than being defined by it. Under-valuing, or under-pricing ecological resources not only affects resource allocation, but can bias the direction of technological development (Dasgupta and Mäler, 2004).

3a. The special case of water

A variety of physical, social, cultural, political and economic factors make water a special case with regard to economic valuation. This in turn presents special challenges in selecting appropriate governance mechanisms. Because of its variable physical nature – it flows, seeps and evaporates, but sometimes is a stationary solid – it is difficult or costly to establish and enforce property rights over water. Another critical aspect is its variability, and increasing uncertainty, in terms of space, time and quality. Because of the many environmental services provided by water and aquatic ecosystems, water is considered a 'public good' – the two defining characteristics being 'non-rivalry in consumption' and 'non-excludability' of non-paying users. However, as shown in **Table 12.2**, it also has many private uses. The general perception of water as 'different' has political as well as economic consequences, all tending to dissuade the private sector from investing in the water sector.

The importance of water for public health means that governments generally attempt to provide some minimal level of water supply and sanitation services, whether or not the full cost of these services can be recovered from users. The water sector is exceptionally capital-intensive – not only as compared to manufacturing and industry generally, but to other public utilities – with assets that are fixed, non-malleable and very long-lived (Hanemann, 2005). An industry typified by economies of scale[4], where a single supplier can provide the least-cost service, is considered a 'natural' monopoly and public ownership or regulation is customary. Such is the case for the water sector, which makes economic analysis a particular challenge. Some argue that the poor appreciation of the complexities of water as an economic commodity contributes to the current water crisis (Hanemann, 2005).

Due to the special nature of water and its particular role in human life and development, there are certain socially and politically determined principles that must be taken into consideration when assessing the value of any policy or programme. These principles reflect collective social values that set the tone for the debate and, in some cases, determine the boundary parameters and influence the

consideration of various different water policies and programmes. Three of the most important of these are the principles of economic efficiency, user-pays and water security.

- **Economic Efficiency**: Efficiency speaks to the importance of maximizing returns for the money, manpower and materials invested – more 'crop per drop'. Given the growing competition for freshwater, making more efficient use of the resources available is critical. Efficiency contributes to equity to the extent that if some users are discouraged from wasteful use, more water will be available for sharing with other users. Economic efficiency is summed up in the measure 'net present value'[5] or 'present value of net benefits'. Equivalent alternative measures include the 'benefit-cost ratio'[6] and the 'internal rate of return'[7]. The formulas for all of these measures involve generating monetary estimates of the costs (including damage costs and opportunity costs[8]) and benefits of the proposed programme for each year and, with the application of a social interest rate, discounting the stream of expected annual net benefits to a single number in current value terms. The goal of economic efficiency can and should serve the parallel goals of social equity and environmental sustainability – the three pillars of IWRM.

- **User Pays**: The 'user pays' principle contends that consumers should pay an amount equivalent to the burden (i.e. the full social cost) that their consumption places on society. Full social costs include both the capital, operating and maintenance expenditures to keep the system operating, and also the opportunity costs. They would also include the costs of damage resulting from the water pollution imposed on the society – the 'polluter pays' principle.

- **Water Security**: Often perceived to conflict with the above 'user pays' principle, the notion of water security holds that resources should be distributed according to 'need'. Accordingly, all individuals have the right to an adequate, reliable and affordable supply of potable water. Because of the importance of potable water for human health and well-being, it is usually regarded as a 'merit good', meaning that in certain cases, people deserve more water than they are willing or able to pay for. Thus 'security' in the context of domestic water supply refers to governance policies designed to help the poor satisfy basic water needs. This may involve setting tariffs according to the

4. Economies of scale are the savings achieved in the cost of production by larger enterprises because the cost of initial investment can be defrayed across a large number of producing units.

5. Net present value (NPV) is the discounted or present value of an annual or periodic stream of benefits minus costs over the life of a project (OMB 1992).

6. The benefit-cost ratio (BCR) is the ratio of the present value of periodic benefits to the present value of periodic costs over the life of the project (Boardman et al., 2000).

7. Internal rate of return (IRR) is the discount rate that will render the present value of a future stream of net benefits equal to zero (OMB, 1992).

8. Opportunity cost is the maximum worth of a good or input among possible alternative uses (OMB, 1992).

criterion of ability to pay, for example, a basic charge less than a certain fraction of net consumer income. A recent OECD (2003) report suggests a 'macro-affordability indicator', or 'water charges as proportion of income or expenditure' be considered. In a given network, lower income households will generally pay a larger proportion of income for water services than the average household.

Numerous additional social and cultural considerations may influence the selection and implementation of water governance policies. The valuation of water resources needs to take into account traditional cultural values that affect how water is perceived and used. Religious and ethical teachings regarding the appropriate use of water can have an important influence on public water management activities. Experience has shown that cultural perceptions of the role of water and perceived rights associated with its use strongly affect social acceptance of government attempts to effect policy changes in this sector. As noted earlier, differential effects with regard to gender also need to be considered in assessing policy initiatives.

Environmental considerations are also receiving growing emphasis worldwide. Protests against the environmental impacts of water development projects have been known in the United States since the early twentieth century, although at that time there were few theoretical tools to aid in water policy analysis. Models for evaluating the environmental as well as the economic impacts of water projects only emerged in the post-Second World War era with the work of Arthur Maass and Maynard Hufschmidt at Harvard University. In the United States, The Omnibus Water Resources Act of 1970 mandated that water project planning consider a range of objectives including social and environmental concerns. The 1983 guidelines of the Water Resources Council strongly influence such analyses still today.

The OECD's *Management of Water Projects* (1985) takes a similar approach, advocating measurement of incremental environmental impacts. A dozen general categories of impacts are suggested for consideration in the evaluation process. Among these are: water quality (including as appropriate for the given case, specific pollutants such as dissolved oxygen, temperature, biochemical oxygen demand, pH, bacteria etc.), amenity and recreation values (e.g., clean water, turbidity; colour and, odour), natural hazard exposure, aquatic ecosystems

and aesthetics (loss of valued scenery, or historical or archaeological sites). The Millennium Ecosystem Assessment (2005) report forcefully reiterates the importance of considering environmental and ecosystem values. The economic significance of anticipating and avoiding environmental damages becomes apparent when one considers the costs of remediation, not to mention the social costs. **Table 12.1** illustrates the high cost of restoring a clean water source in a selected sample of cities across the United States.

3b. Non-market economic valuation

Economic evaluation of resource allocation requires some means of estimating resource values. When markets operate appropriately, a set of market values (prices) appear which serve to allocate resources and commodities in a manner consistent with the objectives of producers and consumers. In the case of freshwater, however, markets seldom operate effectively, or, more likely, are absent. Non-market economic valuation can be defined as the study of actual and hypothetical human behaviour to derive estimates of the economic value (often called shadow prices or accounting prices) of goods and services in situations where market prices are absent or distorted. Such estimated prices are an essential component of economic assessment of public water policy choices. **Table 12.2** illustrates some of the various types of water-related goods, services and impacts that might be measured by non-market valuation techniques.[9]

Most applied methods of water valuation fall into one of two broad categories depending on mathematical procedures and types of data employed: deductive and inductive approaches. Deductive methods involve logical processes to reason from general premises to particular conclusions. Applied to water valuation, the deductive methods commence with abstract models of human behaviour that are fleshed out with data that is appropriate to projected future policy, economic or technological scenarios. Assumptions can be varied and the sensitivity of the results to varying assumptions can be determined. The advantages of deductive models are simplicity, flexibility and the ability to analyse a hypothesized future. In principle, they can incorporate alternative assumptions about prices, interest rates and technology, thus testing the projections about unknown, future conditions.

Inductive methods, on the other hand, involve a process of reasoning from the particular to the general, that is,

Fishery, India

Economic evaluation of resource allocation requires some means of estimating resource values

---

9. See Freeman, 2003 for a state-of-the-art review of the theory of non-market economic valuation of environmental goods and services.

Table 12.1: The high cost of coping with source water pollution in selected communities in the US

Community	Type of problem	Response	Costs (USA)
Perryton, Texas	Carbon tetrachloride[1] in groundwater	Remediation	$250,000 (estimated)
Camden-Rockland, Maine	Excess phosphorus in Lake Chickawaukie	Advanced treatment	$6 million (projected)
Moses Lake, Washington	Trichloroethylene[2] in groundwater	Blend water, public education	$1.8 million (to date)
Mililani, Hawaii	Pesticides, solvents in groundwater	Build and run treatment plant	$2.5 million + $154,000/yr
Tallahassee, Florida	Tetrachloroethylene[3] in groundwater	Enhanced treatment	$2.5 million + $110,000/yr
Pittsfield, Maine	Land-fill leachate in groundwater	Replace supply, remediation	$1.5 million (replace supply)
Rouseville, Pennsylvania	Petroleum, chlorides in groundwater	Replace supply	> $300,000 (to date)
Atlanta, Missouri	Volatile Organic Compounds (VOCs)[4] in groundwater	Replace supply	$500,000 to $600,000
Montgomery County, Maryland	Solvents, freon[5] in groundwater	Install water lines, provide free water	$3 million + $45,000/yr for 50 years
Milwaukee, Wisconsin	Cryptosporidium[6] in river water	Upgrade water system; immediate water utility; Health Department costs	$89 million to upgrade system; millions in immediate costs
Hereford, Texas	Fuel oil in groundwater	Replace supply	$180,000
Coeur d'Alene, Idaho	Trichloroethylene[2] in groundwater	Replace supply	$500,000
Orange County, California	Nitrates, salts, selenium, VOCs in groundwater	Remediation, enhanced treatment, replace supply	$54 million (capital costs only)

Note: This table features a sampling of localities of various sizes that have borne high, readily quantifiable costs due to source water pollution. It attempts to isolate community costs, excluding state, federal and private industry funding. Not included here are the costs to individuals, such as lost wages and medical costs, reduced property values, higher water bills, and, in extreme cases, death.

1. A manufactured chemical most often found in the air as a colourless gas, used in the production of refrigeration fluid and propellants for aerosol cans, as a pesticide, as a cleaning fluid and degreasing agent, in fire extinguishers, and in spot removers; soluble in water.

2. A colourless or blue organic liquid with a chloroform-like odour used as a solvent to remove grease from fabricated metal parts and some textiles.

3. A manufactured chemical used for dry cleaning and metal degreasing. Exposure to very high concentrations can cause dizziness, headaches, sleepiness, confusion, nausea, difficulty in speaking and walking, unconsciousness,   and death. See also: www.atsdr.cdc.gov/tfacts18.html

4. Volatile organic compounds; for more information see: glossary.eea.eu.int/EEAGlossary/N/non-methane_volatile_organic_compound

5. FREON (trade name) is any one of a special class of chemical compounds that are used as refrigerants, aerosol propellants and solvents.

6. A parasitic protozoa found in soil, food, water, or surfaces that have been contaminated with infected human or animal faeces.

Source: NCSC, n.d. ca. 2000.

Table 12.2: Classifying goods and services provided by water

Commodity (or Private) Goods		Non-Commodity (or Public) Goods	
Producers' Goods	Consumers' Goods	Use Values	Non-use Values (Existence and bequest values)
1. Agricultural Producers ■ Crop Irrigation ■ Aquaculture	1. Residential water supply	1. Enhancing beneficial effects ■ Ecosystem services ■ Recreation ■ Aesthetics ■ Wildlife habitat ■ Fish habitat	1. Protection of aquatic environment
2. Off-stream industries ■ Manufacturing ■ Commercial services	2. Residential sanitation	2. Reducing adverse effects ■ Pollution abatement ■ Flood risk reduction	2. Protection of wild lands
3. In-stream industries ■ Hydropower ■ Transportation ■ Fisheries			3. Protection of biodiversity and endangered species

Source: Young, 2005.

from real-world data to general relationships. Observations on water-user behaviour are tabulated and subjected to formal statistical analysis to control for external factors influencing willingness to pay. Use may also be made of surveys of expressed preferences for recreation or environmental improvements, observations of prices on various elements from water rights to land transactions, responses to survey questionnaires, and secondary data from government reports. Because inductive techniques are based on observations of actual behaviour and real-life situations, they are preferred by many analysts. With data from published sources or surveys, these can be readily used to analyse the outcome of previous policies. One limitation in the evaluation of future or hypothesized policies is that analysis may involve assumptions outside the range of available historical observations. The accuracy of inductive techniques depends on several factors, including the representativeness and validity of the observed data used in the inference, the set of variables and the functional form used in fitting the data, and the appropriateness of the assumed statistical distribution. For readers interested in detailed discussion of these methods and the specific contexts in which they may be useful, there are numerous texts available for consultation.[10]

Examples of the application of different methods are provided in **Boxes 12.3** to **12.5**. In **Box 12.3** we see how contingent valuation was used to evaluate the willingness to pay for a proposed improved wastewater disposal

program in Cairo, Egypt. **Box 12.4** summarizes several statistical (inductive) studies using historical government data to develop estimates of investment returns to irrigation in Asia. **Box 12.5** provides an example of a major benefit-cost analysis of global improvements of water supply and sanitation services, using an alternative cost (deductive) approach.

Economic valuation has been criticized for its lack of transparency and difficulty of use. Certainly, competence with survey research and other data collecting methods and complex mathematical and statistical skills, plus significant time and resources are necessary to perform valid economic valuations. A related problem is the dependence of techniques, based on consumer surveys and choice experiments, on public knowledge of the subject in question (Powe et al., 2004). One major criticism is that many stakeholders feel that economic valuation fails to capture all relevant value, especially social, cultural and environmental aspects, of typical water and environmental policy issues. Finally, non-specialists may find it difficult to understand and interpret study results.

Over the last several decades, the theory and practice of non-market economic valuation of water and environmental resource policies have been greatly improved, and those with the skills, time and resources can effectively derive conceptually consistent and empirically valid measures for the value of water and the benefits of water governance strategies. The primary

10. See, for example, Young, 2005.

BOX 12.3: WILLINGNESS TO PAY FOR IMPROVED WATER AND WASTEWATER SERVICES: CAIRO, EGYPT

A review of the status of wastewater disposal in Cairo, Egypt – one of the world's largest and fastest growing urban areas – showed inadequate water supply and wastewater conveyance, treatment and disposal capacities. Contingent valuation methods employing a referendum-type questionnaire to collect information on consumer preferences were used to develop estimates of the willingness to pay and economic net benefits for each of four potential investment programmes: (a) opportunity to connect to the water supply network; (b) improved reliability: provision of all-day water service; (c) wastewater maintenance to eliminate sewer overflows; and (d) an in-home connection to the wastewater disposal network. A separate sample of about 1,000 households was interviewed for each of the four programmes. Statistical analysis of responses showed willingness to pay for a water connection was US $7.70 per household per month compared to an estimated cost of US $2.50 per household per month for a net economic benefit of US $5.40 per household per month (evaluated with a 10 percent discount rate in 1995 US dollars). Respondents also showed a positive net willingness to pay for reliability of US $0.50 per household per month. For the wastewater programmes, an estimate of willingness to pay of US $2.20 per household per month was derived for the programme to eliminate sewer flooding which compared to a cost of US $0.20 per household per month, while a connection to the wastewater network was valued at US $7.60 versus a cost of US $6.30 for a net benefit of US $1.30 per household per month. However, if the wastewater investment programme was designed so as to require the household to pay for treatment in addition to disposal, willingness to pay was insufficient to justify that form of investment programme. And, although the mean net willingness to pay was positive for most of these plans, the distribution of responses showed that some lower income households would be unwilling to participate if charges were set at full costs.

Source: Hoehn and Krieger, 2000.

advantage of applying these techniques is that they generate information on different beneficial and adverse effects in a common denominator – money. This, in turn, enables policy-makers and stakeholders to better understand the trade-offs to be made and be better prepared to make the necessary decisions regarding the most appropriate water development strategies.

3c. Application of valuation techniques in evaluating alternative governance strategies

Public policies addressing water resources allocation and management must serve societal goals, such as equity and environmental sustainability, as well as financial feasibility, administrative practicality and economic efficiency. Thus, choosing the best governance strategy is a challenging process. In the context of water governance, the application of non-market valuation methods to estimate benefits (reflecting demand or willingness to pay) for water services have been used to assess and compare alternative proposed policies and programmes for the management and allocation of water resources.

Governance strategies should be selected to optimize the achievement of societal goals. In this context, valuation can be viewed as a fairly neutral and objective process by which social goals and trade-offs can be identified and debated and the optimal governance strategies chosen. In benefit-cost analysis (BCA) (e.g. Boardman et al., 2000), policy outcomes are assumed to be quantifiable and values are assigned to them in the single monetary metric. Although many impacts, positive and negative, are not properly or fully measured, if at all, by market prices, non-market economic valuation techniques can be used to assign monetary values to these impacts.

In BCA, monetary values must be assigned to each physical unit of input and product. The primary sources of these values are the observable market prices. However, in the case of water, as discussed above, market prices do not adequately reflect all the goods and services associated with water resources nor the true social value, for example, when agricultural commodity prices are controlled by government regulation or when minimum wage rates are set above market clearing prices. In such cases, prices must be adjusted to reflect the full costs and benefits. In many cases where market prices do not exist at all, it is necessary to construct surrogate prices. Whatever the source, the prices used in BCA are interpreted as expressions of willingness to pay (WTP) for, or willingness to accept compensation (WTA) for going without, a particular good or service by individual consumers, producers or units of government. For market prices, this presumption is straightforward, since the market price represents the willingness to pay at the margin for the potential buyers of the good or service. For non-market economic goods, WTP is also the theoretical basis on which surrogate (or shadow) prices are calculated. The assertion that willingness to pay is an

BOX 12.4: ECONOMIC VALUATION OF RETURNS TO IRRIGATION WATER INVESTMENTS IN ASIA

Several recent studies to assess the relative contribution of policies designed to enhance rural economic growth and reduce poverty in Asia have been reported by the International Food Policy Research Institute (IFPRI). Each study used inductive methods to analyse state or provincial time series data on public expenditures and economic output measures. Evenson et al. (1999), having analysed data from nearly all districts of thirteen states in India from 1956 to 1987, found that the marginal internal rate of return (IRR) to investments in irrigation was only about 5 percent. In contrast, public expenditures on agricultural research and extension were reported to yield marginal IRR of 58 percent and 45 percent, respectively. Also for India, Fan et al. (1999), using 1970–93 state data in a cross-sectional, time-series

econometric study, found that irrigation development distantly trailed road building, agricultural research and development, and education in its impact on poverty reduction, although irrigation showed a somewhat more favourable impact on productivity. Fan et al. (2002) also compared returns to irrigation investments in China with other rural development programmes from provincial data for 1970–97. For the nation as a whole and for each of three economic zones, the authors reported high returns to investment in education, agricultural research and development, and rural infrastructure, but '[i]nvestments in irrigation had only modest impact on agricultural production, and even less impact on poverty reduction, even after trickle-down benefits were allowed.' Similarly, for

Thailand, Fan et al. (2004) report that for 1970–2000, estimated economic benefits of irrigation investment failed to exceed costs, while agricultural research and development, electricity and education yielded quite generous rates of return. Overall then, one can infer that rates of economic return to investments in irrigated agriculture have been low in recent decades, particularly when compared to the opportunity cost of capital or to the return on alternative programmes aimed at improving the livelihoods of the rural poor. These conclusions suggest that the deductive methods typically used to evaluate proposed investments in irrigation may have been overly favourable to such programmes.

Source: Young, 2005.

appropriate measure of value or cost follows from the postulate that public policy should be based on the aggregation of individual consumer preferences.

Not all agree with this approach, however. Major opponents challenge the view that the economic efficiency impacts measured by benefit-cost analysis fully reflect society's goals. Sagoff (2004), the well-known and outspoken advocate of the widely held position that benefit-cost analysis has little role to play in environmental planning, argues that political resolution of value conflicts is the most appropriate approach. The basic argument is that other important goals relevant to decisions on appropriate water governance strategies cannot be reduced to the common denominator of money. When values conflict, as they often do, a dilemma arises. Some solution must be found which reconciles the disparate, competing perspectives. A widely used method of reconciliation is called the 'approved process approach' (Morgan and Henrion, 1990). This approach, roughly speaking, requires all relevant parties to apply the concept of 'due process', or to observe a specified set of procedures, to estimate a proposed policy's impacts on all relevant indicators of value. According to this method, any decision reached after a mutually acceptable mediator balances the competing values according to the specified procedures is deemed acceptable.

One variant of the approved process approach often applied in water resources analyses is called 'multi-objective planning'. Multi-objective assessment of water projects and policies has been promoted for some time in both the US and Europe with two well-known manuals by the Water Resources Council (1983) and by OECD (1985). Both emphasize a similar framework of analysis with three goals in common: economic efficiency, environmental quality and social well-being. Each provides advice on developing indicators to reflect the degree of goal attainment by particular strategies focusing on indicators both for beneficial as well as adverse effects for the *with* versus the *without* policy situation. Both approaches call for environmental impacts to be balanced against economic and social welfare considerations. In each case, the manuals emphasize the point that the task of technical analysts is not to come to a final decision on governance strategy, but to illustrate the expected impacts in the appropriate metrics. With many values or goals considered incommensurate (i.e. not reducible to a common denominator), it is assumed that the ultimate resolution or weighting of conflicting values will be referred to the political arena.

Watering ramp in a field, Senegal

BOX 12.5 BENEFITS VERSUS COSTS OF IMPROVED WATER AND SANITATION SERVICES

Adoption of the Millennium Development Goals (MDGs) that deal with extending the availability of water and sanitation services has prompted interest in assessing the net economic benefits of such programmes. Hutton and Haller (2004) evaluated five different scenarios with different levels of intervention for seventeen World Health Organization (WHO) sub-regions. The five levels of intervention were:

1. Water improvements required to meet the MDG for water supply (halving by 2015 the proportion of those without safe drinking water).
2. Water improvements to meet the water MGD for water supply *plus* the MDG for sanitation (halving by 2015 the proportion of those without access to adequate sanitation).
3. Increasing access to improved water and sanitation for everyone.

4. Providing disinfectant at point-of-use over and above increasing access to improved water supply and sanitation.
5. Providing regulated piped water supply in house and sewerage connection with partial sewerage connection for everyone.

Costs were determined to be the annualized equivalent of the full capital cost of the intervention. Benefits were measured in terms of several variables: the time saving associated with

estimated benefit-cost ratios for selected regions. Economic benefits were found to greatly exceed the costs for all interventions, particularly level (4), a result that was robust for all regions and under alternative intervention scenarios.

Source: Hutton and Haller, 2004.
www.who.int/water_sanitation_health/
wsh0404.pdf

WHO Sub-Region	Population (million)	Benefit-cost ratio by intervention level				
		1	2	3	4	5
Sub-Saharan Africa (E)	481	11.50	12.54	11.71	15.02	4.84
Americas (D)	93	10.01	10.21	10.59	13.77	3.88
Europe (C)	223	6.03	3.40	6.55	5.82	1.27
South East Asia (D)	1689	7.81	3.16	7.88	9.41	2.90
Western Pacific (B)	1488	5.24	3.36	6.63	7.89	1.93

Note: The parenthetical letters identify WHO sub-regions as classified by epidemiological (health risk) indicators. See source for definitions.

Part 4. Charging for Water Services

For both municipal and irrigation water services in developing countries, performance, efficiency and conditions of water delivery systems tend to fall far short of normal standards. Many people, but mostly the poor, lack access to safe water supplies and/or sanitation facilities, and for many others, the only access may be via water vendors or public latrines. Often over one-third of water transmission is lost to leakages or to unregulated access. The World Water Council's report 'Financing Water for All' (commonly known as the Camdessus Report), addressed the issue of mustering financial resources to meet internationally agreed water supply and sanitation goals, concluding that currently available sources will be insufficient to maintain and expand coverage (Winpenny, 2003). As the financing of water services is becoming ever more urgent, recovering costs is seen to be central to improving the conditions of water services. In this context, charging for water services is increasingly being promoted as an appropriate response.

An increasingly important aspect of water governance is the regulation of water quality

Criteria applied to tariff-setting

Multiple criteria influence policy decisions on how to finance water services and how much revenue to collect from beneficiaries (cf. Herrington, 1987, 1999; Hanemann, 1997). In addition to the goals of safe and affordable water for all and maximum net social benefits, two key criteria are:

■ **financial sustainability**, requiring the collection of sufficient revenue to meet present and future financial obligations, that is, operating costs as well as the capital costs of facilities and infrastructure, and the

■ **user pays principle**, which holds that consumers should pay an amount equivalent to the burden of their consumption on society. This implies that charges should attempt to recover full costs, including not only operation, maintenance and capital replacement, but taking into account foregone benefits (opportunity costs), as well as any externalities (damages to third parties) (see **Figure 12.1**).

Other characteristics important in the successful implementation of any charging plan are:

■ **simplicity**, which means that the selected tariff plan should be open, understandable and straightforward with users able to see how usage patterns affect the amount payable

■ **transparency**, enabling consumers to understand how their own tariffs and those of other user classes are set, and

■ **predictability**, permitting customers to reasonably anticipate and plan for their water-related expenses.

These criteria often come into conflict. For example, assuring that the less fortunate members of society are charged an affordable rate is likely to clash with both the user pays principle (recovering full costs) and maximizing net social benefits (pricing at marginal social cost). As shown in **Figure 12.1**, tariff-setting must balance both cost and value considerations as the upper level of charges is limited by user willingness to pay. Resolving the conflicts of rate-setting is inherently a political process. Any assessment of the various charging options must consider carefully the incidence of all costs and benefits, if charging is to be equitable as well as efficient.

Structuring user charges

For most marketed goods and services, units are obvious and the price per unit is easily understood. The case of water tends to be more complex. Water users may pay only a charge for access to the delivery network, but not for water itself. Charges may include a fixed periodic (e.g. monthly) access fee as well as a variable charge based on volume used. Many utilities require an initial connection fee. Hence, there is no single 'price'. In general the tariff structure for water services can be described in two dimensions: form and level.

The form refers to if and how the charge relates to the quantity used, while the level refers to the proportion of the cost of service to be recovered from users. Flat rates are more or less independent of the quantity used or may be linked to the projected level of use, according to, say, the number of family members or size of pipe connection. Conversely, charges may vary directly with the quantity of water used. Rate structures are changing because of the falling costs of metering, the increasing tendency to define water as a commodity (rather than a

Figure 12.1: Comparing the cost and value of water

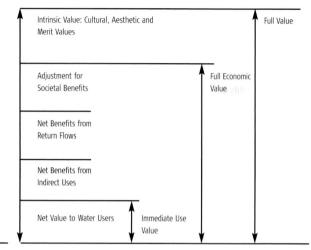

a. Components for cost of water supply

b. Components of water use value

Source: Derived from Rogers et al., 1998, Savenije, H. H. G. and van der Zaag, P. 2001, and Matthews et al., 2001.

public good), and the perceived need to use charging to restrain water use as well as to recover costs.

The level of charges refers to the proportion of costs to be recovered from users and how these costs are divided among user groups. Although previously, water was widely regarded as a public good to be made available to all without charge and financed by from general public revenues, increasingly, policy is changing to one of full cost recovery, except where poverty is an issue. Charging each customer according to the cost each imposes on the system is seen to be consistent with both the criterion of the sustainability and the principle of fairness. Because of the importance of water for health and well-being, less well-off customers may be charged according to ability to pay, rather than full cost. In the residential sector, affordability is often measured by the fraction of household income spent on water. Although the determination of this fraction may be subjective and policies vary by country, commonly the aim is for household expenditures on water to be below 3 percent of annual income (OECD, 2003). Where the balance of costs must be raised elsewhere, one common solution has been cross-subsidization, that is, revenues from better-off residential or industrial users or from city, state or national government coffers covering the cost of less affluent users.

One major problem stemming from the politicized nature of rate setting is that subsidized rates are inadvertently

made available to too large a proportion of the service base. As a result, revenues are insufficient to operate the utility efficiently and to extend service to potential new users. Over time infrastructure deteriorates or other sources of financing must be sought.

4a. Charging for municipal water services

Two conceptual positions vie for the main charging formula: the average historical cost method and the incremental future cost (or long-run marginal cost) method. Critics of the first model point out that only a small portion of the charge to consumers varies with the amount used. However, they contend that with a low marginal charge, customers have insufficient incentive to restrain water use and to invest in water-saving appliances, xeric landscaping and other conservation measures. Sceptics also observe that the historic costs model with its annual depreciation of historic capital costs ignores price inflation and current value or the replacement cost of assets. More generally, critics hold that with low marginal charges, historical cost models in practice encourage profligate water use, and stimulate construction of increasingly expensive supply systems ahead of need, which leads to calls for rate hikes only to support excess capacity. Finally, in practice, the historical cost method ignores social costs, such as the opportunity costs and detrimental environmental impacts.

Advocating an alternative approach, public utility economics literature (e.g. Hall, 1996, 2000) proposes that the relevant costs for determining municipal water charges is long-run incremental social cost. The purpose of charging at long-run incremental social cost is to produce price signals that induce water-use levels that maximize net benefits of the water utility's services. This concept, it is argued, reflects the true scarcity of the resources required to deliver water. A rate structure based on incremental cost, in theory, provides incentives for users to reduce water use, the value of which to them is less than the cost of provision. It would also, it is argued, encourage water users to make investments in plumbing fixtures and appliances for which the cost savings in water is less than the supplier's incremental cost of provision. In contrast to the historical cost method, social costs would include not only the costs of inputs and services acquired on the market, but non-market opportunity costs of the potential value of the water in alternative uses, and the unintended side effects on third parties. One difficulty of adopting the incremental future cost approach is the relative complexity of determining future as opposed to past costs, particularly opportunity and social costs. Another is the large rate hikes that could result unless tariff increases were phased in.[11]

4b. Charging for irrigation water

Around the world, it is rare that water users are charged the full cost of water services. Subsidized irrigation water is justified not only on users' limited ability to pay but on the (disputed) economic grounds of secondary economic benefits, for example, the boost in agricultural production due to increased availability of irrigation water. For public irrigation systems, the aim is to recover only operating and maintenance costs from users, with initial capital costs covered by the general public budget or donor agency contributions. Even in self-supplied pumping systems, typically no account is made for the opportunity cost of water or for the cost of damages to third parties. With cost recovery difficult in practice, inadequate revenues generally force some higher level of government to cover shortfalls or lead to a deterioration of the system.[12]

Of the several potential methods of charging for irrigation water, the most common is area-based charging, set according to the area irrigated, but also possibly according to the season, irrigation method or crop grown (Tsur et al., 2004). Area-based charges are criticized, however, for lacking any incentive to conserve water, for instance by reducing the number of irrigations, by limiting

the amount of water applied per irrigation or by shifting to less water-intensive crops. The main alternative to an area-based charge is volumetric charging, which requires some agreed-upon method of measuring volume, such as length of time of water delivery based on a stable and known rate of flow. In fact, many different types of tariffs exist. Typically, the organization and configuration of irrigation water supply influence the feasibility of alternative methods of charging for irrigation water.

A review of policy reforms to increase farmers' cost shares and apply volumetrically based charges reveals several problems. One is that within the irrigation sector, non-economic goals tend to be more influential than economic efficiency, so in practice, full cost recovery and incremental cost pricing are less important. Moreover, with irrigation charges designed both to signal scarcity and reduce the taxpayer burden, adversely affecting farmers' incomes, governments find such polices neither in the public interest nor in their own political interest. Complicating the issue is the fact that governments often undertake irrigation projects to foster economic development in disadvantaged regions. Moreover, there is the perceived issue of national food security and the belief that production from irrigated land is more stable.[13]

Both cost and benefit considerations suggest that volumetric pricing of irrigation water may not be as desirable as it might first appear. The extra costs of assessing volumetric-based charges are often judged not worth the cost of implementation, thus volumetric-based charges are even less common in agriculture than in municipal and industrial water systems. From the perspective of benefits, the issue of volumetric pricing to encourage water saving is further confounded by the distinction between water withdrawn and water consumed. Water leaked from permeable canals, ditches and fields returns to the hydrologic system (surface or groundwater) hence becoming available from streams or wells. Only when leaked water is permanently degraded and/or lost to future use is it true that water saving at the farm or the district level is important. Thus, the evaluation of the technical efficiency of water use in the agricultural sector must be addressed on a case-by-case basis.

Some observers (e.g. Young, 2005) contend that governments and donor agencies tend to overestimate potential economic returns to irrigation and, as a result, farmers' actual ability to pay for water (see **Box 12.4**). As a result, projects have experienced lower than

A farmers' son uses a motorized pump to irrigate a greenhouse in Mborucuya, Argentina. This small farm has received credit from a support project for small scale producers to finance both the greenhouse and the irrigation pump

11. Recent texts with in-depth discussion of this topic include Shaw, 2005 and Griffin, 2006.

12. Further detailed discussion can be found in Tsur et al., 2004 and Cornish et al., 2004.

13. A more cynical view holds that these below-cost charging policies are merely the result of successful political efforts to obtain government subsidies on behalf of political supporters.

BOX 12.6: IRRIGATION MANAGEMENT TRANSFER (IMT) AS A COST RECOVERY TOOL

Many developing countries (aided by international donors) have in the past several decades invested large sums in irrigation systems with the expectation of increasing agricultural productivity and improving incomes for poor farmers. It was assumed that most such schemes would be economically and financially self-sufficient under reasonable management. However, most developing countries have not implemented charging programmes to recover actual operating and maintenance costs, let alone to pay for the capital costs of the investments. As governments have been unable or unwilling to adopt cost-recovery policies that keep pace with inflation or the need for periodic system rehabilitation, they have found that budgetary demands of the irrigation sector increasingly compete with other public needs. Policy reforms to transfer more of the irrigation costs to water users have come as

part of a package called 'Irrigation Management Transfer' (IMT). These programmes assume that farmer management of public irrigation systems would make the system more responsive to members, and thereby encourage water users to be more receptive to paying costs. Expectations were that local control would not only improve the cost-effectiveness but by transferring costs to users reduce costs to the public exchequer. Results of such reforms have been, at best, mixed. While IMT programmes have been somewhat successful in more developed countries (US, New Zealand and Mexico), elsewhere the results are less promising. In many cases, charges to farmers did increase, but farmer-managed systems have tended to under-invest, thus necessitating public rescue. Little evidence of an overall increase in agricultural productivity or farm incomes has been observed. In large systems with

many smallholders, costs of administration and revenue collection are necessarily high, and the users have ended up with lower productivity and income. In some cases, the systems have collapsed. The conclusion seems to be that IMT can work in cases where irrigation is essential to high-performing agriculture, and farmers are not too numerous, better educated and behave as businessmen. Furthermore, the cost of operating and administering the irrigation system must represent a modest proportion of the increment in farmers' income expected from irrigation. Where the system serves numerous, small farms producing low-value staple crops (such as in the rice-producing regions of Asia), in terms of cost recovery, system efficiency and productivity, IMT has not produced the expected results.

Source: Shah et al., 2002.

expected net returns to water. With subsidies to irrigation capitalized into higher land prices, governments find that levying higher user charges may not only depress farm income, but risk imposing significant capital losses on landowners. Nevertheless, many countries are moving toward collecting a larger proportion of irrigation costs from farmers. As part of a larger reform and decentralization effort, this trend aims not only to reduce public subsidies, but also to increase efficiency and the responsiveness of irrigation delivery. Such policies, often called irrigation management transfer (IMT), seek to shift the administration of all or part of irrigation water delivery to associations of water users, thus sharing the responsibility of water management. **Box 12.6** reviews the experience in various developing countries for transferring responsibility for irrigation water delivery to user groups.

4c. Charging for discharge of industrial effluent

An increasingly important aspect of water governance is the regulation of water quality. Water's solvent properties and widespread availability provide both producers and consumers with an inexpensive means of waste disposal. With public expectations of near zero effluent discharge, policy-makers face a paradoxical situation with regard to water use and quality. In many countries minimum waste

disposal would be enormously expensive, even impossible, unless some important industries were closed altogether. Assessing the costs and benefits in such cases demands careful consideration of the relative effectiveness and desirability of the alternatives, not only from an economic perspective, but also in terms of the distribution of costs and benefits, the ease of monitoring and enforcement, and industry flexibility, among other factors. Although direct regulation has been the main tool of water quality management in the past, water pollution is increasingly being addressed by decentralized systems of incentives and disincentives, such as effluent charges (see **Chapter 8**).

The effluent charge, also called an emissions or pollution tax and essentially a fee levied on each unit of contaminant discharged, is based on the principle of 'polluter pays'. Initially this principle was intended to 'suggest' to governments that they should refrain from subsidizing investments required to comply with pollution-control regulations. A more recent interpretation holds that emission charges should be set so that the costs, or the economic value, of the damages inflicted by polluters on third parties are borne by the polluters themselves, in effect 'internalizing' the previously externalized costs of production. With the unit charge set to rise with

Figure 12.2: Actual and planned water pollution charges in the River Narva and Lake Peipsi Catchment, 1993-2005

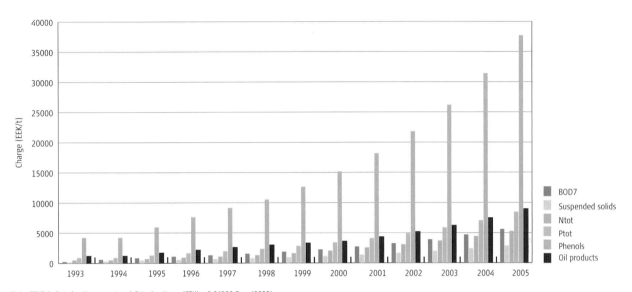

Note: EEK/t is Estonian Kroon per ton. 1 Estonian Kroon (EEK) = 0.06390 Euro (2005).

Source: Environmental Information Center, Tallinn, Estonia.

increased levels of discharge, polluters may respond as they choose, that is, reduce effluent or pay the charges. Firms facing low pollution reduction costs relative to the charges imposed would presumably move to reduce discharges. Others might find it cheaper to pay the tax than to make the necessary pollution control expenditures. Such charges should provide incentives for pollution discharge to be reduced by the least cost methods available. All firms would find it in their interest to seek changes in processes, technologies and/or in discharge treatments that reduce the cost of coping with the problem of residuals disposal.

Criticisms have come from all sides but most prominently from polluters, who complain of potential impacts on profits and hence, over the longer term, on net worth and share value. Public officials, on the other hand, are concerned producers may be forced into reducing output and employment with corresponding negative effects on tax revenues. From the viewpoint of regulatory agencies, effluent charges present challenges of monitoring and enforcement. Environmental groups object to effluent charges on the grounds that they convert the environment into a commodity. Surveys of pollution control strategies in OECD nations show that environmental charges for the most part were not applied to induce less polluting behaviour nor to compensate damaged parties, but to fund specific environmental expenditures. Despite all criticism, effluent charges for water pollution management are seeing increasing application (see **Figure 12.2**).

Chemical outflow, Germany

Part 5. Responding to the Challenge of Valuing Water

In this section, we introduce some of the issues at the forefront of the debate on valuing water. The subjects addressed include public-private partnerships, virtual water and payments for environmental services. Private sector participation in water resources development can assist not only in meeting financial and management needs of this sector but in tailoring water services to better address local concerns and values. The concept of 'virtual water' – recognizing the value of water embedded, directly or indirectly, in various products and services – has come to influence both production and trade policy in an attempt to maximize returns to water as a scarce factor of production. Payment for environmental services, that is, paying upland farmers for land husbandry that preserves the output of clean and regular water supplies, directly acknowledges the value of the water to downstream users.

5a. A shared management approach: Public-private partnerships

Public-private partnerships (PPPs) are essentially a management tool designed to bring the strengths of both public and private sectors to water utilities. They combine the social responsibility, environmental awareness and public accountability of government with the technology and management skills and finance of the private sector (UNDP, 2004). Depending on the extent of private sector participation (PSP), public-private partnerships are also characterized by the state's changing role, that is, from one of sole service provider to monopoly regulator in charge of controlling tariffs and service quality (World Bank, 1994; Estache et al., 2005).

Sharing skills and resources

Despite the wide acknowledgement that the public sector in many regions lacks not only the economic resources, but also the technical and management skills required to meet water services demand in an efficient and environmentally sustainable manner, private participation in water services remains controversial because of water's essentiality for life (Cosgrove and Rijsberman, 2000; Gresham, 2001). In countries where the political and institutional climate make it difficult for governments to involve the private sector, contracting-out services, operation and management of the water supply, allows the public sector to take advantage of private sector technology and skills, while maintaining ownership of

---

BOX 12.7: WATER AND SEWERAGE SERVICES IN YEREVAN, ARMENIA

The World Bank made two loans worth US $80 million to Yerevan for improving water and sewerage services, in particular in poorer areas. In 2000, ACEA (Italy), C. Lotti & Associati and WRc (UK) undertook an operations and management contract for Yerevan. At that time just 21 percent of billed accounts were paid. Billing collection was revived through the introduction of metering from 2002. With an increase in the number of registered customers from 275,500 in 2002 to 311,056 by April 2004, 245,000 of these were metered and 28,000 filed as non-active accounts (empty apartments, etc).

Service indicators	1999	2003	2004
Water provision (hours/day)	6	13	16
Percentage of metered apartments	56	–	95
Percentage of revenue collected	21	87	100

Instead of charging domestic customers a nominal per capita consumption of 250 litres/day, customers are now being billed for actual usage, on average about 100–120 litres/capita/day.

Some 30 percent of Yerevan's population lives below the poverty line. The introduction of

metering has improved service affordability for these people. In 2002, the bottom quintile spent 8.1 percent of their income on water services. This fell to 5.0 percent in 2003 and is expected to reach about 4 percent in 2005, despite a 50 percent overall tariff increase in April 2004.

Sources: OECD, 2005; World Bank, 2005.

Figure 12.3: Share of private sector participation in water supply and sanitation by region

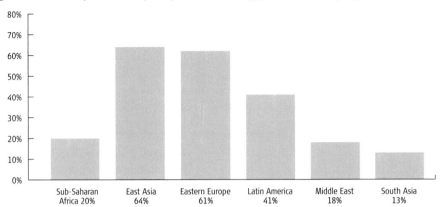

Source: Estache et al., 2005.

key assets (see for example **Box 12. 7**) (World Bank, 1997; Gresham, 2001; Estache et al., 2005). Private corporations are now involved in some dimension of large-scale water supply in almost half the countries of the world, particularly in the developed world, but also increasingly in the developing world as shown in **Figure 12.3**. The private sector's proportion of the water and sewerage sectors in developing countries comprises, on average, only 35 percent, whereas in developed countries it constitutes 80 percent of the market – in particular because of already high coverage rates and an institutional climate conducive to private investment (Estache and Goicoechea, 2004).

Reconciling cost recovery and affordability

Population growth and burgeoning water demand have convinced most policy-makers that the cost of water system development will increasingly have to be met by users, especially if the Millennium Development Goals are to be achieved. Meeting the financial challenge of water supply means the involvement of all stakeholders, with funds from governments, financial markets, international aid and users. However, with private sector participation – ranging from small water vendors to large private utilities – projected to increase in the next decades, the issue of pricing is critical, not only to improve access and quality of service and discourage theft and wasteful use, but to ensure affordability and fairness to all customers (Whittington, et al. 2002).

Recent problematic PSP experiments in some developing countries, such as Bolivia (see **Box 12.8**) and Ghana, highlight the need to ensure the availability of affordable

water supply and sanitation for poor households (Finnegan, 2002). PSP arrangements – particularly those transferring the responsibility for capital investment to the private sector – can increase tariffs to levels often unaffordable for the poor. Accordingly, there is an urgent need to better understand consumers' conditions and to improve subsidy mechanisms in PSP schemes. Research has revealed that these benefits have been captured by middle-income and rich consumers rather than the poor and that the poor in many cases are willing to pay for improved water supplies. In sprawling cities of the developing world, poorer populations typically lack access to formal water systems and may pay more than ten times per cubic metre of water than people with household connections (Raghupati and Foster, 2002).

An inclusive approach

Technological change and more cost-effective, smaller-scale systems continue to alter market structures in water delivery, actively engaging civil society both through community-level initiatives as well as through large-scale water supply schemes (Estache et al., 2005). Public-private community partnerships incorporate innovative grassroots mechanisms to enable service to poor populations in small regions at more affordable levels. The 'private' side of these partnerships refers to a range of different actors, from households to community organizations, NGOs and small businesses. The 'public' side of the partnerships involves not only the public utility and the independent regulator, but local governments committed to facilitating grassroots initiatives (Franceys and Weitz, 2003). As in traditional public-private partnerships, each of these options may

Population growth and burgeoning water demand have convinced most policy-makers that the cost of water system development will increasingly have to be met by users

BOX 12.8: THE 'WATER WAR' IN COCHABAMBA, BOLIVIA

The city of Cochabamba, Bolivia, the third-largest city in Bolivia, has a chronic water shortage. A sprawling city of 800,000 people, whose population has exploded during the last decades with immigrant workers from the countryside, it has many poor neighbourhoods lacking connection to municipal water supplies. In recent years, residents in peri-urban areas pushed for workable community initiatives with the help of foreign aid. Small-scale water companies built electric pumps to access well water and distribute it throughout these neighbourhoods, at a total cost of US $2 to US $5 per month. In 1997, conditions on the World Bank US $600 million loan for debt relief included the privatization of the water supply in Cochabamba, and in 1999, a private operator was granted a 40-year concession contract to rehabilitate and operate the municipal water supply system, as well as the smaller ones. The contract provided for exclusive rights to all the water in the city, including the aquifers used by the water cooperatives. Billing and metering was implemented, with the cost of these services, as well as of connections, being reflected in the tariffs.

Within weeks of taking control of the city's water supply, prices were raised to unaffordable levels, effectively leaving the poor in marginal areas without access to any water as they were no longer permitted to draw water from their community wells. Workers living on the local minimum wage of US $60 per month suddenly had to pay US $15 for the water bill. In 2000, a coalition of workers, farmers and environmental groups, 'Coalition for Defence of Water and Life', organized a general strike and massive protests in opposition of the rate hikes. Bolivians blocked highways, and the city was shut down. Police forces and the military were sent to take control of the city, and martial law was declared. As protests grew stronger despite being suppressed, the private operator withdrew from the city and the government rescinded the concession contract. This experience led the government to reconsider private sector participation, and to enact a law granting legal recognition to traditional communal practices, under which small independent water systems shall be protected.

Source: Finnegan, 2002.

allocate ownership rights and responsibilities for investments and management differently.

In larger-scale initiatives private corporations can also partner with local governments and NGOs. NGOs can provide local governments with information on the specific needs of poor areas, which then can be better addressed in negotiating concession contracts, for example, by defining specific connection targets or obligations for expansion into peri-urban areas. NGOs and communities can also participate in tariff collection on behalf of the private utility in exchange for deferred payment of connection fees. Likewise, municipal governments can facilitate connections by, for example, waiving the land title requirements for slum dwellers. To reduce connection costs, NGOs can help by providing transportation and materials, while the community contributes labour, for instance carrying pipes, digging trenches and laying lines (Franceys and Weitz, 2003). As described in **Box 12.9**, researchers have discovered such innovative approaches in Manila in the Philippines.

The value of public-private partnerships

Both the value and economic valuation of water are important in assessing water supply and sanitation alternatives. While privatization may not be suitable in all cases, neither are underfinanced public utilities a sustainable solution given burgeoning water demand.

Likewise, the global replication of community-driven arrangements is not viable on a large-scale. Experience with both public and private delivery of water services over the past decade has taught us that ownership of water infrastructure, whether public or private, has no significant effect on efficiency nor on the selection of the public versus private sector as service provider (Estache and Rossi, 2002; Wallsten and Kosec, 2005). Indeed, ownership has proved less significant than governance, and thus a good institutional climate is important, not only for private sector investment, but for the transfer of relevant technical knowledge and management skills (Estache and Kouassi, 2002; Bitrán and Valenzuela, 2003). Similarly, institutional mechanisms that enable various degrees of engagement by consumers must be put in place in order for efficiency-oriented water supply schemes to be successful. Ultimately, the decision as to whether to involve the private sector, civil society and government is political and influences the kinds of governance mechanisms needed to ensure efficient and equitable service.

Provided that mechanisms to ensure affordable access by those without ability to pay are put into place, the potential economic and social benefits of improved access to water services are great. In addition to the considerable health benefits gained from connection to the official

BOX 12.9: TARGETING THE POOR THROUGH GRASSROOTS MECHANISMS IN MANILA, THE PHILIPPINES

In 1997, a twenty-five-year concession contract for water supply and sewerage in the city of Metro Manila, Philippines, was granted to two companies: Manila Water Company to supply the east side, and Maynilad Water Services to provide the west side of the city, with an aim at having spare capacity in case of failures. In order to increase access to the poor, the concession agreement provided for public standpipes for every 475 customers in 'depressed' areas. Instead of implementing this conventional solution, both companies have devised innovative approaches to extend service to poor areas.

Manila Water has a programme which relaxes some application requirements in order to enable water connections for poor customers. Group taps are designed for every two to five households where users get together to apply for a single connection. The group is given a 'mother meter' and thus, share the cost of their

usage. Each group chooses a representative, who is in charge of collecting and paying the bill to Manila Water. Besides group taps, Manila Water has a programme of community-managed water connections whereby a metered master connection is provided, and a community association acts as water distributor through individual or shared connections, which allows local residents to manage water according to their needs.

Maynilad Water Services favours individual to shared connections. Under its 'Water for the Community' programme, the land title requirement for connections is waived and payment of connection fees is deferred over a period of six to twelve months, and in some cases twenty-four months. NGOs were crucial in providing information to the private utilities, as well as in information campaigns aimed at community mobilization. They helped with the provision of materials while the community

contributed labour for carrying pipes into the city, which helped decrease connection costs. The number of connections has increased dramatically, and poor consumers, who now pay less for water than under their previous informal supply arrangements, are able to enjoy the same kind of services provided to other sectors of society.

In specific focus group discussions, several residents stated that connecting to the urban water supply had greatly decreased their water bills. In the Liwang Area of Manila, one resident related that, after being hooked up to the network, her monthly bills came to on average between 25 and 50 pesos per month, in contrast with 40 pesos per day spent on informal water vendors. Another resident, who used to pay a flat fee of 300 pesos per month to a neighbour with access to the system, now pays 60 pesos per month for a larger amount of water.

Source: Franceys and Weitz, 2003.

network, poor people freed from the burden of water collection can expect to have more time to engage in productive poverty-alleviating pursuits. Similarly, the public sector can expect to benefit from a reduction in unaccounted-for water losses, enabling them to price water more efficiently and potentially reduce subsidy mechanisms. Finally, participation of all kinds, from information-sharing, to consultation in PSP arrangements, to having a voice in decision-making and management in public-private community partnerships, is crucial for the long-term success of improved water supply and sanitation.

The choice of public-private partnership depends on the political, institutional, social and cultural features of the area where the service is to be provided. An assessment of the capability of governments to provide service in the target areas plus an analysis of the costs and benefits of different options and associated tariffs – including their potential impact on different sectors of society – will enable policymakers to make more informed choices as to which management tool can provide water services that best meet the societal goals of equity, efficiency and environmental sustainability.

5b. Virtual water trade

Virtual water, a concept that emerged more than a decade ago, is defined as the volume of water required to produce a given commodity or service. Allan proposed the term 'virtual water' to describe a phenomenon he observed in countries of the Middle East. They were using imports in the form of water-intensive products, such as food, to create a 'virtual' inflow of water as a means of relieving pressure on scarce domestic water resources (Allan, 1997). Several Middle Eastern nations, most notably Jordan and Israel, have altered their trade and development policies to promote the import of water-intensive products, generally agricultural crops, and the export of crops of high water productivity, that is, high income per unit of water consumed in production (Hofwegen, 2003). The adoption of such policies, in effect, recognizes the value of water.

As Allan (1997) noted, 'It requires about 1000 cubic metres of water to produce a ton of grain. If the ton of grain is conveyed to ... [an] economy short of freshwater and/or soil water, then that economy is spared the economic, and more importantly the political stress of

Wetlands in Amboseli Reserve, Kenya. These wetlands are fed by the Kilimanjaro mountain glaciers

Figure 12.4: Estimated annual water savings attributed to trade in wheat, Egypt, 1997–2001

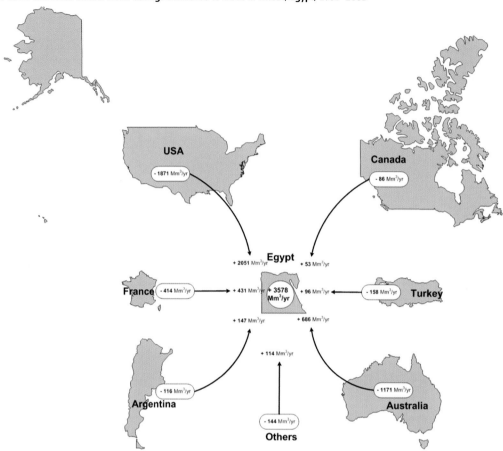

Note: Negative figures indicate the amount of water consumed in the production of the quantity of wheat exported, whereas positive figures indicate the amount of water savings by the importing country. Conversion formulas vary by country depending on various factors, including seed stock, type of technology used, and water management efficiency in the different countries.

Source: Chapagain and Hoekstra, 2005.

mobilizing about 1000 cubic meters of water.' These 'water savings' can be used to produce alternative, higher-value agricultural crops, to support environmental services, or to serve growing domestic needs. As seen in **Figure 12.4**, the imported goods may require more water during production in the alternative source country, but presumably this would be a country suffering from less water stress so that, overall, the efficiency of water use is promoted. Thus, 'virtual water' embedded in products is increasingly perceived as an alternative source of water for some water-stressed nations (see **Chapter 11**).

Recent research has revealed that the flow of such virtual water between nations is substantial (Hoekstra and Hung, 2002; Chapagain and Hoekstra, 2004; Chapagain et al., 2005). As observed by Allan (1997) 'more water flows

into the Middle East each year as virtual water than flows down the Nile into Egypt for agriculture'. Worldwide, the virtual water in international trade totals 1,625 Gm^3 annually. This amounts to about one-fifth of total world trade with approximately 80 percent of virtual water flows as trade in agricultural products, and the remainder as industrial products (Chapagain and Hoekstra, 2004) (see **Chapters 7 and 8**).

Increased trade in 'virtual water' has been proposed as a means to increase 'global water use efficiency', improve 'water security' for water stressed regions, and alleviate environmental impacts due to growing water demand (Turton, 2000). Emerging from the apartheid era, South Africa realized the potential benefits of adopting policy supporting virtual water imports as opposed to an ambitious programme of inter-basin water transfers (Allan, 2003). However, the concept of virtual water has not been accepted worldwide. Attempting to link agricultural imports directly to water dependency is difficult as numerous factors besides water availability affect farmers' planting decisions and production methods. Special trade arrangements, access to foreign exchange credits, market advantage – all affect the market and thus, decisions in the agricultural sector. Changing patterns of trade based on water conservation concerns need to be examined in the context of larger, national issues, including food security, food sovereignty[14], employment, foreign exchange requirements and perceived vulnerability to external political pressure. Additional research needs to be conducted on the social, economic, political as well as environmental 'implications of using virtual water trade as a strategic instrument in water policy' (Hofwegen, 2003). In the current period of political, economic and environmental uncertainty and instability, it is unlikely that societies will soon abandon the goal of food security. The new concept of food sovereignty introduced in recent years reflects the concerns of the small-scale agriculturalists. Llamas (2003) has argued that adopting 'virtual water trade' as a policy would require that the World Trade Organization or another international institution guarantee a prohibition on food embargos.

It should be noted that the concept of virtual water is still in developmental stages and several computational difficulties remain to be overcome. Figures on virtual water trade must be viewed cautiously as considerable uncertainty is associated with some of the underlying assumptions. Given the significant spatial and temporal variability in crop productivity and irrigation efficiency, extrapolations across geographic area and cultures could be problematic. Distinction should also be made with regard to the origin of the 'virtual water' in question (Llamas, 2003). Groundwater and surface water (see **Chapter 4**) have many alternative uses whereas options for soil moisture are more limited. One key question raised is: does adopting the concept of virtual water in designing trade policy contribute to an improvement in water availability?

Virtual water is an interesting concept, especially where water is in critically and chronically short supply and it will undoubtedly play an important role in influencing production and trade strategy for some nations. However, water is not the only factor of production and other

factors, such as energy costs, may come to play an increasingly important role in determining water resources allocation and use. It has been suggested that the concept of 'virtual water' trade is most applicable to the developed or high-income countries and that policies that might work for the relatively rich Middle East countries, however, may not work for the poorer economies of sub-Saharan Africa. This raises the question as to whether this approach will serve to alleviate or accentuate the differences between the rich and the poor countries.

The concept of virtual water could be valuable in promoting the production and trade in goods most suited to local environmental conditions and the development and adoption of water efficient technology. Adopting this approach, however, requires a thorough understanding of the impacts of such policies on socio-cultural, economic and environmental conditions, from local to national and regional levels. As the rigour of the analytical tools improves, undoubtedly so too will the usefulness of the concept of virtual water in terms of integrating the concerns of different sectors. However, as Allan (2003) observed, it may be that 'the virtual water remedy to local water scarcity will be shown to depend more on political processes rather than the scientific authority of the idea or precision with which it is defined'. In effect then, the success of the concept of virtual water may well turn on the achievements of global trade negotiations.

5c. Payments for environmental services

'Valuing' water is increasingly being extended to include an appreciation of human activities in upper catchment areas that contribute to maintaining the regular flow of clean water for downstream users (Pagiola and Platais, 2002; FAO, 2004). As increasingly recognized, land use and land cover management in the uplands affects water cycling through the Earth's natural systems. Healthy, intact ecosystems and their geologic substrate facilitate the hydrological cycle, filtering percolating water, distributing nutrients, providing a nurturing habitat for a wide diversity of wild animal and plant populations (biodiversity), and storing carbon. These and other functions, known as 'environmental services', endure only if fragile areas are protected, over exploitation of resources is avoided, pollution limited, and human intervention mediated by natural buffers. With the increasing recognition of the value of environmental services over the past decade, a variety of schemes have emerged that propose 'paying' for ecosystem services, or rewarding human actions contributing to preserving these

Local community installing water pipes for their village in Kinhare, Tanzania. This project will bring safe drinking water to the region for the first time. Previously people had to collect often unsafe drinking water from distant streams and carry it in buckets

14. Food sovereignty has been defined as the right of peoples, communities and countries to define their own agricultural, labour, fishing, food and land policies so that they are ecologically, socially, economically and culturally appropriate to their unique circumstances. It includes the right to food and to produce food, which means that all people have the right to safe, nutritious and culturally appropriate food and to food-producing resources and the ability to sustain themselves and societies. (*Source*: 'Food Sovereignty: A Right for All', a political statement of the NGO/CSO Forum for Food Sovereignty. Issued on 13 June 2002, in Rome). www.foodfirst.org/ progs/global/food/ finaldeclaration.html

functions. In effect, these schemes attempt to link the 'benefits' enjoyed by downstream users to the 'costs' incurred by the *de facto* catchment managers.

In one sense, paying for environmental services is an extension of the concept of 'cost recovery' discussed above. Payments for environmental services would fall into the category of indirect operating costs (see **Figure 12.1**). Acknowledging and compensating those individuals who actually manage the environment by those who benefit from these services attempts not only to reward good land husbandry, but in formalizing the relationship between the two groups of users, also enhances the long-term security of these ecosystem functions and the downstream benefit flows. Formal legislative support for payments for environmental services is recognized as one means of developing a sustainable source of environmental funding perhaps less vulnerable to political vagaries.

Payment for environmental services (PES)[15] has been commonly considered in the context of watershed management, biodiversity conservation, and more recently, carbon sequestration. Watershed-based PESs have a longer history and thus seem more straightforward. They avoid many of the constraints inherent in some of the newer schemes focusing on biodiversity conservation and carbon sequestration. As Scherr et al. (2004) observed, [m]arkets for watershed services are site and use specific and currently are limited to situations where the downstream beneficiaries – such as hydroelectricity power generation, irrigators, municipal water supply systems and industry – are directly and significantly impacted by upstream land-use'. Most payment schemes are in their infancy with analysts still learning from pilot projects both in developed and developing countries. The wide diversity of approaches reflects the variety of services supplied, participants' concerns, and physical and cultural environments. Ongoing programmes show promising results, however, with good prospects for scaling up to the basin, regional or national level (Scherr et al., 2004; Gouyon, 2003).

Typically, we find that watershed-based payment schemes fall into one of three categories: public payment programmes, self-organized private arrangements and open trading. Latin America and several of the developed countries have most experience with such schemes (FAO, 2004). The types of payment mechanisms associated with catchment protection include: best-management practices contracts, protection contracts, water quality credits,

stream flow reduction licences and reforestation contracts. Compensation generally comes in a variety of formats but mainly occurs as: direct financial remuneration; payment in kind, for example, infrastructure and equipment; and/or privileged access to resources or markets, for example, land-use rights. Case studies show that compensation programmes can have significant positive impacts on local livelihoods (InfoResources, 2004). In Costa Rica, landholders in watersheds designated critical are paid between US $30 and US $50 per hectare per year for good land management practices (Scherr et al., 2004). In Mindanao in the Philippines, regular payments to residents in the catchment area of the Mount Apo geothermal facility have been ongoing for many years (Warner et al., 2004), while in Europe a new PES project is being planned for the lower Danube River Basin (see **Chapter 14**).

Whereas typically, development programmes channel funds to or through local governments, it is proposed that new programmes could more efficiently and effectively target funds directly to the environmental stewards themselves. Transparent processes and multi-stakeholder involvement with emphasis on locally determined priorities and participatory planning, implementation and monitoring are needed to bridge the gap between proposed and existing programmes. Given the close links between poverty and resource dependency, designing PES programmes that reward the poor for good natural resources stewardship is key to effective conservation. While the vulnerability of hydroelectric facilities to poor land management in catchments is of growing concern, for many countries the not unrelated issue of poverty alleviation, particularly in rural areas, has become a national priority, as has the decentralization of government services. The potential of PES schemes to address these several issues is increasingly recognized. Indeed, paying for environmental services, in particular as related to hydroelectric production, responds to the demands for clean energy – hydroelectricity being a well-known example (see **Chapter 9**) – for better catchment area management, for greater local responsibility, and for poverty reduction. Lessons have already been learned, as indicated in **Box 12.10**. In building on royalty programmes traditionally associated with resource extraction, PES seems to have the potential to serve as a new paradigm for sustainable watershed management, integrating the concerns of users throughout the basin.

15. Another related term is 'green water credits', which are payments proposed as a mechanism for the transfer of cash to rural people in return for better management of 'green water' (see **Chapter 4**) resources (Dent, 2005).

BOX 12.10: LESSONS LEARNED IN PAYING FOR ENVIRONMENTAL SERVICES

A review of programmes in which payments for environmental services have been made concluded that these schemes are most successful when:

■ financial sustainability is promoted by independence from long-term external financial support

■ locally defined best management practices are taken into consideration

■ transaction costs are minimized

■ rights and responsibilities of all parties including intermediaries are clearly defined

■ payment is linked to performance monitored regularly

■ resource rights and tenure are secured

■ legal and institutional frameworks create an enabling environment

■ mechanisms for fees assessment, collection and disbursement are locally determined, clearly defined, and transparent

■ poverty reduction is explicitly addressed, especially providing women and disadvantaged groups with opportunities to participate in planning, implementation and monitoring, for example, targeting small-holders as service providers.

Source: Warner et al., 2004.

According to many analysts[16], '[m]arkets for forest ecosystem services are expected to grow in both developed and developing countries over the next two decades' (Scherr et al., 2004). Currently nearly one-third of the world's largest cities depend on forested watersheds for their water supplies (see **Chapter 3**). Water demand, projected to double if not triple in the coming half century, will grow fastest in the developing world. It is increasingly apparent to water providers and water users that investments in watershed protection can be far more economical than investments in costly engineered solutions, such as water treatment plants or long-distance canals. For governments – principal purchasers of many ecosystem services but also catalysts for many private-sector payment schemes – incorporating PES schemes in basin-wide integrated water resources management programmes has the potential to deliver rural poverty alleviation as well as environmental conservation and enhanced water security. Thus,

recognizing the value of environmental services in the very real sense of financial compensation may be an attractive alternative to governments facing growing rural-to-urban migration and increasing pressure on already overstretched urban water supply systems.

The assessment of strictly anthropogenic impacts is tricky, especially as the timing and scale of the impacts of different land-use and vegetation management practices on hydrological function and resources vary according to local environmental conditions – which are sometimes confounded by natural phenomena. Experience has shown that although the effects of human actions are more directly observed in smaller catchments, they are also visible on larger scales. More extensive monitoring and evaluation is needed in order to better understand land-use and water linkages so as to refine the diverse mechanisms emerging for compensation for environmental services (Fauré, 2004).

16. Founded in 2000, the Katoomba Group is a collection of government officials, private sector professionals, academic researchers and NGO representatives devoted to sharing information and experience on development of financial markets for ecosystem services. www.katoombagroup.com/

Part 6. Indicators under Development

Economic valuation has been defined as the task of assigning a money metric to the benefits and costs associated with different policies so that different governance options can be compared and ranked. To enable the comparison of actual achievements against targets or projected outcomes, we must look to selected variables for quantitative measures that we could employ as 'indicators'. The indicators noted below are those that could be used to monitor progress towards society valuing water in a manner likely to realize societal objectives, including efficiency, equity and environmental sustainability. The indicators noted are still in developmental stages and clearly more research and experience is needed to assess their ease of use, robustness, and reliability with regard to understanding their utility for cross-country comparisons.

Water Sector Share in Public Spending: In highlighting public-sector spending in the water sector, this indicator illustrates the level of political commitment to meeting the Millennium Development Goals (MDGs) on water. Expressed as a percentage, this indicator shows the proportion of the total public budget allocated to water systems development. Data on annual investments by sector are generally available from national statistical yearbooks, country-wide economic reviews, and the government office responsible for water sector development. Widely applicable, this indicator could be used at any level where statistics are available.

Ratio of Actual to Desired Level of Investment: This indicator illustrates the extent to which investments required to meet water-related MDGs are on track. Computed as AL/DL, where AL is the actual level and DL refers to the desired level of investment, this ratio indicates the degree to which planned investment is realized. Although not an indicator of actual hook-ups, the allocation of funds for the installation of the necessary infrastructure is a crucial first step. The data required for this calculation should be available from the national budget documents. Data on desired level of investment could be obtained from the project documents and feasibility studies of relevant infrastructure development projects, or perhaps government offices in charge of water resources planning and infrastructure development.

Rate of Cost Recovery: This indicator measures the total amount of fees actually collected as a proportion of total

revenues scheduled to be collected. It reflects as well the effectiveness of fiscal administration and institutional governance in the water sector. Data required to prepare this indicator include: total water charges to be collected and those actually collected. These are usually available from the published annual reports of water utilities and national budget documents. The rate of cost recovery could also be viewed as an indicator of the population's willingness to pay for water services. The effectiveness of the fee collection system will have a direct influence on private sector willingness to invest in this sector and the ability of public water systems, which depend on cost recovery through charging, to meet projected expansion plans as well as maintenance obligations.

Water Charges as a Percentage of Household Income/Expenditure: Water charges are seen as an important instrument for improving cost recovery in the water utilities sector. Expressed as a proportion of household income or household expenditure, water charges illustrate the pressure of this expense on household finances (see **Figure 12.5**). Indirectly, this figure may also serve as an indicator of household willingness to conserve and use water efficiently. A very low rate would indicate little incentive to conserve or use water efficiently. The potential for introducing effective demand management measures would depend on this figure. Data required to estimate this indicator are generally collected through household income and expenditure surveys conducted by governments at regular intervals.

Figure 12.5: Affordability of utility services in Eastern Europe and Central Asia, 2003–04

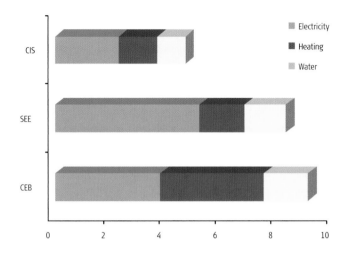

Legend:
■ Electricity
■ Heating
■ Water

Percentage of total household expenditures

Key: CIS = Commonwealth of Independent States: Armenia, Azerbaijan, Belarus, Georgia, Kazakhstan, Kyrgyz Republic Moldova, Russia, Tajikistan, Turkmenistan, Ukraine and Uzbekistan.
   SEE = South Eastern Europe: Albania, Bosnia and Herzegovina, Bulgaria, Croatia, FYR Macedonia, Romania and Serbia and Montenegro.
   CEB = Central Eastern Europe and Baltic States: Czech Republic, Estonia, Hungary, Latvia, Lithuania, Poland, Slovak Republic and Slovenia.
Note: Affordability estimates are unweighted averages. Data on district heating were not available for Albania and Georgia, where heat networks are not functioning.

Source: Fankhauser and Tepic, 2005.

Part 7. Conclusions and Recommendations

The supply and affordability of water is of growing political and economic concern as it is increasingly recognized that safe water is not only essential for health, but also for social and economic development. As the world's population grows in numbers and wealth, the demand for more and better water supply and sanitation services increases, as does the competition between sectors serving other societal needs, such as food, manufactured goods and environmental services. Understanding the value of water is essential if this ever more scarce resource is to be more effectively and efficiently applied to meeting societal goals.

Valuation is the process of assessing the impact of various policies and initiatives then assigning weights to various policy outcomes based on the importance of various policy objectives or criteria. Value, in this sense, is not assigned specifically to water but to consequences of a change in governance or policy initiatives. Values vary depending on the services in question, the location, the policy context and other circumstances. Economic valuation assesses outcomes based on willingness to pay and willingness to accept compensation. Other considerations include social values, such as rights to clean water and adequate sanitation irrespective of ability

to pay, gender equity, and respect for religious and cultural beliefs and environmental concerns, including concern for biodiversity preservation and wetland protection.

Economic valuation can be useful for assessing the potential net benefits of proposed public policy initiatives as well as the realized benefit of previous policies. For example, research indicates that the economic returns to public investments in irrigation in Asia over the past three decades have been quite modest as compared to returns from alternative investments (e.g. research, rural roads and education) or even the cost of capital. High benefit-cost ratios for investments in water supply and sanitation in areas where such facilities are lacking suggest that a reallocation of resources toward domestic water services would improve social welfare.

Though considerable effort has gone into expanding and refining the analytical methods for measuring water-related values, results are only as reliable as the assumptions and data upon which the analyses are based. More work needs to be done on refining these tools and improving data collection. Economic valuation is rarely seen as a wholly acceptable solution. Although helpful in elucidating the trade-offs between different objectives in alternative scenarios, it is often necessary to enter the political realm or formalized negotiations in order to resolve the dissonance inherent in situations containing multiple conflicting objectives. Greater attention needs to be focused on understanding more clearly '*who benefits*' and '*who bears the costs*' in any water resources policy or development initiative. Seeming inconsistencies in governments' stated objectives and activities can often be understood by analysing the distributional aspects of government investments.

Charging, as a governance policy, aims to balance multiple competing objectives. Most water professionals now feel that the reform of charging policies is critical to improving the performance of the water services sector. Revised charging structures need to be more widely implemented to improve cost recovery, to facilitate adequate maintenance and expansion of water supply systems, and to provide incentives for conservation, while making water services affordable and available to all. The political unpopularity of increased charges will need to be overcome with phased tariffs in some areas but also programmes to help consumers understand the true costs and value of regular, reliable water and sanitation services. Given that willingness to pay, the limit to charging, is a function of information, better informed, and better served, customers should facilitate cost recovery and thus the development of water services. For the moment, however, many poor find full-cost charging, unaffordable, so subsidies will probably need to be maintained in many areas if MDGs on water supply and sanitation are to be met in the near term.

Several factors, but especially the shortfall of funds for infrastructure development, have led many national and local governments to look to the private sector for assistance in water systems management and

development. The experience of private sector involvement in the water sector, however, has been decidedly mixed. Indeed, dissatisfaction with water services after private sector involvement brought consumers onto the streets in Bolivia. Given the nature and role of the resource, the nature of associated infrastructure investment and the social sensitivity regarding water supply, it is almost impossible to depoliticize water. However, an increasing number of examples of successful public-private partnerships should serve to enlighten future developments. Governments, for their part, need to take more seriously their regulatory responsibilities to ensure quality service and socially equitable access.

Virtual water, that is, the water embedded in various goods, has become a topic of increasing interest as water-stressed countries reassess their production priorities. Many nations have realized that they can in effect import 'virtual water' in the form of goods requiring significant water for their production, for example, food. Thus, countries experiencing severe and persistent water stress may opt for trade policies focused on importing water-intensive goods while exporting more high-value water-efficient products. Similarly, the growing interest in payment for environmental services reflects societies' and governments' heightened appreciation of the value – including cost savings – of ecosystem functions especially as related to water supply. These include filtering water, regularizing water flows, and buffering against flood and tidal surges.

Throughout this chapter the concept of measurement has dominated the discussion. Although we acknowledge that some values of water are difficult if not impossible to measure, informed decision relies upon such information developed largely through regular monitoring and data collection. Indicators which focus on critical aspects of water resources management and allocation have an important role to play in developing efficient and effective systems of water governance. Continued work is needed to further refine the art and science of developing indicators, particularly with regard to the social and environmental dimensions of value, and at the local and national level. Both theoretical and real-world studies are needed.

As freshwater becomes ever scarcer and increasingly contested, the importance of understanding the diverse values of water increases. Recognizing the distinction between 'valuing' and 'valuation' is critical. Valuing water is not solely a matter of applying sophisticated economic techniques and calibrating various water-related goods and services in terms of a money metric. Rather, it means involving all stakeholders in a process of determining priorities and making informed decisions on specific courses of action that will enable society to better meet its water-related goals. Economic valuation is a tool that can assist in this process and charging is but one strategy among many possibilities. It is important not only that more individuals involved in policy making and planning are made aware of the strengths and weaknesses of the various economic techniques that can be applied in assessing governance strategies, but that stakeholders become better capable of understanding and articulating the wide spectrum of values that water holds for them.

References and Websites

Allan, J. A. 2003. Virtual Water – the Water, Food, and Trade Nexus: Useful Concept or Misleading Metaphor? *Water International* Vol.28, No.1, pp. 4–11.

——. 1997. Virtual Water: A Long-term Solution for Water-Short Middle Eastern Economies? Paper delivered at British Association Festival of Science, 6 Sept. Leeds, UK.

Bitrán, G. and Valenzuela, E. 2003. Water Services in Chile: Comparing Public and Private Performance. (Public Policy for the Private Sector, Note No. 255). March. Washington DC, World Bank.

Boardman, A. E., Greenberg, D. H., Vining, A. R. and Weimer, D. L. 2000. *Cost-Benefit Analysis: Concepts and Practice* (2nd edn). Upper Saddle River, NJ, Prentice Hall.

BBC (British Broadcasting Corporation) 2005. *Recycling Around the World.* news.bbc.co.uk/1/hi/world/europe/4620041.stm news.bbc.co.uk/1/hi/magazine/4373350.stm

Chapagain, A.K. and Hoekstra, A. Y. 2004. Water Footprints of Nations. (Value of Water Research Report Series No. 16). Delft, UNESCO-IHE.

Chapagain, A. K., Hoekstra, A. Y. and Savenije, H. H. G. 2005. *Saving Water through Global Trade.* (Value of Water Research Report Series No.17). Delft, UNESCO-IHE.

Cornish, G., Perry, C. and Bosworth, B. 2004. *Water Charging in Irrigated Agriculture: an Analysis of International Experience.* Rome, FAO.

Cosgrove, W. J. and Rijsberman, F. R. 2000. *World Water Vision: Making Water Everybody's Business.* London, Earthscan.

Dasgupta, P. and Mäler, K. G. 2004. *Environmental and Resource Economics: Some Recent Developments.* Beijer International Institute of Ecological Economics, Stockholm.

Dent, D. 2005. Green Water Credits. Paper delivered at FAO International Conference on Water for Food and Ecosystems: Make it happen! The Hague, 31 January – 4 February.

Estache, A. and Goicoechea. 2005. A Research Database on Infrastructure Economic Performance. (Policy Research Working Paper 3643). Washington DC, World Bank.

Estache, A. and Kouassi, E. 2002. Sector Organization, Governance, and the Inefficiency of African Water Utilities. (Policy Research Working Paper 2890). Washington DC, World Bank.

Estache, A. and Rossi, M. 2002. How Different is the Efficiency of Public and Private Water Companies in Asia? *World Bank Economic Review* Vol.16, No.1.

Estache, A., Perelman, S. and Trujillo, L. 2005. Infrastructure Performance and Reform in Developing and Transition Economies: Evidence from a Survey of Productivity Measures. (Policy Research Working Paper 3514).Washington DC, World Bank.

Evenson, R. E., Pray, C. and Rosegrant, M. W. 1999. Agricultural Research and Productivity Growth in India. (Research Report 109). Washington DC, International Food Policy Research Institute (IFPRI).

Fan, S., Hazell, P. and Thorat, S. 1999. Linkages between Government Spending, Growth and Poverty in Rural India. (Research Report 110). Washington DC, IFPRI.

Fan, S., Jitsuchon, S. and Methakunnavut, N. 2004. The Importance of Public Investment for Reducing Rural Poverty: Thailand. (Discussion Paper 7). Washington DC, IFPRI.

Fan, S., Zhang, L. and Zhang, X. 2002. Growth, Inequality and Poverty in Rural China: The Role of Public Investments. (Research Report 125). Washington DC, IFPRI.

Fankhauser, S. and Tepic, S. 2005. Can Poor Consumers Pay for Energy and Water? An Affordability Analysis for Transition Countries. (Working Paper No. 92). London, European Bank for Reconstruction and Development (EBRD).

Fauré, J.-M. 2004. Land-water linkages in rural watersheds: implications for payment schemes and environmental services. Payment Schemes for Environmental Services in Watersheds, Land and Water. (Discussion Paper 3). Rome, FAO.

Finnegan, W. 2002. Letter from Bolivia: Leasing the Rain, *New Yorker,* 8 April.

FAO (Food and Agriculture Organization). 2004. Payment Schemes for Environmental Services in Watersheds, Land and Water. (Discussion Paper 3). Rome, FAO.

Franceys, R. and Weitz, A. 2003. Public-Private Community Partnerships in Infrastructure for the Poor. *Journal of International Development* Vol.15, No.8.

Freeman III, A. M. 2003. *The Measurement of Environmental and Resource Values: Theory and Methods* (2nd edn). Washington DC, Resources for the Future.

Gleick, P. H., Cain, N., Haasz, D., Henges-Jeck, C., Hunt, C., Kiparsky, M., Moench, M., Palaniappan, M., Srinivasan, V. and Wolff, G. 2004. *The World's Water 2004-2005.* Washington DC, Island Press.

Gouyon, A. 2003. *Rewarding the Upland Poor for Environmental Services: A Review of Initiatives from Developed Countries.* Rome, World Agroforestry Centre, IFAD. www.worldagroforestrycentre.org/sea/Networks/ RUPES/download/paper/Agouyon_RUPES.pdf

Gresham, Z. 2001. *Lessons from the Field: Private Sector Involvement in Water Projects.* Morrison and Foerster, www.mofo.com/news/updates/files/update545.html

Griffin, R. C. 2006. *Water Resource Economics: The Analysis of Scarcity, Policies, and Projects.* Cambridge, MA, MIT Press.

Hall, D. C. 2000. Public Choice and Water Rate Design. A. Dinar, (ed.) *The Political Economy of Water Pricing Reforms.* New York, Oxford University Press.

——. (ed.) 1996. *Marginal Cost Rate Design and Wholesale Water Marketing.* Greenwich, CT, JAI Press.

Hanemann, W. M. 2005. The Economic Conception of Water. (Working paper No. 1005). Berkeley, California, California Agricultural Experiment Station.

——. 1997. Prices and Rate Structures. D. D. Baumann, J. J. Boland and W. M. Hanemann (eds), *Urban Water Demand Management and Planning.* New York, McGraw Hill, pp. 31–95.

Herrington, P. 1999. *Household Water Pricing in OECD Countries.* Paris, OECD.

——. 1987. *Pricing of Water Services.* Paris, OECD.

Hoehn, J. P. and Krieger, D. P. 2000. An Economic Analysis of Water and Wastewater Investments in Cairo, Egypt. *Evaluation Review* Vol.24, No.6, pp. 579–608.

Hoekstra, A. Y. and Hung, P. Q. 2002. Virtual Water Trade: A Quantification of Virtual Water Flows Between Nations in Relation to International Crop Trade. (Research Report no. 11). Delft, UNESCO-IHE. www.ihe.nl/downloads/projects/ report11-hoekstra-hung.pdf

Hutton, G. and Haller, L. 2004. *Evaluation of the Costs and Benefits of Water and Sanitation Improvements at the Global Level.* Geneva, WHO.

InfoResources. 2004. Compensation for Ecosystem Services (CES): A Catalyst for Ecosystem Conservation and Poverty Alleviation? *Focus,* 3/04. Geneva, InfoResources.

International Bottled Water Association (IBWA). 2005. 2005 Market Report Findings. www.bottledwater.org/public/BWFactsHome_main.htm

Llamas, R. 2003. Online Conference on virtual water at the World Water Council, Marseilles.

MacRae Jr, D. and Whittington, D. 1997. *Expert Advice for Policy Choice: Analysis and Discourse.* Washington DC, Georgetown University Press.

Matthews, O., Brookshire, D. S. and Campana, M. E. 2001. The Economic Value of Water: Results of a Workshop in Caracas, Venezuela, August. Water Resources Program, University of New Mexico, Albuquerque.

Millennium Ecosystem Assessment. 2005. *Ecosystems and Human Well-Being: A Framework for Assessment.* Washington DC, Island Press.

Morgan, M.G. and Henrion, M. 1990. *Uncertainty: A Guide to Dealing with Uncertainty in Quantitative Risk and Policy Analysis.* Cambridge, UK, Cambridge University Press.

National Center for Small Communities (NCSC). n.d. (ca. 2000). *Action Guide for Source Water Funding: Small Town and Rural County Strategies for Protecting Critical Water Supplies.* Washington DC, NCSC and Environmental Protection Agency (EPA).

Natural Resources Defense Council (NRDC). 1999. Bottled Water: Pure Drink or Pure Hype? www.nrdc.org/water/drinking/nbw.asp

Office of Management and Budget (OMB). 1992. Guidelines and Discount Rates for Benefit-Cost Analysis of Federal Programs. (Circular No. A-94 [Revised]). Washington DC, US Government.

Organisation for Economic Co-operation and Development (OECD). 2005. Financing Strategy for Urban Wastewater Collection and Treatment Infrastructure in Armenia. Paris, OECD.

——. Environment Directorate. 2003. Social Issues in the Provision and Pricing of Water Services. Paris, OECD.

——. 1985. *Management of Water Projects: Decision-making and Investment Appraisal.* Paris, OECD.

Pagiola, S. and Platais, G. 2002. Payments for Environmental Services. (Environmental Strategy Notes, No.3). Washington DC, World Bank.

Powe, N. A., Garrod, G. D., McMahon, P. L. and Willis, K. G. 2004. Assessing customer preferences for water supply options using mixed methodology choice experiments. *Water Policy* Vol.6, No.5, pp. 427–41.

Raghupati, U. and Foster, V. 2002. Water Tariffs and Subsidies in South Asia: Understanding the Basics, A Scorecard for India. (Water and Sanitation Program, Paper No. 2). Washington DC, World Bank.

Rogers, P., Bhatia, R and Huber, A 1998. *Water as a Social and Economic Good: How to Put the Principle into Practice.* Stockholm, Global Water Partnership/Swedish International Development Agency.

Rogers, P., de Silva, R. and Bhatia, R. 2002. Water is an Economic Good: How to Use Prices to Promote Equity, Efficiency and Sustainability. *Water Policy* Vol.4, No.1, pp. 1–17.

Sagoff, M. 2004. *Price, Principle and the Environment.* Cambridge, UK, Cambridge University Press.

Savenije, H. H. G. and van der Zaag, P. 2001. 'Demand Management' and 'Water as an Economic Good': Paradigm with Pitfalls. (Value of Water Research Report series No. 8). Delft, UNESCO-IHE.

Scherr, S., White, A. and Khare, A. 2004. *For Services Rendered: The Current Status and Future Potential of Markets for Ecosystems Services Provided by Tropical Forest.* (ITTO TS-21). Yokohama, International Tropical Timber Organization.

Shah, T., van Koppen, B., Merrey, D., de Lange, M. and Samad, M. 2002. Institutional Alternatives in African Smallholder Irrigation: Lessons from International Experience with Irrigation Management Transfer. (Research Report 60). Colombo, Sri Lanka, International Water Management Institute (IWMI).

Shaw, W. D. 2005. *Water Resource Economics and Policy: An Introduction.* Northampton, MA, Edward Elgar.

Tsur, Y., Roe, T. Doukkali, R. and Dinar, A. 2004. *Pricing Irrigation Water: Principles and Cases from Developing Countries.* Washington DC, Resources for the Future Press.

Turton, A. R. 2000. Precipitation, people, pipelines and power: Towards a political ecology discourse of water in Southern Africa, P. Stott and S. Sullivan (eds), *Political Ecology: Science, Myth and Power,* London: Edward Arnold, pp. 132–53.

UNDP (United Nations Development Programme). 2004. *What are Public-Private Partnerships?* pppue.undp.org/index.cfm?module=ActiveWeb&page=WebPage&s=what

UNICEF/WHO. 2004. *Meeting the MDG Drinking Water and Sanitation Target.* Geneva, WHO/UNICEF.

van Hofwegen, P. 2003. Virtual Water: Conscious Choices. *Stockholm Water Front,* 2 June. (4).

Wallsten, S. and Kosec, K. 2005. Public or Private Drinking Water? The Effects of Ownership and Benchmark Competition on U.S. Water System Regulatory Compliance and Household Water Expenditures. (Working Paper 05–05). Washington DC, AEI-Brookings Joint Center for Regulatory Studies.

Warner, K, Huang, M and Middleton, D. 2004. *Financial Incentives to Communities for Stewardship of Environmental Resources.* Washington DC, Winrock International and US Agency for International Development (USAID).

Water Resources Council (WRC). 1983. *Economic and Environmental Principles and Guidelines for Water and Related Land Resources Implementation Studies.* Washington DC, Superintendent of Documents (US).

Whittington, D., Boland J. and Foster, V. 2002. Water Tariffs and Subsidies in South Asia: Understanding the Basics. Public Private Infrastructure Advisory Facility (PPIAF) and the Water and Sanitation Program (WSP), (Paper No. 1). Washington DC, World Bank. www.wsp.org

Winpenny, James. 2003. Financing Water for All: Report of the World Panel on Financing Water Infrastructure [a.k.a. The Camdessus Report]. Marseilles, World Water Council.

World Bank. 2005. *Yerevan Water and Wastewater Project: Europe and Central Asia Region.* (Project Appraisal Document 30251). Washington DC, World Bank.

——. 2004. Reforming the Water Sector. *Reforming Infrastructure: Privatization, Regulation and Competition.* Washington DC, World Bank.

——. 1997. *Selecting an Option for Private Sector Participation: Toolkits for Private Sector Participation in Water Supply and Sanitation.* Washington DC, World Bank.

——. 1994. *World Development Report: Infrastructure for Development.* Oxford, UK, Oxford University Press.

World Health Organization (WHO). 2000. Bottled Drinking Water. w3.whosea.org/en/Section260/Section484/Section487_7866.htm

Yardley, J. 2005. China's Next Big Boom Could be Foul Air. *New York Times,* 30 October.

Young, R. A. 2005. *Determining the Economic Value of Water: Concepts and Methods.* Washington DC, Resources for the Future Press.

Beijer Institute website on environmental economics: www.beijer.kva.se/

European Commission on the Environment on water policy: europa.eu.int/comm/environment/water/index.html

IUCN's Water and Nature Initiative on valuing water: www.iucn.org/themes/wani/value/index.html

OECD website on water financing: www.oecd.org/document/7/0,2340,en_2649_201185_33719751_1_1_1_1,00.html

United Nations Development Programme (UNDP) on public-private partnerships (medium-sized): pppue.undp.org/index.cfm

UNESCO-IHE website on virtual water: www.waterfootprint.org/

WWF Freshwater programme: www.panda.org/about_wwf/what_we_do/freshwater/index.cfm

World Bank Water Supply and Sanitation programme website: www.worldbank.org/watsan

World Bank's Public-Private Infrastructure Advisory Facility (PPIAF) regarding large scale PPPs: www.ppiaf.org/

World Water Council on virtual water: www.worldwatercouncil.org/index.php?id=866

For definitions:

www.ecosystemvaluation.org

Knowledge has to shuttle between the local and the global level, taking account of the retroactive effect of the global on the particular.

Edgar Morin

CHAPTER 13

Enhancing Knowledge and Capacity

By

UNESCO-IHE
*(Institute
for Water Education)*

*Madhukari Ganokendra (People's Centre), in
Rajapur village, western Bangladesh, holds monthly
meetings to discuss primary school attendance and
other important issues for the community to take
action*

Key messages:

Financial investments made in the last decades in the water sector have often failed to bring about the expected outcomes, largely due to lack of attention given to enhancing knowledge and capacity. While infrastructure is needed, it is doomed to deteriorate if not properly maintained by adequate human resources and institutional capacity within an enabling environment. In a time of climate change and declining hydrological data collection systems, all countries need to take seriously the threat to their water resources and invest in capacity development.

■ Self-assessments of knowledge and capacity needs are urgently required to assist water resources managers in all challenge areas in setting priorities, identifying gaps and improving the effectiveness with which they can respond to a continuously changing environment.

■ It is essential that the knowledge base of capacity development be enhanced through case studies, best practices, twinning organizations and shared experiences and that the capabilities of national statistical agencies to deal with water sector data be improved.

■ Increased access to education at all levels through information and communication technologies is a cornerstone for development, and efforts to broaden individual capacities through education should be actively pursued.

■ Knowledge requires continuous investment to enable society to adapt to an uncertain future generated by climate change. In particular, increased investments in the hydrological data network and remote sensing are needed to provide the information necessary for modelling future scenarios.

■ The capacity of water management institutions should be increased to ensure that they have a clear mandate, an effective organizational system, and improved decision support through lessons learned and indigenous knowledge.

An education programme provides free of cost relevant life skills ranging from reading, writing, simple calculating to tailoring, furniture making, etc. to out of school youth and adults in Bhutan Katha public school provides an education to 5–16 years olds in Govinpuri slums, south Delhi, India

Part 1. Assessing Knowledge and Capacity

Spurred by the Millennium Development Goals (MDGs), many nations are now intensifying their actions to improve water services and infrastructure development. Over the past two decades, developing countries have invested hundreds of billions of dollars in water services and water resources, a substantial portion of which has failed to bring about the desired outcomes and impacts. The operations assessments by the development banks and other donors attribute this in many cases to inadequate knowledge bases and weak capacities.

As our understanding of the interactions between water management and society develops, it becomes increasingly evident that the past focus on developing infrastructure has overlooked the need for a strong knowledge base and capacity to plan, manage and use that infrastructure and enable proper governance of the water sector. Today, there is a growing consensus that knowledge and capacity in the water sector is a primary condition for sustainable development and management of water services.

Knowledge development and accessibility lie at the heart of this concern. Knowledge takes a variety of forms: as databases; as the competence to integrate and interpret data and create meaningful information that can inform decisions; as capacity to generate new data and information, to identify gaps, to learn from past experiences and to explore the future; and educational and dissemination mechanisms. A knowledge system extends well beyond data pertaining to physical and technical parameters. Involving civil society and increased community participation foster a greater understanding of the interactions of the complex social and environmental processes involved in water management, which enables the rethinking of approaches to effective water development.

The knowledge base is made up of databases, documents, models, procedures, tools and products. It also includes knowledge that may not be explicitly available because it is contextual, cultural and relates to skills, heuristics, experience and natural talents (such as local or indigenous knowledge). This implicit knowledge leads the way for capacity-to-act or a competence to solve problems, but describing and communicating such implicit knowledge remains challenging (Snowden, 2003).

The support of a strong knowledge base can greatly improve capacity development and spur the kind of informed decision-making that drives policy directives, which enable

local institutions to be better equipped to direct their own self-sufficient and sustainable futures in the face of change. As such, research, assessment, know-how and communication are not simply components of a development initiative that compete with other components: they are primary targets in any effort towards effective and sustainable development in water-related sectors.

1a. From knowledge to capacity development

Capacity development is the process by which individuals, organizations, institutions and societies develop abilities (individually and collectively) to perform functions, solve problems and set and achieve objectives (UNDP, 1997; Lopes and Theisohn, 2003). A country's capacity to address water-related issues is not just the sum total of individual capacities, but rather a broad holistic view of the central concerns of management, namely how to resolve conflict, manage change and institutional pluralism, enhance coordination, foster communication, and ensure that data and information are collected, analysed and shared. This involves not only individual capacities (human resources), but also the effectiveness, flexibility and adaptability of organizational processes (institutional capacity) and an enabling and stimulating management framework (the enabling environment). These three levels of capacity development are presented in **Figure 13.1** with its associated activities, outputs and goals. A detailed description of these three levels is included in Part 3 of this chapter.

Sustainable development increasingly requires countries to have the capacity to put in place effective knowledge generation and learning mechanisms. This capacity-to-learn or 'adaptive capacity' is the potential or capability of a system to adjust or change its characteristics or behaviour, so as to better cope with existing and future stresses. More specifically, adaptive capacity refers to 'the ability of a socio-ecological system to cope with novelty without losing options for the future' (Folke et al., 2002)

The knowledge base is seen to be of a higher order than a database... it relates to how such collected explicit knowledge on the world's water resources and their use is archived and analysed

In order to achieve sustained progress, knowledge building and capacity development must be viewed as development objectives in and of themselves...

Figure 13.1: Capacity development: Levels, activities, outputs and goals

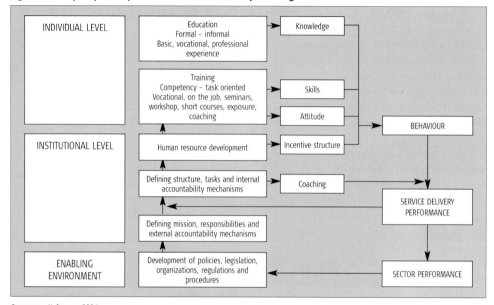

Source: van Hofwegen, 2004.

Section 4: MANAGEMENT RESPONSES & STEWARDSHIP

and 'is an aspect of resilience that reflects learning, flexibility to experiment and adopt novel solutions, and development of generalized responses to broad classes of challenges' (Walker et al., 2002). There is a need therefore to build into capacity development a concern that individuals have the skills to innovate when faced with a non-standard problem and a structural flexibility that does not penalize, but rather rewards and capitalizes on such innovation.

A new paradigm for water development has begun to emerge. It stresses the importance of country ownership and shifts the focus from passive knowledge transfer (e.g., from the North to the South) to knowledge acquisition and integration within the developing countries themselves. It does this by supporting home-grown processes for knowledge development – often using existing local and indigenous capacities – while also specifically including local participatory processes. In order to achieve sustained progress, knowledge building and capacity development must be viewed as specific development objectives, which command their own resources, management attention and evaluation standards, much along the lines of gender, poverty or environmental issues (Morgan, 2000).

The concept of capacity development implies that improved water services delivery and sustainable development are to be achieved as much through

improving the enabling environment, the institutional frameworks and human resources as through the technocratic approach of investments in infrastructure. Capacities must be developed at all three levels while acknowledging that these layers of capacity are mutually interdependent – if one is pursued in isolation, development still remains skewed and inefficient (Fukuda-Parr et al., 2002). The right combination of actions depends on the local situation, which calls for extensive prior analysis and priority setting, for instance, by region or by river catchment (Alaerts et al., 1999).

1b. Identifying socio-economic benefits

While high-income countries have been able to couple large investments in infrastructure with human and institutional knowledge building many middle- and low-income nations lag behind in their ability to adapt to the ever-increasing pace of change in a complex world (Alaerts et al., 1999). Industrialized countries, for example, can afford to invest in better understanding and preparedness for the effects of climate change. Middle-income countries are generally characterized as having built sufficient infrastructure assets to provide adequate water services and prepare for the 'conventional' larger water-related risks such as floods. They may still however lack the necessary institutional and human knowledge base that is needed to reap a greater benefit from water resources development for more sustainable growth. Lower-income countries typically have not yet been able to invest in a minimum stock of water infrastructure, and often do not have the capacity to govern and manage these investments effectively once they are made. They thus have a strong incentive to invest their scarce resources in infrastructure that brings rapid returns. However, past experience shows that heavily investing in infrastructure without enhancing existing local capacities can result in dilapidated infrastructure, exasperated water problems and increased debt.

Even though there exists in the world at large the know-how and knowledge to solve many, if not most, of the world's pressing water problems, this knowledge is often slow to make an impact. National governments, often with overburdened and underpaid staff, possess only limited capability to acquire and interpret that knowledge, and turn it into practical action and realistic proposals. Vested interests often prevent the adoption of new approaches, and staff are forced to respond to short-term priorities.

While it is agreed that good governance and management require local government involvement, the devolution of responsibility for managing a range of water services from national to lower levels further raises the need to strengthen capacities. This is because local-government staff tend to have benefited less from proper education than their colleagues in central-government agencies, and because the local administrative procedures are even less geared to appreciate the value of sound knowledge. Similarly, better governance hinges on users and communities that are informed and have the capacity to access and use information with which they can hold government accountable.

Each country needs a development strategy that recognizes the balance between knowledge, capacity and infrastructure in order to adopt the most suitable governance strategy and utilize its water resources in line with sustainable development. Such a development strategy must acknowledge that radical social, environmental and technological changes are taking place at an increasing rate. As discussed throughout this Report, these include the burgeoning population growth in lower-income countries, the global consequences of climate change, the pervasive influences of globalization and the exponential growth in Internet-based communication.

The ability to predict the trends, measures and potential consequences of such complex systems depends on our capacity to understand and integrate information and knowledge, as well as on our assessment of the

effectiveness of the knowledge economy both of which are fuelled by the new information and communication technologies (ICTs) that facilitate the collection, storage and sharing of data and information globally.

More than ever before, our increasingly interconnected world can enable more societies to identify opportunities and means for determining their own path to sustainable development. However, while the communication of lessons learned, and the sharing of experiences have allowed the international community to better articulate the objectives of water management in various sectors, the knowledge base and the development of capacity to implement and effectively achieve these objectives remain very much 'work in progress'. Major constraints include the large sizes of the funds required to build these knowledge bases and capacity; the low sense of urgency at political levels; and, perhaps most importantly, the fact that people must first recognize the value of better knowledge and capacity, and that capacity-building is inevitably a long-term and continuous process.[1]

Yet because of the complexities faced in turning the social and economic benefits of research and development into knowledge generation and building capacity, economic returns are often overlooked, and there remains a strong reluctance to invest the necessary resources as a sustained initiative (see **Box 13.1**). Understanding and appreciating the need to change the approach to water development is the first step in overcoming deficiencies. The private sector has long since recognized the difficulties involved in designing and managing programmes of intentional change in corporations, and it has accepted occasional failures as normal episodes on a learning curve, spurring efforts to master change as a process (Pasmore, 1994; Senge et al., 1999; Kotter and Cohen, 2002). The response from development agencies, in contrast, has typically been to minimize risks and boost the apparent benefits to be achieved (Morgan et al., 2005). Without further intensive efforts to understand the dynamics of the complex processes of institutional change in international development, initiatives to enhance knowledge and capacity will fail to be properly targeted and will not produce the desired outcomes. Indeed, compared with other sectors, the water sector has been slow to seek out and internalize knowledge from other sectors, affecting issues such as climate change, and to investigate more deeply the longer-term scenarios needed for proper governance.

...the devolution of responsibility for management of a range of water services from national to lower levels further raises the need to strengthen capacities...

1. These long-term goals are gradually becoming recognized, and development banks like the World Bank, the Inter-American Development Bank, the Asian Development Bank, the African Development Bank and the international donor community are providing increasing support for capacity-building.

BOX 13.1: ENHANCING EDUCATION AND CAPACITY: AN ECONOMIC PROPOSITION

Conventional financial analyses of investment projects tend to calculate the rate of return based only on investment in physical assets. The capacity development component is typically treated as an appendix, to which no separate economic relevance is allocated. However, the methodological difficulty in determining the correlation between investment in capacity and improved sectoral performance is not proof that there is no return.

The Indonesian Government has gone through a period of intense institutional change since 1998. To improve its irrigation performance, it

launched large-scale pilot programmes across the country where Farmer/Water User Associations (WUAs) were empowered through capacity-building and appropriate regulatory changes that created an enabling environment. As part of this, the local staff of the Irrigation Services were trained as 'facilitators'. Because of the availability of comparative data, it was possible for the first time to separate out the value of investments in capacity-building as opposed to investments in physical assets. The analysis showed that conventional rehabilitation projects (to repair irrigation schemes after the recent economic crisis) would bring an economic

rate of return (ERR) of 10 to 18 percent, depending on the state of the assets and the productivity of the scheme. However, when the enhanced knowledge and capacity of the WUAs was factored in, the ERR rose to 30 to 40 percent. The ERR over the incremental investment for the capacity-building component was approximately 32 percent. Thus, the larger benefits were created by the investment in the empowerment and training of the users, increasing the 'social capital' of the local communities and strengthening governance.

Source: World Bank, 2003.

Part 2. Enhancing the Knowledge Base

An adequate knowledge base must be available to the water sector worldwide in order to understand and deal with current changes. Besides data describing the state of water resources and their management, there is an urgent need for good applied research to generate knowledge on the current challenges facing the water sector and to collect and share the existing experiences of communities as they develop capacity. Because the knowledge base must also address the socio-cultural and economic processes that feed into all three levels of capacity, factors related to collective learning processes, and democratic participation and empowerment must also be taken into account, requiring knowledge acquisition covering areas far beyond those concerned solely with the state of the resource.

2a. Data acquisition

Faced with climate change and population growth, it is now more urgent than ever that measures be undertaken to improve the state of knowledge on water services and water resources for better management. The knowledge base of the water sector is very broad; it touches on health, agriculture/aquaculture, industry, energy and ecosystem issues and draws on skills and knowledge from technological, scientific, medical, economic, legal and social realms. In order to appreciate the complexity of the interaction between these different issues it is essential to have relevant and reliable data that relate to them and their connection to the water sector. The World Meteorological Organization (WMO) Global Runoff Data Centre reported that there has been a significant reduction in the data collected since mid-1980s. Agenda 21 (UN, 1992) states that the lack of data 'seriously impairs the capacities of countries to make informed decisions concerning the environment and development'. Indeed, raw data from

monitoring the physical and socio-economic systems of our water resources is the foundation on which we assess their state. Because of the high cost of data acquisition, data collection should be targeted at what is critically important.

An examination of the current state of the data in the knowledge base on integrated water resources management (IWRM) reveals some distinct trends. On the one hand, due to political, institutional and economic instability, there has been a severe decline in basic hydrological monitoring for estimating sustainable water supplies (IAHS Ad Hoc Committee, 2001; Grabs, 2003). On the other hand, there have been considerable advances in the acquisition of data on water resources from remote sensing (Vörösmarty et al., 2005). These advances are due in part to the governments' expectations of the availability and capabilities of remote sensing technology and its spatial coverage, in addition to the large costs and uncertainties involved in sustaining ground-based hydrological monitoring networks.

Ground monitoring systems

Ground-based monitoring systems are essential for characterizing a country's water resources. Despite the spatial coverage of remote sensing, such data is generally still less accurate than that of ground-based monitoring, which is needed for confirming remote sensing data and for measuring parameters such as precipitation, discharges and sediment transport in rivers and groundwater levels. Today, however, there is a lack of ground-based hydrological station networks. In large parts of the world, basic networks have been seriously threatened in the last fifteen years as pressure on funding agencies and governmental organizations to reduce their size has often resulted in the elimination of the technical staff responsible for their operation (**Box 13.2**). It was mistakenly believed that the introduction of automation, among other reasons, would justify such staff reductions, resulting in widespread personnel removal, extremely low remuneration and little political support – the effects of which will only be felt in the mid to long term. Political decision-making has sometimes been in conflict with scientific research (see **Chapter 1**), even in countries with an advanced scientific capability, like the US: 'There is a deep disconnection in American politics between scientific knowledge and political decisions' (Sachs, 2005).[2]

WMO's World Hydrological Cycle Observing System[3] (WHYCOS) is making an important contribution to an overall assessment of the world's water resources (see **Map 13.1**) by strengthening the technical and institutional capacities of the hydrological services to collect, transmit and store hydrological data and produce information responding to the needs of the end users. However, it is ironic that the dilapidation of ground-based monitoring networks is happening at a time when remote sensing coupled with geographic information systems can truly complement these traditional labour-intensive methods of data collection with better archiving, access and analysis of the data. Due to the uncertainties in their formulation and application, even sophisticated modelling systems are dependent on good ground-based data.

Today, in large part due to vast technological monitoring advances, the global and regional water balance (as well as water use statistics) has been estimated to a level of detail not achieved previously. However, in these estimates there are still significant uncertainties that need to be addressed by both the new technologies and the ground-based monitoring systems. It is a paradox that governments and agencies are willing to invest many millions of dollars in projects that have such fragile hydrological data foundations and may not be sustainable, but are unwilling to spend the much smaller sums needed to ensure that data are collected and processed to meet current and future needs and demonstrate the sustainability of projects (WMO/UNESCO, 1997a).

Preparations of a remote sensing test in Delft, Netherlands

2. See also UNESCO's World Commission on the Ethics of Scientific Knowledge and Technology (COMEST).

3. See www.wmo.ch/web/homs/ projects/whycos.html for more details.

BOX 13.2: HYDROLOGICAL NETWORKS: THE CASES OF INDIA, MEXICO AND VENEZUELA

In 1998, approximately 250 hydro-meteorological observer jobs (the technicians responsible for the operation and maintenance of the hydro-meteorological network) were eliminated in the National Hydrological Service of Venezuela, as part of the downsizing of the Ministry of Environment and Natural Resources. The plan, supported and encouraged by several development banks and financing institutions, was to fulfil responsibilities through a number of micro-enterprises that would be constituted by the same personnel who would be contracted on a need basis. Unfortunately, almost six years later, these enterprises have not yet been created, and as a result most of the stations of the original network are no longer operational.

In the case of Mexico, the river discharge network was traditionally operated by approximately 800 technicians, each living near one of the measuring stations (see **Chapter 14**). In the last six years, the policy of downsizing the public administration has meant that no new recruitment to replace retiring personnel has been authorized by the competent authorities. As a consequence, approximately 200 discharge stations were inoperative as of 2003, and more were expected to cease operating in the future.

In contrast to these cases, India built a huge Hydrology Information System covering nine states (1.7 million square kilometres) between 1996 and 2003, of which the main activities were to improve institutional and organizational arrangements, technical capabilities and physical facilities. Among others, 265 new river gauging stations were built and 650 stations upgraded, with 2,239 purpose built piezometres and thousands of digital water level recorders for groundwater monitoring, 14 water quality laboratories were upgraded with modern equipment and 9,000 employees were trained with 27,000 training units. In 2004, an additional phase was approved including enlarging the geographic scope to four additional states until 2011. The implementation of the project provides a quantum leap in the understanding of the state of the water resources in India.

Sources: Misión de Evaluación OMM/PROMMA, 2003; WL Delft Hydraulics, 2004; World Bank, 2004.

Map 13.1: WMO'S World Hydrological Cycle Observing System (WHYCOS)

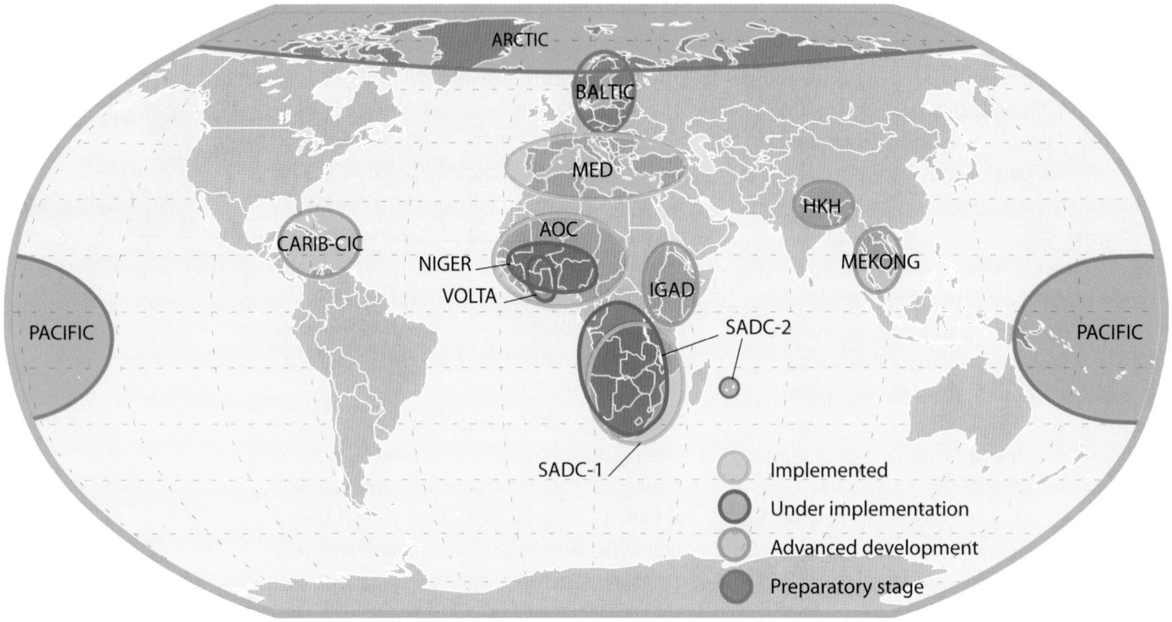

Note: The WHYCOS programme is implemented through various regional HYCOS components, as shown in the coloured areas of the map above. Each component concerns either a transboundary basin or a community of countries. As for 2006, three components have been implemented:

- MED-HYCOS (Mediterranean see medhycos.mpl.ird.fr/ for more details)
- AOC-HYCOS (West and Central Africa see aochycos.ird.ne/ for more details)
- SADC-HYCOS (South African Development Community).

Moreover, three components are still under implementation (in blue):

- Niger-HYCOS
- Volta-HYCOS
- SADC-2HYCOS.

The main activities of each project include updating the observing network, developing regional databases (see www.r-hydronet.sr.unh.edu/), establishing websites for easy data access and dissemination, and training personnel. Data collected through the HYCOS components also contribute to a better understanding of the global water cycle and its variability.

Source: WMO-WHYCOS, 2005.

India is regarded as a world leader in using satellite data techniques for managing its natural resources and supporting rural development

Section 4: MANAGEMENT RESPONSES & STEWARDSHIP

Remote sensing

In recent years, water resources management has benefited from the powerful assessment tools provided by remote sensing. Since the Rio Declaration on Environment and Development in 1992, a number of major developments have occurred. Over 100 new satellite sensors for sustainable development have been put into operation, and advanced warning for extreme storms and floods has increased in some instances to over 100 hours (UNESCAP, 2003). Remote sensing is used for the provision of simple qualitative observations, the mapping/detecting features of hydrological importance and the direct estimation of hydrological parameters and water quality (see **Box 13.3**).

India is regarded as a world leader in using satellite data techniques for managing its natural resources and supporting rural development. However, most countries, including relatively developed ones, do not yet use these techniques on a day-to-day basis to support decision-making in water resources management. Because of this, the United Nations (UN) has made the enhancement of the capacity of countries to use and benefit from remote-sensing technologies a key focus for many space-related activities (UN, 2004). Of particular note is the TIGER Initiative led by the European Space Agency (ESA) in partnership with the United Nations Educational, Scientific and Cultural Organization (UNESCO), the United Nations Office for Outer Space Affairs (UNOOSA) and others, which

BOX 13.3: ADVANCES IN THE PRACTICAL USE OF SATELLITE REMOTE SENSING FOR WATER RESOURCES

Considerable improvements in the assessment of hydrological parameters for water resources management have been made during the last two decades using remote sensing from satellites (Schultz and Engman, 2000). Using a combination of radar and thermal sensors from weather satellites, the accuracy of precipitation estimates for crop forecasting, flooding and river flows over large areas and basins has improved considerably, as has the extent of snow cover and water equivalents. In addition, satellite data provide a unique means of assessing separately the actual evaporation over different areas, such as river basins, irrigated areas and wetlands, using the surface energy balance equation. This has led to methods for determining crop water efficiencies, water use by groundwater irrigation, and wetland water requirements. Another important hydrological parameter that is

monitored using active or passive radar is the moisture of the uppermost soil layer.

Important progress has also been made in surveying the land surface. The Shuttle Radar mission has made freely available a worldwide coverage of digital terrain models, required for example, by rainfall-runoff modelling. Satellites, through radar altimetry, are now surveying water levels in lakes and large rivers within a few centimetres accuracy. This is particularly important for remote water bodies. Satellite images with resolution of 1 or 2 metres can be purchased, enabling the rapid preparation of maps through digital photogrammetry and showing terrain heights of floodplains or coastal areas, which are required for assessing flood risks and the propagation of floods. Land subsidence, often due to groundwater extraction,

can also be measured with high precision by radar interferometry.

Imaging spectrometry (or hyperspectral remote sensing) provides information about the water quality of optically deep-water bodies. The first operational applications from airborne platforms were reported in the 1990s, and the first imaging spectrometry satellites were launched in 2000. The most successfully monitored water quality parameters are chlorophyll, a blue-green (or cyannobacterial) pigment, total suspended matter, vertical light attenuation coefficient and turbidity. The technique can be used in coastal waters for the assessment of the health of coral reefs and for bathymetric mapping.

Sources: Schultz and Engman, 2000; Dekker et al., 2001.

BOX 13.4: TIGER INITIATIVE: IMPROVING WATER SYSTEMS OBSERVATION IN AFRICA

Established in 2003, the European Space Agency's (ESA) TIGER Initiative aims to make earth observation services more accessible for developing countries, with particular focus on Africa. In 2005, there were four separate ESA projects operating under the TIGER umbrella:

■ GlobWetland: provides land cover and land-use change maps on fifteen African wetland sites to support reporting obligations for the Ramsar Convention on Wetlands.

■ Global Monitoring for Food Security (GMFS): maintains a continental-scale overview of sub-Saharan Africa in order to produce sub-national and selected high-resolution crop production forecasts.

■ Epidemio: uses satellites to provide environmental information in the service of epidemiology, including the charting of water bodies in order to prepare malaria risk maps.

■ Aquifer: generates land-use cover and land-use change charts, digital terrain maps, soil moisture

mapping and subsidence monitoring, so that new aquifers can be identified and existing aquifers exploited in a sustainable manner.

The Envisat environmental satellite and European Remote Sensing satellite data are freely available for African hydrology research. TIGER also enhances capacities in space technologies in African regions, while supporting its integration within the user's traditional working procedures to improve the sustainability of water resources management.

Sources: ESA, 2004; earth.esa.int/tiger/

aims to provide earth observation data, capacity-building and technical support services for IWRM in developing countries with a particular focus on Africa (**Box 13.4**).

The advantages of remote sensing lie in its ability to map conditions across regional, continental and even global scales on a repetitive basis at a relatively low cost compared to ground-based monitoring. The coupling

of biophysical, socio-economic, hydrometric and remote sensing data with modelling now leads to the emergence of valuable new information on water stress at global, regional and local levels (see **Box 13.5**). The United States National Aeronautics and Space Administration (NASA) and ESA plan to launch special water management and hydrology-related satellites (see, for example, Alsdorf and Rodriguez, 2005), optimized for

BOX 13.5: ADVANCES IN REMOTE SENSING TECHNOLOGIES

Research undertaken at the University of New Hampshire by Vörösmarty et al. has pioneered a means of coupling a range of different data types in order to generate new information, such as spatially discrete, high-resolution remote sensing data, model-generated climate change data, population density, growth and migration, and industrial development indicators **(see global maps at the head of each section).** Based on the integration of such datasets with appropriate modelling, the researchers have developed indices for local relative water use and reuse and water stress in order to assess the current state and future trends. They conclude that in 1995, 1.76 billion people were under severe water stress and that 'rising water demands greatly outweigh greenhouse warming in defining the state of global water systems to 2025'. Such an analysis is carried out globally at a resolution of about 50 kilometres (km) but has also been done for Africa at 8 km resolution and for the river basins of Lake Victoria down to about 2.5 km resolution. Results show that chronic water stress is high for about 25 percent of the African population; 13 percent of the population experiences drought-related stress once each generation, 17 percent are without a renewable supply of water, and many are dependent on highly variable hydrological runoff from a far distant source.

Sources: Vörösmarty et al., 2000, 2005; wwdrii.sr.unh.edu/

Countries require organizations with individuals who are able to collect, store and analyse data in order to generate knowledge...

Section 4: MANAGEMENT RESPONSES & STEWARDSHIP

4. Those countries include Angola, Belize, Bhutan, Bosnia and Herzegovina, Botswana, Brunei, Cambodia, Djibouti, Eritrea, Gambia, Guinea Bissau, Haiti, Lebanon, Libya, Namibia, Qatar, Samoa and Yemen.

5. See www.iwmbd.org/ for more information.

special measurements of water-related state variables, such as water levels and discharges in rivers. Research efforts have recently concentrated on developing methods that need limited *in situ* measurements, but a large number of applications still depend on correlation and/or combination with measurements from land-based monitoring systems. The usefulness of information will ultimately depend on its suitability for assimilation within hydrological forecasting.

Databases

Thanks to the establishment of water databases and monitoring guidelines, many countries have made strides since 2004 to set up and maintain national water databases.[4] Still, 61 out of 239 countries have not supplied the necessary data to international repository centres, making it extremely difficult to establish the overall state of the world's water resources, and in particular those of a particular region or river basin in a given year, month or day (GRDC, 2005). Such data is needed for inter-country comparisons in order to identify similarities, differences, strengths and weaknesses, to alert better governance and to improve management. The willingness to share data among countries remains a key obstacle to effective transboundary river management (see **Chapter 11**).

However, even when data is openly shared, differences in data characterization and record duration make it difficult to compare data from different databases. Often data collection is initiated by and limited to the duration of specific projects, and ongoing data collection is either ignored or grossly under-funded. These important problems need international attention. While a full assessment of global water database initiatives is beyond the scope of this report, given the scope and scale of the entire field of water data, the list of useful websites at the end of the chapter provides some key listings and some of the better examples of current databases in the water sector.

Modelling systems

It is one thing to acquire data from the real world, but it is another to interpret and utilize that data to better understand water-based processes and systems. The ability to extract, understand and interpret such data and information is crucial for taking advantage of the processes and systems in water management. Computer simulation models are now commonplace tools for assisting in such understanding and interpretation. Such models encapsulate scientific knowledge in their development and application, while operating with data to replicate real-world phenomena. In addition, when possible, the sophisticated use of graphics and video with Geographic Information System (GIS) provides opportunities for visualizing and anticipating new information and knowledge in order to understand complex phenomena.

Models are extremely important for effective water resources management, and the use of computer-based models has increased considerably in the last ten years; indeed, no major water-related project is undertaken without the use of models and corresponding decision-support tools. Centres such as the Institute of Water Modelling in Bangladesh[5] have been established in order to capitalize on local expertise, while using advanced modelling tools for better water management.

A very important trend in modelling is the linking of hydrologic and hydrodynamic models to GIS, ground-based monitoring systems, remote sensing data, numerical weather prediction models and quantitative precipitation forecasts. Such links, properly implemented using information and communication technologies, contribute to the building of integrated hydroinformatics systems leading to considerable improvements in predictive accuracy, especially important in the context of managing extreme events (see **Chapter 10**) and water resources management in transboundary river basins. OpenMI is a European project that is committed to developing a protocol that enables hydrological modelling software from different suppliers, to be linked so that the integrated software will support and assist the strategic planning and integrated catchment management required by the European Water Framework Directive.[6] This opens up opportunities for a better management of transboundary rivers.

2b. Knowledge sharing and accessibility

Access to and sharing of knowledge by individuals and groups are critical to addressing water-related problems. In many developing countries, however, the water resources and services knowledge base has often been limited due to budget constraints, the emphasis on developing new infrastructure, and a lack of professional education and language barriers. This results in decreased capacity to translate available data into usable knowledge. Consequently, data collection has all but ceased in some countries. Countries require organizations with individuals who are able to collect, store and analyse data in order to generate knowledge, which requires first and foremost better education of these individuals. In addition, the willingness to share information and the importance of building trust between parties is critical in the development of a shared vision for water management (UN, 2003b; **Chapter 11**).

New ICTs have facilitated the mechanisms and practices of sharing knowledge.[7] ICTs have made an incomparable contribution to development in water-related sectors, from the rapid collection and exchange of hydrological

information and the integration of advanced warning systems to virtual classrooms, video conferencing (see **Box 13.6**) and networking with GIS (see **Box 13.10**). While telecommunications still favour urban areas, the coverage in rural areas is growing rapidly in many developing countries. In Cambodia, for example, one year after the introduction of a mobile cellular phone network, mobile subscribers had already surpassed the number of fixed telephone lines in the country (UNESCAP, 2004). The media (newspapers, radio, television, films) also has an important role in disseminating information to the general public.

Better information management has repercussions at every level of society, whether on the development of national and international knowledge resources, the efficiency and effectiveness of government services, the focus of donor community programmes or project ownership by local stakeholders through participatory decision-making. Uganda provides an excellent example of this as indicated in the Uganda National Water Development Report (see **Chapter 14**). The Directorate of Water Development

Analysis of water samples for trace elements in Athens, Greece

6. www.harmonit.org/

7. The UN Millennium Declaration includes a specific commitment to ensure that the benefits of new technologies, especially ICTs, notably the telephone, mobile phone, Internet and broadcast networks, are available to all.

BOX 13.6: **THE GLOBAL DEVELOPMENT LEARNING NETWORK (GDLN)**

The Global Development Learning Network (GDLN) is a global network of learning centres that uses advanced information and communication technologies to connect people around the world working in development. Initiated by the World Bank in June 2000, this network has grown from eleven to over ninety affiliated centres, most of which are located in developing countries. In the future, the network is expected to grow dynamically. In middle-income countries across Latin America and Asia, emerging in-country networks are further increasing GDLN's reach and use on a remarkable scale. GDLN facilitated more than 850 videoconference-based activities between July 2003 and June 2004, connecting an estimated 26,000 people around the world. During 2004, the GDLN centre located at UNESCO-IHE Institute for Water Education hosted fifty water-related videoconferences, connecting 1,600 water experts worldwide.

Source: www.gdln.org

Networks of all kinds, representing all sectors, such as professional associations, are powerful tools for knowledge sharing and distribution

Map 13.2: The Knowledge Index, 2005

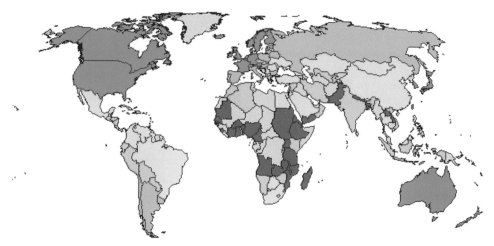

Map Legend (0 is the lowest score and 10 is the maximum score)

| 0 <= KI <= 2 | 2 < KI <= 4 | 4 < KI <= 6 | 6 < KI <= 8 | 8 < KI <= 10 | No data |

Note: The Knowledge Index (KI) benchmarks a country's position vis-à-vis others in the global knowledge base. It is the average of the performance of a country or region in three knowledge economy pillars (education, innovation and information and communications technology) and is calculated considering the following variables: adult literacy rate, secondary enrolment, tertiary enrolment, researchers in research and development per million population, patent applications per million population, scientific and technical journal articles per million population, telephones per 1,000 people, computers per 1,000 people, and Internet users per 10,000 people.

Source: World Bank Institute, 2005.

Arsenic Removal Family Filter being set up in a Bangladeshi home

8. See www.dwd.co.ug

9. www.worldbank.org/kam

(DWD) established a Management Information System for the water sector in 1998 in order to improve management and facilitate decision-making processes. With the purpose of monitoring financial and physical activities, it has since produced a series of design reports that currently constitute the archives for the sector, and it has developed data collection procedures that are used in all the districts of the country. Data are collected, processed and shared with other stakeholders through reports, Intranet and the DWD website.[8]

However, despite considerable progress in recent years, access to ICTs remains unequally distributed. There are, for example, more computers in Brazil, more fixed line telephones in Italy, more mobile phones in Japan and more Internet users in France, than in the whole continent of Africa. Yet the population of Africa and the needs of its people greatly exceed those of these other countries (ITU, 2004). In response, the World Bank Institute's Knowledge for Development Programme has developed a user-friendly tool designed to assist decision-makers in understanding and benchmarking their strengths and weaknesses in terms of their ability to compete in the global knowledge economy. While not specific to water-related fields, the Knowledge

Assessment Methodology uses a series of relevant and widely available measures that can allow for a preliminary country comparison and the identification of strengths and weaknesses in order to focus policy attention or future investments in making the transition to the knowledge economy. The assessment tool is available online.[9] The state of the global knowledge economy, weighted by population, is presented in **Map 13.2**.

The language barrier and quality assurance and control

Today, due to the ever-increasing speed with which technology can collect, store and disseminate data, we are possibly encountering for the first time a situation in which human individuals and their capacities are the primary bottleneck in the chain of information processing, making people the limiting factor for further understanding (Maurer, 2003). Knowledge has to be presented in a way that people can assimilate it. One barrier is that information and knowledge on water management and water use often uses terminology that only academicians, theoreticians and technical people can understand, or worse, in a language foreign to the end user. Language barriers constitute a critical obstacle to local information for literacy and education as well as a level playing field in the global digital knowledge economy. This is unfortunate, considering the potential value that this knowledge could contribute to alleviating the water crisis by increasing public engagement in the process.

With roughly 7,000 living languages in the world (Gordon, 2005), participatory approaches to water management immediately become more complex. SIL International, for example, works to develop community-level capacities to enable communities to carry out their own research, translation, and production of literature in their native tongue. In addition, with more than 90 percent of Internet content today existing in just twelve languages, UNESCO's Initiative B@bel[10] uses ICTs to support linguistic and cultural diversity, protect and preserve languages in danger of disappearing and facilitate access to this important communication medium. To further facilitate the use of software products and websites across multiple platforms, languages and countries, the Unicode Consortium has developed a standardized computer language.[11] This could support the increasing trend in the development and use of online water information networks that can provide another means of surmounting the language barrier. This is particularly appropriate for the translation of technical terms that are peculiar to water development.[12]

Another major problem that arises in using information or knowledge from elsewhere is assuring its quality. Information or knowledge can originate from a reputable source, such as a peer-reviewed journal or the website of a trusted organization, but this hardly accounts for the majority of information found on the Internet. Almost anyone can put anything online. In doing so, they bypass many of the benefits of traditional publications – issuance by an authoritative source, editorial or peer review, evaluation by experts, etc. Quality is still a matter of trust by the recipient in the trustworthiness of the supplier. Third-party confirmation of any information and knowledge is generally recognized as one of the best ways of assuring quality. Responsibility for the use and application of the data provided remains largely in the hands of the user, who has to rely on his or her education and experience to exercise discernment. Networking and involvement in professional associations largely stimulate quality assurance through continuous peer review, as in the case, for example, of the peer-reviewed web-based information service provided by FAO's International Programme for Technology and Research in Irrigation and Drainage (IPTRID).[13]

Knowledge networks

Networks of all kinds, representing all sectors, such as professional associations, are powerful tools for knowledge sharing and distribution. They offer a framework for resource optimization and knowledge combination, saving valuable financial and time resources, in addition to providing an excellent platform for peer discussions (see **Box 13.7**).

Networks for capacity-building in integrated water management are a relatively new phenomenon. The advantages of networking for scaling up capacity-building to reach the MDGs are gaining recognition in the international water community. The advantages are predominantly in providing a more coherent and coordinated approach to capacity-building, increased impact, relevance and sustainability from working with local institutions, improved sharing of knowledge and expertise and a platform for cross-disciplinary and cross-regional discussions.

There are more computers in Brazil, more fixed line telephones in Italy, more mobile phones in Japan and more Internet users in France, than in the whole continent of Africa

10. See www.unesco.org/ webworld/multilingualism for more details.

11. See www.unicode.org/ standard/whatisunicode.html for more details.

12. See water.usgs.gov/wsc/ glossary.html, for example.

13. See www.fao.org/iptrid for more details.

BOX 13.7: FARMNET – FARMER INFORMATION NETWORK FOR RURAL DEVELOPMENT

Since the early 1990s, the UN Food and Agriculture Organization (FAO) has assisted in the development of networks among rural farmers and supported intermediary organizations using ICTs and conventional communication media for capacity development. Operated by farmers, these FarmNets disseminate locally relevant information that is needed to improve livelihoods. FAO adopted a participatory approach to performing the preliminary assessment of needs and then provided the electronic network designs, some basic equipment, logistical support, coordination, technical backstopping and training to local extension and farm organization personnel. The impact of FarmNet has been significant. Transmitting price and market information through computer-based networks cost 40 percent less than using traditional extension methods. In one case, by using the market information provided by the network, a farmer association was able to sell cotton for US $82 per metric quintal as opposed to US $72, which was the price local buyers were trying to impose. Vegetable producers reported that the information on meteorological conditions informed them of climatic conditions faced by competitors in other regions and countries. This enabled them to plan their irrigation strategies and market their produce more successfully.

Source: FAO, 2000.

Local knowledge can lead to a better understanding of the water cycle and could play a vital role in providing solutions to the world's water crisis

Map 13.3: Cap-Net

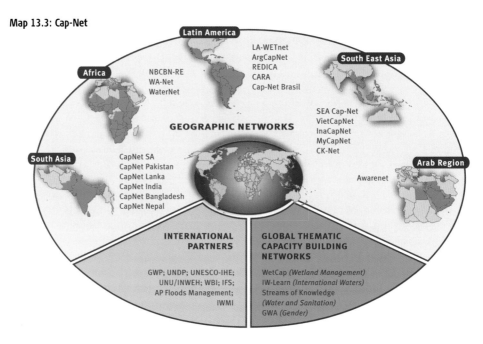

Note: Cap-Net is an international network made up of autonomous international, regional and country networks and institutions committed to integrated water resources management (IWRM) capacity-building. The networking concept is being used by Cap-Net to bring cooperation and coherence to scale up capacity-building in water management.

To date, the programme has trained 550 trainers, who have in turn impacted thousands of decision-makers, water managers, and fellow capacity builders – exponentially increasing capacity in IWRM. The programme has addressed issues such as legal and institutional reform, conflict resolution, IWRM planning, gender and water, and other aspects of IWRM are to follow.

- Twenty geographic networks and four thematic networks affiliated with Cap-Net.
- Over 1,000 member institutions organized in regional and country networks.
- Fifty-five planned network training events and education programmes.
- Nine operational topic or geographic e-discussion groups on capacity-building in IWRM.

Source: Cap-Net, 2005.

Literacy project in Cheikh Anta Diop village, Senegal

When Cap-Net, a United Nations Development Programme (UNDP) initiative, was described in the first *UN World Water Development Report* (2003), their international network committed to capacity-building in IWRM consisted of a partnership of four regional networks at their early stages of development. Today, twenty geographic capacity-building networks, with special attention given to IWRM, and four international thematic networks, with expertise in specific thematic areas relevant to IWRM (water supply and sanitation, wetland management, gender mainstreaming in water, and international waters), form the global network (see **Map 13.3**).

The UNESCO-IHE PoWER Programme, a partnership between seventeen member institutions throughout the world, has also made substantial progress in initiating the development of joint education and training packages (see **Map 13.4**).

Map 13.4: PoWER

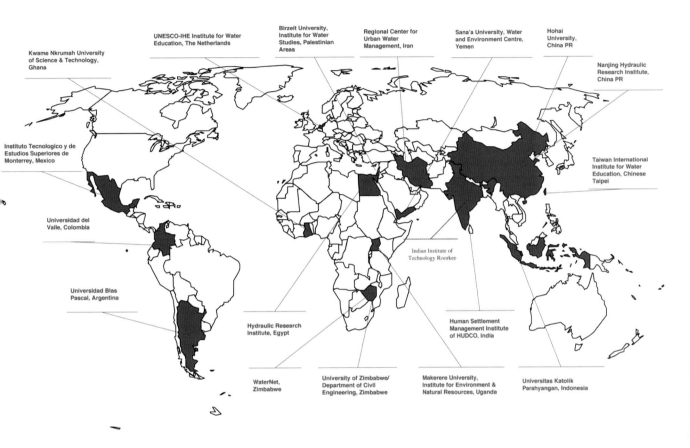

Networks stimulate regional collaboration by addressing and solving water-related problems, and, as such, contribute to an increase in trust and stability throughout the world.

Local and indigenous knowledge

Local and indigenous knowledge refers to the cumulative and complex bodies of knowledge, know-how, practices and representations (including language, attachment to place, spirituality and worldview) as maintained and developed by peoples with extended histories of interaction with their natural environment. Despite the fact that such peoples number about 300 million, representing over 5,000 languages and living in more than 70 countries in all of the world's regions (including 17 countries that are home to more than two-thirds of the Earth's biological resources), they must still struggle for their rights over the water resources they have been

using and protecting for generations (UNESCO, 2003a, 2003b). In fact this knowledge is typically dismissed by planners and still far from internationally recognized as vital to sustainable development and biodiversity management, particularly as it relates to the holistic approach so sought after in IWRM. Worse still, while knowledge development efforts today aim to empower local stakeholders in efforts to move towards effective IWRM (see **Chapter 2**) changing values, globalization and the drive for commercialism have all but extinguished these indigenous practices despite the vital role they could play in providing solutions to the global community's water crisis (**Box 13.8**). And yet such knowledge and habits (constituting the 'social capital' of a community) can lead to a better understanding of the water cycle, of local seasonal effects and of their relationship with nature and agriculture. They also encompass social skills such as the resolution of water

BOX 13.8: LOCAL AND INDIGENOUS KNOWLEDGE FOR SUSTAINABLE IWRM

Indian communities have a long tradition of coping with the recurrent hot and dry seasons by constructing small to medium-sized reservoirs (or tanks). Villages in the past would each have their own tank(s), maintained by the villagers themselves. With the advent of modern technology, planners preferred to construct very large reservoirs that would serve a whole region and needed to be constructed and operated by a dedicated bureaucracy. As a consequence, many of the local tanks went out of use. Over time, however, it has become clear that the centralizing technocratic approach cannot address all local water shortages and a new

movement has started to re-introduce the local tank system.

The island of Bali, Indonesia, has a sophisticated Hinduist culture in which water and irrigation are central. Long before the government started to train and empower irrigators, and indeed started to construct large irrigation systems, the Balinese operated complex irrigation on hillside terraces (*subak*), including water allocation by rotational schedules, and sustainable maintenance procedures.

Many countries have learned how to cope with seasonal floods. In some countries houses are built on stilts; in others, such as Bangladesh, villages are built on artificial mounds for which the soil was taken from 'borrow pits' that in the dry season were neatly maintained to provide water for the household and for cattle. Transportation in the flood season took and still takes place by canoe. This approach proved effective, causing minimum disruption of daily life, allowing the floodwater to deposit fresh fertilizing silt on the fields, and sustaining large fisheries.

Source: Agarwal and Narain, 1997.

---

14. See UNESCO's Local and Indigenous Knowledge Systems (LINKS) project, www.unesco.org/links

15. See www.nuffic.nl/ik-pages

16. See www.unesco.org/most/

17. See www.ik-pages.net

conflicts and water allocation, as well as technologies to harvest, store and canalize water.

UNESCO's Local and Indigenous Knowledge Systems (LINKS) project focuses on the interface between local and indigenous knowledge, and the MDGs of poverty eradication and environmental sustainability.[14] It addresses the different ways that indigenous knowledge, practices and worldviews are drawn into development and resource management processes. It also considers the implications

this may have for building equity in governance, enhancing cultural pluralism and sustaining biodiversity. In addition, the Netherlands Organization for International Cooperation in Higher Education/Indigenous Knowledge (NUFFIC/IK-Unit)[15] in cooperation with UNESCO's Management of Social Transformations (MOST) Programme[16] established in 1999 a database of best practices on indigenous knowledge, which initially contained twenty-seven best practices and gained an additional twenty-two cases during the second phase of work (2001–2002). Recently, the International Institute of Rural Reconstruction (IIRR), a rural development organization working in developing countries, was chosen as a strategic partner in a phased transfer of published indigenous knowledge materials to the global 'south'. Forty documents related to indigenous knowledge for water development are available online.[17] Overall, there remains a serious challenge to properly understand and appreciate the value of such traditional approaches, and to turn them into local tools to achieve better water management.

Section 4: MANAGEMENT RESPONSES & STEWARDSHIP

Literacy course for women in Praia, Cape Verde

Part 3. Enhancing Local Capacities

When challenge areas discussed in this report describe deficiencies and propose goals for a better tomorrow, it is the enhancement of knowledge and local capacities that are needed to fill the gap between the current state and the desired sustainable solution. Success in water development can only be achieved when local capacities have been enhanced to address the water-related problem.

3a. Human resources development

Human resources development is a continuous process aimed at imparting knowledge, developing skills and changing attitudes and behaviours, which allow the maximization of the benefits of knowledge sharing and participatory processes. While all levels of capacity development – human resources, institutional capacity and the enabling environment – are ultimately of equal importance (see **Box 13.9**), adequately skilled staff to develop policy and legal frameworks and the necessary institutions are the starting point of any successful venture. In all capacity-building efforts, attitude, behaviour, education and training, as well as workforce incentives, career paths, and accountability mechanisms in the workplace all influence the capacity to make knowledge-based decisions, and are critical components of human resources development. Developing knowledge through effective sharing at and between each level of capacity development is one of the most important challenges in the water sector.

The many technocrats within the water sector generally do not have the skills to deal effectively with governance issues, such as conflict mediation, mobilization of communities, managing processes of stakeholder participation, etc. 'Social facilitators' can be introduced to complement the technocrats with skills to manage diverse and dynamic social and political processes that typically did not receive much attention in the past.

Basic education

People's effectiveness in managing and using water is only brought about with the provision of a basic education on water, sanitation and hygiene. If children are taught proper hygiene, primary schooling can transform them into health educators for their families, thereby passing on vital information and skills that can reduce household vulnerability to deadly diarrhoeal diseases by at least 40 percent (see **Chapter 6**). This is particularly true for women and girls who are responsible for household hygiene, food and water, and who are, with the elderly, the most at risk from water-related hazards. Increased and safe access to primary education for girls will pave the way to more gender-balanced decision-making processes for water management (see **Chapter 1**), in line with MDG Goal 3 to promote gender equality and empower women.

If children are taught proper hygiene, primary schooling can transform them into health educators for their families...

Higher education and training has an important role in knowledge transfer, not just as a one-time diploma course but also throughout the active professional life of the recipient

As described in the Ethiopia National Water Development Report (see **Chapter 14**), 30 to 40 percent fewer girls than boys attended school at the national level. UNICEF's report on gender parity and primary education (2005) indicated that girls make up most of the 115 million children currently out of school, and 80 percent of children not attending primary school in West/Central Africa, South Asia and Middle East/North Africa had mothers with no formal education at all (UNICEF, 2005a). Reasons for this include the need for girls and women to walk long distances to bring water to the home and/or lack of sanitation facilities in schools, forcing girls to seek refuge in the woods and risk sexual attack, ridicule and shame.

A recent study in Bangladesh indicated that a separate toilet could increase the number of girls in school by as much as 15 percent (UNICEF, 2005b). As Carol Bellamy, the former executive director of UNICEF, aptly noted: 'We will only reap the rewards of investment in education if we safeguard children's health while they learn' (UNICEF, 2005b).

As part of its Water for African Cities Programme, UN-HABITAT has since 2003 embarked on an innovative Value-based Water Education campaign that seeks to impart information on water, sanitation and hygiene to children and communities, all the while inspiring and motivating learners to change their behaviour and adopt attitudes that promote hygienic living and wise and sustainable use of water.

Another interesting initiative is The Focusing Resources on Effective School Health (FRESH), an inter-agency collaboration between UNESCO, UNICEF, WHO, and the World Bank, which has produced a toolkit that offers information, resources, and tools that provide support to the preparation of Hygiene, Sanitation, and Water in Schools policies and projects (World Bank, 2000).[18]

18. See www.freshschools.org for more information on the FRESH Institute and Toolkit on Hygiene, Sanitation and Water in Schools.

Educational systems in general need to strengthen education and training delivery mechanisms through networking and the use of professional associations. Curricula, in addition to ensuring high scientific standards, must constantly adapt to concrete problems. An integrated and multidisciplinary competence in general problem solving rather than in purely technical subjects will prove valuable in many new fields. The use of GIS in support of mapping the sustainability of schools is a growing trend around the world, including education sectors in developing countries. In some cases, building geo-spatial databases and using GIS 'is becoming a standard and/or a requirement for funding (Al-Hanbali et al., 2004) (see **Box 13.10**).

Higher education/training

Higher education and training has an important role in knowledge transfer, not just as a one-time diploma course but also throughout the active professional life of the recipient. The value of such education and training is that appropriate knowledge on water can be packaged and even tailored for professionals and other stakeholders. New blended learning approaches encourage active and participatory learning. Peer-to-peer technology that links users in two-way communication enables collaborative working and distance learning. Whereas face-to-face communication is preferable, the increasing availability of distance learning facilities enlarges the possibilities and opportunities for lifelong learning and continued development of the knowledge and expertise of professionals. Online education has increased tremendously since the UN Conference on Environment and Development in 1992, but the challenge of educating the world's population has never been greater, particularly at a time when over 40 percent of the global population is under twenty years old (van der Beken, 2004).

BOX 13.10: GIS FOR SCHOOL SANITATION AND HYGIENE EDUCATION: TAMIL NADU, INDIA

GIS has been applied innovatively in Tamil Nadu for improving the school sanitation programme. With UNICEF assistance, GIS was used for the first time in India in the village of Panchayat to create water and sanitation facility mapping for schools focusing on five indicators: availability of drinking water, toilets, water for toilets, washing and school sanitation and hygiene education (SSHE).

This triggered significant changes in planning for SSHE, especially the use of spatial data. When the first GIS maps were displayed during a regional workshop, they shocked the district officials, as they did not have any idea about the coverage of water and sanitation facilities in schools. They were drawn towards the GIS maps and started comparing coverage levels between different

districts and decided to take up joint planning and use pooled resources. The data has been used to prepare district action plans for SSHE, jointly owned by the local governments. Higher officials in offices far away also became sensitized to local problems after looking at the GIS data.

Source: DDWS, 2004.

Preliminary needs assessments indicate that nearly all countries lack the numbers of adequately trained individuals in order to meet the MDGs, and large numbers of technical and scientific personnel lack a holistic perspective on water management and use. Appropriate training programmes need to be designed and delivered in order to tackle the problems that affect people in their local realities. Adapted training and awareness raising must be provided to the myriad of professional perspectives that play a role in water sector development. This situation was made clear in a needs assessment carried out in 2001 in Bolivia (Lake Titicaca) (see **Chapter 14** on case studies). The assessment showed how many managers already identified the need to combine knowledge and tools from social sciences with traditional engineering skills, and in a related survey, water resources management emerged as the most needed postgraduate programme, leading to the creation a Masters-level programme on sustainable water resources management. In Egypt, the activities of the Regional Centre for Training and Water Studies of Arid and Semi-arid Zones (RCTWS), working under the auspices of UNESCO as an international knowledge institution since January 2002, help to contribute to capacity-building in the field of Integrated Water Resources Management specifically for the Arab and Nile Basin countries.

3b. Strengthening institutional capacity

Institutional capacity relates to the overall performance of the organization and its capability to function properly, as well as its ability to forecast and adapt to change. An organization's personnel, facilities, technology, knowledge and funding constitute its 'resource base'. Procedures and processes for managing its resources, programmes and external relationships determine its 'management capacity'. Together, this resource base and the management capacity make up the overall capacity of the organization (Horton et al., 2003). Such capacities must be operational (day-to-day activities) and adaptive (response to changing circumstances). If the organizational structure within the water sector is conducive to efficient, effective and expedient decision making, it is likely that the country's capacity to address water-related problems is strong. To ensure effectiveness in services delivery and efficiency in water use, the public sector needs to establish partnerships with local communities and user groups. By empowering and strengthening their capabilities, they can assume part of the management responsibility and authority over the infrastructure and the resource itself. Directly affected sectors or user groups, like water user associations, industrial estates, municipalities or wards, and irrigators,

can be empowered by establishing and formalizing a platform that allows all interested water users to present their interests and have 'voice and choice' in the decision-making and management process of water services. This enhances transparency and accountability and fosters the local sense of ownership, while partially relieving the financial burden on the agencies. However, such user groups need also to have their capacities strengthened through training and access to information to allow them to take informed decisions and play their roles.

At the institutional level, three capacity development needs stand out in particular:

- a clear mandate for managing agencies, water providers and policy-making bodies that promote and enhance the institutionalization of good water management and water use throughout all levels of society
- an organizational system conducive to effective and efficient management decisions
- improved decision support mechanisms through research on lessons learned and indigenous knowledge.

Water management agencies continue to close the door on opportunities for effective integrated water management if they do not provide a voice to relatively powerless groups, such as women (**Box 13.11**), who are central in providing, managing and safeguarding water, and indigenous people (**Box 13.8**) who are custodians of sound, ancestral water management practices. As many ministries and non-governmental institutions deal with water within a country, a national apex body[19] can play a crucial role in coordinating activities and disseminating responsibilities in the network of organizations operating on the ground within a river basin. By considering all stakeholders, a national apex body may adopt policies and laws, carry out institutional reforms and formulate a national water agenda (Asian Development Bank, 2004).

It is crucial to deal with employment issues in order to improve organizational performance. Both public and private sectors need to provide adequate salaries, as well as professional development opportunities and incentives. If the incentives for the staff as individuals and as an organization point in the wrong direction, the possession of other capacities is of little value (Alaerts et al., 1999). This is illustrated by a study of retention rates in a poorly performing public sector in Africa. It was found that the retention rates of trained individuals were much higher than expected for the perceivably low

Visiting South African children playing the World Water Game, aimed at increasing awareness and knowledge about water resources, at the Waterdome during the World Summit on Sustainable Development in Johannesburg

...the public sector needs to establish partnerships with local communities...

19. An apex body is a national organization that guides the water sector in water services and resource management.

BOX 13.11: THE NEED FOR GENDER BALANCE

Women produce between 60 and 80 percent of the food in most developing countries and provide up to 90 percent of the rural poor's food intake. Women are major stakeholders in all development issues related to water. Yet they often remain on the periphery of management decisions and planning for water resources.

To overcome this deficiency, an inter-agency Task Force on Gender and Water was inaugurated to work towards the implementation of gender-sensitive water and sanitation

activities. In addition, the Gender and Water Alliance (GWA) was established at the World Water Forum in 2000.

The GWA has developed a training methodology geared towards building capacity to mainstream gender equity in integrated water resources management. The Vice Ministry of Basic Services and the Ministry of Agriculture of Bolivia have undertaken gender audits in both institutions, turning this unique research and analytical initiative into a 'learning by doing gender'

experience. Although the audits indicated that the approach to gender equity is not reflected in sector policies – nor is the impact of programmes and projects on local women and men systematized with feedback to decision-making levels – issues related to gender responsiveness are becoming increasingly important.

Sources: See www.un.org/esa/sustdev/sdissues/ water/Interagency_activities.htm#taskforce_water for more information; UN, 2003a; Arce, 2005; www.genderandwateralliance.org

Mid-career professionals receive on-the-job training by local experts in Indonesia

salaries, lack of equipment, demoralized environment and the bureaucratically inefficient management. As it turned out, the trained personnel stayed in the public sector but were underperforming because 'the opportunity to use office hours and equipment to significantly augment official salaries through private-income earning activities provided a major incentive to stay in the civil service' (Cohen and Wheeler, 1997). The institution's performance therefore suffered not from a lack of skilled human resources, but from lack of incentives coupled with poor accountability and management structure.

Sound demand-driven research on water-related issues enhances the ability for more rational decision-making on costs, impacts, and benefits of alternative policy options and institutional arrangements. Investing in research and development and its associated infrastructure, equipment and human resources means that conventional approaches to recurring problems can be challenged and new ways of addressing local engineering, social, economic and environmental issues may begin to flourish.

3c. Creating an enabling environment

The enabling environment consists of the broader political, policy, legal, regulatory and administrative frameworks that set the boundary conditions for the execution of the organizational and operational functions of the agencies and institutions entrusted with the development and management of water resources and services. A truly enabling environment is created primarily by policies that focus on sustainable development, consider water as a social and economic good, and are supported by legal and financial frameworks that ensure the policies are

implemented. A proper enabling framework will also emphasize the need for sector agencies to continuously improve their performance, through knowledge creation and acquisition, and through reform. For a broader sector reform to run its course, governments must be able to rely on realistic fiscal and monetary policies in the water sector, including adequate cost-recovery mechanisms, and transparent and equitable judicial systems (van Hofwegen and Jaspers, 1999). Civil society has an important role in developing the enabling environment. Well-informed civil groups and the media can enhance the awareness of the public at large of the need for particular actions, and can at the same time provide the information that empowers and motivates them to change their attitudes (social learning).

As a consequence of their decentralization policies, many governments such as Indonesia and Pakistan are now discussing and facilitating the possibilities of sub-sovereign financing and decentralized funding. Enhancing availability and access to finance especially is an essential element of the institutional capacity and was identified as one of the main recommendations of the Camdessus Panel, an initiative of the World Water Council, the Global Water Partnership and the Secretariat of the 3rd World Water Forum (Winpenny and Camdessus, 2003). To support the investment requirements at sub-sovereign level, various innovative initiatives and financing mechanisms have been recommended and some of them have been launched. These initiatives enhance the amount of financing available as they mobilize local capital markets by provision of guarantees for especially local political and currency risks and they enable financing at sub-sovereign level. Some of them, like Output Based Aid and ADB-Water for All initiatives, focus particularly on reaching the poor (see **Box 13.12**).

Capacity development is dependent on the government's political will to change the existing policy, legal, management and economic frameworks and to implement reforms, as well as on the introduction of new governance systems and the familiarization of decision-makers and implementers with improved ways of managing water. This implies that capacity development actions have to include the political, social, economic and administrative dimensions of systems that may affect management of water resources and delivery of water services. This can go as far as the inclusion of policies in relation to organization of government, delegation of authority, career planning, salary and reward systems within the civil service and the creation of incentive mechanisms to enhance effective governance. It also implies development of policies that enhance access to finance for development and management of water infrastructure and services.

Until now capacity development has often been focused at the level of new utility management, communities or basin and water users associations. However, the decentralization and management transfer policies in many countries add a new dimension to capacity development: the development of new regulatory and governance systems at the decentralized levels. Unfortunately, the changing roles of government are not always accompanied with the associated capacity development and incentive systems required to effectuate the change. This is often due to a combination of a lack of knowledge on the implementation of these new roles, inherent resistance to innovation and a lack of appreciation of local capacity, knowledge and experience.

It follows that an important aspect of good governance in stimulating capacity development is related to research and education. Governments have to provide incentives and mechanisms that stimulate education and research institutions to address the real societal issues and demand. This can be done through applied-oriented research funds and through activation of professional, commercial, civil society and political institutions in the development of education and research programmes. The European Union has provided instruments through its fifth and sixth framework programmes that support the development of such linkages in society and among education and research institutions.

A country may even have to modify its national laws and regulations to enable education institutions to adjust their curriculum in response to demands from society. One such example is the new law on higher education in Indonesia (2003). This law has opened up the possibility to include private education institutions with their linkages to professional organizations and the private sector. Moreover, the accreditation system has been changed from pre-approval of the curricula to post-approval by an accreditation board. This is a big step towards the development of dynamic and society responsive education and research environment.

Similarly, supply-oriented programmes for education and training of water user associations and community organizations have to be changed into demand-driven programmes with a menu structured delivery system of training services where the communities or user organizations can match their needs and priorities. This will facilitate a better assimilation of new knowledge and put it into practice more quickly.

National and international meetings are another way of providing a platform for local authorities, politicians, water

...retention rates of trained individuals were much higher than expected for the perceivably low salaries, lack of equipment, demoralized environment and the bureaucratically inefficient management

Ensuring access to public information is an aspect of good governance

sector institutions and education and research institutions to get exposure and exchange knowledge and experiences. Such interactions among different stakeholder groups provide new insights and ideas on priorities and approaches to capacity development and the development of governance structures. The 4th World Water Forum in Mexico in March 2006, which focuses on strengthening local action, is an example of such a platform.

Ensuring access to public information is yet another aspect of good governance. Unfortunately, the financial support to generate basic information especially in the domain of basic water quality and hydrological and meteorological data has been dwindling. A new discussion is needed on the meaning and implications of property rights, especially in the public domain, and the ethics around charging for information in the public domain, which should also include the international organizations.

Part 4. Identifying Knowledge and Capacity Needs

It is widely acknowledged that greater efforts are required to understand the complex processes of change within all levels of development. While case studies, working papers, reports, manuals, best practices, guidelines, and the like are all valuable sources of knowledge, comprehensive cross-country comparisons (regional, national and/or basin wide assessments) of existing knowledge and capacity are needed as are analyses of capacity development initiatives undertaken in the past. Today, little data is available to allow the identification of national capacities to address development problems specific to water.

...little data is available to allow the identification of national capacities to address development problems specific to water

Identifying methods to measure capacity and monitor development initiatives has proven to be difficult. National statistical agencies must improve and strengthen the collection, storage and analysis of information conducive to the knowledge base of water management. At the international level, such needs are being carried out in order to move the process of water assessments along, and the UN Statistical Commission's Intersecretariat Working Group on Environment Statistics (IWG-ENV) is currently reviewing measures to improve the contribution of statistical work to the management of development in water-related sectors. Other organizations involved with enhancing capacity within national statistics agencies include Paris21, the World Bank, and UNESCO Statistics. Through discussions between the international statistics community, water experts and development agencies, it is expected that the existing sustainable development questionnaires could be adapted and new survey material created in order to segregate and better measure the capacity-enhancing initiatives specific to water development *en route* to meeting the MDGs. Thinking in terms of a non-traditional approach to capacity-development assessment will help to overcome limited international agreement on how to adequately measure such complex qualitative processes.

4a. Indicators
To date, indicators for measuring capacity are very much in their infancy. Rather than a final, easily measurable outcome or output, capacity development is a complex *process* that leads to more varying degrees of capacity in an ever-changing environment. Traditional measurement methods have often concentrated on 'harder' issues (e.g. systems improvements, equipment supply, training, organizational structuring, etc.), with a focus on a specific entity (societal, organizational or individual), rather than reflecting the broader system or environment in which they function and the 'softer' issues of learning, adaptation and changes in attitude. Under these circumstances, when output-based performance indicators conclude that the majority of capacity development projects implemented during the 1980s failed, we learn whether the capacities were 'adequate' or 'good' overall but gain little information about which aspects of the process were failing, where partial successes exist or whether or not the initiative was doomed to fail from the start (Mizrahi, 2004).

By only considering the barriers to capacity development without better assessing the causes of good and bad performance, the opportunity to identify and confirm strengths is forfeited. Because a number of indirect and unrelated inputs can contribute to the good performance of an institution, an excellent agricultural output in one country might be the result of a market collapse in another without any institutional reform or capacity enhancements being made.

Potential indicators for a global assessment might include the identification of users of the growing number of water-related databases or the human resource needs to meet specific water-related goals on the global agenda. However, while aggregated global indicators can make an important contribution to the awareness of global capacity gaps, they can oversimplify the complex processes involved in addressing the myriad of water-related problems and would therefore provide little information on accomplishments, needs or failures. Worse, such aggregated indicators could degenerate into a form of conditionality designed mainly for the reporting needs of the international community, thereby undermining a nation's efforts to address the water-related challenges being assessed.

An effective global assessment demands that a bottom-up approach be undertaken, which endorses ownership, promotes participation and considers the contextual aspects of a sector's objectives. The design and implementation of suitable strategies for change must adopt a focused approach based on region/nation/basin/ community ownership. There is no panacea for assessing and enhancing capacity. When we ask ourselves, 'What are the crucial *capacity* gaps?' We must also ask, 'For what?' and 'For whom?' The capacity for a nation to meet the MDGs water supply and sanitation targets will be different from that same nation's capacity to monitor the resource for better risk management. The capacity required for a government to enable better agricultural trade will be different from the Farmers Association's capacity needs for increasing pressure on the government to do so.

Capacity development is, by definition, a process *leading* to outcomes. As such, benchmarking is required to measure the degrees of capacity attained. However, identifying benchmarks is particularly difficult because a common understanding of the abstract notions of human knowledge, institutional performance and cultural change across the community, basin, national and international levels of society must be achieved among the various actors. Therefore, a participatory, locally oriented approach must be adopted when designing capacity assessment programmes. The ability to prioritize goals and improve self-evaluation methodologies can be

strengthened through negotiation from the beginning of the capacity development initiative.

As each country faces its own water-related problems within the context of its own national priorities and agendas, it must choose its own methodological approach to identifying capacity needs while taking varying political, socio-economic and technical factors into account. It is essential that these assessments be entirely country-driven, undertaken by national institutions and experts to the fullest feasible extent, and responding to national situations and priorities. External agencies cannot effectively assess the capacity needs of a country, nor should they play any role other than that of facilitating the process of analysis and developing capacities to manage and implement change. It is therefore important that countries be self-sufficient in their ability to undertake capacity-needs assessments in water-related sectors, and it should be incumbent on those countries to be forthright in sharing the assessment information with the international community in order to gain from lessons learned, prove institutional strength or direct attention to their capacity needs. By this approach, better clarity of the global capacity needs to address water-related problems will begin to be realized, and actions to address these needs can be undertaken. To enable some consistency in the regional and sector assessments, the general framework for capacity outlined in **Table 13.1** could be used as guidance for indicator development. The framework provides a conceptual map that can be adapted and used to assess capacity related to specific development objectives and then reported within each of the challenge areas of the next UN World Water Development Report. This way, future assessments of the various challenge areas within this Report will inherently incorporate knowledge and capacity needs into their cross-country comparisons. For example, by viewing **Chapter 7** on water for food through the lens of knowledge and capacity, future indicators may include the number of irrigation associations per capita or farmers' access to meteorological information. These sector-based capacity assessments will contribute to the much-needed statistical data on existing capacity and associated initiatives. In addition, they will promote awareness of capacity gaps and be themselves a capacity-enhancing process to those involved.

The readiness of governments to undertake capacity-needs assessments in various water-related sectors would

It is important that countries be self-sufficient in their ability to undertake capacity-needs assessments in water-related sectors...

Lawyers and water managers from Southern Africa playing a negotiation game during a short course on conflict prevention in Maputo, Mozambique

Table 13.1: General framework for capacity development

Human Resources

Refers to the process of changing attitudes and behaviours – imparting knowledge and developing skills while maximizing the benefits of participation, knowledge exchange and ownership.

Job requirements and skill levels	Are jobs correctly defined and are the required skills available?
Training/Retraining	Is the appropriate learning taking place?
Career Progression	Are individuals able to advance and develop professionally?
Accountability/Ethics	Is responsibility effectively delegated and are individuals held accountable?
Access to Information	Is there adequate access to needed information?
Personal/Professional Networking	Are individuals in contact and exchanging knowledge with peers?
Performance/Conduct	Is performance effectively measured?
Incentives/Security	Are these sufficient to promote excellence?
Values, Integrity and Attitudes	Are these in place and maintained?
Morale and motivation	Are these adequately maintained?
Work Redeployment and Job Sharing	Are there alternatives to the existing arrangements?
Inter-relationships and teamwork	Do individuals interact effectively and form functional teams?
Interdependencies	Are there appropriate levels of interdependence?
Communications skills	Are these effective?

Institutional Capacity

Focuses on the overall organizational performance and functioning capabilities as well as the ability of an organization to adapt to change.

Mission and strategy	Do the institutions have clearly defined and understood missions and mandates?
Culture/Structure/Competencies	Are institutions effectively structured and managed?
Process	Do institutional processes such as planning, quality management, monitoring and evaluation work effectively?
Human resources	Are the human resources adequate, sufficiently skilled and appropriately deployed?
Financial resources	Are financial resources managed effectively and allocated appropriately to enable effective operation?
Information resources	Is required information available and effectively distributed and managed?
Infrastructure	Are material requirements such as building, offices, vehicles and computers allocated appropriately and managed effectively?

The Enabling Environment

Focuses on the overall policy framework in which individuals and organizations operate and interact with the external environment.

Policy framework	What are the strengths, weakness, opportunities and threats according to the socio-political, government/public sector, economic/technological and physical environment factors operating at the societal level?
Legal/Regulatory framework	Is the appropriate legislation in place, and are these laws effectively enforced?
Management/Accountability framework	Are institutional responsibilities clearly defined, and are responsible institutions held publicly accountable?
Economic framework	Do markets function effectively and efficiently?
Systems-level framework	Are the required human, financial and information resources available?
Process and relationships	Do the different institutions and processes interact and work together effectively?

Source: Lopes and Theisohn, 2003.

provide an indication of a government's own capacity to appreciate and to adapt to change. The degree to which a participatory process is incorporated in performing the assessment and the utility of the assessment to identify capacity gaps and priorities could also be considered as a global indicator to assess countries in future editions of the UN World Water Development Report. Such information could be compiled in a map on countries' readiness to meet the Johannesburg implementation plan target on IWRM Plan preparation by 2005. As presented in the case studies of this Report (**Chapter 14**), governments of some countries such as Ethiopia and Uganda have taken a positive first step in performing preliminary assessments of needs to operationalize IWRM and their foresight should be recognized in global country comparisons (see **Chapter 14**).

4b. The distribution of knowledge and capacity needs

The knowledge base that supports decision-making and the capacities of individuals, institutions and societies to perform functions, solve problems and set and achieve objectives pertain to all the challenge areas discussed in the preceding chapters. Lopes and Theisohn (2003) present a simple indication of the relative significance of different aspects of capacity development. Their work aims to arrest the ambivalence about capacity issues that remains within many international development agencies and partner countries by illustrating capacity development that does not only depend on individual training courses, but also on every aspect of a country's ability to address water-related problems within all sectors of its development.

4c. Assessment tools and challenges

In recent years, a few examples of national capacity self-assessment handbooks have been created for various development objectives. Of particular note is WMO's and UNESCO's *Water Resources Assessment - Handbook for Review of National Capabilities* (WMO/UNESCO 1997b), which is designed to assess a country's ability to measure and monitor effectively its water resources. The handbook specifically aims to provide guidance for reviewing levels of activity of the basic water resources assessment in the whole or part of a country or region. The activities are compared, when possible, with minimum acceptable requirements in terms of installation density, degree of computerization, skilled human-power and related management structure and education, training and research programmes. The comparisons then provide a

basis for proposing actions considered necessary for achieving minimum requirements.

This methodology has been applied in Latin America and further expansion to other regions is desirable. The guidelines are not prescriptive in any sense and are not meant as a standard methodology that can be applied to all countries at all times under varying political, socio-economic and water resource conditions. Countries should use the water resource and capacity needs self-assessment guidelines to the extent that they consider them to be feasible, or they may choose other methodologies better suited to their national situations and preferences.

Though not widely available, assessments have been undertaken addressing specific sector needs that make them difficult to compare. Moreover, assessments undertaken have followed approaches intended to specifically address different levels of capacity development. At the regional level, assessments have been undertaken in Asia and Latin America to make inventories of the human capacities required and available to address the Millennium Development Goals in water and sanitation (Mejia-Velez and Rodic-Wiersma, 2005a; 2005b; Rodic-Wiersma and Sah, 2005). At the country level, some examples are included in Chapter 14 on case studies. Additional examples are the *Mali Long Term Training and Capacity Building Needs Assessment* (Academy for Educational Development, 2003), and the Capacity Building Project in the Water Resources Sector in Indonesia (Asian Development Bank, 2005). The first example concerns the individual human resources and the second the nature of the institutions. A brief description of the multilevel assessment and implementation of capacity development initiatives in Mexico is described in *Capacity Building for the Water Sector in Mexico: An Analysis of Recent Efforts* (Tortajada, 2001). Some key sources for information on how to perform capacity assessments are included in **Table 13.2**.

In addition, the Global Environment Facility (GEF) Secretariat and UNDP launched the Capacity Development Initiative (CDI), which produced a National Capacity Self-Assessment (NCSA) process. The primary goal of the NCSA is to identify, through a country-driven consultative process, priorities and needs for capacity-building, in order to identify, confirm or review priority issues for action within the GEF's thematic areas of biodiversity, climate change and desertification/land degradation.

Countries should use the water resource and capacity needs self-assessment guidelines to the extent that they consider them to be feasible...

...the true test for any country will be its formulation of a strategic plan of action and the successful implementation of appropriate capacity development initiatives

Table 13.2: Some capacity assessment tools

The UNDP's Capacity Development Resource Book (available at magnet.undp.org/cdrb/Default.htm) is a collection of electronic documents assembled for capacity development practitioners.

UNDP website on capacity development (www.capacity.undp.org/) includes key sources for generic information on how to perform capacity assessments. Additionally, it includes initiatives, networks, resources and tools. It offers access to the Capacity 2015 initiative developed to operationalize the MDGs.

The South African Capacity Initiative (SACI) (www.undp-saci.co.za) developed a Capacity Mobilization Toolkit for Southern African countries, which takes into consideration the particularly complex human capacity challenges associated with the impacts of HIV and AIDS, poverty and recurring disasters on sustaining basic social services to the public at all levels of the Millennium Development Goals.

The World Bank provides an online **Capacity Development Resource Centre**, which provides an overview of case studies, lessons learned, 'how to' approaches and good practices pertaining to capacity development. It is available at www.worldbank.org/capacity

The Canadian International Development Agency (CIDA) has developed a Tool-Kit for Capacity Development, available at www.acdi-cida.gc.ca/ that includes reference documents for capacity development.

The European Centre for Development Policy Management's Capacity Development website (www.capacity.org) aims to look at policy and practice of capacity development within international development cooperation and provides a newsletter and a comprehensive material related to capacity development in all sectors.

The International Development Research Centre (IDRC), the International Institute of Rural Reconstruction (IIRR) and the International Service for National Agricultural Research (ISNAR) implemented a project to better understand how capacity development takes place and how its results can be evaluated. Further information is available at www.idrc.ca/en/ev-31556-201-1-00-TOPIC.html

A team of **the German Technical Cooperation** (GTZ) supported the Indonesian Government in preparing guidelines on how to organize and manage a capacity-building needs assessment process in the regions, resulting in a medium-term regional capacity-building action plan. These very well structured guidelines, field-tested before they became final, are available at www.gtzsfdm.or.id/cb_nat_fr_work.htm.

Country capacity needs and priorities to achieve goals related to conventions in these three areas are then documented globally (excluding high-income countries). This methodology provides a good direction for the implementation of global assessments in the various water-related sectors of the next World Water Development Report.

When assessing capacity needs, the true test for any country will be its formulation of a strategic plan of action and the successful implementation of appropriate capacity development initiatives. This assessment can help identify an entry point to initiate a capacity development programme under conditions of resistance to change (see **Chapter 2**). If the capacity development

process starts through legal reforms, institutional strengthening or awareness-building is necessary, which entails the identification of a suitable entry point among the different levels of capacity. Principles summarizing the implementation of capacity development initiatives are beginning to be formulated (Lopes and Theisohn, 2003). An effective ongoing monitoring programme must accompany the process to improve the capacities of individuals and institutions, such that they can develop a culture of self-assessment and establish an approach to thinking strategically about capacity and performance. Since capacity development is a process, a monitoring system is necessary to understand changes in the process and to feed this information back to those managing it, so that they can enhance the efficiency and effectiveness of the intervention.

Part 5. The Way Forward

This chapter highlights the need for knowledge acquisition and integration to become the responsibility of the country concerned, supported on the one hand by participatory processes for knowledge development that capitalize on existing local and indigenous knowledge and on the other hand by the unprecedented access to the global knowledge base and professional networks in order to adopt a holistic and integrated approach to enhance all levels of existing capacities. As solutions are only found when problems are understood, comprehensive capacity assessments are urgently needed to identify where understanding is deficient and to meet water development objectives. To date, such assessments have not yet been done in developing countries. This trend should be reversed.

If the Millennium Development Goals are to be met... it is incumbent upon donor nations to embrace the new paradigm for water development...

Comprehensive capacity assessments will set the baseline to gauge progress on locally owned strategies to address country-specific water-related problems. The information gained will stimulate the adaptive capacity of countries to anticipate and manage change by fostering a culture of self-assessment and by establishing a strategic approach to sustainable development. In this way the goal of sustainable development of water resources and services and the capacity to manage them effectively and efficiently can be achieved through a continuous renewal process that is at the heart of good governance.

It follows that there should be increased international commitment to statistical capacity-building in water with a focus on the three levels of capacity and their sub-components. Similarly, there should be greater emphasis on sharing knowledge, particularly amongst riparian nations of a transboundary basin, in order to build regional capacity to address water-related problems together.

That part of the knowledge base that concerns the data from monitoring the water resources and services performance is essential for both global comparison and local assessment. Further investment and better management of ground-based networks and remote sensing facilities are needed to ensure that adequate information is available to facilitate sound decision-making at all levels of a country's capacity. At the same time emphasis should be placed on improving the human component of the knowledge base. This acknowledges that all staff and stakeholders, including the general public, require education and training about the pressing issues of water development in the region. Although there is a need for countries to create apex bodies at the basin level, the decentralization of responsibilities will require attention directed at strengthening local institutional capacities. In doing so, the roles of the local

institutions must be clearly defined, and a culture of innovation should be encouraged that appreciates and takes advantage of local and indigenous knowledge and experience. Only by building upon the existing foundation can water development be effective.

If the Millennium Development Goals are to be met, in particular Goal 8 to develop a global partnership for development, it is incumbent upon donor nations to embrace the new paradigm for water development by providing the necessary support (increased aid and debt relief, opening of trade, accelerated transfer of technology and improved employment opportunities) to allow developing nations to expand their knowledge base and enhance their existing local capacities rather than transferring short-term solutions as was common in the past. Likewise, it is the responsibility of the leaders of developing countries to create an enabling environment to enhance the existing local capacities and the knowledge base on water, by setting policies, ensuring adequate funding, and empowering local institutions and stakeholders with decision-making responsibilities, and to monitor performance to ensure good governance and transparency.

In viewing the assessments and country comparisons for each of the challenge areas presented throughout this Report, we can identify the gap between the existing situation and the desired situation. These identified gaps *are* the deficiencies in knowledge and capacity within each of the associated sectors (i.e. food, health, energy, etc.). To continue to be ambivalent about enhancing the knowledge base, or to ignore the holistic approach to capacity development and view it simply as a one-off course for individual training would mean that countries will go on pouring untold resources and time into unsustainable solutions and these gaps will see little reduction in future World Water Development Reports.

References and Websites

Academy for Educational Development. 2003. *Mali Long Term Training and Capacity Building Needs Assessment: Africa Agriculture Capacity Development Training Initiative Strategic Technical Assistance for Results with Training (START)*. Report submitted to the United States Agency for International Development. October 2003. Washington, DC. www.usaid.gov/our_work/agriculture/bifad/mali_training_assessment_oct_03.pdf

Agarwal A. and Narain, S. 1997. *Dying Wisdom: Rise, Fall and Potential of India's Traditional Water Harvesting Systems*. New Delhi, Centre for Science and Environment.

Al-Hanbali, N., Al-Kharouf, R. and Bilal Alzoubi, M. 2004. Integration of geo imagery and vector data into school mapping GIS data-model for educational decision support system in Jordan. ISPRS Commission II, WG II/5 – Design and Operation of Spatial Decision Support Systems, Istanbul, Turkey.

Alaerts, G. J., Hartvelt, F. J. A. and Patorni, G. M. (eds). 1999. Water sector capacity-building: Concepts and instruments. Proceedings of the Second UNDP Symposium on Water Sector Capacity-building, Delft, The Netherlands.

Alsdorf, D. and Rodriguez, E. 2005. WatER: The Water Elevation Recovery Satellite Mission. First Mission Document for the WatER Mission. www.geology.ohio-state.edu/water/WatER_Document.pdf

Arce, M. 2005. Personal interview. Executive Secretary of Gender and Water Alliance. Delft, The Netherlands.

ADB (Asian Development Bank). 2005. *Project Completion Report on the Capacity Building Project in the Water Resources Sector in Indonesia*. Manila, The Philippines. www.adb.org/Documents/PCRs/INO/pcr-ino-26190.pdf

——. 2004. First Regional Meeting of National Water Sector Apex Bodies: Leadership in Water Governance. Report of the meeting, Hanoi, Vietnam, 18–21 May 2004.

Cap-Net. 2005. *Annual Report*. Project No. GLO/02/115 2004. United Nations Development Programme. www.cap-net.org/file_aboutCapnet/3_1_Annual_Report_2004.doc_

Cohen, J. M. and Wheeler, J. R. 1997. Training and Retention in African Public Sectors: Capacity-Building Lessons from Kenya. M. S. Grindle (ed.), *Getting Good Government: Capacity-building in the Public Sectors of Developing Countries*. Cambridge, Mass., Harvard University Press.

Dekker, A. G., Brando, V. E., Anstee, J. M., Pinnel, N., Kutser, T., Hoogenboom, E. J., Peters, S. W. M., Pasterkamp, R., Vos, R., Olbert, C. and Malthus, T. J. M. 2001. Imaging spectrometry of water. F. van der Meer and S. M. de Jong. (eds), *Imaging Spectrometry: Basic Principles and Prospective Applications*. Dordrecht, The Netherlands, Kluwer Academic Publishers.

DDWS (Department of Drinking Water Supply). 2004. School sanitation and hygiene education in India: Investment in building children's future. DDWS, Government of India. Presented at the SSHE Global Symposium, 'Construction is Not Enough'. Delft, The Netherlands, 8–10 June 2004.

ESA (European Space Agency). 2004. TIGER Workshop puts focus on space for African water management. *ESA News*.

3 Nov 2004. European Space Agency. www.esa.int/esaEO/SEMUHV0A90E_economy_2.html

——. 2000. FarmNet: Farmer information network for agricultural and rural development. Rome, FAO Research, Extension and Training Division, SDR and World Agriculture Information Centre. ftp://ftp.fao.org/sd/farmnet.pdf

Folke, C., Carpenter, S., Elmqvist, T., Gunderson, L., Holling, C. S. and Walker, B. 2002. Resilience and sustainable development: Building adaptive capacity in a world of transformations. *Ambio*, Vol. 31, No. 5, pp. 437–40.

Fukuda-Parr, S., Lopes, C. and Malik, K. (eds). 2002. *Capacity for Development: New Solutions to Old Problems*. London, UK and Sterling, VA, Earthscan and United Nations Development Program. www.undp.org/dpa/publications/CapforDevelopment.pdf

Gordon Jr., R.G. 2005. *Ethnologue: Languages of the World*, 15th Edition. Dallas, SIL International. www.ethnologue.com

Grabs, W. 2003. Networks, availability and access to hydrological data. Hydrology and Water Resources Department. Geneva, World Meteorological Organization. grdc.bafg.de/servlet/is/9921/

GRDC (Global Runoff Data Centre). 2005. GRDC Station Catalogue. Global Runoff Data Centre. Koblenz, Germany. grdc.bafg.de/servlet/is/910/

Horton, D., Alexaki, A., Bennett-Lartey, S., Noële Brice, K., Campilan, D., Carden, F., de Souza Silva, J., Thanh Duong, L., Khadar, I., Maestrey Boza, A., Kayes Muniruzzaman, I., Perez, J., Somarriba Chang, M., Vernooy, R. and Watts, J. 2003. *Evaluating Capacity Development: Experiences From Research and Development Organizations Around the World*. ISNAR (International Service for National Agricultural Research); IDRC (International Development Research Centre), ACP-EU Technical Centre for Agricultural and Rural Cooperation. www.isnar.cgiar.org/publications/ecd-book.htm

IAHS Ad Hoc Committee. 2001. Global water data: A newly endangered species. *AGU EOS-Transactions*, Vol. 82, No. 5, pp. 54–58.

ITU (International Telecommunication Union). 2004. *World Telecommunication Indicators Database*. ITU. Geneva. www.itu.int/ITU-D/ict/statistics/

Kotter, J. P. and Cohen, D. S. 2002. *The Heart of Change: Real-Life Stories of How People Change their Organizations*. Boston, Harvard Business School Press.

Lopes, C. and Theisohn, T. 2003. *Ownership, Leadership and Transformation: Can We Do Better for Capacity Development?* London, UNDP and Earthscan Publishing.

Maurer, T. 2003. Intergovernmental arrangements and problems of data sharing. Contribution to Monitoring Tailor-Made IV Conference, Information to support sustainable water management: From local to global levels, St. Michielsgestel, Netherlands, 15–18 September 2003.

Mejia-Velez, D. and Rodic-Wiersma, Lj. 2005a. Higher education in water and sanitation: A preliminary needs

assessment for the achievement of the Millennium Development Goals in Latin America. International IWA Conference AGUA 2005, From Local Action to Global Targets. Cali, Colombia, 31 October – 4 November 2005.

——. 2005b. Higher education in water and sanitation: An assessment in selected African countries. (draft paper)

Milburn, A. 2004. E-mail communication. Kingston-upon Thames, UK, December, 2004

Misión de Evaluación OMM/PROMMA. 2003. *Evaluación técnica del PROMMA 2003*. Proyecto de Modernización de Manejo del Agua (PROMMA) No. 160. Mexico.

Mizrahi, Y. 2004. *Capacity Enhancement Indicators: Review of the literature*. Washington DC, World Bank Institute. siteresources.worldbank.org/WBI/Resources/wbi37232Mizrahi.pdf

Morgan, P., Land, T. and Baser, H. 2005. Study on Capacity, Change and Performance - Interim Report. Discussion Paper, 59A, Maastricht, The Netherlands, ECDPM.

Morgan, P. 2000. Some observations and lessons on capacity-building. I. Grunberg and S. Khan (eds), *Globalization: The United Nations Development Dialogue: Finance, Trade, Poverty, Peace-Building*. (UNU Policy Perspectives 4.) New York, United Nations University Press.

Pasmore, W. 1994. *Creating Strategic Change: Designing the Flexible High-Performing Organization*. Chichester, UK, Wiley.

Rawls, W. J., Kustas, W. P., Schmugge, T. J., Ritchie, J. C., Jackson, T. J., Rango, A. and Doraiswamy, P. 2003. Remote sensing in watershed scale hydrology. Proceedings of the First Interagency Conference on Research in Watersheds, 27–30 October, 2003, Benson, Arizona. pp. 580–85.

Rodic-Wiersma, Lj. and Sah, R. D. 2005. Professional capacity needs assessment for Millennium Development Goal on water supply and sanitation in Asia. XII World Water Congress of IWRA - Water for Sustainable Development Towards Innovative Solutions, 22–25 November 2005, New Delhi, India

Sachs, J. 2005. Bush inherits the wind. Project Syndicate. 20 September 2005. www.project-syndicate.org

Schultz, G. A., and E. T. Engman (eds) 2000. *Remote Sensing in Hydrology and Water Management*. Springer-Verlag, Berlin.

Senge, P., Kleiner, A., Roberts, C., Ross, R., Roth, G. and Smith, B. 1999. *The Dance of Change: The Challenges of Sustaining Momentum in Learning Organizations*. New York, Doubleday/Currency.

Snowden, D. 2003. Complex Knowledge. Presentation at the Gurteen Knowledge Conference, June 2003, Cynefin Centre for Organizational Complexity IBM, UK.

Spicer, M. 2005 Encouraging private investment in water and sanitation: new and traditional approaches. Presentation of Municipal Fund during CSD 13, New York, April 2005.

Tortajada, C. 2001. Capacity building for the water sector in Mexico: An analysis of recent efforts. *Water International*, Vol. 26, No. 4, pp. 490–98, December 2001. www.thirdworldcentre.org/waterinternational.pdf

UN (United Nations). 2004. Water for the world: Space solutions for water management. UN/Austria/ESA Symposium on Space Applications for Sustainable Development to Support the Plan of Implementation of the World Summit on Sustainable Development, Graz, Austria.

——. 2003a. New inter-agency gender and water task force established. Press release for the International Year of Freshwater 2003, 15 September 2003.

——. 2003b. *Shared Natural Resources: First Report on Outlines*, First report of the special rapporteur on Shared Natural Resources, Amb Chusei Yamada, International Law Commission, 55th Session, UN Doc. A/CN.4/533.

UNDP (United Nations Development Programme). 1997. Capacity development, Technical advisory paper 2. New York, UNDP – Management Development and Governance Division, Bureau for Policy Development.

UNESCAP (United Nations Economic and Social Commission for Asia and the Pacific). 2004. *Trade and Investment Policies for the Development of the Information and Communication Technology Sector of the Greater Mekong Subregion.* (Studies in Trade and Investment, 52.) New York, UNESCAP, Trade and Investment Division.

——. 2003. *Use of Space Technology Applications for Poverty Alleviation: Trends, Strategies and Policy Frameworks.* New York, United Nations Economic and Social Commission for Asia and the Pacific.

UNESCO (United Nations Educational Scientific, Cultural Organization). 2003a. Best Practices on Indigenous Knowledge. MOST/NUFFIC (IK-Unit) Database. UNESCO. www.unesco.org/most/bpikreg.htm

——. 2003b. International Year of Freshwater Website. Indigenous Peoples, Paris. www.wateryear2003.org/facts

UNESCO-IHE Institute for Water Education. 2002. Capacity Building: Methods and Instruments. Delft, UNESCO-IHE Institute for Water Education.

UNICEF (United Nations Children's Fund). 2005a. *Progress for Children: A Report Card on Gender Parity and Primary Education.* New York, UNICEF.

——. 2005b. Lack of safe water and sanitation in schools jeopardizes quality education. Press release, New York, UNICEF.

UN-WWAP (United Nations World Water Assessment Programme). 2003. *UN World Water Development Report: Water for People, Water for Life.* Paris/London, UNESCO/Berghahn Books.

van der Beken, A. (ed.). 2004. Water-related education, training and technology transfer. *Encyclopaedia of Life Support Systems*, Oxford, EOLSS Publisher. etnet.vub.ac.be/ePUBLICATIONS21/watereducation.pdf

van Hofwegen, P. J. M. 2005. E-mail communication. Marseilles, France. September, 2005

——. 2004. Capacity-building for water and irrigation sector management with application in Indonesia. *Capacity Development in Irrigation and Drainage Issues, Challenges and the Way Ahead.* (FAO Water Reports, 26), Rome, FAO.

van Hofwegen, P. J. M. and Jaspers, F. G. W. 1999. *Analytical Framework for Integrated Water Resources Management: Guidelines for Assessment of Institutional Frameworks.* IHE Monograph 2. Rotterdam, The Netherlands, A.A. Balkema.

Veevers-Carter, P. 2005. Output-based aid and its use in water and sanitation programmes. Stockholm Meetings on the EU Water Initiative ACP-EU Water facility session, 24 August 2005.

Vörösmarty, C. J., Douglas, E. M., Green, P.A. and Revenga, C. 2005. Geospatial indicators of emerging water stress: An application to Africa. *Ambio*, Vol. 34, No. 3, pp. 230–36.

Vörösmarty, C. J., Green, P., Salisbury, J. and Lammers, R. B. 2000. Global water resources: vulnerability from climate change and population growth. *Science*, Vol. 289, pp. 284–88.

Walker, B., Carpenter, S., Anderies, J., Abel, N., Cummings, G., Janssen, M., Lebel, L., Norberg, J., Peterson, G. D., and Pritchard, R. 2002. Resilience management in social-ecological systems: A working hypothesis for a participatory approach. *Conservation Ecology* Vol. 6, No. 1, p. 14. www.consecol.org/vol6/iss1/art14/main.html.

WHO, UNICEF (World Health Organization, United Nations Children's Fund). 2003. *Meeting The MDG Drinking Water and Sanitation Target: A Mid-Term Assessment of Progress.* www.unicef.org/wes/mdgreport/index.php

WL | Delft Hydraulics. 2004. *The Hydrology Project, India: Development of a Hydrological Information System (HIS).* Project Description Q1990. Delft, The Netherlands. www.wldelft.nl/proj/pdf/3uk00227.scherm.pdf

Winpenny, J. T. and Camdessus, M. 2003. *Financing Water for All: Report of the Global Panel on Financing Water for Infrastructure* (The Camdessus Panel). World Water Council, Global Water Partnership, Secretariat of the 3rd World Water Forum.

WMO/UNESCO (World Meteorological Organization/United Nations Educational, Scientific and Cultural Organization). 1997a. *The World's Water – Is there enough?* Geneva, WMO.

——. 1997b. *Water Resources Assessment: A Handbook for Review of National Capabilities.* www.wmo.ch/web/homs/documents/english/handbook.pdf

WMO/WHYCOS. 2005. *Guidelines for Development Implementation and Governance (2005)* WMO/TD No. 1282. Geneva.

World Bank. 2004. Implementation completion report (Ida-27740 Tf-28729) Report No. 28775–IN, 12 May 2004. Washington DC, World Bank.

——. 2003. Water resources and irrigation sector management project. Appraisal Document, Washington, DC, The World Bank.

——. 2000. FRESH Initiative. School health toolkit. Washington, DC. www.schoolsanitation.org/Resources/Readings/Fresh%20School%20Health%20toolkit.doc

United Nations water-related portals

Earthwatch: earthwatch.unep.net
FAO Water Portals: www.fao.org/ag/agl/portals.stm
GEO-3 Data Portal: geodata.grid.unep.ch
International Year of Freshwater: www.wateryear2003.org
UNEP Freshwater Portal: freshwater.unep.net
UNESCO Water Portal: www.unesco.org/water

For a comprehensive list of Water-related UN programmes, portals, and databases see www.unesco.org/water/water_links/Type_of_Organization/United_Nations_System_Programmes_and_Agencies/

Some global water-related databases

AQUASTAT: www.fao.org/waicent/faoinfo/agricult/agl/aglw/aquastat/main/index.stm
FAOSTAT: apps.fao.org
Global International Waters Assessment (GIWA): www.giwa.net
Global Online Research in Agriculture (AGORA): www.aginternetwork.org
Global Resource Information Database (UNEP GRID): www.grida.no/
Global Runoff Data Centre (GRDC): grdc.bafg.de
International Disasters Database and **Complex Emergencies Database:** www.cred.be/cred/index.htm.
International Groundwater Resources Assessment Centre (IGRAC): igrac.nitg.tno.nl/homepage.html
Transboundary Freshwater Dispute Database: www.transboundarywaters.orst.edu

UNEP GEMS/Water Programme: www.gemswater.org

US National Academies' Water Information Centre: water.nationalacademies.org

Water and Sanitation International Benchmarking Network (IBNET): www.ib-net.org

Water and Sanitation Programme: www.wsp.org

Water Research Network: water.nml.uib.no/

WCA infoNET: www.wca-infonet.org

Some international water networks and professional associations

American Institute of Hydrology (AIH): www.aihydro.org/

Freshwater Action Network: www.freshwateraction.net

International Association of Hydraulic Engineering and Research (IAHR): www.iahr.org/

International Association of Hydrogeologists (IAH): www.iah.org/

International Association of Hydrological Sciences (IAHS): www.cig.ensmp.fr/~iahs/

International Commission on Irrigation and Drainage (ICID): www.icid.org

International Network of Basin Organizations: www.inbo-news.org/

International Water Association (IWA): www.iawq.org.uk/

International Water Resources Association: www.iwra.siu.edu/

International Waters Learning Exchange and Resource Network (IW:LEARN): www.iwlearn.org/

Latin American Network for Water Education and Training (LA WETnet): www.la-wetnet.org/

Nile Basin Capacity-Building Network for River Engineering (NBCBN-RE): www.nbcbn.com

Streams of Knowledge: www.streams.net

Water Environment Federation (WEF): www.wef.org/

WaterNet: www.waternetonline.ihe.nl/

World Meteorological Organization – World Hydrological Cycle Observing System (WHYCOS): www.wmo.ch/web/homs/projects/whycos.html

For additional professional associations see www.unesco.org/water/water_links/Type_of_Organization/Professional_Organizations/

Environmental & Water Resources Institute (EWRI): www.ewrinstitute.org/

Global Development Learning Network (GLDN): www.gdln.org

International Water Management Institute (IWMI): www.iwmi.cgiar.org and www.iwmidsp.org/iwmi/info/main.asp

Research Institute for Development (IRD): www.ird.fr

Some international institutions for water education and research

UNESCO Centre for Water Hazard and Risk Management: www.unesco.pwri.go.jp/

UNESCO-IHE Institute for Water Education: www.unesco-ihe.org

Water Virtual Learning Centre: www.inweh.unu.edu/inweh/Training/WVLC.htm

For additional information on institutions for water-related training, education and research, visit www.unesco.org/water/water_links/Type_of_Organization/Educational_Training_and_Research_Institutions/.

SECTION 5
Sharing Responsibilities

Local-level actions and on-the-ground insights are the starting point of the global strategy to improve the overall quality and quantity of the world's water resources. Lessons learned – successes and failures – are invaluable sources of information and, if properly shared, will help us to solve some of the world's most pressing freshwater-related problems.

Improving water management and stewardship means meeting basic needs, reducing vulnerabilities, improving and securing access to water and empowering the poverty-stricken to manage the water upon which they depend.

Chapter 14 – **Case Studies: Moving Towards an Integrated Approach**

These 16 case studies from around the world examine water resource challenges and provide valuable on-the-ground insights into the facets of the water crisis and different management responses: The Autonomous Community of the Basque Country (Spain), Danube River Basin (Albania, Austria, Bosnia-Herzogovina, Bulgaria, Croatia, the Czech Republic, Germany, Hungary, the Former Yugoslav Republic of Macedonia, Moldova, Poland, Romania, Serbia and Montenegro, Slovak Republic, Slovenia, Switzerland, Ukraine), Ethiopia, France, Japan, Kenya, Lake Peipsi (Estonia, Russian Federation), Lake Titicaca (Bolivia, Peru), Mali, the State of Mexico, Mongolia (Tuul Basin), La Plata Basin (Argentina, Bolivia, Brazil, Paraguay, Uruguay), South Africa, Sri Lanka, Thailand, Uganda.

Chapter 15 – **Conclusions and Recommendations for Action**

Drawing on the essential points and key messages presented throughout the Report, this chapter weaves together a set of conclusions and recommendations to guide future action and enhance the sustainable use, productivity and management of the world's increasingly scarce and polluted freshwater resources.

Water links us to our neighbour in a way more profound and complex than any other.

John Thomson

CHAPTER 14
Case Studies

14

Case Studies: An Overview

As explained in the first *World Water Development Report* (WWDR1), many countries lack the institutions, legislation and financial means to assess the state of their water resources. Self-assessment and recognition of problems are the first and most important steps towards addressing these problems. WWDR1 noted, 'Water professionals need a better understanding of the broader social, economic and political context, while politicians need to be better informed about water resource issues', pointing to the urgent need for capacity enhancement at all levels (UN-WWAP, 2003). One of the key objectives of the World Water Assessment Programme (WWAP) is to help countries improve their self-assessment capability by building on existing strengths and experiences. WWAP fulfils this mission by assisting in the preparation of case studies in countries around the world in order to highlight the state of water resources where different physical, climatic and socio-economic conditions prevail. In this regard, case studies show the diversity of circumstances and different human needs. The second purpose of the case studies is to highlight the challenges that need to be addressed in the water resources sector. In the process, the skills and experience of both local water professionals and policy-makers are engaged and enhanced.

The World Water Assessment Programme is both global and local in scale, for it must check the accuracy of the big picture on the basis of snapshots of water in the field. In the global strategy to improve the overall quality of water resources, local actions often present the starting point to the most fruitful efforts. The WWAP case studies aim to provide a snapshot of those efforts while showing the significance of the decisions taken at local, subnational and national levels. The lessons learned, from both successes and failures, may be shared with other countries interested in addressing such issues. For WWDR1, seven pilot case studies involving twelve countries were prepared. In WWDR2, this has increased to sixteen case studies and thirty-eight countries (see **Map 14.1**). Due to this increase, the case study chapter in the present Report includes an executive summary of each case study report, briefly highlighting its water-related challenges. Given that each case study report is an important point of reference in time they will be available in their entirety on WWAP's website as they become available.[1]

The case studies included in this report were conducted at three different scales – the subnational (regions or basins), national and at international levels – with a focus on transboundary river basins and lakes. Together, they illustrate how water-related challenges are experienced at various scales. For example, transboundary basin studies focus on the challenges of sharing water resources in an international setting, while national and subnational studies are aimed at assessing the state of water resources and progress towards the Millennium Development Goals (MDGs). As a long-term goal, WWAP aims to achieve global coverage by adding a number of new case studies in each edition as more countries undertake the important task of water resources assessment (WRA). WWAP support is important in assisting participating countries in this crucial task while the WWDR and the WWAP website serve as valuable platforms for sharing ideas and stimulating discussion on water issues worldwide.

Case studies are conducted within a multi-objective framework. The overarching goal is to help enhance national institutions, but one of the main objectives and the point of departure for this exercise is bringing key stakeholders – intergovernmental organizations (IGOs), non-governmental organizations (NGOs), universities, the private sector, etc. –

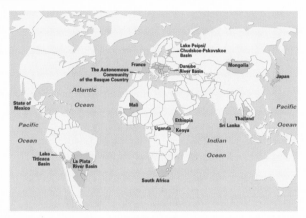

Map 14.1: Case study overview

in the water sector together through a series of national workshops to promote dialogue (not only between IGOs and NGOs but also between IGOs themselves while emphasizing the need for vertical integration between levels of government), identify priorities and create guidelines for the compilation and exchange of information. This step also serves to raise public awareness of existing and emerging problems. In this context, WWAP case study methodology brings bottom-up and top-down approaches together. The most important aspect of the case studies conducted in Africa, for example, was setting up national working groups, which in turn followed up on the group meetings and directed the overall process of national case study preparation. Although the composition of such committees varied, most included governmental institutions, universities, research centres and NGOs. The common challenge faced by nearly all case study working groups was bringing together institutions and agencies that have traditionally been accustomed to working in isolation, and helping these groups to communicate better and work together. This enhanced dialogue contributes to a more voluntary and open exchange of information among the various institutions.

---

1. See www.unesco.org/water/wwap/case_studies

This interaction also permitted the identification of institutional and technical deficiencies in the water sector as a whole. Some of the reported problems recognized to be hindering progress were as follows:

- lack of adequate or reliable data, training and qualified manpower

- lack of communication between the organizations responsible for water management

- lagging implementation of integrated water resources management (IWRM) due to financial constraints and lack of appropriate tools and policies, fragmented institutional structures and an abundance of government agencies without clear-cut definitions of their responsibilities.

In sum, through WWAP's process of case study preparation, case study partners were able to reach a consensus on the challenges to be addressed in the water sector, thus making the first step towards addressing them.

The WWAP case studies are also intended to benchmark the current situation in order to provide a basis on which to analyse change taking place in the water sector over time. In this context, and with its capacity-building directive in mind, WWAP not only facilitated the testing of indicators suggested in WWDR1, but also assisted various countries in developing their own sets of indicators. These indicators are recognized to be critical building blocks of a larger monitoring, evaluation and reporting (MER) system and a key element of good water governance.

WWAP's case study partners have widely varying development profiles. For example, Japan and France have fully functional MER systems in place based on a very comprehensive set of indicators and are also in the process of developing additional indicators to suit their special needs. On the other hand, some countries have either no MER system or are only in the very early stages of implementation. Research evaluating the effectiveness of proposed WWAP indicators by South Africa concluded that there was a high level of relevance and linkage between the WWAP challenge areas and the core components of the South African water resources management policy. Furthermore, our South African partners felt that WWAP indicators highlighted potential priority issues of interest to decision-makers and stakeholders more broadly (IWMI, 2004).

Progress in the identification of a comprehensive set of sectoral and national indicators is promising but slow. The biggest roadblocks in this process are a lack of data, which stems from the only partial coverage of established hydrological monitoring systems, and outdated and unreliable information currently available. Experience has shown that indicator development is a lengthy and iterative process, which requires significant and sustained financial and technical support.

Currently, indicator development is the responsibility of the specialized United Nations (UN) agencies. With regard to water issues, these efforts are coordinated by a core group of experts in cooperation with WWAP with the aim of bringing all the relevant expertise and experience of UN agencies together in preparation of indicators for the WWDR. Given that indicator development should be an inclusive process open to the benefit of all external expertise, case study countries may play a valuable role in this process. Broadly relevant indicators that have been developed by our case study parties can be evaluated by the other case study partners and considered for wider utilization or further refinement. WWAP plays a catalytic role in bringing the know-how and information accrued by our case study partners to a wider audience. Involvement of UN Member States in indicator development could also facilitate UN agencies' access to local knowledge and information while improving the data availability problem that has so often plagued indicator development in the past. It is important to remember that enhanced country involvement in indicator development is a complex, long-term process that needs sustained support.

Case study development should be considered as the fusion of many parallel but complementary WWAP activities conducted at the local, subnational and national levels. They address water issues from a very practical perspective and thus are critical to developing policy-relevant information for decision-makers committed to better water governance.

WWAP case study findings

The current case studies illustrate the complexity of water resources management in different geographic settings of the world with varying water-related stress, socio-economic circumstances and human needs, complementing the picture drawn in WWDR1. The findings are alarming. In many developing countries with abundant water resources, socio-economic problems present an obstacle for better access to safe water and sanitation services, whereas in other countries, water scarcity continues to be a limiting factor for development.

The disparities in water resources are blatant. For example, available per capita water resources in Peru are more than 60,000 cubic metres per year (m^3/y), but only 1,000 m^3/y in South Africa. Yet the incidence of poverty in Peru is greater than in South Africa, which means that a large amount of available water resources is not sufficient in and of itself to combat poverty; good management of these resources is also necessary. Mali's rich water resources cannot be fully put to use, due to economic difficulties and the spatial distribution of water resources. In many African countries, the livelihoods of rural dwellers are dependant on rainfed agriculture. However, due to the absence of large-scale irrigation schemes, droughts severely affect harvest and undermine food security. Furthermore, access to food for the poor and other marginalized people of society continues to be a great challenge. Local poverty and the hope of finding better living conditions often drive people from rural areas to migrate to urban settlements. The urban population in the La Plata River Basin, for example, has increased from an average of 45 percent in the early 1960s, to almost 87 percent. Here, as in other regions, the accelerating rate of urbanization often hampers the extension of safe water and sanitation service provision, and general water coverage statistics do little to elucidate the fact that though coverage is often greater in urban settlements, water supply and sanitation facilities are generally insufficient in the peripheral areas of these centres where many of the poor and marginalized groups of society live. However, as demonstrated in the case of Brazil and the PROSANEAR project, when the poor become organized and involved in decision-making processes, this greatly contributes to the provision of better services. Such projects have demonstrated that even the poor are willing to pay as long as they are well-informed of what they are paying for and what kind of services to expect in return.

Water-borne diseases continue to be among the major causes of morbidity in developing countries. Millions of people are suffering from diseases that could be prevented by better hygienic practices such as handwashing. In Thailand, for example, despite the development of water supply and sanitation infrastructures, hygiene-related problems, such as acute diarrhoea and food poisoning are still on the rise, whereas other water-related diseases (such as enteric fever, dysentery and helminths) have decreased. Some case study partners, such as Argentina, Mexico, Peru, Sri Lanka and Thailand, have reported on the dangers of groundwater resources with naturally high arsenic content.

Meeting the growing need for energy is a challenge for our developing partners. Although hydropower is considered among the cheapest source of energy and its development is promoted, its environmental and social impacts are also given careful consideration. Energy is the number one requirement for a better quality of life and industrial development. The hydroelectric potential of our case study partners is not being utilized to its full extent. For example, in Ethiopia, the contribution of hydropower to annual energy production is approximately 1 percent. While an estimated 30,000 megawatts (MW) of hydropower could be generated using available water resources, a mere 670 MW (approximately 2 percent) of hydropower potential is actually used, due to economic difficulties. Instead, fossil fuels and firewood are often used. Likewise, in Mali, 90 percent of basic energy needs are met through firewood and charcoal. The environmental implications (e.g. deforestation and the emission of greenhouse gases) stemming from these sources are serious. In addition, while the electricity grid is being expanded to serve most of the previously marginalized rural areas, energy consumption in residential and commercial sectors seems to have risen only gradually, due to the low income of households served.

Industrial development is closely linked to urbanization. As industry creates jobs, the expectation of finding a job often results in mass migrations to industrial centres. The continuation of the current industrial development and population migration trends will likely place increased pressures on the scarce water resources and limited service infrastructure in these locations, such as in the case of Gauteng in South Africa.

Gender equity in education continues to be a problem in developing countries. The main reason for this is that girls and women are considered as a source of cheap or free labour. In Sri Lanka, although 40 percent of agricultural workers are women and almost 80 percent of them are involved in straining physical activities, more than 70 percent of the women workers render their services without receiving anything in return. Furthermore, in many developing countries, carrying water home, sometimes from considerable distances, is seen as the responsibility of women and girls. These kinds of hard living conditions unfortunately put them at an educational disadvantage. Funds that are devoted to improving educational systems have so far been unable to address these issues and increase the enrolment of girls. Furthermore, higher learning and research institutions specialized in the technical and management aspect of water resources are insufficient in quantity and variable in quality. This in turn limits the institutional capacities of national organizations due to the absence of trained manpower.

MER systems are at different stages of implementation in the countries of our case study partners. In Europe, for example, the implementation of the European Union Water Framework Directive has been a positive step towards a comprehensive MER system (see **Box 14.1**). WWAP's efforts to help countries improve their MER systems have been very well received by our case study partners. For example in Uganda, the national process initiated by the WWAP Secretariat facilitated the bringing together of several institutions and agencies to better communicate and work together.

Agriculture is the most water-demanding sector, in addition to being a major source of employment and a major contributor of the national gross domestic product (GDP) of many of our case study partners in Africa. Agriculture in Ethiopia, for example, provides 86 percent of employment and generates 57 percent of its GDP. However, the industrial sector is becoming increasingly important to many of our case study partners. In South Africa, it has developed into the fastest growing economic sector, generating 29 percent of South Africa's GDP and employing over 25 percent of the work force. Mining also has a big share in international trade.

The challenge of meeting the increasing water needs of industry while preserving the ecosystem remains a prominent challenge. The mining industry in Mongolia contributes approximately 20 percent of national GDP and accounts for over 50 percent of the country's overall exports, but has caused a detrimental effect on the well-being of ecosystems. On the other hand, however, with almost 80 percent recycling efficiency, Japan is a prime example of how high-level technology and up-to-date regulations can contribute to the effective use of water in industry.

Regardless of their level of development, managing risks is a concern for all our case study partners. Water-related disasters continue to claim the lives of many and cause considerable socio-economic damages. In many countries, both structural and non-structural measures have been implemented to mitigate the negative effects of droughts and floods. However, experience has shown that defensive action against water-related hazards alone is insufficient. The action programme, recently adopted by the members of the International Commission for the Protection of the Danube River, recognizes floods as a natural part of the hydrologic cycle and emphasizes the need to be cautious when planning development activities in flood-risk areas and manage risk through a basin approach, with the participation of governments, municipalities and stakeholders. Within the framework of the programme, which primarily refers to United Nations Economic Commission for Europe (UNECE) Guidelines on Sustainable Flood Prevention and adopts EU Best Practices on Flood Prevention, Protection and Mitigation, conservation and improvement of water-related ecosystems are a high priority, since wetlands act as a buffer against floods, thereby reducing their intensity.

In several regions, both surface and groundwater resources are under varying levels of stress due to growing population and continuing industrialization. In many instances, water resources are not utilized in a sustainable manner. In the State of Mexico, it is estimated that groundwater resources are used at twice their rate of natural recharge which causes land subsidence up to 40 centimetres (cm) per year due to the shrinking of aquifers as water is drawn out and leads to the disruption of water and sanitation infrastructure and increases vulnerability to floods.

Droughts are also a part of the water cycle and take place with varying frequency and severity. The outcomes of our case studies show how drought aggravates the level of poverty and famine, especially in Africa. In Ethiopia, one of the poorest countries in the world, there have been about thirty major drought episodes over the past nine centuries, thirteen of which were severe at the national scale and put millions of Ethiopians in dire need for basic food assistance. Even countries with abundant rainfall are prone to droughts. In Sri Lanka, twenty-three droughts were reported between 1947 and 1992, severely disrupting the Sri Lankan economy. During the 2001 drought, for example, the country faced power cuts for up to eight hours per day. In 2004, over fifty thousand hectares of crops were damaged, and the government had to appeal for assistance to provide food rations for approximately one million people for a six-month period. Flood and drought forecasting systems are necessary to take precautionary measures and to reduce the socio-economic impacts of such natural disasters. However, lack of funding or limited funding slows down the effective implementation of such systems. Consequently, flood-warning systems are missing in a great number of flood-prone basins. Flood-forecasting models in Sri Lanka fail to simulate real-life situations due to the poor mathematical algorithms employed. In Kenya, disaster management has not been viewed as an integral part of development planning, and water-related disasters were responded to in an ad hoc manner whenever they occurred. As a result, the variation in rainfall has had a significant effect on rainfed agriculture, upon which Kenya's economy heavily relies.

Climatic variations affect livelihoods of urban and rural dwellers. For example, in the Lake Titicaca Basin (Peru and Bolivia), glaciers, which are the major source of water for drinking and irrigation, are receding and reducing volumetrically. This trend will spell disaster for small and medium-scale irrigation, causing an increase in water prices, possibly aggravating poverty and triggering social movements.

BOX 14.1: THE EUROPEAN UNION WATER FRAMEWORK DIRECTIVE

Abundant and clean water is a given for most of the people living in the European Union (EU). However, many human activities put a pressure on both water quality and quantity. Polluted water from industry, agriculture and household use causes damage to the environment and affects the health of those using the same water resources. The EU Water Framework Directive (WFD) came into force on 22 December 2000 and aims to establish a framework for the protection of surface and groundwater, as well as coastal waters.

This directive requires all inland and coastal waters to reach 'good status'[1] by 2015. The definition of the good water status includes the chemical composition of water and the ecological elements. In order to reach this goal, a river basin structure is established within which certain environmental targets are set. The most important aspects of the WFD is that it calls for sustainable development, requires the adoption of integrated river basin management and links and coordinates all previous water policies, such as the directives on urban waste water treatment, nitrates, bathing or drinking water into a common framework. Finally, the integration of water policy with other major EU policies (like agriculture, hydropower and navigation, for example) is a prerequisite for successful protection of the aquatic environment.

In 2009, measurement programmes will be established in each river basin district for delivering environmental objectives (article 11). The first river basin management plan for each river basin district, including environmental objectives for each body of surface or groundwater and summaries of programmes of measures (article 13) will also be published.

Recognizing that water management must respond to local conditions and needs, the WFD, has strong public information and consultation components that encourage all interested parties to become involved in the production, reviewing and updating of river basin management plans.

1. The values of the biological quality elements for the surface water body type show low levels of distortion resulting from human activity, but deviate only slightly from those normally associated with the surface water body type under undisturbed conditions.

Source: EC, 2000.

The importance of Integrated Water Resources Management (IWRM) is becoming increasingly recognized throughout the world and the legislative and regulatory frameworks needed for putting IWRM tools into use are being created and revised. The involvement of stakeholders is encouraged through the establishment of community councils and river basin organizations, which share the responsibility of water management with national institutions. However, the World Summit on Sustainable Development (WSSD) target for the preparation of IWRM and efficiency plans in all countries by 2005 has not been fully met. Furthermore, although water management laws, policies, programmes and regulations do exist, their enforcement and implementation remain problematic. Implementation has proven to be particularly difficult in cases where there has been little public involvement. Hence, facilitating the participation of water users and stakeholders in the management and allocation of water resources remains an important challenge.

The major problem plaguing many of our case study partners is the lack of coordination between institutions and agencies responsible for drafting and implementing policy. This is especially critical for multi-state

countries, such as Mexico where decisions taken at the federal level need to be implemented at the state level. In the State of Mexico, the legal framework has been revised to allow the creation of the Secretariat of Water, Public Works and Infrastructure for Development (SAOPID), which is single-handedly responsible for preparing and implementing State policy guidelines concerning public works and infrastructure development. This secretariat, which reports back to National Water Commission at the federal level, is the first of its kind in Mexico.

Lastly, and perhaps most importantly, the case studies demonstrate that where gross inadequacies exist in the provision of water and sanitation facilities, a lack of financial and human resources capacity can clearly be seen. Human resources capacity is not only essential to the implementation of policies and programmes, but to the proposal of innovative solutions overall. Furthermore, a lack of synergy and an unclear division of responsibility among institutions often exacerbates these problems and inhibits reforms from reaching the local level. Until these issues can be addressed, they will likely remain the most outstanding problems challenging the water sector of developing countries in the near future.

1. The Autonomous Community of the Basque Country

The Autonomous Community of the Basque Country (ACB) is one of seventeen autonomous bodies in Spain. It is densely populated, with 5 percent of the overall population of Spain (over 2 million people) living in 1.4 percent, or 7,234 square kilometres (km^2), of the total surface area of Spain (EUStat, 2005). Accordingly, the population density was 292 inhabitants per km^2 as of 2003. The surface area of exclusive internal basins is around 2,200 km^2 with a population density of over 600 inhabitants per km^2.

The ACB is a highly mountainous territory located across the western end of the Pyrenees and the eastern part of the Cantabrian Mountains. The Cantabrian-Mediterranean water divide formed by mountain ranges of modest altitude (1,000 to 1,600 m) divides the territory. A great portion of the ACB lies in the Bay of Biscay-Mediterranean watershed. However, on both sides of this basin, there are a series of small catchments, generally characterized by a high level of rainfall and extremely uneven terrain. Rainfall is abundant throughout the ACB, with an annual average of over 1,000 mm and a long-term variability of about 20 percent. Despite its relatively constant levels of rainfall, the region has experienced serious flooding and a number of droughts. The region's rugged surface conditions and high rainfall have prompted ACB to establish an extremely dense hydro-meteorological monitoring network, with over 330 control stations currently in operation.

Urban settlements are the biggest user of water resources. In fact, 72 percent of the overall water demand is utilized for urban consumption, whereas 14 percent is utilized by industry, and the remaining 14 percent

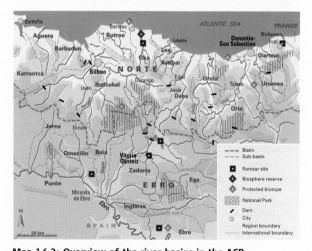

Map 14.2: Overview of the river basins in the ACB
Source: Prepared for the World Water Assessment Programme by AFDEC, 2006.

is used by agriculture. Although non-consumptive demands like hydroelectric energy production and aquaculture exert considerable local pressure on the movement of water in the region, these activities do not constitute an important part of the regional economy.

In parallel to industrial and urban development, the quality of the region's water resources and aquatic ecosystems has constantly degraded. In response to this situation, a network with 360 operational sampling points has been set up in order to survey the environmental status of all aquatic ecosystems and regional water bodies (rivers, lakes, reservoirs, transitional waters, coastal waters and groundwater). The data collected from these points is used to assess the current condition of all water bodies in accordance with the EU's Water Framework Directive (WFD) which entered into force in 2000 (see **Box 14.1**). In order to comply with the WFD, the Basque Government *carried out* a detailed study exclusively on its internal basins, comprising 122 rivers, 4 lakes, 14 transitional water bodies, 44 aquifers and 4 coastal waters, in an effort to characterize the freshwater resources and their associated ecosystems from an environmental and socio-economic perspective.

The Hydraulic Administration of the Autonomous Community of the Basque Country submitted a detailed study to the EU, in which the economic aspects of water use and the environmental impact of humans were analysed for each water body and all protected areas were registered.

The results of these studies have indicated that improving water quality and curbing the destruction of ecosystems remain challenges for the region. While the reclamation and restoration of rivers and estuary banks are, to a degree, feasible and currently underway in the ACB, the

likelihood of their success will depend upon the degree of damage that has previously occurred. Though such projects may not be able to completely restore water resources to their prior conditions, they can nevertheless help to improve their current state. These projects, combined with an increase in the level of public awareness, new laws and directives concerning the region's water resources, have been key in initiating a trend towards the restoration and better preservation of fragile ecosystems.

Conclusion

In order to promote the sustainability of water resources while meeting the water demands of various sectors, IWRM policies are actively being implemented. Accordingly, the varying water needs of Basque society are fully met, and the full cost of providing these services is recovered through the current water management scheme. The central challenge for the future is to define and successfully implement a series of case-specific and efficient programmes to protect and improve the status of valuable water resources and the associated ecosystems.

2. The Danube River Basin

The Danube River Basin (DRB) covers a vast area of 801,463 km^2, making it the second largest river basin in Europe, after the Volga. It is also the basin that covers the greatest number of countries in the world, with a total of eighteen states. The DRB lies to the west of the Black Sea in Central and Southeast Europe (see Map 14.2). It discharges into the Black Sea via the Danube Delta, which lies in Romania and Ukraine. With an average discharge of about 6,500 m^3/sec, the Danube is the Black Sea's largest tributary.

Due to its large surface and diverse relief, the Danube River Basin has a varied climate and a multiplicity of habitats. The upper regions in the west have high precipitation, whereas the eastern regions have lower precipitation and cold winters. Depending on the region, precipitation can range from less than 500 mm to over 2,000 mm per year, which strongly affects surface run off and discharge levels in streams.

Transboundary and regional aquifers are common in the DRB region. In some cases, groundwater resources represent as much as 30 percent of the countries' total internal renewable water resources. Although aquifers are the main sources of drinking and industrial water in the DRB region, there is little information concerning the availability of groundwater or potential extraction capacity in many countries.

There are 26 major tributaries of the Danube River, all of which have their own sub-basins. The Tysa (also called Tisza or Tisa) River Basin is the largest sub-basin in the DRB (157,186 km^2). It is also the Danube's longest tributary (966 km). By flow volume, it is the second largest after the Sava River. The Sava River is the largest Danube tributary by discharge (average 1,564 m^3/s) and the second largest by catchment area (95,419 km^2). The Inn is the third largest by discharge and the seventh longest Danube tributary.

In the DRB, there are several freshwater lakes of varying sizes. The most prominent are the 'Balaton' in Hungary (605 km^2) and the 'Neusiedlersee' (also called Fertö tó), which is shared by Austria and Hungary (315 km^2). Furthermore, the Razim-Sinoe Lake System is composed of several interlinking large brackish lagoons that are separated from the sea by a sandbar.

Some countries such as Austria, Hungary, Romania, Serbia and Montenegro and the Slovak Republic are almost completely situated within the DRB, whereas less than 5 percent of the territories of Albania, Italy, Macedonia, Poland and Switzerland lie in the basin. More than 26 percent of the overall basin population is Romanian. This is by far the largest population group in the DRB, followed by populations from Germany, Hungary, and Serbia and Montenegro.

The International Commission for the Protection of the Danube River (ICPDR) was established in 1998 to promote and coordinate sustainable and equitable water management practices, including conservation, improvement and rational water use. The ICPDR, with thirteen cooperating states[2] and the EU, pursues its mission by making recommendations for the improvement of water quality, developing mechanisms for flood and industrial accident control, agreeing on

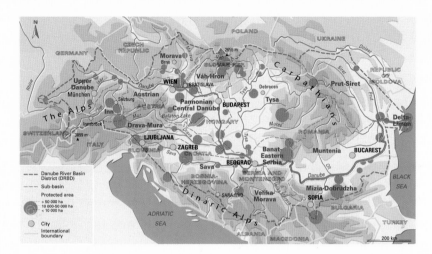

Map 14.3: Overview of the Danube River Basin
Source: Prepared for the World Water Assessment Programme by AFDEC, 2006.

standards for emissions and by ensuring that these measures are reflected in the Contracting Parties' national legislations and applied in their policies (see **Box 14.2**).

Ecosystems and transportation

Floodplain forests, marshlands, deltas, floodplain corridors, lakeshores and other wetlands form the basis of the rich biodiversity in the DRB. In fact, the Danube River Basin extends into five of the eight biogeographical regions of Europe, each with its own particular characteristics. However, in those regions, industrialization, population growth and agriculture have had a negative impact on the size and biodiversity of wetlands.

Regulation works for navigation in the Upper Danube region started as early as the nineteenth century. Navigation is now possible in the Danube River itself and in some tributaries in the lower portions of the basin. In order to make the river navigable, the meanders were cut off in several places, the main channel was straightened and lateral dams were built to narrow the river's width. Consequently, in some parts of the river, the length of the watercourse was shortened considerably. Additional artificial waterways were also built along the Danube River for transport purposes. These include the Main-Danube Canal in Germany, which provides a link to the Rhine and the North Sea, the Danube-Tysa-Danube Canal System in Serbia and Montenegro, and the Danube-Black Sea Canal in Romania. The Danube itself is now regulated along more than 80 percent of its length. The canals, in some areas, also serve as flood protection measures while providing recreational areas and tourist attractions.

The hydraulic works for navigation improvement have had a major impact on natural floodplains and their ecosystems. In many places along the

river, the floodplains and meanders were cut off from the river system. As a result, 80 percent of the historical floodplain on the large rivers of the Danube River Basin has been lost during the last 150 years. Some of the remaining areas have either received protection status under different national or European legislation or international conventions (such as the Ramsar Convention on Wetlands) while other areas remain vulnerable (e.g. the Middle and Lower Drava-Mura wetlands in Croatia, Hungary and Slovenia). Large dikes and disconnected meanders also suppressed the exchange of surface and groundwater, which reduced the recharge of groundwater utilized for the drinking supply.

Given the ecological and hydrological importance of wetlands (preserving a diversity of species, recharging groundwater aquifers, flood buffering, etc.), the protected areas within the DRB are being compiled into an inventory. This inventory will help to provide crucial input for the preparation of the DRB Management Plan (see **Box 14.2**). The timetable for completion of the inventory is based on the European Commission's progress in the establishment of 'Natura 2000', which will represent an authoritative network of protected sites in the European Community with the highest value for natural habitats and rare, endangered or vulnerable plant and animal species.

Managing pollution and floods

Six EU Member States (Austria, Czech Republic, Germany, Hungary and the Slovak Republic) and three accession countries (Bulgaria, Croatia and Romania) are working towards implementing WFD requirements in the DRB. Parallel efforts are also underway in the DRB, within the framework of the ICPDR, to record all the significant point sources of pollution (municipal, industrial and agro-industrial) and specific discharges. The ICPDR has prepared inventories for point source emissions for the years 2000 and 2002, which is becoming more complete as states continue to provide more detailed information.[3] In addition, chemical and biological variables are measured in thirteen ICPDR member countries at seventy-nine monitoring stations on the Danube and its major tributaries to

---

2. The cooperating states are Austria, Bosnia and Herzegovina, Bulgaria, Croatia, Czech Republic, Germany, Hungary, Moldova, Romania, Serbia and Montenegro, Slovak Republic, Slovenia and Ukraine.

3. To learn more about this please refer to the ICPDR website: www.icpdr.org/

BOX 14.2: THE DANUBE RIVER BASIN MANAGEMENT PLAN: CONVERGING WITH EU-WFD

The International Commission for the Protection of the Danube River (ICPDR) serves as the platform for coordinating the development of the Danube River Basin Management Plan, which is to be implemented by 2009. Preparation of basin management plans by this date is compulsory for all European Union (EU) countries as per the requirements of Water Framework Directive (WFD).

In the ICPDR, all Contracting Parties support the implementation of the WFD in their territories and cooperate in the framework of the ICPDR to achieve a single basin-wide coordinated Danube River Basin Management Plan. To this end, each country is in the process of preparing national reports and 'roof reports', which give an overview of WFD issues such as the pressures on the Danube River Basin (DRB) main surface and groundwater resources and related impacts exerted on the environment. The 'roof reports' will be the basis for the preparation of the Danube River Basin Management Plan.

The ICPDR has also requested that other DRB countries cooperate in order to achieve a basin-wide coordinated management plan. Albania, The FYR of Macedonia, Poland and Switzerland have offered their support. On the operational level, it is the obligation of the Contracting Parties to ensure the necessary coordination with their DRB neighbours.

Source: Modified from Danube Basin Analysis (WFD Roof Report, 2004). For further examples of the implementation of the WFD, see the case studies for France and Lake Peipsi/Chudskoe-Pskovskoe.

---

analyse the impact of organic pollution. In fact, based on the results of the biological impact assessment, the Danube is classified as 'moderately polluted' to 'critically polluted' in the Saprobic system – a method used to detect biodegradable organic pollution by measuring presence or absence of certain indicator species in water.[4] The major cause of organic pollution is insufficient urban wastewater treatment, due to a lack of wastewater treatment plants. The construction of wastewater treatment plants is therefore expected to be included as a priority action within the programme of the ICPDR's Basin Management Plan by the end of 2009 (see **Box 14.2**).

In addition to biological pollutants, pollution from other hazardous substances are also significant in the DRB. Unfortunately, other than the data available for heavy metals and pesticides, the full extent of contamination by hazardous substances cannot be evaluated to date. Cadmium, lead and pp'-DDT (a derivative of DDT) levels are substantially higher than current ICPDR standards. The risk of industrial accidents is also high in several parts of the DRB. In order to minimize the occurrence of such incidents, the Accident Emergency Warning System (AEWS) was put into place in 1997. Still, the cyanide accident in the Tysa River Basin in January 2000 caused massive harm to the environment and had a substantial impact on the economy of the entire region. This clearly demonstrated the need for better preparedness against such events.

Agricultural activities also exert pressure on water resources. Although some DRB states are highly industrialized, other countries like Bulgaria, Croatia and Romania rely on agricultural activities which generate around 10 percent of their GDPs. This share is between 1 and 3.7 percent in the remaining countries of the basin. Overall 47.4 percent of land resources in the DRB is used for agriculture.

Many large floods have occurred in the DRB, causing numerous human casualties and material damage. While floods are naturally occurring events of the water cycle, human impacts increase the risk of flooding through inappropriate land-use in high-risk areas and by interfering with natural processes. The extreme and devastating floods along the Morava and Odra rivers in 1997, the Tysa River between 1998 and 2001 and the Elbe and Danube rivers in 2002 have clearly demonstrated the destructive and unpredictable nature of floods and the need for careful planning for development projects in flood-sensitive areas.

Climatic variations are expected to further increase the risk of flood-related damages. As a response, in 2004, ICPDR adopted the long-term Action Programme for Sustainable Flood Prevention in the Danube River Basin. This initiative is based on the sustainable flood protection programmes developed in the various Danube countries and utilizes existing networking structures. The most important point of the action programme is that it deviates from the common practice of mainly taking defensive action against water-related hazards and recognizes floods as a natural part of the hydrologic cycle. It emphasizes the need to learn how to live with floods and manage risk through a basin approach, with governments, municipalities and stakeholder participation. Conserving and restoring water-related ecosystems is also given high priority within the terms of the action programme, as wetlands play an important role by buffering and reducing the intensity of floods. The action programme primarily refers to UNECE Guidelines on Sustainable Flood Prevention and adopts EU Best Practices on Flood Prevention, Protection and Mitigation.

Energy in DRB countries

In DRB countries, various technologies are used for energy production. Austria relies on hydropower to generate almost two-thirds of its overall energy production. On the other hand, some other basin countries rely heavily on conventional thermal power, such as the Croatia (67 percent), Czech Republic (53 percent), Hungary (57 percent), Romania (61 percent)

---

4. To find out more about Saprobic Indices in Water Quality Assessments see:
www.who.int/docstore/water_sanitation_health/wqassess/ch10.htm

and Serbia and Montenegro (67 percent), whereas Bulgaria and the Slovak Republic have mostly invested in nuclear power (over 50 percent).

Conclusion
The main problem in the DRB is the water quality rather than quantity. Nine countries (six EU members and three concession countries) are at different stages of implementation of the WFD. The other contracting parties of the ICPDR are also working towards the common goal of improving the quality of water resources. However, marked differences in economy, sociology and topography complicate the tasks of the states. For this reason, neither WFD nor ICPDR goals are yet to be implemented uniformly throughout the region, and there is still a substantial amount of work to be done at the national level. However, members of ICPDR consider the sustainable utilization of water resources as the overriding priority and work together to this end.

3. Ethiopia

Ethiopia is located in East Africa and constitutes a major portion of the Horn of Africa. Its terrain consists mostly of a huge central plateau and surrounding lowland plains, producing three climatic zones: tropical in the south and southwest, cold to temperate in the highlands and arid to semi-arid in the northeastern and southeastern lowlands. As a result, the amount of rainfall and surface run off is highly variable and depends on location and altitude. In fact, four basins located on the western part of the country contribute 83 percent of the national surface water potential, while other areas produce very low surface runoff.

Ethiopia has 71 million inhabitants, half of which live at around 2,200 metres above mean sea level (m.a.s.l), in the areas with cooler temperatures, higher rainfall and fewer instances of malaria. Another 40 percent lives between 1,400 and 2,200 m.a.s.l. The remaining population lives at altitudes below 1,400 m.a.s.l. Thus, Ethiopia's population is also unevenly distributed, with nearly 80 percent of the 71 million inhabitants living in only 37 percent of the total area of the country.

Ethiopia has seven transboundary basins that carry over 95 percent of annual runoff. However, there is no comprehensive agreement binding riparian states. The riparian countries of one of these basins, the Nile Basin, taking into consideration the challenges of meeting their growing water needs in a sustainable manner, have launched the Nile Basin Initiative (NBI) in 1999. While basin countries are currently engaged in negotiations, it is hoped that the NBI will provide the basis for a permanent legal and institutional framework.

With a per capita gross national product (GNP) of US $100 in 1994, Ethiopia ranks as one of the poorest countries in the world. In 1994, the per capita GNP income in Ethiopia was less than half of those in sub-Saharan Africa as a whole, where per capita GNP was approximately US $259. Overall, it is estimated that nearly 52 percent of the population is below the national poverty line,[5] with poverty in urban and rural areas estimated at 58 percent and 48 percent respectively.

Water resources
Ethiopia has nine major rivers and twelve big lakes. Lake Tana, for example, in the north is the source of the Blue Nile. However, apart from the big rivers and major tributaries, there is hardly any perennial flow in areas below 1,500 m. While the country's annual renewable freshwater potential is 122 billion m³, only 3 percent of this amount remains in the country. It is estimated that 54.4 billion m³ of surface runoff and 2.6 billion m³ of

groundwater can be developed for utilization. Currently less than 5 percent of surface water potential is used for consumptive purposes.

Challenges to life and well-being
Ethiopia is largely dependent on the agricultural sector, which provides 86 percent of the country's employment and 57 percent of its GDP. Rainfed crop cultivation is the principal activity and is practised over an area of 27.9 million hectares (ha), or approximately 23 percent of

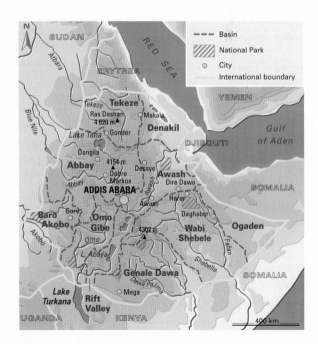

Map 14.4: Overview of the river basins in Ethiopia

Source: Prepared for the World Water Assessment Programme by AFDEC, 2006.

potentially arable land. Frequent and severe droughts cause serious decreases in the incomes of rural inhabitants who tend to rely heavily on agriculture. While estimates have shown that up to 3.7 million ha can be irrigated, a mere 300,000 ha of irrigation has been developed. To make matters worse, projected large- and medium-scale irrigation schemes will likely do little to secure the food supply for the rapidly growing population.

Wetlands in Ehiopia are very valuable areas for rural communities. They contribute directly to food security by providing vegetables in the early rainy season when the supply of food from the upland fields is running out for many families. Moreover, many rural inhabitants obtain drinking water from the springs around the wetlands. However, wetlands are being degraded due to human-related activities, such as draining for agriculture, cattle grazing, industrial pollution and unsustainable utilization of resources. Although there are some policies that specifically address wetlands, there is, at the national level, an overall lack of wetland policy.

Ethiopia's urban population is close to 10 million people, 25 percent of whom are located in Addis Ababa. Half of the urban population lives in towns smaller than 30,000 inhabitants. Although Ethiopia is rural-centred (85 percent of the population lives in rural areas), the rate of urban growth is increasing at a yearly average rate of 5 percent, which is much higher than the average national growth rate of 2.9 percent per year.

The status of water and sanitation infrastructure is very poor in Ethiopia: only 10 percent of Ethiopians have access to proper sanitation facilities and 31 percent to safe water. Service coverage is considerably higher in urban areas than in rural areas (74.4 percent and 23.1 percent respectively). Furthermore, almost 25 percent of water installations in rural areas are not functional at any given time. Central Statistic Authority (CSA) results from 1998 showed that 63.8 percent of people living in rural areas had to fetch water from a source within a distance of 1 km. The situation gets worse during dry periods, as water carriers have to walk longer distances for even smaller quantities of lower quality water. Accordingly, the incidences of diseases related to unsafe water supply and inadequate sanitation are very high. High population growth rates, low education levels and high rates of illiteracy have also contributed to the burden of ill health.

The major causes of morbidity among patients seeking treatment in health facilities include respiratory infections, malaria, skin infections, diarrhoeal diseases and intestinal parasitic infections. About three-quarters of Ethiopia is known to be a breeding ground for mosquitoes, the vector of malaria. Malaria is especially endemic in hot lowlands, which prompts many Ethiopians to live in the highlands. Diarrhoea, the most prevalent water-related disease, accounts for 46 percent of the under-five child

mortality rate. The five illnesses mentioned above account for over 63 percent of all reported cases of child morbidity. Women and girls are particularly vulnerable to water-borne and water-related diseases as they more frequently come into contact with contaminated water (they are usually responsible for fetching water for the family).

The main source of energy production in Ethiopia (about 93 percent) is biomass (fuel wood, coal, agricultural residues, animal wastes, etc.), which leads to rapid forest depletion. The contribution of hydropower to annual energy production is approximately 1 percent. It is estimated that 30,000 MW of hydropower can be generated using available water resources. However, merely 670 MW of hydropower potential is currently developed. Household consumption accounts for 87.5 percent of total energy consumption, while industry accounts for 5 percent. As Ethiopia's economy depends almost entirely on subsistence agriculture, the need for electricity has been quite low. However, this situation is changing, as urbanization and industrialization increase energy demands. The Ethiopian Electric Power Corporation aims to set up a variety of hydroelectric, oil and gas resource development schemes in order to improve access to electricity from 15 percent to 20 percent by 2010.

The effects of water-related natural disasters

As most of the rivers in Ethiopia flow in deep gorges, floods have not traditionally been a common phenomenon. However, due to massive deforestation and loss in surface vegetation, flooding now annually occurs in some areas, such as in the banks of the Blue Nile River and in the vast plains of the Baro Akobo Basin in the country's southwestern region. Although sometimes associated with economic and social damages, floods provide much needed water to ensure the fertility of grazelands, making them anticipated events, especially for nomads, whose incomes are dependant on animal husbandry.

Drought is a frequent natural disaster in Ethiopia. Recent observations have shown that the frequency of droughts have increased over the last few decades. There have been about thirty major drought episodes over the past nine centuries. Of these drought episodes, thirteen were very severe at the national level. **Table 14.1** shows the number of people affected by droughts and the population which required basic food assistance between 1990 and 2004.

The contingency plans for water-related natural disasters are prepared by the Disaster Prevention and Preparedness Commission (DPPC), which brings together all concerned stakeholders to draft a plan of action to be implemented by all relevant organizations.

Water policy implementation

The Federal Water Resources Management Policy, issued in 1999, elaborates on the water supply and sanitation, irrigation and hydropower sectors. It promotes the sustainable development of water resources for equitable social and economic benefits through public participation and IWRM. In order to implement the requirements set forth in the policy, various legal and

---

5. National poverty line is deemed appropriate for a country by its authorities. For this reason, the national poverty line should not be used for comparison between other countries as it varies significantly (Human Development Report, 2005).

Table 14.1: Number of people affected by recent droughts

Year	Population affected	Food assistance requirements (number of people)
1990	3,429,900	374,400
1991	1,850,000	838,974
1992	5,228,530	1,288,737
1993	1,644,040	739,280
1994	889,000	577,586
1995	3,994,000	492,460
1996	3,153,000	253,118
1997	1,932,000	199,846
1998	5,820,415	572,834
1999	2,157,080	1,138,994
2000	7,732,335	836,800
2001	6,242,300	639,246
2002	5,181,700	557,204
2003	14,490,318	1,461,679
2004	9,369,702	964,690

institutional capacity-building efforts are currently underway. For example, the fifteen-year Water Sector Development Programme (WSDP) was put into effect in 2002, and the Water Resources Management Proclamation was issued the same year to provide legal ground for the implementation of the Water Policy. The Water Sector Development Programme is composed of five programmes and sets the targets on water supply and sewerage, irrigation

and drainage, hydropower development, general water resources programme and institutions/capacity-building. Furthermore, vocational and technical training centres, operational since 2003, have been established to train technicians on irrigation development schemes and water supply and sanitation services. In addition, the government has taken the initiative to establish basin institutions. For this purpose, with the financial and technical aid of international donors, an institutional study has been initiated for the Blue Nile (Abbay) Basin as a pilot project. Upon the successful completion of this project, the establishment of similar institutions in other basins are foreseen. However, awareness-raising activities to disseminate existing plans and policies at various levels (public and national institutions) are lacking. Furthermore, due to the absence of a functioning monitoring and evaluation system, the rate of implementation and the effectiveness of policies have not yet been assessed.

Conclusion

most of the Ehiopians do not have access to safe water and sanitation. The Water Sector Development Programme (WSDP) prepared for 2002-2016 aims to improve the existing situation; however, the investment required for the implementation of this programme cannot be financed by national funds alone. Attracting international donors will therefore likely remain a priority in order to alleviate the heavy burden of disease, poverty and hunger that the country currently faces.

4. France

Excluding its overseas territories, over 60 million people live in France, within a surface area of 551,695 km².

There are six major river basins in France: the Adour-Garonne, the Artois-Picardy, the Loire-Britanny, the Rhine-Meuse, the Rhone-Mediterranean and the Seine-Normandy. These basins are managed by separate basin agencies that were established by the 1964 Water Law and further reinforced by the 1992 Water Act.

The WFD is similar to the French institutional system in that it requires the implementation of IWRM at the basin level. The most recent French water law (passed in 2003) takes the WFD into account, calling on all EU Member States to achieve 'good status' for all of their water bodies by 2015 (see **Box 14.1** on the WFD and the case studies for the Danube River Basin and the Lake Peipsi/Chudskoe-Pskovskoe Basin for further information on the implementation of the WFD).

France's six major river basins have different climatic, hydrological and socio-economic characteristics. Consequently, six basin agencies have been set up to address the differing challenges of each basin. The specific basin challenges are briefly summarized below.

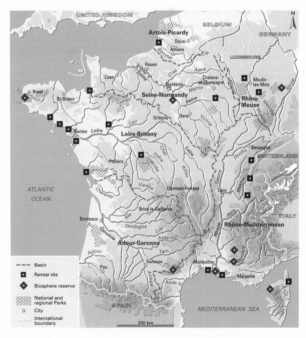

Map 14.5: Overview of the river basins in France
Source: Prepared for the World Water Assessment Programme by AFDEC, 2006.

The Adour-Garonne Basin

The Adour-Garonne Basin covers 116,000 km^2, or 21 percent of France. In this basin, 35,000 farmers irrigate 645,000 ha of land, approximately 40 percent of the total irrigated surface area in France. Although there is a dense network of tributaries, there are no major rivers. Low rainfall in the summers results in severe low-water levels from the end of spring. Normally, irrigation water accounts for 35 percent of the water abstracted throughout the year; however, this ratio increases to 80 percent during low-water-level periods. In order to cope with the adverse affects of such conditions, planning tools like strict low-water target flow (DOE, Débit Objectif d'Etiage) and low-water management scheme (PGE, Plans de Gestion d'Etiage) were put into practice. DOEs are the fixed flow rates at strategic points of the basin during low water periods. PGEs involve all relevant stakeholders and set the rules concerning how to allocate limited water resources at the basin scale and specifically in water deficit areas. Those tools have had overall positive results, such as the establishment of better dialogue among stakeholders and reduced frequency of low water crises. Furthermore, the basin administration constantly provides sound advice for promoting rational water use and equipment for monitoring water abstraction. However, irrigation charges are still highly subsidized, and as a result, the revenues collected for irrigation water are still far from adequate for meeting the real cost of providing services (€3.83 million collected in 2002 versus a full cost of €107 million) (see **Chapters 7** and **3**).

The Artois-Picardy Basin

The Artois-Picardy Basin covers 3.6 percent of the national territory. As an area previously dominated by the coal and steel industries, both surface water and groundwater resources have been highly contaminated by various hydrocarbons and toxic metal salts. Furthermore, the abundant water resources of the region were previously used by factories in an unsustainable fashion, which resulted in a considerable lowering of the water table. In order to preserve water resources, groundwater abstraction and pollution charges were implemented and have been kept consistently high since the 1970s. As a result, pollution has been considerably reduced. Discharges of organic matter went down from 440 to 74 tons a day. Furthermore, underground water abstraction has decreased from 300 million tons in 1971 to 100 million tons in 2003. Undoubtedly, the adoption of technical solutions, such as wastewater treatment plants, recycling of water and utilization of advanced manufacturing processes, to reduce or in some cases omit water usage has played an important role in reducing the damage caused to the basin's water resources (see **Chapter 8**).

The Loire-Brittany Basin

The Loire-Brittany Basin is the biggest basin in France, covering 28 percent of the country: 58 percent of the total number of farms and 65 percent of the livestock production in France is located in this basin. The surface area utilized for agricultural purposes covers 64 percent (100,000 km^2) of the basin and produces 50 percent of national cereal production. Following the end of Second World War, policies were adopted towards ensuring food for all and creating jobs. Although this resulted in a boost in the productivity of livestock-raising and cereal production, it resulted in excessive nitrate pollution in surface and underground waters. After the reform of the Common Agricultural Policy (CAP) in 1992, appropriate measures were taken, such as agro-environmental measures and a nitrogen absorption programme at the European level, to reduce the impact of agricultural activities on water quality. In addition, the farm pollution management programme (PMPOA, Programme de Maîtrise des Pollutions d'Origine Agricole) was introduced at the national level to monitor pesticides and fertilizer pollution, which provides financial incentives through subsidies for the farmers to upgrade their livestock effluent management. Despite positive signs emerging in some sub-basins concerning nitrate and pesticide content, the overall progress on water quality remains modest thus far. However, given that most developments have been undertaken recently and agro-environmental policy is based on voluntary participation, it will take some time to observe the real outcomes.

The Rhine-Meuse Basin

The Rhine-Meuse is a transboundary basin that encompasses nine countries: Austria, Belgium, France, Germany, Italy, Liechtenstein, Luxembourg, the Netherlands and Switzerland. The Rhine River is 1,320 km long and the size of its catchment area is 186,765 km^2. The biggest part of the basin lies in Germany (106,000 km^2), followed by Switzerland (28,000 km^2), France (23,000 km^2) and the Netherlands (22,700 km^2). The total population of the catchment area is 78 million inhabitants, 1.7 million of which live in France. In order to create a transboundary cooperation forum among the Rhine Basin countries, the International Commission for the Protection of the Rhine (ICPR) was established in 1950. The ICPR was given the task of determining pollution levels and adopting appropriate measures for the protection of the Rhine. In 1986, an industrial accident in Basel, Switzerland caused approximately 20 tons of highly toxic pesticides to flow into the Rhine. This has had a devastating impact on the ecosystem. Following this incident, the Rhine Action Plan (RAP) was put into effect in 1987 and completed in 2000. Within the RAP framework, a warning network with six international warning centres has been set up to notify downstream states and riverside inhabitants in case of accidents. Based on the achievements of the RAP, the Rhine 2020 Programme on Sustainable Development of the Rhine was initiated in 2001. In addition, an action plan on flood defence was adopted in 1998. Combined, they focus on flood protection, prevention and improving water quality through waste discharge control, industrial accident prevention and the ecological restoration of the Rhine. The adoption of the WFD is likely to have a positive effect on the quality of the Rhine River, as by 2015 all the rivers in EU states will be required to reach 'good water' status (see **Chapter 11** and **Box 14.1**).

The Rhone-Mediterranean Basin

The Rhone-Mediterranean Basin covers 25 percent of the surface area of France. The basin is characterized by a dense network of rivers of varying length, 6,500 of which are longer than 2 km. The Rhone River, the biggest river in the basin, is shared with Switzerland. In order to meet the energy requirements of developing industry, the construction of hydroelectric power plants (HEPP) was started in 1946, and by 1986, eighteen HEPPs were installed on the Rhone River. Overall, the dams installed in this basin generate 64 percent of the national hydroelectric production and 8 percent of the total national energy production. Today, hydropower ranks as the second biggest (after nuclear) source of energy production in France. The dams built for energy production also serve different purposes, such as flow regulation and water supply for drinking, irrigation, navigation and recreational activities. However, the dams divert more than 80 percent of the river flow and so have a direct impact on the aquatic environment by preventing fish migration and altering the natural flow regime. These problems are being addressed by specific action plans that aim to increase water flow in the bypassed sections of the river. Consequently, a reduction of eutrophication and an increase of flora and fauna diversity has occurred. However, the measures taken to restore the free movement of fish have not been successful, due to a lack of monitoring and enforcement.

The Seine-Normandy River Basin

The Seine-Normandy River Basin accommodates 17.5 million people, which corresponds to 30 percent of the overall population of France.

The capital city, Paris, and other big urban settlements such as Rouen, Caen, Le Havre, Reims and Troyes, are also located in this basin. Of the 1.5 billion m^3 water used in the basin, 40 percent comes from surface waters and 60 percent from groundwater resources. The main problem in the basin remains improving water quality under the strain of increasing pollutant concentrations, particularly nitrates and pesticides. Given this problem, it is anticipated that despite the current action plans and high capacity wastewater treatment plants, meeting the targets required by the WFD will not be possible for many years to come (see **CD-ROM**).

Conclusion

France's great productivity in agricultural and industrial products has caused complex environmental problems, stemming from the pollution of surface and groundwater resources by agricultural, domestic and industrial wastes. Water legislation reform of 1992 laid out the principles for a balanced management of water resources with the aim of keeping the needs of humans and the environment in balance. Furthermore, the WFD has already been integrated into French law. However, finding a compromise between the needs of ecosystems and other water uses continues to be a real challenge for the six basin agencies.

5. Japan

Located off the East Asian coast in the North Pacific Ocean, Japan is comprised of a chain of 6,852 islands. The four largest islands – Hokkaido, Kyushu, Honshu and Shikoku – make up 98 percent of Japan's total land area of 377,899 km² (see CD-ROM for a discussion of the water challenges of the Greater Tokyo region).

Japan receives abundant precipitation, due to regular monsoons. Nevertheless, water shortages are frequent, due to the spatial and temporal variation of rainfall, marked topographic differences, small river catchments and sudden drops in altitude causing short and swift rivers. This situation is further aggravated by severe droughts. The amount of available water resources per capita is 3,300 m^3/year.

Total annual water use is approximately 85.2 billion m^3, 88 percent of which is obtained from rivers. The agriculture sector makes up more than 65 percent of annual water abstraction, followed by domestic and industrial uses (20 and 15 percent respectively).

Ensuring drinking water supply and access to sanitation

Based on the 1961 Water Resources Development Promotion Law, comprehensive water resources development (including infrastructure like water supply reservoirs) and efficient use of water resources have been advanced in order to ensure a stable supply of water resources over a wide area to respond to the rapid development of industry and increase in urban population. Nearly 100 percent of the population in Japan is connected to safe drinking water supplies. The average per capita daily

water consumption of 320 litres (L) has remained unchanged since the 1990s. The total population connected to public sewerage was estimated to be about 68 percent in 2004, whereas the rate in towns and villages with population less than 50,000 is only 36 percent. The government's target for 2007 is to expand the coverage of the public sewer system to 72 percent and increase the proportion of population served with advanced wastewater treatment from 13 to 17 percent. Thanks to the adoption of proper waste management techniques, water-borne diseases have been drastically decreased.

Safeguarding ecosystems

Japan's varied landscape and climate provide a rich but fragile natural environment for thousands of different plant and animal species, which has been deteriorated by industrialization and urbanization. In order to prevent further degradation of freshwater sources and the surrounding environment, the government strictly regulates effluent from the industrial and public sectors and imposes regulations on agricultural chemicals. Accordingly, the environmental quality parameters (e.g. biological and chemical oxygen demands) of rivers, lakes and reservoirs are improving.

Map 14.6: Overview of the river basins in Japan

This map shows major river basins in each nine regions in Japan including Greater Tokyo. The major hydroelectric power plants on the map are those whose power generation is ranked in the top ten in Japan.

Source: Prepared for the World Water Assessment Programme by AFDEC, 2006.

The Environmental Impact Assessment (EIA) Law was fully put into force in 1999. The purpose of this law is to ensure that environmental considerations are taken into account when implementing large-scale projects by conducting EIAs and reflecting on the results in decision-making. For this purpose, the EIA Law sets forth the procedure and defines the responsibilities of government regarding the EIA.

The River Law, which was originally enacted in 1896 for flood control, was comprehensively amended in 1964 and 1997. The latest amendment promotes the establishment of a comprehensive river administration system for flood control, water use and environmental conservation, which has resulted in an increase in the number of river restoration programmes throughout the country. It has also put more emphasis on public involvement in the planning process. In 2003, the Promotion of the Nature Restoration Act came into force, and the New Biodiversity National Strategy was put into place by the government as a comprehensive and systematic plan to protect natural biodiversity and restoration of the fluvial environment and ecosystems. A practical

implementation of this strategy is the National Census on River Environment, which aims to periodically survey the status of rivers and dam reservoirs from an environmental perspective.

Water for agriculture, industry and energy

Water for flooded rice paddy fields and fish culture comprises most of Japan's agricultural water use. For the last two decades, agricultural surface area has been decreasing. However, due to increasing requirements for higher crop productivity and measures on water quality, the amount of water utilized for irrigation has not changed significantly. Due to its limited amount of cultivatable surface area, Japan imports many products, especially grains, crops and meat.

Starting in the 1960s, a period of high economic growth, the water requirement of Japanese industry increased. The industrial water consumption has been stabilized since 1975 in response to the regulations on drainage and efficient water use. Currently, with an average water-recycling rate of around 79 percent, industry is the sector that consumes the least amount of water in Japan.

The share of hydropower energy accounts for about 9.5 percent of the total electrical energy generated. As of 2004, there were over 1,800 hydropower plants across the country, generating about 47 billion watts, or 17 percent of the maximum generating capacity of all electric power plants in Japan.

Coping with water-related disasters

Given Japan's unstable geology, mountainous topography and small amount of inhabitable land, there have been significant variations in land-use and population concentrations over the centuries. Currently, 50 percent of the nation's total population and 75 percent of its total physical assets are concentrated on the alluvial plains, which account for only 10 percent of Japan's total land area. The natural and social constraints exacerbated flood damage and prompted the creation of a national programme for comprehensive flood disaster control measures. The programme promotes the holistic integration of structural measures such as river improvement schemes (e.g. levees, channel improvement, dredging, floodway construction, etc.) and non-structural measures, such as controlling basin land development, the creation of a warning system, the establishment of a community flood fighting corps and the dissemination of flood hazard maps. Yet despite these measures, the flooding of a number of small rivers in 2004 caused a reported 275 deaths and about US$ 17 billion in damage. As a response, the Flood Fighting Act was revised in 2005 to expand the scope of flood forecasting activities to include smaller rivers in order to promote a more responsive and timely emergency evacuation. The amended act also aims to enhance and adopt extensive flood information and communication systems for an additional set of rivers, while improving already existing flood forecast systems to disseminate information necessary for ensuring a smooth evacuation operation.

The Law for Enhancing Motivation on Environmental Conservation and Promoting Environmental Education was established in July 2003 and enforced in October 2004 to promote environmental education in schools and community workplaces and enhance public awareness and education about the different issues of environmental protection and the conservation of natural water bodies. Some of the programmes and campaigns aimed at public

education are National Water Day on August 1, the Annual Forum on the Water Environment and the Disaster Prevention Poster Competition, among other local activities.

In addition to universities, there are several high-level research institutes and centres that focus on issues concerning water resources, environmental protection and disaster prevention. These institutions not only pursue

scientific research and development but also actively make policy recommendations for better management. The National Institute for Land and Infrastructure Management (NILIM) and the Public Works Research Institute (PWRI) in Tsukuba City are the leading water-related research institutes in the country, the latter of which hosts the International Centre for Water Hazard and Risk Management (ICHARM) under the auspices of UNESCO.

The Sediment-Related Disaster Prevention Law was enacted in 2000 with the aim of implementing comprehensive non-structural measures that would protect people from sediment-related disasters. Measures included raising public awareness on high-risk areas prone to sediment-related disaster, the development of a warning and evacuation system, the restriction of new land development for housing and other purposes and the promotion of relocation for some existing houses. After the revision of the Law in 2005, new regulations to prevent housing development in hazardous areas were introduced and the preparation and dissemination of hazard maps for smooth evacuation mechanisms made obligatory.

The development of advanced forecasting and warning systems in Japan is backed by the dense network of rain gauges and water-level telemeters. These observation points, in combination with twenty-six radar systems, provide high precision information concerning the spatial and temporal distribution of rainfall (see **Chapter 10**).

Conclusion

Risk reduction and the mitigation of water-related disasters is considered as one of Japan's main challenges. Accordingly, it has revised and amended disaster-related legislation. To ensure the sustainability of water resources, comprehensive water resource development plans have been put into action and the efficient use of water resources have been promoted in all sectors. In the sanitation sector, the government is attempting to expand the coverage of the public sewer system. In order to combat environmental degradation caused by human activities new legislation to regulate the use and discharge of effluent has been brought into action. Water-related decisions and policies aim to increase public prosperity by integrating the needs of modern life into a well-functioning healthy ecosystem. In the light of these facts, the Government of Japan is continuously striving to overcome current and future water challenges.

---

6. Kenya

Kenya is a water-scarce country. Located in East Africa, Kenya sits on the coast of the Indian Ocean, which serves as an important outlet. Surface waters cover only 2 percent of Kenya's total surface area. The climate varies from tropical along the coast of the Indian Ocean to arid in the interior, and two-thirds of the country is covered by semi-desert or desert land. As a result, only about 160,000 km^2 of land, most of which is situated in the wetter southwest area, is suitable for the current population of approximately 33 million. Per capita available water is about 650 m^3/year. Future projections show that by the year 2020, per capita water availability will drop to 359 m^3 as a result of population growth.

The uneven distribution of rainfall in addition to temporal and spatial variations often lead to recurring droughts in the north and east and flooding during rainy seasons. More than 50 percent of annual water abstraction is used for domestic purposes and livestock production, and the remainder is used for irrigated agriculture. The demand management strategies are lacking, and water resources allocation decisions related to surface and groundwater abstractions are made without adequate data. It is estimated that more than 50 percent of water abstractions are illegal. Water metering systems are used in few projects; as a result, revenue

collection is very low and corresponds to just 55 percent of the total operation and maintenance costs.

Major challenges: Poverty, access to safe water and sanitation, food and energy

Due to a steady decline in economic performance during the last two decades, the level of poverty in Kenya is steadily increasing, especially in semi-arid and arid areas. The welfare monitoring survey indicated that between 1994 and 1997 the poverty level rose from 47 to 53 percent in rural areas and from 29 to 49 percent in urban areas. As of 2005, approximately 42 percent of the population is below the national poverty line (UNDP, 2005). The poverty line for urban settlements is about US $35 per adult per month and US $16 for rural settlements.

In order to alleviate poverty levels, the Kenyan Government proposed the Economic Recovery Strategy for Wealth and Employment Creation (ERS), which charts the country's economic course from 2003 to 2007 and asserts that past institutional arrangements were simply insufficient to win the battle against poverty. The ERS promotes initiatives that would facilitate the achievement of MDGs, recognizes water as a pivotal element in poverty reduction and emphasizes the importance of providing services to the poor while ensuring adequate water for competing demands. It suggests undertaking comprehensive institutional reforms to facilitate 'pro-poverty water and sanitation programmes'. In this context, Kenya's poverty reduction strategy programme, initiated in 2000, commits the government to providing water and sanitation services to the majority of the poor at a reasonable distance (less than 2 km). The proposed strategy is to involve communities and local authorities more actively in the management of water and sewerage systems and services.

Over 70 percent of the population, about 24 million people, live in rural areas. However, half of the urban population is settled in informal settlements. The percentage of people with access to safe water is 68 percent in urban areas and 49 percent in rural settlements, according to the most recent data from 2003. In urban areas, almost 40 percent of water goes unaccounted for, lost through either leakage or illegal connections. Access to sanitation in urban areas is at 65 percent compared to 40 percent in rural areas. Accordingly, water-borne or sanitation-related diseases make up the majority of Kenya's morbidity rate and are responsible for over 60 percent of premature deaths. The most common instances of disease in Kenya are malaria (32.6 percent), respiratory system infections (24.6 percent) and diarrhoea and intestinal worms (17 percent).

Agriculture is the leading sector of the national economy, employing about 80 percent of the population and accounting for 26 percent of Kenya's GDP. However, Kenya has not yet put its available land resources to full use. Out of 9.4 million ha of potentially cultivable land, only 2.8 million ha are devoted to agriculture, which heavily relies on rainfed production with very little irrigation. The irrigation potential for the country is estimated at approximately 550,000 ha, but only about 109,000 ha has been put to use. Irrigation is the only way to ensure food security considering the variation in rainfall patterns and recurring droughts (**Box 14.4**). Kenya has been struggling to achieve food security for the last two decades; however, recent surveys reveal that the situation is getting worse. For example in 2004, the 'food poor', those who cannot meet the daily necessary minimum of 2,250 kilocalories, stood at 15 million people, up from 7.3 million in 1973. Of these, 3 million are in constant need of relief, and the number of malnourished children is also mounting.

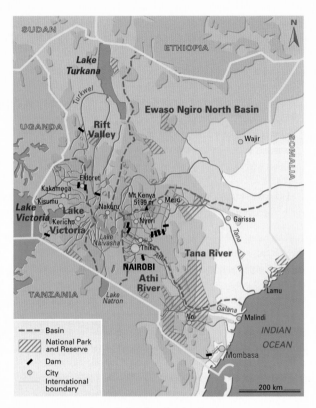

Map 14.7: Overview of the river basins in Kenya

Source: Prepared for the World Water Assessment Programme by AFDEC, 2006.

Kenya is mainly dependant on biomass for energy. Both fuelwood and charcoal accounted for 66 percent of the total energy consumption in 1996. In the same year, petroleum ranked second in energy production (24 percent) and electricity (hydropower combined with geothermal energy) produced the lowest amount of energy (9 percent). Seventy percent of the electricity supply comes from hydropower generation. Official technically feasible hydro potential is 2,023 MW, of which 677 MW (about 33 percent) is put into use. Generally, a heavy dependence on hydropower increases vulnerability to droughts, as low water supplies can cause power shortages. However, new hydropower projects have been implemented less frequently since 1996, decreasing the percentage of hydroelectricity in energy production. Instead, Kenya is becoming more dependent on fossils fuels, which emit higher amounts of greenhouse gases and other pollutants.

While less than 10 percent of the population is connected to the national grid, demand for power is increasing at the rate of 6 percent per year. This lag in the development of energy supply has had a negative impact on urban and industrial development. In 1997, reforms initiated in the power sector led to the creation of the Electricity Regulatory Board (ERB), which formulates policy and regulates the energy sector. The objectives of the energy sector are to enhance the energy supply and delivery capacity to all sectors of the economy, institutionalize environmental impact assessments in energy development, promote energy conservation through the use of efficient and cost-effective technologies and create an enabling environment for private sector participation in the supply of energy including electricity. Currently, the Kenya Generating Company (KenGen) is the major supplier of energy and provides almost 90 percent of national power. The Kenya Power and Lighting Company (KPLC) is the only organization responsible for the transmission and distribution of power.

Water sector reform

The level of water scarcity in some regions of Kenya has become a serious limiting factor for development activities. Consequently, the need to change the scattered structure and functioning of the water management system has arisen. In 2002, major reforms were initiated with the revision of the Water Act, which defines clear roles for the different actors involved in the decentralized institutional framework that separates policy formulation from regulation and services provision. When possible, the participation of stakeholders in the decision-making process is promoted by involving communities and other actors such as NGOs, community-based organizations (CBOs) and the private sector.

Under the revised system, the Ministry for Water Resource Management and Development (MWRMD) is responsible for formulating the National Water Policy and for carrying out reforms by bringing together all the

stakeholders in the water sector. This is achieved through transferring the responsibility of water management to basin organizations. Furthermore, since 2004, the provision of water and sanitation services are being transferred to private companies as a part of the decentralization process.

National legislation like the Environmental Management Coordination Act from 1999 aims to ensure the sound management of the environment. All projects that might have a potential impact on water bodies must complete an Environmental Impact Assessment. In addition, there are approved standards for drinking water quality and effluent discharges; however, the relevant rules and regulations are not strictly enforced due to a lack of skilled personnel and limited funds. As a result, water pollution from urban and industrial wastes continues to degrade water quality; the heavy use of pesticides and fertilizers in agriculture leads to deterioration of surface water and underground resources; deforestation for firewood production continues at an increasing pace; and the overall exploitation of the country's resources remains an imminent threat to ecosystems.

Enhancing water sector capacity

Water education in Kenya is carried out through university degree programmes at both the undergraduate and graduate levels. In addition, the Kenya Water Institute provides short-term courses tailored to meet the specific needs of clients in the water and sanitation sectors. These courses train approximately 600 candidates per year on topics like water

---

BOX 14.4: DISASTERS AND INCOME

The effects of natural disasters on the national income

In the past, disaster management has not been viewed as an integral part of development planning, and water-related disasters have been responded to in an ad hoc manner whenever they occurred. Similarly, the important elements of disaster management, such as prevention, mitigation, disaster preparedness, recovery and rehabilitation have been either ignored or haphazardly dealt with. As Kenya's economy heavily relies on rainfed agriculture, the variation in rainfall has a significant affect on the gross domestic product (GDP). The following figure shows the rainfall variability between 1979 and 2000. During drought years, the agricultural GDP shows a massive deficit with the overall GDP following it.

The effects of drought have become more pronounced in recent decades: in the 1990s,

there were three major droughts. The effect of the 1991-1992 drought in the arid districts led to livestock losses of up to 70 percent and high rates of child malnutrition of up to 50 percent. During this drought, 1.5 million people in seventeen arid and semi-arid districts of four provinces received relief food assistance. The second major drought occurred in 1995-1996

and affected an estimated 1.41 million people. The third and worst drought affected Central, Eastern, Rift Valley, Coast and North Eastern Provinces, with 4.4 million people requiring food assistance in the year 2000. The energy sector, which suffered huge financial losses, and rice production, which dropped by 40 percent, were particularly impacted.

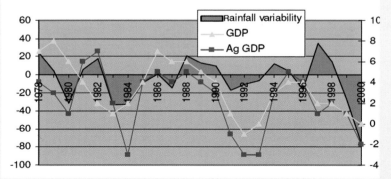

Rainfall variability, Ag GDP and GDP

meter installation, servicing and reading; the operation and maintenance of pumping and treatment plants; and water pollution control assistants. Although these efforts are a good way to build up the necessary technical human resources base, a detailed analysis of the water sector has not yet been done to identify current existing gaps in capacity (i.e. required skills, levels of competency and experience), which makes it difficult to estimate the impact of higher-level water education and short courses for technician training.

Conclusion

The food deficit in Kenya is a major problem, resulting in millions of chronically undernourished people. National food policies, while aiming to boost productivity, do not address access to food and quality. Access to safe water and sanitation services have not caught up with the needs of the growing population. Inadequate funding curbs the rehabilitation and expansion of the water supply and sewerage systems, and as a result, many diseases claim the lives of poor people every year. The need for domestic, industrial and agricultural water supply is growing, but the absence of demand-management strategies means that the increase in demand will likely outstrip the available supply. The construction of new dams is essential for providing the energy needed for development and meeting the increasing demand for drinking and irrigation water. The absence of international funding, however, remains a major obstacle for development efforts.

7. Lake Peipsi/Chudskoe-Pskovskoe

Lake Peipsi/Chudskoe-Pskovskoe (referred to here as Lake Peipsi) is the fourth largest and the biggest transboundary lake in Europe. It consists of three unequal parts: the biggest northern Lake Peipsi sensu stricto (s.s.)/Chudskoe, the southern Lake Pihkva/Pskovskoe and the narrow strait-like Lake Lämmijärv/Teploe connecting Lake Peipsi s.s. and Lake Pihkva/Pskovskoe.

The Lake Peipsi Basin is situated in Russia (59 percent), Estonia (33 percent) and Latvia (8 percent), although the last has a negligible effect on the basin. It is officially managed by a joint commission of Estonia and Russia; Latvia is not a member of any agreements or commissions concerning the lake.

The case study report published in WWDR1 (see **CD-ROM**) concluded that lake pollution (see **Chapter 12**), eutrophication and economic growth were the most critical problems facing the region. The fishing industry, which has been the lake's major economic activity, has suffered from environmental damage inflicted by pollution and overfishing, both of which have led to a depletion of fish populations. Although some improvements have taken place, the prior trends have not significantly changed.

Changing climatic and socio-economic contexts

Although there has been no specific research conducted on the effects of climate change in the region, the analysis of data collected over the last fifty-four years shows a slight decrease, approximately 3 cm, in the average thickness of ice cover.

There has been a slight change in the demographic characteristics of the region (the birth rate in Russia has started to increase). However due to aging populations in both countries the departure of younger and educated people for big urban areas continues to be a problem. On the other hand, economic activities in the region are improving and diversifying in both Estonia and the Russian Federation. This change is due to several factors, including the accession of Estonia into the

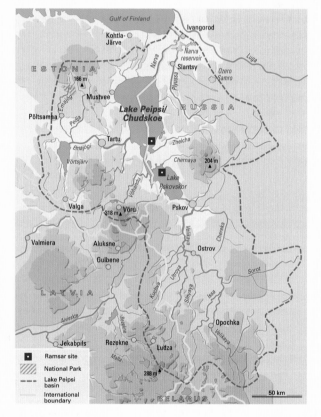

Map 14.8: Overview of the Lake Peipsi/Chudskoe-Pskovskoe Basin

Source: Prepared for the World Water Assessment Programme by AFDEC, 2002.

European Union, the economic recovery of the Russian Federation from its recent crisis and an increased profit from oil exportation because of record high prices. Furthermore, both countries are eager to develop joint economic activities and gain more access to neighbouring markets. Accordingly, economic difficulties are easing, and the high levels of unemployment are decreasing.

Reforms underway

The Joint Lake Peipsi Management Programme, an integrated water resources management tool, is being developed by both countries in accordance with the requirements of the WFD. When completed, the Estonian-Russian Transboundary Water Commission will be responsible for its implementation and updating. The programme will take into consideration both the surface and groundwater resources of the basin as a unity.

At the institutional level, there have been changes in both countries. As a result of the administrative reform in the Ministry of Natural Resources of the Russian Federation, the Water Resources Agency performs all activities related to the utilization, protection and rehabilitation of water resources. In Estonia, WFD rules and norms are being implemented (see **Box 14.5**), including the development of water basin administrations. As a result, the West Estonia, East Estonia and Koiva River Basin Administrations were established. The East-Estonian River Basin includes Lake Peipsi and the Narva River, which is shared with the Russian Federation.

In order to promote better communication among different stakeholders and build partnership in the region, the Peipsi Council was elaborated as a joint effort.

A bilateral agreement concerning water transportation in Lake Peipsi, Lake Lämmijärv and Lake Pskovskoe was signed in March 2002. One objective that has been facilitated by the agreement and is being currently pursued is the opening of a ferry line between the Russian city Pskov and the Estonian city Tartu. The ferry line is meant to support the local and bilateral development activities of the two countries.

As elaborated in WWDR1, Lake Peipsi is a relatively new transboundary water basin (the control line was formed when Estonia separated from the Soviet Union in 1991). Therefore, the border between the Republic of Estonia and the Russian Federation, a big portion of which runs along Lake Peipsi, has not yet been officially determined, since the official border has not been finalized.

Environmental issues and meeting the MDGs

In the last few years, considerable investments have been made in both countries to improve water supply facilities and wastewater treatment plants. For example, recent improvements in wastewater treatment facilities in Tartu has helped reduce the basin's pollution load. Due to its unique characteristics, several international and national projects and programmes have been launched with the aim of preserving the biodiversity of the lake and its protected areas. The promotion of regional ecotourism is also considered as a way to improve the regional economy while contributing to current reclamation efforts in the lake.

Progress towards the MDGs in the Russian Federation is reflected in the reporting under the state development programmes, which correspond partially to MDGs and incorporate other state-specific aims and objectives. Estonia, on the other hand, as a member of the EU, addresses water supply and sanitation issues as part of the implementation stage of the WFD. Both countries are on track for achieving the water-related MDGs (UNDP, 2005).

Conclusion

The region's economy is improving and Estonia and Russia are looking to further increase their economic relations through the establishment of new ferry lines. The water sector in both countries is also improving to face current challenges: Estonia, as a relatively new member of the European Union, has adopted the WFD and is in the process of implementing it, while the Russian Federation is going through administrative reform and revamping its water code. Furthermore the Joint Lake Peipsi Management Programme is in the process of elaboration which would promote implementation of IWRM. Consequently, the changes in both countries will help to create a sustainable utilization of lake resources while protecting the ecosystems.

BOX 14.5: BUILDING A COMMUNITY FRAMEWORK

A prime example of the implementation of the EU Water Framework Directive (WFD) is the preparation of the Viru-Peipsi water management plan. In November 2002, the project implementation unit, in cooperation with the environmental departments of four counties (Tartu, Põlva, West-Viru and East-Viru), organized four seminars for ten counties. The main target groups were local authorities (county governments, city governments and rural municipalities), environmental departments, land reclamation bureaus, health protection departments and water companies. In April and July 2004, two seminars were organized for the participation of local stakeholders in Jõgevamaa, Tartumaa, Võrumaa and Põlvamaa counties.

On the Russian side, such consultations took place within the context of a UNDP/GEF (United Nations Development Programme/Global Environment Fund) project. The goal of the project was to facilitate the planning processes of the Joint Lake Peipsi Management Programme, taking into account the requirements of both the European Union (EU) and the Russian Federation. Meetings and seminars were carried out mainly in Pskov region and resulted in a number of proposals and recommendations for the future Management Plan (see **Boxes 14.1** and **14.2**).

8. Lake Titicaca Basin

The Lake Titicaca Basin is composed of four major basins: Lake Titicaca, Desaguadero River, Lake Poopó and Coipasa Salt Lake. These four basins form the TDPS System, the main element being Lake Titicaca, the largest lake in South America and the highest navigable lake in the world. The TDPS System stretches approximately 140,000 km^2 and is located between 3,600 and 4,500 m.a.s.l.

Poverty and conflict: Persistent challenges

The initial case study report presented in WWDR1 (see **CD-ROM**) concluded that poverty was the most critical social problem in the TDPS system, affecting both rural and urban populations and undermining attempts to implement solutions to various problems. Unfortunately, in the past three years, no significant progress has been made to improve the situation.

In January 2005, the inhabitants of El Alto, Bolivia (located near La Paz), the main city of the TDPS System with 800,000 inhabitants (Instituto Nacional de Estadistica, 2005), protested the contract with Aguas del Illimani (Waters of Illimani), a subsidiary of French Suez Lyonnaisse des Eaux that was running a thirty-year concession for the water and sewage services in La Paz and El Alto. A week of civil disturbances finally came to an end with the resignation of the Constitutional President of Bolivia and the government's unilateral decision to end the water concession with Aguas del Illimani. The political transitions occurring in some Latin American countries since the 1980s have further added to the complexity of finding a solution to poverty. Peru was among the first Latin American countries to shift to a democratic regime. However, an increasingly authoritarian regime led to public outrage and caused the president to flee the country in 2000.

These events can be linked to structural poverty (see WWDR1 for more information), which stems from the combination of several socio-economic factors. Some of these factors are land property fragmentation (causing the under-utilization of land resources and thus low productivity) and indigenous cultural patterns leading to social exclusion. The effects of these factors are more pronounced in rural areas. Consequently, migration to urban settlements becomes the only choice for the rural poor, who hope to find better living conditions and mostly end up in crowded degraded districts. These migrants, the inhabitants of the Bolivian urban TDPS System, were the real actors of the social upheaval that took place in October 2003.

The impact of climate change on glaciers

During dry seasons, glaciers are the main source of drinking and irrigation water for many urban dwellers and farmers living in Peru and Bolivia. However, the climate variability and associated changes in ambient temperatures have started affecting the tropical glaciers of the region. The loss in volume of these unique tropical glaciers is alarming, and continuing melting trends will translate

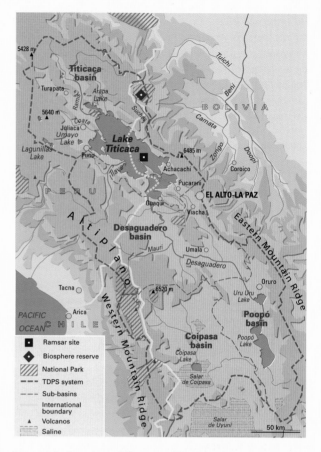

Map 14.9: Overview of the Lake Titicaca Basin
Source: Prepared for the World Water Assessment Programme by AFDEC, 2006.

into drought for thousands of people. **Figure 14.1** illustrates the impact of climate change on the availability of water resources in the TDPS System.

The consequences of glacial melting for local populations are serious. Acting as reservoirs, glaciers regulate stream flow and diminish seasonal discharge variation. This effect is vital, especially between September and November, when ice melting (and water demand) is at its maximum. Discharges in glacier basins are important during those months, since the flows of other rivers in the Altiplano Basins reach minimum levels.

To counterbalance the negative effects of glacial melting, more dams and reservoirs will have to be constructed, increasing the cost of the water supply to Andean cities. It can be expected that the additional cost will be transferred to urban users by means of tariff increases, particularly in El Alto and La Paz, where the urban water supply is under private administration. Judging from recent social movements, any tariff increase would likely trigger potential conflicts, particularly in the poorest areas of El Alto. The additional costs of flow regulation in glacier basins could also be hard to afford for small and medium-sized irrigation systems, rendering rural poor more vulnerable.

Conclusion

Poverty remains the underlying cause of many social problems experienced by both rural and urban populations. Since the first WWAP case study was conducted in 2003, there has unfortunately been no improvement in living conditions. The poor are still struggling to meet the most basic of food and water needs. The expectation of better living conditions tempts young people to migrate to the cities; however, most of these people find themselves living in degraded crowded informal settlements, which lack even the most basic of utilities. The poor, even if they have physical access to water and health services, can only marginally take advantage of them due to poverty. In this context, the water-related problems of basin countries cannot be isolated; they must be addressed within the greater social framework. Better management of these countries' land, water and gas resources is the only means to break the vicious circle of poverty.

Figure 14.1: Areal and volumetric variation of the Chacaltaya Glacier

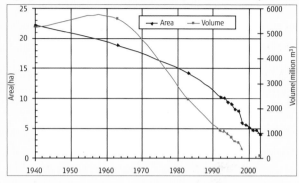

The data collected in the TDPS System shows the receding trend of tropical glaciers. Between 1991 and 2003, Zongo and Chacaltaya glaciers suffered both areal and volumetric losses. In fact, the accumulated mass balance, expressed as water depth, was -11.02 m for Zongo and -15.06 m for Chacaltaya. Chacaltaya glacier, a small glacier located at a medium altitude, lost 97 percent of its mass between 1960 and 2003 and is expected to disappear completely by 2010. This figure clearly shows that the receding trend started in the 1960s and has accelerated in the last twenty years.

9. Mali

Located in the heart of western Africa, Mali has a surface area of 1,241,000 km², over 50 percent of which is located in the Sahara Desert. More than 1,000 km away from the sea, the country is completely landlocked. Mali's location means that the country's climate can sometimes be quite unpredictable: years of abundant rainfall and years of extreme drought.

Three climatic groups can be discerned: arid desert in the northern region, arid to semi-arid in the centre and savannah in the south. The Sahara region, in the northwest tip of Mali, covers up to 57 percent of the national territory with an arid and semi-arid desert climate (rainfall usually does not exceed 200 mm per year). At its centre, the country's climate is characterized by the Sahel, encompassing about 18 percent of the land. The humid rainy season (June to October) usually brings between 200 and 700 mm of rainfall per year. The Niger River is an important part of this region, as the annual flooding of the river makes the surrounding land fertile for agricultural production. In the southern region of Mali, the rainy season generally brings over 1,200 mm of rain per year. This region and climate covers approximately 25 percent of the country. It is by far the most fertile area, where the majority of the population resides and where most agricultural activities take place.

Despite its northern desert, Mali has a number of important water resources. Two major rivers – the Niger River and the Senegal River – run through Mali. These two rivers constitute the majority of Mali's perennial surface water resources, providing the country with 56 billion m³ of water. Important non-perennial surface waters are estimated at a volume 15 billion m³. Mali also has seventeen large lakes situated near

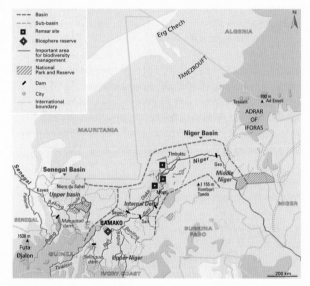

Map 14.10: Overview of the river basins in Mali

Source: Prepared for the World Water Assessment Programme by AFDEC, 2006.

the Niger River, and renewable groundwater resources from aquifers have been assessed at 66 billion m³. The volume of renewable water resources per capita per year is 10,000 m³.

However, these water resources are geographically dispersed and not always available when needed, greatly limiting their exploitation and economic development: overall, only 0.2 percent of Mali's potential water resources is put into use. Furthermore, the country has had many droughts in the past, compounding problems of water shortage issues.

Poverty, increased irrigation, access to safe water and environmental degradation

Mali is one of the world's poorest countries, with a per capita GDP of US $296, where over 90 percent of the population lives on less than US $2 a day. Like many other African countries, Mali's economy is heavily dependent on agricultural production, as well as on herding and fishing, with more than 80 percent of the population working in agro-pastoral activities. The agricultural sector represents 40 percent of the country's GDP, whereas the industrial sector represents 16 percent and the service sector 40 percent. Mali's agriculture is largely rainfed, but irrigation also plays a major role for some crops, such as rice. A number of droughts have devastated agricultural production and livelihoods in the past.

As of 1998, Mali had approximately 9.8 million inhabitants and over 10,000 villages with a population growth rate of 2.2 percent. Population density was around 8 inhabitants per km², with notable disparities between regions. Ninety percent of the villages are situated in five regions that occupy approximately 38 percent of the country's surface area. Although all these regions have at least one major urban centre, the areas are still largely rural. In fact, almost 70 percent of all Malians live in rural areas.

Approximately 30 percent of the population lives in urban settlements. There are seven major cities in Mali, the largest being the capital city Bamako, with a population of about 1 million inhabitants. In 1992, only nineteen urban centres were equipped with water facilities, whereas today twenty-seven centres out of thirty-three have been set up. Recently, much water infrastructure work has been implemented to improve access to safe water. National studies indicate that the percentage of rural and urban populations with at least one point of access to water has risen from 55 percent in 1998 to about 84 percent in 2002, based on one modern point of access to water per 400 inhabitants.

As these statistics indicate, a concerted effort has been made to provide drinking water to cities with populations of 10,000 or more inhabitants. However, the disorderly development of housing settlements has influenced the availability of water infrastructure in urban areas over the last two decades. This situation is further aggravated by the rapid growth of drinking water needs. The shortage of functioning infrastructure also continues to be highly problematic in rural areas. Additionally, mounting pollution combines with these factors to seriously impact the quantity and quality of water available to residents, dramatic affecting Malians' health.

Mali's main environmental challenge is the continual degradation of natural resources and the environment as a whole. Desertification and deforestation are two particularly menacing environmental problems for the country. Population growth, increasing desertification, soil degradation, intense firewood and charcoal production as well as a lack of a waste treatment system for the industrial and other sectors have seriously contributed to growth of environmental problems. In addition,

deforestation and desertification have decreased the area of the natural habitats of numerous plant and animal species and contributed to an increase in human migration further south. An increase in population growth in these areas has quickly led to an over-cultivation of the soil and increase in pollution. After the United Nations Conference on Environment and Development in Rio de Janeiro in 1992, legislative and regulatory measures were set up as guidelines for the protection of water resources in Mali. However, few of these guidelines have been implemented, and the country's water resources are being increasingly polluted by industrial urban and agricultural pollution.

Challenge areas: Health, food, and energy

Water-related diseases, such as cholera, diarrhoea and the guinea worm disease, represent more than 80 percent of all illnesses in Mali. However, other water-related diseases stemming from poor hygiene and parasites also frequently occur. For very young children, malaria and diarrhoeal problems are very common. Furthermore, 11 percent of children under the age of 5 suffer from acute malnutrition and emaciation, and 33 percent of all children are underweight – the latter being more pronounced among children aged twelve to twenty-three months (48 percent), children born with a frail build (48 percent) and children living in a rural environment (37 percent). Though access to water is, on the whole, increasing, access to adequate quantities of quality water remains highly problematic in both rural and urban areas.

The ability to ensure the continual provision of food for the population of Mali remains a major challenge. Droughts in the 1970s and 1980s were particularly harmful to food production. However, more recently, a number of diverse actions, including enlarged irrigation schemes, have been carried out to address the challenge of food security. Accordingly, annual grain production reached an average of 2.26 million tons (between 1990 and 2002) compared to an annual average of 1 million (between the years 1964 and 1990), signalling a two-fold increase in production in approximately twenty years time.

Currently, 270,000 ha of land is irrigated. Water abstraction for irrigation is about 4.5 billion m³, 98 percent of which is obtained from surface water resources. However, it is estimated that in order to ensure food security, the amounts of irrigated land and water allotted for irrigation both need to be expanded at least two-fold. The current shortage in food supply is mainly due to the inability of agricultural production to keep up with the rapid growth of the population and greater climatic volatility. However, the water sector suffers from a lack of coherent national pricing policies and causing serious difficulties in collecting the amounts owed by consumers leading to a poor track record for attracting private investments.

According to the data presented in 2002 by the National Energy Commission, Mali has an energy potential of 1,119 MW, which could allow it to produce 4,849 gigawatt hours (GWh) per year. Out of this identified potential, 378 MW could come from the Niger River and 740 MW from the Senegal River. Currently, two large dams (the Selingue and Manantali) provide 980 GW per year, which represents 20 percent of the identified potential and 98 percent of what is actually produced. Despite the country's sizeable energy potential, hydroelectricity only represents 1 percent of total energy consumption at the national level, whereas 90 percent of basic energy needs are met through firewood and charcoal. The dependence on firewood is one of the main causes of deforestation, which contributes to the process of desertification in fragile environmental zones. Until the economic advantages of pursuing alternative energy sources become clearer in practice, firewood will continue to be the main source of energy for households.

Management responses and stewardship

Since the early 1990s, Mali has been managing its water resources according to the Water Resources Development Framework. This strategy focuses on decentralizing water and sanitation administrative bodies between the central government and local communities involving a multitude of government agencies in water resource management. At the national level, the water sector falls under the responsibility of the Ministry of Mining, Energy and Water, which operates under the structure of the National Water Directorate (DNH, Direction Nationale de l'Hydraulique). Sanitation is the responsibility of the DNH and the ministries of Environment and Health. At the local level, over 700 communal councils have been created to share the responsibility of water management and infrastructure maintenance. These communal councils are funded by national authorities and help to ensure that implemented water infrastructure and service costs are recoverable through fee collection.

Increasing the knowledge base and technical expertise of water resources remains a major challenge in Mali. There has been limited progress made in the development of strong assessment indicators, namely the density of hydrologic and hydro-geologic stations, the quality of the information available about the water sector and the quality of the training and research institutions operating in the sector. Still, some knowledge has been accumulated and monitoring processes have been established and implemented in several projects. Unfortunately, however, the overall development of indicators is still fairly limited. Measures are being taken to correct this, but it will take time and money before they produce concrete results.

Conclusion

Many of Mali's water problems can be characterized as problems of access, largely provoked by an uneven temporal and geographical distribution of water combined with an under-exploitation of available water resources. In recent decades, the Government of Mali has taken a number of steps to ameliorate the situation in an attempt to meet the population's basic needs. However, a great deal of work remains, notably in the provision of infrastructure for safe drinking water and sanitation. Other enormous challenges for the country include controlling the level of pollution, developing alternative energy sources and decreasing deforestation and desertification. Mali's ability to address these issues will depend on a number of factors: namely, the country's capacity to raise the level of national technical expertise through increased educational programmes and research institutions, to develop strategies to better utilize available national water resources, to decrease the negative impact of urban population growth on water resources, as well as to attract investment for sustainable future water schemes.

BOX 14.6: SHARED WATER MANAGEMENT

Mali shares two large transboundary rivers, the Niger and Senegal rivers, with many other countries and is highly dependent upon these two large basins, particularly the Niger River Basin, as it is where most of the country's economic activity is centred and where more than half of its population resides. These basins are managed by two basin organizations, in which Mali participates. Since the United Nations Conference on Environment and Development in Rio de Janeiro in 1992, these two organizations have functioned on the principles of integrated water resources management. Continued cooperation in both organizations seems crucial

to the development of future sustainable water projects.

Mali's Niger River Agency aims to safeguard the Niger River, its tributaries and catchments, as well as the integrated management of its resources. Under the aegis of the organization, Mali has recently participated in a project to reverse damage done to the Niger and its surrounding land area. Pollution of the Niger has been a significant problem, since large amounts of wastewater stemming from the capital city, Bamako, flows back into the Niger.

Mali also participates in the Organization for the Development of the Senegal River. A 1972 Convention and 2002 Charter established the organization's legal and regulatory framework and clearly state that river water must be allocated to each of the various sectors. There is no agreement allocating the river's water to riparian states in terms of volumes of water to be withdrawn, but rather to use as a function of possibilities (i.e. agriculture, livestock-raising, hydroelectric energy production, drinking water supply, navigation, environment, etc.). The Senegal River Basin and its organization was presented in WWDR1 (see **CD-ROM**).

10. State of Mexico

Mexico's total surface area is slightly less than 2 million km². Annual runoff from its rivers is 399 km³, of which 87 percent comes from the thirty-nine main rivers of the country, whose basins occupy 58 percent of the country's total surface area. Average annual per capita water availability is 4,547 m³, with great variation between the southeast (13,566 m³) and the north, centre and northeast of the country (1,897 m³) (CNA, 2004). This uneven distribution of water resources causes water shortages in densely populated areas. The north, centre and west of the country, where only 32 percent of the runoff takes place, is home to 77 percent of the country's population and 85 percent of Mexico's GDP (CNA, 2004).

In so far as water stress is concerned, one of the country's most critical cases is the State of Mexico, with a population of almost 15 million inhabitants corresponding to approximately 14 percent of the nation's total population but only 1 percent of the country's total surface area. The State of Mexico is an industrial centre with a wide range of economic activities; it is ranked second in the nation in terms of its GDP contribution, about 9.5 percent.

Water and land resources

The State of Mexico is located within the geographical area of three main basins, namely the Valley of Mexico, the Lerma River and the Balsas River basins. The Valley of Mexico Basin lies in the north and northeast of the State and includes the Federal District[6] and parts of the states of Hidalgo and Tlaxcala. There are 22 million inhabitants living in these states, 10 million of which reside in the State of Mexico's part of the basin (CAEM, 2004). Combined, they represent 20 percent of the national population and contribute 31.5 percent of the total GDP. Conversely, the amount of available water resources in the basin is only 3.9 km³, or only 0.9 percent of the country's total water availability (CNA, 2004).

The Lerma River Basin is the lifeline of the city of Toluca, the capital of the State of Mexico, with approximately 1.5 million inhabitants living in the metropolitan area and 2.5 million living throughout the basin (CAEM, 2004). This region is also a centre of heavy economic activity. The high water demand in this basin has necessitated the implementation of water transfer schemes from the Balsas River Basin, located 130 km from Mexico City.

As for groundwater resources, there are nine aquifers in the State of Mexico, six of which are shared with Mexico City[7] (CAEM, 2004). Since these aquifers are the main source of water supply for the State of Mexico and Mexico City, they are exploited well beyond their renewal capacity. In general, it is estimated that underground water resources are overexploited at a rate of 100 percent or more, with the Texcoco aquifer in the Basin of the Valley of Mexico being overexploited at a rate of more than 850 percent (CAEM, 2004). As a direct consequence, in many aquifers the hydrostatic pressure has been lost, some springs have dried, and the ground is sinking up to 40 cm per year in some areas

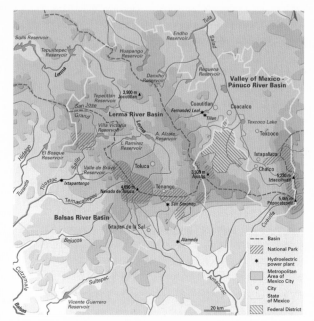

Map 14.11: Overview of the river basins in the State of Mexico
Source: Prepared for the World Water Assessment Programme by AFDEC, 2006.

of the Valley of Mexico. The intense overexploitation is further aggravated by the fact that the clayey topsoil in both the Valley of Mexico and the Lerma Valley enhances the runoff of rainwater and substantially reduces the natural recharge of aquifers. In order to curb the destruction of aquifers, the Federal Government has forbidden further development. However, unauthorized usage remains a problem.

Water and land uses

Of the available water resources in the State of Mexico, 48 percent is utilized for domestic purposes, 34 percent for irrigation and 5 percent for industry. The remaining 13 percent is transferred to the Federal District for consumption.

Agriculture is the main economic activity in the state, practised over approximately 50 percent of the overall surface area. Irrigation for agriculture is practised on a smaller scale, covering only 7 percent of the State's land surface. Almost 80 percent of the water used for irrigation is pumped from aquifers.

---

6. The capital of Mexico.

7. The Metropolitan Area of Mexico City (henceforth Mexico City) includes the Federal District and some boroughs of the States of Mexico and Hidalgo.

Water transfer

Water shortage in the State of Mexico is already at an alarmingly critical level. The situation is expected to worsen due to increasing domestic, industrial and agricultural water demands. Although the state government is constantly pursuing new mechanisms to slow down urban growth and promote efficient water use, water transfer from other water basins remains necessary to meet growing demand. Currently, water is transferred from both surface and groundwater resources to meet the demands of Mexico City and, to a lesser degree, the State of Mexico itself. For example, water is transferred from the Balsas River Basin to the Lerma River Basin and the Valley of Mexico, mainly to provide the Federal District's potable water supply. Underground water resources of the Alto Lerma System are also channelled to the Mexico City, causing their overexploitation. The extent of these transfer schemes (i.e. the distance from which the water is diverted), is also likely to grow, which could trigger disputes over water resources.

Water and health

In 1990, a cholera outbreak initiated in Peru and expanded throughout the continent. The first sign of cholera was seen in the State of Mexico in June 1991, signalling deficiencies in proper chlorination. Since then, the Government of the State of Mexico has assumed responsibility for the production of the chlorine supply and the maintenance of chlorination equipment. As a result, chlorination effectiveness has increased by 300 percent in the past decade.

The coverage of water and sanitation services in the State of Mexico is above the national average. Over 90 percent of the population has access to safe water and approximately 80 percent to sanitation services. However, water scarcity is still the major factor behind water-borne diseases. Through state social programmes, public awareness is being raised concerning preventive health measures, such as handwashing and cleaning water storage tanks.

Since underground water resources are exploited on a regular basis, the water wells are sealed to protect the naturally high quality of groundwater by preventing direct contamination by pollutant leakage. However, human activities pose a constant threat to groundwater quality. For example, in the State, wastewater is generated approximately at the rate of 30 m^3 per second (m^3/s), about 19 percent of which is directly discharged without any kind of treatment. Solid wastes are disposed of into open pits or partially controlled waste disposal sites. In addition to this is agricultural pollution, caused by the utilization of wastewater for irrigation and the use of fertilizers and insecticides. There is no exact data concerning the health consequences of such activities.

Water management

The water policies adopted in the State of Mexico (see **Box 14.7**) complement the National Development Plan for 2001-2006, which correctly identifies water as a scarce resource and puts forward the

common policy of integrated management for the sustainable utilization of water resources. In the State of Mexico, the Water Commission of the State of Mexico (CAEM, Comisión del Agua del Estado de Mexico) is an independent decentralized government institution authorized to manage the country's water resources, as well as to formulate national water policy. The Secretariat of Water, Public Works and Infrastructure for Development (SAOPID, Secretaría de Agua, Obra Pública e Infraestructura para el Desarrollo), the Water Consulting Council and the Water Commission of the State of Mexico are the coordinating bodies for the planning and programming of the state's water sector and for the surveillance of compliance with policies, strategies, plans and programmes. Laws, standards and regulations have been issued at federal, state and municipal levels. However, the full enforcement of such regulations is missing. Furthermore, a lack of consolidation of the bodies in charge of providing and maintaining the water and sanitation services has led to the inefficient use of water supply. In order to minimize this problem, in 2004, the Federal government started PROMAGUA, a programme aimed at modernizing the water operating bodies, and established an independent body for the regulation of the supply of potable water, sewer systems and wastewater treatment utilities. In addition to national funds allocated for the modernization and extension of utility services, the collaboration of the private sector is also being sought through public-private partnerships. Unfortunately, low levels of revenue collection coupled with a lack of respect for efficient water use might pose a problem for private sector involvement.

Water and ecosystems

The State of Mexico is ranked fourth in the world in terms of diversity of flora and fauna, after Brazil, Indonesia and Colombia (CNA, 2004). However, as a consequence of human activities, many species have become extinct, and wetlands and forests have either decreased in size or have been completely destroyed. Early records from the seventeenth century show that more than 58 percent of the State of Mexico's territory was forest, covering more than 1.3 million ha, and wetlands of the Lerma Swamp and the lakes of the Basin of the Valley of Mexico extended over 88,000 ha. Today, in the valleys of Mexico and Alto Lerma, the natural water bodies have been reduced by more than 80 percent overall; some of them having totally disappeared with their ecosystems. It is estimated that the forest area was reduced to one-third of its original area. The main causes of deforestation are stockbreeding, human settlements, road systems and firewood production for domestic use. Deforestation causes the surface run off to carry greater amount of debris, silting up dams, rivers and channels, reducing the capacity of storage and evacuation of storm water. The loss of vegetal cover also enhances surface run off, thus reducing the amount of water of infiltration and severely affecting the recharge of aquifers.

The National Development Plan of Environment and Natural Resources for 2001-2006 (PND, Plan Nacional de Desarrollo) was formulated to pursue economic development while reversing environmental degradation as

---

BOX 14.7: WATER RESOURCES DEVELOPMENT IN THE STATE OF MEXICO

In the State of Mexico, the following criteria set the main framework of water resources development:

- Implementation of IWRM throughout the State

- Giving priority to enhancing the quality and the extent of water utilities for raising the quality of life of citizens

- Assisting in the consolidation and efficiency of bodies in charge of rendering services

- Establishing a 'water culture' that prioritizes the efficient and sustainable utilization of water resources

- Involving the private sector in the financing, construction, operation, maintenance and administration of infrastructure

- Promoting modernization of the legal framework

---

much as possible. In this regard, the State of Mexico's Secretariat of Ecology is responsible for executing state policy and evaluating its effectiveness.

Risk management

Urban settlements in the state have experienced high growth rates accompanied by the rapid expansion of informal settlements. Consequently, people living on settlements that are constructed on slopes, old lacustrine areas, river and stream banks and beds are highly vulnerable to water and mud floods. Furthermore, the overexploitation of aquifers has caused differential ground sinkage and impeded the surface run off of storm water. Water and sewer services have also been either interrupted or completely disconnected due to ground sinkage.

Flood risk is especially high in the plains of the Valley of Mexico and Alto Lerma. Due to the lacustrine origin of the land, the natural drainage is very limited. The situation is worsened by the fact that the largest percentage of the State's population lives within this area. In order to reduce the risk of catastrophic floods, large-scale drainage systems have been implemented. However, as mentioned previously, the sinking of the ground and the extreme siltation caused by the loss of vegetal cover lower the discharge capacity of the drainage systems.

During the last eleven years, the State Government has compiled a flood atlas that gathers information on the social and economic impacts of water-related extreme events. The latest flood event in 2004 affected over 35,000 people in diverse municipalities of the State of Mexico.

The National Development Plan of 2001-2006 defined water-related risk reduction as a priority. In this context, the structural measures against flood prevention will continue to be financed at the federal and state levels.

Conclusion

Although Mexico has sufficient water resources, the State of Mexico is under severe water shortage stemming from a very dense population coupled with an accelerated growth of approximately 380,000 thousand inhabitants per year. The situation is even more critical in the Valley of Mexico, where the Metropolitan Area of Mexico City contains approximately 20 million inhabitants. The increasing water demands of various sectors have led to 100 percent or more overexploitation of underground water resources. The effects of the overuse of aquifers are striking: ground has been sinking up to 40 cm per year; piezometric levels have dropped significantly; aquifers have lost their hydrostatic pressure; and some springs have dried up. Water and sewer infrastructure has been either disconnected or become unusable due to sinking ground. This further complicates the challenge of providing the public with safe water and sanitation services. Although further abstraction of underground water resources is forbidden, illegal utilization continues to grow. Population growth has also had a toll on the vegetation cover and ecosystems. Forests have decreased by one-third and natural water bodies have been reduced by more than 80 percent in area; the associated ecosystems have shrunk dramatically or disappeared altogether. Because of deforestation, topsoil has lost its ability to retard surface run off, which in turn has reduced the infiltration rate and recharge of aquifers. Intra-basin water transfer schemes have been implemented to cope with the growing demand for water, but this unfortunately caused disputes between user groups. The quality of surface and groundwater resources is decreasing due to domestic, industrial and agricultural pollution. Although the construction of treatment plants is underway, financial problems have hindered their full realization. The capacity of existing wastewater treatment facilities cannot cope with the sheer volume of discharge.

The National Development Plan has underlined the importance of water resources as well as the necessity for integrated basin-wide management and stakeholder participation in decision-making. This plan is further backed by specialized agencies and coordinating bodies that are responsible for implementing projects and surveying the compliance with rules and regulations. However, the enforcement of such regulations has not been effective.

Hydrological and economic difficulties are unfortunately compounded by the lack of social awareness towards efficient use of water resources. Raising public awareness will facilitate the sustainable utilization of water resources in an environmentally sound fashion.

11. Mongolia with special reference to the Tuul River Basin

Mongolia has an annual precipitation of 361 km³, about 90 percent of which is lost to evapotranspiration. Of the remaining 10 percent, 37 percent infiltrates into the soil while 63 percent turns into surface runoff. Almost 95 percent of the surface runoff component flows out of the country (Box 14.8). Consequently, only 6 percent of Mongolia's annual precipitation is transformed into available water resources in surface water bodies (Altansukh, 1995). The total surface water resource of Mongolia is estimated as 599 km³/year and is composed of water stored in lakes (500 km³/year), glaciers (63 km³/year) and rivers (36 km³/year) (Myagmarjav and Davaa, 1999). The amount of renewable groundwater resources has been estimated at 10.8 km³/year. Groundwater resources continue to be a major source of water, especially during winter when many surface water resources are frozen.

There are approximately 3,500 lakes in Mongolia with a total surface area of about 15,600 km², about 54 percent of which is located in the Gobi region, mainly in the form of small shallow or salty lakes. As a consequence of human activities, many of these lakes are now severely depleted or dry (Altansukh, 1995).

Large rivers originate in the country's mountainous northern and western areas, while very few surface streams are found in the south. The country's largest watershed is the Selenge River Basin in the north with its major sub-basins the E'Gyin, Ider, Orhon and Tuul Rivers.

The height of the Tuul River Basin varies from 1,200 to 2,700 m.a.s.l. The Tuul River, the main river in the basin, is formed by the confluence of the Namiya and Nergui streams at the southwestern slope of the Khentei Mountain, which is a world watershed divide that separates the Arctic and Pacific ocean basins and the internal drainage basin in Central Asia. The Tuul River is 704 km long, with a catchment area of 49,840 km². It drains into one of the main tributaries of the Selenge River, which is the main artery of the Lake Baikal, the world's largest freshwater lake by volume.

The Tuul River Basin covers only 3.19 percent of the country's territory, but is home to more than half of the country's population.

Current contexts

Administratively, the country is divided into provinces (*aimags*), each of which is divided into *sum* (territorial administrative unit subordinate to district) and *bag* (the smallest administrative unit in rural district).

The capital city Ulan Bator is located in the Tuul River Basin and is home to 772,000 inhabitants, or about 32 percent of the country's population. About 60 percent of the nation's population is classified as urban, and more than half of this urban population lives in the capital city Ulan Bator.

Forty percent of the population lacks access to safe water resources and only 25 percent of the population has adequate access to sanitation facilities. Clearly, poverty is one of the main reasons. Official figures suggest that almost one-third of the population lives below the national poverty level, defined as the inability to afford a basket of basic food and non-food items. Urban poverty is on the rise due to increased migration from rural

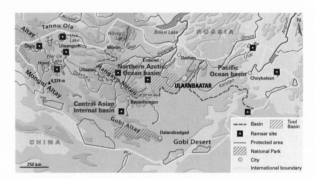

Map 14.12: Overview of the river basins in Mongolia

Source: Prepared for the World Water Assessment Programme by AFDEC, 2006.

areas; almost half of the poor live in urban areas and one-quarter of the urban poor is located in Ulan Bator. Recent statistics show that the depth of poverty and disparity has increased following several calamities in 1999, which caused a loss of livestock and a sharp decline in agricultural production in 2000 and 2001. If this trend is not reversed, the MDG target of halving the poverty headcount by 2015 will not be met (MFA, 2004).

Water and ecosystems

Growing urbanization and the mining industry have significantly polluted surface and underground water resources, which has had a significant impact on associated ecosystems. Furthermore, overuse of groundwater resources has led to lowering of the groundwater table, which has consequently caused some springs, lakes and their associated ecosystems to dry up. Increasing numbers of livestock and uncontrolled grazing practices are also affecting the balance of ecosystems.

Currently, there are forty protected areas, twelve of which include lacustrine ecosystems. The Government has declared its intention to raise the surface area of protected areas. The National Development Plan, adopted by the Mongolian Parliament in 1994, assigned a high priority to ecosystem protection. The basic guidelines for environmental protection were further identified in the Ecological Policy of the Mongolian State directive in 1997. These initial steps were followed by more than twenty laws on conservation. Practically speaking, the legal basis for sound environmental management is in place. Furthermore, the Ministry of Education, Culture and Science has included an environmental education

---

BOX 14.8: TRANSBOUNDARY WATER RESOURCES IN MONGOLIA

There are about 210 rivers flowing through Mongolia into Russia and China. Mongolia aims for international cooperation concerning the equitable utilization of transboundary waters with its neighbours. The first international agreement on transboundary water resources was between the governments of Mongolia and the USSR in 1974 on the use of water and protection of the Selenge River Basin, which plays an important role for the economic and industrial development of both countries. The agreement made between

the governments of Mongolia and the Russian Federation in 1995 on the protection of transboundary water resources focuses on over 100 small rivers and streams located in the western part of the country. In general, the drainage basins of transboundary rivers between Mongolia and the Russian Federation cover about 31.4 percent of the Mongolia's territory.

In 1994, an agreement was signed between China and Mongolia on the protection of transboundary

water resources concerning Lake Buir, the Kherlen, Bulgan, Khalkh rivers, and eighty-seven small lakes and rivers located near the border. Transboundary water resources shared with China include surface water bodies in Dornod, Khovd, and Bayan-Ulgii provinces and groundwater resources in Gobi-Altai, Umnugobi, Bayankhongor, Sukhbaatar and Dornogobi provinces.

---

programme into the secondary school curriculum. However, due to the competing interests of different sectors and a lack of incentives for environmental protection, the rate of implementation of rules and regulations has been weak (MFA, 2004).

Challenges to well-being and development

Average per capita water consumption in Mongolia is very low. The average water consumption of populations living in yurt (the traditional tent-like structures used by nomads) districts of big settlements is around 10 litres per person per day, far from being enough to meet sanitary requirements. There are 10,000 cases of diarrhoea every year in Mongolia and almost 70 percent of these cases occur in Ulan Bator. Dysentery and hepatitis are also common. These infections stem from a lack of access to safe water and sanitation infrastructure.

Water for food

Nomadic livestock husbandry has long been the dominant economic activity in Mongolia. This sector employs 47 percent of the total population, produces 34.6 percent of agricultural gross production and accounts for 30 percent of the country's exports. Until recently, crop production was not considered a significant economic activity in Mongolia. Intensive land cultivation only began in 1958. Currently, about 130 million ha of land is used for agriculture. Almost 98 percent of this surface area is utilized as pastureland whereas farmland occupies less than 1 percent (806,000 ha) of this land (UNEP, 2002). As of 2000, agriculture employed 48 percent of the total work force, made up about 35 percent of Mongolia's GDP and 30 percent of total export products. Until 1990, crop production was sufficient to surpass the total domestic demand for flour, and surpluses of flour, potatoes and vegetables were exported. However, after the collapse of Soviet Union, both cropping area and yield have declined, due to a lack of funding and technical and managerial problems. Today, wheat production satisfies only 50 percent of domestic demand, and potato and vegetable production barely meets 40 percent of demand. Yet irrigation continues to be the most water-

demanding sector. Approximately 43 percent of annual water abstraction is used for agriculture.

In recent years, climate changes have caused groundwater levels to fall, which has resulted in the drying up of some wells and springs (NSO, 2000). This has a great impact on animal herders living in remote areas of Mongolia. Consequently, the risk of livestock losses during the dry periods has increased enormously, and pastures near abundant water sources have become overused. The increasing number of livestock (from 25 million in 1990 to 30 million in 2000) clearly indicates that the problem is likely to get worse.

Water and industry

The mining industry contributes approximately 20 percent of national GDP and accounts for over 50 percent of overall exports. While mining is the largest industry in Mongolia, traditional industries such as fur and leather processing have also caused water pollution and affected ecosystems. Industrial water demand corresponds to 26 percent of annual supply. This rate of utilization is expected to increase in parallel to economic growth: since the 1990s, many new enterprises have been established, but environmental problems have increased due to lack of adequate environmental precautions.

Water and energy

Mongolia experiences an extremely cold climate for eight months of the year, making energy for heat generation crucial for survival. The large geographical area of the country and its low population density makes the provision of energy services a very difficult task. Wood and coal are commonly used for heating and cooking purposes. During the last decade, however, deforestation caused by firewood production has become one of the most serious and urgent environmental concerns in the country. Currently, only about 8 percent of Mongolia's territory (mostly in the north) is covered by forest. Using coal and wood for heat generation leads to serious air pollution.

People living in steppe, Gobi and desert areas face serious fuel shortages. The government of Mongolia has given top priority to developing the energy sector as the main electricity grid covers only 30 percent of the total land area, supplying power to about 1 million people.

Mongolia's hydropower potential is stagnant, due to a lack of funds for the implementation of large-scale hydropower projects. Currently, hydroelectricity is produced at five small hydropower plants in the western region of Mongolia.

Risk management and responses

The central and northern parts of the country are prone to floods during the periods of heavy rain. The inhabitants of yurt settlements are the most affected, as they are usually located in flood-prone areas. Floods cause greater economic damages when they take place in densely populated areas. For example, in July 1966, the water level of the Tuul River increased by 3 m. This flooded the industrial region of Ulan Bator, claiming the lives of 130 people and causing US $7.5 million in economic damages (UNEP, 2002).

Due to low average rainfall, drought is very common, especially in the desert-steppe zone of country, where droughts up to three consecutive years have been recorded. The biggest impact of drought is definitely on the agriculture sector, including animal husbandry. For example, in the central and southern regions of the country, droughts are frequently observed during the first stage of the growing period (UNEP, 2002). As a result, crop cultivation is becoming more and more dependent on large-scale irrigation schemes.

Unfortunately, neither flood nor drought prevention measures are organized in a systematic manner. In the case of floods, communities lack the advantage of early warning systems. Furthermore, there is a definite lack of public awareness.

Water resources management

The Government recognizes that conservation of water resources is of primary importance for the long-term development of the economy. This is reflected in the terms of reference of the National Water Programme, which aims to ensure sustainable development of the country by the efficient use and protection of water resources. In 2000, the National Water Committee (NWC) was established with the purpose of coordinating and monitoring the National Water Programme's implementation. It serves as the coordinating body of a number of ministries and local governments. However, there are no resources allocated for the realization of the National Water Programme. Furthermore, no specific milestones were identified. As a result, the NWC struggles to coordinate the actions of several ministries within the fragmented management scheme of water sector.

The legislative and regulatory framework for the use of water resources is in place and updated when necessary. For example, the Water Law, which was adopted in 1995, was amended in 2004 to integrate river basin management practices (including the establishment of enhanced water resources information systems, the development of river basins management plans and the establishment of river basin organizations) with the goal of better utilizing water resources while protecting ecosystems. The Water Law also recognizes the economic value of water, requires capacity-building in the water sector, focuses on the decentralization of water management, puts forward the need for environmental impact assessments and sets new penalties for violating water legislation. However, the provisions of the law are vague and open to interpretations by different sectors. Furthermore, although the newly amended law foresees provisions for IWRM, public involvement at the local level is missing. Therefore, developed policies and programmes lack any public ownership. Facilitating the involvement of water users and stakeholders in managing the allocation of water resources remains a challenge.

Water-related policies and programmes developed at the national level often do not reach the local level. Policy implementation and monitoring mechanisms are also strained. At the institutional level, financial and human resource capacity is limited. The coordination of numerous institutions at national and local levels is missing, and the division of responsibility is not clear. Due to financial limits, laws and regulations are not adequately enforced.

Ulan Bator and the surrounding settlements located upstream of the Tuul River Basin are the biggest water users. However, no management plan currently exists for the water resources of the Tuul River Basin.

Mongolia's pricing policy is decentralized; local authorities are entitled to set up and revise the water tariffs. Although in theory, the Mongolian Government gives priority to the interests and water needs of the poor and marginalized, in practice, the current pricing scheme has become pro-industry and pro-wealthy due to weak regulations. Water tariffs for the mining industry are about US $0.006 per 1,000 L, whereas small businesses pay about US $0.48 per 1,000 L (eighty times more). For metered apartment users, a fixed rate of between US $1.5 and $7.5 per month is charged per inhabitant. The rate for yurt consumers, similar to small businesses, is eighty-four times higher than for industries and mining companies. As a result, those with the lowest income pay the highest and consume the least.

Conclusion

After the fall of the Soviet Union, Mongolia has been going through a profound economic and political transition period. Poverty is on the rise, only a limited portion of the population has access to safe water, sanitation facilities are poor, the quality of water resources are decaying, water-related diseases are common, and health services are out of reach for the poor. These problems are further accentuated by water scarcity, a very cold climate and recent disasters. The Government of Mongolia is committed to implementing reforms in water resources management and environmental protection, but due to lack of financial resources and the limited number of trained personnel, policies cannot be implemented, and laws and regulations cannot be enforced. Improving the implementation of

legal frameworks and policy coordination in the water sector are dire necessities. Sectoral interests have prevented the adequate protection of water resources and the environment. The decentralization of water pricing has promoted economic growth by providing low-cost water to business and industry but has disregarded the needs of the poor.

12. La Plata River Basin

The La Plata River Basin is the fifth largest river basin in the world, extending over 3.1 million km², and its surface area is second only to the Amazon River Basin in South America. It covers an extensive part of central and northern Argentina, southeast Bolivia, almost all the southern part of Brazil, the whole of Paraguay and a large part of Uruguay (see Map 14.12).

With over 100 million inhabitants, close to fifty major cities, seventy-five large dams and an economy that represents 70 percent of the per capita GDP of five countries, the basin has enormous economic and social importance for the region overall.

The La Plata River Basin has four main sub-basins: the Paraná, Paraguay and Uruguay River systems and the La Plata River sub-basin itself. The Paraná River system is the biggest of the three, constituting 48.7 percent of the basin's overall surface area. The Paraguay and Uruguay River systems respectively comprise 35.3 percent and 11.8 percent of the basin. The remaining 4.2 percent corresponds to the La Plata River sub-basin itself.

In terms of discharge, the Paraná River System is the most important in the basin, with a mean annual flow of about 17,100 cubic metres per second (m³/s) at Corrientes.[8] The Uruguay River system has a mean annual flow of about 4,300 m³/s, while the Paraguay River System has the lowest capacity with a mean annual flow of approximately 3,800 m³/s at Puerto Pilcomayo.[9]

Long-term measurements over a large part of the La Plata Basin show certain trends in climate and rainfall patterns. For example, annual minimum temperatures are increasing by about 1°C per century. Furthermore, hydrological records show evidence of an increase both in rainfall and runoff in the La Plata Basin after 1970. El Niño has also had an impact on stream flows in the basin. For example, in the middle section of the Paraná River, the four largest discharges on record followed the four El Niño events of 1905, 1982–1983, 1992 and 1998. In 1982 and 1983, more than 40,000 people were affected in more than seventy towns along the reach of the

Map 14.13: Overview of La Plata River Basin

Source: Prepared for the World Water Assessment Programme by AFDEC, 2006.

Uruguay River within the Brazilian state of Rio Grande do Sul. Severe flooding, with extensive damage to infrastructure and economic production, are frequent occurrences, especially in the Paraná and Uruguay sub-basins. The Paraná River and its tributaries have many riverside towns that are frequently flooded. This is the case in the Argentinean cities of Resistencia, Corrientes, Rosario, and Santa Fe. In the La Plata Basin as a whole, losses associated with El Niño events were estimated at more than US $1 billion.

Water and environment

Thanks to climatic conditions, rainfed agriculture is common in the basin. In fact, the proportion of irrigated land to the overall agricultural area is relatively low, varying between 0.3 percent (in Paraguay) to 16.8 percent (in Uruguay). Soybean, maize and wheat are widely produced in the basin, and animal husbandry and fisheries are other important sources of food and income.

However, soil loss from agricultural areas and organic and chemical contamination stemming from agriculture and animal husbandry are also

---

8. Located on the left bank of the Paraná River (Argentina), after its confluence with the Paraguay River.
9. Located on the right bank of the Paraguay River (Argentina), after its confluence with the Pilcomayo River.

sources of pollution. Furthermore, pollutants and heavy metals from mining operations and inadequately treated urban sewage are other causes of environmental concerns in the basin. Increased tourism is also leading to overfishing, damage to flora and fauna and the illegal exportation of endangered species. In addition, it is argued that the Hidrovía project, which has been proposed as a means of facilitating the transport of agricultural products, might negatively affect the extremely rich array of wildlife in Pantanal ecosystem (see **Box 14.9**).

With the support of the World Bank, the Government of Brazil initiated the Pantanal Project in 1991. The resulting Plan for the Conservation of the Upper Paraguay River Basin (PCBAP, Plano de Conservação da Bacia do Alto Paraguai) employed an environmental zoning approach to delineate general and site-specific guidelines for the conservation, rehabilitation and preservation of degraded lands; created a geographic information system (GIS) to facilitate the dissemination of available physical, biological, social, legal and economic information; and proposed the operation of a real-time flood warning system.

Environmental problems are not only limited to Pantanal. In the last few decades, rapid population growth, road development, expanding agricultural frontiers, mining and large-scale hydraulic engineering (including dams, waterways, and irrigation projects) have resulted in a decrease in the overall quality of the basin environment and created ongoing problems, such as siltation of waterways and resevoirs, intense deforestation and degradation (e.g. in the Chaco semi-arid woodland), tropical forest loss and fragmentation of the rainforests of Argentina, Brazil and Paraguay.

Severe erosion on the eastern slopes of the Andes has caused agricultural land loss in Bolivia and Argentina, as well as a devastating siltation process on the Bermejo and Paraguay rivers, which extends into the Paraguay, Paraná and La Plata rivers. A large part of fertilizers and pesticides used in farming are carried by runoff into watercourses. This toxic pollution puts the populations that depend on the rivers' productivity for their livelihoods at risk and threatens the biodiversity of the maritime front of the La Plata River.

Levels of poverty

Important economic crises at the beginning of this decade have affected all basin countries and had a negative impact on the success of poverty reduction strategies. The poverty rates in Argentina and Uruguay decreased rapidly in the 1990s and then increased again between 1999 and 2002, more than doubling in Argentina. On the other hand, poverty decreased rapidly in Brazil between 1992 and 1995 and has remained more or less stagnant since then. Lately, poverty has also been on the rise in the urban areas of Bolivia and Paraguay.

People with low incomes often live in informal settlements established in marginal areas where safe water and sanitation infrastructure is either insufficient or non-existent, increasing the percentage of people suffering from diseases that stem from a lack of water and sanitation. Problems related to informal settlements in the region have aggravated in the last twenty years.

Table 14.2: Percentage of urban and rural populations with access to drinking water and sanitation services

Countries	Safe water (%)		Sanitation (%)	
	Urban areas	Rural areas	Urban areas	Rural areas
Argentina	85	30	89	48
Bolivia	93	44	82	35
Brazil	96	65	94	53
Paraguay	70	13	85	47
Uruguay	99	93	95	85

Access to safe water and sanitation

Access to safe water and sanitation varies significantly between urban and rural areas in the La Plata Basin. In all the countries, urban areas have better access to safe water and sanitation services than rural areas. In fact, between 70 and 99 percent of the urban population has access to safe water and 82 to 95 percent has access to sanitation services. Meanwhile, in rural areas between 13 and 93 percent of the population have access to safe water and 35 to 85 percent have access to sanitation services (**Table 14.2**).

Water and health

Biological contamination stemming from a lack of proper sanitation infrastructure and inadequate wastewater treatment facilities constitutes a severe problem in several urban and rural settlements in the basin. Consequently, occurrences of waterborne diseases such as diarrhoea, cholera, malaria and dengue are quite common in certain regions. Other diseases of fewer occurrences are leptospirosis, leishmaniasis and yellow fever. Diarrhoea is by far the most common waterborne disease, affecting children especially. In 2003, in Argentina alone, over 900,000 people suffered from diarrhoea.

In different zones of Argentina, among several other Latin American countries, the population has to constantly utilize water resources with naturally high arsenic content, surpassing acceptable limits for drinking water standards. Arsenic is a naturally occurring element found in the earth's crust (see **Chapter 6**). Water resources that are in contact with rock layers that are rich in arsenic get polluted by this carcinogenic mineral. Arsenic is also utilized in some industrial processes and can leak into water bodies if not handled with care. Significant efforts are made to minimize or eliminate arsenic contamination in urban drinking water by chemical treatment. However, although low-cost methodologies for removal of arsenic at home level have been tested with success, some rural communities continue using groundwater resources that are contaminated with arsenic.

Water and industry

The La Plata River Basin comprises a great potential for economic activities. In this regard, there are several industrial centres in the five countries situated along numerous tributaries. However, the most significant industrial centres are located in Brazil, in the São Paulo metropolitan area, and in Argentina, along the industrial strip on the Paraná River and in the metropolitan area of Buenos Aires. Mining sector is prominent at the upper part of Paraguay River sub-basin and in Bolivia close to the Pilcomayo River tributaries.

The greatest industrial water demand occurs in the Paraná River System (20 percent), where the industrial sector is mostly concentrated. In the La Plata River sub-Basin, almost 98 percent of the water abstracted is utilized along the Argentinean bank by factories located in the Buenos Aires metropolitan area.

The industrial centres, although providing employment and contributing to the national GDP, are a source of pollution in the basin. Consequently, depending on the type and extent of industrialization and the absorption capacity of rivers, the level of contamination varies in the four river sub-systems. However, there are localized cases of contamination where big urban settlements, industrial zones and mining pits are located. In order to curb industrial pollution, the five countries of the La Plata River Basin are drawing policy guidelines and implementing programmes for promoting cleaner industry.

Water and energy

Growing population and industrialization necessitate an increase in energy production. Given an estimated potential of 92,000 MW, the production of hydropower has thus become a regional priority. So far, approximately 60 percent of this potential has been put to use.

More than 90 percent of the energy used by Brazil comes from hydropower, the greater part of which is generated by dams on the Paraná River and its tributaries. By taking into consideration the environmental and social impacts that the dams cause, a financial compensation tax has been placed on the hydroelectric sector for the utilization of water resources. By law, 6 percent of the value of the electric energy produced is channelled back to the areas where the facilities for energy production are located or areas that have been flooded due to the dam reservoirs. Furthermore, a certain percentage of these funds is allocated to the Ministry of Environment for the implementation of the National Water Resources Management System.

Hydropower development projects are not only national in character. Riparian countries have implemented joint projects such as the Salto Grande (Argentina and Uruguay), Itaipú (Brazil and Paraguay) and Yacyretá (Argentina and Paraguay) dams to further improve their energy production.

Sharing water resources

There are substantial underground water resources in the La Plata River Basin. The Guaraní Aquifer System (GAS), for example, is one of the world's most important fresh groundwater reservoirs, due to its extent and volume. It underlies portions of Argentina, Brazil, Paraguay and Uruguay, stretching over an area of approximately 1.2 million km^2, with almost 15 million inhabitants. The capacity of the GAS is estimated at around 40,000 km^3.

BOX 14.9: THE PARAGUAY-PARANÁ WATERWAY (HIDROVÍA) PROJECT

The waterways of the La Plata Basin have been navigated since the early sixteenth century. The Paraguay and Paraná Rivers are natural transport corridors extending in a north-south direction, connecting the heart of South America to the Atlantic Ocean. Although they remain an important transportation artery linking the five riparian countries, the continuous maintenance of those waterways poses a problem to riparian countries.

The Hidroviá waterway, as planned, runs from Puerto Cáceres (Brazil) in the north to Nueva Palmira (Uruguay) in the south, following the Paraguay and Paraná rivers over 3,000 km. The aim of the project is to expand the five countries' navigation possibilities, promote development of the region by reducing the cost

of transport of goods and improve links with commercial centres, while granting an outlet to the sea to landlocked Bolivia and Paraguay.

However, the project's construction and operation may have a number of severe and complex impacts on the environment of the region, particularly the Pantanal, an immense plain located in the Upper Paraguay River Basin, considered to be one of the world's largest wetlands (with an estimated area of 140,000 km^2). This large and rich ecosystem, which has so far remained relatively untouched, might be seriously damaged due to a significantly modified flow regime, whose repercussions might not only be limited to a decrease in biodiversity but might also lead to significant

changes in water levels at the confluence of the Paraná and Paraguay rivers. Other concerns include the alteration of natural aquifer systems, increased water contamination due to expected growth in local populations and increased commerce, industry and irrigation.

As a result, this project is under debate by scientists and conservation organizations. A more in-depth environmental impact assessment is necessary to address the various social, environmental and economic aspects of this development project.

Sources: Modified from Bucher and Huszar, 1995; Gottgens et al., 1998; and Petrella and Ayuso, 1996.

A joint project is currently under way to support GAS countries in implementing a common institutional framework for managing and preserving the GAS. The project also aims to expand and consolidate the current knowledge base through monitoring and evaluating water resources, in order to promote stakeholder participation in decision-making and control pollution.

An international legal framework for the management of transboundary groundwater resources currently does not exist. However, transboundary groundwater management is necessary in regions that are subject to water scarcity and fierce competition among users (see **Chapter 11**). In the case of the GAS, helping to shape an institutional framework regarding transboundary groundwater can make a contribution that could serve as a potential model for other countries and regions.

With respect to surface water, the main concern is the sustainability of the resources in the long term. For this purpose, many bi- or multilateral projects are currently in progress. Some examples are the integrated management and master plan of the Pilcomayo River Basin (Argentina, Bolivia, and Paraguay) and the strategic action programme for the Bermejo River Binational Basin (Argentina and Bolivia). Through these projects, basin countries aim to promote better utilization of water and land resources while conserving and rehabilitating ecosystems. These projects also facilitate information exchange in addition to providing a basis for strengthening regional information systems.

Managing the resource: Institutional frameworks

The first step towards the initiation of a comprehensive basin-wide study was taken in 1967, as a decision of the Ministers of Foreign Affairs of the five countries. The direct outcome of this decision was the establishment of the Intergovernmental Coordinating Committee of the Countries of La Plata Basin (CIC). In 1968, the committee was entrusted to draw up a treaty in order to enforce the institutionalization of the basin. Approved in 1969, this treaty provides the basis for further bilateral and multilateral agreements concerning jurisdictional matters, navigation, fishing, pollution prevention, scientific research, etc.

Currently, the CIC has a new 'Program of Action' and is preparing a Framework Programme, with the support of the Global Environment Facility (GEF), through the United Nations Environment Programme (UNEP) and the Organization of American States (OAS), in order to implement the environmentally and socially sustainable economic development of the La Plata Basin, specifically by protecting and managing its water resources and adapting to climatic change and variability.

Conclusion

The La Plata River Basin is the fifth largest river system in the world and has enormous economic and social importance for the region.

Due to a series of economic crises affecting the countries of the basin, rising poverty remains the most important social issue to be addressed. Given limited funds allocated, meeting safe water and sanitation needs of the people, and especially the poor, is a big challenge. As a result, water-borne diseases stemming from lack of water and sanitation continue to be among one of the major causes of morbidity in the basin.

The basin is blessed with a rich array of wildlife and extensive ecosystems, however, rapid population growth, expanding industrial, agricultural and mining activities and large-scale hydraulic engineering have caused extensive environmental deterioration in the basin.

Basin countries share the common vision of sustainable development through bilateral and multilateral cooperation in utilising the extensive surface and groundwater resources in an optimal fashion. For this purpose, many joint projects are currently in progress. through these projects, basin countries aim to promote better utilisation of water and land resources while conserving and rehabilitating ecosystems. The La Plata Treaty, based on a basin-wide institutional framework, provides the foundation for such efforts towards the protection and integrated management of water resources and adaption to climate change and variability.

13. South Africa

South Africa is the fourth largest country in Africa, with a surface area of 1,219,090 km^2 and an estimated population of 48 million. Average rainfall is 450 mm per year, but can vary significantly from less than 100 mm along the west coast to more than 1,000 mm on the east coast and along the escarpment. In the interior, seasonal rivers generate 27 percent of the runoff from 54 percent of the surface area, while in the west 24 percent of the surface area is drained by episodic rivers without any significant contribution to the runoff. The natural mean annual runoff (MAR) is about 49,000 million m^3 per year, of which only 27 percent is currently available as reliable yield. Due to the high temporal and spatial variation in rainfall, high evaporation and the location of water users, the remaining economic development potential is only 5.4 million km^3 per year (11 percent of MAR). Accordingly, per capita water availability is approximately 1,060 m^3 per year (based on MAR), of which the utilizable portion is only 300 m^3 per person per year. South Africa's existing water resource availability comprises 77 percent surface water, 9 percent groundwater and 14 percent re-use of return flows.

Water and Ecosystems

South Africa is home to six world heritage sites, sixteen international Ramsar sites, twenty national parks and over 500 terrestrial reserves. However, ecosystems are under the threat of extensive land use, urbanisation, industrialisation and water resources development. Although over 50 percent of wetlands have been destroyed for land-use changes, such as a reduction of 40 percent of the area of Mfolozi Swamp (South Africa's largest fluvial plain), South Africa has successfully managed and protected numerous other wetland conservation areas – many with increased biodiversity thanks to environmental and natural resource management policies and practices. For example, the National Water Act of 1998 includes formal provisions for the protection of the aquatic ecosystems, including the classification of water resources, determination of the Reserve and setting of resource quality objectives. Consequently, no water use is licensed without first determining the Reserve (environmental requirement) and its possible impacts on the functioning of ecosystems.

Map 14.14: Overview of the river basins of South Africa

Source: Prepared for the World Water Assessment Programme by AFDEC, 2006.

Water and settlements

South Africa is an urbanizing nation. Approximately 28 million people (59 percent of the overall population) live in more than 3,000 urban communities, including informal settlements. The nine largest cities are home to 16 million inhabitants (37 percent of the national population) and provide 50 percent of the nation's work force. These cities cover only two percent of the overall surface area of the country.

The rapid growth of informal urban settlements presents a major challenge. According to recent statistics, approximately 5 million people (28 percent of the urban population) live in such settlements without proper water services infrastructure. Some informal settlements are located along river reaches, which exposes them to waterborne diseases and makes them vulnerable to flooding. The fast growth of settlement development poses major challenges to municipalities and service providers. Extensive effort and funding is directed to water services, housing and integrated programmes to address urbanization and the creation of sustainable human settlements. This goes hand-in-hand with social development programmes and associated job creation.

The rural population (about 20 million people or 41 percent of the total population) also presents a major challenge for ensuring sustainable livelihoods. Although groundwater represents only about 9 percent of available water resources, 74 percent of South African rural communities are dependent entirely on groundwater, while another 14 percent depend partially on it.

South Africa has initiated a comprehensive basic water services programme to provide effective, affordable and equitable water services to all (see **Chapter 6**). South Africa successfully achieved the MDG target of halving the proportion of people lacking access to safe water by 2015 (reduced from 40 percent to 19 percent since 1994). The remaining challenge is to address the remaining 9 million people who still lack access to water supply, 64 percent of which live in rural settlements. In addition, 16 million people are still without acceptable basic sanitation facilities, 56 percent of which are rural inhabitants. The scattered nature of rural settlements presents major challenges for providing sustainable services.

Poverty is a profound socio-economic challenge in South Africa, which particularly affects female-headed households and rural inhabitants. Over 34 percent of the population live on less than US $2 per day and 70 percent of the country's poorest households live in rural areas.

Water and health

Nineteen percent of the population still lack access to safe water, and 33 percent lack basic sanitation services. Public institutions also suffer from a lack of access to safe water and sanitation services: 59 percent of all schools (over 16,000) and clinics (over 2,500) lack access to acceptable sanitation facilities, while 27 percent of all schools (over 7,500) and 48 percent of all clinics (over 2,000) lack access to safe water supply.

Water-related epidemics and diseases such as diarrhoea, cholera, dysentery, hepatitis and schistosomiasis occur in South Africa. In 2004, there were 2,780 cholera infections with 35 fatalities and 9,503 hepatitis A infections with 49 fatalities. Malaria is also a common disease; between 2000 and 2004, there were 77,854 reported incidences of malarial infection, 875 of which ended with fatality. Each year, some 2.5 million people fall ill with bilharzia (schistosomiasis), of whom about 10 percent are 'severely infected', although few die of the disease. Schistosomiasis infections are highest (up to 70 percent) amongst children living in lower-lying areas of Limpopo Province, Mpumalanga and KwaZulu-Natal. Infections of *shigella dysenteriae* vary significantly; there was a total of 894 cases recorded in 2004. Diarrhoea is a result of various types of bacteriological, viral and parasitic infections, affecting more than 3 million patients each year and causing up to 3 percent of annual deaths (over 15,000 deaths), of which not all are directly related to water. The high prevalence of HIV/AIDS is also a major concern (see **Chapter 6**). These diseases significantly affect the economic productivity and social activities of affected households.

The free basic water (FBW) programme to ensure access to effective water supply and sanitation services, in association with access to health facilities and services, plays a major role in addressing water-related diseases and improving the health and quality of life of all people. The South African Government is also promoting improved hygiene practices through national programmes, campaigns and education awareness at schools and in communities. Furthermore, the provision of free basic water services to the poor has become a national policy since 2000. This programme aims to ensure that poor households receive 6,000 litres of FBW per month. Beyond this basic allowance, users must pay for their consumption. The programme is progressively implemented by the Water Services Authorities, and over 76 percent of the population already receives access to FBW. Progress is also being made to ensure access to water supply and sanitation services at all schools and clinics.

Water for food

South Africa's agricultural production is limited by the availability of natural resources (soil, climate and water), as only 15 percent of the country is suited for conventional cultivation. Approximately 10 percent of the surface area is currently cultivated. In spite of the semi-arid climate, over 70 percent of crop production is rainfed, and less than 30 percent is produced with irrigation. In total, agriculture accounts for over 60 percent of total water utilization. Primary and processed agriculture currently contribute 15 percent to the GDP. Furthermore, agriculture is the main economic activity of rural areas and a major supplier of jobs, with almost 40 percent of poor households involved in agriculture for food or cash crops.

South Africa is self-sufficient in most major food crops and a main exporter of food to neighbouring states, largely through production on large commercial farms. In addition to this, approximately 1.6 million households (35 percent of rural households) depend on their own farming and food production abilities to meet their nutritional needs. The Government is giving increasing attention to social programmes in poverty-stricken areas to reduce the vulnerability of poor households. An example of this is the development of food plots and vegetable gardens for poor communities. Furthermore, institutional reform is underway to facilitate equitable access to water resources and representation in water management institutions.

Water and energy

South Africa is the world's sixth largest producer of coal, which constitutes about 74 percent of the country's total energy consumption. Coal-fired thermal plants supply 93 percent of the country's total energy requirements. The energy sector has a considerable share in the economy as it employs about 240,000 people, which corresponds to approximately 1.5 percent of the economically active population and contributes 13 percent to the GDP. South Africa is the largest energy consumer on the continent, with an electricity consumption equivalent to two-thirds of the overall electricity use in Africa. To ensure the security of electricity supply, the energy sector receives a preferential allocation of water resources, while it only accounts for 2 percent of national water use.

Hydropower plants contribute only 0.4 percent to current electricity generation. This is mainly due to limiting factors like the quantity and variability of surface water. However, South Africa has prioritized hydropower development for managing electricity demand peaks and as a development potential within the Southern African Development Community (SADC). The SADC is a regional economic community that aims to combat poverty, ensure food security and promote industrial development through the integration of regional economies. Within the context of the SADC, the hydroelectric potential of the Zambezi and Congo rivers are explored in collaboration with Mozambique, Zimbabwe and Democratic Republic of Congo. Given the variability of surface water runoff, South Africa relies on dams and transfer schemes to ensure access to water in locations of high economic activity. In fact, the total storage capacity currently represents 66 percent of the total MAR.

To serve growing energy demands, South Africa is currently planning additional nuclear and thermal power generation schemes as well as some smaller solar and hydropower plants. In line with the national water resource strategy, water conservation measures have been introduced in the energy sector, resulting in savings of up to 40 percent of water consumption per energy unit generated.

Water and industry

The industrial sector is the most prominent and the highest growing economic sector in South Africa. It generates 29 percent of the GDP and 54 percent (including mining) of all exports and employs over 25 percent of the total work force. Together, industry and mining account for 11 percent of the total water use in South Africa. Many industries and mines are located far from available water resources, which necessitates extensive water infrastructure developments to transfer water from other basins, sometimes in neighbouring countries.

These sectors are major impactors on water resources, and as a result, pollution control and water resource quality management receive high priority. Specific attention is given to effective use and conservation measures through various legislations such as the 1998 National Water Act.

Water management and risk mitigation

In South Africa, water is governed by the National Water Act and the Water Services Act, supported by a dedicated Minister, a National Department of Water Affairs and Forestry, as well as various water institutions at various levels. Legislation in South Africa recognizes water as a national asset and a strategic resource for economic and social development. It also recognizes the need to protect the environment and ensure quality of life. IWRM has been adopted to achieve these goals. To date, four of the nineteen planned Catchment Management Agencies have been established. A further 170 municipalities have been given the responsibility to act as Water Services Authorities. This is supported by various infrastructure, finance, capacity-building and management programmes. Currently almost 3 percent of the national budget is allocated to water governance, and additional funds are provided for specific water-related programmes and infrastructure development.

Disasters like drought, floods, fires and epidemic outbreaks of diseases are common in South Africa. To manage these risks, South Africa has adopted a proactive planning and management approach. Through the national disaster management policy, institutional arrangements and early warning systems have been established.

Water resources and water services management is guided by a National Water Resource Strategy, catchment management strategies, integrated strategic perspectives and water services development plans. These are supported by institutional reform and the development of comprehensive regulatory frameworks. A key challenge is the development of appropriate skills and the capacity-building of newly established water institutions. South Africa is also in the process of establishing a comprehensive integrated monitoring framework for water resources and water services. These various governance initiatives instill an effective, participative and sustainable water management culture in South Africa.

Sharing water

There are four major transboundary river basins in South Africa, encompassing 65 percent of the country's surface area, 72 percent of the population and 40 percent of available water resources. These are the Limpopo (South Africa, Botswana, Zimbabwe, and Mozambique), the Komati (South Africa, Swaziland, and Mozambique), the Maputu/Usuthu (South Africa, Swaziland and Mozambique) and the Orange basins (South Africa, Lesotho and Namibia).

South Africa subscribes to IWRM and therefore promotes equitable allocation of water resources among uses and users at both the national and international level. Accordingly, South Africa has entered into nineteen international water agreements and treaties with its neighbouring countries, such as the revised protocol on shared water resources of the SADC.

South Africa is also sharing water across national river basins. The uneven distribution of water availability in South Africa in relation to the location of the country's economic growth centres necessitated the building of various inter-basin transfer infrastructure to facilitate optimal water utilization and sharing between economic sectors and stakeholders.

Ensuring the knowledge base

South Africa's strong science base is well reflected in the water sector. However, the science and technology disciplines have a much skewed human resource base, with only 20 percent of the country's science, engineering and technology capacity constituted in the previously disadvantaged population groups. The knowledge base of the water sector, at present, reflects a similar disparity. For this purpose, various capacity-building programmes have been directed towards local authorities and water institutions. These initiatives include IWRM programmes, as well as other specific training, support and capacity-building programmes. This is further promoted by a participatory and cooperative governance approach, as well as various information, knowledge and advisory systems.

Water research plays a major role in establishing and maintaining the knowledge base. The Water Research Act of 1971 established the Water Research Commission (WRC), mandating it to coordinate and support water research, using funds from a dedicated Water Research Fund. Besides the direct impact on water resources, governance, management and development, these research projects also play a major role in capacity-building: more than 100 Masters and thirty PhD degrees were awarded in 2004.

Research is also carried out by universities and other institutions, as well as government-subsidized science councils, including the Council for Scientific and Industrial Research, the Agricultural Research Council and the Human Sciences Research Council.

Conclusion

In South Africa, water scarcity is a limiting factor for development. The value of water is therefore high in all aspects of society, the economy and the environment.

South Africa is a country emerging from a history of political oppression to become a nation of democratic values for human dignity, equality and freedom. Poverty is the foremost social concern, and the government aims to address the needs of the poorest in society by ensuring access to basic services through dedicated programmes for infrastructure and free basic water services. The social value of water is founded in the desperate need of the 3.6 million people (8 percent of the population) who currently do not have access to any water supply infrastructure, and the 9 million people (39 percent of the population) who do not receive minimum basic water supply services.

Securing household food security is a common concern, as many families live a subsistence lifestyle and depend on rainfed irrigation to produce their own food. Although irrigation plays a strategic role in providing food security during dry years, water scarcity impedes irrigation on a wider scale.

The industrial and mining sectors are major contributors to the overall wealth of South Africa. To support economic growth and development, South Africa must reconcile the growing demands of the different uses with limited water availability, while ensuring the sustainability of ecosystems.

To address these challenges, South Africa has undertaken a comprehensive policy and legislative reform and is in the process of implementing these through various national programmes. This goes hand-in-hand with institutional reform and capacity-building programmes in order to ensure that IWRM is implemented and sustainable effective service delivery is ensured.

14. Sri Lanka

Over a surface area of 65,600 km², 19.5 million people live in Sri Lanka. Water bodies, a considerable portion of which are man-made, cover about 4 percent of the land. The terrain of the island is mostly made up of coastal plains, with mountains rising only in the south central part.

Sri Lanka has more than 100 water basins, varying from 10 km² to over 10,000 km² in size. The Ruhuna Basins,[10] which are located at the southern part of the island, were featured in WWDR1 (see **CD-ROM**).

The climate of Sri Lanka is tropical and heavily influenced by monsoons that bring rain throughout the year. The mean annual rainfall volume is approximately 120 km³. Rainfall totals range from under 1,000 mm to over 5,000 mm. Sri Lanka's groundwater resources are considered minor compared to its surface water resources. The estimated groundwater potential in Sri Lanka is 7.8 km³ per annum and is widely used for domestic, small-scale irrigation, industrial and other uses. However, in recent years, due to increased irrigation and population growth, both shallow and deep aquifers have been subject to over-extraction. Consequently, the drying up of domestic wells during dry periods has become more common.

Water and ecosystems

There is a rich diversity of ecosystems in Sri Lanka, including wetlands, natural forests and marine and coastal ecosystems. Sri Lanka is considered one of the world's twenty-five 'Biodiversity Hotspots' (i.e. very rich in biodiversity). Overall, there are three Biosphere reserves, one World Heritage site, three Ramsar sites (see **Chapters 5** and **12**) and

forty-one wetland sites included in the Asian Wetland Directory. Coastal ecosystems are diverse, but their fragmentation, in addition to that of forests, is extremely high (UN, 2002). In 1999, the Government imposed a ban on logging in all natural forests in order to curb deforestation.

Sixteen of Sri Lanka's coastal lagoons are classified as threatened and constitute nearly half of the country's threatened wetlands. Environmental degradation of the coastal zone is a major hazard faced by Sri Lanka as an island state. During the last two decades, there has been increasing pressure for development in the coastal zone, particularly for tourism and recreational purposes, near shore fisheries, fish farming, industrial development and housing. Communities have exploited the use of natural resources, such as sand and coral, on a commercial basis. Development pressures have also led to the reclamation of estuarial, lagoon and marsh waters and the unrestricted disposal of untreated sewage, leading to major pollution problems. The main threat to natural ecosystems, however, is population growth and migration, reducing the available habitat for ecosystems to thrive. Some other threats to the island's biodiversity are natural disasters, soil erosion, sedimentation and large-scale sand mining.

10. See www.unesco.org/water/wwap/case_studies/ruhuna_basins/ for more information.

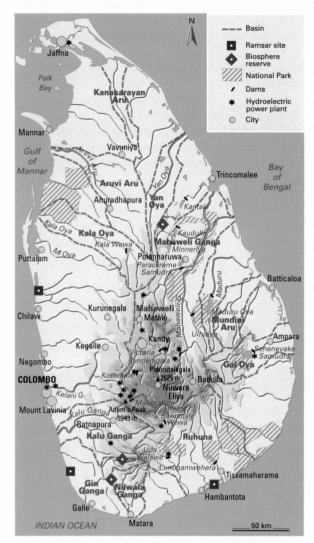

Map 14.15: Overview of the river basins in Sri Lanka
Source: Prepared for the World Water Assessment Programme by AFDEC, 2006.

In order to address these concerns, the Ministry of Environment was established in 1990. This was followed by the creation of two legal instruments, namely the Environmental Impact Assessment (EIA) and Environment Protection Licence (EPL). All approved development projects are required to obtain EIA and EPL clearance. These legal instruments ensure the integration of environment protection measures into development projects at the early stages of planning. Environmental concerns are being addressed through ongoing government programmes

with active support from NGOs. However, the limited annual budget of the Ministry of Environment necessitates external sources of funding for any substantial progress to be made (UN, 2002).

Poverty

Over 45 percent of the population lives on less than US $2 per day, and the percentage of the population living below the national poverty line is 19 percent. The estimated number of preschool children suffering from stunting[11] is 330,000, while 210,000 are wasted[12] and 540,000 are underweight.[13] Furthermore, 620,000 preschool children and about 36 percent of pregnant women were found to be anaemic. Although the nutritional status shows a slight improvement from the 1990 levels, it is necessary to enhance food production in the country and improve the accessibility of food for the poor and other vulnerable populations at the household level. This can be done through the efficient mobilization and equitable allocation of land and water for food production.

Water and health

Approximately 78 percent of the population in Sri Lanka is rural. Although the availability of safe drinking water varies, on average, 75 percent of the rural population and 95 percent of the urban population has access to safe water resources. Furthermore, the percentage of households with safe sanitation is between 85 and 90 percent. However, it should be noted that there are considerable population groups in the peripheral areas of urban centres where water supply and sanitation facilities are generally poor.

About 70 percent of the urban population is served by piped water systems, whereas drinking water for rural populations are mainly supplied by dug and tube wells. However, studies have shown that deep wells in fractured crystalline rock yield water containing excessive amounts of fluoride, which can lead to dental problems (fluorosis) among children (see **Chapter 6**). The Government aims to provide 100 percent of the urban population with a piped water supply by 2010 and all the major urban areas with piped sewerage systems by 2015. The target is to provide safe drinking water for the entire population by 2025. Diarrhoeal diseases are still one of the ten leading causes of hospitalization in Sri Lanka. In recent years, there were several outbreaks of diarrhoeal diseases, but better medical practices have helped to decrease the diarrhoea-related mortality rate. Vector-related diseases, such as malaria and Japanese encephalitis, still continue to be a major public health problem and socio-economic burden in Sri Lanka. However, effective vaccination programmes and spraying of host areas have reduced the number of incidences.

Water and food

For the past 2,500 years, Sri Lanka has remained mainly rural and agriculture-based. Agriculture is practised over 1,8 million ha of land, or 28 percent of overall surface area. Rice is the major agricultural crop, and paddy land, 80 percent of which is under irrigation, covers 40 percent of agricultural land. It is estimated that approximately 85 percent of the overall water extraction is used for agriculture.

---

11. A child is stunted if his or her height-for-age is two standard deviations or more below the median of the reference population.
12. A child is wasted if his or her weight-for-height is two standard deviations or more below the median of the reference population.
13. A child is underweight if his or her weight-for-age is two or more standard deviations below the median of the reference population.

Women make up 40 percent of the agricultural workforce, but are rarely allowed to take part at the decision-making level, serving instead mostly as cheap labour. Studies in minor irrigation systems show that 75 to 85 percent of women are involved in considerable physical activities, such as land preparation, on-farm water management, sowing, transplanting, harvesting and selling products. More than 70 percent of the women agricultural workers render their services without receiving anything in return. On the other hand, there are few female landowners (4 to 8 percent) who play major roles in agricultural activities.

National capital investment in irrigation and drinking water supplies, as a ratio of total capital investment, declined from 7.7 percent in 1993 to 2.2 percent in 2003. Since the 1980s, the irrigation sector has stressed better water resources management and planning, which partially justifies the declining trend of the proportion of public investment, especially in the construction and restoration of irrigation infrastructure. However, there is still need for investment in the water sector.

Considering the limited nature of Sri Lanka's water resources, demand management has become a necessity. Demand management in the irrigation sector involves the adoption of a cultivation calendar and an irrigation schedule. Active community participation in decision-making is customary in Sri Lanka, and farmers, through their institutions, participate in the planning processes with formal irrigation officials.

Water for energy

There are over 10,000 dams of varying sizes in Sri Lanka (of which eighty are classified as large dams), but hydropower contributes only 9 percent to annual energy production. The country's major energy sources are petroleum (41 percent) and fuel wood (50 percent). The contribution of hydropower to electricity generation has decreased from about 100 percent in 1990 to about 42 percent in 2000. This drop was mainly due to the fact that there was an inability to produce enough energy during frequent droughts, especially in the late 1990s and early 2000s. However, dams and reservoirs have helped to compensate for water deficiencies in dry areas, and the Government plans to pursue the development of all possible hydropower options in the future.

Almost half of the electricity consumed in Sri Lanka is used for domestic purposes. At the beginning of the year 2002, only 65 percent of the population had access to electricity from the national grid, although this ratio is scheduled to be increased to 77 percent by the end of 2006.

Water and industry

The pollution of surface and groundwater resources by industries is a grave concern. For example, Kelani River, which is the main source of drinking water for over 2 million inhabitants in the capital city Colombo, is polluted by industries. Groundwater pollution has also been detected in well-water in mixed residential and industrial areas. The Central Environmental Authority and local government authorities are given the responsibility for controlling industry-related water pollution.

Management responses

Demand management in the irrigation sector involves the adoption of a cultivation calendar and an irrigation schedule to optimize water use. Both irrigation officials and farmers take part in this joint effort. The utilization of different kinds of seeds (like the introduction of short-term rice paddy varieties) is another method for maximizing productivity. Community participation in irrigated agriculture is now common in Sri Lanka, although this was accepted as government policy only after the mid-1980s. Since then, farmer institutions have been included in the formal institutional structure. In the water supply sector, public participation in rural water supply schemes is substantial, and communities contribute by sharing costs and management roles.

Demand management in other sectors is mainly affected through the tariff structure. For example, per unit of industrial water use is priced up to six times higher than domestic water use charges. In allocating water among different sectors, water for drinking, sanitation, irrigation, ecology, environment and hydropower generation has priority over industrial and commercial water requirements.

Risk management

Floods, droughts and landslides are the most common and destructive types of natural disasters in Sri Lanka, with twenty-three droughts reported to the Department of Social Services between 1947 and 1992.

More recent droughts occurred in 1995-1996, 2001, 2002 and 2003. The disruption of livelihoods and productivity losses that occurred as a result of these droughts severely disrupted the Sri Lankan economy. During the 2001 drought, for example, the country faced power cuts for up to eight hours per day. In 2004, an estimated 52,000 ha of crops were damaged in seven districts, and the government had to appeal for assistance to provide food rations for over 1 million people during a six-month period (Ministry of Social Welfare, 2004).

Several basins in Sri Lanka are flood-prone. The most recent flood, which took place in May 2003, was one of the worst since 1947. It affected 139,000 families, completely destroyed 9,500 houses and claimed the lives of 250 people. The cost of the damage was estimated at US $76.8 million.

For drought and flood mitigation, a number of structural and non-structural measures have been taken. Reservoirs built for irrigation purposes also serve as flood protection and drought mitigation. Flood warning systems are unfortunately missing in a great number of basins. Furthermore, flood-forecasting models fail to simulate real-life situations, due to the poor mathematical algorithms employed. Efforts to minimize the possible damages of water-related disasters include raising public awareness and arranging insurance schemes for those who are frequently subjected to floods and other natural disasters. However, inadequate resources for data collection and dissemination and the inadequacy of early warning systems remain major constraints for effective disaster

BOX 14.10: THE SOCIAL, ECONOMIC AND ENVIRONMENTAL COSTS OF THE TSUNAMI IN SRI LANKA

The 26 December 2004 tsunami was the biggest natural disaster to strike Sri Lanka. It resulted in 38,900 deaths and displaced about 443,000 people on the eastern and southern coasts. The damage to homes, infrastructure, ecosystems and agricultural lands was enormous. The cost of the damage to assets is estimated at US $1 billion, while the cost of full recovery is about twice that.

Coastal populations earning their livelihoods from fishing and tourism have been deeply affected. Infrastructure, such as roads, bridges, railways, harbours, schools and telecommunications networks was highly damaged as well as fishing vessels, water supplies and many dwellings, including hotels..

However, cost estimates alone do not fully reveal the human loss and suffering that occurred as a result of the tsunami. It is estimated that more than 900 children were orphaned or separated from parents, and many parents lost all of their children. Clearly, women and children were the most affected groups during the disaster as well as in the aftermath, and ensuring their security remains a continuing challenge and a grave responsibility even today.

Ongoing research has revealed that environmental damage to coastal ecosystems, such as coral reefs and mangroves, reduced the capacity of the natural barriers to mitigate the tsunami force and thus intensified the destruction. Therefore, the Government and non-governmental organizations are planning to restore these natural barriers where possible and preserve the ecosystems with the participation of the community (for a discussion of early warning systems, see **Chapters 1** and **10**).

management (see **Chapters 1 and 10**). **Box 14.10** gives an overview of the dramatic damages caused by the 2004 tsunami.

Ensuring the knowledge base

Sri Lankan universities do not offer undergraduate courses specifically on water resources, but the curricula of civil engineering, agricultural science and some other science programmes contain water-related courses. However, the scopes of these courses vary depending on the nature of the degree. Because water-related courses are taught as optional subjects, in-depth water topics usually do not exceed 15 to 20 percent of the total course content.

Currently, universities and a few government agencies carry out scientific research on water resources and related topics. On the other hand, several government agencies carry out research relevant to their primary responsibilities, which include pollution control, irrigation, agriculture and sanitation.

Conclusion

About half of the Sri Lankan population struggles to survive on an income of less than US $2 per day. Hundreds of thousands of children suffer from malnutrition. However, the use of high-yielding crops, more fertilizers, better pest management practices and improved irrigation infrastructure have contributed to an increase in food production. As a result, Sri Lanka is on track to achieving the hunger-related MDGs. Industrialization and poor agricultural practices are threatening the quality of surface and groundwater. Although water-related legislation exists, the fragmentation of institutions, unclear responsibility, a lack of accountability and inadequate resources prevent the satisfactory implementation of provisions for controlling water pollution. Water-related disasters also present a great concern, as their financial and social damages put a heavy burden on the island's already fragile economy. Early warning systems are lacking in many basins, and forecasting models fall short of making accurate predictions.

The great challenge lying ahead is to improve the quality of life for inhabitants without jeopardizing ecosystems. The establishment of river basins organizations for the better management of valuable resources will help to alleviate poverty and environmental degradation.

15. Thailand

The Kingdom of Thailand's population is estimated to be around 63 million inhabitants, and its surface area is approximately 513,000 km^2, with elevations ranging from sea level in the south to high mountains in the northwest. The country's centre is dominated by the predominantly flat Chao Phraya River valley, which was studied in the case study chapter of WWDR1 (see **CD-ROM**). Thailand has a tropical monsoonal climate with a distinct wet season and a long hot dry season. The monsoon season is from mid-May to early September, during which time most parts of the country receive over four-fifths of their annual rainfall. Thailand possesses abundant water resources; total internal renewable water resources are estimated to be approximately 210 km^3 (FAO, 2000), 20 percent of which is used for agriculture.

Water and ecosystems

Thailand has over 40,000 wetlands, which are important at the local, national and international levels. As a contracting party of the RAMSAR convention (see **Chapters 5 and 12**), Thailand has twelve designated sites. Nonetheless, investment in agriculture has caused the overexploitation of forests. Between 1960 and 1990, the area devoted to agriculture doubled, while forest area was reduced by more than half, consequently producing widespread watershed degradation. Industry has recently become Thailand's main source of GDP. Though this has, in part, helped to decrease rural agricultural production and its related deforestation pressure, industrial areas have polluted wetlands with toxic chemicals. In an effort to lessen these negative trends, environmental concerns have been included in national socio-economic development plans.

Thailand's diverse coastal and marine areas and tropical and subtropical mountain ranges have made tourism a major part of the Thai economy. Unfortunately, tourism has also contributed to the clearance of coastal mangrove forests, the pollution of near-shore marine environments and the destruction of coral reefs. These environments are vital to sustained tourism revenues as well as the nation's important commercial fisheries (ICEM, 2003).

Natural resources management

The National, Economic and Social Development Plan (NESDP), prepared every five years, is the main mechanism for policy development and planning for the improved utilization of natural resources and environmental protection. Since the eighth NESDP (1997 to 2001), a participatory approach has been adapted to include important civil society organizations in its preparation. This was followed by an institutional reform and a restructuring of agencies responsible for the conservation and management of protected areas, which resulted in the establishment of the Ministry of Natural Resources and Environment in 2002.

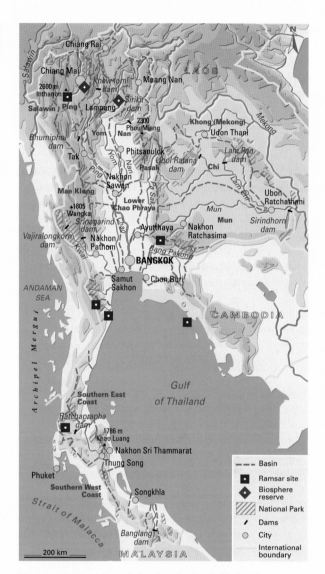

Map 14.16: Overview of the river basins in Thailand
Source: Prepared for the World Water Assessment Programme by AFDEC, 2006.

A framework for environmental conservation has been established under the Policy and Prospective Plan for National Environmental Quality Enhancement and Protection (1997-2016) in addition to an Environmental Management Plan (2002-2006). However, despite of the reforms and changes in the institutional system, the management of natural resources has not been fully integrated into sectoral planning. Different ministries have varying responsibilities, objectives and priorities for managing water resources and the environment. The efforts of individual ministries responsible for different sectors are isolated and not well-coordinated. This lack of coordination renders IWRM implementation difficult, in addition to impeding the Government in reaching its objectives in water management (ICEM, 2003).

Poverty and progress towards MDGs

Thailand's economy has undergone rapid development in the last three decades, and the level of poverty has greatly reduced. The number of people living in poverty dropped from 3.4 million in 1975 to fewer than 500,000 in 1995 (Ahuja et al., 1997). During the same period, however, income inequities between urban and rural areas increased. The proportion of poor living in rural areas rose to 92 percent in 1992, and the Asian economic crisis of 1997 and 1998 exacerbated the situation. Nevertheless, under 10 percent of the overall population currently lives below the poverty level. Thailand has made good progress towards achieving several of the MDGs, including those related to water. For example, access to safe water and basic sanitation is above 90 percent in both rural and urban areas. The current challenge is to address the opportunity disparities that exist between different regions.

Water and health

The most serious diseases in Thai people are helminthes, diarrhoea, dysentery and enteric fever (typhoid and paratyphoid fever), which are mainly caused by poor sanitation and insufficient hygiene (see **Chapter 6**).

Despite the development of water supply and sanitation infrastructures, morbidity rates remain high due to the high bacterial contamination of water. The Bureau of Epidemiology (2001) found that acute diarrhoea and food poisoning are still increasing, whereas between 1983 and 2001, the incidences of enteric fevers, dysentery and helminthes decreased. The main reasons for increasing diarrhoea and food poisoning are considered to be unhygienic food handling and a lack of awareness concerning the protection of water resources. Also alarming is the increasing number of diseases caused by chemical and toxic substances contaminating water resources. These contaminants are of domestic, industrial and agricultural origin. For example, lead and tin poisoning has been linked to improper mining practices, and high concentrations of fluoride in groundwater resources have also caused dental problems.

Management conflicts in local authority wastewater treatment systems often result in ineffective and non-continuous performance, which causes high bacterial contamination of receiving water bodies. The major obstacles preventing effective functioning of wastewater treatment facilities are insufficient financing for system operation and a lack of regular maintenance.

Water for food

Its fertile and well-watered central plains have helped to make Thailand a major international exporter of agricultural crops (rice in particular) and processed agricultural products. The cultivated area in Thailand is 28 million ha (54.5 percent of the total land area), approximately half of which is used for growing rice. At present, water used for irrigation is

equivalent to approximately 70 percent of the total water storage capacity of all reservoirs and structures. Agricultural production, although it comprises only 10 percent of the national GDP, is the major source of income for the rural poor. Accordingly, food security remains the primary issue on the political agenda. Thailand allocates about 10 percent of its national budget to irrigated agriculture, and its water policy calls for the nation-wide distribution of water for subsistence irrigation (up to the capacity limits of the river basins). In the mid-twentieth century, government policy promoted conversion of forests and swampland for agriculture, leading to significant deforestation. However, in recent decades, increasing urban migration and employment has reduced Thailand's rural farming population (from around 90 percent in 1950 to 40 percent), creating opportunities for reforestation.

Water for energy and industry

Thailand has a growing energy demand due to rapid industrial development and an increase in domestic demands. Imported fossil fuels and especially natural gas is the main source for energy production. In 2002, fossil fuels provided over 90 percent of national electricity production whereas hydropower was in the vicinity of 3 percent. To be able to meet growing energy demand, Thailand also imports hydro-electricity from neighbouring countries. In order to reduce external dependency on energy and to curb pollution, the government has plans to increase the share of renewable energy through the utilization of solar, wind, biomass and hydropower. Micro-hydropower development schemes, in particular, are considered as good choices given the topographic conditions and ecological sustainability.

The main source of water in industry is aquifers. However, over-utilization of these water resources, especially around Bangkok region, has caused serious land subsidence. For this reason, the government promotes the utilization of alternative sources of water supply and water-efficient processes in industry. The availability of freshwater can thus be a constraint to industrial development in the future. At present, the Thai government encourages the private sector to provide water to industry as well as to the municipalities. For example, Provincial Waterworks Authority of Thailand (PWA) has engaged a private company (East Water Company) to supply water to the industrial sector in the eastern part of Thailand.

Water allocation

In the dry season, the water stored in reservoirs is distributed according to priorities. The first allocation priority is water for household consumption, followed by other sectors, such as agriculture. The allocation system also takes into account the provision of water for the ecosystems and for the prevention of sea water intrusion.

Water allocation for agriculture is conducted at two levels:

■ From its main storage to secondary or sometimes tertiary canals, water is allocated by government agencies, which are fully responsible for

BOX 14.11: THE IMPACT OF THE TSUNAMI IN THAILAND

The tsunami created by an earthquake in the Indian Ocean on 26 December 2004 caused heavy social and economic damages in Thailand. According to official figures, there were over 5,000 dead, 8,400 injured and 3,000 missing persons. The sectors most badly damaged have been tourism and fishing. The beach resorts along the Andaman Sea coast have been extensively damaged and large amounts of investment are needed to enable

the private sector to recover. Furthermore, thousands of low-income Thais who are dependent on tourism-related industries have lost their jobs.

The extensive destruction of fishing boats has also affected food security in Thailand. It has led to a loss in livelihoods for individual fishing families who also lost their homes and cannot afford to replace their fishing equipment.

The tsunami has also caused severe damage on near shore aquifers as well as ecosystems in coastal wetlands and coral reefs. Furthermore, the spread of various kinds of wastes and industrial chemical solids further threatens the environment (see **Chapter 1** and **Box 14.10** for discussions on the 2004 tsunami).

operation and maintenance.
■ At the farm level, it is allocated by farmers and water user groups/organizations.

Thailand increasingly encourages water user groups/organizations to more actively participate in the management and allocation of water in secondary canals.

Water-related disasters
In Thailand floods of varying severity, size and duration are associated with tropical typhoons and usually occur between May and October. However, no matter how small or how large, floods cause serious social and economic damage. In 1995, a flood in the Chao Phraya River basin caused about US $290,000 in damage and was the costliest in the last seventeen years. In order to prevent flood damage, many structural and non-structural measures are being implemented, including the utilization of GIS and the creation of a flood risk map covering twenty-five basins. Furthermore, the installation of warning systems, such as supervisory control and data acquisition (SCADA), are being suggested.

As Thailand is located in a monsoon-risk area, with low precipitation from December until May or June, drought becomes a problem in various areas of the country. Although the impact of droughts is not as violent as that of floods, from 1989 to 2003, the cumulative damage from drought was in the vicinity of US $112 million. The droughts especially affect agricultural production, since rainfed irrigation is quite common, and only 23.7 percent of farmland is irrigated.

To prevent and mitigate flood and drought in Thailand, the Department of Water Resources has established the Water Crisis Prevention Center for collecting data, monitoring and formulating disaster policy responses. Unfortunately, at this stage, risk management activities remain poorly implemented.

Transboundary water resources
The Mekong River is the twelfth longest river in the world and has a water basin with nearly 60 million inhabitants distributed over 800,000 km². It encompasses six countries, including Cambodia, China, Lao People's Democratic Republic (PDR), Myanmar, Thailand, and Viet Nam. The Mekong River Commission (MRC) aims to protect the rich and diverse resources of the Mekong River Basin as a combined effort of Cambodia, Lao PDR, Thailand and Viet Nam. The Cooperation for the Sustainable Development of the Mekong River Basin Agreement, signed in 1995, provides the framework of the MRC and promotes basin-wide cooperation. The issue that is given the foremost importance in the 1995 agreement and its strategic plan is ensuring the sustainability of water resources and the environment while promoting overall basin development. Fisheries, agriculture and navigation are only some of the important issues that are also central to the agreement. The programmes established under the MRC strive for capacity enhancement and focus on the current and future needs of riparian countries and are meant to complement and support national and bilateral development initiatives.[14]

Conclusion
While Thailand possesses abundant water resources, growing demand coupled with pollution puts an increasing pressure on these resources. For many rural communities, cultivation has been the main source of livelihood and survival. Although the significant expansion of agriculture has contributed to ensuring food security and reducing poverty in Thailand, it has caused significant deforestation and widespread watershed degradation. While industry has become the major source of GDP, toxic wastes have polluted surface and groundwater supplies and damaged aquatic ecosystems. As a country that has achieved most of the MDGs, effective systems for conserving and protecting natural resources have become central to national development projects.

14. More information concerning the MRC can be found at www.mrcmekong.org

16. Uganda

Situated southeast of Uganda is Lake Victoria, the principal source of the White Nile and the second largest freshwater lake in the world. Uganda's rivers and lakes, including wetlands, cover about 18 percent of the total surface area of the country.

Lake Victoria is very significant for the Ugandan economy, since it is the source of almost all of the country's hydropower and provides the domestic and industrial water supply for the three biggest towns in Uganda: Kampala, Jinja and Entebbe. It is also an important location for the fishery and horticulture industries. Additionally, the lake serves as a key transport link between Uganda, Kenya and Tanzania.

Uganda's total annual renewable water resources are estimated to be 66 km^3. With an annual average of 2,800 m^3 of water available per capita, Uganda is better off than many other African countries. However, rapid population growth, increased urbanization and industrialization, uncontrolled environmental degradation and pollution are placing increasing pressure on the utilization of freshwater resources.

Water and Ecosystems

With 13 percent of its total surface area covered by wetlands, Uganda is very rich in biodiversity. In spite of the existence of national policies and laws for the conservation of ecosystems, there has recently been an observed decline in aquatic biodiversity in most of Uganda's water bodies. This has mainly been attributed to destructive fishing habits, increasing eutrophication as a result of pollution, degradation of riparian watersheds and deforestation (see **Chapter 5** for a discussion of the alarming loss of biodiversity in Lake Victoria).

Rural areas

The percentage of rural inhabitants with access to improved sanitation increased from 68 percent in 1991 to 85 percent in 2002. However, access to clean and safe water is still far from universal (see **Chapter 6**). In 2003, only 59 percent of rural inhabitants had such access. Frequently, people have to collect water from distant locations. This burden falls mainly on women and children, who are the most vulnerable members of society. The long distances they travel significantly reduce their productive time and subsequent contribution to the economic development of the country. Furthermore, the amount of water that can generally be collected is insufficient to meet drinking, cooking and hygiene needs. According to National Surveys conducted in 1996 and 1999, average rural per capita water consumption was found to be about 13 litres per day. Though the sanitation coverage has increased significantly, in some rural areas, basic sanitation still remains elusive, due to poverty and low hygiene and sanitation awareness.

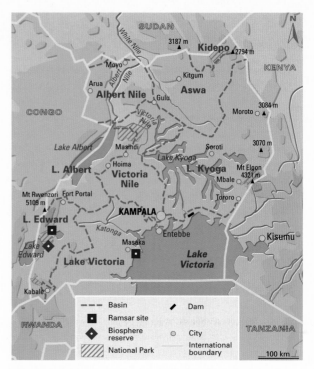

Map 14.17: Overview of the river basins in Uganda

Source: Prepared for the World Water Assessment Programme by AFDEC, 2006.

Urban settlements

In Uganda, urban areas are defined as settlements with over 5,000 inhabitants. Towns of 5,000 to 15,000 inhabitants are classified as small, and those with more than 15,000 inhabitants as large. Based on these criteria, there were 106 small towns and 43 large towns in Uganda in 2004. The current urban population is estimated to be 3.7 million out of a total population of 25 million. The urban population has been growing faster than that in rural areas – the overall population growth rate is 3.4 percent, while that in urban areas is 4.1 percent. The percentage of the population residing in urban areas increased from 12 percent in 1993 to 14 percent in 2003. National urban water coverage is an estimated 65 percent, up from 54 percent in 2000. The sanitation coverage is equally low, 65 percent.

Cost recovery

The current tariff structure of Ugandan water resource systems is aimed towards covering only operation and maintenance costs. Full cost recovery (operation and maintenance, depreciation and investment) would require a significant increase in tariffs. Therefore, major investments in system improvement and extension are financed separately through grants from the Government and international donors. The collection efficiency of revenues, although variable, is about 79 percent on average. Although funding levels are increasing, significant investment is still required to raise the safe water and sanitation coverage to meet the national targets and MDGs.

Water and health

Low access to clean water has had many health implications in Uganda. According to a study carried out in 2002, diarrhoea alone accounted for approximately 19 percent of infant mortalities in the country. Furthermore, statistics from the Ministry of Health indicate that malaria is the leading cause of child morbidity. Approximately 70,000 to 100,000 children in Uganda die every year from malaria. This represents 30 percent of the country's child mortality rates (between the ages of 2 and 4), and accounts for 23 percent of total disability-adjusted life years (DALYs) lost and 25 percent of all instances of illness in Uganda. Estimates from the Ministry of Health indicate that the average expenditure on malaria-related treatments are as high as US $300 million annually. AIDS is the leading cause of death for people between the ages of 15 and 49 and is responsible for 12 percent of all annual deaths (see **Chapter 6**).

Food security

The total potential irrigable area in Uganda is approximately 202,000 ha (FAO, 1995). However, a recent study by JICA (2004) revealed that about 14,000 ha of the potential irrigable area is under official irrigation and 6,000 ha under unofficial irrigation, particularly for rice production. The total amount of water used for irrigation is 12 km^3 per year, whereas the annual total renewable water resources are 66 km^3. These figures reveal the high potential for irrigated agriculture in Uganda. Currently, most of Uganda's agriculture is rainfed and thus more vulnerable during climatic variations. Food shortages and nutritional deficiencies are common in many parts of the country: 40 percent of deaths among children in Uganda are due to malnutrition. According to the 2002 Uganda Population and Housing Census, the country's annual population growth rate was 3.4 percent, while the annual growth rate of food production was only about 1.5 percent. If food production levels do not increase, food shortages will become more acute in the near future.

Livestock production is concentrated along 'the cattle corridor' which runs southwest to northeast across Uganda, encompassing twenty-nine

districts. Animal husbandry is a considerable source of income. It represents 7.5 percent of the GDP and 17 percent of the agricultural GDP. However, water scarcity in the cattle corridor reduces productivity and triggers conflict among herders.

Fisheries also contribute to food security in Uganda and are crucial to populations living along rivers, lakes and islands as well as the disadvantaged rural poor. Current annual fish consumption is estimated to be 10 kg per capita. In the past, Uganda's fishing industry boasted over 300 endemic fish species, but unsustainable fishing practices and a deterioration in the quality of local water bodies have greatly reduced the number of commercial fish species. Today, only twenty-three remain. The Ugandan Government is also promoting aquaculture to boost fisheries production to better meet the increasing fish demand in both the domestic and international markets (see **Chapter 5**).

Poverty

As of 2002, close to 40 percent of Uganda's population lives below the poverty line, giving Uganda a rank of 142 out of 162 countries in terms of poverty. Poverty reduction has been a leading objective of Uganda's development strategy since the early 1990s. The Government, in its combat against poverty, prepared a Poverty Eradication Action Plan (PEAP) in 1997. The plan, which has been revised twice, employs a multi-sectoral approach that takes into consideration the multi-dimensional nature of poverty and the inter-linkages between influencing factors. In this regard, the government is making continuous efforts for development in the areas of agricultural modernization, land management, rural credit and microfinance, rural electrification, primary health care, primary education and water supply and sanitation. Of all these, perhaps the PEAP's most critical intervention is the modernization of agriculture. Considering that the agricultural sector employs 82 percent of Uganda's labour force and is the mainstay of the economy, these efforts have the potential of improving the living standards of most Ugandans. Furthermore, through the Plan for the Modernisation of

---

BOX 14.12: THE IMPACT OF RISING TEMPERATURES

The continent of Africa's temperature has risen by 0.5°C in the past century. The five warmest years in Africa's recorded history all occurred after 1988. Recent studies have shown that the glaciers and ice fields on Rwenzoris, one of a few of permanently ice-capped mountains in Africa, have decreased markedly both in number and size and that the greatest rate of shrinkage has been after 1990.

Malaria has for long been the leading cause of illness in Uganda and accounted for almost 39 percent of all mortality cases in 2002. Today, malaria incidences in the highlands (1,500 to 1,800 m a.s.l.) are thirty times higher than at the beginning of the twentieth century. Rising temperatures in addition to heavy El Niño rains, local climate changes arising from wetland drainage, population growth and human migrations are thought to be some of the most important factors contributing to this increase.

Rising temperatures will have a detrimental effect on the agriculture sector of Uganda. For example, if the current trend continues, a further 2°C rise in temperature would lead to an 85 percent shrinkage in the area suitable for growing rubusta coffee, which constitutes a significant portion of Uganda's export (see **Chapters 4 and 10** for discussions on climate change).

Agriculture (PMA), the Government has initiated programmes to boost agricultural production and the marketing and processing of agricultural goods.

In recent decades, Uganda's elaborate plans and investments to combat poverty have started to pay off. Income poverty levels declined from 50 percent in 1992 to 35 percent in 2000. However, the economic recession that hit many parts of the world after 2000 slowed down the steady progress and returned poverty levels to around 40 percent in 2002.

In recognition of the progress made by Uganda in implementing economic reforms and poverty reduction, the international community, through the International Monetary Fund's (IMF) Heavily Indebted Poor Countries (HIPC) Initiative, has cancelled a large part of Uganda's external debt – between 1998 and 2000, approximately US $2 billion or 60 percent of Uganda's external debt. These measures have enabled the country to rechannel its financial resources to the fight against poverty.

Water and industry

The major industries in Uganda are agro-processing oriented, mainly fish processing, sugar, tea, cooking oil, diary processing, breweries and soft drinks. Factories for textile, paper products and tobacco processing are also fairly common.

The current low level of economic development in Uganda is partly attributable to an inadequate power supply, which cannot support large-scale manufacturing industries and agro-processing factories. Industry employs only 5 percent of the total labour force, whereas agriculture employs 82 percent and the service sector about 13 percent. The Ugandan Government is working to steer the country away from an over-dependence on agriculture by increasing the importance of the industrial and service sectors.

The industrial sector is a source of pollution due to the discharge of untreated or partially treated industrial effluent into nearby water bodies. Pollution stemming from mining activities, on the other hand, is still low and does not yet threaten the general quality of surface and groundwater. Localized pollution, however, exists in the areas where mining activities take place.

Water and energy

Biomass, principally firewood and charcoal, is the most important energy source in Uganda, constituting about 93 percent of the country's consumed energy. Petroleum products contribute only about 6 percent and electricity about 1 percent of annual energy demand.

Hydropower is the major source of electrical power in Uganda. Most of Uganda's hydropower potential is concentrated along the White Nile, with a total estimated potential of 2,000 MW. In addition, there are also several small rivers in different parts of the country, with a potential for mini- and micro-hydropower development (see **Chapter 9**). Currently,

only about 15 percent of of existing hydropower potential (300 MW) is utilized, and power demand, which is growing at a rate of 8 percent per year, exceeds available supply. The shortage in generation capacity limits growth in many sectors of the Ugandan economy. The Government formulated a Hydropower Development Master Plan to guide the hydropower planning and development process in Uganda. The Master Plan includes a comprehensive study of all the potential large- and small-scale hydropower schemes in the country and outlines the energy development strategy based on criteria such as power demand forecast, project generation potential, environmental effects and cost criteria.

Overall, only 9 percent of Uganda's population is supplied with grid electricity (20 percent in urban areas compared to only 3 percent in rural areas), and 70 percent of these customers reside in the three major towns of Kampala, Entebbe and Jinja. Official records show that there are about 230,000 grid electricity users. The national average annual per capita electricity consumption is about 44 kilowatts per hour (kWh), compared with an average of 170 kWh for the major urban areas and 10 kWh for rural areas.

Rural electrification forms an integral part of the Government's wider rural transformation and poverty eradication agenda. In this context, the most noticeable programme is the Uganda Photovoltaic Pilot Project for Rural Electrification, which aims to expand the access to electricity using solar technology in isolated and dispersed rural areas that will not have access to electricity grid in the near future and have both the ability and the will to pay the unsubsidized cost of the systems. As a result, more people in rural areas are switching from kerosene to solar lighting.

Legal framework and reform of the water sector

In order to meet the emerging challenges of the water sector, a Water Action Plan (WAP) was prepared in 1993–94, which recognized that water is an economic good with an economic value. The WAP principles were followed by a set of policies and laws throughout the 1990s.

In order to ensure efficiency and cost effectiveness in water resources management, government-initiated reforms in the water sector were established in 1997. As part of the reform process, a comprehensive Water Sector Strategy, detailed sub-sectoral investment plans and a clear definition of national targets for the sector were prepared. One of the key strategic outcomes of the reforms is the adoption of a Sector Wide Approach to Planning (SWAP). The SWAP framework, which has been embraced by both the Government and water sector development partners, promotes the participation of all stakeholders in the planning and implementation of water sector activities. This transparency has resulted in increased confidence from development partners who have agreed to finance water sector programmes through the regular government budget, contrary to the project-specific funding of the past. This is an important step, as 75 percent of the sector's funding comes from donors.

In addition, the water sector is also implementing a comprehensive sector-wide capacity-building and sensitization programme. The sector-wide approach to capacity-building mainly focuses on equipping sector personnel with relevant skills and knowledge in the management of water and sanitation programmes through specific tailor-made training courses and formal graduate training.

Water-related disasters

Water-related disasters, such as droughts, floods, landslides, windstorms and hailstorms contribute to well over 70 percent of natural disasters in the country and destroy an average of 800,000 ha of crops annually, causing economic losses in excess of approximately US $65 million. Large-scale atmospheric events, such as El Ninõ and La Ninã, are identified as the principal causes of the most severe water-related disasters in Uganda.

Disaster Preparedness and Management Strategy is designed to establish and improve national and local capabilities to minimize the damages caused by natural hazards and ensure that they do not result in disasters. The fundamental principle underlying the strategy is that the costs of responding to disasters once they strike far exceed those of disaster prevention and risk reduction activities. Further efforts are also being made to strengthen legal and institutional frameworks and ensure involvement of all relevant sectors. Raising public awareness has proven to be a pivotal point for effective hazard mitigation.

Conclusion

Uganda is on track for meeting the MDG targets for safe water and sanitation. However, the funds required for achieving those targets are in the vicinity of US $1.5 billion, an amount too high for the Ugandan national budget. Therefore, raising funds remains a critical issue. The country would also greatly benefit from an improved technical capacity within governmental institutions and a greater exchange of information among water-related agencies.

References and Websites

Case Studies: An overview

EC (European Commission). 2000. Directive 2000/60/EC of the European Parliament and of the Council of 23 October 2000 establishing a framework for Community action in the field of water policy. Bruxelles. europa.eu.int/comm/environment/water-framework/index_en.html

Walmsley, R.D., Havenga, T., Braune, E., Schmidt, C., Prasad, K. and van Koppen, B. 2004. *An Evaluation of Proposed World Water Programe Indicators for Use in South Africa.* Working Paper 90. Colombo, Sri Lanka, Internaitonal Water Management Institute.

1. The Autonomous Community of the Basque Country
Unless otherwise noted, all information is from the Executive Summary of the Case Study Report of the Autonomous Community of the Basque Country.

Basque Government Department of the Environment: www.ingurumena.ejgv.euskadi.net/
EUStat, Basque Statistics Office: www.eustat.es/about/a_euskadi-general_i.html and www.eustat.es/document/datos/1.medio_físico_i.pdf

2. The Danube River Basin
Unless otherwise noted, all information is from ICPDR (The International Commission for the Protection of the Danube River). 2004. Danube Basin Analysis (WFD Roof Report,

2004).www.icpdr.org/pls/danubis/danubis_db.dyn_navigator.show
Chapman, D. (ed.). 1996. *Water Quality Assessments: A Guide to Use of Biota, Sediments and Water in Environmental Monitoring.* Cambridge, Cambridge University Press with UNESO/WHO/UNEP.
ICPDR (The International Commission for the Protection of the Danube River). 2004. *Action Programme for Sustainable Flood Protection in the Danube River Basin.* www.ecologicevents.de/danube/en/documents/INFOFloodActionPlanEN_000.pdf
UNECE (United Nations Economic Commission for Europe) Draft Guidelines on Sustainable Flood Prevention:www.unece.org/env/water/publications/documents/guidelinesfloode.pdf

3. Ethiopia
Unless otherwise noted, all information is from the preliminary version of the Ethiopia Case Study Report.

UNDP(United Nations Development Programme). 2005. *Human Development Report: International Cooperation at a Crossroads, Aid, Trade and Security in an Unequal World.* Nairobi, UNDP.

4. France
All information was provided by the Ministère de l'écologie et du développement durable (Ministry of Ecology and Sustainable Development).

5. Japan
All references are from the information provided by the Ministry of Land, Infrastructure and Transportation (MLIT).

6. Kenya
Unless otherwise noted, all information is from the preliminary version of the Kenya Case Study Report.

UNDP(United Nations Development Programme). 2005. *Human Development Report: International Cooperation at a Crossroads, Aid, Trade and Security in an Unequal World.* Nairobi, UNDP.

7. Lake Peipsi/Chudskoe-Pskovskoe
All information is from the preliminary version of the Lake Peipsi/Chudskoe-Pskovskoe Case Study Report.

8. Lake Titicaca Basin
All information is from the preliminary version of the Lake Titicaca Case Study Report and the National Institute of Statistics.

9. Mali
All information is from the preliminary version of Mali Case Study Report.

10. Mexico
Unless otherwise noted, all information is from the preliminary version of the State of Mexico Case Study Report, SAOPID (Secretaría de Agua, Obra Pública e

Infraestructura para el Desarrollo). 2005. Damages Caused by the Overexploitation of Aquifers in the State of Mexico, Mexico. Mexico City. 2005.

CAEM (Comisión de Agua del Estado de México. 2005. *Atlas de Inundaciones No. 11.*

—— 2004a. *Prontuario de Información Hidráulica del Estado de México.* Mexico City.

—— 2004b. *Situación Actual y expectativas del Subsector Agua y Saneamiento en el Estado de México.* Mexico City.

CNA (Comisión Nacional del Agua). 2004. *Estadísticas del Agua en México.* Mexico City.

11. Mongolia

Unless otherwise noted, all information is from the preliminary version of the Mongolia Case Study Report.

Altansukh, N. 1995. *Country Report to the FAO International Techincal Conference on Plant Genetic Resources.* Ulan Bator, National Plant Genetic Resources Research and extension Programme.

Myagmarjav, B and Davaa, G. (eds). 1999. *Surface waters of Mongolia.* Ulan Bator (in Mongolian).

UNEP (United Nations Environment Programme). 2002. *State of the Environment, Mongolia.* Ulan Bator, UNDP.

NSO (National Statistical Office). 2000. *Child and Development Survey.* Ulan Bator.

MFA (Ministry of Foreign Affairs). 2004. *Millennium Development Goals: The 2004 National Report on the Status of Implementation in Mongolia.* Ulan Bator.

12. La Plata River Basin

Unless otherwise noted, all information is from the executive summary of the La Plata River Basin Case Study Report.

Bucher, E. and Huszar, P. 1995. Critical Environmental Costs of the Paraná-Paraguay Waterway Project in South America. *Ecological Economics*, Vol. 15, No. 1, pp. 3-9.

Gottgens, J., Fortney, R., Meyer, J, Perry, J. and Rood, B. 1998. The Case of the Paraguay-Paraná Waterway (Hidrovía) and its Impact on the Pantanal of Brazil: A Summary Report to the Society of Wetlands Scientists. *Wetlands Bulletin* pp. 12-18.

Petrella, F. and Ayuso, A. 1996. *The Paraguay-Paraná Waterway: Towards Convergence with the Plata Regime, a Personal Approach.* Proceedings of an International Conference, Harvard University, David Rockefeller Center for Latin American Studies, Cambridge, Massachusetts, 3-4 April 1996.

13. South Africa

All information is from the preliminary version of the South Africa Case Study Report.

14. Sri Lanka

Unless otherwise noted, all information is from the executive summary of Sri Lanka Case Study Report.

UN (United Nations). 2002. Sri Lanka Country Profile. *The 2002 Country Profiles Series.* World Summit on Sustainable Development, Johannesburg, 2002.

Ministry of Social Welfare. 2004. *Request for Drought Relief Assistance, Initial Assessment of Emergency Requirement (Revised).* Colombo.

15. Thailand

Unless otherwise noted, all information is from the executive summary of the Thailand National Case Study Report.

ICEM (International Centre for Environmental Management). 2003. Review of protected areas and development in the Lower Mekong River Region, Indooroopilly, Queensland, Australia.

Ahuja, V., Bidani, B., Ferreira, F. and Walton, M. 1997. *Everyone's Miracle? Revisiting Poverty and Inequality in East Asia.* New York, World Bank.

FAO (Food and Agriculture Organization). 2000. Irrigation water use per country in the year 2000, *Aquastat 2000.* www.fao.org/ag/agl/aglw/aquastat/water_use/index.stm

Mekong River Commission:www.mrcmekong.org/

16. Uganda

All information is from the preliminary version of the Uganda Case Study Report.

All know the way, but few actually walk it.
Bodhidharma, 6th Century

CHAPTER 15

Conclusions and Recommendations

15

Flooded slum on the edge of Pasig River, Manila, Philippines

Key Recommendations:

1. **We need to recognize that access to clean water is a fundamental right.** In 2002, the UN Committee on Economic, Social and Cultural Rights affirmed that 'sufficient, affordable, physically accessible, safe and acceptable water for personal and domestic uses' is a fundamental human right of all people and a pre-requisite to the realization of all other human rights. Although not legally binding for the more than 140 countries ratifying the International Covenant on Economic, Social and Cultural Rights, this decision carries the weight of a moral obligation on the signatories to progressively ensure that all the world has access to safe and secure drinking water and sanitation facilities, equitably and without discrimination. As the world is currently falling short of meeting the targets set to ensure adequate water services for all, it is our shared responsibility to maintain vigilance and continue to monitor our progress towards this goal.

2. **Poverty remains the biggest problem facing the world today.** The Millennium Development Goals (MDGs) adopted at the 2000 United Nations Millennium Summit focused world attention on this issue. Inadequate drinking water and sanitation services are key aspects of poverty with serious implications – death, disease and delayed development in the populations immediately affected. Broader ramifications include wasted economic opportunities, social and political unrest and environmental pollution. These problems are especially acute in the fast-growing human settlements of the developing world, particularly in shanty towns and slums with little if any water services. Rapid industrial growth and pollution in this context exacerbate the competition for water resources, often without providing the needed jobs for migrants. **We need to focus on better water governance that embraces all stakeholders and civil society, in both the public and private sectors, with strong support from the international community as the only plausible solution to these expanding problems.**

3. Climate change with increased variability exacerbates the spatial and temporal variability of water resources and intensifies the urgent need for the sound management of water resources. Given the finite quantity of freshwater, the current business-as-usual approach to development can only limit usable water resources as a result of continual and widespread physical and chemical pollution from virtually all sectors. Inadequate data collection, poor reliability of existing data and our limited understanding of the functioning of hydrological systems are a serious handicap to good planning and management. **We need to better understand complex environmental systems and the impacts of human activities, if society is to anticipate, mitigate and adapt to environmental changes and changing circumstances.**

4. **We need to recognize that sectorally and geographically, water problems and challenges are neither independent nor isolated.** Their solutions thus need to be addressed in a comprehensive and holistic manner, taking into account a variety of circumstances with solutions tailored to the situation. Hence we see the emergence and broad acceptance of the concept of Integrated Water Resources Management (IWRM). While IWRM may vary in different socio-economic settings and should be flexible enough to fit the attitudes and principles of local people, its core principles of equity, efficiency and environmental sustainability are invariable. Yet only a very few countries were able to meet the Johannesburg target calling for IWRM to be incorporated into national plans by 2005.

5. It is increasingly recognized that healthy ecosystems have importance far beyond their amenity or biodiversity preservation value. Healthy ecosystems are integral to the proper functioning of the hydrological cycle, thus environmental preservation must be at the heart of IWRM. Environmental pollution and the disruption of natural flows from all sectors (municipal, agricultural, industrial, energy, transport, etc.) must be addressed both in terms of detrimental impacts on aquatic habitats and the broader implications for the sustainable availability of clean freshwater resources. **We need to understand that water moves within natural limits, which usually do not correspond to the administrative units within which societies organize themselves.** Addressing water management issues from the perspective of natural boundaries, rather than political administrative units, will facilitate the consideration of environmental issues in IWRM. The urgent need for the integration of environmental and socio-economic concerns must be overcome with the greater collection and use of geo-referenced data.

6. **With growing demand and decreasing supply, competition among different sectors and users is increasing, requiring greater wisdom in the allocation of the resource and greater efficiency of water use.** The implementation of an IWRM approach with transparent mechanisms, e.g. tariffs, for allocating water among competing sectors is necessary to ensure sustainable widespread availability of limited freshwater resources. More efficient water use must be accomplished not only through the adoption of a variety of new technologies and the application of proven traditional knowledge, but also, and most importantly, through better water governance and the recognition that demand management must be a shared responsibility across all sectors.

7. **With fast changing socio-economic conditions occurring against the backdrop of unprecedented environmental change, water crises in many parts of the world are becoming increasingly severe.** Whether the problem is too much water or not enough, extreme pollution or excessive diversion, water remains a critical necessity for social and economic development. The solution to many, if not most, water-related problems lies first and foremost in better governance. Regardless of the particular mix of characteristics within a society or polity, the principles of transparency and accountability should prevail. Monitoring and indicator development at all levels are critical to supporting these critical aspects of good governance and informing important policy decisions.

8. **We must increase focus on the governance aspects of water management.** Good governance, although increasingly recognized as the key to more equitable, efficient and sustainable resource management, continues to encounter problems on the ground. Inadequate institutions, weak and ineffective legal frameworks, and limited human and financial resources continue to plague implementation and impede reform in water resources management. We need to enhance capacity across all sectors, and build awareness so that citizens and policy-makers may be better informed about water issues in order to encourage responsible decision-making at all levels.

Chapter 1 sets out a wide range of the issues at stake in the water sector and the global contexts in which they are taking place. These are further discussed, as appropriate, in the different water challenge areas covered by subsequent chapters. The many aspects of water, as they relate to poverty and environmental degradation, are briefly reintroduced together here, before moving on to recommendations for the future, such as ensuring that water-related factors do not contribute to the prolongation of poverty and environmental decay; facilitating the productive and sustainable use of water to reach the MDGs' aims of socio-economic development and environmental protection; and ensuring the use of a holistic approach to water and land-use management, which embraces Integrated Water Resources Management (IWRM) as its central principle.

As we have seen throughout the Report, even though there is plenty of freshwater at the global scale, it is unevenly distributed over time and space. For example, many of the relatively rich and sparsely populated countries in temperate zones have bountiful freshwater resources and rainfall throughout the year in addition to low evaporation, whereas poorer, more densely populated areas in tropical zones generally have less water per capita, and the bulk of their precipitation often occurs during a period of several weeks each year. In the tropics, water evaporation is high due to warmer climate, and dry spells occur frequently in some areas. It is expected that climate change and variability will have the most dramatic effect on tropical zones. Floods and hurricanes are already increasing in number and severity in many countries in the tropics. The hydrology and climatology of these regions differ from those of the rest of the world and thus may require completely different institutional and technological means of providing sufficient water throughout the year.

Uneven distribution of water resources also occurs between regions, communities and income groups within countries. In many cases, large- and small-scale infrastructure development, such as irrigation canals, water reservoirs and water transfer canals, has made it possible to distribute water more evenly both over time and space, benefiting households and various productive uses of water. However, many regions, such as the Middle East and North Africa, have reached a point where it is too expensive or logistically unfeasible from a hydrological point of view to respond to water crises by increasing the supplies of water.

Increases in water shortages and stress are a pressing problem in many countries. Although in absolute terms, water in many countries is not scarce, many people still lack sufficient and reliable access to clean water and sufficient water for food production and other productive uses. For example, water is wasted by inefficient irrigation and the poor operation and maintenance of water works. In many places of the world, a staggering 30 to 40 percent of water or more goes unaccounted for due to water leakages and illegal tapping. The Food and Agriculture Organization (FAO) has estimated that the overall water-use efficiency for irrigated agriculture in developing countries averages 38 percent. A basic insight – which has not yet attracted enough attention – is that the insufficiency of water, particularly drinking water supply and sanitation, is primarily driven by an inefficient supply of services rather than by water shortages (see **Chapter 2**).

A lack of basic services is often due to mismanagement, corruption, lack of appropriate institutions, bureaucratic inertia and a shortage of new investments in building human capacity as well as physical infrastructure. Water supply and sanitation have recently received more international attention than water for food production, despite the fact that in most developing countries, agriculture accounts for 80 percent of total water use. It is increasingly agreed that water shortages and growing pollution levels are to a large extent socially and politically induced challenges, which means that they are issues that can be addressed by changes of water demand and uses through, for example, increased awareness, education and water policy reforms. The water crisis is thus increasingly about how we, as individuals, and as part of a collective society, govern the access to and control over water resources and their benefits.

This final chapter draws on some of the key issues identified in the previous chapters and references of the Report, looking at them through the lenses of poverty, the environment and governance.

The water crisis is thus increasingly about how we, as individuals, and as part of a collective society, govern the access to and control over water resources and their benefits

Freshwater ecosystem, Cambodia

Many poor families suffer from housing insecurity, because they are in rented property or occupying land illegally

Part 1. Water and Poverty

The lifestyle of the extreme poor is almost literally hand-to-mouth; what they earn, in an urban area on a good day, will buy food and water for the family for that day. In rural areas, the food and water needed by families is taken largely from the natural environment: water carried from a distant spring or water body and some not very nutritious food grown on or gathered from marginally productive land, insufficient to satisfy hunger and provide needed nourishment. Very poor people struggle to pay for adequate food and water, rent for housing, medicines and drugs to treat sick family members, transport to get to places of work or carry sick family members to treatment centres, the education of their children and so forth. Very often, the quantity of water needed for good personal and domestic hygiene, laundry, etc. is too expensive to buy from street water vendors, too far to carry in the case of distant water sources, or necessitates the use of polluted water from nearby, heavily used rivers and streams. Rarely do they have access to improved sanitation, and where this may be available from a public facility in towns and cities, the cost to the whole family may be prohibitive.

The payment structure for many utility services like water and electricity, including up-front connection and monthly consumption charges, are often too expensive for the poor (see **Chapter 12**). Water-related disease (**Chapter 6**) and threats from water-related hazards (floods, landslides, droughts, etc., see **Chapter 10**) add to the precarious nature of their environments and lifestyles. Very poor farmers, working marginal farms, cannot afford the soil nutrition additives (agricultural chemicals) and the irrigation services needed to improve the reliability and productive capacity of their land (**Chapter 7**). On top of all of this, indebtedness frequently adds to the burdens of poor households.

1a. Insecure and overcrowded housing
Many poor families occupy land over which they have no formal legal rights – in a squatter community or slum, or farming on marginal lands owned by others with limited access to reliable water (see **Chapter 3**). They lack the savings and stores of surplus food and water to tide them over during lean times. In fact, most official statistics probably over estimate the savings of the poor, because they make no allowance for the depletion of natural

capital by poor communities – excessive cutting down of trees for fuel wood, exhausting the nutrition of soils, over fishing, among many others (Sachs, 2005). Most of the extremely poor are illiterate and unskilled. Women and girls in particular often have the least entitlement to household or family assets (see **Chapters 12** and **13**).

In urban areas, the land occupied by the poor is mainly the most marginal, prone to flooding, steep hillsides, etc. Often, they live right alongside streams and rivers that are grossly polluted, frequently by small-scale industries, involved in metal finishing, textiles, tanning, etc., using older chemical-based processes inherited from the industrialized countries and for which there is little or no affordable treatment technology. Chronic overcrowding is common, and the close proximity of households provides opportunities for the transmission of a range of infectious diseases. Many poor families suffer from housing insecurity, because they live in rented property or occupy land illegally (see **Chapter 3**).

1b. Inadequate access to public infrastructure and basic community services
Very poor households are rarely connected to urban infrastructure – piped water and sanitation, electricity supply, etc. The latter is a significant problem; in many parts of the world, access to electricity lags far behind access to improved water supplies. Drainage systems for urban rain and storm water are frequently inadequate; no formal systems for solid waste collection are provided; and there is a lack of paved areas – footpaths, roads etc. The latter are important not just for movement, they also provide a location for the installation, operation and maintenance of network utility services, such as water, drainage and electricity. There is a lack of flood protection infrastructure. All of this creates an ideal environment for disease transmission, vulnerability to loss of housing and possessions and, overall, a low quality of life.

For the rural poor, the lack of paved roads makes access to markets and health services very difficult. Water plays a very big role in transportation – many waterways, large and small, provide essential transport corridors, while many key roads and bridges are washed away in the rainy season. The rural poor are often at the end of irrigation systems, and dependent on richer upstream users for water, or pushed out onto land dependent totally on what may receive increasingly erratic rainfall as a result of growing climate variability.

Whether because of inadequate provision of basic community services by local authorities – health care, transport, education and training, emergency services, law enforcement, etc. – or their inability to pay for some of these, the poor are excluded from many vital opportunities. Confronting water-related disease – including malaria, which causes 300 to 500 million episodes of sickness and 1.6 to 2.5 million deaths each year – must also be done. While the urban poor may be close to many of these services, the rural poor often face the added burden of distance and transport costs. All of this increases vulnerability and prevents the development of much needed livelihood skills and, collectively, the capacity needed for greater self sufficiency and enhanced resilience.

1c. Lack of safety nets and adequate legal protection of rights

Poor families find it difficult to accumulate any surpluses, be they nutritional or financial, which means that it is hard to maintain consumption when their incomes are interrupted or their crops fail. In such circumstances, it can be a big problem to find ways to ensure access to water, food, healthcare, education, essential transportation and other necessities. The insurance provisions that are part of the way of life in higher-income countries are almost always denied to poor people.

In addition, there is a common lack of protection, while laws, regulations and procedures that concern legal and political rights, environmental health and protection, occupational health and safety, crime prevention and safeguarding from exploitation and discrimination are often limited or unenforced. Many rural poor suffer from limited rights to land, water and other natural resources. Indigenous people frequently have to struggle for rights to the water they have been using and protecting for generations. Deep well water abstractions by richer farmers and industries can lower water tables to the extent that poorer families and communities cannot then access groundwater. Untreated municipal and industrial effluents pollute the surface and groundwater sources relied on by the poor for their water supplies. Unbridled competition from richer farmers and industrial concerns for water, productive land and fisheries often puts the poor at a serious disadvantage. The implementation of national food policies (through subsidies, taxes, tariffs,

food aid, etc.) can distort markets and marginalize the rural poor, and inadequately organized and non pro-poor international trade liberalization can exacerbate this.

1d. Lack of voice and power within political and administrative systems

It is often very difficult for the poor to assert their rights and needs in order to receive a fair entitlement to public goods and services and hold service providers, NGOs and bureaucracies accountable. Local authorities fail to identify and put in place measures to protect poor communities from water-related hazards and disease. Indigenous communities find their detailed understanding of local water resources and its management is often ignored, while gaining access to information on water-related hazards and water resources is difficult. Many local authorities have little experience in dealing with poor community groups and may indeed be afraid to enter slum and squatter areas. Corrupt practices, in all aspects and levels of society, further complicate matters. These matters affect the approximately 1 billion people worldwide (one-sixth of the total population) in extreme poverty, who through sickness, hunger, thirst, destitution and marginalization find it nearly impossible to climb out of the cycle of poverty (Bass et al., 2005).

Other water-related factors further inhibiting economic growth, such as landlocked countries with poor transportation links, inadequate roads prone to water-related damage, lack of navigable waterways and good natural harbours. Widespread poverty and a lack of savings mean that governments cannot borrow from domestic sources or collect tax to provide essential public goods and services, so they borrow heavily, creating international debt burdens, which they cannot then service. Governments may fail to create the environment favourable to private business investment, both foreign and domestic.

Cultural barriers that discourage an active role for women marginalize an invaluable productive resource and prolong the demographic transition from high to low fertility. This exacerbates the problems of affordable education for all the children in a poor family and providing enough food and healthcare as well as reducing the amount of land per person to be passed on to the next generation. Poor countries tend to be very low on innovation, as they do not have the financial resources for the research and development needed to support economic growth. Trade sanctions put barriers in the way of trade by poor countries, often, remarkably, between adjacent poor countries (Sachs, 2005).

Slum in Jakarta, Indonesia

Public water pump in Amboceli Reserve, Kenya

Untreated municipal and industrial effluents pollute the surface and groundwater sources relied on by the poor for their water supplies

Part 2. Water and the Environment

The time when countries could industrialize with no regard for pollution and then invest in a massive clean-up, once and if they could afford to do so, is long gone

2a. The worrying deterioration of natural capital

Chapter 6 points to an alarming deterioration in freshwater aquatic ecosystems and species. In addition, the Millennium Ecosystem Assessment (MEA), in a review of some twenty-four ecosystem services (including a stable climate, freshwater replenishment, fresh air, soil fertility, pollination of crops, waste clean-up and nutrient recycling) highlighted the fact that fifteen of these are being degraded or used unsustainably.

In 2000, the total forest area of the planet was about 3.9 billion hectares (ha), or around 30 percent of the world's total area. During the 1990s, FAO estimated that some 94,000 square kilometres of forest (km²), an area roughly the size of Portugal, was lost to clearances of one sort or another each year. Forests contribute directly to the lives and livelihoods of over a billion people living in great poverty, providing them with freshwater, food, meat, medicines and building materials. As a result of such deterioration, poor communities face reduced levels of food protein, clean water and income-generating potential, which undermines poverty reduction strategies and is causing unprecedented rates of species extinction rates. Normal aquatic biodiversity is very rich, with high levels of endemic species. As species decline, biodiversity is reduced and essential ecosystem resilience diminishes.

The UN Millennium Project has made plain that long-term success in meeting the MDGs depends on environmental sustainability. Without it, any gains will be short-lived and inequitable. Yet, apart from climate change and warning systems for natural hazards, very little is being done on an international scale. Part of the problem is the very modest political effort devoted to sustainable development, compared with global economic growth. Although specific action programmes for forests, climate change and biological diversity were put into place in the latter part of the last century, they were mostly agreed before the MDGs and retain rather limited backing now (Concern/*Guardian*, 2005).

2b. The growing pressures on natural systems

As this Report points out, there has been a significant growth in freshwater-related disasters since the turn of the century in rich and poor countries alike, with over 400,000 lives lost and 1.5 billion people affected.

Some 13 percent of the world's population, over 800 million people, do not have enough food to live healthy and productive lives. Providing the water needed to feed a growing population and balancing this with all the other demands on water, is one of the great challenges of this century. Providing water for environmental flows and industry will tax water resources even more. Extending water services to the 1.1 billion unserved with improved water supply and the 2.6 billion lacking improved sanitation, will enlarge the challenge even further. Energy's water needs also need to be recognized: whereas some 90 percent of the urban world has access to improved supplies, only 37 percent has access to energy.

Water pollution worldwide is of huge concern, but has not received adequate attention. **Chapter 4** indicates the vast amount of water used to dilute and transport wastes. It has now become clear that dilution is not a viable solution to managing pollution. **Chapter 1** explains that the world's sinks for pollution are filling up fast – rivers, seas, atmosphere. The water sector has done little long-term forecasting or scenario development, but what has been done suggests that 'the problem of water is the most important global scale issue of the present century'(Simovic, 2002). In particular, the current use of clean water for the dilution and transport of wastes is not sustainable.

Yet the world has a vast knowledge of a wide variety of wastewater treatment systems – at all scales, many different degrees of ease of use and a wide range of affordability. We can treat household and industrial wastes, we have farm management practices which inhibit the polluting run-off from the use of agricultural chemicals into rivers, streams, the sea and groundwaters. Urgent steps must be taken worldwide to begin to implement tried and tested methods of wastewater treatment before the situation gets out of control. The time when countries could industrialize with no regard for pollution and then invest in a massive clean-up, once and if they could afford to do so, is long gone. The world no longer has the sinks for pollution that can accept this approach.

The net effect of this broad range of increasing pressures on water is a serious worsening of global water quality and a steady reduction of available per capita quantities of clean freshwater.

Part 3. Water and Governance

As made plain in this report, water is absolutely central to alleviating poverty, protecting the environment, promoting socio-economic development and achieving the MDGs. Yet despite this, not very many lower-income countries include water as a key feature of their national planning and budget processes, according to the Poverty Reduction Strategy papers and the outputs from the UN World Summit in 2005. Furthermore, the evidence suggests a widespread mismanagement of water in many countries, characterized by a lack of integration, sectoral approaches and institutional resistance to change by large public agencies in a context of increasing competition.

The available information alarmingly suggests that very few of the world's many significant and often transboundary rivers (of which there are 264, with 40 percent of humanity living in shared river basins) have well resourced, competent basin management commissions. The same is true for many of the world's important aquifers. Along the same lines, too few competent, properly resourced, independent basin regulatory agencies are operating with the needed powers. Too few water and electricity utilities in lower-income countries function even moderately well or are even close to recovering their full operating costs, not to mention depreciation. Much of their infrastructure is run down and degraded, and additional capacity – infrastructure and human and other resources – needed to meet the MDGs is lacking.

In many countries there is a huge deficit of water storage and flood protection infrastructure at all levels and scales, which will be aggravated, especially in the light of increased climate variability and volatility.

Only a minority of local authorities and water associations have the resources needed to carry out the delegated responsibilities they have inherited from central governments. Yet it is at the local level that authorities can empower community groups to self-manage the installation and operation of water supply and sanitation systems and safely collect and dispose of solid waste.

Just what is the total demand for community health and agricultural extension services in lower-income countries worldwide? Given that the knowledge to solve many, if not most, of the world's pressing water-related problems

exists, how well is this knowledge shared with those who really need it? Again, it is difficult to be precise, but general demand for advice, expertise and skills vastly outstrips the resources to provide it.

Previous chapters provide clear and convincing evidence that data on almost every subject related to water issues is usually lacking, unreliable, incomplete or inconsistent. We have learned that merely collecting data is not enough. It must be brought together, analysed and converted into information and knowledge (see **Chapter 13**), then shared widely within and between countries and stakeholders to focus attention on water problems at all scales. It is only when the data has been collected and analysed that we can properly understand the many systems that affect water (hydrological, socio-economic, financial, institutional and political alike), which have to be factored into water governance.

To facilitate understanding, advocacy and access to needed resources, the sector's many challenges need to be summarized and presented in simple but realistic terms. Trends must be discerned and progress monitored so that those who are falling behind can be helped, successful experiences can be identified and the lessons shared. The key to this is good, robust indicators, an ongoing iterative process that is impeded by a lack of good, reliable and consistent data, which needs input from sources external to the UN system to expedite the process.

3a. Awareness and advocacy

Each of the Report's chapters point to the challenges facing the water sector in order to raise awareness and advocate early action to tackle the world's outstanding water problem – poor water governance – by reminding the world that its water problems are not going away.

The lack of coverage of water in the Poverty Reduction Strategy Papers and the UN World Summit in 2005 is a matter of serious concern to the water sector, which must rigorously investigate the reasons for these omissions and set out systematically to change perceptions about water, while making clear that it occupies a central place at the head of the development agenda.

Lack of such understanding has contributed to serious under-investment and inadequate donor aid to the sector.

...there seems to be a huge deficit of water storage and flood protection infrastructure at all levels and scales...

Children playing in the river, Cambodia

Investment in improved water supply and sanitation has a strong potential for yielding three to thirty-four times the original investment...

Private investors are deterred because they feel that the water sector offers higher risks, but longer and lower returns on investment, than other sectors. Both public and private sector investors are deterred by what they perceive to be inadequate governance. Yet the cost-effectiveness of water investments is plain to see, as discussed below. This evidence and their supporting arguments need to be better organized and more forcefully projected to secure needed resources.

3b. The cost-effectiveness of water investments

Recently, an impressive range of information has become available relating to the cost-effectiveness of investments in water resources, water supply and sanitation service delivery, and in ecosystem protection.[1]

Investment in improved water supply and sanitation has a strong potential for yielding three to thirty-four times the original investment, depending on the local circumstances. By adding rapidly deployable interventions targeted at the poor, such as improved household water treatment and storage, returns can go up to sixty times the original investment. It has been estimated that 322 million working days per year, with an annual value of US $750 million, would accrue from meeting the MDG water and sanitation targets. Furthermore, the World Health Organization (WHO) has estimated that meeting these targets would yield time and convenience savings of US $7 billion and a further US $340 million in savings due to the costs avoided in seeking treatment, including the costs of care, drugs and transport and the opportunity costs of time spent in seeking medical attention.

By comparison, the annual per capita costs of meeting the MDG water and sanitation targets are extremely low: somewhere between US $4 and $7 in countries like Bangladesh, Cambodia, Ghana, Tanzania and Uganda. Illustrating the effect this can have on a country, those with improved water supply and sanitation enjoyed an annual growth rate of 3.7 percent of GDP, whereas those without grew at a paltry 0.1 percent.

For the irrigation sector, drip irrigation and treadle pumps (see **Chapters 1** and **7**) are two ways in which access to small-scale water technology can be provided to poor farmers. Research has shown that the direct total net benefit of promoting small-scale water technologies to 100 million poor farmers would mean gains estimated at US $100 to $200 billion.

Well-managed ecosystems more than pay for themselves, while providing a wide range of services, as indicated earlier. Yet many lower-income countries are losing a staggering 4 to 8 percent of their GDP through environmental degradation. Industrial income lost to water pollution in China in 1992 alone amounted to US $1.7 billion. On the other hand, an investment in watershed protection can save anywhere from 7.5 to 200 times the original investment in costs of waster treatment saved. The annual benefits of protecting a wetland in Cambodia, for example, have been estimated at US $3,200 per household.

With respect to climate variability, for example, it has been estimated that improved resilience to floods and drought could help Kenya's GDP to grow at an annual rate of 5 to 6 percent – the amount needed to start effectively reducing poverty – rather than its current 2.4 percent.

3c. Integrated Water Resources Management

Change is virtually the only constant of modern times, as emphasized by **Chapter 1**, with globalization, urbanization, climate variability, hydrological variability, cooperation and conflict all vying for attention within the water management setting. All of which emphasizes the necessity for societies and their socio-economic systems to be adaptive and resilient.

The political systems of the world vary greatly, driven by different underlying cultures, attitudes, relationships and natural environments. Relationships between different levels of government also vary, within their institutions, legislative, regulatory and socio-economic settings. Watershed and basin boundaries often do not coincide with administrative boundaries, causing many overlaps to occur. Strategic basin and watershed issues cannot always be dealt with at the local level. Growing demands for water, availability reduced by pollution, competition from the various sectors and the many and growing number of users, is another constant. Globally, regionally, nationally and locally, ecosystems are under growing threats.

All of this emphasizes that the various water issues are not independent of each other and focuses on the need for greater foresight in the allocation and management of water resources. At both strategic and local levels, a flexible approach is essential. The answer to all of this, including meeting the MDGs, lies in a holistic, ecosystem-based approach, known as Integrated Water Resources Management (IWRM).

The different chapters of the Report, which address the challenge areas associated with meeting the MDGs, set out what needs to be done in the different water-using sectors. Each stresses how the IWRM approach is essential to an optimum and efficient response to the challenges. But there is no panacea for implementing IWRM; it must be tailored to prevailing conditions and flexible enough to permit this. Local circumstances can put obstacles in its way:

- lack of appropriate governance
- lack of proper coordination of management activities
- lack of appropriate management tools
- institutional fragmentation
- insufficiently trained or qualified manpower
- shortfalls in funding
- inadequate public awareness
- limited involvement by communities, NGOs and the private sector.

Probably because of these and other difficulties, very few countries have met the Johannesburg Plan of Implementation (JPOI) target that IWRM should be incorporated into national water resources plans by the end of 2005. Thus, it is clear that more analysis of the practical means of moving from a fragmented, sector-by-sector approach to IWRM needs to be carried out for lower-income countries, and these experiences need to be shared widely (see **Chapter 14**).

3d. The need for international and national cooperation

The necessary overhauls of water governance and the challenge of meeting the MDGs are closely linked. In order to succeed in both, action at all levels of society is required. Individuals must take greater responsibility, both for their families and their communities. Provincial and national governments, with full transparency and accountability, must take steps towards making resources available and creating enabling environments for beneficial change, while ensuring that water policies and plans are set firmly within the context of regional and national development plans and budgets. Action to increase progress on the MDGs must be initiated within countries themselves, involving the whole country and maximizing the capacity for community self-organization.

At the international level, industrialized countries must play their part. **Chapter 1** makes clear that the MDGs are a joint project. The first seven goals are directed at alleviating poverty; while the eighth goal is to create the partnerships of rich and poor countries to meet the first seven. Lower-income countries are tasked with delivering promised policy changes and improvements to governance, and the industrialized countries must follow through with their long-standing commitments to increase Overseas Development Assistance (ODA) and technical assistance.

Globally, regionally, nationally and locally, ecosystems are under growing threats

Women's group for microfinance, Andra Pradesh, India

Part 4. Cautious Optimism

'In economic terms, the human race has never been richer, or better armed with the medical knowledge, technical prowess and intellectual firepower to make poverty history'

2. It is not certain how many microfinance organizations there are worldwide, but the number is thought to be large and growing. Indonesia alone claims 600,000 and other countries in Asia and in Africa claim many thousands more. Today, some of the world's biggest banks (Deutsche Bank, Citigroup, HSBC and ABN Amro) are showing an interest. Credit rating agencies are beginning to provide affordable services to microfinance organizations, and big banks are finding their way into the sector through the remittances that overseas workers send to their families back home.

4a. Economic progress is being made

As previous chapters indicate, there should be cause for cautious optimism. We know that economic development can and does work in many parts of the world. Despite the sometimes daunting data and statistics on the extent of poverty, at least five-sixths of the world's population is at least one step above extreme poverty. Nearly 5 billion people are living in countries where average incomes rose in the 20 years between 1980 and 2000. Over a similar timeframe, life expectancy increased in a range of countries in which some 5.7 billion people live. Out of the total world population of 6.3 billion people, nearly 5 billion have managed to advance into at least the first stages of social and economic development. The truth is that, thankfully, the scale of extreme poverty is lessening, both in terms of the total numbers affected and as a proportion of the total world population (Sachs, 2005).

In addition, the growth of microfinance is proceeding and has great potential for speeding up poverty alleviation.[2] Microfinance is a system of providing small loans to the very poor, which can then be used by local communities to build a well, for example. Microfinance is presently receiving a lot of attention from policy-makers, with its proponents asserting that it has enormous potential in the fight against poverty – sweet music indeed for those who fight for the rights of the poor and the abolition of poverty (*Economist*, 2005). It is showing signs that it may be about to expand substantially by providing financial services that can be made available to the very poor, or low-cost insurance to protect them against the risks and setbacks to which they are particularly vulnerable – water-related hazards, crop and livestock losses, death of the family breadwinner and others.

Among the barriers to providing financial services to the very poor are inflation, incompetent governments (which allow corruption and fail to provide an enabling environment for financial services) and property laws that make it impossible for homes (for those who own their own) to be used as security for loans. Funding for microfinance organizations must go beyond governments, aid agencies and charities, and the cost of operations must be brought down, because as it is presently organized, microfinance is very labour-intensive.

4b. Reform is underway in the water sector

Although evidence of its effectiveness is hard to come by, a tremendous worldwide revolution is ongoing in the reform of the water sector's many institutions. Progress is patchy, sometimes slow and not as synchronized with national development planning and budgeting as it should be. Many local initiatives, often by poor communities, are underway but usually under-reported. Rapidly growing lower-income countries, such as Brazil, China and India, are coming up with a wide range of novel initiatives to deal with their water governance and water service delivery challenges, which are robust and could be adapted by other countries. The UN Millennium Project (see **Chapter 1**) has produced a range of plans and ideas to meet the MDG targets on time, much of which involves activities within and directly related to the water sector. Universities training water managers have shown that they understand the issues and challenges of contemporary water management and are responding positively to them.

This report covers a lot of ground, reflecting the breadth and scope of what needs to be done in the water sector worldwide and indeed what is being done. It has made plain just how central water is to poverty alleviation and development and how little this fact has been recognized and acknowledged. There are other important sectors that would wish to claim priority for scarce resources and investments, yet water is *primus inter pares* – the first among equals. No matter how many mobile phones are in circulation, how many new drugs and new seed varieties are produced, without access to secure water supplies, development will stall and the MDG targets will fall short. The same can be said about the environment, which is also dependent upon good quality water to sustain it.

It is not that the world does not have the resources to do what is needed, both for water and the MDGs. The Millennium Project has made clear that the world does have the wealth and the tools to do what is needed, an idea that *The Economist* (2004) has succinctly summarized: 'Optimism is certainly justified. In economic terms, the human race has never been richer, or better armed with the medical knowledge and technical prowess and intellectual firepower ... to make poverty history'.

Given the nature of the challenges, the thoughts of the 2005 Stockholm Water Prize winner, Ms Sunita Narain, seem particularly appropriate. She noted that water is about more than just water; it is about building people's institutions and their power to take control over decisions. However, water cannot be made everyone's business without fundamental changes in the way we do business with water: 'Humanity must realise, policy makers and public alike, that water management which involves communities and households needs to become the greatest cooperative enterprise in the world.'

Given the state of water around the world, and the challenges facing contemporary water managers, there has probably never been a more exhilarating time to be in the profession. Certainly, managing water today is a tough proposition, but the rewards are immense: world poverty alleviation and environmental sustainability.

In some of the concluding remarks to WWDR1, it was said that, given the evidence available at that time, the prospects for hundreds of millions of poor people in lower-income countries, as well as the natural environment, did not look good. Has the situation improved since then, in the intervening years? Yes, it has. Certainly the major water-related challenges have not changed very much, but a worldwide process of reform of water is underway. We have convincing evidence of the very positive cost-effectiveness of investments in water. The growth in microfinance has the potential to provide essential capital for the extension of water service provision, through a much enhanced availability of funds to the very poor, while also contributing to lessening their insecurity in many other ways. The MDG review has been carried out; we know what has to be done to meet the targets, and a plan to do this has been produced. We know that there has been and continues to be progress in poverty alleviation and socio-economic development. We know, in fact, how to bring to an end the exclusion of the poor from their fair share of the Earth's resources. With determination and political will, the levels of international cooperation agreed on in the Millennium Declaration, and reconfirmed at the 2005 UN World Summit, the MDGs can be achieved, and the water sector can be reformed.

There is a danger of complacency however. The fast changing context of today's world, especially the accelerated pace of climate change, can only heighten the urgency with which we must address our water-related challenges. We all share the responsibility to ensure that water – critical to every aspect of our life – remains at the forefront of the political agenda.

Jiuzhaigou valley, Sichuan, China

References

Bass, S., Reid, H., Satterwaite, D. and Steeple, P. 2005. *Reducing Poverty and Sustaining the Environment: The Policies of Local Engagement.* London, Earthscan Publications

Concern/*Guardian.* 2005. *Look into the Future: Are the MDGs a Ray of Hope for the Poorest People?* Booklet prepared for the 2005 UN World Summit by Concern Worldwide and *The Guardian,* London/Manchester, England.

Economist. 2005. The hidden wealth of the poor: A survey of microfinance. 5 November 2005. London, England.

——. 2004. Making poverty history. 18 December 2004. London, Economist Group.

Sachs, J. 2005. *The End of Poverty: How We Can Make it Happen in Our Lifetime.* London, Penguin Books.

Simonovic, S. 2002. Global water dynamics: Issues for the 21st century. Paper presented at the 2001 Stockholm Water Symposium. *Water Science and Technology.* Vol. 45 No. 8. pp. 53-64. London, IWA Publishing.

SIWI (Stockholm International Water Institute). 2005. Stockholm Waterfront. 2005. Forum for Global Water Issues. Stockholm, Sweden.

SIWI/WHO (Stockholm International Water Institute/World Health organization). 2005. Making Water a Part of Economic Development, Stockholm/Geneva.

Boxes by chapter

Boxes by region

Maps

Figures

Tables

Acronyms

ACB: Autonomous Community of the Basque Country

ACHR: Asian Coalition for Housing Rights

ACIA: Arctic Climate Impact Assessment

ACP-EUWF: European Union Water Facility for the African, Caribbean and Pacific Countries

ADB: Asian Development Bank

ADR: Alternative Dispute Resolution

AEWA: Agreement on the conservation of African Eurasian Migratory Waterbirds

AEWS: Accident Emergency Warning System

AGORA: Access to Global Online Research in Agriculture

AIH: American Institute of Hydrology

APFM: Associated Programme for Flood Management

AQUASTAT: Country Information on Water and Agriculture

ARC: Agricultural Research Council

ASCE: American Society of Civil Engineers

ASR: Artificial Storage and Recovery

BAT: Best Available Technology

BCA: Benefit-Cost Analysis

BEP: **Best Environmental Practice**

BOD: Biological Oxygen Demand

BOO: Build-Own-Operate

BOT: Build-Operate-Transfer

CAP: Common Agricultural Policy

CAPNET: International Network for Capacity Building for Integrated Water Resources Management

CAZALAC: Regional Water centre for Arid and Semi-arid Zones of Latin America and the Caribbean (Chile)

CBD: Convention on Biological Diversity

CCs: Collaborating Centres

CDI: Capacity Development Initiative

CDM: Clean Development Mechanism

CEH: Centre for Ecology and Hydrology

CES: Compensation for Ecosystem Services

CESCR: Covenant on Economic, Social and Cultural Rights

CGIAR: Consultative Group on International Agricultural Research

CI: Conservation International

CIC: The Intergovernmental Coordination Committee of the Plata Basin Countries

CRED: Centre for Research on the Epidemiology of Disaster

CO2: Carbon Dioxide

COD: Chemical oxygen demand

CODI: Community Organizations Development Institute (ex-UCDO: Urban Community Development Office)

COMEST: World Commission on the Ethics of Scientific Knowledge and Technology

CPS: Cleaner Production Assessment

CRC: Convention on the Rights of the Child

CSA: Central Statistic Authority

CSD: Commission on Sustainable Development

CSIR: Council for Scientific and Industrial Research

CVI: Climate Vulnerability Index

CWS: Cities Without Slums

DALY: Disability-Adjusted Life Years

DDT: Dichlorodiphenyltrichloroethane (toxic Chemical)

DEFRA: Department for Environment, Food and Rural Affairs (United Kingdom)

DES: Dietary Energy Supply

DEWA: Division of Early Warning and Assessment

DFID: Department For International Development

DHF: Dengue Hemorrhagic Fever

DHS: Demographic Health Surveys

DMCs: Drought Monitoring Centres

DNH: National Water Directorate (Direction Nationale de l'Hydraulique)

DPPC: Disaster Prevention and Preparedness Commission

DPSIR: Driving forces, Pressure, State, Impact, Response

DRB: The Danube River Basin

DRBMP: The Danube River Basin Management Plan

DRI: Disaster Risk Index

DWD: The Directorate of Water Development

EC: European Commission

EDSS: Educational Decision Support Systems

EIA: Environmental Impact Assessment

EMAS: Eco-Management and Audit Scheme

EMIS: Educational Management Information Systems

EMS: Environmental Management Systems

EPA: Environmental Protection Agency

EPL: Environment Protection Licence

ERR: Economic Rate of Return

ESA: European Space Agency

ETIC: The Euphrates-Tigris Initiative for Cooperation

EU:	European Union	**IAHS:**	International Association of Hydrological Sciences
EWRI:	Environmental & Water Resources Institute	**IAS:**	Invasive Alien Species
FAO:	Food and Agriculture Organization	**IBNET:**	Water and Sanitation International Benchmarking Network
FBW:	Free Basic Water		
FDA:	Food and Drug Administration	**ICARM:**	Integrated Coastal Area and River Basin Management
FEMIP:	Facility for Euro-Mediterranean Investment and Partnership		
		ICESCR:	International Covenant on Economic, Social and Cultural Rights
FEWS:	Famine Early Warning Systems Network		
FFEWS:	International Workshop on Flood Forecasting and Early Warning Systems	**ICHARM:**	International Centre for Water Hazard and Risk Management (Japan)
FRESH:	Focusing Resources on Effective School Health	**ICID:**	International Commission on Irrigation and Drainage
FRIEND:	Flow Regimes from International Experimental and Network Data		
		ICOLD:	International Commission on Large Dams
GAS:	Guaraní Aquifer System	**ICPDR:**	International Commission for the Protection of the Danube River
GDLN:	The Global Development Learning Network		
GDP:	Gross Domestic Product	**ICT:**	Information and Communication Technology
GEF:	Global Environment Facility	**ICZM:**	Integrated Coastal Zone Management
GEMS/WATER:	Global Environmental Monitoring System, Freshwater Quality Programme	**IDA:**	International Desalination Association
		IDNDR:	International Decade for Natural Disaster Reduction (1990–2000)
GEO:	Global Environment Outlook		
GIS:	Geographic Information Systems	**IDRC:**	International Development Research Centre
GIWA:	Global International Waters Assessment	**IETC:**	International Environmental Technology Centre
GLIMS:	Global Land Ice Measurements from Space		
GMFS:	Global Monitoring for Food Security	**IFAD:**	International Fund for Agriculture and Development
GMOs:	Genetically Modified Organisms		
GNIP:	Global Network for Isotopes in Precipitation	**IFC:**	International Finance Corporation
GNP:	Gross National Product	**IFI:**	International Flood Initiative
GPA:	Global Programme of Action	**IFM:**	Integrated Flood Management concept
GPA:	Global Plan of Action for the Protection of the Marine Environment from Land Based Activities	**IGRAC:**	International Groundwater Resources Assessment Centre
		IHA:	International Hydropower Association
GRDC:	Global Runoff Data Centre	**IHE:**	Institute for Infrastructural, Hydraulic and Environmental Engineering
GRID:	Global Resource Information Database		
GSLRP:	Gash Sustainable Livelihood Regeneration Project	**IHP:**	International Hydrological Programme
		IIED:	International Institute for Environment and Development
GTZ:	German Technical Cooperation		
GWP:	Global Water Partnership	**IIRR:**	International Institute of Rural Reconstruction
HDI:	Human Development Index		
HDR:	Human Development Report	**ILA:**	International Law Association
HELP:	Hydrology, Environment, Life and Policy	**ILC:**	International Law Commission
HIA:	Health Impact Assessment	**IMF:**	International Monetary Fund
HIPC:	Heavily Indebted Poor Countries	**IMT:**	Irrigation Management Transfer
HOMS:	Hydrological Operational Multipurpose System	**INBO:**	International Network of Basin Organizations
		IGOs:	Intergovernmental Organizations
HSRC:	Human Sciences Research Council	**IPCC:**	Intergovernmental Panel on Climate Change
IAEA:	International Atomic Energy Agency		
IAH:	International Association of Hydrogeologists	**IPTRID:**	International Programme for Technology and Research in Irrigation and Drainage
IAHR:	International Association of Hydraulic Engineering and Research	**IRBM:**	Integrated River Basin Management

IRC:	International Water and Sanitation Centre
IRD:	Research Institute for Development (Institut de Recherche pour le Développement)
ISARM:	International Shared Aquifer Resource Management
ISDR:	International Strategy for Disaster Reduction
ISI:	International Sediment Initiative
ISO:	International Standards Organization
ITNs:	Insecticide-treated nets
IUCN:	World Conservation Union
IVA:	Industrial Value Added
IWA:	International Water Association
IWG-ENV:	Intersecretariat Working Group on Environment Statistics
IWMI:	International Water Management Institute
IWRM:	Integrated Water Resources Management
IWT:	Inland Water Transport
JMP:	Joint Monitoring Programme
JPOI:	Johannesburg Plan of Implementation
JRC:	Joint River Commission
KenGen:	Kenya Generating Company
KI:	Knowledge Index
KPLC:	Kenya Power and Lighting Company
LA WETnet:	The Latin American Network for Water Education and Training
LIMCOM:	Limpopo Watercourse Commission
LINKS:	Local and Indigenous Knowledge Systems
LPI:	Living Planet Index
LRS:	Large River System
MAP:	Mediterranean Action Plan
MAR:	Managed Aquifer Recharge
MAR:	Natural Mean Annual Runoff
MCK:	Municipal Council of Kisumu (Kenya)
MDGs:	Millennium Development Goals
MEAs:	Multilateral Environmental Agreements
MER:	Monitoring, Evaluation and Reporting
MICS:	Multiple Indicator Cluster Surveys
MIS:	Management Information System
MLIT:	Ministry of Land, Infrastructure and Transport (Japan)
MOST:	Management of Social Transformations
MRC:	Mekong River Commission
MRET:	Mandatory Renewable Energy Targets (Australian Government's)
MWRMD:	Ministry for Water Resource Management and Development
NASA:	The United States' National Aeronautics and Space Administration
NBCBN-RE:	Nile Basin Capacity-building Network for River Engineering

NBI:	Nile Basin Initiative
NCSA:	National Capacity Self-Assessment
NCSC:	National Center for Small Communities
NEHAP:	National Environment Health Action Plan
NEPAD:	New Partnership for Africa's Development
NGO:	Non-governmental Organization
NHMSs:	National Hydrological and Meteorological Services
NOx:	Nitrogen Oxide
NPs:	National Platforms for Disaster Risk Reduction
NPP:	Net Primary Productivity
NSIDC:	National Snow and Ice Data Centre
NUFFIC/ IK-Unit:	Netherlands Organization for International Cooperation in Higher Education/Indigenous Knowledge
OAS:	Organisation of American States
OBA:	Output-Based Aid
ODA:	Overseas Development Assistance
ODI:	Overseas Development Institute
OECD:	Organization for Economic Cooperation and Development
OFDA:	Office of Foreign Disaster Assistance (United States)
OMVS:	Organization for the Development of the Senegal River (Organisation pour la mise en valeur du fleuve Sénégal)
OPP:	Orangi Pilot Project
OPP-RTI:	Orangi Pilot Project's Research and Training Institute
ORT:	Oral Rehydration Therapy
OTCA:	Amazonian Cooperation Treaty Organization (Brazil)
PAR:	Pressure and Release model
PCBAP:	Plan for the Conservation of the Upper Paraguay River Basin, (Plano de Conservação da Bacia do Alto Paraguai)
PCCP:	From Potential Conflict to Cooperation Potential
PEAP:	Poverty Eradication Action Plan
PEEM:	Panel of Experts on Environmental Management for Vector Control
PES:	Payment for Environmental Services
PIM:	Participatory Irrigation Management
PMA:	Plan for the Modernisation of Agriculture
PMPOA:	Farm pollution management programme (Programme de Maîtrise des Pollutions d'Origine Agricole)
PPP:	Public-Private Partnership
PPPUE:	Public-Private Partnerships for the Urban Environment

PRA:	Participatory Rapid Appraisal
PROSANEAR:	Water and Sanitation Project for the Low-income Urban Population (Brazil)
PRSP:	Poverty Reduction Strategy Papers
PSP:	Private Sector Participation
PUB:	Prediction in Ungauged Basins
PWRI:	Public Works Research Institute
QA/QC:	Quality Assurance/Quality Control
RBI:	River Basin Initiative
RBM:	Roll Back Malaria Initiative
RCMs:	Regional Circulation Models
RCMRD:	Regional Centre for Mapping of Resources for Development
RCTWS:	Regional Centre for Training and Water Studies of Arid and Semi-arid Zones
RCUWM:	Regional Centre on Urban Water Management (Iran)
SACI:	South African Capacity Initiative
SADC:	Southern African Development Community
SAFE:	Integrated strategy involving Surgery, Antibiotic treatment, promotion of Facial cleanliness and the initiation of Environmental changes
SAOPID:	Secretariat of Water, Public Works and Infrastructure for Development (Mexico)
SAR:	Sodium Absorption Ratio
SCADA :	Supervisory Control and Data Acquisition
SDAGE:	Master Plan for Water Management (Schémas directeurs d'aménagement et de gestion des eaux)
SDC:	Swiss Agency for Development and Cooperation
SHP:	Small Hydropower
SIDA:	Swedish International Development Cooperation Agency
SIWI	Stockholm International Water Institute
SSHE:	School Sanitation and Hygiene Education
SWAP:	Sector Wide Approach to Planning
TAI:	The Access Initiative
TI:	Transparency International
TEST:	Transfer of Environmentally Sound Technology
TDPS:	Titicaca, Desaguadero, Poopó, Coipasa Salt Lake
UCDO:	See CODI
UN:	United Nations
UNCED:	United Nations Conference on Environment and Development
UNDESA:	United Nations Department of Economic and Social Affairs
UNDP:	United Nations Development Programme
UNDP/BCPR:	United Nations Development Programme's Bureau for Crisis Prevention and Recovery
UNECA:	United Nations Economic Commission for Africa
UNECE:	United Nations Economic Commission for Europe
UNECLAC:	United Nations Economic Commission for Latin America and the Caribbean
UNEP:	United Nations Environmental Programme
UNEP/GRID:	United Nations Environmental Programme's Global Resource Information Database
UNESCAP:	United Nations Economic and Social Commission for Asia and the Pacific
UNESCO:	United Nations Educational, Scientific and Cultural Organization
UNESCWA:	United Nations Economic and Social Commission for Western Asia
UNFCCC:	United Nations Framework Convention on Climate Change
UN/GA:	United Nations General Assembly
UNICEF:	United Nations Children's Fund
UNIDO:	United Nations Industrial Development Organization
UNISDR:	United Nations International Strategy for Disaster Reduction
UNOOSA:	United Nations Office for Outer Space Affairs
UNU:	United Nations University
USAID:	US Agency for International Development
US NGWA:	US National Ground Water Association
UV:	Ultraviolet
VIP:	Ventilated Pit Latrine
VMAD:	Virgin Mean Annual Discharge
WAP:	Water Action Plan
WB:	World Bank
WBI:	World Bank Institute
WCD:	World Commission on Dams
WCDR:	World Conference on Disaster Reduction
WCMC:	World Conservation Monitoring Centre
WCP:	World Climate Programme
WEF:	Water Environment Federation
WEHAB:	Water, Energy, Health, Agriculture, Biodiversity
WERRD:	Water and Environmental Resources in Regional Development
WES:	Water, Environment and Sanitation Programme
WFD:	Water Framework Directive
WGMS:	World Glacier Monitoring Service
WHO:	World Health Organization
WHYCOS:	World Hydrological Cycle Observing System

WHYMAP:	World-wide Hydrogeological Mapping and Assessment Programme
WMO:	World Meteorological Organization
WOC:	Water Operating Center
WRA:	Water Resources Assessment
WRC:	World Research Commission
WRI:	World Resource Institute
WRVI:	Water Resources Vulnerability Index.
WSI:	Water Stress Index
WSP:	Water Safety Plans
WSDP:	Water Sector Development Programme
WSSCC:	Water Supply and Sanitation Collaborative Council
WSSD:	World Summit on Sustainable Development
WSH:	Water, Sanitation and Health Programme
WSS:	Water Supply and Sanitation
WTA:	Willingness to accept compensation
WTP:	Willingness to pay
WUAs:	Water User Associations
WWAP:	World Water Assessment Programme
WWC:	World Water Council
WWDR:	World Water Development Report
WWF:	World Wilde Fund for Nature
ZAMCOM:	Zambezi Watercourse Convention

Main units of measurement

€	euro		**m.a.s.l**	metres above sea level
G	giga		**M**	megamillion
ha	hectare		**s**	second
k	kilo		**T**	tera
L	litre		**t**	ton/tonne (metric)
m	metre		**US $**	United States dollar
m²	square metre		**W**	watt
m³	cubic metre		**Wh**	watt hour

Note: billion = 1,000,000,000

Global Maps: Water Systems Analysis Group
University of New Hampshire, US

The global maps and related charts at the head of Sections 1 through 4 of WWDR2 are provided by the Water System Analysis Group (WSAG) at the University of New Hampshire's Institute for the Study of Earth, Oceans, and Space, located in Durham, New Hampshire. WSAG was founded in 1999, and its mission is to serve as a research and advanced training facility for analysing the global water system and the impacts of human activities and a changing water system on the natural environment and society. WSAG research integrates hydrology, biogeochemistry, and human-water interactions in analysing the full dimension of global change at local, regional and global scales. These include studies of the biogeochemistry of coastal watersheds in the northeast US, pan-Tropical and pan-Arctic water cycle studies, global biogeochemical modelling, and world water resource assessment. WSAG personnel actively participate in UNESCO's International Hydrological Programme (IHP), the United Nations World Water Assessment Programme (WWAP), Global Environmental System of Systems (GEOSS), International Council of Scientific Unions, International Association of Hydrological Sciences, and the newly-formed Global Water System Project. Additional information can be found at www.watsys.unh.edu.

Funding for the research on which the maps are based was provided mainly by the National Aeronautics and Space Administration (NASA) and the US National Science Foundation. Support from UNESCO's IHP, WWAP and the University of New Hampshire helped to produce the maps. Charles Vörösmarty, Ellen Douglas and Stanley Glidden from WSAG designed and assembled the compendia of indicator maps and created the website. Datasets and maps are available for downloading at http://wwdrii.sr.unh.edu/. Graphical material for the box on population was kindly provided by Deborah Balk, Bridget Anderson, and Marc Levy at the Center for International Earth Science Information Network (CIESIN), the Earth Institute, Columbia University, Palisades, New York. Population data are available at http://beta.sedac.ciesin.columbia.edu/gpw/index.jsp.

References and Websites:

CIESIN, 2005. Global Urban-Rural Mapping Project, Center for International Earth Science Information Network (CIESIN), Columbia University, http://beta.sedac.ciesin.columbia.edu/gpw/index.jsp.

Douglas, E. M., Githui, F., Mtafya, A., Green, P., Glidden S. and Vörösmarty, C. J. 2006. The application of water scarcity indicators at different scales in Africa. *Journal of Environmental Management.* (in press).

Ericson, J. P., Vörösmarty, C. J., Dingman, S. L., Ward, L.G. and Meybeck, M. 2006. Effective sea-level rise in deltas: Sources of change and human-dimension implications. Global and Planetary Change. (in press).

Galloway, J. N., Dentener, F. J., Capone, D. G., Boyer, E.W., Howarth, R. W., Seitzinger, S.P. Asner, G., Cleveland, C., Green, P., Holland, E., Karl, D.M., Michaels, A.F., Porter, J., Townsend, A. and Vörösmarty, C. J. 2004. Global and regional nitrogen cycles: Past, present and future. *Biogeochemistry.* Vol. 70, pp. 153-226.

Green, P., Vörösmarty, C.J., Meybeck, M., Galloway, J., and Peterson, B.J. 2004. Pre-industrial and contemporary fluxes of nitrogen through rivers: A

global assessment based on typology. *Biogeochemistry.* Vol. 68, pp.71-105.

McCully, P., 1996. Silenced Rivers: *The Ecology and Politics of Large Dams.* Zed Books, London, UK.

Postel, S. L, Daily, G.C. and Ehrlich, P.R. 1996. Human appropriation of renewable fresh water. *Science* Vol. 271 pp. 785-88.

Shiklomanov, I. (ed.). 1996. *Assessment of Water Resources and Water Availability in the World: Scientific and Technical Report.* St. Petersburg, Russia, State Hydrological Institute.

Syvitski, J.P.M., Vörösmarty, C. J., Kettner, A. J. and Green, P. 2005. Impact of humans on the flux of terrestrial sediment to the global coastal ocean. *Science,* Vol. 308 pp. 376-80.

UN (United Nations). 2003. *World Urbanizations Prospects: The 2003 Revision.* http://www.un.org/esa/population/publications/wup 2003/WUP2003Report.pdf.

Vörösmarty, C. J., Douglas, E. M., Green P. A., and Revenga, C. 2005. Geospatial indicators of emerging water stress: An application in Africa, *Ambio,* Vol. 34, No. 3, pp. 230-36.

Vörösmarty, C.J., Leveque, C., Revenga C., Caudill, C., Chilton, J., Douglas, E. M., Meybeck, M. and Prager, D. 2005. Fresh Water. *Millennium Ecosystem Assessment, Volume 1: Conditions and Trends Working Group Report.* Island Press. (in press).

Vörösmarty, C. J., Meybeck, M., Fekete, B., Sharma, K., Green, P. and Syvitski, J. 2003. Anthropogenic sediment retention: Major global-scale impact from the population of registered impoundments. *Global and Planetary Change.* Vol. 39, pp. 169-90.

Vörösmarty, C. J., Green, P., Salisbury, J. and Lammers, R. 2000. Global water resources: Vulnerability from climate change and population growth. *Science.* Vol. 289, pp. 284-88.

Walling, D.E., and Fang, D. 2003. Recent trends in the suspended sediment loads of the World's Rivers. *Global and Planetary Change.* Vol. 39, pp. 111-26.

Willmott, C.J., and Feddema, J. J. 1992. A more rational climatic moisture index. *The Professional Geographer* Vol. 44, pp. 84-7.

WRI (World Resources Institute). 1998. *World Resources: A Guide to the Global Environment 1998-99.* Washington, DC.

Photography

The World Water Assessment Programme would like to thank Bastien Affeltranger, the Ankara Fotoğraf Sanatçıları Derneği (AFSAD), Yann Arthus-Bertrand, the Australian Water Partnership, Robert Bos, Thomas Cluzel, Deanna Donovan, FAO, Richard Franceys, the GAP Bölge Kalkınma İdaresi Başkanlığı Arşivi (GAP-BKİ), the Ministry of Water and Agriculture of Kenya, IFAD, Christian Lambrechts, ICHARM, Alexander Otte, José María Sanz de Galdeano Equiza, Andras Szöllösi-Nagy, the Secretariat of Water, Public Works and Infrastructure for Development of the Government of the State of Mexico (SAOPID), Surapol Pattanee, UNESCO, UNESCO-IHE, UN-HABITAT, UNHCR and Sajith Wijesuriya for generously providing photographs.

cover
© SAOPID Mexico
© UNESCO – Andes / CZAP / ASA
© UNESCO – I. Forbes
© Sven Torfinn / Panos
© UNESCO – J. W. Thorsell
© Yann Arthus-Bertrand/La Terre vue du Ciel
© Surapol Pattanee
© Chris Stowers / Panos
© Australian Water Partnership

Front matter
IV: © Wim Van Cappellen / Still Pictures
VI: © Thomas Cluzel
VII: © Thoma Cluzel

SECTION 1
ii: © Thomas Cluzel
1: © Yann Arthus-Bertrand/La Terre vue du Ciel, © Sean Sprague / Still Pictures, © Yann Arthus-Bertrand/La Terre vue du Ciel

Chapter 01
8: © Ron Giling / Still Pictures
9: ©UN-HABITAT
10: © UNHCR/D. Shrestha
12: © UNESCO
13: © Mark Edwards / Still Pictures
16: © UNESCO – O. Brendan
17: © AFSAD / Selim Aytac
19: © Thomas Cluzel
20: © UNESCO
21: © Yann Arthus-Bertrand/La Terre vue du Ciel
27: © UNESCO - Evan Schneider
29: © UNESCO – Ines Forbes
34: © UNESCO – Niamh Burke
40: © Chris Stowers / Panos

Chapter 2
42: © Sean Sprague / Still Pictures
44: © Yann Arthus-Bertrand/La Terre vue du Ciel © Richard Franceys, © Mark Edwards / Still Pictures
45: © Julio Etchart / Still Pictures

52: © Thomas Cluzel
53: © Jorgen Schytte / Still Pictures
54: © Sean Sprague / Still Pictures
60: © Ton Koene / Still Pictures
74: © UNESCO / O. Brendan
77: © Hartmut Schwarzbach / Still Pictures
80: © Dirk R Frans / Still Pictures
81: © UNESCO / O. Brendan
83: © Thomas Cluzel

Chapter 3
86: © Ron Giling / Still Pictures
88: © UNESCO – Alexis Vorontzoff, © UN-HABITAT, © UN-HABITAT
89: © John Maier, Jr / Still Pictures
93: © Yann Arthus-Bertrand/La Terre vue du Ciel
94: © Andras Szöllösi-Nagy
96: © Andras Szöllösi-Nagy
97: © Mikkel Ostergaard / Panos
99: © Neil Cooper / Still Pictures
103: © UNESCO - Alexis Vorontzoff
105: © UN-HABITAT, © UN-HABITAT
107: © UNHCR/D. Shrestha
108: © UN-HABITAT
111: © Alexander Otte, © Alexander Otte / Veolia, © UNESCO – Dominique Roger

SECTION2
114: © Voltchev/UNEP / Still Pictures

Chapter 4
120: © Thomas Cluzel, © FAO/17287/ J. Holmes, © Mitchell Rogers/UNEP / Still Pictures
123: © Manit Larpluechai / UNEP / Still Pictures
125: © Thomas Cluzel
127: © UNESCO – A. de Crepy
136: © Yann Arthus-Bertrand/La Terre vue du Ciel
142: © Thomas Cluzel
143: © UNESCO - G. Boccardi
146: © UNESCO - A. Wheeler
147: © Ron Giling / Still Pictures
157: © AFSAD / Serpil Yıldız

Chapter 5
158: C Johnson /WWI / Still Pictures
160: © SOAPID Mexico, © UNESCO – I. Forbes, © Yann Arthus-Bertrand/La Terre vue du Ciel
161: © FAO/17121/M. Marzot
163: © C. Zöckler
164: © Nicolas Granier / Still Pictures
166: © SAOPID Mexico, © Sajith Wijesuriya
167: © SAOPID Mexico
168: © UNESCO – I. Forbes
171: © UNESCO
173: © Paul Glendell / Still Pictures
175: © Sajith Wijesuriya
184: © UNESCO – E. Timpe
191: © Alexander Otte/Veolia
192: © Christopher Uglow/UNEP / Still Pictures
197: © UNESCO – Peter Coles

SECTION 3
198: © Marcia Zoet / UNEP / Still Pictures

Chapter 6
202: © Julio Etchart / Still Pictures
204: © UN-HABITAT, © Andras Szöllösi-Nagy, © Shehzad Noorani / Still Pictures
205: © FAO/19526/G. Bizzarri
208: © UNESCO – O. Brendan
210: © Yann Arthus-Bertrand/La Terre vue du Ciel
212: © Jorgen Schytte / Still Pictures
221: © Jorgen Schytte / Still Pictures
223: © UNESCO – Dominique Roger
225: © UNESCO - Henry Bernard
227: © UNESCO/IHE – Fred Kruis
235: © Jacob Silberberg / Panos
236: © Robert Bos
241: © Mark Edwards / Still Pictures

Chapter 7
242: © UNEP/Still Pictures
244: © FAO/17346/R. Faidutti, © Mark Edwards / Still Pictures, © Yann Arthus-Bertrand/La Terre vue du Ciel, FAO/18992/R. Faidutti

245: © Glen Christian / Still Pictures
247: © SAOPID Mexico
248: © FAO/17268/ C. Sanchez
250: © FAO/15157/A. Conti
252: © FAO/17343/R. Faidutti
254: © AFSAD / Serpil Yıldız
255: © FAO/22404/ R. Faidutti, © Jinda Uthaipanumas/UNEP / Still Pictures
256: © FAO/22375/R. Messori
257: © FAO/19756/G. Bizzarri
259: © FAO/17086/M. Marzot
261: © Peter Frischmuth / Still Pictures
262: © Jeremy Horner/Panos
263: © Yann Arthus-Bertrand/La Terre vue du Ciel
264: © Joerg Boethling / Still Pictures
265: © Julio Etchart / Still Pictures
270: ©Alexander Otte/Veolia
271: ©Alexander Otte/Veolia, © FAO/13504/I. de Borhegyi

Chapter 8
274: © Jim Wark / Still Pictures
276: © UNESCO / I. Forbes, © MARK EDWARDS / Still Pictures
277: © Agence de l'eau Artois Picardie
279: © Adrian Arbib / Still Pictures, © Ron Giling / Still Pictures
284: © Agence de l'eau Artois Picardie
288: © UNESCO – Dominique Roger
290: © Jochen Tack / Still Pictures
292: © William Campbell / Still Pictures
294: © Yu Qiu/UNEP / Still Pictures
299: © Agence de l'eau Artois Picardie
300: © Agence de l'eau Artois Picardie

Chapter 9
304: © William Campbell / Still Pictures
306: © UNESCO, © Sean Sprague / Panos
307: © Yann Arthus-Bertrand/La Terre vue du Ciel
308: © Martin Bond / Still Pictures
309: © Sean Sprague / Still Pictures
310: © Jorgen Schytte / Still Pictures

312: © Ray Pfortner / Still Pictures
313: © GAP-BKİ
315: © KLEIN / Still Pictures
319: © GAP-BKİ
320: © Valérie SANTINI
321: © Hartmut Schwarzbach / Still Pictures
324: © Mike Powles / Still Pictures
325: © Kent Wood / Still Pictures
329: © Sean Sprague / Still Pictures
335: © Gilles POUSSARD

SECTION 4
336: © Thomas Cluzel

Chapter 10
340: © UNESCO - E. Schneider
342: © Andras Szöllösi-Nagy, © Bastien Affeltranger, © Bastien Affeltranger, © Bastien Affeltranger, © AFSAD – Aydan Adsaz, © Yann Arthus-Bertrand/La Terre vue du Ciel
343: © A.Ishokon/UNEP / Still Pictures
347: © G. Griffith / Still Pictures
348: © Peter Frischmuth / Still Pictures
349: © UNESCO – Nigel Swann, © UNESCO – Felipe Alcoceba
353: © UNESCO - Hameed A. Hakeem
358: © UNESCO - B. Bisson
360: © Thomas Cluzel
361: © UNESCO - E. Schneider, © Jim Wark / Still Pictures
362: © Bastien Affeltranger
365: © Ken Kerr / Still Pictures, © UNESCO - E. Schneider
367: © UNESCO - Michel Ravassard

Chapter 11
370: © Jose Roig Vallespir / UNEP / Still Pictures
372: © Secret Sea Visions / Still Pictures, © Julio Etchart / Still Pictures, © C.Garroni Parisi / Still Pictures
382: © Jim Wark / Still Pictures
388: © John Isaac / Still Pictures
392: © UNESCO - J.C. Simon
395: © UNESCO – Dominique Roger

Chapter 12
398: © Jorgen Schytte / Still Pictures
400: © Thomas Cluzel, © IFAD / Radhika Chalasani, © Deanna Donovan
401: © IFAD / Radhika Chalasani
402: © Hartmut Schwarzbach / Still Pictures
403: © UNESCO – André Abbe, © UNESCO - Tang Chhin, © UNESCO (Beijing)
404: Map 12.1: 1: © UNESCO – Alexis N. Vorontzoff, 2: © UNESCO - Hans de

Vaal, 3: © UNESCO – Anthony Lacoudre, 4: © UNESCO – Georges Malempré, 5: © UNESCO – Dominique Roger, 6: © UNESCO - Niamh Burke, 7: © UNESCO – T. Margoles
Map 12.2: 1: © UNESCO, 2: © UNESCO - Bruno Cottacorda, 3: © UNESCO - Peter Coles, 4: © UNESCO, 5: © Christian Lambrechts/UNEP (2004), 6: © UNESCO – Raoul Russo
405: © Thomas Cluzel, © Thomas Cluzel
407: © Thomas Cluzel
411: © UNESCO – Dominique Roger
412: © Wolfgang Schmidt / Still Pictures
413: © Ron Giling / Still Pictures,
415: © IFAD / Giuseppe Bizzarri
417: © B.Blume/UNEP / Still Pictures
418: © Julio Etchart / Still Pictures
421: © Deanna Donovan
423: © Sean Sprague / Still Pictures
424: © Thomas Cluzel
426: © Acharya / UNEP/Still Pictures
427: © Wang Fu-Chun/UNEP / Still Pictures

Chapter 13
432: © UNESCO – O. Brendan
434: © UNESCO - Yannick Jooris, © UNESCO – O. Brendan, © UNESCO - Yannick Jooris © UNESCO – O. Brendan, © UNESCO - O. Brendan
439: © UNESCO-IHE
443: © UNESCO – N. Burke
444: © UNESCO/IHE – Fred Kruis
446: © UNESCO – Ines Forbes
448: © UNESCO – Dominique Roger
449: © UNESCO-IHE
451: © UNESCO-IHE
452: © UNESCO-IHE
454: © UNESCO-IHE
455: © UNESCO-IHE
459: © UNESCO-IHE
463:© Shouli Lin / UNEP / Still Pictures

SECTION 5
465: © Tanya Wangniwiatkul/UNEP / Still Pictures

Chapter 14
466: © José María Sanz de Galdeano Equiza, © UNESCO - Bruno Cottacorda, © FAO/17067/ M. Marzot, © Gilles POUSSARD, © ICHARM, © Ministry of Water and Irrigation of Kenya, © Ago Jaani, © Thomas Cluzel, © FAO/13702/John Isaac, © SAOPID Mexico, © UNESCO - Michel Setboun, © UNESCO - Dominique Roger, © UNESCO – J.W. Thorsell, © Sajith

Wijesuriya, © Surapol Pattanee, © UNESCO
517: © UNESCO – Ariane Bailey

Chapter 15
518: © A. Appelbe/UNEP / Still Pictures
520: © Thomas Cluzel, © Roger De La Harpe / Still Pictures
521: © Thomas Cluzel
523: © Jusuf Jeremiah/UNEP / Still Pictures, © Deanna Donovan
524: © Deanna Donovan
525: © Dan Porges / Still Pictures
526: © Thomas Cluzel
527: © Sean Sprague / Still Pictures
546: © Neil Cooper / Still Pictures

Back cover
© UNESCO-IHE
© AFSAD / Doganay Sevindik
© Christopher Uglow / UNEP / Still Pictures
© AFSAD / Selim Aytac

YANN ARTHUS-BERTRAND
Yann Arthus-Bertrand's photographs of *The Earth from Above* are meant to show that, now more than ever, levels and modes of consumption and exploitation of natural resources are not sustainable in the long term. Whereas world production of goods and services has multiplied by 7 since 1950, 20 percent of the world population has no access to drinking water, 25 percent has no electricity, and 40 percent has no sanitary installation. In other words, a fifth of the world's population lives in industrialized countries, consuming and producing in excess and generating massive pollution. The remaining four-fifths live in developing countries and, for the most part, in poverty. To provide for their needs, they make heavy demands upon the Earth's natural resources, causing a constant degradation of our planet's ecosystem and limited supplies of fresh water, ocean water, forests, air, arable land, and open spaces.

At this critical stage, the alternative offered by a sustainable development policy should help in bringing about the necessary changes in order to 'meet the needs of the present without compromising the ability of future generations to meet their own needs.'* Inseparable from the accompanying text

commentaries, *Earth from Above* images invite each one of us to reflect upon the planet's evolution and the future of its inhabitants.
We can and must act individually on a daily basis for the future of our children.

Earth from Above team

* Quoted from the Brundtland report, The world commission on environment and development: *Our common future*, Oxford University Press 1987.

Cover and page 21
CENTRE-PIVOT IRRIGATION, Ma'an, Jordan (N 29°43' E 35°33'). This self-propelled, centre-pivot irrigation machine, invented by the American Frank Zybach in 1948 and patented in 1952, drills for water in the deep strata 30 to 400 m below the surface. A pivoting pipeline with sprinklers, extending about 500 m is mounted on tractor wheels, and irrigates 78 hectares of land. The countries of the Middle East and North Africa experienced the world's most rapid increase in grain imports in the 1990s. Production of 1 tonne of grain requires about 1,000 tonnes of water, and these countries prefer to import grain to meet their growing needs rather than produce it domestically because of the scarcity of water. In fact, at the current rate of use in Jordan, subterranean water reserves could dry up before 2010. Underground water is already overexploited in the United States, India and China. Watering technologies, however, waste less and respond better to plant needs, saving 20 to 50 percent of the water used in agriculture.

Page 1
REFUSE DUMP IN MEXICO CITY, Mexico (N 19°24' W 99°01'). Household refuse is piling up on all continents and poses a critical problem for major urban centres, like the problem of air pollution resulting from vehicular traffic and industrial pollutants. With some 20 million residents, Mexico City produces nearly 20,000 tonnes of household refuse a day. As in many countries, half of this debris is directed to open dumping. The volume of refuse is increasing on our planet following population growth and economic growth. Thus,

an American produces more than 700 kg of domestic refuse each year, about four times more than a resident of a developing country and twice as much as a Mexican. The volume of debris per capita in industrialized nations has tripled in the past twenty years. Recycling, reuse and reduction of packaging offer solutions to the pollution problems caused by dumping and incineration, which still absorb 50 percent and 35 percent, respectively, of the annual volume of household refuse in France.

Pages 44 and 92
***Favelas* in Rio de Janeiro, Brazil (22°55′ S, 43°15′ W).** Nearly one-quarter of the 10 million cariocas – residents of Rio de Janeiro – live in the city's 500 shantytowns, known as *favelas*, which have grown rapidly since the turn of the twentieth century and are wracked by crime. Primarily perched on hillsides, these poor, under-equipped neighbourhoods regularly experience fatal landslides during heavy rains. Downhill from the favelas, the comfortable middle classes of the city (18 percent of cariocas) occupy the residential districts along the oceanfront. This social contrast marks all of Brazil, where 10 percent of the population controls the majority of the wealth while nearly half of the country lives below the poverty level. As a result of urban growth, approximately 25 million people in Brazil, and 600 million in the world, inhabit the slums of great metropolitan areas, where overpopulation and poor conditions threaten their health and their lives.

Page 136 and 263
FIELDS NEAR QUITO, Sierra region, Ecuador (N 0°17′ W 78°41′). Between the Western and the Royal cordilleras, or chains, of the Andes, the plateaus of Quito benefit from the humid, gentle climate of the sierra, which favours the cultivation of cereals (corn, wheat, barley) and potatoes. Agriculture, which makes up only 12 percent of the gross national product, still supports nearly half of the population of Ecuador. Cultivated land comprises nearly one-third of the national territory, and is crucial in the country's history: the agrarian reforms

of the 1960s and 1970s, which eradicated the dominance of the great haciendas of Spanish colonists, failed to solve the problem of unequal distribution of farmland. The best lots, those in the valleys and on the coasts, devoted to the export crops (bananas, sugarcane, cacao), remain in the hands of rich landholders, whereas small farmers share the land on upper plateaus, barely subsisting from their produce. In a country of growing pauperization (65 percent of the population live in poverty and 19 percent are unemployed), such inequalities are providing fertile ground for a possible future crisis.

Page 160
CONFLUENCE OF THE RIO URUGUAY AND A TRIBUTARY, Misiones, Argentina (S 27°15′ W 54°03′). Drastically cleared to make way for farming, the Argentine tropical forest is today a less effective defence against erosion than it was in the past. Heavy rains falling in the province of Misiones (2,000 mm per year) wash the soil and carry off significant quantities of ferruginous earth into the Río Uruguay, turning the waters a dark reddish ochre. Swollen by tributaries bearing vegetal debris, the Río Uruguay (1,612 km long) empties into the Atlantic Ocean in the area of the Río de la Plata – forming the Earth's largest estuary (200 km wide) – where the river dumps the sediment it has carried. The sediment accumulates in the access channels to the port of Buenos Aires, which must be dredged regularly to remain navigable. Deposits built up at the mouths of rivers can change landscapes by forming deltas or extending land into the sea.

Page 210
Washing laundry in a creek, Adjamé district in Abidjan, Côte-d'Ivoire. (5°19′ N, 4°02′ W). In the neighborhood of Adjamé in northern Abidjan, hundreds of professional launderers, *fanicos*, do their wash every day in the creek located at the entrance of the tropical forest of Le Banco (designated a national park in 1953). They use rocks and tyres filled with sand to rub and wring the

laundry, washing by hand thousands of articles of clothing. Formerly a fishing village, Adjamé has been absorbed gradually by the metropolis of Abidjan, and it is now a working-class district. Abidjan is the economic and cultural centre of the country, yet some parts of it are without running water or electricity. It has undergone staggering urban growth: its population has increased fifty-fold since 1950 and today it has more than 3 million residents, one-fifth of the national population. The city has seen a proliferation of dozens of small trades, such as these *fanicos*, which offer the only means of subsistence for the poorest groups.

Page 244
Rendille enclosure between Lake Turkana and Marsabit, Kenya (N 2°20′-E 37°10′). In northwestern Kenya, the Rendilles, who are said to descend from the Samburu (with whom they have close kinship and economic links) and the Somali people, comprise some 22,000 camel herders. Their space is organized into large, semi-permanent camps inhabited by married men, women and children, and mobile encampments entirely made up of young men looking after flocks and searching for new pastures. Every evening the Rendille livestock are rounded up and placed in enclosures of spiny plants to prevent them from wandering and to protect them against predators. The tribe's girls are responsible for the goats and sheep, which they take out to pasture during the day. After the rainy season (June and July), the young herdsmen are able to find pastures closer to the main family encampments and attend the ceremony known as Almhata, a ritual feast during which they drink large amounts of milk. The tribe then moves off to a new site.

Page 307
BLUE LAGOON, near Grindavík, Reykjanes Peninsula, Iceland (N 63°54′ W 22°25′). The volcanic region of Reykjanes Peninsula, Iceland, has numerous natural hot springs. The Blue Lagoon (*Bláa Lónid*, in the Icelandic language) is an artificial lake fed by the surplus water

drawn from the geothermic power station at Svartsengi. Captured at 2,000 m below ground, the water is raised to 240°C by the molten magma and reaches the surface at a temperature of 70°C, at which point it is used to heat neighbouring cities. The milky blue colour of the lagoon results from the mineral mixture of silica and chalk from the basin combined with the presence of decomposing algae. Rich in mineral salts and organic matter, the hot waters (about 40°C) of the Blue Lagoon are known for their curative properties in the treatment of skin ailments. The use of geothermy, a renewable, clean, and inexpensive energy source, is relatively recent, but it is being used with growing frequency. In Iceland, in 1960 less than 25 percent of the population benefited from this source of heat, whereas today it meets the needs of 85 percent of Icelanders and provides heating for pools and greenhouses.

Page 342
TORNADO DAMAGE IN OSCEOLA COUNTY, Florida, United States (N 28°17′ W 81°24′). On February 22, 1998, a force-4 tornado (with winds of 300 to 400 km per hour) finished its course in Osceola County, after having devastated three other counties in central Florida. Several hundred homes were destroyed in its whirlwind, and thirty-eight people were killed. This type of violent tornado, rare in Florida, is generally linked to the climatic phenomenon of El Niño, which causes strong meteorological disturbances all over the world about every five years. Major natural catastrophes are more frequent and devastating than ever before. Human activities have significantly disturbed natural sites, reducing their resistance and their ability to withstand the effects of extreme climatic events. People also aggravate these consequences by living in areas exposed to risks. The decade of the 1990s saw four times more natural catastrophes than occurred in the 1950s, and the economic losses thus caused in that decade totalled US $608 billion, more than the costs for the four previous decades combined.

Index

Notes: text in boxes is indicated by **bold** page numbers and in figures and tables by *italics*.